William Latham Bevan

The student's manual of ancient geography

William Latham Bevan

The student's manual of ancient geography

ISBN/EAN: 9783742808370

Manufactured in Europe, USA, Canada, Australia, Japa

Cover: Foto ©Klaus-Uwe Gerhardt /pixelio.de

Manufactured and distributed by brebook publishing software
(www.brebook.com)

William Latham Bevan

The student's manual of ancient geography

NEW LATIN-ENGLISH DICTION-
WORDS OF FORCELLINI AND FREUND.
(...). Medium 8vo. 21s.

Dictionary the best representation of the
is undoubtedly that of Dr. Wm. Smith."—

S SMALLER LATIN-ENGLISH DIC-
With Proper Names, Calendar, Weights and
Abridged from the above work. Crown 8vo.

DR. WM. SMITH'S LATIN-ENGLISH VOCABULARY.
ARRANGED ACCORDING TO SUBJECTS AND ETYMOLOGY; INCLUD-
ING FIRST LATIN-ENGLISH DICTIONARY TO PHÆDRUS, CORNELIUS
NEPOS, AND CÆSAR'S GALLIC WAR. 12mo. 3s. 6d.

DR. WM. SMITH'S NEW CLASSICAL DICTIONARY
OF MYTHOLOGY, BIOGRAPHY, and GEOGRAPHY. For
the Higher Forms in Schools. Compiled from his larger
works. With 750 Woodcuts. 8vo. 18s.

DR. WM. SMITH'S SMALLER CLASSICAL DICTION-
ARY OF MYTHOLOGY, BIOGRAPHY, and GEOGRAPHY.
Abridged from the above. With 200 Woodcuts. Crown 8vo.
7s. 6d.

DR. WM. SMITH'S SMALLER DICTIONARY OF GREEK
AND ROMAN ANTIQUITIES. Abridged from his larger work.
With 200 Woodcuts. Crown 8vo. 7s. 6d.

THE STUDENT'S GREEK GRAMMAR. For the Use
of Colleges and the Upper Forms in Schools. By Professor
Curtius. Edited by Dr. Wm. Smith. New and Cheaper Edition.
Post 8vo. 6s.

A SMALLER GREEK GRAMMAR. For the Use of
the Middle and Lower Forms. Abridged from the above.
12mo. 3s. 6d.

INITIA GRÆCA, A First Greek Course. Containing
a Delectus, Exercise Book, and Vocabularies. By Dr. Wm.
Smith. 12mo. 3s. 6d.

MATTHIÆ'S SHORTER GREEK GRAMMAR. For
the Use of Schools. Abridged by Blomfield, revised by
Edwards. 10th Edition. 12mo. 4s. 6d.

HUTTON'S PRINCIPIA GRÆCA: An Introduction to
the Study of Greek. Comprehending Grammar, Delectus,
Exercise Book, Vocabularies, &c. 12mo. 3s. 6d.

Continued.

STANDARD SCHOOL BOOKS

(Continued).

BUTTMAN'S LEXILOGUS; a Critical Examination of the Meaning and Etymology of Passages in Greek Writers. Translated, with Notes, by FISHLAKE. *Fifth Edition*. 8vo. 12s.

BUTTMAN'S CATALOGUE OF IRREGULAR GREEK VERBS. With all the Tenses extant—their Formation, Meaning, and Usage. Translated, with Notes, by FISHLAKE. *Third Edition*. Post 8vo. 6s.

THE STUDENT'S LATIN GRAMMAR. FOR THE USE OF COLLEGES AND THE UPPER FORMS IN SCHOOLS. By WM. SMITH, LL.D. *New and Cheaper Edition*. Post 8vo. 6s.

DR. WM. SMITH'S SMALLER LATIN GRAMMAR. FOR THE USE OF THE MIDDLE AND LOWER FORMS. Abridged from the above work. 12mo. 3s. 6d.

KING EDWARD VI.'s LATIN GRAMMAR: LATINÆ GRAMMATICÆ RUDIMENTA, or an Introduction to the Latin Tongue. *Nineteenth Edition*. 12mo. 3s. 6d.

KING EDWARD VI.'s FIRST LATIN BOOK. THE LATIN ACCIDENCE; Including a short Syntax and Prosody, with an ENGLISH TRANSLATION. *Fifth Edition*. 12mo. 2s. 6d.

OXENHAM'S ENGLISH NOTES FOR LATIN ELEGIACS, designed for early Proficients in the Art of Latin Versification, with Rules of Composition in Elegiac Metre. *4th Edition*. 12mo. 3s. 6d.

DR. WM. SMITH'S PRINCIPIA LATINA. Part I. A FIRST LATIN COURSE. Comprehending Grammar, Delectus, and Exercise Book, with Vocabularies. 12mo. 3s. 6d.

DR. WM. SMITH'S PRINCIPIA LATINA. Part II. LATIN PROSE READING BOOK. Serving as an Introduction to Ancient Mythology and Geography, Roman Antiquities, and History. With Notes and a Dictionary. 12mo. 3s. 6d.

DR. WM. SMITH'S PRINCIPIA LATINA. Part III. LATIN POETRY. Consisting: Easy Hexameters and Pentameters; Eclogæ Ovidianæ; Prosody and Metre; First Latin Verse Book. 12mo. 3s. 6d.

DR. WM. SMITH'S PRINCIPIA LATINA. Part IV. LATIN PROSE COMPOSITION. Containing the Rules of Syntax, with copious Examples, Explanations of Synonyms, and a Systematic Course of Exercises on the Syntax. 12mo. 3s. 6d.

DR. WM. SMITH'S PRINCIPIA LATINA. Part V. SHORT TALES AND ANECDOTES FROM ANCIENT HISTORY, for Translation into Latin Prose. 12mo. 3s.

JOHN MURRAY, ALBEMARLE STREET.

January, 1867.

EPHESUS RESTORED

FROM ABOVE THE STADIUM

a. Mt. Coressus
b. Mt. Prion. { *c. Theatre.*
d. Upper Agora.
e. Odeum.

f. Temple of Diana.
g. Circular Temple, unknown.

THE FOLLOWING ARE NOW READY.

PREFACE.

THE following Manual is based upon the 'Dictionary of Greek and Roman Geography.' The original work contains a great mass of information derived from the researches of modern travellers and scholars, which have not yet been made available for the purposes of Instruction in our colleges and schools. It has therefore been thought that a Manual, giving, in a systematic form and in a moderate compass, the most important results embodied in the Dictionary would prove an acceptable addition to our school and college literature.

It would, however, be doing injustice to Mr. Bevan's labours to represent them as only systematizing the larger work. Besides adapting it for a different class of readers, he has likewise made many valuable additions, of which the most important are :—

1. A history of Geography in Antiquity, containing an account of the views of the Hebrews, as well as of the Greeks and Romans, and tracing the progress of the science from the mythical accounts of the poets through the progressive systems of Herodotus, Eratosthenes, Strabo, Ptolemy, and intermediate writers. This portion of the work is illustrated by maps of the world as known to the poets, historians, and geographers. It concludes with a chapter upon the Mathematical and Physical Geography of the Ancients.

2. As full an account of Scriptural Geography as was consistent with the limits of the work. Not only is considerable space devoted to Palestine and the adjacent countries, but information is given upon all other Scriptural subjects, such as the Travels of St. Paul, which can be illustrated by a knowledge of Geography. In this part of the work important assistance has been derived from the recently published 'Dictionary of the Bible.'

ANC. GEOG.   b

vi   PREFACE.

3. Numerous quotations from the Greek and Roman poets, which either illustrate or are illustrated by the statements in the text.

These are the principal additions made to the original work. In arranging the materials in a systematic form, great pains have been taken to make the book as interesting as the nature of the subject would allow. The tedium naturally produced by an enumeration of political boundaries and topographical notices is relieved by historical and ethnographical discussions, while the numerous maps, plans, and other illustrations, give life and reality to the descriptions. The Retreat of the Ten Thousand Greeks, the Expedition of Alexander the Great, and similar subjects, are discussed and explained. It has been an especial object to supply information, on all points required by the upper classes in the public schools and by students in the universities.

As regards the arrangement of the materials, the plan adopted has been to descend by a series of gradations from the *general* to the *particular* description of each country, commencing with the boundaries, character, climate, and productions; proceeding next to the physical features, such as mountains and rivers; then describing the inhabitants, political divisions, and principal towns; and concluding with a brief notice of the less important places, of the roads, and of the political history. This arrangement, which has been uniformly followed, will enable a student to arrive at both the kind and the amount of information he may require. Should he wish to study the physical features alone, he will find them brought together as a separate branch of the subject : should he, on the other hand, desire topographical particulars, he will know at once where to turn for them, both by the order observed in the treatment of the subject, and by the alteration in the type.'

A Manual of Modern Geography on a similar plan is in course of preparation.

WILLIAM SMITH.

November 1868.

CONTENTS.

BOOK I.

HISTORY OF ANCIENT GEOGRAPHY.

BOOK II.

ASIA.

CONTENTS.

LIST OF ILLUSTRATIONS.

ANCIENT GEOGRAPHY.

Mount Ararat.

BOOK I

HISTORY OF ANCIENT GEOGRAPHY.

CHAPTER I.

THE WORLD AS KNOWN TO THE HEBREWS.

§ 1. Original abode of man; rivers of Eden.   § 2. Ante-diluvian era.
§ 3. Ararat; Armenia. § 4. Shinar. § 5. Tripartite division of the
human race. § 6. Limits of the world as known to the Hebrews.
§ 7. Egypt. § 8. Ethiopia. § 9. Arabia. § 10. Syria. § 11.
Phoenicia. § 12. Mesopotamia. § 13. Babylonia and Assyria.
§ 14. Geographical Ideas of the Hebrews. § 15. Biblical nomen-
clature.

§ 1. The Bible contains the earliest geographical notices, commencing with the description of the original abode of man and
carrying us through a period long anterior to the rise of classical
literature. The primæval abode of the human race was situated on

one of the plateaus of Western Asia, but its precise position cannot be fixed. The "garden of Eden" in which the first man dwelt, is described (Gen. ii. 10-14) as having been situated in some central and lofty district, whence four rivers issued in various directions, viz. the Pison, Gihon, Hiddekel, and Euphrates. With regard to the two latter rivers, there can be no doubt that they are identical with the Tigris and Euphrates; with regard to the two former a great variety of opinion exists.

Rivers of Eden.—Many ancient writers, as Josephus, identified the Pison with the Ganges, and the Gihon with the Nile. Others, guided by the position of the two known rivers, identify the two unknown ones with the Phasis and Araxes, which also have their sources in the highlands of Armenia. Others, again, have transferred the site to the sources of the Oxus and Jaxartes, and place it in Bactria; others, again, in the valley of Cashmere. Such speculations may be multiplied *ad infinitum*, and have sometimes assumed the wildest character.

§ 2. So long as the position of Eden remains undecided, so long will it be futile to attempt any settlement of the other questions of ante-diluvian geography. The human race appears to have been divided into two great branches—the Cainites and Sethites—each having their distinct abodes and characteristics. The Cainites went eastward (Gen. iv. 16) from Eden, and settled in the land of Nod (= "exile"), which has been identified variously with Susiana, Arabia, Parthia, Tartary, and India: their first capital was Enoch, of equally uncertain position. The Sethites, we may infer, went westward, descending to the districts with which the Hebrews were afterwards best acquainted. The Cainites were agriculturists ; the Sethites adopted the pastoral life. To the former are attributed the establishment of towns, and the discovery of various useful and ornamental arts ; the latter, we may assume, retained their habits of primitive simplicity with the tenacity which, even to the present day, characterises the pastoral nations of the Eastern world.

§ 3. With the subsidence of the deluge we enter upon a new era in geography : the names of well-known localities appear in history. The ark "rested upon the mountains of Ararat" (Gen. viii. 4), meaning the mountains of Armenia, for Ararat in Biblical geography (2 K. xix. 37 ; Jer. li. 27) is not the name of a mountain, but of a district—the central region, to which the name of Araratia is assigned by the native geographer Moses of Chorene. This being the case, we are not called upon to decide a point which the sacred writer himself leaves undecided, namely, the particular mountain on which the ark rested.

Mount Ararat.—In a matter of such deep interest as the narrative of the Deluge, we cannot be surprised that attempts should have been made to fix on the precise spot among "the mountains of Ararat" where the ark rested, and Noah stepped forth on the regenerated world.

Nicolaus of Damascus assigned a mountain named Baris, beyond the district of Minyas (the Minni of Scripture), as the scene of that event. Berosus, who lived at Babylon, fixed on the lofty ridge of the Carduchian or *Kúrdish* range, which overlooks the plain of Mesopotamia in the neighbourhood of the Tigris: his opinion was followed by a large portion of the eastern world, so much so that in several ancient versions the name "Kardu" is substituted for Ararat, while the Koran gives the modern name "*Al-Judi*." The belief that the remains of the ark exist amid the lofty summits of that range is still cherished by the inhabitants of the surrounding district. Josephus, who notices these opinions (*Ant.* i. 3, § 6), further informs us, that the Armenians had fixed on the spot where Noah descended from the ark, and had given it a name which he translates *Apobaterium*, i. e. "landing-place:" he is supposed to refer to the place now called *Nackchivan*, which bears a similar meaning, in the valley of the Araxes. Nothing would be more natural than that the scene of the event should in due course of time be transferred to the loftiest of the mountains of Armenia, and that the name of Ararat should be specially affixed to that one: accordingly all the associations connected with the ark now centre in the magnificent mountain which the native Armenians name *Macis*, and the Turks *Aghri-Tágh*. This is the culminating point of the central range of Armenia, the Abus of the ancients. It rises majestically out of the valley of the Araxes to an elevation of 17,260 feet above the level of the sea, and about 14,350 above the valley, and terminates in a double conical peak, the lower or Lesser Ararat being about 400 feet below the other. The mountain is very steep, as implied in the Turkish name, and the summit is covered with eternal snow. Until recently it was believed to be inaccessible, but the summit was gained by Parrot in 1829, and the ascent has been effected since his time. A terrible earthquake occurred in the year 1840, which shattered the northern side of the mountain and carried vast masses of rock into the valley, doing immense damage.

It is important to observe how admirably Armenia is adapted by its geographical position to be the central spot whence the streams of population should pour forth on all sides of the world. The plateau of Armenia is the most elevated region of Western Asia, some of the plains standing at an elevation of 7000 feet above the level of the sea. It is equidistant from the Caspian and Euxine seas in the N., and from the Mediterranean and the Persian Gulf in the S. Around those seas the earliest settlements of civilised man were made, and they became the high roads of commerce and colonization. Armenia had communication with them by means of the rivers which rise in its central district, the Euphrates opening the path to Syria and the Mediterranean in one direction, as well as to the Persian Gulf in the other; the Tigris leading down to Assyria and Susiana; the Araxes and Cyrus descending to the Caspian, the latter also furnishing ready access to the Euxine by the commercial route which connected its valley with that of the Phasis. Westward the plateau of Armenia merges into that of Asia Minor, and eastward it is connected with the large plateau of Iran, the ancient Persia. If we add to these considerations, that in all directions the contrasts of climate, soil, and natural productions, were such as to invite emigration, we shall see how fitly the scene of the first dispersion of the human race is assigned to Armenia.

§ 4. The earliest settlements of any importance in the ancient world were in "the plain of Shinar" (Gen. xi. 2), the later Chaldæa, about the lower course of the Euphrates, and the shores of the Persian Gulf. In connexion with these settlements the Biblical narrative transports us back to a time when "the whole earth was of one language and of one speech " (Gen. xi. 1), and assigns to that region the development of those distinctive features of race and language which are embodied in the tripartite division of Noah's descendants, Shemites, Hamites, and Japhetites.

§ 5. The earliest and indeed the only systematic statement that we possess as to the distribution of these three great divisions over the face of the earth is contained in the 10th chapter of Genesis. That statement assumes the form of a genealogy: but a large admixture of geographical information is contained in it, the intention of the writer being to specify not only the nations, but the localities wherein they lived, and thus to present to his readers a map of the world as it existed in his time. Some of the names are purely geographical designations: Aram, for instance, means "high 'ands;" Canaan, "low lands;" Eber, the land "across" the river Euphrates; Sidon, " fishing station;" Madai, "central land;" Mizraim, in the dual number, the "two Egypts;" Ophir, "rich" land. Indeed it is not improbable that the three great divisions of the human race had originally a geographical meaning: Japheth, the " widely extended" region of the north; Ham, the "black" soil of Egypt; and Shem, the "mountainous" country.

The Mosaic world.—The world appears to have been divided into three zones, northern, central, and southern, which were occupied respectively by the descendants of Japheth, Shem, and Ham. The names of the nations may be in most cases identified with the classical names either of races or places.

(1.) The *Japhetites*—Javan, *Ionians*, in Greece and Asia Minor; Elishah, *Ælians*, in the same countries; Dodanim, *Dardani*, in Illyricum and Troy; Tiras in *Thrace*; Riphath, *Rhipæi Montes*, more to the north; Kittim, *Citium*, in Cyprus; Ashkenaz, near the *Axinus*, or Euxine Pontus, in Phrygia; Gomer, *Cimmerii*, in Cappadocia and in the *Crimea*; Tarshish, *Tarsus* (?), in Cilicia, but at a later age undoubtedly *Tartessus* in Spain; Tubal, *Tibareni*, in Pontus; Meshech, *Moschi*, in Colchis; Magog, *Gogaræus*, in northern Armenia, the Biblical name for the Scythians; Togarmah, in Armenia; and Madai in *Media*.

(2.) The *Shemites*—Elam, *Elymäis*, in Susiāna; Asshur, in *Assyria*; Arphaxad, *Arrupachitis*, in northern Assyria; Lud, *Lydia*; Aram, in Syria and Mesopotamia; the descendants of Joktan, in Arabia.

(3.) The *Hamites*—Cush, an appellation for the dark races, like the Greek *Æthiopia*; Mizraim in Egypt; Phut in Libya; Naphtuhim and

Lehâbim on the coast of the Mediterranean, west of Egypt; Caphtôrim in Crete; Casluhim from the Nile to the border of Palestine; Pathrúsim in the Thebäis; Seba in Meroë; Sabtah on the western coast of Bab-el-Mandeb; Havilah still more to the south; Sabtechah in the Somauli country; the various tribes of the Canaanites in Palestine and Phœnicia; Nimrod in Babylonia; Raamah and Dedan, on the southwestern coast of the Persian Gulf.

Map of the Distribution of the Human Race, according to the 10th chapter of Genesis.

§ 6. The limits of the known world in the Mosaic age may be fixed at the following points : in the N. the Euxine Sea ; in the S. the Indian Ocean, and Ethiopia ; in the E. the range of Zagros, which bounds the Mesopotamian plain ; and in the W. the Libyan

Desert and Ægean Sea. The knowledge of the Hebrews did not
extend much beyond these limits at any period of the Old Testament history; even within those limits, some districts, as Asia
Minor, were wholly unknown; while others, as Armenia and Assyria,
were but partly known. The only countries with which the Hebrews had intimate acquaintance were those immediately adjacent
to them—Egypt, and (in connexion with Egypt) Ethiopia, the
northern part of Arabia, Syria, Phœnicia, Mesopotamia, Assyria,
and Babylonia.

§ 7. Egypt was the land with which the Hebrews were best
acquainted : it was at the earliest period of the Bible history the seat
of a powerful empire, high civilization, and extended commerce.
Active communication was maintained between Canaan and Egypt
in the time of the Patriarchs, as evidenced by Abraham's visit (Gen.
xii. 10), the journey of the Ishmaelites (Gen. xxxvii. 25), and the
trade in corn (Gen. xlii. 1). The lengthened residence of the ancestors of the Hebrews in Egypt before the Exodus, the alliance which
subsisted between the two countries in the time of Solomon, and
the asylum which was afforded to a vast number of the Jews at the
time of the Babylonish captivity—all combined to establish an intimate relation with it, and account for the numerous references to it
in the Bible.

(1.) *Names.*—The Scriptural name "Ham" seems to be identical
with the indigenous name of Egypt, as it appears in hieroglyphics,
"Khemmi," and refers to the black colour of the soil: the name was
retained in that of the town Chemmis. The special name in Scriptural
geography was "Mizraim," a noun in the dual number signifying the
two (i.e. the Upper and Lower) *Misr*, the name by which Egypt is still
designated by the Arabs: it means "red mud." Occasionally the name
occurs in the singular number, "Mazor," in which case it is more
strictly appropriate to Lower Egypt (Is. xix. 6; 2 K. xix. 24, "besieged
places," A. V.). "Mizraim" is occasionally used in the same restricted
sense (Is. xi. 11; Jer. xliv. 15). We must also notice the poetical
name, "Rahab" (Ps. lxxxvii. 4, lxxxix. 10; Is. li. 9), an image of the
strength (comp. Is. xxx. 7) or violence of the nation.

(2.) *Divisions; the Nile.*—On this subject our information at an
early period is scanty. The name "Mizraim" implies that the same
twofold division, which existed in later historical times, existed in the
earliest period, being based on the natural features of the country.
These divisions were named by the Hebrews "Pathros" and "Mazor,"
the former representing the Thebaid, or Upper Egypt, which the
Hebrews regarded as the "land of birth," i.e. the *mother country* of the
Egyptians (Ez. xxix. 14): it was the abode of the Pathrusim (Gen. x. 14).
The Nile is occasionally named "Shihor" (Is. xxiii. 3; Jer. ii. 18); but
more commonly "Yeor" (Gen. xli. 1; Ex. i. 22), after the Coptic *iaro*,
"river;" the Hebrews also applied to it sometimes the term *yom*,
"sea" (Is. xix. 5; Ex. xxxii. 2; Nah. iii. 8).

(3.) *Towns and Districts noticed in the Bible.*—The district of Goshen
or Rameses (Gen. xlvii. 11), in which the Israelites were located, was

situated between the Delta and the Arabian Desert, on the eastern side of the Pelusiac branch of the Nile: the valley now called *Wadi-Tumeylah* appears to be the exact locality: Rameses may be the name of the *nome* in which Goshen was situated. The towns noticed are – Migdol (Ex. xiv. 2), *Magdolum*, on the border of the desert, the most northerly, as Syene was the most southerly of the towns of Egypt (Ez. xxix. 10, margin): Sin, *Pelusium*, well described as the "strength of Egypt" (Ez. xxx. 15), not only from its natural position and fortifications, but as commanding the entrance into Egypt from the north; it was situated at the mouth of the Pelusiac branch of the Nile: Tahapanes (Jer. ii. 16), Tahpanhes (Jer. xliv. 1), or Tehaphnehes (Ez. xxx. 18), *Daphne*, in the same neighbourhood, possessing a royal palace (Jer. xliii. 9), and evidently a place of importance (Ez. xxx. 18): Zoan (Num. xiii. 22), *Tanis*, on the Tanitic branch of the Nile, surrounded by a fine alluvial plain, "the field of Zoan" (Ps. lxxviii. 12), the residence of the 21st and 23rd dynasties, and regarded in the time of the Prophets as the capital of Lower Egypt (Ex. xxx. 14): Pi-beseth (Ez. xxx. 17), *Bubastis*, higher up the course of the river: Pithom, *Patumus*, and Raamses, *Heroopolis* (Ex. i. 11), on the eastern side of the Pelusiac arm, which were built by the Israelites as treasure-cities, probably for Rameses II.: On (Gen. xli. 45), or Aven (Ez. xxx. 17), "Ein-Re" in hieroglyphics, meaning "abode of the sun," and hence rendered Beth-shemesh (Jer. xliii. 13) by the Hebrews, and *Heliopolis* by the Greeks: the magnificent Temple of the Sun, of which Poti-pherah was priest (Gen. xli. 45), was approached by an avenue of sphinxes, terminated by two fine obelisks, the "images" or rather *columns* to which Jeremiah refers (xliii. 13): Moph (Hos. ix. 6), or Noph (Jer. ii. 16), *Memphis*, the city of "princes" (Is. xix. 13), as being the capital of Lower Egypt; it was situated on the left bank of the Nile, near the head of the Delta; the "idols and images," with which it was once lavishly adorned, have now utterly disappeared (Ez. xxx. 13): Hanes (Is. xxx. 4), probably another form of the name Tahpanhes: No (Ez. xxx. 14; Jer. xlvi. 25), or No Ammon ("populous," Nah. iii. 8), *Thebæ*, the capital of Upper Egypt, "situate among the rivers" (Nah. iii. 8), being probably surrounded by artificial canals communicating with the Nile: lastly, Syene (Ez. xxix. 10, xxx. 6), on the borders of Ethiopia. Of the above-mentioned towns, Migdol, Tahpanhes, Noph, and No were the chief abodes of the Jewish exiles (Jer. xliv. 1).

§ 8. To the south of Egypt, the kingdom of Cush, or Ethiopia, was one of high antiquity, possessing two capitals, Meroë (near *Dankalah*) in the south, and Napâta (*Gebel Birkel*) in the north? which owed its importance to its proximity to the border of Egypt. Active intercourse between Egypt and Ethiopia was maintained from the earliest ages. A large portion of the caravan-trade, from Libya on the one side, and the Red Sea on the other, converged to the banks of the Nile in this district, and was thence conveyed to Egypt. The two nations were frequently united under one sovereign: Herodotus (ii. 100) records that eighteen Ethiopian kings ruled Egypt before the time of Sesortasen; and we have undoubted evidence that in the latter part of the 8th century B.C. an Ethiopian dynasty held sway over Egypt. Two of the kings of this dynasty are well known to us from Scripture: So, or Sebichus, the ally of Hoshea king of Israel (2 K. xvii. 4), and Tirhakah, or Tarachus, who created a diversion in favour of Judæa when Sennacherib was besieging Jerusalem (2 K. xix. 9): the latter appears not to have held undivided sway, Sethos being contemporaneously the ruler of Lower Egypt.

The Notices of Cush in the Bible.—These are numerous, but it is difficult to apply them all to the Ethiopia of classical geography. In the Prophets, indeed, the African Ethiopia is distinctly defined as to the south of Syene (Ez. xxix. 10), the district intended being that which surrounded the northern capital of Napata, while the more southern territory of Meroë is described as "beyond the rivers of Ethlopia" (Is. xviii. 1). The African Ethiopia is undoubtedly referred to in 2 K. xix. 9; Ps. lxviii. 31; Is. xx. 4; Ez. xxx. 4, 6. In other passages, however, the term is extended to all the dark races of the south (Jer. xiii. 23); and in some the Asiatic or Arabian Cush seems more particularly intended (Gen. ii. 13; Job xxviii. 19; Hab. iii..7).

§ 9. Arabia bounded Palestine on two of its sides, viz. the south and east. Its inhabitants were in some instances connected with the Hebrews by the ties of a common descent, and in others by the commercial relations which from an early period existed between the two countries. The character both of the country and of the inhabitants prevented the Hebrews from penetrating into the country, and making themselves acquainted with the localities: still they must have known much relating to its physical features, its natural productions, and its wandering tribes.

(1.) *Name.*—The name of "Arabia" does not occur until the time of Solomon, and even then refers only to a few wandering tribes in the northern districts. The special name applied by the Hebrews to the northern part of the country was *Eretz-Kedem,* i. e. "Land of the East" (Gen. xxv. 6; Matt. ii. 1), while the remainder of the country was broadly described as "the South" (Matt. xii. 42). The district immediately S. of Palestine was named Edom or *Idumæa.*

(2.) *Places and Towns.*—The notices in the Bible are chiefly confined to the commercial districts of Arabia. Active trade was carried on between Tyre and the tribes on the shores of the Persian Gulf, Dedan und Raamah, as well as with Sheba and Uzal in the S. (Ez. xxvii. 15, 19, 20, 22); the "travelling companies of Dedanim" (Is. xxi. 13) were evidently the carriers who monopolized the caravan trade of Central Arabia: their trade consisted in ivory and ebony, which were Indian productions, and embroidered stuffs, which they probably manufactured themselves. The notices of Sheba are numerous: its productions were spices, frankincense, "the sweet cane from a far country" (Jer. vi. 20), gold and precious stones (1 K. x. 2; Ps. lxxii. 15; Is. lx. 6; Ez. xxvii. 22). The queen who visited Solomon was undoubtedly from this country: "the companies of Sheba" (Job vi. 19) traded northwards as far as Petra. Uzal is probably noticed in Ez. xxvii. 19, as trading with Tyre from its port Javan in "bright iron (i. e. steel), cassia, and calamus;" the same Javan is noticed in Joel iii. 6 as import-

ing slaves from the N. Ophir is mentioned in connexion with the
commerce of Solomon; if it was on the coast of Arabia, as seems to
be implied in Gen. x. 29, it was probably in the neighbourhood of the
modern *Aden*. The positions of Mesha and Sephar, which are given as
the limits of Arabia (Gen. x. 30), are uncertain; the former may be
identical with *Musa*, near the entrance of the Red Sea, and the latter
with *Saphar*, the modern *Daphar*, on the southern coast. The Midianites were active traders in the N. of Arabia; they were the merchantmen who took Joseph into Egypt (Gen. xxxvii. 28): their "camels and
dromedaries" (Is. lx. 6) were the means by which the northern trade
was carried on: their wealth is noticed in Judg. viii. 26. Other tribes
adopted the pastoral nomadic life which still prevails throughout the
greater part of Arabia: the "flocks of Kedar and the rams of Nebaioth"
(Is. lx. 7) wandered over the deserts to the E. of Palestine, and supplied
the markets of Tyre: the dark tents of the former people were so
familiar to the Jews (Ps. cxx. 5; Cant. i. 5), that the name seems to
have been adopted for the whole of Arabia (Is. xxi. 17), or perhaps
rather for the nomadic tribes (the Bedouins) as distinct from the
dwellers in villages, whose districts were named Hazor (Jer. xlix. 28).
The Nebaioth seem to have roamed as far as the Euphrates, for they
are noticed in the Assyrian inscriptions of Sennacherib, under the name
Nabatu, as having been defeated by him. At a later period they became
active traders, and seem to have transferred their residence to the
neighbourhood of Petra (Strab. xvi. p. 779; Diod. Sic. li. 48). The
Hagarites (1 Chron. v. 10), or Hagarenes (Ps. lxxxiii. 6), the *Agræi* of the
geographers, were a roaming tribe of Ishmaelites occupying a portion of
Northern Arabia to the E. of Palestine; they are noticed in the Assyrian inscriptions, under the name *Hagaranu*, as having been defeated
by Sennacherib. The towns that deserve notice are few. Elath, *Ælanu*,
stood at the head of the Ælanitic Gulf; David secured it (2 Sam. viii.
14), and Solomon thence fitted out his fleet for Ophir (1 K. ix. 26): it
was subsequently lost to the kingdom of Judah in the reign of Joram
(2 K. viii. 20), regained by Uzziah (2 K. xiv. 22), and again lost through
its conquest by Rezin (2 K. xvi. 6). Ezion-Geber, on the other side of
the channel, was the port whence the fleet actually sailed. Petra is
undoubtedly noticed under the name of Selah, each of these names
meaning "rock;" it was taken by Amaziah (2 K. xiv. 7), and afterwards by the Moabites (Is. xvi. 1); its position and its natural strength
rendered it an important acquisition for military purposes; equally
great was its commercial importance, as the central spot whither the
routes from Babylon, the Persian Gulf, Southern Arabia, Egypt, and
Tyre converged. Bozrah was another important town of the Edomites
(Gen. xxxvi. 33), whose destruction was frequently predicted by the
Prophets (Is. xxxiv. 6, lxiii. 1; Am. i. 12): it was situated to the N.
of Petra, at *Busairah*. The positions of the other ancient capitals of
the kings of Edom, Dinhabah, Avith, Rehoboth, and Pau (Gen. xxxvi.
32, 35, 37, 39), cannot be identified.

§ 10. Syria was contiguous to Palestine on its northern and north-
eastern border. The Hebrews were familiar with it from an early
period: the patriarchs had passed through it on their journeys to
and from the land of Mesopotamia, and Abraham had a native of
Damascus as his steward. At a later period, in the early days of
the monarchy, David extended his dominion over the whole of

Syria to the banks of the Euphrates: Solomon retained it for the greater part of his reign, and carried on an active trade along its southern frontier with Babylon and the East. Still later, the Syrians were constantly engaged in wars with the Hebrews, until they were themselves carried into captivity by the Assyrians.

(1.) *Name.*—The Biblical name of this district was "Aram," which extended to the "highlands" on both sides of the Euphrates. The name "Syria" appears to be an abbreviation of Assyria, introduced by Greek writers.

(2.) *Districts and Towns.*—Syria was divided into several districts, of which we may notice Aram-Maachah (1 Chron. xix. 6), between Palestine and Damascus; Aram of Damascus (2 Sam. viii. 5; Is. vii. 8, xvii. 3), the district surrounding the town of that name; and Zobah (1 Sam. xiv. 47; 2 Sam. viii. 3), an extensive district to the north of Damascus, reaching from Phœnicia to the Euphrates. Of the towns, Damascus and Hamath were the most important. The first was beautifully situated on the banks of the Abana (*Barrada*) and Pharpar (2 K. v. 12), and is noticed as early as the time of Abraham (Gen. xiv. 15, xv. 2). Hamath was situated on the Orontes, and commanded the pass into Palestine between the ranges of Lebanon and Anti-Lebanon: "the entering in of Hamath" (2 K. xiv. 25; 2 Chron. vii. 8) was the key of Palestine on the north; hence Hamath, with Riblah, which was in its territory, is frequently noticed in connexion with military operations (2 K. xiv. 28, xxiii. 33, xxv. 21), and its conquest was a subject of pride to the Assyrian monarchs (2 K. xviii. 34, xix. 13). The district of Hamath was regarded as the extreme northerly limit of the promised land (Num. xxxiv. 8; Ez. xlvii. 17). In addition to these we may notice Tiphsah (1 K. iv. 24), *Thapsacus*, an important point, as commanding one of the fords of the Euphrates; Helbon (Ez. xxvii. 18), near Damascus, famed for its wine; Tadmor, *Palmyra*, built, or, more probably, enlarged, by Solomon (1 K. ix. 18), as a commercial entrepôt for the caravan-trade between Palestine and Babylon; and Berothai (2 Sam. viii. 8), or Chun (1 Chron. xviii. 8)—perhaps *Birtha* on the Euphrates.

§ 11. Phœnicia was contiguous to Palestine on its northern frontier along the sea coast, and was familiar to the Hebrews partly from the enterprise of its merchants, and partly from the alliance which existed between the two countries in the reigns of David and Solomon. Wars occasionally occurred at a subsequent period, and numerous prophecies were directed against the capital, Tyre.

(1.) *Name.*—No general name for this country appears in the Bible: it was regarded as a portion of the land of Canaan, as being a maritime district.

(2.) *Towns and Districts.*—The following places may be regarded as the abodes of the tribes noticed in the Mosaic table (Gen. x. 15-18), in their order from N. to S.:—*Arādus*, of the Arvadites, whose skill in seamanship is mentioned by Ezekiel (xxvii. 8, 11); *Sinna*, a mountain fortress of no historical note, of the Sinites; *Simyra*, at the mouth of the Eleutherus, of the Zemarites; *Arca*, of the Arkites; and Sidon, which may, perhaps, be intended as the name of a district

rather than of the town, in the sense in which Homer uses Sidonia
(*Od.* xiii. 285). Sidon is frequently noticed; it was in the earliest
ages regarded as the "border of the Canaanites" (Gen. x. 19); a
little later Jacob speaks of it as "the haven of the sea, the haven of
ships" (Gen. xlix. 13). Although nominally within the limits of the
promised land, it was never conquered by the Israelites (Judg. i. 31).
It was emphatically the "great Sidon" (Josh. xi. 8), whose merchants "passed over the sea" (Is. xxiii. 2). At a later period we have
notice of *Byblus* as the abode of the Giblites (Josh. xiii. 5), the best
shipbuilders in Phœnicia (Ez. xxvii. 9), and the "stone-squarers"
employed in the building of Solomon's temple (1 K. v. 18). Zarephath,
or Sarepta (1 K. xvii. 9; Obad. 20; comp. Luke iv. 26), was a small
town about midway between Sidon and Tyre. Tyre is not noticed
until the time of Joshua (xix. 29), though probably an older town than
Sidon, and, subsequently, of much more importance in relation to
Palestine; the prophets expatiate upon its "perfect beauty" (Ez.
xxvii. 3; comp. Hos. ix. 13) and its commercial greatness—"the city
whose merchants are princes, whose traffickers are the honourable of
the earth" (Is. xxiii. 8); Ezekiel (xxvii.) in particular gives a detailed
account of the countries with which it interchanged its wares. Achzib,
the later *Ecdippa*, was on the sea-coast (Josh. xix. 29); Acco (Judg. I. 31),
afterwards called Ptolemais (Acts xxi. 7), a little to the N. of Carmel;
and Dor, or *Dora*, to the S. of it (Josh. xi. 2, xvii. 11).

§ 12. Mesopotamia was situated eastward of Syria between the
Euphrates and Tigris. The close connexion between the Hebrews
and the Aramæans of this district is marked by several circumstances : here Abraham sojourned on his passage to Canaan (Gen.
xi. 31); here Isaac's wife, Rebecca, spent her early days (Gen.
xxiv. 10); here Jacob served Laban (Gen. xxviii. 5); and here the
ancestors of the Israelitish tribes, with the exception of Benjamin,
were born.

(1.) *Name.*—The Biblical name of this country is "Aram-naharaim,"
i. e. "Aram of the two rivers" (Tigris and Euphrates) (Gen. xxiv. 10).
The term "Aram," *i. e.* "highlands," would restrict the original application of the name to the mountainous district about the upper courses
of the rivers. A portion of it was called "Padan-Aram," *i. e.* "the
cultivated land of the highlands" (Gen. xxv. 20, xxviii 2), being probably
the district immediately adjacent to the Euphrates; and another portion
"Aram Beth-rehob" (2 Sam. x. 6), the position of which is uncertain.

(2.) *Towns and Places.*—These are connected either with the history
of Abraham or with the Assyrian wars. Haran (Gen. xi. 31) was
situated in the N.W., on the river Bilias; it was identical with the
classical *Charræ;* it appears to have been a place of considerable trade
in Ezekiel's time (Ez. xxvii. 23). "Ur of the Chaldees" is by many
supposed to be at *Edessa*, in the same neighbourhood; by others it
has been placed to the S.W. of Nineveh; it was probably a *district*,
and not a town, and we can only say with certainty that it was to the
E. of Haran (Gen. xi. 31). The district of Gozan (2 K. xix. 12),
whither a colony of Israelites was transplanted (2 K. xvii. 6 ; 1 Chron.
v. 26), lay about the upper course of the Habor (2 K. xvii. 6), the
Aborras or *Chabóras* of classical geography. Along the course of the
Euphrates we have notice of Carchemish (Jer. xlvi. 2 ; 2 Chron. xxxv.

20), *Circesium*, at the junction of the Chaboras, the scene of the great battle between Necho and Nebuchadnezzar; Hena, lower down the river at *Anatha*; and Sepharvaim, *Sippara*, on the borders of Babylonia, the capture of which is noticed in the Assyrian inscriptions (2 K. xvii. 24, xix. 13). The positions of Hezeph and Thelassar (2 K. xix. 12) are uncertain: the former is supposed to be *Risapha*, on the W. of the Euphrates, S.W. of Thapsacus, and the latter, *Teleda*, in the same direction.

§ 13. Babylonia and Assyria were at different periods the seats of the most powerful empires of Western Asia. Their early importance is testified by the notice of their capitals in the Mosaic ethnological table (Gen. x. 10-12). In the time of Abraham a powerful confederacy issued from those regions, which extended its conquests for a while almost to the shores of the Mediterranean (Gen. xiv.). At a still later period the Assyrian armies overran Palestine, carried the ten tribes captive, and threatened the destruction of Jerusalem itself. This, however, was reserved for the Babylonian dynasty, which succeeded to the supremacy of the west after the overthrow of Nineveh by Cyaxares. The remnant of the Jewish nation was carried into captivity, and passed a lengthened period in the territories of the king of Babylon.

(1.) *Names.*—The southern district of Babylonia was known as "Shinar," and sometimes as the "land of the Chaldæans;" Assyria was designated "Asshur," after the original occupant of that district.

(2.) *Capitals of Babylonia.*—The Bible gives the names of four cities as having been originally founded by Nimrod in the plain of Shinar—Babel, Erech, Accad, and Calneh (Gen. x. 10): in addition to these, we have notice of Ellasar (Gen. xiv. 1). The sites of these towns have not been identified with certainty. (i.) It is doubtful whether the Babel of Nimrod's kingdom is the same as the Babylon of history, which was of comparatively recent date. The name "Babel" is supposed to mean "gate of Belus," and we may perhaps identify it with a town which was dedicated to Belus, and probably bore the name of Belus, the site of which is marked by the mound of *Niffer*, about 50 miles to the S.E. of Babylon. (ii.) Erech, the residence of the Archevites, may be identified with the modern *Warka*, situated near the left bank of the Euphrates, about 80 miles S.E. of Babylon: (iii.) Accad with the remains at *Akker-kuf*, near *Baghdad*; (iv.) Calneh with the classical *Ctesiphon*: (v.) Ellasar with *Senkereh*, about 15 miles S.E. of *Warka*. The fame of these cities, however, was wholly eclipsed by the rise of the later capital on the banks of the Euphrates—the Babylon of history, to which the name of Babel was transferred—the ruins of which at *Hillah* still strike the beholder with astonishment. This city is described at length in a future chapter.

(3.) *Capitals of Assyria.*—These are described in the following terms in the Bible:—"Out of that land went forth Asshur, and builded Nineveh, and the city Rehoboth, and Calah, and Resen between Nineveh and Calah: the same is a great city" (Gen. x. 11, 12). The identification of these places is not yet satisfactorily settled. The mounds opposite *Mosul*, named *Kouyunjik* and *Nebbi Yunus*, no doubt represent Nineveh, or a portion of it: it has been further conjectured that the

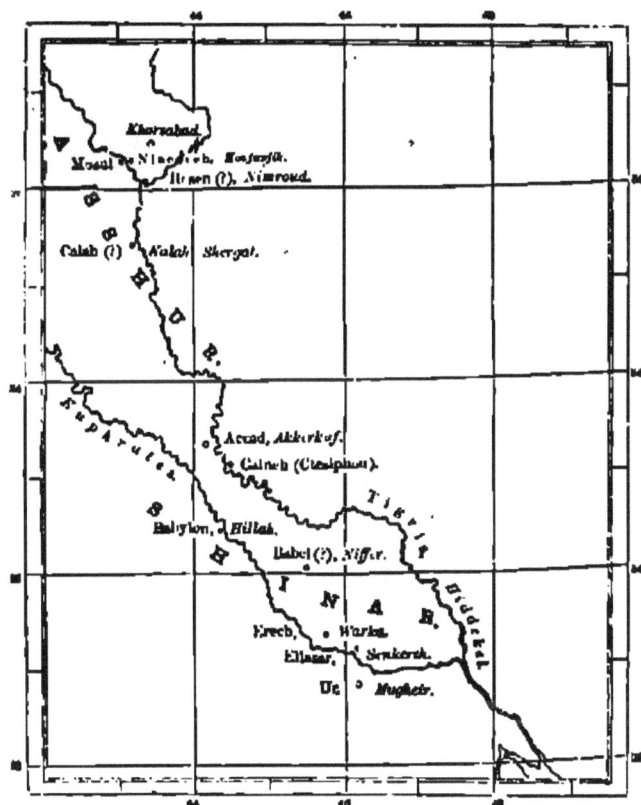

Map to Illustrate the Capitals of Babylonia and Assyria.

city may have extended over the whole quadrangular space inclosed between the four points, *Kouyunjik*, *Nimroud*, *Khorsabad*, and *Karamles*, in which case Jonah's description of it as "a city of three days' journey" would be strictly verified: this, however, is not decided. If Calah be identified with *Kalah-Shergat*, as the name suggests, then *Nimroud* would naturally represent the "great" city of Resen, which, according to the Bible, was between Calah and Nineveh. Rehoboth or Rehoboth Ir cannot be fixed at any place: the name describes the "broad, open streets" of an Oriental town.

§ 14. With regard to the opinions of the Hebrews as to the form, the size, and divisions of the earth, our information is but scanty, being derived wholly from scattered notices, many of which occur

In the poetical books of the Bible, and do not admit of being construed too rigidly.

(1.) The earth was circular (Is. xl. 22), with Jerusalem as its centre (Ez. v. 5) or navel (Judg. ix. 37; Ez. xxxviii. 12), and bounded on all sides by the ocean (Deut. xxx. 13; Job xxvi. 10; Ps. cxxxix. 9; Prov. viii. 27). The passages we have quoted cannot indeed be considered as conclusive; for a place may be described as centrally situated, without any idea of a circle entering into our minds, and Jerusalem was undoubtedly so situated with regard to the great seats of power, Egypt and Mesopotamia. Still the view, derived primâ facie from the words in Ez. v. 5, harmonizes with what experience would lead us to expect, and it was retained on the strength of that passage by a large section of the Christian world even so late as the 14th century, as instanced in the map of the world still existing in Hereford cathedral.

(2.) The earth was divided into four quarters, corresponding to the four points of the compass: the most usual method of describing these was by their position relatively to a person looking towards the east, in which case the terms "before," "behind," "the right hand," and "the left hand," would represent respectively E., W., S., and N. (Job xxiii. 8, 9). Occasionally they were described relatively to the sun's course, "the rising," "the setting." "the brilliant quarter" (Ez. xl. 24), and "the dark quarter" (Ez. xxvi. 20), representing the four points in the same order. The north appears to have been regarded as the highest, and so the heaviest, portion of the earth's surface (Job xxvi. 7).

(3.) The Hebrews, as other primitive nations, gave an undue importance to the earth, in comparison with the other parts of the universe. It was the central body, to which sun, moon, and stars were strictly subordinate. The heaven was regarded as the roof of man's abode — the curtain of the tent stretched out for his protection (Ps. civ. 2; Is. xl. 22): it was supposed to rest on the edges of the earth's circle, where it had its "foundations" (2 Sam. xxii. 8) and its massive pillars (Job xxvi. 11). It was the "firmament" for the support of the reservoirs of the rain (Gen. i. 7; Ps. cxlviii. 4), which descended through its windows (Gen. vii. 11; Is. xxiv. 18) and doors (Ps. lxxviii. 23). The sun, moon, and stars were fixed in this heaven, and had their respective offices assigned with an exclusive regard to the wants and convenience of the earth (Gen. i. 14-18; Ps. civ. 19-23). Beneath the earth was sheol, "hell," which extended beneath the sea (Job xxvi. 5, 6), and was thus supposed to be conterminous with the upper world: it had in poetical language its gates (Is. xxxviii. 10) and bars (Job xvii. 16), and was the abode of departed spirits, "the house appointed for the living" (Job xxx. 23).

§ 15. Before quitting the subject of early Biblical geography, it would be well to remind the reader that the Hebrew names are retained as the designations of the tribes or the countries inhabited by them throughout the whole of the Old Testament. Our translators have unfortunately adopted the classical names instead, and thus we have "Mesopotamia" for Aram-Naharaim; "Ethiopia" for Cush; "Chaldæa" for Chasdim; "Græcia" for Javan; "Armenia" for Ararat; and "Assyria" for Asshur.

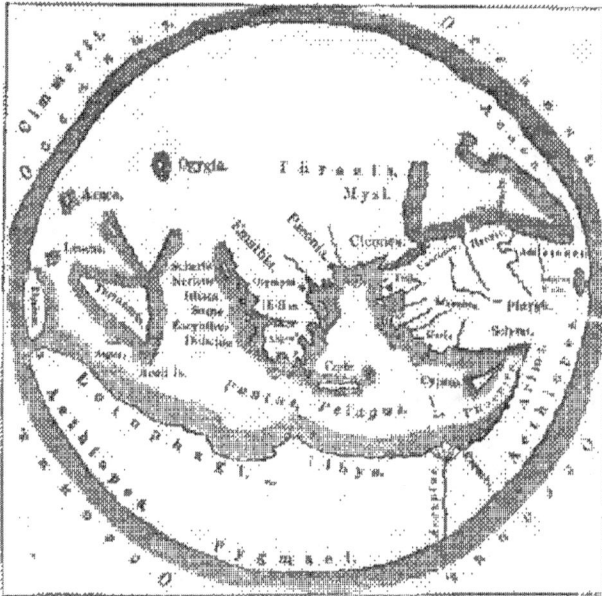

Map of the World, according to Homer

CHAPTER II.

§ 1. The earliest description of the world in classical literature is found in the Homeric poems. Without fixing the date of their composition, we may safely assume that they represent the views of the Greeks from about the 10th to the 8th century B.C. Homer is supposed to have been a native of Smyrna: however this may be, there is abundant evidence in the poems themselves that he had lived for some time in Greece ; his descriptions are those of an eye-witness : he must have been acquainted with all that lies southwards of the Ambracian Gulf on the western coast, and of Olympus on the eastern, though more intimately with some parts than others. The western coast of Asia Minor was also known to him. Beyond these limits his information was evidently derived from vague reports, and it becomes an interesting question whence these reports were obtained. In order to ascertain this, we must cast a glance at the progress of early maritime discovery. The Greeks themselves were not a seafaring race in that age : a voyage from Greece to Troy was regarded as a hazardous undertaking ; to Africa or Egypt, a terrible affair (*Od.* iii. 318); to Phœnicia no less so (*Il.* vi. 291). Even the seafaring Phæacians considered a voyage round the coast of Greece from Scheria to Eubœa a long one (*Od.* vii. 321). The Greeks must therefore have heard of distant lands from other more enterprising nations—among which we may notice firstly the Phœnicians, and secondly the Carians and Cretans.

(1.) *The Phœnicians.*—The Phœnicians carried on a most extended commerce long before the age of Homer: the coasts of Spain (Tarshish) and of Northern Africa were familiar to them; in short, the Mediterranean was a Phœnician lake. From their colonies about the Bosporus they carried on trade with the Euxine, and in other directions (as we know from Scripture) with Syria, Armenia, Southern Arabia, Africa, and India. They had settled on the islands of the Ægæan, and even on the mainland of Greece, and Homer speaks of them (*Od.* xv. 415, 458; *Il.* xxiii. 743) in terms which prove that the Phœnicians carried on an active trade in those parts; Corinth in particular had risen to wealth (*Il.* ii. 570) through their presence. Their influence is strongly marked in Homeric geography : there can be no doubt that the more distant points noticed, such as the Ocean, the Cimmerians, the Ocean mouth, Atlas, the land of Ææa, &c., were known to the Greeks only through the reports, designedly obscured and invested with terror, of the Phœnician traders.

(2.) *The Carians.*—The Carians appear to have been the earliest race connected with the Greeks, who established themselves as a naval power in the Ægæan sea. They were the "corsairs" of antiquity (Thuc. i. 8), and had stations on most of the islands as well as on the mainland of Asia Minor. They also possessed Cius on the Propontis, whence they traded with the shores of the Euxine Sea.

(3.) *The Cretans.*—The Cretans succeeded the Carians in their naval supremacy : to Minos was assigned the credit of having swept away piracy from the waters of the Mediterranean (Thuc. i. 4), reducing the Carians to peaceable submission, and prosecuting naval expeditions as far as Phœnicia in one direction (Herod. i. 2) and Sicily in the other (Herod. vii. 170). The period of Cretan supremacy is placed before the Trojan War, at which time it had declined (*Il.* ii. 852).

(4.) *The Argonautic Expedition.*—The legend of this expedition was probably founded on the accounts, which some of these seafaring nations communicated, about the commercial wealth of the Euxine Sea and the dangers that attended its navigation. That the Greeks themselves undertook such an expedition we think highly improbable ; but we see no grounds for doubting that the Phœnicians carried on an active trade from Pronectus, and the Carians from Clus; and that the commercial route, which was known to exist in later times between Central Asia and Europe, by the Oxus to the Caspian, and thence by the courses of the Cyrus and the Phasis to the Euxine, was established as early as the period we are now describing. The story of the Argonauts, as it comes before us, is evidently the fabrication of many generations. Homer (*Od.* xii. 69 ff.) merely notices the passage of the Argo between the whirling rocks on its return from Æaa. The golden fleece is first noticed by a writer of Solon's age (Strab. i. p. 46), and the earliest detailed account now extant is that of Pindar (*Pyth.* iv.) The position of Æaa—the route which the Argonauts pursued—and the extent of their voyage—were altered and enlarged from time to time to suit the geographical knowledge of the day.

§ 2. Homer is styled by Strabo the "author of geographical experimental science,"[1] in reference to the particular knowledge of places and institutions displayed in his poems. In as far as the actual experience of Homer or his countrymen is concerned, he fully merits the praise bestowed upon him by Strabo; but beyond this range his geography is involved in inextricable confusion. Homer had no idea of the spherical form of the earth: he conceived it to be the upper surface of a body of great thickness, which was as round as the shield of Achilles (*Il.* xviii. 607), and so flat that a god could look across it from Lycia to Scheria (*Od.* v. 282). This circular surface was edged by a river named Oceanus, just as a shield is bordered by its rim. On either side of this body, he conceived a domed covering to rest, the firmament of heaven on the upper side, and on the lower surface Tartarus, the counterpart of heaven, and equi-distant from the earth. In the interior of the earth's body was situated Hades, the abode of the dead. The earth's surface was divided between the masses of land and water, the latter occupying the largest space. Oceanus was regarded as the parent of all other bodies of water, the "sea," *i. e.* the Mediterranean, being connected with it at its western extremity, and the rivers by subterranean channels. The sea (θάλασσα, πόντος, πέλαγος, ἅλς) was supposed to extend indefinitely to the north, and perhaps to be connected with the Euxine in that direction; in the N.W. lay the fabled island of Ogygia, "the navel of the sea," the centre of an unlimited expanse.

§ 3. The land was regarded as a single undivided body—the names Europe, Asia, and Libya marking, not the continental divisions, but particular regions, Europe (which first appears in one of the hymns) the northern part of Greece, Asia the alluvial plain about the Cayster, and Libya a maritime tract west of Egypt. The usual division of the earth into quarters is not recognised by Homer, but instead of this we have it divided into *halves*, the eastern and

---

[1] Ἀρχηγέτης τῆς γεωγραφικῆς ἐμπειρίας, i. p. 2.

western, the former being described as the sunny side of the earth (πρὸς ἠῶ τ' ἠέλιόν τε), and the latter as the dark side (πρὸς ζόφον). Sunrise and sunset were, therefore, the cardinal points in Homeric geography, and had their features of similarity and contrast. As the sun apparently approached the earth at those points, its power was held to be greater there than elsewhere, and accordingly the people who lived in the adjacent regions, whether in the E. or W., were named Æthiopians, "dark complexioned:" at each too there was a country called Ææa, which seems to be an appellative for an extremely distant land. In the E. was the "Lake of the Sun," whence he arose, as a "giant refreshed," to take his daily course ; in the W. was the "glittering rock" Leucas, which formed the portal of his chamber. The W., as being the side of darkness, was naturally connected with the subject of death : there, consequently, Homer placed Elysium, the abode of the blessed, and the entrance to Hades—the former on this side, the latter on the other side of the stream of Ocean.

§ 4. In considering the special localities noticed by Homer, we have to distinguish the real or historical from the fanciful or mythical. It is difficult to draw an accurate line of demarcation, as there is a certain substratum of truth in many of the descriptions, which yet cannot be reconciled with fact. Generally speaking it will be found that all the notices of peoples and places in the E. and S. are reconcilable with fact, while the greater part of the notices in the W. and N. fall within the range of fiction, so that if a straight line were drawn through Corcyra in the direction of N.E. and S.W., it would divide the Homeric world with tolerable accuracy into the regions of fact and fiction. In the former district would be included the southern coast of the Euxine, the Ægæan Sea, and the coasts of the Mediterranean eastward of Greece ; while in the latter we should have the confused notices of Sicily and Italy, and the fabulous voyages from the Mediterranean to the Euxine and the western coast of Greece. The notices of special localities are, as might be supposed, very unequally dispersed, Greece and the western coast of Asia Minor being tolerably well filled up, while the more distant 'countries are but indefinitely described.

Details of the Homeric Geography.—Most of the important rivers and mountains of Greece have a place in Homer. Of the former, Acheloüs, "the king of rivers," Cephisus, Asopus, Alpheus, Spercheus, Enipeus, Titaresius ; of the latter, Olympus, the abode of the gods, Ossa and Pelion, Parnassus, Taygetus, and Erymanthus. The lakes Bœbeis and Cephisis, and the promontories Sunium and Malea are also noticed. Homer knew no general name for Greece : Hellas is with him but a small district in the south of Thessaly, and the Hellenes the inhabitants of that district : Peloponnesus is first noticed in one of the Hymns; in the earlier poems it is described by the term Middle Argos.

Of the names of provinces in northern Greece, afterwards familiar to us, only Ætolia, the Locri, Boeotia, and Phocis appear; Acarnania is named Epirus; the plain of Thessaly, Pelasgic Argos; Epirus may, perchance, be referred to under the name Apeira (*Od.* vii. 8). In Peloponnesus, Elis, Messenia, and Arcadia are named, while Argolis appears under the name Argos, and Laconia as Lacedæmon. The names of the occupants of these provinces are, in many instances, different from those of later times. Homer describes the general mass of the nation under the three names, Danaans, Argives, and Achæans. Among the special names we may notice the Curetes in Ætolia, the Cadmeans about Thebes, the Minyans about Orchomenus in Boeotia, and northwards of the Pagasæan gulf; the Æthices in the N.W. of Thessaly, the Selli about Dodona, the Epeans in Elis, and the Caucones in Triphylia. At this period the northern coast of the Peloponnesus was inhabited by Ionians, Argos and Laconia by Achæans, and Corinth by Æolians. Achæans were also settled in southern Thessaly. The towns are generally described as we afterwards know them: it should be noted, however, that there are two Dodonas, one in Thessaly (*Il.* ii. 750), and the other in Epirus (*Il.* xvi. 234): Delphi appears under the name Pytho: Corinth is also described as Ephyre (*Il.* vi. 152): Pylus, Nestor's capital, is probably the Messenian town of that name, though those in Triphylia and Elis contested the honour.

In Asia Minor we have on the western coast the rivers, Æsepus, Granicus, Simois, Scamander (or Xanthus), Hermus, Caÿster, Mæander, and several lesser streams; and the mountains, Ida with the peak Gargarus, Placus, Tmolus and its offset Sipylus, and Mycale: on the northern coast only the rivers Sangarius and Parthenius: on the southern, the river Xanthus, and perchance a reference to Mount Chimæra with its jets of inflammable gas in *Il.* vi. 179; beyond this limit, the Aleïan field in Cilicia is the only object. The inhabitants of the peninsula were arranged thus: on the western coast, the Dardani in Troas; the Mysians, Celeans, and Cilicians, in Mysia; the Mæonians in Lydia; and the Carians in Caria: on the northern coast, the Amazons about the Parthenius, the Halizones and Henети in Paphlagonia, and the Caucones in Bithynia: on the southern, the Lycians in Lycia, and the Solymi more to the east: in the interior, the Phrygians, and the Paphlagonians. Of the places on the coast, Ilium will be hereafter described: Thebe, the residence of the Cilicians, was near Placus; Larissa was a Pelasgic town in Æolis; Milêtus was in existence; several towns are noticed in Paphlagonia (*Il.* ii. 853), but there is some doubt whether the passage is not interpolated.

Proceeding to countries less known to Homer, we find the Syrians noticed under the name Arimi, connected with the Biblical Aram; then, the Phœnicians and especially the Sidonians; and the Erembi, another form of the name Arabians, at the S.E. angle of the Mediterranean. In Africa, the Nile is described as Ægyptus, with the isle Pharos at a day's sail distance from its mouth, and the hundred-gated Thebes on its banks. West of Egypt was Libya, and still more to the westward the Lotophagi, while in the extreme south, by the Ocean, were the Pigmies. Both of the two last mentioned peoples had a real existence: the Lotophagi are noticed by Herodotus (iv. 177) as living on the shore of the Lesser Syrtis, and both eating and extracting an intoxicating liquor from the lotus or jujube: the same writer (ii. 32) also notices dwarf races in the interior of Africa: the *lotus* is still eaten in *Tripoli*, and a dwarfish race, the *Dokos*, are known to exist in the S.W.

of Abyssinia. Atlas, in Homer, is not the mountain range of that
name, but rather a deity, the personification of the power which
sustained the vault of heaven.

North of the Ægean Sea, the mountains Athos and Nyseïum, and
the countries Pieria, Pæonia, Emathia in Macedonia, the Cicônes on the
coast of Thrace, the Mysi on the western coast of the Euxine, Thrace
in the interior, and in the extreme north the Scythian tribes Hippemolgi ("mare-milkers") and Abii are mentioned.

Many of the islands of the Ægean and Ionian seas are mentioned :—
Delos is occasionally named Ortygia; Euboea appears as the residence
of the Abantes ; the Calydnian isles (Il. ii. 677) were a group off the
coast of Caria ; Carpathus is named Crapathus ; Crete was occupied
by a variety of tribes, Eteocrêtans, Cydonians, Dorians, Achæans, and
Pelasgians, and possessed ninety (Od. xiv. 174), or according to Il. ii.
649, a hundred cities; the inhabitants of Lemnos are named Sintians,
a Thracian tribe of "robbers" (σίνομαι) ; Samothrace is given in its
resolved form "the Thracian Samos:" Tenêsa, whither the Taphians
traded for copper, was probably in Cyprus, but it has also been
identified with Tempsa in Italy. In the Ionian Sea, the group off the
coast of Acarnania is frequently referred to ; the occupants are named
Cephallenians, the island afterwards called after them being named
Samos or Same ; Lecucadia or Leucas is described as a promontory
of the mainland ; Ithaca is fully and accurately described. The
Echinâdes lie opposite the mouth of the Achelous ; Dulichium is
generally supposed to have been the largest of the group, but it
may have been situated on the mainland and hence is described as
"grassy" and "abounding in wheat" (Od. xvi. 396): the Taphians
occupied a small group of islands between Leucas and Acarnania.
Lastly, Corcyra is perhaps referred to under the name Scheria, the
residence of the seafaring Phæacians, though these may perhaps be regarded as a poetical fiction.

§ 5. The poetical geography of Homer is involved in inextricable
difficulties : it seems as though the poet had received certain scraps
of intelligence from Phœnician navigators as to the western and
northern districts of Europe, and had worked them up into a narrative without any regard to the true position of the localities.
Thus we have the Cimmerians, who really lived in the Crimea,
transported to the extreme west, and again the Planctæ, which
probably represent the Symplegâdes at the mouth of the Thracian
Bosporus, placed near Sicily ; the Argonauts are brought round
from the western Ææa to the eastern land of Æetes : Ulysses is
carried northward an immense distance inside the Ocean mouth,
and returns from Ogygia straight to the shores of Greece. It is
difficult to form any theory which will reduce the narrative to anything like consistency with geographical facts : it has been suggested that Homer had received reports of two ocean mouths, one
in the E. (the Straits of Yenikale), and one in the W. (the Straits
of Gibraltar), and that he transferred to each of them features that
belonged to the other (Gladstone's Homeric Studies, iii. 263): but
even this theory fails to reduce the narrative to consistency. We

therefore restrict ourselves to a brief notice of the localities described in the wanderings of Ulysses with a notice of anything that serves · to account for the narrative.

Wanderings of Ulysses.—Leaving Troy, he passed by the Cicones in Thrace, Cape Malea, and the island of Cythéra, to the land of the Lotophagi in Africa. Henceforth we enter on the realm of fiction: he first reaches the island Ægûsa, a reference to the Ægätes, but erroneously placed to the S. instead of W. of Sicily; he then passes to the land of the Cyclopes, either on the southern coast of Sicily or in Italy; it is termed the "continent" (ἤπειρος), which, however, is occasionally applied to large islands; Æolia (a reference to the Æolian group with volcanic *Stromboli*) was next visited, and then Læstrygonia, a city in a land of perpetual day (in reference to the long summer days of northern climates), generally placed on the northern coast of Sicily: the island of Ææa lay near the Ocean mouth, and thence he reaches the banks of Ocean stream, the land of the Cimmerians, and the entrance to Hades: he returns to Ææa, passes by the isle of Sirens, the Planctæ "wandering rocks," [1] to the W. of Sicily, Scylla, and Charybdis, and reaches Thrinacria, which must from its meaning "triangular," apply to Sicily; thence he is carried far to the northward to Ogygia, the "navel" of the sea, the residence of Calypso "the hidden one," and returns in a south-easterly course by Scheria to the shores of Greece.

§ 6. In the poems of Hesiod (about B.C. 735) we find the same general views as to the earth's form maintained with but slight deviation. The stream of Ocean still surrounds the earth's disk, its sources being placed in the extreme west. The vault of heaven still rests on the edge of the earth, upborne by Atlas, and as far removed from the earth in height as Tartarus in depth. Tartarus is represented as co-extensive with the earth and heaven, and as having its entrance in the west: the earth was rooted in its unfathomable depths. Hades is, generally speaking, placed on the surface of the earth in the extreme west, although occasionally the idea of a subterranean Hades is still expressed. In experimental knowledge a considerable advance had been made in the knowledge of the western countries of Europe. We have notice in Italy of the Tyrrhenians and of their king Latinus; of Ætna and the town of Ortygia (the later Syracuse) in Sicily, and of the Ligyans in Gaul. The gardens of the Hesperides, with their golden apples, are located opposite Atlas, with evident reference to the groves of oranges and citrons in Spain. In the extreme west are the "islands of the blest," and in the place of Homer's Elysium the fabled Isle of Erytheia. Hesiod knows nothing of the Cimmerians, but notices, according to Herodotus (iv. 32), the Hyperboreans who spent a happy life in the extreme north-western regions.

---

[1] In the later books of the *Odyssey* the names of Sicania (xxiv. 307), and of the Siceli, its inhabitants (xx. 383, xxiv. 211), first appear. Both the Sicani and Siceli lived at one period on the mainland of Italy, but they had probably crossed into Sicily before these books were composed.

Details of Hesiod's Geography.—Hesiod further notices the rivers
Eridanus, on whose banks were the amber-distilling trees, the Ister
in the N., the Phasis in the E., and the Nile in the S., which Homer
had named Ægyptus. The Ethiopians are correctly placed in the S. ;
and the name of Scythians is applied to the Hippemolgi of Homer, one
tribe of whom, named the Galactophagi, are described as a nomad race.
In Greece itself the names of various localities first appear, among
which we may notice Hellopia, the district about Dodona (*Fr.* v. 112),
and Abantis, an ancient name of Euboea : he also notices the alluvial
deposit which connected the Echinades with the mainland of
Acarnania (Strab. i. p. 59).

§ 7. In the poems of Æschylus we find some advance : the
three continents are noticed, Europe being divided from Asia either
by the Phasis, by which he probably means the later Hypanis, or
by the Cimmerian Bosporus, and from Libya or Africa by the
Straits of Hercules. The four quarters of the heavens are recognised, east, south, west, and north. The mythical element still
appears in the notices of the fountains of the ocean ; of Delphi as
the centre of the earth ; of the ocean encircling the world ; and of
the Ethiopians, both in the extreme east and also in the extreme
west, where he also placed a second Lake of the Sun.

The Wanderings of Io.—These cannot be reconciled with real geography : it is clear indeed from the writings of Æschylus (*Suppl.* 548 ;
comp. *Prom.* 705) that he was not careful to give even a consistent
story. We will therefore only observe that the Chalybes were probably the Cimmerians of the Crimea ; that the Hybristes may be
either the Don or the Kuban ; that the Amazons must be placed in
Colchis ; and that the Salmydessian Rock refers to the rocks near the
Thracian Bosporus. According to these notices, Io followed the line
of the Euxine along its eastern and southern coasts ; she then crossed
the Thracian Bosporus from Asia to Europe, and followed the Euxine
back to the Cimmerian Bosporus. She crossed the Palus Mæotis into
Asia, and arrived after some wanderings at the Gorgonæan plains of
Cisthênes in Ethiopia. The Bosporus mentioned in this part of her
course is the so-called Indian Bosporus, at the spot where Asia and
Africa were supposed to be contiguous at their southern extremities.
The Arimaspi of the north are transplanted to this district. From the
Indian Bosporus Io reached the river Æthiops, probably the upper
part of the Nile, and descended that river by the cataracts down to
the Delta. A considerable advance was made in the knowledge of
eastern countries, as might be expected from the historical events of
the poet's time. We have notice In Asia of the Indians, Susa, Cissia,
Babylon, Ecbatana, Bactria, Syria, and Tyre ; and in Egypt, of the
cataracts, the Delta, and the towns Memphis and Canôpus.

§ 8. In the writings of Pindar (B.C. 522-442) the same views still
prevail ; he recognises the three continents, and seems to make the
Phasis and the Nile the divisions. Cyrêne in Africa, Gadeira in
Spain, Cyme in Italy, and various Greek towns in Sicily, are first
noticed in his poems.

Map of the World, according to Hecatæus.

CHAPTER III.

THE WORLD AS KNOWN TO THE GREEK HISTORIANS.

§ 1. Causes which led to advanced knowledge: commerce and colonization ; voyages of discovery ; intellectual activity ; historical events. § 2. Hecatæus. § 3. Herodotus ; his life and travels. § 4. His character as a geographer. § 5. General views as to the earth's form, &c. § 6. Physical features. § 7. Political divisions and topography. § 8. Xenophon : the Anabasis. § 9. Ctesias. § 10. Alexander the Great. § 11. Extent of his discoveries. § 12. Arrian : histories of Alexander's life.

§ 1. Geographical knowledge made immense progress during the centuries that elapsed between Homer the first of the poets, and Herodotus the first of the historians. Nor was it confined simply to the increased extent of the earth's surface laid open to civilization : contemporaneously with this there sprung a spirit of scientific inquiry, which, not satisfied with the simple creed of an earlier age, sought out the physical nature of the earth, and of the phenomena connected with its economy. Among the various causes which led to these results, the following may be enumerated as most prominent :—(1.) The advance of commerce and colonization ; (2.) voyages of discovery ; (3.) the spirit of intellectual activity ; (4.) historical events.

(1.) *Advance of Commerce and Colonization.*—The spirit of commercial adventure was at an early period developed in the inhabitants of the isles and towns of the Ægean Sea. The Æginetans, and at a later period the Rhodians distinguished themselves for their bold seamanship ; the latter are said to have planted colonies in Iberia and among the Opicans and Daunians of Italy. The foundation of Metapontum in Italy by the Pylians on their return from Troy, and of Cumæ by Eubœans of Chalcis, cannot be regarded as well authenticated ; but there can be no doubt that from the eighth century the coasts of Magna Græcia in Italy and of Sicily were constantly visited by the Greeks, who planted the following colonies on them : Naxos (735 B.C.) ; Syracuse, Mybla, and Thapsus (734) ; Sylaris (720) ; Croton (710) ; Tarentum (708) ; Locri Epizephyrii (683) ; Rhegium (668) ; Himéra (648) ; and Selinus (about 628). The Phocæans of Ionia explored the coasts of Spain, Gaul, Western Italy, and the Adriatic : they were reputed to be the founders of Massilia, *Marseilles* (B.C. 600), and certainly settled at Alalia, in Corsica, about B.C. 564. The Samians under Colæus (about 650 B.C.) had penetrated beyond the *Straits of Gibraltar* to Tartessus : they were followed by the Phocæans, who settled there, in the year C30 B.C.

In another direction, the Milesians had thoroughly explored the Euxine, and are said to have changed its name from Axinus "inhospitable" to the more propitious name of Euxinus "hospitable." They lined its coasts with flourishing colonies during the eighth and two following centuries, B.C., other commercial towns following their example, but not to the same extent. Of these colonies we may notice, on the southern coast, Heraclea, Sinôpe, Amisus, Trapezum ; on the eastern, Phasis, Dioscurias, and Phanagoria ; in the Tauric Chersonese, Panticapæum ; on the northern coast, Tanis and Olbia ; and on the western, Istria, Tomi, Callâtis, Odessus, Apollonia, and Salmydessus.

Lastly, by the foundation of Cyrène by the Theræans (B.C. 630), and by the liberal policy of the Egyptian king Psammetichus, who gave to the Greeks Naucrâtis as a commercial station, the continent of Africa, hitherto a sealed book to European nations, was opened to them.

It should be remembered that each colony was a fresh starting point for more extended discoveries, the limits of which cannot be fixed with any precision. Herodotus (iv. 24) informs us that the Greek merchants penetrated to the extreme north of Scythia, and even beyond this to the mountain range of Ural. Tartessus again was undoubtedly an entrepôt for the prosecution of the northern trade in tin and other articles. From Naucratis the Greeks not only penetrated into Egypt, but learnt very much regarding the interior of Africa.

(2.) *Voyages of Discovery.*—Foremost among these we must mention Necho's expedition for the circumnavigation of Africa, about 600 B.C. Herodotus, who records it (iv. 42), expresses his doubts as to the account the Phœnician navigators gave, "that the sun was on their right hand;" this particular, however, forms the strongest argument in support of the real accomplishment of the undertaking, and it is the opinion of many distinguished geographers that the Cape of Good Hope was doubled more than 2000 years before the time of Vasco de Gama's discovery. It is important to observe that the Phœnicians started from the Red Sea and returned through the Straits of Gibraltar, thus gaining the advantage of the northern monsoon to carry them southwards to the tropic, thence a strong current setting to the south along the coast of Africa, and after doubling the Cape, the southern trade-wind to carry them home.

Sataspes undertook an expedition with a similar object, by the command of Xerxes, which proved a failure; the result is attributable to his having taken the opposite course, starting through the Straits of Gibraltar, in consequence of which he found himself baffled when he reached the coast of Guinea (Herod. iv. 43). The course at present taken by sailing ships is to cross over to the coast of South America, in order to avail themselves of the trade-wind.

An expedition into the interior of Africa was undertaken by some Nasamonians, as related by Herodotus (ii. 32); they reached a large river flowing from west to east (probably the *Niger*), and a town occupied by negroes (perhaps *Timbuctoo*).

Lastly, Scylax of Caryanda explored the Indus at the command of Darius Hystaspis; he started from Caspatyrus, descended the river to the sea, and thence returned by the Indian Ocean and the Red Sea (Herod. iv. 44).[1]

(3.) *Intellectual Activity.*—The spirit of commercial activity thus developed among the Greeks of Asia Minor, awakened a corresponding degree of intellectual excitement. The earliest school of physical science arose in that district under the guidance of the celebrated teachers Thales (B.C. 640-550), Anaximander (B.C. 610-547), and Anaximenes, who flourished about 530 B.C. The opinions entertained by these philosophers will be hereafter noticed. Of these, Anaximander conferred the most direct benefit on practical geography, by the introduction of maps of the world.

The Ionian School was succeeded by the Eleatic, founded by Xenophānes of Colophon about the year 536 B.C., and the Atomic School of Leucippus about 500 B.C., and lastly by that founded by Pythagōras, who flourished about 540-550 B.C.; to the latter is assigned the credit of having discovered the spherical form of the earth, a doctrine which did not gain general acceptance until the time of Plato. Some of the philosophers contributed to the advance of practical geography: we may instance Democritus of Abdēra, who composed several works, "Periplus of the Ocean," "Periplus of the Earth," &c., containing the results of his own observations; and Heraclitus of Ephesus (B.C. 500) who undertook and described a journey to the Ocean.

Another class of writers, the logographers, gave to the world descriptions, partly historical, partly geographical, of the various countries laid open. Of the majority of these, only the titles and a few fragments remain; yet these are interesting as showing the increased range of knowledge and the lively interest felt by the public on this subject.

---

[1] The expedition of Hanno occurred about this same period, but the notice of it is postponed, as it does not appear to have been known to Herodotus.

[2] The following is a list of the names and dates of the authors, with the titles—

The most important of these writers was Hecataeus, of whom, as the more immediate predecessor of Herodotus, we shall give a special notice. Of the others it may be observed that Hellanicus is supposed to have mentioned "Rome," and Damastes certainly did so: the latter writer and Pherecydes exhibited a very advanced knowledge of the western districts of Europe.

(4.) *Historical events* had their influence on the knowledge of geography. The growth of the Persian empire had excited curiosity as to the interior of Asia, and had opened fresh sources of information regarding the distant regions of the east. The expedition of Darius against Scythia, which he penetrated perhaps as far as the Volga, and his conquest of upper India, drew attention to both of those quarters. The disputes with the Ionian Greeks, and the subsequent invasions of Greece, led to the valuable information preserved to us in the pages of Herodotus. Nor should we omit notice of the facilities offered for travelling throughout the vast extent of the Persian empire. Herodotus gives a detailed account (v. 52) of the royal road from Sardis to Susa, which was furnished with stations at regular intervals.

§ 2. Hecataeus of Miletus flourished about 500 B.C., and took an active part in the political events of the day, particularly in the Ionian revolt. Previously to this he had travelled extensively, visiting Egypt, Persia, the coast of the Euxine, Thrace, Greece, Italy, Spain, and Africa; and he embodied the results of his observations in two works, the one geographical, the other historical. The former was named a "Survey of the World," and consisted of descriptions of the different districts of the then known world. His opinions are frequently referred to, indirectly, by Herodotus. Hecataeus supposed the habitable world to be an exact circle, surrounded by the Ocean, with which the Nile was connected at its source. He divided the land into two continents, the northern being Europe, and the southern Asia; these were separated by the Straits of Gibraltar in the W., and the Tanais, or more probably the Araxes and Caucasus, in the E. Libya he considered as a part of Asia: he describes the western parts of Europe at greater length than even Herodotus himself, and added much to the previous knowledge of Thrace, the coasts of the Euxine and Caspian seas, and the inhabitants of Caucasus, Persia, and India.[*]

of their works:—'Miletus and Ionia,' by Cadmus of Miletus (B.C. 520); 'Description of the World,' 'Persia, Troas, &c.,' by Dionysius of Miletus (B.C. 510); 'Description of the World,' containing special chapters on Asia, Europe, Africa, &c., by Hecataeus of Miletus (B.C. 549–486); 'Ethiopia, Libya, and Persia,' and a 'Periplus of the Lands outside the Pillars of Hercules,' by Charon of Lampsacus (B.C. 480); 'Lydia,' by Xanthus (B.C. 480); 'Sicily,' by Hippys of Rhegium (B.C. 495); 'Troas, Persia, Egypt, and the Greek States,' by Hellanicus of Mytilene; a 'Periplus,' 'Catalogue of Nations and Cities,' 'Greek Chronicles,' by Damastes of Sigeum, or of Citium in Cyprus; 'Antiquities of Attica,' by Pherecydes of Leros (about 500 B.C.).

[*] The Fragments which remain are remarkable for the number of names which appear in no other writer: some of these admit of identification with other forms, *e.g.* Darsians (Dersaeans, Herod. vii. 110); Dartilepilans (Danthaletians, Strab. vii.

Details of the Geography of Hecatæus.—Among the names of interest which first appear in his writings we may notice, in Europe—Pyrène (Pyrenees); the Celts of Gaul, with their town Narbo; Massilia; the Œnotri and Ausonians of Italy; Nola, Inpygia, Syracuse, and various other towns of Sicily; Cyrnus (Corsica); the Illyrians and Liburnians, and the Melanchlæni of Scythia : in Asia—Pontus Euxinus; the Hyrcanian (Caspian) sea; the Colchians, Moschians, Armenians, and Tibarenians; the Caspian gates; the Parthians and Chorasmians; the Indians, with the river Indus and the town Caspapyrus; the Persian Gulf; Canytis in Syria (Gaza); and Chna (Canaan, *i.e.* Phœnicia) with Gabala (Gebal); in Africa—Magdôlus (Migdol) and Chembis (Chemmis), towns of Egypt; the Psyllians, Mazyans, Zauecians (compare the Roman *Zeugitana*), and Carthage on the northern coast, and the river Lixas, perhaps the Lixus of Hanno, on the western. It may be noticed that he names certain islands in the Nile, Ephesus, Chios, Lesbos, Cyprus, and Samos ; this may be perhaps regarded as an indication that Greek colonies were planted on them. Whether the name Amalchium Mare (= " frozen sea ") applied to the Northern Ocean originated with Hecatæus, is doubtful ; it may be due to Hecatæus of Abdera. Lastly, he improved the map of Anaximander, and it has been supposed that it was his which Aristagoras showed to Cleomenes, as related by Herodotus (v. 49).

§ 3. Herodotus was born at Halicarnassus, B.C. 484, and probably died at Thurii in Italy.[4] At an early age he entered upon a course of travel, and in his great historical work he has recorded much that he saw. Great difference of opinion exists as to the extent of his travels ; we have positive evidence that he visited Egypt (ii. 29), Cyrene (ii. 181), Babylon (i. 181-3), Ardericca in Susiana (vi. 119), Colchis (ii. 104), Scythia (iv. 81), Thrace (iv. 90), Dodona (ii. 52), Zacynthus (iv. 195), and Magna Græcia (iv. 15, v. 45). Some of these countries, particularly Egypt, Greece, Asia Minor, and the islands of the Ægean Sea, he knew intimately : or others his narrative shows only a partial knowledge. He seems to have visited only the coast of Scythia, between the Danube and Dniepr; the same may be said of Phœnicia, Syria, and Thrace, while in Magna Græcia he notices only some few of the Greek towns. The dates of the chief events of his life may be fixed with some probability as follows : Egyptian travels, B.C. 460-455 ; visit to Thrace, about B.C. 452 ; removal from Halicarnassus to Greece, B.C. 447 ; removal to Thurii, B.C. 443.

§ 4. As a geographer Herodotus has both merits and defects. Among the former we may notice the fidelity with which he records what he had himself seen, and the candour with which he relates

---

818); Mazyes (Mazykes, Ptol. iv. 8; Mazyes, Herod. iv. 191); Caspapyrus (Caspatyrus, Herod. iii. 102); Ellbyrge (Illiberis); Canytis (Cadytis, Herod. ii. 159); Zygantes (Gyzantes, Herod. iv. 194); others are wholly unknown.

4 The date of the death of Herodotus has been a subject of much dispute. Some writers place it in B.C. 430, and others not earlier than B.C. 408.

the statements of others, even when he himself attached no credit
to them. To this latter quality we owe some of the most interesting
notices in the whole of his work : since most of the statements which
he regarded as incredible, some of which indeed are incredible in
the form in which they appear, are nevertheless found to have a
large substratum of truth, which, by the light of modern research,
can be separated from the fiction mixed up with them. Among
his defects we may notice the very unscientific and unmethodical way
in which he treats his subject, and the indistinctness of his statements whenever he attempts a general sketch either of a land or of
a continent. The first of these defects may be partly excused on
the ground that his work was rather historical than geographical :
the second, though not admitting of the same plea, may nevertheless
be explained as resulting in many instances from a laboured attempt
to be distinct, without a sufficient regard to the facts with which he
deals : hence he adopts a symmetrical arrangement in cases where his
subject does not admit of it. We may instance his account of the
continent of Asia with its two actai (iv. 37 ff), which is apparently
simple enough, but becomes more and more complicated as he goes
on ; for he seems not to have observed that the four nations selected
as occupying the heart of the continent, did not live due north of one
another, nor yet that, according to his theory, the whole of Africa
became merely an appendage of one of the actai. Again, his idea
of the relative positions of Egypt, Cilicia, Sinope, and the mouth of
the Danube, as being in the same meridian (ii. 34), can only be
regarded as an instance of false symmetry. Lastly, his description
of Scythia as a four-sided figure (iv. 101), has been quite a vexata
quæstio to his commentators.

§ 5. With regard to his general views as to the form, boundaries,
and divisions of the world, Herodotus had gained sufficient knowledge to lead him to reject the theory of Hecatæus, that the world
(i.e. the habitable world, the land) was "an exact circle as if
described by a pair of compasses" (iv. 36), the projections of Arabia
and Libya disproving this to his mind. He had not, however,
sufficient knowledge to enable him to propound any theory of his
own ; the boundaries of Europe, and consequently of the northern,
eastern, and western parts of the world were unknown (iii. 115, iv.
45), and it was therefore ridiculous in his eyes to attempt a defini-
tion of its form. As far as we can gather from his description, he
supposed the world to be oval rather than circular, Greece holding
a central position (iii. 106). He rejected the idea of the "river" of
Ocean as a poetical fancy (ii. 23), and doubted whether the world
was surrounded by the Ocean at all (iv. 8, 36, 45); though he
does not expressly reject, yet he shows his extreme distrust of the
report of a northern sea, which had evidently been reached (iii. 115).

He knew that the western shores of Europe and Africa were washed by the Atlantic ocean (i. 203), which was connected with the Mediterranean at the Pillars of Hercules (iv. 42); and he further knew that the Atlantic was connected with the great southern Ocean that surrounded Libya and Asia, which he names the Erythraean Sea (i. 203, iv. 40).   With regard to the division of the world into continents, he adopts, without approving of, the recognised divisions into Europe, Asia, and Libya : his own view was that the earth formed but one continent in reality ($\mu\iota\tilde{\eta}$ $\dot{\epsilon}o\dot{v}\sigma\eta$ $\gamma\tilde{\eta}$, iv. 45), and he disliked the ordinary division, partly because it was unsymmetrical, Europe being as large as the other two put together (iv. 42), and partly because there was no well-defined boundary between Asia and Libya, the Nile, which was usually regarded as the boundary, dividing in its lower course, so that the Delta was neither in Asia or Africa (ii. 16).   Herodotus evidently considered Africa below the dignity of a continent : it is only a portion of the great southern projection of Asia (iv. 41), separated from Asia by Egypt (ii. 17, iv. 41), in short a *district* and not a continent; at the same time he occasionally falls into the usual phraseology, and u-es Libya as inclusive of Egypt (iv. 42).   Herodotus justly notes the awkwardness of dividing a country like Egypt between the two continents (ii. 17), and insists that the land of the Egyptians must be regarded as one : it is singular that he did not see the way of meeting the difficulty by constituting the Red Sea the boundary. He regarded Europe as co-extensive with Asia in the east (iv. 42), and therefore he included northern Asia in it : the boundary between the two thus ran east and west, and consisted of the Mediterranean Sea, the Euxine, the Phasis, the Caspian Sea, and the Araxes (Jaxartes), as we gather from detached notices (iv. 37, 40). His view of the contour of Europe is defective in the west, for he supposes the land to stretch out beyond the Pillars of Hercules to a great extent (ii. 33).   His knowledge of this continent did not go beyond the Danube, except in the neighbourhood of the Euxine Sea.   Asia was known only as far as the Indus eastward : from the direction which he gives to the course of that river (iv. 44), it would naturally be inferred that he carried the Ocean round towards the north before reaching its mouth.   The peninsula of Arabia is duly accounted for in his description, but Asia Minor is disfigured by the undue contraction of its eastern side, which is represented as one hundred miles too little (i. 72).   Africa was known as far south as Abyssinia, from which point Herodotus probably supposed the sea at once to trend round to the W.   The form of the northern coast is modified by the notice of only one Syrtis (ii. 32).

§ 6. The most important physical features in the world of Herodotus are the seas, rivers, and mountains, the 'ast being but

Inadequately noticed as compared with the two former. The seas
are the Mediterranean, Euxine, Caspian, and Red Sea. The rivers
are the Nile, Danube, Halys, Tigris, Euphrates, Indus, Tyras,
Borysthenes, Araxes, and several other Scythian rivers. The
mountains are Caucasus, the Matienian mountains, Atlas, Hæmus,
and several of the ranges in Greece. Of these objects a more
particular account is given in the following paragraphs.

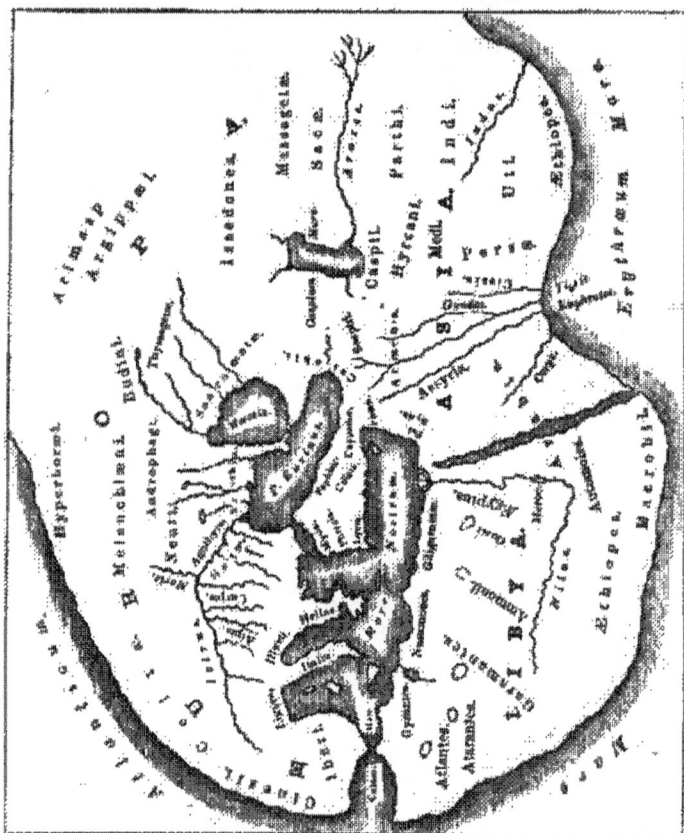

Geography of Herodotus — Physical Features. (1.) Seas.—The Mediterranean was the only sea to which Herodotus applies the term
θάλασσα: he describes it as "our" sea (ἥδε ἡ θάλασσα, i. 1, 185, 'v.

41), and the "northern" sea in reference to Africa (ii. 11, 32, 158, iv. 42)—a name which it still retains among the Arabs: it was divided into the following subordinate seas, to which he applied the terms πόλπος, πόντος, and πέλαγος—the Adriatic (ὁ Ἀδρίης, i. 163); the Ionian Gulf, which is another term for the Adriatic (vi. 127, vii. 20), at all events for the eastern coast of the Adriatic; the Ægæan, of which he notices the width (χάσμα πελάγεος, iv. 85); the Icarian (vi. 95), off the coast of Caria; the Sardinian (i. 166); the Egyptian (ii. 113); and the parts about the islands Carpathus (iii. 45), and Rhodes (i. 174). The Euxine is named "Pontus," as being the largest inland sea with which the Greeks of that day were acquainted; in reference to Asia it is the "northern" sea (iv. 37), in reference to Scythia, the "southern" (iv. 13). Herodotus exaggerates its size (iv. 85); its length, between the points which he incorrectly regards as extreme, being 630 miles, instead of 1280, and its breadth 270, instead of 380: its greatest length in reality is through the middle of the sea, and the greatest width between the mouths of the *Telegul* and *Sakhariyeh*. The Palus Mæotis (*Sea of Azov*) Herodotus describes as nearly as large as the Euxine (iv. 86); in this he exaggerates; it is highly probable, however, that it extended eastward along the course of the *Manytch* for a considerable distance, as he implies (iv. 116), and, from the present rate of its decrease, we may well suppose it to have been four or five times as large as it is now. Great changes have undoubtedly taken place in the levels of the land north of the Euxine, by which some of the rivers have altered their courses, and others have altogether disappeared: Herodotus' description of the *Crimea* as an *acte* (projecting tract of land) similar to Attica (iv. 99), would lead us to suppose that the *Putrid Sea* had come into existence since his time. The Bosporus, Propontis, and Hellespont, are described with tolerable accuracy (iv. 85). The Caspian is more accurately described by Herodotus than by many of his successors: he knew it to be a distinct sea (i. 203), whereas it was generally believed after his time to be connected with the northern Ocean: the proportions which he assigns to its length and width (750 and 400; i. 203) are very nearly correct; nor is there any reason to infer that he reversed the position of the lake as is occasionally represented in Herodotean maps. The *Sea of Aral* is not noticed: it has been conjectured, by many eminent geographers, that the Caspian extended very much to the eastward so as to include *Aral*, and the appearance of the country favours this idea. Geologists, however, have come to the conclusion that the elevation, which separates these two seas, occurred at a period anterior to the creation of man, and even before the separation of the Caspian from the Euxine by the elevation of Caucasus. The Caspian has, nevertheless, undergone great changes even in historical times; not improbably the *Gulf of Kuli Deryu*, on its eastern coast, extended far over the alluvial flats to the eastward, receiving the *Oxus (Jyhun)* by a course which may still be traced. The Red Sea is described as the Arabian Gulf (ii. 11); Herodotus probably supposed that the breadth which he had seen at the *Gulf of Suez* (about twenty-five miles) continued through its whole course; for he gives it as half a day's journey in a row-boat, whereas it is in reality 175 miles.

(2.) *Rivers.*—In Europe the Ister (*Danube*) was the largest river known to Herodotus: he placed its sources very much too far westward near Pyrēne, a city of the Celts beyond the Pillars of Hercules (ii. 33), and supposed it thus to intersect Europe in its whole length

Its tributaries are described at length (iv. 49), but cannot be wholly
identified : on the right bank, the Alpis and Carpis must represent
streams that rise on the Alps, either the Save and Drave, or the Salia
and Jan ; the Angrus, which flows through the Triballian plain, may
be the Ibar ; it was a tributary of the Brungus, Morava ; the Scius is
probably the Isker ; the other six which he enumerates are unimportant streams between the Isker and the sea ; Herodotus is mistaken
in describing them as large: on the left bank of the Ister, the Maris is
the Marosch, which falls not immediately into the Danube but into
the Theiss; the Tiarantus is the Schyl, the Ararus the Aluta, the Naparis
the Ardisch, the Ordessus the Berith, and the Porata the Pruth. The
lower course of the Ister is awkwardly described: it is said to discharge
itself into the Euxine in the same meridian as the Nile, opposite Sinope
(ii. 34), and near Istria (ii. 33): neither of these statements can be reconciled with the facts : Istria was sixty miles from the present mouth
of the river, ground of considerable elevation intervening; and in what
sense Herodotus supposed the Danube to be opposite Sinope is a mystery:
we may perhaps attribute his remarks to his love of symmetry. Of
the other rivers of Europe he notices—in Scythia, the Tyras, Dniestr ;
the Hypanis, Boug ; the Borysthenes, Dnieper ; the Panticapes, which
cannot be identified, flowing into the Borysthenes, and having its course
in a more easterly direction; the Hypacyris, which is described as
reaching the sea near Kalantchak, after having received a branch of the
Borysthenes, named the Gerrhus; and the Tanais, Don (iv. 51-57) :
beyond Scythia, the Ilyrgis (iv. 57) or Syrgis (iv. 123), perhaps the
Donets, a tributary of the Don ; the Oarus, perhaps the Volga, which,
however, is described as flowing into the Palus Maeotis (iv. 123) ; and
the Lycus : lastly the Phasis in Colchis (i. 2), which formed the
boundary between Europe and Asia. The Eridanus is noticed as
flowing (according to report) into the Northern Ocean : Herodotus
discredited the report (iii. 115) ; but without doubt the shores of the
Baltic were visited for the sake of procuring amber, and the name
Eridanus may still survive in the Rhodaune which flows by Dantaic.

Of the rivers in Asia Herodotus notices the Halys as rising in the
mountains of Armenia (i. 72) and flowing (in its lower course) in a
northerly direction between Syria (i. e. Cappadocia) and Paphlagonia
(i. 6): the Thermodon and the Parthenius, about which the Syrians
(Cappadocians) lived (ii. 104); the latter is probably not the Bartan,
but some other river of the same name east of the Halys: the
Euphrates, as dividing Cilicia and Armenia (v. 52), and flowing by
Babylon (i. 185); the Tigris, as flowing into the Erythraean Sea (i.
189), after having received two rivers having the same name (the two
Zabs), and the Gyndes, probably the Diala (v. 52): the story of the
division of the latter into 360 channels (i. 189), may be founded upon
the extensive hydraulic works for irrigation which were carried out on
that river : the Choaspes, Kerkhah, on the banks of which Susa stood
(i. 188 ; v. 49, 52); the river is now 1½ mile from the site of the
city, but not improbably it formerly bifurcated and sent a branch
by the town : the Aces, which is described as splitting into five
channels (iii. 117), perhaps in reference to the waters of the Herirud,
which admits of being carried through the Elburz range in the manner
described ; the Indus, to which Herodotus assigns an easterly course
(iv. 44), perhaps under the impression that the Cabul was the main
stream ; and the Corys, in Arabia, represented as a large river (iii. 9),
but probably identical with the small torrent of Corr. The Araxes

cannot be identified with any single river: the name was probably an appellative for a river, and was applied, like our *Avon*, to several streams, which Herodotus supposed to be identical: the Araxes which he describes as rising in the Matienian mountains (i. 202), is the river usually so called, flowing into the Cyrus ; the Araxes which separated the Massagetæ from Cyrus's empire is either the Oxus or the Jaxartes (i. 201) ; the Araxes which the Scythians crossed into Cimmeria is probably the *Volga* (iv. 11).

In Africa the Nile is described as of about the same length as the Danube (ii. 33) ; its sources were entirely unknown (ii. 28, 34), nor does Herodotus notice the division into the Blue and White Nile, but the easterly course described in ii. 31, and the supposed course as described in caps. 32 and 33, would apply (if at all) to the latter of the two branches. The periodical rise of the Nile is attributed to the unequal force of the sun's attraction (ii. 25). The cataracts (*Katadupi*) are noticed (ii. 17, 29): the windings of the river below the 23rd par. of lat. are transferred to the district near Elephantine (ii. 29). The division of the main stream at the head of the Delta into three large and four smaller channels is noticed (ii. 17). The other rivers noticed in Africa are the Triton (iv. 178) described as a large river flowing into Lake Tritonis: no large river, however, exists in the district referred to: the lake probably includes the *Shibk-el-loudeah* and the Lesser Syrtis, the Triton being one of the streams flowing into the lake: the Cinyps (iv. 175), near Leptis, was a mere torrent not easily identified. The Niger is probably the river which the Nasamonians reached (ii. 32).

(3.) *Mountain Chains.*—Herodotus is peculiarly defective in his notices of mountains. Caucasus is correctly described as the loftiest chain known in his day (i. 203): Atlas is described, not as a chain, but a peaked mountain, somewhere in south-eastern Algeria (iv. 184): the great range of Taurus is not noticed at all: the mountains of Armenia are generally noticed (i. 72) ; the Matienian mountains, which contained the sources both of the Araxes, *Aras*, and the Gyndes, *Diala* (i. 189, 202), answer to the Abus range and the northern part of Zagrus; the *names* alone of the European ranges were known to him, but were transferred to other objects, Pyrēne (the *Pyrenees*) to a town (ii. 33), Alpis and Carpis (the *Alps* and *Carpathians*) to rivers (iv. 49): the *Ural* range is referred to in the account of the Ægipōdes, and as forming the boundary between the Issedonians and Argippæans (iv. 23, 25), and the gold mines of the *Altai* are probably referred to in iv. 27. Of the ranges nearer Greece he notices Hæmus (iv. 49), Rhodōpe (iv. 49), Pangæum (vii. 112), and Orbēlus (v. 16) in Thrace.

§ 7. The political and topographical notices are very unequally distributed over the map of the world as Herodotus would have delineated it. In the west of Europe, we have not nearly so many notices as Hecatæus gives us. Scythia, on the other hand, is very fully described ; so also is the sea-coast of Thrace, in connexion with the Persian expeditions: the notices of spots in Greece are, of course, very numerous. In Asia, the political divisions are fully and correctly given, according to the system of satrapies established in the Persian empire: the topographical notices of the western coast of Asia Minor are numerous, as might be expected: in other

quarters they are scanty. In Africa, Egypt is fully described; so
also is the sea-coast as far as Carthage westward, and the tribes
occupying the interior at a short distance from the coast. We sub-
join a brief review of each continent.

Geography of Herodotus — Political Divisions. (1). *Europe.* —
Commencing from the W., we have notice of Iberia (Spain) (i. 163),
with the towns Tartessus (iv. 152) and Gadeira, *Cadiz* (iv. 8), and
the island Erytheia (iv. 8), either *Troondero* or an island which
has been absorbed into the mainland by the deposits of the
Guadalquiver. Beyond the Pillars of Hercules, in the extreme W.,
were the Cynesians (ii. 33), or Cynétes (iv. 49), a people but seldom
afterwards noticed. Next to these came the Celts, with the town
Pyrene and the sources of the Danube (ii. 33, iv. 40). In Gaul we
have notice of the Ligyans (Ligurians) as living above Massalia, *Marseilles* (v. 9), and of the Elisyci (vii. 165). In Italy—a name which first
appears in Herodotus (i. 24, iv. 15), as applicable, however, only to the
southern district of Magna Graecia—we have notice of Tyrrhenia
(i. 94, 163) on the western coast, the Ombrici (iv. 49), or Umbrians,
Iapygia (iv. 99) at the heel, Œnotria to the S.W., and various well-
known towns, of which we need only observe that Vela and Posidonia
(i. 167) are the same as Elea and Pæstum. Of the islands off the
coast of Italy, we hear of Sardo, *Sardinia,* which he correctly describes
as the largest in the then known world (i. 170, v. 106, vi. 2) ; Cyrnus,
Corsica, with the Phocæan colony of Alalia (i. 165, vii. 165); and
Sicily (vii. 170), in which he notices the majority of the Greek
colonies, Messana appearing under the name Zancle (vi. 22). The name
"Hellas" appears as an ethnological title applying to any country
where Hellenes were settled, and thus including spots in Italy, Asia
Minor, and Africa (i. 92, ii. 182, iii. 39, vii. 158). The country of
Greece receives no general title; but the southern peninsula is described
as Peloponnesus (viii. 73), and the land of Pelops (vii. 8). The notices
of places and peoples are very numerous, but contain little that is
peculiar; the omission of the name Epirus may be noticed. The name
Macedonia applies in Herodotus only to the district south of the
Haliacmon, the remainder being described according to the names of
the separate tribes. In Illyria, the Enêti, *Venetians* (i. 196), and the
Encheleans (v. 61, ix. 43) on the coast above Epidamnus, are noticed;
the Triballian plain is probably *Servia* (iv. 49), and the Sigynnæ
(v. 9), north of the Danube, may be placed in *Hungary* and the adjacent countries; beyond this the country was deemed uninhabitable
from the bees (probably the mosquitoes) about the river (v. 10). The
Thracians are noticed as a very powerful race, divided into a great
number of tribes, of which the Getæ (of Dacia) were the most powerful (v. 3, iv. 93); there is little of special interest in his notices of
the other tribes. The northern coast of the Ægean, together with
the towns upon it, is described at length, and in a manner that
does not call for observation; the eastern district is also noticed in describing the Thracian expedition of Darius (iv. 89-93) ; the route that
he followed is not clearly marked out; he struck into the interior to
the western side of the *Little Balkan,* passing by the sources of the
Teärus, *Simerdere,* whose 38 fountains can still be numbered, a
tributary of the Contadesdus, *Karishtiran,* and this of the Agriânes,
Erkene, which joins the Hebrus; then he met with the Artiscus, gene-

rally identified with the *Arda*, but more probably the *Tekaderch* more
to the E.; he crossed the Balkan in the neighbourhood of *Burghas*,
and thence followed the defiles that skirt the sea coast. Scythia and the
countries adjacent to it are described at considerable length in Book iv.
(17-20, 99-117); his account of the shape of the country in cap. 101 has
been variously understood, but may be most simply explained in the
following manner: Herodotus regarded the coast from the mouth of
the Danube to that of the Tanais as a straight line, the interruption
caused by the Crimea being overlooked; this line formed one side of
his quadrilateral figure, which thus touched two seas, the Euxine and
the Palus Mæotis. The position of the other sides was regulated by
this: the western boundary joined the sea at the mouth of the Ister,
which thus touched Scythia obliquely (cap. 49) without forming the
boundary throughout its course; the eastern boundary was in great
measure formed by the Tanais; and the northern was an imaginary
line drawn from the upper course of the Tanais at the distance
of 4000 stades from its mouth to the upper course of the Tyras, at a
similar distance. The inhabitants of this district were partly Scyth-
ized Greeks, but mainly Scythians; the tribes on the northern and
eastern frontiers were not Scythians, but still resembled the Scythians
in many respects. The position of the various tribes referred to
may be described thus: the Callipidæ and the Alazônes between the
Hypanis and Tyras, the former on the sea-coast, in the modern *Kher-
son*; the Agathyrsi in *Transylvania*; the agricultural Scythians between the Hypanis and the Pantiœpes, which was probably somewhat
eastward of the Borysthenes, in *Ekaterinoslav*; the Neuri in *Volhynia*
and *Lithuania*; the Androphagi ("cannibals") in *Koursk* and *Tchernigov*;
the nomad Scythians, eastward of the Panticapes in the eastern parts
of *Ekaterinoslav* and in *Kharkov*; the Royal Scythians in *Taurida* and
the steppes of the *Don Cossacks*; the Budini and Gelôni in part of
Tambov; the Melanchleni ("black-coats") between the Tanais and the
Desna in *Orlov* and *Toula*; and the Sauromâtæ on the steppe between
the *Don* and the *Volga*. The positions of the other tribes can only
be conjectured; the Thyssagêtæ, W. of the *Volga*, about *Simbirsk*;
the Iyrcæ on the opposite bank of that river; the Revolted Scythians
on the left bank of the *Kaama*; the Argippæi on the western slopes
of the *Ural* range, about the sources of the river just noticed; the
Issedônes on the opposite side of *Ural*; and the Arimaspi perhaps about
the western ranges of the *Altai*. Within the limits of Herodotus's
Europe, we must also include the Massagêtæ, who occupied the steppes
of the *Kirghiz Tartars* between the *Volga* and the *Sirr*, the latter being
probably the Araxes intended by Herodotus (i. 201). The only places
noticed in this wide district are—Olbia or Borysthenes (iv. 18), at the
mouth of the Hypanis; Prom. Hippolaus, on the opposite, *i.e.* the left
bank of that river; the Course of Achilles, the *Cosa Tendra*, and *Cosa
Djarilgatch*; Carcinitis, probably *Kalantchak* (55); and Cremni on the
northern coast of the Mæotis (20). The *Crimea* is described under the
name Taurica, the eastern part being named the "Rugged Chersonese" (99), which was separated from the rest of the country by
the slaves' dyke (20). With regard to the northern districts of Europe
Herodotus appears to have heard a rumour of the large lakes of *Ladoga*
and *Onega*, as he describes the Tanais as rising in a large lake (57).
The more western districts he supposed to be utterly unknown, and
therefore rejects the reports of the amber brought from the coasts of
the Baltic and the tin from the Cassiterides (iii. 115).

(2). *Asia.*—Asia Minor was occupied, according to Herodotus, by 15 races, arranged thus: four on the southern coast from E. to W., the Cilicians, Pamphylians, Lycians, and Caunians; four on the western coast from S. to N., the Carians, Lydians, Mysians, and Greeks; four on the shores of the Euxine, the Thracians, Mariandynians, Paphlagonians, and Cappadocians; and three on the central plateau, the Phrygians, Chalybes, and Matiēni. The divisions occupied by these tribes varied considerably from those of a later period; the Cilicians crossed the ranges of Taurus and Antitaurus, and occupied the upper valley of the Halys (i. 72), extending eastward to the Euphrates and the border of Armenia (v. 52). Pamphylia probably included the southern part of Pisidia, which is nowhere named by Herodotus, the northern portion falling to Phrygia. Lycia extended westward to the river Calbis; it was divided into three districts, Lycia Proper along the coast, occupied by the Termilæ and the Trees; Milyas, the eastern half of the inland plain, on the borders of Pisidia; and Cabalis, *Salala,* the western half, to the Calbis, which was occupied by the Cabalians and Lasonians, remnants of the old Mæonian people. The Caunians occupied the coast from the Calbis to the Ceramian bay, which was afterwards known as Perea. Caria included the western coast from the Ceramian bay to the mouth of the Mæander; Lydia thence to the bay of Elæa, while to Mysia its usual limits were assigned; the Greeks were dispersed over the western coast—the Dorians in the peninsula of Cnidus and along the northern shore of the Ceramian bay; the Ionians from the bay of Iassus to the river Hermus, with the Phocæan peninsula to the north of it; and the Æolians from Smyrna to the bay of Adramyttium, though he occasionally implies that they extended above this point over the whole of Troas (i. 151, v. 122). On the northern coast, Thrace corresponds with the later Bithynia, as far as the river Sangarius; this district was occupied by two tribes which immigrated from Europe, the Thynians and the Bithynians, the former occupying the coast, the latter the interior. The Mariandynians held the coast between the river Sangarius and Prom. Posidium (*C. Baba*), and the Paphlagonians thence to the Halys, while the Cappadocians occupied the remainder of the coast to Armenia, and the northern portion of the table-land, including a part of Galatia. In the interior the Matiēni occupied the table-land about the upper course of the Halys (the later Cappadocia), while the Phrygians held the whole of the remainder, including Lycaonia, Phrygia, and parts of Galatia, Pisidia, and even Lydia, the Catacecaumēné being considered as part of it. The Chalybes dwelt in the mountain ridges that run parallel to the Euxine in the neighbourhood of Sinope. The Hygennians (iii. 90) are not noticed by any other writer; perhaps the reading should be Hytennians, the people of Etenna in Pisidia. Proceeding eastward we come to Armenia, separated from Cilicia by the Euphrates (v. 52), and extending over a considerable portion of Mesopotamia, which is nowhere noticed by Herodotus as a separate district. Contiguous to Armenia on the E. was a district named Pactyica (iii. 93), distinct from the one noticed in iv. 44. Northward of Armenia lay Colchis, whose inhabitants, dark-complexioned and woolly-haired, were believed by Herodotus to be of Egyptian extraction (ii. 104); the mythical Æa was placed in this country (i. 2, vii. 193). South-west of Armenia, and conterminous with Cilicia, was Syria, which commenced at Posidium, *Bosyt,* about 12 miles S. of the Orontes (iii. 91), and extended along the coast of the Mediterranean sea to the borders of Egypt, with the exception of a small interval between Cadytis, *Gaza*, and a place named Ienysus, which belonged to the Arabians (iii. 5); the southern portion was termed Syria Palæstina and the northern Phœnicia (iii. 91). The towns Ascalon (i. 105), Azotus (ii. 157), Cadytis (ii. 159), and Agbatana (iii. 62), are noticed in the former; Agbatana may perhaps have reference to *Batanæa* (Bashan), the first syllable representing the Arabic article *el*; Cadytis has been taken either for Jerusalem, the "holy" city (*Kedesh*), or for Gaza; the notices favour the latter opinion; in Phœnicia, Tyre (ii. 44) and Sidon (ii. 116) are noticed. South of Syria was Arabia, which according to Herodotus touched the Mediterranean Sea between Cadytis and Ienysus, the exact position of which is unknown; it was on the coast near Egypt (iii. 5); the productions of the country are described at length (iii, 107-113); the term "Arabian" is used somewhat indefinitely by Herodotus; Sennacherib is termed king of the "Arabians," and his army the "Arabian" host (ii. 141). Contiguous to Arabia on the E. was Assyria, of which Babylonia formed a portion (iii. 92), with the towns Babylon (i. 178), Is, the modern *Hit* (i. 179), Ardericca, probably *Akkerkuf*(i. 185), Opis, probably *Khafaii*, near the confluence of the *Diala* and Tigris (i. 189), and Ampe, near the former mouth of the Tigris (vi. 20); the advance of the coast prevents any identification of its site. Eastward of Assyria came Cissa (iii. 91), the Susians of later geographers, with the town Susa (v. 52), and a second Ardericca (vi. 119), perhaps at *Kir-ab*, 35 miles N.E. of Susa. In Persia no places are noticed; but the habits of the people are described at length (i. 131-140), and the tribes, which were of three classes—(1) three dominant races, the Pasargadæ, Maraphians, and Maspians; (2) three agricultural, Panthialæans, Derusiæans, and Germanians (probably from *Carmania*); and (3) four nomad, Daans (*i.e. rural*; the Dahavites of Ezra iv. 9), the Mardians (*i.e. Aeroes*), the Dropicans, and the Sagartians (i. 125). North of Persia were the Medes, divided into six tribes (i. 101), with the town Agbatana, *Takht i-Soleiman*, in Atropatene (i. 98); westward, in the ranges of Zagrus, the Matienians; and north of these, in the upper valleys of the Cyrus, the Saspirians (i. 104, 110, iv. 37), perhaps the same as the later Iberians, with the Alarodians, about Lake Lychnitis; and the Caspians on the western shore of the Caspian Sea. The positions of many of the nations enumerated in the account of the satrapies can only be conjectured; their probable localities are as follows: the Daritæ* and Pausicæ to the S. of the Caspian Sea; the Pantimathi,* Paricanii, and Hyrcanii, at its south-eastern angle; the Sagartii in the desert of later Parthia; the Parthi more to the N., about Nisæa; the Chorasmii, Arii, Bactri, and Sogdi in their later districts; the Ægli among the Sogdi, near Alexandria Ultima; the Sacæ between the head waters of the Oxus and Iaxartes; the Dadicæ and Aparytæ,* in the southern part of Bactria; the Gandarii on the banks of the *Cabul*; the Sattagydæ* (the *Thataguah* of the Assyrian inscriptions), about the upper course of the Etymander; the Sarangæ about Aria lacus, and the Thamanæi* to the N. of the same, the Paricanii* in the northern part of *Baluchistan*, and the Æthiopians on the sea-coast; the Myci* (the *Maka* of the inscriptions) about the neck of the Persian Gulf; and the Orthocorybantes* perhaps in Media. The India of Herodotus is confined to the upper valley of the Indus, the *Punjab*; he notices a

---

* The names thus marked do not appear in any other writer.

second district named Pactyica with the town Caspatyrus (iii. 102), which has been identified with Cabul and with Cashmere, neither of which however agree with the notice of its being on the Indus (iv. 44); the Padæi (iii. 99), who were regarded even in the age of Tibullus[1] as living in the extreme east; and the Callatians (iii. 38), or Calantians (iii. 97); the abodes of these tribes are uncertain. Eastward of India (i.e. to the north of the Himalayan range), stretched the vast sandy desert (iii. 98), which reaches to the confines of China.

(3.) *Africa.*—The description of Egypt as an "acquired country, the gift of the Nile" (ii. 5), is, geologically speaking, incorrect. The level of Egypt has undoubtedly been raised by the alluvial deposit of the Nile, but the land has not gained upon the sea within historical times; the line of the coast remains very nearly what it was in the age of Herodotus. Still more incorrect is his notion of the influence of the Nile on the depth of the Mediterranean (ii. 5); the depth described (11 fathoms) is not found until within about 12 miles of the coast. His measurements are, as usual, exaggerated; the length of the coast is 300 miles, and not 400 (ii. 6), and that of the Delta from the coast to the apex about 100 instead of 173 miles (ii. 7). His description of the Nile valley is not altogether reconcilable with the facts; its breadth above the Delta is about 13 miles, instead of 23 (ii. 8); nor does the valley widen in the place described mid-way between Heliopolis and Thebes. The distance between these two places is 421 instead of 552 miles, and between Thebes and Elephantine, 124 instead of 206 (ii. 9). Herodotus divides Egypt into two portions, the Delta (ii. 15), and Upper Egypt, which, however, he refers to but once (ii. 10); he notices 18 nomes only out of the 36 usually enumerated (ii. 165 ff.); and he describes most of the great works of art, particularly the Pyramids (ii. 124-134), the Labyrinth, and Lake Mœris (ii. 148), and the canal which connected the Nile with the Red Sea (ii. 158, iv. 39). The notices of the towns are very numerous, and belong to the general geography of Egypt. To the S. of Egypt lived the Ethiopians, divided into two tribes—the Nomads (probably the "Nobatæ" are intended), and the other Ethiopians (ii. 29); the capital of the latter was Meroë; the northern capital, Napata, is not noticed. Beyond the Ethiopians were the Automöli in *Abyssinia*; on the coast of the Red Sea, the Ichthyophagi ("fish-eaters"), whom Herodotus describes as being met with at Elephantine (iii. 19), and the Macrobii near *Cape Guardafui*, in the extreme S. (iii. 17). West of the valley of the Nile, seven days' journey from Thebes, was the city Oäsis, the capital of the Great Oasis, of *Khargeh*, "the island of the blessed" (iii. 26), and more to the north the Oasis which contained the temple of Ammon, the modern *Siwah* (ii. 32). The remainder of northern Africa is divided by Herodotus into three zones, the sea-coast, the wild-beast tract, and the sandy ridge (ii. 32, iv. 181); the first of these represents *Barbary* or the states of *Morocco*, *Algeria*, *Tunis*, and *Tripoli*; the second, the hilly district, parts of which are still infested with wild beasts; and the third the *Sahara*, which he elsewhere more distinctly describes as an uninhabitable desert beyond the sandy ridge (iv. 185). The tribes occupying the sea-coast district were divided into two classes, the nomads as far as the Lesser Syrtis, and the agriculturists to the west of that point (186): their residences were as follows:—the Adyrmachidæ from the borders of Egypt to Port Plynus, probably *Port*

---

[1] "Impia nec sævis celebrans convivia mensis
       Ultima vicinus Phœbo tenet arva Padæus."—iv. 1, 144.

Bardeah (iv. 168); the Gilligammæ thence to the Isle Aphrodisias, N. of Cyrene (iv. 169); they are not elsewhere noticed; the Asbystæ, S. of Cyrene (iv. 170); Cyrenaica itself, occupied by a Greek colony, with the towns Cyrene and Barca (iv. 160, 199), and Irasa, probably *El Kubbeh*, near *Derna*, with its beautiful spring (iv. 158); westward of the Gilligammæ, the Auschisæ, touching the sea coast at Euesperides, *Benghazi*, and the Cabalians (compare modern *Cabyles*), near Tauchira, *Taukra* (iv. 171); the Nasamonians to the S. of the above-mentioned tribes, touching the sea at the eastern bend of the Syrtis Major (iv. 172); then the Psylli and the Macæ on the shores of the Syrtis (iv. 173, 175); the Gindanes, nowhere else noticed, on the coast (iv. 176); the Lotophagi in *Tripoli* (iv. 177); the Machlyans about the southern coast of the Lesser Syrtis (iv. 178); the Auseans, nowhere else noticed, on the western coast of the Syrtis (iv. 180); and westward of the Syrtis, the Maxyans (191); the Zaveciana (193), not mentioned elsewhere; and the Gyzantians (194), or Zygantians, off whose coast was the Isle Cyraunis, *Karkenna*; the names of the two latter tribes may be traced in those of the Roman provinces Byzacium and Zeugitania; Carthage fell within the territory of the Gyzantians; the place and its inhabitants are frequently referred to (i. 166, iii. 17, 19, iv. 195); but its position is not defined. Of the more easterly regions of Africa Herodotus knew but little; he rightly describes it as extending beyond the Pillars of Hercules (185), and alludes to the "dumb commerce" carried on between the natives and the Carthaginians (196). 2. In the wild-beast district he notices only the Garamantians, S. of the Nasamonians (174); if the reading is correct, of which there are doubts,* they must be regarded as distinct from the people, afterwards noticed (183). 3. In the sandy zone he places the Oases; that of the Ammonians, *Siwah*, which, however, lies 20 days' journey (400 geog. miles) from Thebes, and not 10 days as described (181); Augila, which is correctly described (172, 182); the Garamantes in *Fezzan* (183), whence was a caravan route to the Lotophagi, coinciding with the present route from *Murzouk* to *Tripoli*; the Atarantians, perhaps the *Tuariks* of the *Western Sahara* (184); and the Atlantes about the range of Atlas (184) in Western Algeria. Below the sandy region in the interior were the Ethiopian Troglodytes (183), the *Tibboos* to the S. of *Fezzan*.

§ 8. The expedition of Cyrus, so graphically described by Xenophon in his ' Anabasis,' abounds with geographical notices of the highest interest, relating to countries with which the Greeks of his day had little more than a general acquaintance. The expedition was undertaken by Cyrus in the year 401 B.C. with the object of dethroning his brother Artaxerxes, then in possession of the throne of Persia. His route may be briefly described as follows: starting from Sardis, he struck across Phrygia and Pisidia until he reached Cilicia; entering that province by the pass over Taurus, named the " Cilician Gates," and leaving it by the " Syrian and Cilician Gates' on the shore of the Bay of Issus, he followed the line of coast to Myriandrus, whence he struck inland, and, crossing the range of Amânus by

---

* Pliny and Mela give the name as Gamphasantians.

the pass of *Deilan*, entered on the plain of Syria, and reached the Euphrates in about 36° lat. He crossed the river at Thapsacus, and descended the left bank of the stream through Mesopotamia to Cunaxa, a place some distance N.W. of Babylon. Cyrus lost his life in the battle that took place there, and the command of the Greeks devolved on Clearchus, and after his death on Xenophon. Returning very nearly on their former course as far as the Median wall, they struck across the plain of Babylonia to the Tigris, and crossing that river followed up its left bank to the borders of Armenia ; their course through the high lands of Armenia cannot be traced with certainty ; they ultimately reached the boundaries of Pontus, and from the range of Theches looked down on the Euxine Sea. They gained the coast at Trapezus, and following it by land as far as Cotyōra, they took ship, and were conveyed to Heraclea in Bithynia, whence they reached home by well-known routes.[7]

§ 9. Ctesias, of Cnidus in Caria, was a contemporary of Xenophon, and was to a certain extent associated with him, if we may receive the statement of Diodorus that he was taken prisoner at the battle of Cunaxa. He passed many years in Persia as physician at the court of Artaxerxes Muemon, and on his return to his native land he recorded what he had seen in several works, of which his treatises on Persia and India were the most important. All that has survived of his writings is contained in an abridgment by Photius and a few. fragments preserved in other writers. His credulity and love of the marvellous deservedly brought him into great discredit.

§ 10. The military expeditions of Alexander the Great form an important epoch in the history of ancient geography. Not only was the extent of the country over which he himself travelled very considerable, but the conquests which he effected had a permanent influence on the future progress of discovery. The establishment of the Græco-Bactrian kingdom constituted a link between the extreme east of Asia and the west ; the subjection of the Punjab led his successors forward to the plains of Central India and to the mouth of the Ganges. A new world was, in short, opened to Greek enterprise, and physical science received a fresh impetus from the discovery of the rich and varied products of the eastern world.

§ 11. The extent of Alexander's discoveries may be briefly described as reaching to the Jaxartes in the N.E., and the Hyphasis, or most easterly river of the Punjab, in the E. Between these limits and the borders of Persia lay a wide extent of country which had hitherto been a *terra incognita* to the Greeks, comprising Parthia, Hyrcania, Aria, Margiāna, Drangiana, Arachosia, Bactriana, Sogdiana,

---

The topographical questions arising out of this narrative are referred to in a future chapter.

the countries lying along the course of the Indus and its tributaries, Gedrosia, and Carmania.

§ 12. The interest excited by these conquests is shown by the number of literary works which were issued at the time—mostly the composition of persons attached to the army of Alexander. To give some idea of the literary zeal displayed, we append the names of the authors and the titles of their works.[a] Most of the works themselves have been lost to us ; but we fortunately possess a very faithful and graphic narrative, written by Arrian in the second century after Christ, the materials of which were gathered from these contemporary sources, particularly from the works of Ptolemy and Aristobulus.

[a] 'The History of the Wars of Alexander,' by Ptolemy, son of Lagus ; 'The Journal' of Nearchus, describing his voyage down the Indus and along the Indian Ocean to the mouth of the Euphrates ; 'The Annals of Alexander,' and other works, by Onesicritus, describing the lands in the interior of Asia—Bogdiana, Bactria, &c., and India : he is the first to notice Taprobane, Ceylon ; 'History of Alexander,' by Cleitarchus, who not only describes India, but portions of the west and north of Europe ; 'Alexander's Campaigns,' 'History of Greece,' by Anaximenes of Lampsacus ; 'Alexander's Campaigns,' by Aristobulus of Cassandria in Macedonia ; 'History of Greece,' and other works, by Callisthenes of Olympus ; 'Alexander's Life,' by Hieronymus of Cardia, the author also of an historical work describing the foundation and antiquities of Rome.

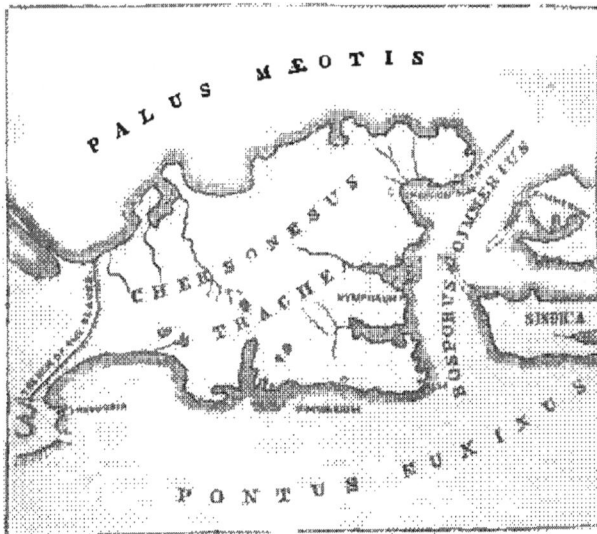

Map of the Chersonesus Trachea, according to Herodotus.

CHAPTER IV.

THE WORLD AS KNOWN TO THE GEOGRAPHERS.

§ 1. Review of the progress of discovery: India; Caspian Sea; China and the East : Western Europe: the Amber Isles : Atlantic Ocean: Phœnician influence : northern discoveries of Himilco and Pytheas: Africa, Hanno, Euthymenes, Periplus of Arrian. § 2. Geographical writers. § 3. Eratosthenes. § 4. Hipparchus. § 5. Polybius. § 6. Minor geographical writers. § 7. Strabo: Posidonius ; Geminus; Marinus. § 8. The discoveries of the Romans: Italy, Illyria, Spain, Africa, Armenia, Gaul and Britain, Asia, Mœsia, &c. § 9. Roman writers: Cæsar, Sallust, Tacitus, Livy. § 10. Mela; Pliny; Arrian; Pausanias. § 11. Ptolemy; Agathemerus, Dionysius, Periegetes, Stephanus Byzantinus. § 12. Peripli and Itineraries.

§ 1. We are now approaching the time when, under the auspices of Eratosthenes, geography was raised to the dignity of a science. Hitherto it had been treated incidentally and superficially : in future we shall see it studied for its own sake and systematically, receiving light and support from the sister sciences of mathematics and astronomy. But, before we enter upon this period, it is desirable to take a review of the position of geographical knowledge and the events which led to its gradual advance during the interval that elapsed between Alexander the Great and the commencement of the Christian era.

(1.) *India.*—The advance had thus far been directed towards the East : the conquests of Alexander may be said to have doubled the area of the world as known to the Greeks of his day. We cannot be surprised that his successors followed in the path which he had so successfully opened, and advanced the frontier of the known world from the Indus to the Ganges. This was achieved by Seleucus Nicator in his war with Sandrocottus, the records of which have been unfortunately lost: the date may have been about 300 B.C. Megasthenes was despatched on an embassy to Palibothra (probably near *Patna*), the residence of Sandrocottus, and on his return he described what he had seen in a work on India in four books. Another ambassador, named Daimachus, spent several years at the court of Allitrochades, the successor of the king just mentioned, and he also gave an account of his experience. Various expeditions were sent into the Indian Ocean. Patrocles, the admiral of Seleucus Nicator, wrote an account of the one placed under his command ; and Euhemerus, who was sent by Cassander, did the same. The latter discovered, or pretended to have discovered, a number of islands, of which he gave a fabulous account. The establishment of a regular commercial intercourse with the shores of India was due to the Egyptian Ptolemies. A navigator, named Hippalus, who had studied the character of the monsoons, ventured in a straight course from the Red Sea to the western coast of India, trading to Limyrice, Mangalore, in the south, and Barygaza, Baroach, in the north. From these points the interior of Hindostan would become more or less known.

(2.) Caspian Sea.—In the north of Asia the progress of discovery was but slow. The Caspian Sea presented in that age the same sort of problem which the "north-west passage" has been in modern days,— the question to be decided being whether any communication existed between it and the northern ocean. Herodotus, as we have already seen, entertained a correct view on this point; but among his successors the opinion gradually gained credence that such a passage did exist. Alexander the Great determined to settle the question, and would doubtless have done so had his life been extended. Patrocles, the admiral of Seleucus Nicator, was fully convinced of the existence of a north-west passage from India into the Caspian; and his ignorance is the more singular from the circumstance that he was fully aware of the commercial route down the Oxus and across the Caspian. Both Eratosthenes and Strabo held to the same false view, and the error was not rectified until the latest period of ancient geography.

(3.) China and the East.—The countries in the extreme east of Asia were to a certain extent known through the commerce carried on by way of Bactria. It is evident that the trade in silk was extensively prosecuted at this period, and that a regular overland route existed between China and the West. The Chinese themselves conveyed the goods as far as the "Stone Tower," a station probably on the western side of the Bolor range: from this point they were transported by Scythians across the passes to the head-waters of the Oxus and Jaxartes, and thence partly by those rivers to the Caspian Sea, partly by an overland route through Parthia to the west of Asia.

(4.) Western Europe.—The progress of discovery in the west was not equally satisfactory: indeed, it presents a remarkable contrast. While the Indian ocean was well known to the Greek writers, the Atlantic and even the Mediterranean Sea were still regions of uncertainty. A few instances will illustrate the extent of this ignorance. The Periplus of Scylax, compiled about 350 B.C., mentions only two towns on the coasts of Italy, Rome and Ancona, in addition to the Greek colonies. Heraclides Ponticus calls Rome a Greek city; Theopompus (about 300 B.C.) describes its position as not far from the ocean. Timæus (280 B.C.), who is supposed to have surpassed his contemporaries in the knowledge of the west, describes Sardinia as being near the ocean, and the Rhone as having an outlet into the Atlantic. Theopompus thought that the Danube discharged itself into the Adriatic as well as into the Euxine; and this is repeated by Dexippus (about 280 B.C.) with the monstrous assertion that there was a mountain near the Danube whence both seas could be seen.

(5.) The Amber Isles.—In no instance is the ignorance of the Greeks more conspicuous than in regard to the amber trade. It is well known that even before the days of Herodotus a considerable traffic in this highly-prized article was carried on from the Eridanus, which, according to the report he had received, flowed into the Northern Ocean. The amber really came from the shores of the Baltic, and was conveyed overland to the head of the Adriatic, which thus became the entrepôt for the trade. Several of the Greek geographers (Dexippus may be

instanced) consequently conceived this to be the locality where the
amber was found, and represented certain islands, which they named
Electrides Insulæ, as existing at the head of the Adriatic. Even when
this error was exploded, the true seat of the trade remained unknown. Timæus places the Amber Island (Raunonia) north of Scythia;
Strabo names it Basilia, but was equally mistaken as to its northern
latitude.

(6.) *Atlantic.*—The Atlantic Ocean was known only by dark rumours: Plato believed it to be so slimy from the effects of a sunken
island, which he names Atlantis, that no vessel could navigate it.
Aristotle believed it to be just as shallow as the Mediterranean was
deep, and so liable to dead calms that sailing was out of the question.

(7.) *Phœnician Influence.*—In all these reports and in the ignorance
which the Greeks display, we can trace the influence of the Phœnicians, who were bent on preserving a monopoly of the ocean-traffic,
and to this end propagated the most exaggerated rumours. Their
determination to keep navigation a secret is well illustrated by a story
related by Strabo, that when a Greek ship followed in the track of a
Carthaginian vessel, the captain of the latter deliberately ran his ship
upon a rock, in order to deter the Greeks from any further attempt at
discovery. Most of the rumours which they propagated appear to have
had some foundation; but the truth was distorted and the dangers
magnified. Thus the opinions both of Plato and Aristotle probably
have reference to the *Sargasso* sea in the neighbourhood of the
Azores. The Phœnicians themselves were undoubtedly acquainted with
the western shores of Europe as far as the British Isles; but, with the
exception of the expedition of Himilco, we hear little of their proceedings. In Europe, Marseilles was most distinguished for maritime
discovery, and produced several distinguished navigators, particularly
Pytheas and Euthymenes.

(8.) *Northern Expeditions.*—There is no contemporary history of Himilco's expedition; we are indebted to Pliny and to Festus Avienus, a
poet of the 4th century A.D., for the information we possess in regard
to it. Himilco is supposed to have lived about 500 B.C., and is reputed
to have been the discoverer of the British Isles. Avienus describes the
Scilly Isles under the name Œstrymnidæ, the *Land's End* as Œstrymnis,
and *Ireland* as Sacra Insula, probably confounding the native "Eri"
with the Greek Ἱερά. Many particulars connected with the voyage are
evidently misplaced: thus the sea-weed which checked his progress must
have been, as already remarked, in the *Sargasso* sea in the neighbourhood
of the Azores, and not to the north of Britain.

The report of the British Isles must have been pretty widely spread,
as Aristotle mentions both Albion and Ierne, and a notice of the latter
occurs in one of the Orphic poems, the date of which, however, is
uncertain.

Pytheas of Massilia, born about 334 B.C., explored the northern and
western ocean, and published a 'Description of the World,' and a
treatise on the Ocean, of which but a few fragments remain. He
followed the coasts of Spain and Gaul to the shores of Britain; he
explored the eastern coast, and, advancing beyond its northern extremity, reached Thule, where he found perpetual daylight. More to
the northward he was stopped by masses of sea-weed. He returned
through the German Ocean to the mouth of the Rhine, and then made
for the amber coast of the Baltic, where he met with the Teutones. A
river which he names the Tanais was the limit of his advance in this

direction. Strabo (li. p. 75) blames him for placing Britain too far
to the north, he himself having committed a greater error in the other
direction. His estimate of the length of the British coast (20,000
stados) was probably intended to include the southern as well as the
eastern coast.

(9.) *Africa.*—Lastly, we have to notice the progress of discovery in
a southerly direction. Here again the Carthaginians were in advance
of other nations. About 500 B.C., as is probable, Hanno undertook a
voyage beyond the Pillars of Hercules for the purpose of establishing
colonies on the western coast of Africa. The account of his expedition
is contained in a Greek translation of a statement which he himself drew
up in the Punic language.

The localities noticed are of doubtful position, but may probably be
identified thus: Prom. Soloeis with *C. Spartel* near *Tangier*; the river
Lixus with the *Alharytch*; the Island of Cyrne with *Arguin*; the river
Chretes with the *St. John*; the river containing crocodiles with the
Senegal; the Western Promontory with *C. Verd*; the mountain Theön
Ochêma with *Sierra Leone*, or with *Sangareah* in 10° N. lat.; the Southern
Promontory with *Sherbro' Sound*, and the island with *Plantain* Island, in
about 6° N. lat. The Gorillæ which he describes in the latter have
been with some probability explained as a species of ape still called
Toorilla. Euthymenes of Marseilles (about 300 B.C.) conducted a
similar expedition outside the Pillars of Hercules, and Eudoxus of
Cyzicus is said to have circumnavigated Africa from Gades to the Red
Sea. We have no detailed account of the eastern coast until the
Periplus of Arrian, compiled probably in the first century A.D., which
gives a survey of the coast down to Rhapta, probably the modern
Quiloa, in 10° N. lat. In the interior no great discoveries were made:
the Ptolemies prosecuted an active trade with Abyssinia from their
ports Berenice, Arsinoë, and Philotra.

§ 2. While a considerable portion of the earth's surface was laid
open by these discoveries, there was a constant supply of geographical works, emanating from authors whose subjects and places
of abode show how widely diffused the taste had become. Most
of these works have been lost, but the titles alone are instructive,
as showing the amount of materials at the command of the later
geographers.

Geographical Works.—'History of Sicily,' by Antiochus of Syracuse (about 400 B.C.), Strabo's chief authority in regard to the Greek
colonies in Italy and Sicily. A large historical work by Ephorus of
Cumæ (about 350), an authority both with Strabo and Diodorus
Siculus. 'History of Greece,' by Theopompus of Chios (about
350 B.C.), praised by Dionysius and Pliny for his knowledge of
Western Europe. 'Description of the World,' by Eudoxus of
Cnidus (about 330 B.C.), a mathematician and astronomer as well as a
practical geographer: he travelled extensively in Egypt, Asia, and
Sicily. A 'Periplus' of Scylax, compiled in the reign of Philip of
Macedon, being a description of the coasts of the Mediterranean,
Propontis, Euxine, and Palus Mæotis, commencing at the Pillars of
Hercules and terminating at the island of Cerne, off the coast of
Africa. 'Periplus' of Phileas, describing the same coasts. 'De-
scription of the World' and other works, by Dicæarchus of Messana

(about 310 B.c.), who was specially devoted to drawing maps. A 'Book of Distances,' by Timosthenes, noticed by Strabo and Pliny, giving the distances between different places about the Mediterranean coasts and elsewhere. 'Treatise on Greece and Sicily,' by Timæus of Tauromenium (B.c. 280), with much information regarding the north and west, and particularly regarding Italy and Sicily; the amber-producing island Basilia is noticed by him. 'Heracleia,' by Herodorus of Heraclea in Pontus, a contemporary of Aristotle, yielding information in regard to Spain in particular. 'Altitude of Mountains,' by Xenophon of Lampsacus, who also refers to the Amber Island under the name Baltia. Lastly, the treatises of Heraclides of Heraclea Pontica, containing various notices of interest.

§ 3. Eratosthenes (B.c. 276-196), a native of Cyrene and educated at Athens, held the post of librarian at Alexandria, under Ptolemy Euergotes. He brought mathematics and astronomy to bear on the subject of geography, and was thus enabled to construct a very much improved chart of the world, which exhibited parallels of latitude and longitude, the tropics, and the arctic circles. His equator divided the earth into two equal halves, and from it he drew eight parallels of latitude through the following points— Taprobane (*Ceylon*), Meroö, Syene, Alexandria, Rhodes, the Hellespont, the mouth of the Borysthenes, and Thule. That which passed through Rhodes (named the διάφραγμα) divided the habitable world into two halves, the northern including Europe, the southern Asia and Libya. These lines were crossed at right angles by seven parallels of longitude passing through the following points —Pillars of Hercules, Carthage, Alexandria, Thapsacus, the Caspian Gates, the mouth of the Indus, and that of the Ganges: the third of these was his main parallel. The circumference of the earth he estimated at 252,000 stades, or about 28,000 miles: the habitable world he conceived to be like a Macedonian *chlamys*, *i. e.* of an oblong shape, the proportions being 77,800 stades in length and 38,000 in breadth, but drawing to a point at each end. In his descriptive geography, he added considerably to the knowledge of the East, which Alexander's campaigns had then opened; in the West a few fresh names appear. The peculiar features in his map are—the mistaken direction given to the British Isles; the undue easterly elongation of Africa below the Straits of Bab-el-Mandeb; the connexion between the Caspian Sea and the Northern Ocean; the Oxus and Jaxartes flowing into that sea, and not into the Sea of Aral; the absence of the peninsula of Hindostan; the Ister communicating with the Adriatic sea through one of its branches; the omission of the Bay of Biscay; the compression of the northern districts of Europe and Asia; and the total omission of the eastern half of Asia and the southern half of Africa. He made numerous calculations of distances, the correctness of which varies considerably, from

Map of the World, according to Eratosthenes.

the circumstance of his having made his meridians of longitude parallel to each other. His great work on geography is unfortunately only known to us from the extracts preserved by Strabo and other writers: it consisted of three books, the first of which contained a review of the progress of geography; the second treated of mathematical, and the third of descriptive geography.

Places, &c., of interest in Eratosthenes' Geography.—In Europe, he notices the Spanish rivers Anas and Tagus, the promontory of Calpe, and the town of Tarraco; off the coast of Gaul, a group of islands, of which Uxisama represents *Ushant*; in Germany, Orkynia, or the Hercynian wood. In Africa, he is the first to notice the two tributaries of the Nile, Astápus and Astaboras; the Cinnamon coast, S. of the straits of Bab-el-Mandeb; the Nubians in the interior of Libya; the town Lixus in Mauretania; and the rock Abylax, the later Abyla, opposite Calpe. Asia he describes as intersected by a continuous range of mountains, consisting of Taurus, Paropamisus, Emodi Montes, and Imaus, which terminated in the promontory of Thinæ on the coast of the Eastern ocean. The southern portion of the continent is divided into four sections—India, Ariana, Persis, and Arabia. The river Ganges, the islands of the Persian Gulf, Tylus, Arádus, &c., the Arabian tribes Nabatæi, Scenitæ, Agræi, and Sabæi, with the towns Petra, Mariaba, and Sabáta, are first noticed by Eratosthenes.

§ 4. Hipparchus of Nicæa in Bithynia (about B.C. 160) improved on Eratosthenes' plan by calculating distances from the observations of eclipses: he thus obtained a method of determining the true position of any locality. In other respects he is famous for his bitter criticisms of his great predecessor, and for his erroneous ideas that Ceylon was the commencement of a great southern continent (which he probably supposed to be connected with Africa at its southern extremity), and that the Danube flowed into the Adriatic as well as into the Euxine Sea.

§ 5. Polybius of Megalopolis in Arcadia (B.C. 205-123) must be ranked as a practical rather than a mathematical geographer, his object, as he himself tells us (iii. 59), being to enlighten his contemporaries in regard to ` foreign lands, especially Rome and Carthage. He differed from his predecessors in subdividing the torrid zone by the equator, thus making six instead of five zones: he believed in the southern connexion of Africa and Asia: he calculated the extent of many of the lands of Europe, and the distances between certain spots. He describes at some length Iberia (Spain), Celtica (Gaul), Italy, and Sicily: but his descriptions are very vague and imperfect. The greater part of his historical work is lost to us: of the forty books in which it was written, only the first five and fragments of the others remain.

§ 6. Between the times of Polybius and Strabo many important works on geography were composed, which have wholly disappeared. The fragments of some few remain, among which we may notice the description of the world by Apollodōrus of Athens (B.C. 140): of the Red Sea by Agatharchǐdes (B.C. 120); the Geography of Artemidōrus of Ephesus (B.C. 100); the description of Europe in iambic verse by Scymnus of Chios (B.C. 100); and the Periplus of the Mediterranean by Menippus (contemporary with Augustus).

The following are authors of less importance. Polemon of Glycea in Troas (about 200 B.C.), the author of a 'Geography of the World,' and various topographical works. Mnaseas of Patara in Lycia (about 150 B.C.), the author of a 'Periplus.' Demetrius of Scepsis (about 140 B.C.), the author of a treatise on the nations engaged in the Trojan war. Nicander of Colophon (150 B.C.) and Alexander of Ephesus, authors of poetical works on geographical subjects. Cornelius Polyhistor, the author of a 'Periplus' in forty books, descriptive of various countries. Apollodorus of Artemita (about 100 B.C.), the author of works on Parthia and the Bactrian kingdom.

§ 7. Strabo, of Amasia in Pontus (B.C. 66—A.D. 24), gave the world the first systematic description of the world, in a work composed in seventeen books. He had travelled extensively, "from Armenia to Tyrrhenia (Western Italy), and from the Euxine to the borders of Ethiopia" (ii. p. 117), and he had studied deeply the writings of earlier geographers. His work was intended, not as a philosophical treatise, but as a manual of useful information for the educated classes; hence he unfortunately omits much that would have added to the intrinsic value of his work, as the exact division of the earth into climates, and the statement of the latitude and longitude of places; he is also deficient in his notices of the physical character and the natural phænomena of the countries which he describes; and he does not show the spirit of true criticism in his undue estimation of Homer and his depreciation of Herodotus. He agrees generally with the views of Eratosthenes: he holds the earth to be spherical, concentric with the outer sphere of the heaven, but immovable. He recognizes five zones, of which the northern was uninhabitable from extreme cold, and the southern from extreme heat: he divides the earth into two hemispheres at the equator; and the habitable world also into two instead of three portions. The map of the world, as Strabo describes it, is defective in many respects: the Bay of Biscay is altogether omitted, and the coast slopes off regularly from Spain towards the N.E., bringing Britain close to the latter country; the Caspian Sea is connected with the Northern Ocean by a channel; the Ganges flows eastward to China; the peninsula of Hindostan is absent; and the coast strikes northward from the eastern extremity of India, to the omission of the Malay peninsula: the southern elongation of the continent of Africa is still unknown.

Posidonius, Geminus, Marinus.—Posidonius of Apamēa in Syria (B.C. 135-51), divided the world into seven zones; he combated the

Map of the World, according to Strabo.

view of Polybius, that the heat was greatest at the equator, on the ground that the level of the land was low in that part ; and he compared the shape of the habitable world to a sling, as being broad in the centre and gradually contracting towards either extremity.

Geminus the Rhodian (about 70 B.C.), a mathematical geographer, is chiefly known for his recognition of the *antipodes*, in whose existence he believed, although he knew nothing of them; he contrasts them with the *antœci*, by whom he means the occupants of the same zone but in the southern hemisphere, and the *synœci* and *periœci* in the same zone and the same hemisphere, the former contiguous to, the latter distant from any given people.

Marinus of Tyre (A.D. 150), the true predecessor of Ptolemy, has the merit of having rectified in a great measure the errors, which appeared on the maps of Eratosthenes and others, by the multiplication of parallels of latitude and longitude. He had a much truer conception of the forms of the continents, extending Asia eastwards, Africa southwards, and describing the northern coast of Europe with tolerable correctness.

§ 8. As we are now entering on the last stage of ancient geography, we must turn aside to consider to what extent Ptolemy and the world at large were indebted to the Romans for contributions to the general stock of information on this subject. It will be found that they did but little for geography as a science; but that they nevertheless advanced practical geography by the extent of their conquests, and by the manner in which the vast dominions under their charge were systematized and consolidated. The portions of the world which were more thoroughly explored by them were Spain, Gaul, Britain, Germany, Dacia, Illyria, and the northern part of Africa. The description of the time when and the manner in which these countries were laid open, involves a brief review of the external history of Rome.

Progress of Geography among the Romans.—The progress of geography among the Romans is coincident with the progress of the Roman Empire.

(1.) *Italy.*—Their knowledge even of the peninsula of Italy was extremely limited down to a comparatively late period. The proposal of Fabius to cross the Ciminian hills in Etruria, in the year 309 B.C., was regarded by the Roman Senate as an act of unwarrantable foolhardiness. At a somewhat later period, 282 B.C., Roman ships first ventured into the Bay of Tarentum. Gradually, however, they established their sway over the whole of the peninsula by 265 B.C.

(2.) *Illyria: Gallia Cisalpina.*—The eastern coasts of the Adriatic were explored in the Illyrian war, 230 B.C., the object of which was to extirpate the hordes of pirates who had, until that time, swept the coasts of Italy and Greece. This was followed by the Gallic war, which led the Romans across the Po, 224 B.C., and opened Northern Italy to the foot of the Alps: It was not, however, until the subsequent reconquest of the Gallic tribes, B.C. 191, and the subjection of the Ligurians, who occupied the Maritime Alps and the Upper Apennines from the mouth of the Rhone to the borders of Etruria, in the year 180 B.C., that the pacification of Northern Italy was complete.

(3.) *Spain.*—The Punic wars resulted in the subjugation of the peninsula of Spain, not, however, without a long and severe contest: during the second Punic war the Roman territory extended along the eastern coast over the modern provinces of *Catalonia, Valencia, Murcia,* and *Andalucia.* The Celtiberians were pacified by Tib. Sempronius Gracchus, B.C. 179, and thus the interior districts of *Castile* and *Aragon* were added. The Lusitanians of Western Spain and Portugal were subdued, B.C. 138, by Dec. Junius Brutus, who was reputed the first man who had seen the sun sink beneath the Atlantic Ocean. Finally, the Numantine war, 143-134 B.C., established the Roman supremacy in Central Spain, and no part of the country remained unexplored except the northern coast of the Cantabri and Astures, who were not subdued until B.C. 25.

(4.) *Greece.*—It is unnecessary to follow in detail the progress of the Roman empire in the East, as no great advance in geographical discovery resulted from it. It will suffice to say that Macedonia became a Roman province in the year 167 B.C.—that Illyria was completely subdued the same year—and that Greece was reduced to a province by the fall of Corinth in the year 146. The arms of Rome had penetrated across the Hellespont, and had decided the fate of Asia Minor in the war with Antiochus, B.C. 192-190.

(5.) *Gallia Transalpina: the Getæ, Cimbrians, and Teutons.*—It was, however, in the west and north that new countries were opened to the world. Southern Gaul was invaded B.C. 125: the Salluvii saw the first Roman colony planted on their soil at Aquæ Sextiæ (*Aix*), B.C. 122: the Allobroges and the Arverni were defeated in the following year, and their territory constituted a Roman province three years later; Narbo (*Narbonne*) was founded to secure the coast-route to Spain. The same period witnessed the first movements of the northern hordes, who ultimately overran the whole of the south. The Getæ had crossed the Danube from Dacia into the districts adjacent to Macedonia; the Roman generals drove them back, and Curio advanced as far as the Danube, but feared to cross the river. The Cimbrians and Teutons penetrated into Gaul and Italy, but were annihilated by Marius, B.C. 102-1.

(6.) *Africa.*—The interior of Africa first became opened through the wars with Jugurtha, Rome having already acquired and organized into a province the coast-district which had previously belonged to Carthage: her armies now penetrated into Numidia, B.C. 109, and southwards into Gætulia in the following year. The history of Sallust contains many geographical notices connected with these campaigns.

(7.) *Armenia and the East.*—The scene of the Mithridatic wars was chiefly laid in Asia Minor: Lucullus, however, penetrated into the interior of Armenia and took Tigranocerta, B.C. 69; and his successor, Pompey, three years later, B.C. 66, advanced as far as the valleys of the Phasis and Cyrus and the southern slopes of Caucasus. After the settlement of Pontus as a Roman province, Pompey subdued Syria and Palestine, B.C. 64. At this period Egypt alone, of all the lands bordering on the Mediterranean, remained unsubdued.

(8.) *Gaul and Britain.*—The Gallic wars of Cæsar first made the Romans acquainted with the countries of Northern Europe, and his own simple narrative furnishes us with almost the whole of the information which we possess relating to Gaul itself. In his first campaign, B.C. 58, after defeating the Helvetii, he passed northwards through Vesontio, *Besançon,* to attack Ariovistus: the battle took place somewhere N. of Bâle. The following year, B.C. 57, he subdued the Belgæ, defeating the Nervii on the banks of the Sabis, *Sambre*, and taking the stronghold of the Aduatici in *South Brabant*; he also received the homage of the various tribes bordering on the Ocean, *i. e.* in *Brittany*, and cleared the valley of the Rhone, in Switzerland, of the chieftains who levied "black mail" on the merchants crossing by the *Great St. Bernard*. In the following campaign (B.C. 56) he defeated the Veneti, of Southern Brittany, who had revolted, subdued the Unelli in *Cotantin*, and the greater part of the Aquitanian tribes between the *Loire* and *Garonne* by his general Crassus, and the Morini and Menapii, the former of whom occupied the coast of the British channel from *Gesoriacum, Boulogne*, to *Cassel*. In the next year (B.C. 55) Cæsar advanced against the German tribes, Usipetes and Tenctheri, who had crossed the Rhine, and defeated them near *Coblents* ("ad confluentem Mosæ (*Moselle*) et Rheni," *Bell. Gall.* iv. 15); crossed the Rhine between *Coblents* and *Andernach*, and after staying eighteen days in Germany returned into Gaul, and made his first expedition to Britain. In B.C. 54 Cæsar first visited the Treviri on the banks of the *Moselle*, and then undertook his second expedition against Britain, in which he advanced westward as far as *Berkshire*, and northward into *Hertfordshire*. In B.C. 53 he crossed the Rhine a second time, and received the submission of the Ubii, and wasted the territory of the Eburones in *Limbourg*. In the winter of 53-52 the Carnutes, Arverni, and other tribes revolted : by a series of decisive movements he took Vellaunodunum, Genabum (*Orleans*), Noriodunum, and Avaricum (*Bourges*); he was himself subsequently defeated at Gergovia, but was again victorious, and succeeded in quelling sedition. In B.C. 51 the pacification of the Gallic tribes was completed by the renewed subjugation of the Carnutes, and the defeat of the Bellovaci who lived on the banks of the *Marne*. This brief review of Cæsar's campaigns will serve to show how wide an extent of country was now for the first time laid open to the civilization of Rome.

(9.) *Asia*.—In the East no great progress was made: the campaigns of Crassus, 53 B.C., and of Antony, 38 B.C., were conducted in countries already well known. The ignorance that prevailed as to the country far east is shown by the hope which Crassus expressed, that after the defeat of the Bactrians and Indians he should stand on the edge of the Ocean. At a somewhat later period, 24 B.C., Augustus sent out an expedition under Ælius Gallus to explore Arabia and Ethiopia; the expedition failed through the treachery of the native guides, and at no time got far from the coast of the Red Sea.

(10.) *Mœsia, &c. : Germany*.—In the north progress was still being made: the important district of Pannonia was first entered by Octavianus, B.C. 35, and its subjugation completed by Tiberius, A.D. 8, and thus the boundaries of the empire were carried to the Danube and the *Save*. Mœsia was permanently subdued by Licinius Crassus, B.C. 29. Thrace was ravaged B.C. 14, and gradually reduced to peaceable subjection, though not made a province until the reign of Vespasian. Rhætia, Vindelicia, and Noricum, yielded to the arms of Drusus and Tiberius, B.C. 15. The German tribes, from the mouth of the Rhine to the Elbe, were invaded by Drusus, B.C. 12-9, and the Roman supremacy was for a time established by Tiberius as far as the *Visurgis* (*Weser*) eastward; the Romans were thenceforward constantly engaged in wars with the German tribes, and acquired considerable information respecting them. Britain became better known

subsequently to the expedition of Aulus Plautius, A.D. 43, and more
particularly by the conquests of Agricola (A.D.78-84), whose fleet sailed
round the island. The coast of Denmark was explored as far as the
northern extremity of Jutland by an expedition sent out under the
auspices of Augustus, and the coasts of the Baltic were visited by
Nero's orders for the purpose of getting amber. Finally, the lower
course of the Danube was more thoroughly made known by the
expeditions of Trajan into Dacia, A.D. 101-106 : he connected the two
banks of the river by a bridge at *Seberin*. The empire of Rome at its
greatest extent stretched eastward to the Caspian Sea and Persian Gulf;
northward to Britain, the Rhine, the Danube as far as its junction with
the Tibiscus (*Theiss*), and thence along the northern boundary of Dacia
to the Tyras (*Dniester*); southward to the interior deserts of Africa and
Arabia; and westward to the Atlantic Ocean.

§ 9. While the Romans thus contributed most materially to the
advance of geographical knowledge by their military successes, they
did but little to forward the subject in a literary or scientific point
of view. Many of their historians, indeed, abound in incidental
notices of countries and places, in which the events they record
occurred. We have already noticed Cæsar's work, 'De Bello
Gallico,' as an authority for the geography of ancient Gaul; Sallust
(B.C. 85-35) iu his 'Jugurthine War' (cap. 17-19), gives a brief
sketch of the state of Africa at the time of his narrative; Tacitus
(A.D. 60 to about 120) describes briefly the geography of Germany
in the early chapters of his 'Germania,' and gives scattered notices
relating to that country in his other works; he has frequent notices
of localities in Britain in his 'Life of Agricola.' Livy (68 B.C.-
19 A.D.) in his great historical work had no occasion to introduce
his readers to new scenes: his deficiencies, as a geographer, are
remarkable in describing countries which he ought to have known
familiarly; his account of Hannibal's march into Etruria, of the
passage of the Alps, of the engagement on the Trasimene Lake, and
of the Caudine Forks, are instances of this.

§ 10. The only Latin writers on geography, whose works have
survived to our day, are Pomponius Mela and the older Pliny.
The former, who flourished about 40 A.D., compiled a useful
manual, entitled 'De Situ Orbis,' in three books. The most remarkable feature in his system is, that he believed in the existence
of a vast southern continent, the inhabitants of which he names
'Antichthons;' he supposed Ceylon to be the commencement of
it. In his description of the world, he takes the sea as his guide,
and surveys the coast-lands of Africa, Europe, and Asia, in order.
His information in regard to Britain was more full than that of any
previous writer : but in his account of the extreme northern, eastern,
and southern parts of the world he revives the long-exploded fables
of sphinxes and other imaginary monsters. Pliny (A.D. 23-79) in
his 'Historia Naturalis,' has devoted four out of the thirty-seven

books, of which that great work consisted, to a sketch of the known
world. His work is a compilation of incongruous materials gathered
from writers of different ages. As a systematic treatise, therefore,
it is comparatively worthless ; but the mere record of ancient names,
and the incidental notices with which his work abounds, render it
valuable to the critical reader.

Arrian, Pausanias.—These writers, though using the Greek language,
may fairly be reckoned as belonging to the age of Latin literature.
Arrian, who, as a Roman citizen, bore the praenomen of Flavius, was
born at Nicomedia towards the end of the 1st cent. A.D., and held high
office under the emperors Hadrian and Antoninus Pius. We have already [1]
referred to his 'History of the Expedition of Alexander:' in addition
to this he was the author of a work on India, and of a 'Periplus
of the Euxine Sea,' which was undertaken at the command of Hadrian,
and in which he describes the coast from Trapezus to Byzantium.
Pausanias, a Lydian by birth, and a contemporary of Arrian, settled
at Rome after a long course of travel, and there compiled a 'Description of Greece,' in 10 books, a work of the highest value for the topography, buildings, and works of art of that country, and containing
occasional notices of other lands which he had visited.

§ 11. Claudius Ptolemy completed the science of geography in a
work which served as the text-book on the subject not only in his
own age, but down to the 15th century, when the progress of
maritime discovery led to its disuse. Of the life of this great
man we know positively nothing beyond the fact that he flourished
at Alexandria about A.D. 150. His work, entitled Γεωγραφικὴ
'Υφήγησις, and drawn up in eight books, is filled with accurate
statements as to the position of places, but is scanty in descriptive materials. In his map of the world the following features
are noticeable: he extends the world southwards to $16\frac{1}{4}°$ S. lat.,
and northwards to Thule somewhere N. of the British Isles: the
eastern limit he unduly extends to a point beyond *China*, and the
western he places at the Insulæ Fortunatæ (*Canaries*). He represents the parallels of latitude in a curved form, as though drawn
from the pole as a centre, and the meridians of longitude as converging towards the poles from the equator. He extends the mass
of land too much in an easterly direction. The Baltic appears as
part of the Northern Ocean ; the Palus Mæotis is unduly elongated
towards the north : the Caspian is restored to its true character as
an inland sea, but its position is reversed, its greatest length being
given as from E. to W. : the peninsula of Hindostan is but faintly
represented, while Ceylon is magnified to four times its real size ;
the Malay peninsula appears on his map, but, instead of carrying
the line of coast northwards from that point, he brings it round the

---

[1] Cap. III. §11.

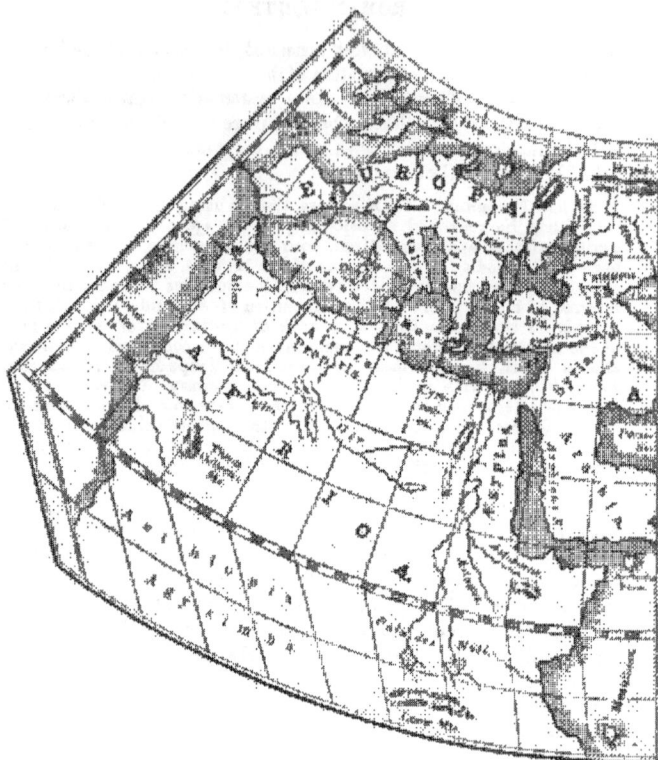

Map of the World.

Sinus Magnus (*Gulf of Siam*) in a southerly direction, and connects
it with the southern extremity of Africa, thus enclosing the Indian
Ocean; the form which he assigns to the western coast of Africa is
also very erroneous, the westerly curve being omitted, and the line
of coast brought straight down from the *Straits of Gibraltar*; the
eastern coast is correct until it reaches the point where he supposed
it to trend eastward to meet Asia. With regard to the new places
noticed, the most interesting are the river Nigir, and the Mountains
of the Moon in the interior of Africa, and a group of 1378 islands
near Ceylon, evidently the *Lacdiva* and *Maldiva* groups.

Agathemerus, Dionysius Periegetes, Stephanus of Byzantium.—Of the
writers who followed Ptolemy, we may notice Agathemerus, the author

according to Ptolemy

of an epitome of Ptolemy's work, in which, however, he renews the error with regard to the Caspian Sea, and describes Britain as reaching from the middle of Spain to the middle of Germany, and Scandia (the Scandinavian peninsula) as an island opposite the Cimbric Chersonese: Ceylon is designated by a name, Salike, which seems to be the prototype of its modern title. Dionysius Periegetes (about A.D. 300) was the author of a poetical manual of geography, in which he follows Eratosthenes and other writers of an earlier age. Lastly, Stephanus Byzantinus (about the commencement of the sixth century) compiled a Geographical Dictionary entitled 'Ethnica,' with articles on countries, peoples, and towns, natural objects being omitted : the work was epitomized by Hermolaus in Justinian's reign: of the original but a few fragments remain, but the quotations from it are numerous.

§ 12. Among the works which contributed materially to the

stock of knowledge with regard to special localities, the Peripli and
the Itineraries deserve particular notice. I. The former consisted
of descriptions of sea-coasts, with the distances of the places from
each other : in addition to those which we have already noticed in the
preceding chapters, we possess portions of six,[1] describing the following seas :—(1.) The Mediterranean ; parts relating to the African and
Asiatic coasts alone survive. (2.) The Indian Ocean; the south coasts
of Arabia, Persia, and India being described. (3.) The Euxine ; for
the most part a mere repetition of the Periplus of Arrian. (4.) The
Euxine and the Palus Mæotis, which is valuable as containing materials borrowed from Scymnus. (5.) The Euxine. (6.) The Ocean,
by Marcian, composed about the commencement of the 5th century,
describing the southern coast of Asia, and the western and northern
coasts of Europe. II. The Itineraries were of two classes, *scripta*
and *picta.* The former were exactly what our old road-books were,
giving directions as to the routes, the distances, the more important
places, and the resting-places. Of this class we have the two so-called
Itineraries of Antonine,[2] giving the routes throughout almost every
province of the Roman empire, the distances from place to place
being given in Roman miles ; and the *Itinerary of Jerusalem* or
Bourdeaux, compiled by a Christian in the 4th century, describing
the route between these two places, as well as between Heraclea and
Milan, with historical notices, and references to all localities connected with sacred events. Of the Itineraria Picta, or illustrated
guide books, only one specimen, or rather copy, has come down to
us, the *Tabula Peutingeriana,* so named after its early possessor
Conrad Peutinger. The original was probably drawn up about
A.D. 230 ; the present copy dates from the 13th century. The
whole of the Roman empire, with the exception of the western
districts, which have been accidentally lost, is depicted in this
itinerary, the roads alone being given, with the names of the provinces and places, the distances, the junction of bye-roads, and the
various objects—woods, towns, castles, &c.—by which they pass.

---

[1] The dates at which the first five of these Periplii were compiled are quite
uncertain : they belong probably to the period of the Roman emperors.

[2] This work was undoubtedly official ; but there has been much controversy
respecting its date. It was probably published in the reign of Caracalla, who also
bore the name of Antoninus ; but it received alterations after his time down to the
reign of Diocletian, subsequently to which we have no evidence of any alterations,
for the passages in which the name "Constantinopolis" occurs are probably
spurious.

Temple of the Winds.

CHAPTER V.

MATHEMATICAL AND PHYSICAL GEOGRAPHY OF THE ANCIENTS.

I. MATHEMATICAL.—§ 1. Formation of the Earth. § 2. Its position in the universe. § 3. Its shape. § 4. Its size. § 5. Tropics, zones, &c. § 6. Parallels of latitude; meridians of longitude. § 7. Climates. § 8. Maps; globes. § 9. Measures of distance. II. PHYSICAL.—§ 10. Divisions; land, sea, air: terms relating to land. § 11. Mountains. § 12. Springs. § 13. Rivers. § 14. Lakes. § 15. Seas. § 16. Winds. § 17. Temperature. § 18. Changes produced by earthquakes, volcanic eruptions, and alluvial deposits.

1. MATHEMATICAL GEOGRAPHY.

§ 1. *Formation of the Earth.*—The Greeks did not hold the same opinion as ourselves on the subject of the formation of the universe. We, on the authority of Scripture, believe that the Almighty "created the heavens and the earth," *i. e.*, not only shaped nature into the forms which it assumes, but brought matter itself into existence. They, on the other hand, held that the universe was constructed out of pre-

existent matter, though they were not agreed as to what the nature
of this matter was. Thales considered water to have been the original
element ; Anaximēnes and Archelāus air, Heraclítus fire, Xenophānes
earth, Anaximander something infinite (τὸ ἄπειρον), meaning probably a mixture of simple unchangeable elements : the opinions, however, which obtained most wide and permanent sway were, either
that the original matter consisted of a mixture of the four elements
(earth, air, fire, water), which was the creed of Empedócles, Plato,
and Aristotle ; or that it was composed of "atoms," i. e. small indi-
visible particles, combined together in various ways, which was the
creed of Leucippus, Democritus, and Epicûrus. Equally various were
the theories as to how this matter came to assume its present form :
the most distinctive views on this subject were, on the one side, that
matter was shaped by the infusion into it of an intelligent principle
(νοῦς) ; on the other hand, that it was the result either of necessity or chance. Lastly, there were various theories as to whether the
world would be destroyed, and by what means: while the Eleatic
school, who held all existing things to be eternal, and the later Stoics,
who held the world to be a development of the Deity, came to the
conclusion that it would never be destroyed, the majority of the
philosophers whom we have above noticed adopted the opposite view,
and supposed that it would be destroyed either by fire, or water, or by
their joint action, or again by a resolution of the forms of matter into
the original atoms.

§ 2. *Position.*—The position of the earth in reference to the universe was another subject on which the Greek philosophers held erroneous views. They did not suppose the earth to be a planet, but a
fixed central body, around which the celestial bodies revolved. The
heaven, in which these bodies were fixed, was of a definite form and
circumscribed within definite limits ; it was generally supposed to be
a large sphere, concentric with the earth, and hence was sometimes
compared to the shell of an egg, the earth representing the yolk
enclosed in it. Whether there were "more worlds than one" was a
question discussed in ancient as in modern days, although in a different sense; the question being, whether, beyond the system of which
the world was supposed to be the centre, other systems might not
exist in the boundless realms of space. It was never supposed that
the stars themselves were the centres of such independent systems.

§ 3. *Form.*—The form of the earth was originally held to be a disk,
i. e. a flat round surface, some difference of opinion existing as to the
precise degree of roundness, whether it was circular or oval. Thales
supposed this body to float, as a cork, on water; Anaximander held
that the earth was of a cylindrical form, suspended in mid air, and surrounded by water, air, and fire, as an onion is by its coats ; Anaximēnes
supposed it to be supported by the compressed air at its lower surface ; and Xenophanes supposed it to be firmly rooted in infinite space.
The true view of the spherical form of the earth originated with the
Pythagoreans, and obtained general belief: its exact form (an oblate
spheroid) was not known, although the revolution of the earth on its
axis, which leads to the compression of the surface at the poles,
appears to have been surmised by Aristarchus, B.C. 280. It was supposed that this spherical body was suspended in space, and kept in
its proper position either by its own equilibrium, or by the pressure
of the air on every side. While the idea that the earth moved round
the sun was confined to a few astronomers of a comparatively late

date, it was, nevertheless, supposed that the earth revolved on one
and the same axis with the universe about it.

§ 4. *Size.*—The size of the earth was variously estimated: Herodotus, who had no notion of its spherical form, probably thought its
length to be from 37,000 to 40,000 stades. When the spherical
theory was received, the size of the earth was unduly magnified;
Aristotle estimated it at 400,000 stades (about 46,200 miles), and
Archimédes at 300,000 (about 34,700 miles), its real circumference
being about 25,000 miles. Eratosthénes calculated it by an ingenious
method[1] at 250,000 stades, or about 28,800 miles; it was afterwards
diminished by Posidonius to 240,000, and again to 180,000 stades.
The latter of these estimates was adopted by Marinus and Ptolemy,
and partly by Agathemerus, though the statements of this writer are
not consistent. The diameter of the earth was estimated at one-third
of the circumference.

§ 5. *Tropics, Zones, &c.*—The mathematical divisions of the earth's
surface were founded on astronomical observation, and were the counterpart of the divisions previously established in the celestial charts.
The most important of these lines were the "equator" (ἰσημερινός,
æquator), which was originally divided by Eudoxus into 60 degrees,
and afterwards subdivided into 360; the summer and winter "tropics"
(θερινὸς, χειμερινὸς τροπικός); and the "arctic" and "antarctic"
circles (ἀρκτικός, ἀνταρκτικὸς κύκλος). The tropics were placed 24
degrees N. and S. of the equator, and the arctic and antarctic circles
36 degrees from the poles, leaving thus an interval of 30 degrees
between these and the tropics. In modern geography the tropical
circles are placed at 23½ degrees from the equator, and the polar circles
at a similar distance from the poles. These lines formed the basis of
the division into "zones" (ζῶναι, *zonæ, plagæ*), of which five were
generally enumerated,[2] viz. the "torrid" (διακεκαυμένη, *torrida*),
two "temperate" (εὔκρατοι, *temperatæ*), and two "frigid," (κατε-
ψυγμέναι, *frigidæ*). Sometimes the torrid zone was subdivided into
two or even three parts.

§ 6. *Latitude and Longitude.*—Parallels of latitude and meridians of
longitude were drawn in the first instance not at equal intervals, but
through certain well known points. Ptolemy was the first to adopt
equal intervals, and further improved the system by drawing the meridians not in parallel but converging lines, and by adding parallels of
latitude south of the equator. To him we owe the introduction of
the terms "latitude" (πλάτος) and "longitude" (μῆκος), to describe
the position of any given place in relation to the *breadth* and *length* of
the world respectively.

§ 7. *Climates.*—The term "climates" (κλίματα) has a totally dif-

---

[1] He ascertained by astronomical observation that the arc between Alexandria
and Syene was 1-50th part of the earth's circumference: he then measured the
distance between these two places, and found it to be 5000 stades; whence the total
circumference would be 250,000. The mode of calculation was correct, but his
observations were not sufficiently nice to ensure an accurate result.

† " Quinque tenent cœlum zonæ: quarum una corusco
    Semper Sole rubens, et torrida semper ab igni;
    Quam circum extremæ dextra lævaque trahuntur
    Cæruleæ glacie concretæ atque imbribus atris:
    Has inter mediamque, duæ mortalibus ægris
    Munere concessæ Divûm."—Virg. *Georg.* l. 233-238.

ferent sense in ancient and modern geography. In the former it
signified parallel belts on the earth's surface, representing equal lengths
of day; in other words, an equal distance from the equator. The
necessity of such a division is entirely superseded by the subdivision
of the earth's surface into regularly marked parallels of latitude, for
each degree represents in reality a "climate" or equal length of day
to all places through which it passes.

§ 8. *Maps.*—The invention of maps for geographical purposes is
attributed to Anaximander, but it is not improbable that maps of
separate countries were used before he drew one of the whole world.
The art of drawing a map is described by the term γεωγραφία in its
special sense; the map itself being called πίναξ, or more fully πίναξ
γεωγραφικός, and occasionally περίοδος τῆς γῆς. Herodotus refers to
Hecatæus's map (iv. 36), and also describes Aristagoras as producing
a bronze tablet on which all the seas and rivers of the earth were depicted (v. 49). The maps of the Greek geographers, Eratosthenes, Strabo,
and Ptolemy, have been reproduced from the descriptions which they
have left, and are given in the preceding chapters: Ptolemy adopted
a more scientific style of projection than his predecessors. The invention of globes is attributed to Crates of Mallus in the second century
B.C. The Romans used maps both for political and educational purposes. Among the important measures which Julius Cæsar originated
may be noticed the survey of the whole Roman empire, with maps of
the several provinces, which was ultimately carried out by Augustus.
Varro (*De Re Rust.*, i. 2, § 4) refers to a map of Italy delineated on a
wall; and at a later date Propertius (iv. 3, 37) complains—

"Cogor et e tabula pictos ediscere mundos."

§ 9. *Measures of Length.*—The methods of ascertaining distances are an
important subject in connexion with ancient geography. The standard
measure among the Greeks was the stadium (στάδιον), among the
Romans the mile (*milliarium*), among the Persians the parasang (παρα-
σάγγης), and among the Egyptians the schœnus (σχοῖνος). The stadium contained 606 feet 9 inches English: about 8⅓ stades, therefore,
equal a mile. In considering the distances as given in stades by Herodotus and other writers, it is important to remember that these were
not *measured*, but simply calculated. Thus a day's journey by land =
200 or 180 stades, or, in the case of an army, 150; the rate of a sailing
ship = 700 stades by day, and 600 by night (Herod. iv. 86, 101, v. 53).
The result of this mode of calculation was that distances were generally over-estimated. The Roman mile = 1618 English yards, and is
thus less than an English mile by 142 yards. The parasang was commonly estimated at 30 stades, but, like the modern *farsakh* of Persia, it
indicated rather the time spent in traversing a certain district, than
the space traversed. The schœnus was estimated as equal to two
parasangs, or 60 stades. The admixture of the idea of time and space
in the same word may be illustrated by the use of the German word
stunde, which in one sense means an "hour," in another a "league."

II. PHYSICAL GEOGRAPHY.

§ 10. The physical geography of the ancients is most conveniently
treated by considering separately the three constituent elements of
land, water, and air.

Land.—The terms descriptive of the various forms which land assumes are as follows—continent (ἤπειρος, *terra continens*), islands (νῆσοι, *insulæ*), isthmuses (ἰσθμοί, *isthmi*), tongues of land (ταινίαι, *linguæ*), peninsulas (χερσόνησοι, *peninsulæ*), plains (πεδία, *campi, planities*), mountains (ὄρη, *montes*), valleys (αὐλῶνες, ἄγκη, *κοιλάδες, valles, convalles*), gorges or ravines (νάπαι, φάραγγες, χαράδραι, *fauces*), and passes (πύλαι, *portæ*).

§ 11. *Mountains.*—These were either isolated hills or chains (ὄρη συνεχῆ, *montes continui*). The heights of mountains were calculated by the Alexandrian geographers, but in a very imperfect way: the loftiest mountains in each continent were reputed to be, in Asia—Caucasus, Paropamisus, and Imāus; in Africa—Atlas and Theôn Ochêma; and in Europe—the Alps and the Sarmatian mountains, and next to them the Pyrenees. The protrusion of mountain-chains into the sea formed promontories (ἀκρωτήρια, *promontoria*). Certain mountains were known as volcanoes, the most famous being Mosychlus in Lemnos, Ætna, Vesuvius, the Æolian and Liparian Isles, Chimæra in Asia Minor, and Theôn Ochêma in Africa: they were reputed to be the residence of Vulcan (whence their title), and the eruptions to be the consequences of the struggles of giants and Titans. Caves (σπήλαια, ἄντρα, *antra, speluncæ*) attracted much notice among the ancients: the largest known were the Corycian caves of Parnassus and Cilicia, and the *Grotto of Posilippo* near Naples: some of those whence mephitic vapours arose, as at Delphi, were the seats of famous oracles; others of a similar nature were reputed the entrances to the nether world (ἀχερόντια, *Plutonia, ostia Ditis*).

§ 12. *Springs* may be noticed in connexion with mountains. Homer supposed all the springs to be united by subterraneous channels with the river of Ocean: later philosophers held views hardly more consonant with truth on this subject: Aristotle, for instance, supposed that rain was formed inside the earth, just as it is outside it, by the compression of the internal air; Seneca went farther, and held that the earth itself turned into water, which, through the pressure of the air, circulated about the earth, as the blood does in the human body. Water was held to be in itself tasteless, inodorous, colourless, and imponderous, the opposite qualities being attributed wholly to the admixture of earthy particles. It was supposed to be cool in proportion to the depth of its source, the phenomenon of hot springs being ascribed to the presence of volcanic action. Mineral springs were resorted to for medicinal purposes; among the most famous may be reckoned those at Baiæ in Campania, the springs at *Aix* (which is merely a corruption of *Aqua*) in France and Prussia, and many others: there is abundant proof that *Bath* (Aquæ Solis) was the fashionable resort of the wealthy Romans in Britain. The various qualities of springs were carefully noted, as the petrifying springs at Tibur, and on the island of Cos; the pitch-springs of Zacynthus; the oily springs of Nyssa, &c. No spring, however, has attained such celebrity as Castalia at Delphi, in which all visitors were ordered to purify themselves, Apollo[1] himself not disdaining to do so.

§ 13. *Water* may be described according to the two principal aspects which it presents, as either *running* in the form of rivers, brooks, &c., or *standing* in the form of lakes, seas, marshes.

Rivers.—Any phenomena connected with rivers were carefully noted;

---

[1] "Qui rore puro Castaliæ lavit
Crines solutos."--Hor. *Carm.* III. 4, 61.

for instance, streams which disappeared for a space beneath the earth,
as the Eulæus, Orontes, Mæander, Acheloüs, and others—a circumstance
on which was founded the poetical idea of the union of distant streams,
as of the Alphéus with the fountain of Arethusa in Sicily, the Mæander
with the Asopus in Sicyonia. Briny streams, such as the Phasis and the
Sicilian Himéra were reputed to be—petrifying streams, as the Silarus—
and again those which brought down gold-dust, as the Pactólus and
the Tagus, were also noticed; as also the not unusual occurrence of
confluent rivers keeping their waters distinct for some distance from
their junction; the Titaresius, for instance, refusing to mingle with the
Peneus, and the Hypanis with the Borysthénes: and lastly, rapids and
cataracts (καταρράκται, dejectus aquæ), as in the Nile, Euphrates, Danube,
and other rivers.

§ 14. *Lakes* not unfrequently possess peculiarities, which were noticed
by classical writers. The vapours of Avernus, the medicinal qualities
of the Lake Velinus, the salt lakes of Phrygia, the asphalt of the Dead
Sea, the naphtha of the Lake of Samosata, the natron-lakes Thonitis
and Ascanius, may be cited as instances. Marshes were held to be
prejudicial to the health; the Pontine Marshes are a well known
example.

§ 15. *The Sea.*—Various opinions were broached as to the origin of
the sea: Anaximander held it to be the surplus moisture which the
fire had failed to consume; Empedocles thought it to be the sweat of
the earth; and so forth. The original view held by Homer was that
the ocean flowed round the earth in a circle, and fed the various seas
and rivers, the Mediterranean being connected with it at its western
extremity. The progress of discovery exploded this view, and the
ocean was recognised to be not a river, but a vast sea covering a large
portion of the earth's surface. The general view held was that all
the different seas (Atlantic, Indian, &c.) were connected together,
though many took the opposite view. The Northern Ocean was invested with many terrors in the eyes of the ancients: navigators
reported the existence of constant darkness, calms, impenetrable masses
of sea-weed; each of these reports had a certain amount of foundation,
though the truth was distorted; the fact of its being frozen was first
discovered in Strabo's time. As to the depth of the sea, the ocean was
held to be unfathomable, but the Mediterranean had been sounded in
various spots. The temperature of the sea was observed to be more
equable than that of the land, being cooler in summer and warmer in
winter. From the circumstance of its not freezing, it was supposed
to have a higher temperature generally than rivers, which was attributed to its constant motion. The specific gravity of sea-water was
observed to exceed that of fresh. The saltness of sea-water was attributed by Anaximander to the constant evaporation of the water, by
which a large residuum of salt and other bitter particles was left
behind. Empedocles, following up his opinion of the earth's sweat,
was at no loss to account for the saltness on the ground that sweat is
salt; while others attributed it to large deposits of salt. The colour
of the sea, when quiet, is expressed in Homer by the term μέλας;
and, when in motion, by πορφύρεοι, οἶνοψ, ἰοειδής, ἠεροειδής, γλαυκός,
πολιός; the Romans described it by the terms cæruleus, viridis, and
purpureus. The constant motion of the sea was usually attributed to
the influence of wind: Strabo and some others, however, conceived
that there was some internal agency at work even during calm weather,
analogous to the heaving of the chest in taking breath. Waves were in

all cases the result of wind: the Greeks believed the *third* wave (τρι κυμία), the Romans the *tenth* to be the strongest and most dangerous. The ebb and flow of the tide (πλημμυρὶς καὶ ἄμπωσις, *aestus et recessus*) was explained in various fanciful ways. The Stoics literally believed that ocean lived, and explained the rise and fall of the water as the panting of the giant's breath: Aristotle supposed it to arise from the pressure of the exhalations raised by the sun acting upon the water and driving it forward: Seleucus attributed it to the influence of the moon, whose motion he supposed to be in a contrary direction to that of the earth, and so to cause conflicting currents of air, which, alternately gaining the supremacy, made the water flow backwards and forwards. The Phœnicians were well acquainted with the ordinary phenomena of the tides, but the early Greeks could have known but little of the matter, as the tides in the Mediterranean are hardly perceptible. The currents in the sea were supposed to originate in the waters seeking a lower level. Whirlpools were caused either by the sudden depressions in the bed of the ocean, by the presence of reefs, or by antagonistic currents of wind. The level of the sea was by some supposed to be everywhere equal; by others a contrary opinion was held, and, in proof of their opinion, it was alleged that the Red Sea was higher than the Mediterranean, an opinion which has been reproduced in modern times, and has only lately been falsified.

§ 16. *Air.*—Of the various phenomena connected with the air, those which have the most direct bearing upon geography are winds and temperature.

Winds (ἄνεμοι, *venti*).—Various terms were used to describe them, according to their violence or their source: thus we hear of landbreezes (ἀπόγειοι, *obogei venti*), sea-breezes (τρόπαιοι, *altani venti*), storms (χειμῶνες, θύελλαι, *procellæ*), hurricanes (ἐννεφίαι, *tempestates fœdæ*), and whirlwinds (τυφῶνες, *turbines*). The most prevalent and important winds proceeded from the four quarters of the heavens, N., S., E., and W., and were termed the cardinal winds (γενιώτατοι, *cardinales*). Their names were (1) *Notus* (Νότος) or *Auster*, the south wind, which prevailed in the early part of the summer, and from the end of the dog-days to the beginning of harvest—a violent, capricious, and unhealthy wind, generally accompanied with wet; (2) *Borras* (Βορέας) or *Septemtrio*, from the north, a clear, cold, but healthy wind; (3) *Zephĝrus* (Ζέφυρος) or *Favonius*, the west wind, which set in with early spring, and was particularly prevalent at the time of the summer solstice; in Greece it brought rain and stormy weather, in Italy it was a mild breeze; (4) *Eurus* (Εὖρος) or *Vulturnus*, the east wind, which prevailed about the winter solstice, and was known for its dry character. We need not assume that these winds proceeded from the exact cardinal points of the compass, but rather that they represent generally the four quarters of the heavens, just as the terms are used by ourselves in ordinary conversation. In addition to these cardinal winds, we meet with others in later writers—viz. (5) *Solānus*, Ἀπηλιώτης, which was substituted for Eurus, to specify *due east* wind; (6) *Aquilo*, Καικίας, from the N.E., very constant at the time of the vernal equinox, bright and cold; (7) *Africus*, Λίψ, from the S.W., moist and violent, prevalent about the autumnal equinox; (8) *Corus*, *Caurus*, Ἀργέστης, Ἰάπυξ, from the N.W., cool and dry. The eight already specified were marked on the Horologium of Andronicus Cyrrhestes, commonly called the Temple of the Winds, at Athens. We may further notice the winds named Μέσης, N.N.E.; Φοινικίας,

S.S.E.; Θραςίας, N.N.W.; and Λιβόνοτοι or Λιβοφοίνιξ, S.S.W.
The Etesian winds blew regularly from the N.W. in the interval
between the spring solstice and the rise of Sirius. It was a favourite
idea of the poets that the winds had their several fixed abodes, whence
they issued; hence it was inferred that the lands beyond these abodes
were not subject to the influence of the winds, and that thus beyond
the abode of Boreas, which was supposed to be in one of the northern
mountain-ranges, there might be a country enjoying a superlatively
mild climate, where the Hyper-boreans passed their tranquil life.

§ 17. *Temperature.*—The temperature of any spot was held to be
mainly dependent upon its proximity to the sun's course, and to be
modified by the presence either of mountain-chains or of bodies of water.
Great mistakes arose, however, as to the degree of proximity to the
sun which certain spots attained. Homer supposed the E. and W. to
be the hottest, as the sun seemed to touch these spots in his rising
and setting, and there accordingly he placed the Ethiopians. This
was found to be an error; but it was succeeded by one hardly less
egregious—that the south pole was the hottest point in the world, as
being opposite to the north, which was known to be cold. The effect
of a chain of mountains shielding a district from the cold north wind,
could not escape notice: the altitude of any spot above the level of
the sea was also known to have its influence.

§ 18. The ancient geographers were observant of the changes that
took place on the surface of the earth. These were attributable to
three causes, earthquakes, volcanic eruptions, and alluvial deposits.

(1.) *Earthquakes.*—The cause of these convulsions was originally
referred to the action of water, whence Neptune was styled the
"earth-shaker" ('Εννοσίγαιος or 'Ενοσίχθων): this was the opinion of
the Ionian philosophers, though they were not agreed as to what was
the disturbing cause—whether heat or air coming in contact with the
water. Aristotle explained earthquakes as arising from the escape of
vapours generated within the earth's bowels. Others, again, attributed
them to the action of subterraneous fire in various ways. Great effects
were assigned to earthquakes, as the separation of Sicily from Italy,
and of Euboea from Boeotia, and the formation of the Vale of Tempe.

(2.) *Volcanic Eruptions.*—The activity of volcanic agency at particular
spots was supposed to arise either from a superabundance of fire in
those spots or from a thinness in the crust of the earth. The ordinary
phenomena attendant on an eruption were closely observed, and one
famous philosopher (Pliny) sacrificed his life to his scientific zeal in
reference to this question. The most striking effect of volcanic action
was the elevation or depression of masses of land, which led occasionally to the sudden appearance of new islands.

(3.) *Alluvial Deposit.*—Great changes were observed to take place on
the sea-coast through the amount of mud and sand brought down by
rivers. Herodotus supposed, though erroneously, that the existence
of Egypt was wholly attributable to the deposits of the Nile: he also
remarks the advance of the coast of Acarnania, by which some of the
Echinades were absorbed into the mainland, and again the changes
that took place in the coast of Asia Minor at the mouth of the
Maeander. The plain of Cilicia is due to the alluvial deposits of the
Sarus and Pyramus. Many districts have been entirely altered since
classical times by the same cause—particularly the pass of Thermo-
pylae, the western coast of the Adriatic, the coast line of the Persian
Gulf, and of the western coast of Asia Minor.

The Mesopotamian Plain.

BOOK II.

ASIA.

—o—

CHAPTER VI.

THE CONTINENT OF ASIA.

§ 1. Boundaries. Name. § 2. Oceans. § 3. Mountains. § 4. Plateaus
and plains. § 5. Rivers. § 6. Climate. § 7. Productions. § 8.
Commerce and commercial routes. § 9., Ethnography.

§ 1. THE continent of Asia was but partially known to the geographers of Greece and Rome. Their acquaintance with it was limited to the western and southern quarters; the north and east were a *terra incognita*. The true boundaries of the continent in the latter directions were consequently unknown : it was surmised, indeed, that the world was bounded on all sides by water, and consequently that Asia, as the most easterly of the three continents, was washed on the E. by an ocean, to which some few geographers assigned the name of Oceanus Eōns, the " Eastern Ocean :" the true position of this ocean was, however, entirely unknown. We have seen that both Eratosthenes and Strabo conceived it to commence on the eastern coast of *Hindostan*, the island of Taprobāne, or *Ceylon*,

being at the extreme S.E. of the world : we have also seen that
Ptolemy, whose information as to the east was more extensive, carried
on the Indian Ocean beyond that point to the coast of *Cochin China*,
but that he supposed the coast then to trend towards the S. instead
of the N., and consequently ignored the existence of an eastern
ocean altogether. We must therefore regard the opinions of those
who notice the ocean as the eastern boundary of Asia as a *surmise*,
rather than an ascertained fact : the boundary was really unknown.
The same observation applies to the northern boundary : the belt
of sandy steppes, which stretches across the continent from the
eastern shores of the Euxine to the confines of *China*, formed an
impassable barrier to the progress of discovery in that direction,
and may be regarded as really the northern boundary of Asia as
known to the ancients. It was, indeed, surmised that an ocean
existed in this direction also : but this surmise seems partly to have
been grounded on the assumption, that so large a sea as the Caspian
must have had a connexion with the ocean, and that as no outlet
existed towards the S., E., or W., it must have been towards the
N. ; accordingly, the geographers who recognized the existence of
such an ocean (as Strabo and Eratosthenes did), placed it a very
short distance N. of the Caspian Sea. Ptolemy, who knew that this
was incorrect, but was unable to supply the true boundary, leaves
out the ocean altogether. The southern boundary was the well-
known Oceanus Indicus. The western boundary was formed partly
by land, and partly by water : the Red Sea, the Mediterranean, the
Euxine, and the chain of intermediate seas connecting the two
latter, have supplied, in all ages, fixed limits, but more to the N.
the limit has varied considerably. The usually recognized boundary was formed by the Palus Mæōtis, *Sea of Azov*, and the Tanais,
Don : it has since been carried eastwards to the Caspian and the
river *Ural*.

Name.—The origin of the name "Asia" is uncertain : most probably
it comes from a Semitic root, and means the "Land of the East," as
distinct from Europe, "the Land of the West." Greek mythology
referred it to Asia, the daughter of Oceanus and Tethys, and the wife
of Prometheus ; or to a hero named Asius. The name first occurs in
Homer, as applicable to the marsh about the Cayster,[1] and was thence
extended over the whole continent. The Romans applied it in a
restricted sense to their province in the W. of Asia Minor.

§ 2. The physical features of the continent first demand our
attention—its oceans, seas, mountains, plains, and rivers : these we
shall describe in the order named, noticing at present only such as
hold an important position on the continent, and reserving the
others to a future occasion.

---

[1] Ἀσίῳ ἐν λειμῶνι, Καϋστρίου ἀμφὶ ῥέεθρα.—*Il.* II. 461.

(1.) The only ocean which requires notice is that which washes the southern coast of Asia, and which was generally named the "Southern Ocean" (*νοτία θάλασσα, μεσήμβρινος ὠκεανός*), occasionally the "Red Sea" (*ἐρυθρὰ θάλασσα*, Herod. ii. 102), and after improved knowledge of India, **Oceanus Indicus**. The coast line of this ocean is regular as compared with that of Europe, and irregular as compared with that of Africa, being, on the one hand, deficient in those numerous inlets and estuaries which characterize the former, and, on the other hand, devoid of that general uniformity which characterizes the latter. The sinuosities, in short, are on a large scale: two extensive bays penetrate deeply into the interior, viz. the **Sinus Gangeticus**, *Bay of Bengal*, and the **Mare Erythraeum**, *Arabian Sea*, divided from each other by the peninsula of *Hindostan*, and bounded, the former on the E. by the Aurea Chersonesus, *Malay Peninsula*, the latter on the W. by the Arabian peninsula. From the latter of these seas, two gulfs penetrate yet more deeply into the interior, viz. the **Persicus Sinus**, *Persian Gulf*, and the **Arabicus Sinus**, *Red Sea*. The Persian gulf occupies the southern portion of the Mesopotamian plain, and, spreading out into a broad sheet, divides the plateau of *Iran* from that of Arabia: the Red Sea seems to occupy a deep narrow valley between the plateaus of Arabia and Africa. The Red Sea is divided at its northern extremity by the mountains of the Sinaitic peninsula into two arms, the western named **Sinus Heroopolites**, *Gulf of Suez*, and the eastern **Sinus Elanites**, *Gulf of Akaba*, after the towns of Heroopolis and Ælana, which stood respectively at the head of each. In addition to these, we may notice the less important seas in the Gangeticus Sin., named **Sabaricus Sin.**, *Gulf of Martaban*, and **Perimulicus Sin.**, *Straits of Malacca*; as also **Magnus Sin.**, *Gulf of Siam*, and **Sinarum Sin.**, *Gulf of Tonquin*, which were regarded as portions of the Indian Ocean.

(2.) The Mediterranean Sea, **Mare Internum** or **Magnum**, which bounds Asia on the W., belongs to the three continents, but more especially to Europe, under which it is described at length.

The parts adjacent to Asia received the following special designations—**Mare Phoenicium**, along the coast of Phoenicia; **M. Cilicium**, between Cilicia and Cyprus; **M. Icarium**, so named after the Island of Icaria, along the S.W. coast of Asia Minor; and **M. Ægæum**, the extensive basin which separates Asia Minor from Greece.

(3.) The **Pontus Euxinus**, *Black Sea*, which in ancient geography belongs rather to Asia than to Europe, was regarded by the ancients as a part of the Mare Internum, being connected with it by a chain of intermediate seas—the **Hellespontus**, *Dardanelles*, on the side of the Ægæan, a strait about a mile in breadth, and probably regarded by

Homer, who gives it the epithet "broad,"[2] as a river ; the **Bosporus
Thracius**, *Straits of Constantinople*, on the side of the Euxine, about
seventeen miles long, and at one point only 600 yards across ; and
the **Propontis**, *Sea of Marmora*, between the two, an extensive sheet
of water, about 120 miles from the entrance of one channel to that
of the other. The shape of the Euxine was compared to that of a
Scythian bow, the north coast from the Bosporus to the Phasis
representing the bow itself, and the southern coast the string.

Names.—The Black Sea is said to have been originally named Axénus,
"inhospitable,"[3] in consequence of the violent storms that sweep over
it ; this name was changed to "Euxinus," when it became better
known to the Greek navigators. The Hellespont was reputed to be
so named from the legend, that Helle, the daughter of Phrixus, was
drowned in attempting to cross[4] its waters : and the Bosporus, from
the legend of Io having crossed it in the form of a heifer. The Propontis owes its name to its relative position, as the "sea before the
Pontus."[5]

(4.) The **Palus Mæotis**,[6] *Sea of Azov*, is a considerable sheet of
water to the N.E. of the Euxine, connected with it by the
Bosporus[7] Cimmerius, *Straits of Yeni-Kalé* ; it is described by the
ancients as of greater extent than it at present has.

(5.) The **Mare Caspium** or **Hyrcanum**, *Caspian Sea*, was but
partially known to the ancients, no vessels being built on its shores,
and the impervious character of the country which surrounded it,
preventing exploration by land. We have already had occasion to
notice the erroneous views entertained by them in regard to this
sea : it was, after all, but natural to suppose that so large a body
of water was connected with the ocean. The Caspian is consider-

---

[2] ἐπὶ πλατεῖ Ἑλλησπόντῳ.—Il. vii. 86.
[3] " Frigida me cohibent Euxini littora Ponti.
      Dictus ab antiquis Axenus ille fuit :
      Nam neque jactantur moderatis æquora ventis,
      Nec placidos portus hospita navis adit."—Ov. *Trist.* iv. 4, 55.
[4] Hence it is termed Ἕλλης πορθμός.—Æsch. *Pers.* 745.
[5] Compare Ovid's expression :
      " Quaque tenent Ponti Byzantia littora fauces."—*Trist.* i. 10, 31.
[6] It was regarded by Æschylus as at the very extremity of the world :
      Καὶ Σκύθην ὅμιλος οἱ γῆς
      Ἔσχατον τόπον ἀμφὶ
      Μαιῶτιν ἔχουσι λίμναν.—*Prom.* 416.
[7] The name was referred to in the legend of Io's wanderings by Æschylus :
      Ἰσθμὸν δ' ἐπ' αὐταῖς στενοπόραις λίμνης πύλαις
      Κιμμερικὸν ἥξεις, ὃν θρασυσπλάγχνως σὲ χρὴ
      Λιποῦσαν αὐλῶν' ἐκπερᾶν Μαιωτικόν·
      Ἔσται δὲ θνητοῖς εἰσαεὶ λόγος μέγας
      Τῆς σῆς πορείας, Βόσπορος δ' ἐπώνυμος
      Κεκλήσεται·—*Prom.* 731-736.

ably more shallow now than formerly, the sea being constantly
reduced by the alluvial deposit of the rivers. Its level is some
eighty feet below that of the Euxine, so that its waters could
never have been drained off into the latter, as some of the ancients
imagined. The steppe E. of the Caspian had altered considerably
within historical times, inasmuch as the Oxus at one time discharged itself into the Caspian.

(6.) Whether the Oxiana Palus of the ancients represents the
Sea of Aral, is doubtful : Ptolemy describes the former as a small sea,
and not as the recipient of the Oxus and Jaxartes : the first undoubted
reference to the latter occurs in Ammianus Marcellinus in the 4th
century A.D. Its waters are also continually decreasing ; its level is
about 110 feet higher than the Caspian Sea.

§ 3. The mountain-system of Asia is regular and clearly defined.
(1.) A series of mountain-ranges traverses the whole length of the
continent, from the shores of the Ægean Sea to those of the Eastern
ocean, dividing the continent into two unequal portions—the
northern, which is by far the most extensive, including the vast
regions N. of the Euxine and Caspian Seas ; and the southern
embracing the peninsulas and plateaus that lie adjacent to the
Indian Ocean. The main links in this great central chain consist of
the ranges of Taurus, Abus, *Ararat*, Caspius Mons, Paropamisus,
Hindú Kúsh, Emodi Montes, *Himalaya*, and Semanthini
Montes. (2.) From this central range depend subordinate,
though still important systems, some of which exhibit great regularity. Thus in Central Asia there are three parallel ranges, now
named *Kuen-lun*, *Thian-shan*, and *Altai*, which are connected
with the more southerly range of *Himalaya* by a series of transverse
ranges, of which *Bolor* is the most important. The regularity of
the mountains in this region is so strongly marked, that Humboldt[a]
has divided them into two classes, viz. those which coincide with
parallels of latitude, and those which coincide with meridians of
longitude. A similar, though not an equal degree of regularity
pervades the mountains of Western Asia, as viewed from the
central highlands of Armenia. (3.) Another marked feature in
the Asiatic mountains, resulting in part from this regularity, is the
tendency to *parallelism*. This feature did not escape the observation of the ancients, and is expressed in the names Taurus and
Antitaurus, Lebanon and *Antilebanon* : it may be noticed on a
larger scale in the ranges of Zagrus which bound the plain of
Mesopotamia on the E., and in the ranges which cross Armenia ;
and on a still larger scale in the lines which form the natural
boundaries of the countries of Western Asia, communicating to

---

[a] *Aspects of Nature*, i. 94.

them their peculiarly regular, we might almost say *geometrical*, forms.

The mountain-system of Western Asia may best be regarded from Armenia as a central point. Turning towards the N., the lofty[*] range of **Caucasus** forms a strong line of demarcation, striking across the neck of land that divides the Euxine and Caspian Seas in a south-easterly direction. Turning westward, three ranges may be traced entering the peninsula of Asia Minor—one skirting the northern coast and connecting with the European system at the Thracian Bosporus, the most important links being **Paryadres** in Pontus, and the Bithynian and Mysian **Olympus**—another, under the name of **Antitaurus**, striking across the plateau of Cappadocia towards the S.W.—and a third, **Taurus**, yet more to the S., skirting the Mediterranean Sea to the very western angle of the peninsula: the second of these forms a connecting link between the first and third, being united with Taurus on the borders of Cilicia, and with Paryadres by an intermediate range named **Scydises**, on the borders of Pontus and Armenia: the range may be traced even beyond the point of its junction with Paryadres, in the **Moschici Montes** on the shores of the Euxine, and the chains which connect these with Caucasus. Turning southward, it will be observed, that, near the N.E. angle of the Mediterranean, Taurus sends out an important offshoot, which skirts the eastern shore of that sea, and is carried down through Syria and Palestine to the peninsula of Sinai, and along the shores of the Red sea to the *Straits of Bab-el-Mandeb*: the most important links in this chain were named, **Amanus** on the borders of Cilicia, **Bargylus** in Syria, **Lebanon** on the borders of Phœnicia, the mountains of Palestine, the **Nigri Montes**, or (as they are more usually called) the **Sinai** group, and the **Arabici Montes**. Lastly, turning eastward, two chains may be traced—one of which, under the name of **Caspius Mons**, skirts the southern coast of the sea of the same name, and after culminating in the lofty height of **Coronus**, proceeds in an easterly direction, under the names of **Labutas** on the borders of Hyrcania and **Sariphi Montes** in Aria, to form a junction with **Paropamisus**, and so with the mountains of Central Asia—the other strikes off towards the S.E. towards the Persian Gulf, and was named **Zagrus** between Media and Assyria; and **Parachoathras** in Susiana and Persia. We must lastly notice the mountain chains of Armenia itself, which form the connecting links between the various ranges already

---

[*] *Æschylus* refers to its great height in the lines,

Πριν ἀν πρὸς αὐτὸν Καύκασον μόλῃς, ὀρῶν
Ὕψιστον . . . . . ἀστρογείτονας δὲ κρη
Κορυφὰς ὑπερβάλλουσαι.—*Prom.* 721, 722.

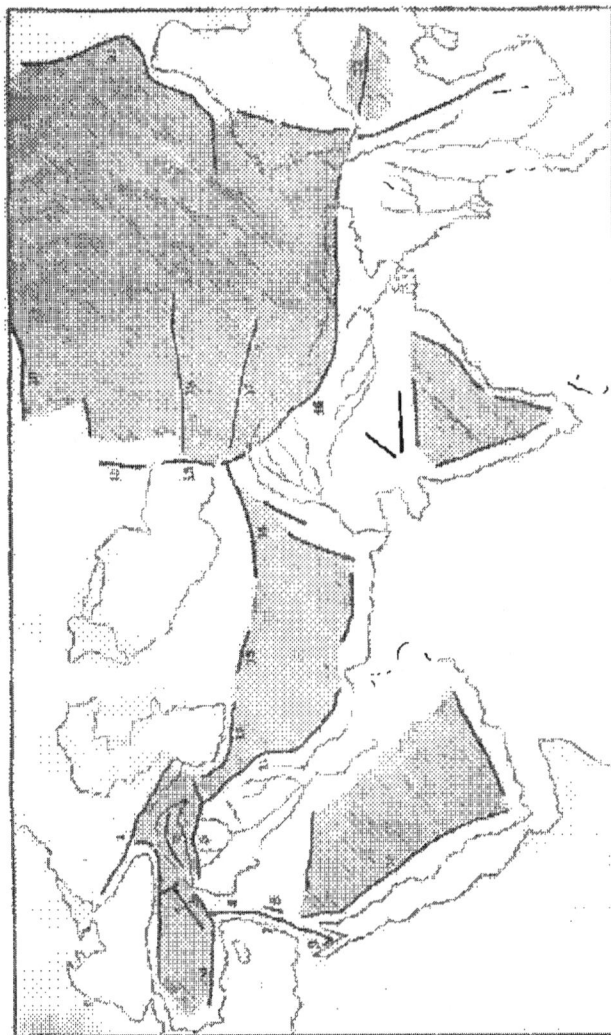

Sketch Map of the Mountain Ranges, Plateaus, and Plains of Asia, as known to the Ancients.   N.B.—The shaded portions represent the plateaus.

described. Two important chains traverse it in nearly parallel
lines from W. to E.; one a continuation of Antitaurus, the other of
Taurus. The former was named Abus, and culminates in the
magnificent heights of the *Greater* and *Less Ararat*, overlooking the
valley of the Araxes: the latter assumed the names of Niphates
in the W., and Caspius Mons in the E., and under the latter
designation connected with the mountains to the S. of the Caspian
Sea: an offset from this range, named Masius, skirts the head of the
Mesopotamian plain, and returns in a northerly direction, under the
name of Gordyæi montes, to the E. of the Tigris.

The ranges of Northern, Central, and Eastern Asia were but little
known to the ancients. In the former direction, the Hyperborei
montes represent the Ural chain; the Rhymnici montes, the mountains
between the rivers *Wolga* and *Ural*, and Noroægus, the chain in which
the latter river has its sources. In Central Asia, the chain of *Bolor*,
which strikes northwards from the junction of Paropamisus and Emodi
montes, was named Imäus, though this was also applied to the
Himalayan range: the yet more northerly range of *Mustag* seems to
have been named Comedärum montes: from this, parallel ranges are
omitted towards the E. and W.—in the former direction, the parallel
ranges previously referred to, and which may be identified in the
following manner, Serici montes with *Kuen-lun*, Ascatancæs with
Thian-Shan, and Auxacii and Annibi montes with the *Greater* and
Less Altai—in the latter direction, the Fogdil and Oxii montes,
between the Oxus and Iaxartes, representing the present *Kara* and *Ak
Tagh*; the Aspisii montes more to the N., in the *Kirghiz* steppe;
and the Anarsi montes, the *Tchingis* range, yet more to the N.
In Eastern Asia, the continuations of *Himalaya* were known to a
certain extent, and were named—Bepyrrus, about the sources of the
Doanas; Damassi montes, about the sources of the Dorins; and
Semanthini montes, in the direction of the *Gulf of Tonquin*. The
range which supports the desert of *Gobi* on the E. may be referred to
under the name Asmirei montes, *Khaigan*.

§ 4. The plateaus and plains of Asia next demand our attention.
The amount of high table-land in this continent is one of its most
striking features: while Europe possesses but one plateau of any
extent, viz. Spain, the greater portion of Western and a large
portion of Central Asia stands at a very high elevation. Not to
speak of the immense plateau of *Gobi*, N. of India, with which
the ancients were but slightly acquainted, we may notice the
plateau of *Iran*, or Persia, which stands at an average elevation of
about 4000 feet; that of Armenia, about 7000 feet; and that of
Asia Minor, at a less elevation. Central Arabia, again, is a plateau;
so also is the peninsula of *Hindostan*. Indeed it may almost be
said, that, with the exception of the strip of low land that skirts the
shore, and the depression between the plateaus of Iran and Arabia
which is occupied by the plain of Mesopotamia, the whole of
Western Asia is elevated ground: even the plain of Syria partakes

of the same character to a certain extent ; for there is a perceptible
difference in its elevation, when compared with Mesopotamia. It
must not be supposed that these plateaus are throughout level :
extensive districts of unbroken plain are indeed one of their
characteristics, but not unfrequently lofty ranges rise out of them
as from a new base, as may be marked particularly in Armenia and
Persia. The plains or lowlands of Asia, though not so extensive,
were important from their position and physical character: they
were the seats of commerce, not unfrequently of empire, and from
peculiarities of soil and climate, were eminently fertile : the well-
watered plain of Mesopotamia was the key-stone of the successive
empires of Nineveh, Babylon, Persia, and Syria : the plains of
Northern India, about the valleys of the Indus and Ganges, have in
all ages held a position of similar importance.

§ 5. The rivers of Asia are comparatively few. It is a necessary
consequence of the structure of plateaus, that few outlets should
exist for the waters of the interior. No river of any importance
attains the sea from the plateaus of Arabia and Persia : the Mediterranean coast is unbroken by the embouchure of any considerable
stream ; the mountain wall that skirts the sea-coast forbids access.
Many of the rivers gather into lakes, or are absorbed in the sands ;
and hence we may institute a classification of them into oceanic and
continental, the former including those which reach the sea, the
latter those which are confined to the interior.

(1.) The rivers of the first class are found, as might be expected,
in the plains. There were but four with which the ancients were
well acquainted, and these retain their classical names to the present
day, viz., the Euphrates, Tigris, Indus, and Ganges.

The **Euphrates** rises in the highlands of Armenia, and consists in its
upper course of a double stream, of which the northern is now named
Kara-su, and the southern *Murad-chai*, the latter being the most
important. These unite, after a westerly course, on the borders of
Asia Minor, and thence pursue a southerly course until the plain of
Mesopotamia is gained. The river then flows towards the S.E., converging to and ultimately uniting with the Tigris. Its lower course
has evidently changed much even in historical times. The Euphrates and
Tigris had originally separate outlets into the Persian Gulf, as also had the
Eulæus : these three unite in a single stream, now named *Shat-el-Arab*.
The Euphrates is navigable as high as Samosata, above which it assumes
the character of a mountain-stream, though its width and depth are
very considerable. It was fordable in several places in its mid-course—
at Samosata, Commagene, Birtha, and Thapsacus. As it issues from
a snowy country, it is liable to periodical floods, which commence in
March, and attain their greatest height towards the end of May. The
Tigris also rises in Armenia, but at a lower point than the Euphrates,
its source being a lake not far from the junction of the *Kara-su* and
Murad-chai. Its direction in its upper course is towards the E. ; and
in this part it drains the extensive district enclosed by Taurus and

E 2

Niphates on the N., Masius on the W. and S., and Gordyæi montes on the E. The latter range gives the Tigris a southerly direction, and after escaping from the deep gorge by which it passes through the lateral ridges of that chain, it enters upon the Mesopotamian plain. Preserving its southerly bearing, it converges to the Euphrates, and above Babylon comes within twenty miles of it, but, again receding, ultimately unites with it in the Shat-el-Arab. The Tigris is shorter than the Euphrates, their respective lengths being 1146 and 1780 miles: it is narrower and swifter, whence its name Hiddekel, "arrow." The Tigris receives numerous tributaries, one of which, rising in Niphates not far from Lake Arsissa, lays claim to the name of Tigris. The Indus (or Sinthus, as some writers call it with a more exact conversion of the native name Sindhu) was comparatively little known to the ancients until the time of Alexander's expedition. Its sources were erroneously placed in Paropamisus, whereas they really are to be found to the north of Himalaya in about 83° long. and 31° lat. Pursuing in this part of its course a westerly direction, until arrested by the transverse chain of Bolor, it bursts through the ranges of Himalaya in a south-westerly direction, and, receiving on its right bank the Cophes or Cophen, Kabul, with its affluent the Choaspes or Choas, Kameh, enters the plain of the Punjab, and receives on its left bank the united waters of the four rivers which water that district, the Acesines, Chenab, the Hydaspes or Bidaspes, Jelum, the Hydraôtes, Ravi, and the Hypanis or Hyphasis, Sutledge or Gharra: it thence pursues an unbroken course to the Indian Ocean, into which it discharges itself by several channels, two of which, named the Buggaur and Sata, are the principal: these channels have been in a constant state of change, but it is probable that the same general features have been preserved in all ages, and that the statement of Strabo and others, that there were two principal outlets, is not really inconsistent with that of Nearchus and Ptolemy, that there were several, according to the latter seven, outlets. The Ganges was not known until a comparatively late period; subsequently to the age of Alexander the Great[1] it was frequently visited, and excited considerable interest among geographers. It rises in the western ranges of Himalaya, and pursues a south-easterly course to the Gangeticus Sinus. Ancient writers vary in their reports of its size, which was, generally speaking, much exaggerated, and of the number of channels through which it reaches the sea. Fifteen of its tributaries are enumerated by Arrian, the names in several cases agreeing with the modern appellations, as in the case of the Jamānes, Jumna, Sonus, Sone, and others. The Dyardānes, Brahmapulra, was regarded as an affluent of the Ganges. The Ganges forms an important feature in the map of Ptolemy, as the intermediate boundary of Eastern and Western India. The names of other important rivers more to the E. were known to the ancients, but cannot be identified with certainty: the Doānas, Irawaddy, the Dorias, Salren, which discharge their waters into Sabaricus Sinus; the Serus, Meinam, flowing into the Magnus Sinus: the Ambastus, the Camboja; the Cottiāris, Si Kinng; and the Bautisus, Hoang-ho.

[1] Ovid refers to the Ganges as a very distant river, in the lines,
"Nec patria est habitata tibi, sed ad usque nivosum
Strymona venisti, Marticolamque Geten :
Persidaque, et lato spatiantem flumine Gangem,
Et quascunque libris decolor Indus aquas,"—Trist. v. 3, 21.

(2.) The chief continental streams are the Jaxartes, the Oxus, the Rha, the Cyrus, and Daix, which were regarded as all flowing into the Caspian, though the two first now join the *Sea of Aral.*

The Jaxartes, *Sir-deria,* rises in the central range of Asia, the Comedarum montes, and pursues a north-westerly course, in length about 900 miles, to the *Sea of Aral.* The Oxus, *Amou* or *Jyhún,* rises more to the S. in Imaus, and pursues a generally parallel course. The upper courses of these rivers were well known, as they watered the fertile districts of Bactriana and Sogdiana: their lower courses crossed a sandy desert. The Cyrus, *Kur,* and its tributary the Araxes, *Aras,* drain a large portion of the district between the Caspian and Euxine Seas. The former rises in the ranges of Scordisca, the latter in Abus, and after a lengthened course through the highlands of Armenia, they converge and unite at a distance of 110 miles from the Caspian. As they are fed by the snows of the high country, their streams are at certain periods very impetuous, and hence the difficulty experienced by the Romans in maintaining bridges.[1] The Rha, *Wolga,* is first noticed by Ptolemy, who describes it as rising in the country of the Hyperborean Sarmatians, and as being divided in its upper course into two arms, one of which is now named the *Kama,* the other the *Wolga.* The Daix, *Ural,* rises in the *Ural* chain, and flows southwards to the Caspian, with a course of about 900 miles.

§ 6. The climate and temperature of Asia is of the most diversified character. While the northern district falls within the arctic circle, the southern extremity very nearly reaches the equator, and in these parts the extremes of cold and heat are experienced. But with the exception of the peninsulas that protrude towards the S., the southern portion of the continent enjoys a fine temperate climate, adapted to the growth of almost every production requisite for the sustenance and comfort of man. The elevation of the plateaus of Western Asia contributes to moderate the heat which would otherwise be excessive, and offers a most agreeable alternation to the inhabitants of the adjacent lowlands. The climate of the central steppes is more severe, from the openness of the country, the absence of foliage, and the small amount of rain that falls there. But even here it is sufficiently warm to mature every species of vegetation, wherover shelter and irrigation exist.

§ 7. The productions of Asia are too numerous to be specified with any degree of minuteness. We shall therefore briefly notice such as entered largely into the commercial arrangements of the continent, and these we shall class under the following heads— I. Metals, Precious Stones, &c. II. Materials of Clothing. III. Spices and Aromatic Drugs.

I. Gold was evidently very abundant in ancient times. The eastern monarchs not only employed it largely in personal decorations, but

---

[1] "Pontem indignatus Araxes."—Virg. *Æn.* viii. 728.

even in furniture and the equipment of their equipages. Gold was procured in some quantities from Mount Tmolus in Asia Minor, whence it was carried down by the rivers Pactolus and Mæander: it was from this source that the Lydian monarchs enriched themselves. But the chief supply was undoubtedly obtained from the mountains of the north. Herodotus (iii. 102) tells us that the Indians collected it for the Persian monarch on a sandy desert: he refers probably to the district of Gobi, the mountains that separate it from Bokhara being to this day auriferous. Yet even this district would hardly supply the amount of gold which appears to have been current. There is good reason for believing that the mines of the Altaic range—the main source at present to the Russian Empire—were worked in ancient times, and that from these arose the report which was current in Herodotus's time (iii. 106), that gold was obtained in large quantities from the extreme east. If this were the case, the gold was in part supplied from the neighbourhood of Lake Baikal and the sources of the Ono, about which are the chief mines at the present time. It was also believed that Arabia yielded gold: this is not the case in the present day, and it is therefore possible that it was one of the articles of commerce introduced through that country; still the very general unanimity of ancient writers on this subject may have had a more substantial ground even than this. Silver is not found in equal abundance in Asia; the main supply is in the Caucasian range, to which Homer[3] perhaps refers in his notice of the Halizonians; there were also silver mines in Bactriana. The amount of silver appears, however, to have exceeded these sources of supply, and it is therefore probable that large quantities were imported by the Phœnicians from Spain. Iron and copper were derived from the mines of Pontus in Asia Minor from the days of Ezekiel (xxvii. 13-14): the latter was also found in Carmania, and was possessed by the Massagetæ, who may have obtained it from the Kirghiz steppes. Precious stones formed another of the valuable productions of Asia. Whether the ancients were acquainted with the diamond mines of Golconda, on the eastern coast of India, is uncertain; but it appears probable, from a passage in Ctesias (India, cap. 5), that they were aware of the productiveness of the mountainous districts of Central Asia, particularly of the range E. of Bactriana, where the jasper, lapis lazuli, and onyx, still abound. Pearls were found in the Persian Gulf, and along the shores of India and Ceylon.

II. In the second class of productions, we have first to notice cotton, described by Herodotus (iii. 106) as "tree-wool" (exactly answering to the German term baumwolle). It was found, according to that author, in India; it also grew on the island Tylus in the Persian Gulf. Silk was not introduced into Western Asia until a comparatively late period. The earliest notice of the silkworm occurs in Aristotle (H. N., v. 19), the term translated "silk" in the Bible being really applicable to a different texture; it was manufactured into robes at Cos, whence the Latin expression Coa vestis. As soon, however, as the Romans became acquainted with the habitat of the silkworm, they named it Sericum after the Seres of China. Flax grew in India and elsewhere. The finest kind of linen was named by the Greeks byssus, after a Hebrew word of the same meaning. Wool of fine quality was produced

[3] Αὐτὰρ Ἀλιζώνων Ὀδίος καὶ Ἐπίστροφος ἦρχον
Τηλόθεν ἐξ Ἀλύβης, ὅθεν ἀργύρου ἐστὶ γενέθλη. — Il. ii. 856.

in many districts, particularly in the neighbourhood of Miletus, in
Syria (according to Ezekiel xxvii. 18), and in Northern India or
Cashmere, the flocks of which country are noticed by Ctesias (Ind.,
cap. 13, 20). The chief manufactories of woollen stuffs were in Babylonia and Phœnicia. The fine goats' hair of Ancyra in Asia Minor was
also highly prized.

III. The chief supply of spices and aromatics was obtained from
Yemen, the southern part of Arabia Felix. Hence was derived frankincense, ladanum (the gum of the *Cistus ladaniferus*), myrrh, gum
storax, balm, and (according to Herodotus, iii. 110, 111) cassia and
cinnamon, though these were more properly the productions of
Ethiopia than of Arabia: perhaps he really referred to a different
production under the name of cinnamon. It is worthy of remark, as
illustrating the origin of spices, that the Greek and in many cases the
English terms are of Semitic origin, and may be referred to Hebrew
roots.

In addition to the productions above enumerated, we may further
notice - the dyes of Phœnicia, some of which were derived from certain kinds of shell-fish, the *buccinum*, and the *murex* or *purpura*, while
the scarlet dye was produced from an insect named the *coccus*, which
is found on the holm oak in Armenia and Persia—indigo, the very name
of which (from *Indicum*) implies the country whence it was obtained—
glass, which was originally invented and afterwards manufactured in
Phœnicia—rice, noticed by Strabo (xv. p. 690, 692) as growing in
India and Syria—and the citron, which was considered as indigenous
in Media, and hence called *Medica*. The cherry was introduced into
Europe from Cerasus (whence the name) in Pontus by the Roman
consul Lucullus: and the pheasant derives its name from the river
Phasis in Colchis.

§ 8. The commerce of Asia was chiefly carried on overland by
caravans—then, as now, the only means adapted to the wide open
plains, the insecure state of society, and the various difficulties and
dangers which attend the lengthened journeys across this vast continent. The merchants engaged in the trade of these parts met
at certain points for the interchange of their wares, and thus the
goods changed hands several times before reaching their final destina-
tion. In ancient times Babylonia formed one of these focuses for the
prosecution chiefly of the Indian trade: Bactriana was another such
entrepôt, as *Bokhara* is at the present day, for the commerce of the
north and east, and particularly of China: Phœnicia, again, was
the mart where the products of Asia and Europe were exchanged
and forwarded to their respective destinations: and on a smaller
scale, Southern Arabia was the entrepôt for the trade of South
Africa and the coasts of the Indian Ocean.

Commercial Routes of Asia.—The points above specified were centres,
to which the great commercial routes converged. Some of these are
minutely described to us by ancient writers; others are not described,
but are known to have existed.

I. From Babylonia the following routes existed:—(1.) To Asia
Minor, by the "Royal Road," which led from Ephesus to Susa: this
road is described by Herodotus (v. 52); it was provided with stations

and caravanserais, and followed very nearly the same line as that of the modern route between *Smyrna* and *Baghdad*, keeping along the central plateau of Asia Minor, crossing the Euphrates probably near Melitene, or perhaps lower down at Samosata, thence crossing northern Mesopotamia to the Tigris at Nineveh, and down the course of the river to Babylon. (2.) To Phœnicia, by the course of the Euphrates as far as Thapsacus, thence across the desert by Palmyra and Damascus to Tyre. (3.) To Mesopotamia, by the same route as far as Thapsacus, and thence across the desert to Edessa. (4.) To India, through Ecbatana to Hecatompylos, E. of the Caspian Gates, thence by Alexandria in Aria, *Herat*, Prophthasia and Arachotus, and the valley of the *Cabul*, to Taxila on the Indus; then either down the course of that river, or across to the valley of the Ganges, and by Palimbothra near *Patna*, to the shores of the *Bay of Bengal*. (5.) To Bactria, by the same route as far as Hecatompylos, and thence towards the N.E. through Antiochia Margiana, *Merv*, to the valley of the Oxus.

II. From Bactria. (1.) To Serica, *China*, across the ranges that intervene between the upper valleys of the Oxus and Jaxartes to where a pass leads across the central range to the desert of *Gobi*: the Chinese merchants came as far as this range, and interchanged their goods at a spot called the Turris Lapidea, "stone tower," probably identical with the Hormeterium, or "merchants' station," to which Ptolemy refers: the position of this spot cannot be accurately made out: the name *Tachkend* means "stone tower," but its position is somewhat too low on the Jaxartes; *Takht-i-Souleiman* stands nearer the western entrance of the pass and was probably the chief mart, while the ruins of an old building now called *Chikel-Sutun*, "the forty columns," not far distant, have been identified with the "stone tower." (2.) To India, by the pass of *Bameean* to Ortospāna, *Cabul*, and thence to the Indus. *Cabul* appears to have been an important trading station, being the spot where three roads converged, and hence termed the Bactrian *Trivium*, one perhaps leading to the Indus, another to Persia, and the third to Bactriana. (3.) To Europe, by the course of the Oxus to the Caspian Sea, which was crossed to the mouth of the Araxes on the opposite shore, and then by that stream to the head-waters of the Phasis, and so down to the Euxine.

III. From Phœnicia the overland routes led—(1.) To Babylonia by Palmyra as already described. (2.) To Gerrha on the Persian Gulf, which was the chief trading station for India. (3.) To southern Arabia, either wholly by land or perhaps by sea as far as the S.E. angle of the Mediterranean, where the "Arabian marts" referred to by Herodotus (iii. 5) were situated, and thence by Petra to the S.

IV. In Arabia, overland routes led—(1.) Northwards from Mariaba, the great commercial capital of the southern district, through Macoraba, *Mecca*, to Petra. (2.) From the same point to Gerrha on the Persian Gulf. (3.) From Gerrha across the country to Petra. (4.) From Petra, westward to Egypt and northward to Palestine: Petra was thus the great entrepôt of Northern Arabia. Lastly, from some point on the southern coast of Arabia, probably *Aden*, an extensive maritime trade was prosecuted with the eastern coast of Africa, and the western coast of India. The commercial route established by Solomon, with the aid of the Phœnicians, from the head of the Red Sea to Ophir (1 Kings, ix. 28; x. 22, 23), was probably directed to some entrepôt on the southern coast of Arabia, where the varied productions of India, South Africa, and Arabia, could be procured.

§ 9. The ethnography of the continent of Asia is a subject of great
interest and importance, but one which in this work we can only
treat incidentally. Asia was, as we have already observed, the
cradle of the human race: there the first family "became fruitful,
and multiplied, and replenished the earth:" there the different
types of language and physical conformation were first developed:
and thence issued the various nations to their respective homes in
the four quarters of the globe. In Asia, therefore, we might expect
to see the greatest diversity of race and language, and to be able to
trace those differences back to the point of their original divergence.
Such a diversity did in point of fact exist, as testified by the trilingual
inscriptions of the Persian Empire: and we are enabled, by the
light of history, and still more by the analysis of language, to arrive
at a probable opinion as to the time when, and the place where, the
divergence commenced. If we refer to the Bible, which furnishes us
with the only historical narrative of these events, we find it stated
that the human race remained "of one language and of one speech"
until a period subsequent to the flood—that the place where the
difference of language originated was in the plain of Shinar, the later
Babylonia—and that a tripartite division was there established,
consisting of the descendants of Shem, Ham, and Japhet.

(1.) Modern philology confirms in a remarkable degree the
statements of Scripture. There are still existing abundant traces
of a language, which, from its simple and unspecific character and
from the wide area over which it prevailed, may be regarded as
the representative of the "one language and one speech" of the
Bible. Ethnologists assign to this language and to the races speaking
it the titles "Turanian," "Allophylian," "Scythic," and "Tâtar."
The Scythians of the ancient world, the Tâtars of the modern, are
the most prominent races of this type.

Turanian or Scythic Branch.—The language in its most ancient form
survives in the Assyrian, Armenian, and Persian inscriptions, which
are for the most part trilingual, one column being in the Scythic
speech. The language and other characteristics of the following ancient
races, viz. the Parthians, Sacæ, Colchians, Asiatic Ethiopians, Saspeiri,
Tibareni, and Moschi, point them out as belonging wholly to this
primitive stock; while the Armenians, Cappadocians, Susianians, and
Chaldæans, contained a large admixture of the same element.

Out of this primitive language were gradually developed more
perfect forms, apparently at considerable intervals of time. The
earliest of these developments was probably the Hamitic language,
which appears to have originated in Egypt (pre-eminently the
"land of Ham"), and to have spread eastward along the shores of
the Arabian Sea to the Persian Gulf and the Indian Ocean. The
extension of Hamitism eastwards to Babylonia is supported by the

Mosaic genealogy, which represents Nimrod as the grandson of Ham (Gen. x. 8), and thus extends the territory of Cush from Abyssinia, which was the proper position of the race, to the eastern Cuthah in Babylonia.

Hamitic or Cushite Branch.—The nations which may be assigned to this family are—the southern Arabs, the early Chaldæans, the early Susianians, the Ethiopians of Asia, and perhaps the early Canaanites.

(2.) The Semitic form of language appears to have emanated from Babylonia. This circumstance appears to be indicated in the notices that Asshur went forth out of Babylonia to Assyria (Gen. x. 11), that a Semitic race settled in Elam (Susiana) (Gen. x. 22), and that the Semitic family of Terah dwelt in Ur of the Chaldees (Gen. xi. 28). The period when this movement originated may be assigned to the earlier part of the 20th century B.C. : the westerly migrations of Abraham to Canaan, of the Joktanidæ to Arabia, and of the Phœnicians to the Mediterranean coast, were connected with this movement.

Semitic Branch.—The nations which may be grouped together in this family are the later Babylonians (as distinct from the Chaldæans), the Assyrians, Syrians, Phœnicians, Canaanites, Jews, Cyprians, the later Cilicians, the Solymi, and the northern Arabians.

(3.) The Japhetic or Indo-European family is the third great division of the human race. Its name implies an ethnical affinity between the Indian and European nations, a fact which has long been established on most indubitable evidence. Hence we must suppose a double migration, eastward and westward, from some central point. Armenia is supposed to have been that point.

Japhetic or Indo-European Branch.—From Armenia issued westward the Thracians, Pelasgians, Celts, Teutons, Phrygians, Bithynians, Lydians, and Lycians; eastward the Getæ of the Caspian steppes and the progenitors of the modern Hindoos, who settled in the upper valley of the Indus, whence one branch appears to have retraced its steps across the *Hindú Kúsh*, and to have settled in Sogdiana, Bactria, Aria, Hyrcania, Arachosia, Media, Persia, Carmania, and Drangiana, while another descended to the plains of *Hindustan*, and took possession of the whole of that peninsula.

Harbour of Alexandria Troas.

CHAPTER VII.

ASIA MINOR.—MYSIA, LYDIA.

§ 1. Asia Minor is the name assigned by geographers to the
large peninsula which stretches westward from the main body of the
continent of Asia, and which is bounded on three of its sides by
water—on the W. by the Ægæan; on the N. by the Euxine, and
the chain of intermediate seas that connect it with the Ægæan,
viz. the Hellespont, Propontis, and Thracian Bosporus ; and on the
S. by the Mediterranean : on the E. it was separated from Syria by
the ranges of Amanus and Taurus, from Armenia by the Euphrates
and one of the ranges of Paryadres, and from Colchis by the river
Phasis.

The Name.—The application of the name "Asia Minor" to this peninsula may be traced as follows:—The name "Asia" originated, as we have already seen, in the alluvial plain of the Caÿster, and seems at all periods to have adhered in a special sense to portions of the peninsula, even after its extension to the whole of the continent. Herodotus, for instance, describes the territory of the Lydian monarchs as "Asia within the Halys;" Strabo and Livy as "Asia within Taurus;" the kings of Pergamus adopted the title of "Kings of Asia," and when the last of these died, and bequeathed his territories to the Romans, they constituted a portion of them into a province named "Asia," partly, perhaps, in imitation of the princes whom they succeeded, and partly because it was the first territory on that continent of which they took formal possession. From the province of Asia, which only included the western district, the name was gradually extended to the whole peninsula, and the addition of "Minor" first appears in Orosius, a writer of the fourth century of our era. It is most important to note, in connexion with classical and even Biblical literature, that the term "Asia" was at no period co-extensive with the whole of the peninsula: it applied either to the continent, or to a portion of the peninsula—in Latin authors frequently, and in the New Testament exclusively, to the latter. But the idea of Asia Minor, as a distinct and united country, was quite foreign to the mind of the ancients. The modern name of the peninsula is *Anadoli, i. e.* "the east."

§ 2. The position and physical character of this peninsula destined it to hold a conspicuous place in the history of the ancient world. Situated at the extreme west of Asia and in close contiguity to Europe, it became, as it were, the bridge to unite the two continents: as such, it was traversed by successive waves of population as they surged westward from Central Asia, and it served as the great high-road on which the contending hosts of the East and West marched to the conflict, and not unfrequently the battle-field on which the question of supremacy was decided between them. In a strategetical point of view, it may be regarded as the outwork of the citadel of Asia: so long as any of its numerous lines of defence were sustained—whether the Hellespont, the Halys, the passes of Taurus and Amanus, the maritime plain of Issus, or the valley of the Euphrates—so long the safety of Europe or of Asia was inviolable. Not less marked was the importance of Asia Minor in the progress of commerce and civilization. In this respect the western district occupies the first place. Holding easy communication by sea with Phœnicia in one direction, with Greece by the isles that stud the Ægean in another, and with the Euxine in a third—with a coast well adapted to early navigation, being broken up into bays and estuaries, and fringed with islands—with a soil fertile in the productions most valued in ancient times—with a brilliant sky and a pure air—it was well calculated to become the nursery of commerce and art. It was here that the activity of the Greek mind was first developed: Miletus and Phocæa were foremost in commercial

enterprise : the first school of philosophy was planted on the soil of
Ionia : both epic and lyric poetry were born and matured in this
favoured district : the earliest historical writers of importance,
Hecatæus, Charon, Hellanicus, and Herodotus, were natives of Asia
Minor. Lastly, in the culture of the fine arts, she was not
behind her contemporaries ; the temples of Diana at Ephesus, and
of Juno at Samos, erected in the sixth century B.C., the monumental sculptures of Xanthus and Halicarnassus, the statuary of
Branchidæ, and the paintings of Phocæa, attested, and in many
instances still attest, the taste and skill of the artists of Asia Minor.

§ 3. The general features of the peninsula of Asia Minor may be
described in the following manner :—In form, it is an irregular
parallelogram, the sides facing the four cardinal points ; in size, it
has a length of about 650, and a breadth of about 350 miles, its
area being about half that of France ; in physical conformation, it
consists of a central plateau, surrounded by a maritime district, the
plateau occupying a length of about 500, and a breadth of about
250 miles, or about one-half of the whole peninsula. The general
fall of the land is towards the N., as indicated by the courses of the
rivers ; the southern part of the plateau is therefore higher than the
northern. The sea-coasts vary in character : while the N. and S.
are regular, the former even more so than the latter, the W. coast
is extremely irregular, the Propontis and the Ægean being deeply
indented with bays and inlets.

Considerable changes have taken place in the coast-line within historical times, through the large amount of alluvium deposited by some
of the rivers. The Elæan Bay has been diminished on its northern side
by the deposits of the Evenus and Caicus ; the Hermæan Bay, which
at one time opened out widely in the direction of Temnos, is now so
contracted at the mouth of the Hermus as to present the appearance
of a double bay ; the port of Ephesus is entirely filled up, and the
general level of the plain, on which the town stood, is raised by the deposits of the Caÿster ; but the greatest change of all is in the neighbourhood of Miletus, where the Mæander has protruded a considerable
plain into the very centre of the Latmian Bay, turning the head of the
bay into an inland lake, swallowing up the islands of Lade and Asteria,
and removing the sea to a considerable distance from the site of ancient
Miletus. On the southern coast a marked change has occurred in the
lower course of the Pyramus, which formerly reached the sea by a
direct channel, but now turns off at right angles to its upper course
near the site of Mopsuestia, and doubling round Mount Parium reaches
the sea in an easterly direction.

§ 4. The mountains which form the framework of the plateau are,
Taurus in the S., Antitaurus and Scydises in the E., Paryadres and
its continuations to the Mysian Olympus in the N., and a series of
subordinate heights that connect the latter with Taurus in the W.

The most important of these mountain-ranges is **Taurus**, which derives its name from the Aramaic word *Tur*, "height." In its western portion it consists of an irregular series of detached mountains, which cover the provinces of Lycia and Pisidia, in the former penetrating to the sea-coast, and terminating in a series of promontories, while, in the latter, they are removed somewhat inland, and leave the comparatively level strip occupied by Pamphylia. The range assumes a more decided form on the borders of Cilicia, and presents the appearance of an unbroken wall throughout the whole length of that province, the only spot where it can be crossed by an army being at the celebrated Portæ Ciliciæ. On the eastern border of Cilicia it throws off a southern limb named **Amānus Mons**, *Almadagh*, which, pressing closely on the Mediterranean shore, presents an almost insurmountable barrier in that direction. Taurus itself continues its easterly course, and forms the boundary of Asia Minor on the border of Cappadocia. **Antitaurus** strikes off from the main chain in a northerly direction from the border of Cilicia, and divides Cappadocia into two parts: the lofty **Argæus**, *Argish Dagh*, whence, according to Strabo, both the Euxine and Mediterranean seas could be seen, forms its culminating point: its height is estimated at 13,000 feet. On the frontier of Cappadocia and Pontus Antitaurus takes an easterly direction, bounding the valley of the Halys, and passes out of Asia Minor into Armenia Minor, where it connects with **Scydises**. This latter throws off a northern offset, which ultimately connects it with the Moschici Montes on the eastern frontier of Pontus. Another offset of Scydises forms the connecting link between the Taurian system and the lofty range of **Paryadres**, *Kutlag*, which runs parallel to the Euxine Sea, and throws off from its central chain numerous spurs, reaching to the neighbourhood of the coast, and enclosing short parallel valleys. Paryadres terminates at the valley of the Iris, and thenceforward the continuity of the northern range is broken, though the system may be traced through the Galatian and Mysian **Olympus** to the very shores of the Propontis. Lastly, a southern range of subordinate height, which leaves the Mysian Olympus and passes near Cotyæum, completes the framework of the country by bounding the plateau on the W. Westward of the line just indicated the table-land breaks up into numerous ridges, which descend towards the Ægæan: of these we may notice—**Messōgis**, *Kestaneh Dagh*, which separates the basins of the Mæander and Cayster—**Tmolus**, *Bouz Dagh*, between the Cayster and Hermus; and **Temnus**, *Ak Dagh*, which divides the upper basin of the Hermus from the Macestus and Rhyndacus, which take a northerly course.

§ 5. The chief rivers of Asia Minor seek the Euxine. Not only is the general slope of the country in that direction, but also more numerous outlets are offered among the broken chains of the north, than along the serried line of Taurus. We may notice, as running in that direction—the **Phasis**, *Rion*, which forms the boundary between Colchis and Asia Minor—the **Acampsis**, *Tchoruk*, in Pontus—the **Iris**, *Kasalmak*, in the same province—the **Halys**, *Kizil Irmak*, i. e. "red river," the most important in the whole country—and the **Sangarius**, *Sakkaryeh*, in Bithynia. The Propontis receives an important feeder in the **Rhyndacus**, *Lupad*. Proceeding southwards along the coast of the Ægæan, we meet with the **Hermus**,

Kodus Chai, in Lydia, and the **Maander**, *Meinder*, in Caria. The streams that fall into the Mediterranean are necessarily short, from the close approach of the Taurus range: from this description, however, we must except the **Saras**, *Sihun*, and the **Pyramus**, *Jihun*, in the eastern part of Cilicia, which rise between the ranges of Taurus and Antitaurus, and thus have longer courses. The rivers above enumerated will be more minutely described in the subsequent accounts of the provinces, with the exception of those which hold an important place in the general geography of Asia Minor.

The **Halys** rises on the borders of Armenia, and traverses Cappadocia in a south-westerly course as far as Mazaca; thence it turns gradually towards the N., and finally towards the N.E., separating in this part of its course Paphlagonia from Galatia and Pontus, and discharging itself into the Euxine: it derives its modern name from the "red" colour of the water when impregnated with the soil of the country. The **Sangarius** rises in the Phrygian mountain Adoreus, and, flowing northwards, receives an important tributary from the neighbourhood of Ancÿra; it afterwards assumes a westerly direction, until its junction with the Thymbres, when it again turns northwards, and in a tortuous course crosses Bithynia to the Euxine: it was navigable in its lower course, and yielded an abundance of fish. The **Phasis** rises in the Moschici Montes, and flows in a semicircular course, with a rapid stream, into the Euxine; in the upper part of its course it was named **Boas**: its water is described as being very cold, and so light that it swam like oil on the Euxine. The **Sarus** rises in Cataonia, and first flows towards the S.E. through Cappadocia, and then towards the S.W. through Cilicia, traversing in its lower course the rich Aleïan plain, and emptying itself into the Mediterranean S. of Tarsus. The **Pyramus** also rises in Cataonia, and has a general S.W. course: for a certain distance it is said to disappear under ground; on its reappearance it becomes a navigable stream, and forces its way through a glen of Taurus, which in some parts is so narrow that a dog can leap across it; it then crosses the eastern part of the Aleïan plain to the sea.

§ 6. The lakes form a conspicuous feature in the map of Asia Minor. The central plateau is not (it should be observed) a dead flat, but intersected by numerous ranges of mountains of varying altitude. In the southern portion of the plateau these ranges form basins in which the waters gather into lakes, no outlet towards the sea existing in any direction. These lakes are for the most part strongly impregnated with salt. The largest of them is **Tatta**, *Tuzla*, on the borders of Lycaonia and Cappadocia, about 75 miles in circumference. **Coralis** and **Trogitis**, in Pisidia, are also of a large size.

§ 7. The soil and the climate of Asia Minor are, as may be supposed, exceedingly variable. The alluvial plains about the lower courses of the rivers of the western district and Cilicia surpass all in fertility. The extent and flatness of these plains is remarkable; the mountains rise out of them at their upper extremities.

"like islands out of the ocean;"[1] they are sheltered from the severe
cold of the upper regions, and are for the most part well watered.
The most extensive of these alluvial plains is in the eastern part of
Cilicia, hence designated Campestris, which is formed by the rivers
Cydnus, Sarus, and Pyramus. Of a similar character are the lands
which surround many of the lakes in the interior; these have at
one period occupied larger beds than at present; the dry margins
are consequently beds of rich alluvial soil. Fertile plains of a
different class are found occasionally on the sea-coast; of these, that
of Attalia on the southern coast was the most extensive. The hills
of the western district are clothed with shrubs and wood, and in
some cases cultivated to their very summits. The climate of the
maritime region is fine, but the heat sometimes excessive. The
western portion of the central plateau consists of extensive barren
plains traversed by deep gullies which the streams have worked
out for themselves. The southern part is subdivided into numerous portions by ranges of considerable height; in the northern part the
hills are of less height, and consequently the plains present a more
unbroken appearance. The same peculiarity, which we have already
noted in regard to the alluvial plains, also characterizes the upper
plains; "they extend without any previous slope to the foot of the
mountains, which rise from them like lofty islands out of the surface
of the ocean."[2] The climate of the central district is severe, the
loftier hills being tipped with snow throughout the greater part of
the year. The northern district along the shores of the Euxine,
from the Iris to the Sangarius, is fertile, the hills being of no great
elevation; on either side of these limits the country is too mountainous to admit of much cultivation.

§ 8. The population of Asia Minor was of a very mixed character: Turanian, Indo-European, and Semitic races are found
there coexisting in different proportions, the predominant element,
however, being the Indo-European. This admixture is indicated in
the Mosaic table, where Lud, the progenitor of the Lydians, is represented as a son of Shem, while the remainder of the northern and
western parts of the world are assigned to the Japhetites—Gomer,
Ashkenaz, and Riphath being (according to the best authorities) the
representatives of the races in the western part of Asia Minor, while
Meschech and Tubal undoubtedly held the eastern part.

(1.) *Turanian Races.*—The most important were the Moschi, the
Meschech of Scripture, and the *Muskai* of the Assyrian inscriptions, the
progenitors of the *Muscovites*; and the Tibareni, the Tubal of Scripture.
These races occupied the later Cappadocia, and were pressed northwards
to the shores of the Euxine by the entrance of the Cappadocians. At a

¹ Fellows's *Asia Minor*, p. 26.   ² Leake's *Asia Minor*, p. 93.

later period Scytho-Thracian tribes recrossed the Bosporus from Europe into Asia, and settled along the northern coast, under the names Thyni, Bithyni, and Mariandyni.

(2.) *Indo-European Races.*—The Phrygians, Trojans, Mysians, Mæonians, Mygdonians, and Dolionians, as well as the Pelasgians, who were closely allied to the Phrygians, belong to this class. The Phrygians (whose name appears under the different forms of Phryges, Bryges, Briges, Breuci, Bebryces, and Berecynthæ) were in early times the dominant race in Asia Minor, and had even crossed over the Hellespont into Europe, whence, however, they were driven back by the advance of the Illyrians and Scytho-Thracians, and resettled on the shores of the Propontis, in the districts named Lesser Phrygia and Mysia. A Celtic race, the Galatians, entered Asia Minor at a comparatively late period.

(3.) *Semitic Races.*—These were chiefly located on the shores of the Mediterranean Sea. The Cilicians were connected by their own traditions with the Phœnicians. The Pisidians and the early inhabitants of Lycia, the Solymi and Termilæ, were undoubtedly of Semitic origin the frequent occurrence of Semitic names in the latter district, as Solymi (Salem), Phœnix (Phœnicia), and Cabalis (Gebal), furnishes a proof of this. The Lydians on the western coast are supposed to be also a Semitic race, but this question can hardly yet be considered as decided. The same may be said of the Cappadocians, who are described as Syrians by Herodotus—a *primâ facie* ground for inferring that they were of Aramæan and thus of Semitic origin. That description may however have been attached to them from their having entered Asia Minor from the side of Syria. The Cappadocians are by some ethnologists supposed to be of the Arian division of Indo-Europeans, an opinion which is favoured by the comparatively late period of their immigration.

§ 9. The territorial divisions of Asia Minor varied considerably in different ages. We have described the positions which the several races were supposed to occupy in the age of Herodotus (p. 30). Subsequently to that time we may note the following changes:—(1.) the introduction of the name "Pontus," which first appears in Xenophon (*Anab.* v. 6, § 15), to describe the province lying along the shore of the Euxine in the N.E.; (2.) the separation of Pisidia from Phrygia and Pamphylia, which was not formally effected until the time of Constantine the Great; (3.) the immigration of the Gauls into the district named Galatia; and (4.) the consequent contraction of the boundaries of Phrygia and Bithynia. The divisions usually recognised in geographical works belong to the period of the Roman empire, and are partly a political, partly of an ethnographical character. They are the following 14: on the western coast, Mysia with Troas and Æolis, Lydia with the northern portion of Ionia, and Caria with southern Ionia and Doris; on the southern coast, Lycia, Pamphylia, and Cilicia; in the interior, Cappadocia with Armenia Minor, Lycaonia with Isauria, Pisidia, Phrygia, and Galatia; and on the northern coast, Bithynia, Paphlagonia, and Pontus.

History — In the earliest historical period Asia Minor was parcelled
out into a number of independent kingdoms, among which the Phrygian appears to have been the most powerful. The Trojan and earlier
Lydian dynasties are also known to us. The last of the Lydian dynasties, the Mermnadæ, extended their sway over the whole of Asia Minor
westward of the Halys from B.C. 720 to 546, when their territory, along
with the rest of the peninsula, was incorporated by Cyrus into the Persian Empire. Asia Minor remained subject to Persia until the time of
Alexander the Great, B.C. 334, when it was transferred to the Macedonian Empire. After the death of the conqueror it fell in the first instance to Antigonus, and after the battle of Ipsus, B.C. 301, to Lysimachus. About 20 years later, Seleucus attached the greater part of it
to Syria, while several provinces, Bithynia, Galatia, Cappadocia, Pontus,
Paphlagonia, and Armenia Minor, and the town of Pergamus, became the
seats of independent monarchies. The battle of Magnesia, B.C. 190, terminated the supremacy of the Seleucidæ, and the Roman conquerors
handed over Lycia and Caria to the Rhodians, Mysia, Lydia, and Phrygia
to the kings of Pergamus. The last of these kings bequeathed his territory to Rome, B.C. 133, and the Roman province of Asia was formed, including a large part of Phrygia, Mysia, Lydia, and Caria, which last
had been taken away from the Rhodians, Lycia being declared independent. By degrees the other portions of Asia Minor fell into the
hands of the Romans ; Bithynia by the bequest of Nicomedes IV., B.C.
75 ; Cilicia by the conquest of Pompey, B.C. 67 ; Pontus partly after
the defeat of Mithridates, and the remainder in the reign of Nero ; Galatia and Lycaonia after the death of the Tetrarch Amyntas, B.C. 25 ;
Cappadocia after the death of Archelaus, A.D. 18 ; and lastly Armenia
Minor, after the death of Tigranes in Vespasian's reign. Asia Minor
was then divided into the following provinces :—Asia, Lycia, Cilicia
with Pamphylia, Cappadocia, Galatia with Lycaonia, Bithynia with
Pontus, and Armenia Minor. In Constantine's division Asia Minor
(with the exception of Cilicia and Isauria, which were added to the
Diocese of the East), was divided into two Dioceses, Asiana and Pontus,
the latter consisting of Pontus, Bithynia, Galatia. and Cappadocia, the
former of the remaining provinces.

Site of Abydos, from the West.

I. Mysia, with Æolis.

§ 10. The province of **Mysia** lay in the north-west of Asia Minor, bounded on the N. by the Propontis and the Hellespont, on the W. by the Ægean, on the S. by Mount Temnus and Lydia, and on the E. by Bithynia and Phrygia, the boundary in this direction being marked by the river Rhyndacus and Mount Olympus. It is generally mountainous, but possesses some plains on the sea-coast. It is also well watered by a number of small rivers. Nevertheless it was not in ancient times so productive as other portions of Asia Minor, and many parts of it were covered with marshes and forests. Besides the ordinary products and the wheat of Assus, Mysia was celebrated for the *lapis assius*, found near Assus, which had the property of quickly consuming the human body, and was hence used for coffins. Near the coasts of the Hellespont there were excellent oyster beds.[3]

Name.—The name Mysia is probably only another form of Mœsia, derived from a Celtic word signifying "a marsh." The Mysians were sometimes distinguished from the Mœsians by the title of "Asiatic."

§ 11. The mountains of Mysia are irregular. The highlands of the central plateau break up into a number of ranges, which seek the sea in various directions, though with a general tendency towards the W. The most important of these ranges are—Olympus on the eastern border—**Temnus** on the southern border—and Ida in Troas near the Ægean.

Olympus, *Ketchich Dagh*, distinguished from other mountains of the same name by the title of "Mysian," is an extensive range between the valleys of the Sangarius and Rhyndacus, and attains the height of 7000 feet. The lower regions are well clad with forests, which in ancient times harboured dangerous bands of robbers; the summit is covered with snow for the greater part of the year. Temnus traverses the province in a north-westerly direction from the angle in which Mysia meets Phrygia and Lydia to the neighbourhood of Ida; it is only noticed by the later geographers, and has no associations of any interest. Ida[4] is an irregular ridge running out into several branches

---

[3] Pontus et *ostriferi* fauces tentantur Abydi.—Virg. Georg. i. 207.

Hellespontia, cæteris *ostreosior* oris.—Catull. xviii. 4.

Pontus et *ostriferam* dirimat Chalcedona cursu.—Luc. ix. 950.

[4] The proximity of Ida to Troy leads to its being frequently noticed by the poets. Virgil describes the meteor as disappearing behind its wooded heights :—

Illam, summa super labentem culmina tecti
Cernimus Idæ claram se condere silvis.—Æn. ii. 695.

So, again, it appears among the ornaments of Æneas's vessel :—

Immicet Ida super, profugis gratissima Teucris.—Æn. x. 158.

Ida was further celebrated in mythology as the birthplace of Cybele :—

Alma parens Idæa deûm.—Virg. Æn. x. 252.

It is also used as a synonym for Trojan ; as in the expressions—*Idæus judex* for

near the Ægean; the highest point, named **Gargūrus**, attains an elevation of 4650 feet above the level of the sea; these ranges are well covered with wood, the haunts in ancient times of wild beasts, and contain the sources of numerous rivers.[5]

The sea-coast is also irregular, particularly in the southern part of the province, where the **Sinus Adramyttēnus**, *Gulf of Adramytti*, advances far inland between Lesbos and the mainland, and is succeeded by a series of sinuosities terminating with the **Sin. Elaïtīcus**, *Gulf of Sandarli*, on the borders of Lydia. The promontories of **Rhœtēum**,[6] *Intepeh*, on the Hellespont—**Sigēum**,[7] *Yenisheri*, at the entrance of the Hellespont—and **Lectum**, *Baba*, the extremity of the range of Ida—are frequently noticed by classical writers.

The less important promontories are—**Abarnus**, near Lampsacus— **Dardânis**, S. of Abydus near Dardanus—and **Cane**, *C. Coloni*, W. of the mouth of the Caïcus.

§ 12. The most important rivers are the **Rhyndācus** and the **Caïcus**. The former rises in Northern Phrygia, and flows in a north-western direction between Mysia and Bithynia through the Lake of Apollonia, and, after receiving the **Macestus** from the S.W., falls into the Propontis. The **Caïcus**,[8] *Ak-su*, rises in Temnus, and consists in its upper course of two streams, which unite near Pergamum: thence it flows into the Bay of Elæa. In addition to these, there are numerous streams, unimportant in point of size, but invested with historical associations, which we will briefly notice.

Paris (Ov. *Fast.* vi. 11), *Idæa naves* (Hor. *Od.* i. 15, 2); or for Phrygian, as *Idæa urbes* (Virg. *Æn.* vii. 207); or lastly for Roman, as being descended from Troy, as *Idæus sanguis* (Sil. Ital. l. 126).

Propertius confounds this Ida with the one in Crete:—

    Idæum Simoenta Jovis cunabula parvi (ill. 1, 27).

[6] 'Ιδην δ' ἵκανεν πολυπίδακα, μητέρα θηρῶν.   Hom. *Il.* viii. 47.

    Concidit : ut quondam cava concidit, aut Erymantho
    Aut Ida in magna, radicibus eruta pinus.—Virg. *Æn.* v. 448.

    Ardua proceris spollantur Gargara silvis :
    Innumerasque mihi longa dat Ida trabes.— Ovid. *Heroid.* xvi. 107.

[6] Rhœteïus is often used as a synonym for "Trojan;" e. g. *Rhœteius ductor* scil. Æneas (Virg. *Æn.* xil. 456) ; *Rhœteia littora* (Luc. vi. 351) ; and by a secondary application, for "Roman," e. g. *Rhœteia regna* (Sil. Ital. vii. 431).

[7] The naval camp of the Greeks was formed near Sigeum : hence it is frequently noticed by Homer and Virgil. The latter alludes to its position just where the Hellespont widens out into the Ægean :—Æn. ll. 312.

    Sigea igni freta lata relucent.—*Æn.* ll. 312.

Sigeus, or Sigeïus, is also used as a synonym for "Trojan;" e. g. *Sigei campi* (*Æn.* vii. 294), *Sigeo in pulvere* (Stat. *Achil.* l. 84) ; and for "Roman :"—

    Seu Laurens ubi Sigeo sulcata colono
    Arridet tellus.   Sil. Ital. ix. 293.

[8]   Mysusque Caïcus.   Virg. *Georg.* iv. 370.

    Et Mysum capitisque sui ripæque prioris
    Pœnituisse ferunt, aliâ nunc ire, Caïcum.— Ovid. *Met.* xv. 277.

The Propontis receives the *Æsēpus*,[2] which rises in Ida, and flows towards the N.E., forming the eastern boundary of the Troad – and the Granicus,[1] the scene of the victory of Alexander the Great over the Persians, B.C. 334, and of Lucullus over Mithridates, B.C. 73; it is probably the same as the *Kodsha-su*. The Hellespont receives the following streams from E. to W.—the Parcoōtes, *Brogaz*, the Practius, *Muskakoi-su*, the Rhodius, the Simois, *Dumbrek-chai*, formerly a tributary of the Scamander, but now an independent stream, and the Scamander or Xanthus, *Mendere-su*, which flowed by the walls of Troy, with its tributary, the Thymbrius, perhaps the *Kamara-su*, which still flows into the *Mendere-su*, though the name *Timbrek* is applied to a stream which has an independent course to the sea. The Satniois, *Tuzla*, in the southern part of Troas, rises in Ida and flows into the Ægean Sea: the Evēnus, *Sandarli*, rises in Temnus, and flows into the Bay of Elæa. Most of these streams owe their celebrity to their connexion with the Homeric poems. The Scamander is described by Homer as having two sources close to Ilium, one of them sending forth hot water, the other cold; he also describes it as a large and deep river;[3] it was named Xanthus from the yellow colour[4] of its water. Pliny describes the Xanthus and Scamander as distinct streams; Ptolemy gives the Simois and Scamander independent courses to the sea. The probability is that even in ancient times considerable changes had taken place in the line of coast by the alluvial deposits carried down by these streams. The Simois crossed the plain of Troy, and was therefore the scene of some of the most striking events in the Trojan war.[5]

§ 13. The inhabitants of Mysia belonged to various races. (1.) The Mysians themselves in the age of Homer appear to have lived on the shores of the Propontis in Mysia Minor; thence they advanced southwards and eastwards, and about the time of the Æolian migration founded the kingdom of Teuthrania. (2.) The Trojans occupied the district of Troas in the Homeric age; they were probably, like the Mysians, an immigrant race from Thrace; they amalgamated with the Phrygians, and hence the terms are used indifferently. (3.) Greek colonists settled at an early period along the western coast; they consisted of Achæans, Bœotians, and Æolians, of whom the latter possessed the chief influence, and communicated their name both to the migration and the district.

---

[2] Ὅι δὲ Ζέλειαν ἔναιον ὑπαὶ πόδα νείατον Ἴδης,
Ἀφνειοὶ, πίνοντες ὕδωρ μέλαν Αἰσήποιο. HOM. *Il.* ii. 824.

[1] Ovid describes it as bifurcating near its mouth:—
Alexirhoē, Granico nata bicorni.—*Met.* xi. 763.

[3] Ὃν Ξάνθον καλέουσι θεοί, ἄνδρες δὲ Σκάμανδρον.—HOM. *Il.* xx. 74.
Xanthus hence appears to have been the more ancient of the two names.

[3] Ξάνθου βαθυδινήεντος.—*Il.* xxi. 15.

[4] Ἀργυρόδίνης.—*Il.* xxi. 8.

[5] Καὶ Σιμόεις, ὅθι πολλὰ βοάγρια καὶ τρυφάλειαι
Κάππεσον ἐν κονίῃσι, καὶ ἡμιθέων γένος ἀνδρῶν.—HOM. *Il.* xii. 22.
Sævus ubi Æacidæ telo jacet Hector, ubi ingens
Sarpedon: ubi tot Simois correpta sub undis
Scuta virum, galeasque, et fortia corpora volvit.—VIRG. *Æn.* i. 99.

Mysia was divided into the following districts :—(1.) **Mysia Minor** or **Hellespontica**, the coast-district along the Hellespont and the Propontis. (2.) **Mysia Major**, the southern portion of the interior of the province, with Pergamum for its capital, and hence sometimes termed Pergaméne. (3.) **Troas**, the northern part of the western coast from the Hellespont to the Bay of Adramyttium. (4.) **Æolis**, the southern part of the coast, though more especially applicable to the portion between the rivers Caïcus and Hermus. (5.) **Teuthrania,**[4] a district on the southern frontier, where the Mysians under Teuthras had settled about the time of the Æolian migration. Under the Persians the western portion of the coast of the Hellespont was named **Phrygia Minor**.

§ 14. The towns of Mysia belonged to various historical eras, and are invested with associations more than usually varied, and extending over a long series of ages. The position of Mysia, in command of the most easy point of crossing the channel that separates Asia from Europe, naturally rendered it the high-road of communication between the two continents. Hence it was visited by Darius both in his Scythian and Greek expeditions, by Xerxes, by Alexander the Great, by Antiochus in his advance into Greece, and by Lucullus in the Mithridatic War. The banks of the Granicus witnessed more than one contest for the empire of the East, and the beach of Abydus was oft-times the parade-ground of hosts gathered from every nation of the known world. The towns of Mysia either lined the sea-coasts of the Propontis and the Ægean, or were situated within easy communication with the seaboard. In the Heroic age, as depicted in the Homeric poems, the towns were the seats of small sovereignties : the far-famed Ilium, Dardánus, Antandrus, Thebe, Scepsis, and many others, belong to this age. The period of Greek immigration followed : most of the towns that were favourably situated received colonies either immediately from Greece or from the Greek colonies on the shores of Asia Minor. The Æolians settled at Cyme and ten[7] other places, and, at a later period, these again sent out colonies to Antandrus, Ilium Novum, and elsewhere : Adramyttium claimed Athens as its founder : the Milesians, accompanied in some instances by other Greek colonists, settled at Cyzīcus, Abydus, Priāpus, Parium, Lampsācus, and Gargāra. Some of the old towns perished from the effects of war or natural

---

[4] Forsitan, ut quondam Teuthrantia regna tenenti,
Sic mihi res eadem vulnus opemque feret.—Ov. Trist. ii. 19.
Teuthranteusque Caïcus.—Id. Met. ii. 243.

[7] The names of the other ten were—Temnos, Larissa, Neon-Tichos, Ægæ, Myrina, Grynium, Cilla, Notium, Ægiroëssa, and Pitāne : Smyrna was originally an Æolian colony, but was afterwards occupied by Ionians.

decay; others from the foundation of new towns and the forcible removal of their inhabitants. The period succeeding Alexander the Great witnessed the rise of Ilium Novum, Alexandria Troas, and Pergämum: each of these owed its prosperity to a different cause— Ilium Novum to its associations with the Troy of Homer, Alexandria Troas to its favourable position on the sea-coast, and Pergamum to the establishment of the monarchy which through the favour of the Romans held sway over the greater part of Asia Minor. After the extension of the Roman Empire over Asia, the towns of Mysia received various boons conducive to their prosperity: Pergamum is described by Pliny as "longe clarissimum Asiae:" Cyzicus and Dardanus became free cities: Parium and Alexandria Roman colonies. The fine air and scenery of Cyzicus rendered it a fashionable resort of the wealthy Romans.[8] These towns are described below in their order from N. to S.

Cyzicus was well situated on the shores of the Propontis, at the inner extremity of an isthmus which connects a peninsula of considerable size with the mainland. The isthmus[9] was severed by an artificial channel, over which two bridges were thrown, and thus the place was easily defensible on the land side. Between the peninsula and the mainland were two roadsteads, one on each side of the isthmus. The Doliones[10] were reputed its earliest inhabitants, but its prosperity was due to the Milesians who settled there. It fell to the Persians after the conquest of Miletus—to the Athenians and Lacedæmonians alternately in the Peloponnesian War—and to the Persians again at the peace of Antalcidas. It was besieged by Mithridates, B.C. 74, but relieved by Lucullus; and, in gratitude for its resistance, it was made a free city by the Romans. Its gold coins, named Cyzicēni, had a very extensive circulation. The

Coin of Cyzicus.

oysters and the marble of Cyzicus were much prized. The ruins of Cyzicus are extensive, and are named Bal Kiz. Lampsacus stood on the Hellespont, near the modern Lamsaki, and nearly opposite to Callipolis, on the Thracian coast: it was named Pityūsa before the Milesians settled there. During the Ionian revolt it was

---

[8] Frigida tam multos placuit tibi Cyzicos annos
Tulle, Propontiaca qua fluit Isthmos aqua.—PROP. III. 22, I.

[9] There is some doubt as to whether the ground on which Cyzicus stood was originally an island or a peninsula. The great length of the isthmus (above a mile) renders it probable that it was made into an artificial island, by a narrow channel dug across, rather than into an artificial peninsula by so long a bridge or mole.

[10] Hence it is termed Hæmonia, i. e. Thessalian:—

Illaeque Propontiacis haerentem Cyzicon oris,
Cyzicon Hæmoniæ nobile gentis opus.—OV. Trist. 1. 10, 29.

taken by the Persians, and remained under their supremacy, though governed by a native tyrant. After the battle of Mycale it joined the side of Athens, and, having revolted from her, was besieged and taken by Strombichides. It was the birthplace of several illustrious men—Charon the historian, Anaximenes the orator, and Metrodorus the disciple of Epicurus.[1] Abydus was situated at the point where the Hellespont is narrowest,[2] being no more than 7 stadia across: on the other side of the strait was Sestos, about 30 stadia distant. Xerxes erected his bridge of boats from a point a little N. of the town, B.C. 480. Under the Romans it became a free town, in gratitude for its sturdy resistance to Philip II. of Macedon. Abydus is well known in mythology as the scene of Leander's exploit of swimming across the strait to visit Hero.[3] Dardanus stood about 8 miles from Abydos, and is supposed to have communicated to the strait its modern appellation, *Dardanelles*: it was regarded as the ancient capital of the Dardanians, and is further known as the spot where Sylla concluded peace with Mithridates, B.C. 84. Further to the S., at the junction of the Hellespont with the Ægean sea,[4] we enter upon the plain of Troy,[5] the stage on which the events of the Iliad were enacted. We have already had occasion to remark that the features of the sea-coast, and of the plain itself, have undergone much alteration, and that the Simois no longer flows into the Scamander. The site of Troy itself is a matter of great uncertainty: some fix it at Ilium Novum, the modern *Kissarlik*, about

---

[1] Lampsacus was the chief seat of the worship of Priapus :—

Et te ruricola, Lampsace, tuta deo.—Ov. *Trist.* i. 10, 26.

Hellespontiaci servet tutela Priapi.—Virg. *Georg.* iv. 111.

Hence Lampsacenus is used as a synonym for "obscene : "—

Nam mea Lampsacio lascivit pagina versu.—Mart. xi. 16.

Quantam Lampsaciae colunt puellae.   Id. xi. 51.

[2] Hence the expression "*fauces* Abydi."—Virg. *Georg.* i. 207.

The junction of the two shores, effected by Xerxes, was regarded as one of the greatest feats of skill and labour :—

Fama canit tumidum super æquora Xerxem
Constravisse vias, multum cum pontibus ausus,
Europamque Asiæ, Sestonque admovit Abydo
Incessitque fretum rapidi super Hellesponti.—Luc. ii. 672.

Tot potuere manus vel jungere Seston Abydo,
Ingesloque solo Phrixenm elidere pontum.—Id. vi. 55.

[3] Vel tua me Sestos vel te mea sumat Abydos.—Ov. *Heroid.* xviii. 127.

Utque rogem de te, et scribam tibi, si quis Abydo
Venerit, aut quæro, si quis Abydon eat.—Id. xix. 30.

[4] Longus in angustum qua clauditur Hellespontus
Ilion ardebat.   Ov. *Met.* xiii. 407.

[5] By Latin writers the place was usually called Troja ; the poets, however, frequently used the names Ilium, Ilion, and Ilios : e. g.

O divûm domus *Ilium*, et inclyta bello
Mœnia Dardanidûm.   Virg. *Æn.* ii. 241.

Ilion aspicies, firmataque turribus altis
Mœnia, Phœbeæ structa canore lyræ.—Ov. *Heroid.* xvi. 179.

Non semel *Ilios*
Vexata.   Hor. *Od.* iv. 9, 18.

CHART
of the
COUNTRY ABOUT TROY.

12 stadia distant from the sea; others at a spot more to the S.E., distant 42 stadia from the sea, now named *Hanarbashi*: the former opinion has in its favour the voice of antiquity, down to the time of Demetrius of Scepsis and Strabo, and must be received as most probably the correct view. The town is described in the Iliad as situated on rising ground[*] between the rivers Simois and Scamander;[7] to the S.E. rose a hill, a spur of Ida, on which stood the acropolis named Pergamum, containing temples and palaces: the city was surrounded with walls, and one of the gates leading to the N.W. was named the Scæan or "left gate." The town was believed to have been destroyed about B.C. 1184, and rebuilt at a later period, with the title of "*New Ilium*," in which Æolian colonists settled. This was probably the place which was visited by Xerxes, Alexander, and Julius Cæsar, and which, as the representative of the ancient Troy,[*] was enlarged and favoured by the Romans. During the Mithridatic war New Ilium was taken by Fimbria, B.C. 85, and suffered severely. In the neighbourhood were several spots associated with the Homeric poems—**Sigéum**, on the sea coast, where the mounds still exist which were reputed to cover the bodies of Achilles and Patroclus; and **Rhœtéum**, on the Hellespont, with the mound of Ajax: near each of these spots towns sprang up, the materials in the case of Sigeum being procured out of the ruins of Troy. **Alexandria Troas**, or, as it was sometimes briefly termed, **Troas**, stood on the coast opposite the S.E. point of the island of Tenedos: it owed its foundation to Antigonus, one of Alexander's generals, and its enlargement to Lysimachus, king of Thrace, who changed the original name of Antigonia into that of Alexandria. Its position rendered it valuable to the Romans, and they did much for it in the way of public works and buildings, of which an aqueduct to bring water from Mount Ida was the most remarkable. Julius Cæsar is said to have designed making it the Roman capital of the East, and

---

[*] The epithets applied to it are αἰπεινή, ὀφρυόεσσα, and ὀφρυόεσσα.

[7] Arisaraci trilcu, quam frigida parvi
Findunt Scamandri flumina
Lubricus et Simois.   Hor. *Ep.* 13, 13.

[*] The site of old Ilium was sought for in the neighbourhood. Cæsar's visit to it is described by Lucan in the following passage:—

Sigeasque petit famæ mirator arenas,
Et Simoentis squas, et Graio nobile busto
Rhœtion, et multum debentes vatibus, umbras.
Circuit exustæ nomen memorabile Trojæ,
Magnaque Phœbei quærit vestigia muri.
Jam sylvæ sterilea, et putres robore trunci
Assaraci pressere domos, et templa deorum
Jam fessa radice tenent, ac tota teguntur
Pergama dumetis: etiam periere ruinæ.
Adspicit Hesiones scopulos, syrisque latentes
Anchisæ thalamos: quo judex sederit antro:
Unde puer raptus cœlo: quo vertice Naïs
Luxerit Œnone: nullum est sine nomine saxum.
Inscius in sicco serpentem pulvere rivum
Transierat, qui Xanthus erat: securus in alto
Gramine ponebat gressus; Phryx incola manes
Hectoreos calcare vetat. Discussa jacebant
Saxa, nec ullius faciem servantis sacri.—ix. 961-976.

Constantine hesitated between this spot and Constantinople. The ruins of Troas supplied a large amount of stone for the erection of Constantinople. The Turks still call the site *Eski Stamboul*, "Old Constantinople." Assus stood on the southern coast of Troas, eastward of Prom. Lectum: it possessed a harbour formed by a mole, and must have been a flourishing place, to judge from the extensive ruins of temples, tombs, and other edifices, still existing on its site at *Beriam Kalesi*. Of these remains the Street of Tombs, a kind of Via Sacra, is the most remarkable. It was the birth-place of Cleanthes, and the temporary residence of Aristotle. Farther along the same coast we meet with Gargara,¹ surrounded by a plain of remarkable fertility, and Antandrus, the Pelasgis of Herodotus (vii. 42), advantageously situated under a spur of Ida, and thus supplied with abundance of timber for ship-building.¹ It was taken by the Persians in the reign of Darius, and, though it for a while shook them off in the time of the Peloponnesian War, it remained generally subject to them. Adramyttium, at the head of the bay named after it, rose to some importance as a seaport,² under the kings of Pergamum, and was the seat of a Conventus Juridicus under the Romans. Pergamum or Pergamus, *Bergamah*, was situated on the banks of the Caicus, near the junction of the streams Selinus and Cetius. Tradition assigned to it a Greek origin, but it remained an unimportant place until it was chosen by Lysimachus, one of Alexander's generals, as the receptacle of his vast treasures. Philetaerus, to whose care these were entrusted, rendered himself independent. The town was enlarged and embellished by one of his successors, Eumenes II., the founder of a magnificent library, second only to Alexandria: the massive substructure of some of the buildings still attests the solidity and splendour of the town. Pergamum remained a remarkably fine town under the Roman empire.³ Elaea, *Kliseli*, was situated on the bay to which it gave name, about 12 stadia S. of the mouth of the Caicus: it served as the port of Pergamum. Cyme,⁴ *Sandarli*, was on the coast, opposite the southern extremity of Lesbos: it was the most flourishing of all the Æolian towns, and has some few historical associations in connexion with the Ionian revolt. Scepsis, *Eski-Upsi*, was the chief town in the interior: it stood on the Æsepus, and was the seat of a school of philosophy: it was here that the works of Aristotle are said to have been buried in a pit after the death of Neleus, who had acquired them from Theophrastus.

We may further give a brief notice of the following towns of less importance :—Priâpus, on the Propontis, a Milesian colony, and the

---

¹ Nullo tantum ac Mysia culta

Jastat, et ipsa suos mirantur Gargara messes.—Virg. *Georg.* L. 102.

¹ Hence Æneas is represented as building his fleet here—

Classemque sub Ipsâ

Antandro et Phrygiæ molimur montibus Idæ.—Virg. *Æn.* III. 5.

¹ "A ship of Adramyttium" conveyed St. Paul from Caesarea (Acts xxvii. 2).

² It was celebrated for its manufacture of parchment, which derived its name (charta *Pergamena*) from this city. It has a still higher interest for us as being the site of one of the Seven Churches of Asia.

⁴ The Italian Cumae is said to have been partly founded by a native of Cyme, Hippocles, and to have derived its name from that circumstance. It was also the birth-place of Hesiod's father, and of the historian Ephorus.

F 2

chief seat of the worship of Priapus; **Parium**, *Kemer*, more to the W., with a good harbour, occupied by a mixed colony of Milesians, Erythraeans and Phocaeans; **Cremaste**, near Abydus, with gold-mines in its neighbourhood; **Sigèum**, the position of which has been already described, an Æolian colony, which was for a long time the source of dispute between Athens and Mitylene, but ultimately fell to the former, and became the residence of the Peisistratidæ; **Larisa**, near Alexandria Troas, an old Pelasgian town, but not regarded as the one to which Homer refers (*Il.* ii. 841); **Hamaxitus** and **Chrysa**, in the southern part of Troas, in both of which Apollo was worshipped under the form of a mouse, with the appellation of Sminthous; **Atarneus**, opposite Lesbos, for some time the residence of Aristotle, and the place where Histiaeus the Milesian was captured by the Persians; **Cane**, opposite the southern point of Lesbos, where the Roman fleet wintered in the war with Antiochus; **Pitane**, on the bay of Elæa, with two harbours; **Grynium**, on the coast S. of Elæa, the seat of a celebrated temple and oracle of Apollo;[5] **Myrina**, S.W. of Grynium, a strong place with a good harbour, occupied for a while by Philip of Macedonia in his wars with the Romans; **Ægæ**, a short distance from the coast, near Cyme; and **Temnos**, S. of the Hermus. The position of the old Homeric town **Lyrnessus**[6] is uncertain: it is usually placed near the sources of the Evenus. Several of the towns on the Bay of Elæa were destroyed by earthquakes in the first century of the Christian era; such was the fate of Temnus, Myrina, Elæa, Pitane, and Ægæ.

History.—The history of Mysia resolves itself into that of the towns which from time to time were dominant, this province having at no

Coin of Lampsacus.

period acquired any specific national existence. In the Heroic age Ilium was the seat of a small sovereignty, which survived the destruction of its capital, B.C. 1184, and was ultimately overthrown by the growing power of the Phrygians. At a later period Mysia formed a part successively of the Persian and Macedonian empires, and after the death of Alexander fell to the lot of the Seleucidæ. Gradually Pergamum became the seat of a petty sovereignty under the management of Philetaerus (B.C. 283-263), Eumenes I. (B.C. 263-241), and Attalus I. (B.C. 241-197), the latter of whom amassed enormous wealth, and established an alliance with Rome. At this period the possession of Mysia

---

[5] Hence Apollo is named Grynaeus:—

  Ille tibi Grynei nemoris dicatur origo.—VIRG. *Ecl.* vi. 72.

  Sed nunc Italiam magnam Grynaeus Apollo.—ID. *Æn.* iv. 345.

[6] It was the birthplace of Briseis:—

  Fertur et abducta Lyrnesside tristis Achilles.—OV. *Trist.* iv. 1. 15.

  Audierat, Lyrnesi, tuos, ablaeta, dolores.—ID. *Art. Am.* ii. 403.

Compare *Il.* II. 690, *Æn.* xii. 547.

was contested between the kings of Pergamum and Bithynia. Eumenes II. (B.C. 197-150) continued the Roman alliance, and received a large portion of Asia Minor for his territory in return for his services. He was succeeded by Attalus II. (b.c. 159-138), and he by Attalus III. (b.c. 138-133); who on his death bequeathed his kingdom to the Romans.

St. Paul's Travels.—Mysia was visited by St. Paul in his second journey. Though it was really a portion of "Asia" in the Biblical sense of the term, the ancient name of Mysia was retained as a territorial designation, as distinct, however, from that of the district of Troas. He entered it on the side of Galatia, and, descending to the coast, probably at Adramyttium, reached the town of Troas, and thence set sail for Macedonia (Acts xvi. 7-11). In his third journey he returned to this same spot from Philippi, and spent a week there: crossed over by land to Assus, following the Roman road which connects these two towns, and there took ship and coasted down the Gulf of Adramyttium to Mitylene, and thence southwards (Acts xx. 6-14). We may infer from 2 Cor. ii. 12 that he had visited previously Troas on his way from Ephesus to Macedonia in this same journey.

§ 15. The following islands lie off the coast of Mysia:—In the Propontis, **Proconnesus**, *Marmora*, which supplied Cyzicus and other places with the fine streaked marble to which it owes its modern appellation, with a town of the same name colonized by the Milesians—in the Ægean, **Tenedos**, *Tenedo*, 40 stadia distant from the coast, about 10 miles in circumference, with a town on its eastern coast which possessed a double harbour; and **Lesbos**, now named *Mitylene* after its ancient capital, situated in the Gulf of Adramyttium, and separated from the mainland by a channel about 7 miles broad. The shape of Lesbos is very irregular: it resembles a triangle, the three angles being formed by the promontories Argennum in the N., Sigrium in the S.W., and Malea in the S.E.: on the side between these two latter, two inlets penetrate deeply into the interior; one near Malea, probably the Portus Hieræus of Pliny, now *Port Hiero*, the other named Euripus Pyrrhæus, *Port Caloni*. The interior is mountainous, Olympus, in the S., attaining an elevation of above 3000 feet. The Pelasgians, Ionians, and Æolians, entered the island in succession; the latter race, however, became dominant, and here they retained a vigour both of intellect and character far beyond the standard of their race elsewhere: Lesbos has been rightly described as "the pearl of the Æolian race."[1] They possessed six cities—Methymna, *Molivo*, and Arisba, on the northern coast; Antissa and Eressus, near Cape Sigrium; Pyrrha, at the head of the Euripus named after it; and Mitylene, which retains its name, on the eastern coast, opposite the mainland. The last of these towns became, from its position and capacities, the natural capital of Lesbos: it was originally built on a small island,

---

[1] Niebuhr's *Lectures*, I. 213.

which was afterwards joined to the main island by a causeway, and thus a double harbour was made, the one N. of the causeway

Coin of Mitylene.

adapted for ships of war, and the southern for merchant-ships. The beauty of the town and the strength of its fortifications are noticed by several classical writers. Its history is involved in that of Lesbos itself, and will be noticed below. The Arginusæ were three small islands between Mitylene and the mainland, off which the ten Athenian generals defeated the Spartans, B.C. 406.

History of Tenedos and Lesbos.—Tenedos was a place of considerable importance so early as the time of the Trojan[1] War, and remained at

Coin of Tenedos.

all periods a valuable acquisition for warlike purposes, as it commanded the entrance of the Hellespont. During the Persian War it was occupied by the Persians: it sided with Athens in the Peloponnesian War, and was consequently ravaged by the Spartans, B.C. 389. Restored to Persia by the peace of Antalcidas,

it revolted on more than one occasion. In the Macedonian wars of the Romans it was held as a maritime station, and in the Mithridatic War was the scene of Lucullus's victory, B.C. 85. In the reign of Justinian it became an entrepôt for the corn-trade between Egypt and Constantinople. Lesbos appears as an important island in the Homeric poems. It joined the revolt of Aristagoras, and suffered severe punishment from the Persians. In the early part of the Peloponnesian War it sided with Athens: in the fourth year of the war, however, Mitylene revolted, and suffered the destruction of her walls and the forfeiture of her fleet : all the island, with the exception of the territory of Methymna, was divided among Athenian settlers. After the peace of Antalcidas it became independent. Alexander the Great made a treaty with it, and in course of time the Macedonian supremacy was established. In the Mithridatic War Mitylene was the last city that held out against the Romans, and was reduced by Minucius Thermus. Pompey made it a free city, and it became the chief

---

[1] Est in conspectu Tenedos, notissima famâ
Insula, dives opum, Priami dum regna manebant :
Nunc tantum sinus, et statio male fida carinis :
Huc se provecti deserto in litore condunt.—Virg. Æn. ii. 21.

town of the province of Asia. In addition to its historical fame,
Lesbos has acquired celebrity as the primitive seat of the music
of the lyre.[9] The lyre of Orpheus was believed to have been carried to
its shore by the waves. It was the birth-place of Lesches, Terpander,
Arion, and, above all, of Alcæus and Sappho. Its women were famed
for their beauty,[1] and, unfortunately, for their profligacy, which passed
into a proverb in the term Λεσβίαζειν. The historians Hellanicus and
Theophanes, and the philosophers Pittacus and Theophrastus, were
also Lesbians. Lastly, we must notice the healthiness of the climate,
justifying Tacitus's encomium, "insula nobilis et amœna;" and its
highly-prized wine.[2]

Ruins of Sardis.

---

[a] Hence the expression "*Lesbio plectro*" (Hor. *Carm.* i. 26, 11), and the
allusion in the lines—

   Age, dic Latinum,
   Barbite, carmen,
Lesbio primum modulate civi.—Id. *Carm.* i. 32, 3.

[1] Homer describes them in the complimentary terms—
   Ἀι κάλλει ἀρίστας φῦλα γυναικῶν.—*Il.* ix. 130, 272.

[2] Non eadem arboribus pendent vindemia nostris,
   Quam Methymnæo carpit de palmite Lesbos.—Virg. *Georg.* ii. 89.

Innocentis pocula Lesbii.—Hor. *Carm.* i. 17, 21.

Tu licet abjectus Tiberina molliter unda
   Lesbia Mentoreo vina bibas opere.—Prop. i. 14, 1.

II. Lydia.

§ 16. **Lydia** was bounded by the Ægæan Sea on the W., Mysia on the N., Phrygia on the E., and Caria on the S. In the latter direction the boundary was carried down by Strabo to the Mæander; the range of Messogis, however, forms the more correct limit. Within these limits is included the northern part of Ionia, which stretches along the sea-coast from the Hermæan Bay in the N.

Lydia is mountainous in its southern and western parts, but it contains extensive plains and valleys between the various ranges. It is one of the most fertile countries in the world, even the sides of the mountains admitting of cultivation; its climate is mild and healthy, and the chief drawback to the country is the frequency of earthquakes. In the eastern portion of the province there are evident traces of volcanic action: numerous extinct volcanos, and particularly three conical hills of scoriæ and ashes, with deep craters, and lava-streams issuing from them, are found on an extensive plain, to which the ancients gave the name Catacecaumēne, i.e. "burnt." The most important productions of Lydia were an excellent kind of wine, saffron, and gold.

§ 17. The chief mountain-ranges are **Tmolus** and **Messogis**, whose general direction has been already described (p. 86). The former ramifies into several subordinate ranges towards the W., viz.: **Draæon** and **Olympus** in the direction of the Hermæan Bay—**Sipylus** more to the N., the fabled scene of Niobe's transformation—the isolated height of **Gallesius**, in the neighbourhood of Ephesus—and the irregular cluster of hills which form the peninsula of Erythræ, named **Corycus** and **Mimas**, and which terminate on the shores of the Ægean in the promontories of Melæna, Argennum opposite Chios, and Corycæum. The slopes of Tmolus were clothed with vines,[4] and it was

---

Νῦν δέ που ἐν κόρησαν, ἐν οὔρεσιν οἰοπόλοισιν
'Εν Σιπύλῳ, ὅθι φασὶ θεάων ἔμμεναι εὐνὰς
Νυμφάων, ——　　　　　　　　　　Hom. Il. xxiv. 614.
Flet tamen, et validi circumdata turbine venti
In patriam rapta est. Ibi fixa cacumine montis
Liquitur, et lacrymas etiamnum marmora manant.—Ov. Met. vi. 310.
The mountain is said from a certain point of view to assume the appearance of a woman weeping.

[4] Virgil praises them in Georg. ii. 98, and Ovid in the following lines:—
Jamque nemus Bacchi Tmoli vineta, tenebat.—Fast. ii. 313.
Cumque choro meliore, sui vineta Tymoli,
Pactolonque petit.　　　　　　　　　Met. xi. 86.
Saffron also grew plentifully upon it:—
Nonne vides crocos ut Tmolus odores.—Virg. Georg. i. 56.
The prominent appearance of Tmolus in the landscape is well described by Ovid:—
Nam freta prospiciens late riget arduus alto
Tmolus in adscensu.　　　　　　　Met. xi. 150.

rich in gold mines. With Messogis is connected the range of
Pactyas, which presses close on the Cayster near Ephesus; and its
westerly continuation Mycale, terminating in the promontory of
Trogillum. St. Marie, opposite Samos : the name of Mycale is rendered illustrious by the battle between the Greeks and Persians,
fought partly on the beach at its foot, partly in the adjacent channel,
B.C. 479. The line of coast is very irregular, two bays penetrating
deeply into the interior on each side of the peninsula of Erythræ,
viz. the Hermæus Sinus, *G. of Smyrna*, on the N., and Caistrianus
Sin., *G. of Scala Nuova*, on the S.

§ 18. The chief river is the Hermus, *Kodus-chai*, which rises in
the Phrygian Mount Dindymus, and flows with a very devious
course, but with a general westerly direction, into the bay to which
it communicates its name, receiving on its right bank the Hyllus
and Lycus, and on its left the Cogamus and Pactolus, *Sarabat*. The
plains through which it flows in its middle course are broad and
fertile : that which stretches from Magnesia to Sardis was specifically
named Hermæus campus, while a more northerly portion was named
Hyrcanus campus. Both the Hermus[5] and the Pactolus[6] are said
to have carried down large quantities of gold-dust from Mount
Tmolus. In the S. of Lydia, between Tmolus and Messogis, is the
river Cayster, *Little Meinder*, which rises on the slope of Tmolus,
and winding about the flat rich plains which border it, falls into the
gulf named after it, near Ephesus. The upper plains of the Cayster
were named Cilbiani campi, and were divided into "upper" and
"lower." The broader plains about its mid course were the proper
Caystriani campi, while near its mouth was a narrow maritime plain
shut off from the central plain by the projecting spurs of Pactyas
and Gallesius. This last was the original Ἄσιος λειμών of Homer
(*Il.* ii. 461), the favourite resort of wild-fowl,[7] particularly swans.

---

[5] Auro turbidus Hermus.—VIRG. *Georg.* II. 137.

Mæonium non ille vadum, non Lydia mallet
Stagna sibi, nec qui rigno perfunditur auro
Campum, atque illatis Hermi flavescit arenis.—SIL. ITAL. i. 137.

Aut quales referunt Baccho sollennia Nymphæ
Mæoniæ, quas Hermus alit, ripasque paternas
Percurrunt auro madidæ : lætatur in antro
Amnis, et undantem declinat prodigus urnam.
                                        CLAUD. *Rapt. Pros.* II. 67.

[6] Pactolusque irrigat auro.—VIRG. *Æn.* x. 142.

Sis pecore et multa dives tellure licebit
Tibique Pactolus fluat.   HOR. *Epod.* xv. 19.

[7] Jam variæ pelagi volucres, et quæ Asia circum
Dulcibus in stagnis rimantur prata Caystri.—VIRG. *Georg.* i. 383.

Ceu quondam nivei liquida inter nubila eycni,
Cum sese e pastu referunt, et longa canoros

§ 19. The earliest inhabitants of this province were the Mæönes, a Tyrrhenian or Pelasgian race. The Lydians, whose name first appears in the poems of Mimnermus, were a kindred race to the Carians and Mysians, and gradually overpowered the Mæonians, probably about the time when the Mermnadæ supplanted the Heracleid dynasty. In addition to these the Hellenic race contributed an important element in the colonies which were planted along the sea-coast at different periods and by various branches of the Hellenic race, among whom the Ionians became dominant, and communicated their name to the district.

§ 20. The towns of Lydia may be arranged into two classes—the Greek towns which lined the coast, and the old Lydian towns of the interior, situated amid the fertile plains of the Hermus and Cayster. The former comprised Phocæa, Smyrna, Clazomenæ, Erythræ, Teos, Lebëdos, Colöphon, and Ephesus, which were members of the Ionic confederation. The sites in most instances had been previously occupied by Carians, Lelegea, and other kindred races; and Smyrna at a later period by Æolians. The Ionians seized them, and their choice justifies the character for taste which Herodotus (i. 142) imputes to this race. Of the Lydian towns we know but little. Sardis is the only one which comes prominently forward. The hostilities which existed between the Lydian monarchs and the Greek cities of the coast bring into early notice Smyrna, Clazomenæ, and Colophon, the first of which was utterly destroyed. Sardis itself, after the death of Crœsus, retained its position as the residence of the Persian governors, but was never a place of commercial importance. The Greek towns succumbed to Persia after the Ionian revolt. Phocæa, which had hitherto been the first in commercial enterprise, sunk at this period, through the withdrawal of its inhabitants. The Alexandrian age witnessed the rebuilding of Smyrna, the ruin of Lebedus and Colophon, whose inhabitants were removed by Lysimachus to Ephesus, and the rise of Ephesus to a state of commercial eminence. Thyatira and Philadelphia belong to a somewhat later period—the former owing its name to Seleucus Nicator, the latter to one of the kings of Pergamum. In the Syrian wars Smyrna, Erythræ, and the Colophonians of Notium, sided with Rome, and received various immunities in return. On the consti-

---

Dant per colla modos : sonat amnis, et Acis longe
Palm palus.   VIRG. Æn. vii. 699.

Hic niger, in ripis errat quum forte Caystri,
  Inter Ledæos ridetur corvus olores.—MART. l. 34.

Utque jacens ripâ deficre Caystrius ales
  Dicitur ore suam deficiente necem,
Sic ego, Sarmaticas longe projectus in oras,
  Efficio, tacitum ne mihi funus est.—OV. Trist. v. 1, 11.

tution of the province of Asia, Ephesus was selected as the capital, and was thenceforth the capital of the whole surrounding district. Most of the cities of Lydia suffered severely from an earthquake in the reign of Tiberius. We shall describe these towns in their order from N. to S.

Site of Ephesus.

Phocæa stood at the head of a small inlet on the peninsula between the bays of Cyme and Hermus. The mouth of the inlet was closed by the island of Bacchium, which contained the chief public buildings, and protected the two harbours of the town. Phocæa became a place of commercial importance, and must have been strongly fortified. It was besieged by Harpagus in the Ionian War, on which occasion the greater part of its population emigrated to Corsica.[2] It revived, however, and was strong enough to sustain a long siege from the Roman fleet under Æmilius in the Syrian War. Its ruins retain the ancient name, Palæo Foggia. Smyrna was originally built on the northern side of the Hermæan bay, near its head. This was destroyed by Alyattes, B.C.

---

[2] Nulla sit hac potior sententia (Phocæorum
     Velut profugit exsecrata civitas
     Agros atque lares patrios, habitandaque fana
     Apris reliquit et rapacibus lupis),
     Ire, pedes quocunque ferent, quocunque per undas
     Notus vocabit, aut protervus Africus.—Hor. Epod. 16, 17.

The Phocæans are said to have founded Massalia on this occasion : but the traditions in regard to this vary. The Latin poets use the term *Phocaïcus* as a synonym for Massilian : e. g.

     Scipio Phocaïcis amne referebat ab oris.—Sil. iv. 52.

See also i. 835, and Luc. iii. 301.

The purple shell-fish was abundant on this part of the coast :—

     Phocaico bibulas tingebat murice lanas.—Ov. Met. vi. 9.

627, and for 400 years the town ceased to exist. A second town, named New Smyrna, was then founded on the southern side of the bay by Antigonus, and completed by Lysimachus. The former was the old colony of the Æolians, and, subsequently to B.C. 688, of the Ionians. The latter was the Smyrna which attracted so much admiration by the beauty of its streets and the excellence of its harbour; and which has a special interest for the Christian as the seat of one of the Seven Churches, and the scene of St. Polycarp's martyrdom. Smyrna, alone of the Ionic towns, retains its ancient importance, and is the chief mart of the Levant. The Cyclopean walls of the Acropolis mark the site of the old town: the stadium and theatre are the most striking remains of the new town. It claimed to be Homer's birth-place,[9] and had a temple erected to him. **Clazomenæ** was on the southern coast of the Hermæan bay, at the entrance of the peninsula on which Erythræ stood.

Coin of Clazomenæ.

Originally on the mainland, the town was transferred to an adjacent island, which was at a later period turned into a peninsula by a causeway connecting it with the coast. It derives its chief interest from being the birth-place of Anaxagoras. **Erythræ** was situated at the head of a capacious bay opposite the island of Chios, the entrance to which was partially closed by a small group of islands named Hippi. Its history is unimportant. The remains at *Ritri* consist of the ancient walls, a theatre scooped out of the solid rock, and traces of aqueducts and terraces. **Teos** stood opposite Clazomenæ, on the southern side of the Erythræan peninsula. Under the Persians its inhabitants emigrated to Abdera in Thrace; and the town, though still existing in the time of the Peloponnesian War, ceased to be of importance. There are interesting remains of a theatre and of a splendid temple of Bacchus at *Sighajik*, one of the ports of the city. It produced two illustrious men, Anacreon[1] and Hecatæus. **Lebedus** stood on the coast about 10 miles E. of *Prom*. Myonnesus, and by its commerce and the fertility of its district it flourished until the removal of the bulk of its inhabitants to Ephesus by Lysimachus. Under the Romans it was a poor deserted place,[2] but attained some celebrity as the head-quarters of the guild of

---

[9] Hence the expressions *Smyrnœus vates* (Luc. ix. 984), and *Smyrnœo plectra* (Sil. viii. 596) : so also Statius,

　　　　　Non si pariter mihi vertice lœto
　　Nectal adorains et Smyrna et Mantua laurus,
　　Digna loquar.　　　　　　*Silv.* iv. 2, 3.

[1] Vitabis œstus, et fide Teia
　　Dices laborantem in uno
　　Penelopen, vitreamque Circen.—Hor. *Carm.* l. 18, 19.
　　Anacreonta Teium.—Ib. *Epod.* xiv. 10.
　　Sit quoque vinosi Teia Musa senis.—Ov. *Art. Am.* iii. 330.

[2] An Lebedum laudas, odio maris atque viarum ?
　　Scis Lebedus quid sit ! Gabiis desertior atque
　　Fidenis vicus.　　　　　　Hor. *Ep.* i. 11, 6.

actors. A few shattered masses of masonry at *Ecclesia* are all that remains of it. Colophon was on the banks of the small river Hales, about 2 miles distant from the shore and from its port of Notium, with respect to which Colophon was designated the "upper city" (Thuc. iii. 34). Its history is almost wholly concerned with the disputes of its own citizens. After the removal of its inhabitants by Lysimachus, it sunk; but Notium still existed, and was unsuccessfully besieged by Antiochus, B.C. 190. It claimed to be the birth-place of Homer, and produced Mimnermus the poet. Ephesus was finely situated near the spot where the Cayster discharges itself into the head of the bay named after it. The original town of Androclus was on the slope of Coressus: thence it gradually spread over the adjacent plain, and was afterwards extended by Lysimachus over the heights of Prion. Down to the Alexandrian age, Ephesus derived its importance almost entirely from its connexion with the worship of Diana: under Lysimachus it became a commercial town, and under the Romans[3] it attained its greatest prosperity as the capital of the province of Asia. The original temple of Diana existed on the spot before the Ionians came there: the first Greek edifice, erected about the 6th century B.C., perished by fire on the night of Alexander's birth. A new one was erected, 425 feet in length, and 220 in width, adorned (according to Pliny) with 127 columns, each 60 feet high. It was the largest of all the Greek temples. This was the temple which existed in St. Paul's time, and survived until Christianity over-spread the land. The trade of Ephesus under the Romans was considerable: it had easy access to the interior of Asia Minor, and possessed an excellent double harbour. It has acquired an especial interest for the Christian from the visit which St.

Coin of Ephesus.

Paul made to it, and the dangers to which he was exposed from the worshippers of Diana. He founded a Church there, to which he addressed an epistle: this was one of the Seven Churches of Asia. Ephesus has perished through the extinction of its port by the deposits of the Cayster. Numerous remains of it exist at *Ayasaluk*, but the site of the great temple has not been made out. The stadium, the theatre (which was the scene of the tumult raised by Demetrius), and the agora, are the most remarkable objects. Sardis, the old capital of the Lydian monarchy, was well situated on the plain between Mount Tmolus and Hermus, on both sides of the small river Pactolus, with its acropolis posted on a precipitous height. It was destroyed by fire on three occasions: by the Cimmerians in the reign of Ardys, by the Ionians at the time of their revolt, and by Antiochus the Great in his war with Achæus. It was the seat of one of the Seven Churches. A small village, named *Sert*, still exists on the site, with the remains of a stadium, theatre, and the walls of the acropolis. Magnesia,

---

[3] Ephesus was much admired by them :—

Laudabunt alii claram Rhodon, aut Mitylenen,
    Aut Epheson.   Hor. Carm. L. 7, 1.

Magnesia, surnamed ad Sipylum, to distinguish it from the town on the Mæander, stood on the left bank of the Hermus, and is celebrated for the victory gained by the Scipios over Antiochus the Great, B.C. 190. Though destroyed by the earthquake in Tiberius's reign, it revived, and existed down to the 5th century. **Philadelphia**, on the Cogamus, was founded by Attalus Philadelphus, king of Pergamum, and derives its interest from having been one of the Seven Churches. Parts of its walls and of its ruined churches, twenty-four in number, exist at *Allahsher*. **Thyatira**, between Sardis and Pergamum, is frequently noticed in the history of the Roman wars with Antiochus. It is better known, however, as one of the Seven Churches, and the abode of Lydia the purple-seller.

We may briefly notice the following less important towns:—**Leuce**, S. of Phocæa, the scene of a battle between Licinius Crassus and Aristonicus,

B.C. 131; **Clarus**,[4] near Colophon, the seat of a famous temple and oracle of Apollo; **Pygela**, S. of Ephesus, with a temple of Diana; and **Metropolis**, N.W. of Ephesus, which produced an excellent kind of wine.

History of the Lydian Empire.—According to Herodotus (i. 7), Lydia was successively governed by three dynasties—the Atyadæ, down to about B.C. 1200; the Heracleidæ, down to about B.C. 700; and the Mermnadæ, down to B.C. 546. The dates are still undecided, the death of Crœsus being sometimes placed in 554. The two first of these dynasties are almost wholly mythical: real history commences with the third. The first of this race, **Gyges**, B.C. 713, instituted an aggressive policy against the Greeks of the sea-coast, waging war with Miletus and Smyrna, and capturing Colophon. His successor Ardys, B.C. 678, carried on the war, and captured Priene. The latter part of his reign was disturbed by the Cimmerian invasion. Alyattes, B.C. 617, expelled the Cimmerians, and extended his dominion as far as the Halys, where he came in contact with Cyaxares: he also conquered most of the Greek cities. The tomb of Alyattes, which Herodotus (i. 93) describes as only inferior to the monuments of Egypt and Babylon, is still extant. It is an immense mound of earth about half a mile in circumference. In the centre a sepulchural chamber has been recently discovered. Crœsus, B.C. 560, raised the power of the Lydian throne to its highest pitch of greatness, his authority on the western side of the Halys being undisputed. He was conquered by Cyrus, and his territories added to the Persian empire; and thenceforth the history of Lydia is involved in that of the peninsula generally.

Coin of Smyrna.

---

[4] Phœbi

Qui tripodas, Clarii lauros, qui sidera sentis.—Virg. Æn. iii. 359.

Mihi Delphica tellus

Et Claros, et Tenedos, Patareaque regia servit.—Ov. Met. i. 515.

Hence Clarius is an appropriate epithet of a poet:—

Nec tantum Clario Lyde dilecta poetæ.—Ov. Trist.-i. 6, 1.

St. Paul's Travels.—St. Paul's first visit to Lydia occurred in the course of his second apostolical journey, when he touched at Ephesus on his return from Greece: on that occasion his stay was but short (Acts xviii. 19-21).   On his third journey he must have traversed Lydia on his way from Phrygia to Ephesus.   The route he pursued is a matter of conjecture: as he probably never visited Colossae, he may have descended the valley of the Hermus, and crossed from Sardis to Ephesus.   He remained in Ephesus three years, during which he appears, from expressions in his Second Epistle to the Corinthians, to have paid a short visit to Corinth.   At the conclusion of his visit he went northwards, probably by sea, to Troas (Acts xix.).

§ 21. Off the coast of Lydia lies the important island of Chios, *Scio,* separated from the mainland by a channel 5 miles in width. Its length from N. to S. is about 32 miles; its width varies from 18 to 8 miles; and its area is 400 square miles, or about thrice the area of the Isle of Wight.   The whole island is rocky and uneven;[5] the mountains of the northern portion rise to a great height, and form a striking object from the coast of Asia Minor.   The most valuable productions were the wine[6] which the Roman writers describe as " vinum arvisium," and the gum-mastic produced from the lentiscus tree.   The Chian women were famed for beauty.   The highest summit in the island was named Pelinæus, *Mount Elias;* the chief promontories were Posidium, *Mastico,* on the S., Phanæ[7] on the W., and Melæna, *S. Nicolo,* on the N.W.   The oldest inhabitants were either Pelasgi or Leleges; settlers from Crete, Euboea, and Caria, afterwards entered.   The chief town, also named Chios, stood on the eastern coast, on the site of the modern capital; no remains of antiquity have been found there.   Delphinium, on the same coast, was a strong position.

---

[5] Ἡ καθύπερθε Χίοιο ποιμαίνω παιπαλοέσσης,
        Νήσου ἐπὶ ψυρίης, αὐτὴν ἐπ' ἀριστέρ' ἔχοντες,
        ῾Ἡ ὑπένερθε Χίοιο, παρ' ἠνεμόεντα Μίμαντα—Hom. *Od.* III. 170

[6] Quo Chium pretio cadum
        Mercemur.   Hor. *Carm.* III. 19, 5.

        At sermo lingua concinnus utrâque
        Suavior: ut Chio nota si commixta Falerni est.—Id. *Sat.* I. x. 23.

  It should be observed that the quantity of the penultimate is different in the substantive and adjective:—

        Quid tibi visa Chios?—Hor. *Ep.* I. 11, 1.

        Capaciores affer huc, puer, scyphos
        Et Chia vina, aut Lesbia.—Id. *Epod.* ix. 33.

  The figs of Chios are celebrated by Martial:—

        Chia seni similis Baccho, quem Setia misit
        Ipsa merum secum portat et ipsa salem.— vIII. 23.

        Nam mihi, quæ novit pungere, Chia sapit.— vII. 25.

[7] The grape of Phanæ was famed :—
        Rex ipse Phanæus.—Virg. *Georg.* II. 98.

History.—Chios was a member of the Ionian confederation, and held a conspicuous place as a maritime power until the Ionian revolt, when

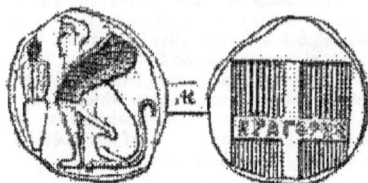

Coin of Chios.

it became subject to Persia, and remained in that position until the battle of Mycale, B.C. 479, when it joined Athens, and remained among its allies until B.C. 412, when it revolted, and was in consequence devastated. It sided with the Romans in the Syrian and Mithridatic wars, and was gifted with freedom in reward for its fidelity. Chios claimed Homer as one of her sons, and gave birth to the historian Theopompus, and the poets Theocritus and Ion.

§ 22. The important island of Samos, *Samo*, is situated just opposite the point where Lydia and Caria meet, and is separated from the mainland by a channel less than a mile in width, which was the scene of the battle of Mycale. Its length from E. to W. is about 25 miles; its breadth is very variable. The island is covered with mountains of great elevation, rendering Samos a very conspicuous object in the landscape. It is to this that it owes its name Samos, "a height." The island was productive to a proverb, and famed for its dried grapes and other fruits. It possessed a stone used for polishing gold, and its earthenware was so prized at Rome that the title "Samian ware" was transferred to the red lustrous pottery of the Roman manufacturers. The general name for the mountain range which traverses the island was Ampělus. It culminates in a height named Cercěteus, *Kerkis*, at an elevation of 4725 feet, and terminates in the promontories of Posidium in the E., and Cantharium in the W. The original inhabitants were Carians and Leleges. Colonies of Æolians from Lesbos, and Ionians from Epidaurus, afterwards settled on it. The principal town, also named Samos, was situated on the S. coast, at the extremity of a plain, at the other end of which stood the famed temple of Juno.[6] Under Polycrates it ranked as the greatest[7] city in the world; its harbour protected by a double mole, and an immense tunnel which formed

---

[6] Hence the affection with which Juno was supposed to regard the island.

Quam Juno fertur terris magis omnibus unam
Posthabită coluisse Samo.   VIRG. Æn. l. 15.

Et jam Junonia lævă
Parte Samos fuerant, Delosque, Parosque relicto.
                                   OV. Met. viii. 220.

[7] Horace characterizes it as "*concinna Samos*" (*Ep.* l. 11, 2): it was among the spots which the Romans most admired :—

Romæ laudetur Samos et Chios et Rhodos absens.—ID. 21.

an aqueduct, were the most remarkable objects in it. The town lay partly on the plain, partly on the slopes of the hills that back it, on one of which, named Astypalæa, the citadel was posted. The theatre and a portion of the walls alone remain. The temple of Juno was of enormous size—346 feet long and 189 broad, of the Ionic order, and decorated with statues and paintings. It was burnt by the Persians; and, after its restoration, plundered by the pirates in the Mithridatic War, by Verres, and by M. Antony.

History.—Samos was at an early period famed for its commercial enterprise, and was an influential member of the Ionian confederacy. Under Polycrates (B.C. 532) it became the greatest Greek maritime power, and entered into commercial relations with Egypt. After his death it became subject to Persia, and remained so until the battle of Mycale, B.C. 479, after which it joined Athens, and adhered to that power through the Peloponnesian War. In the Syrian wars it sided with Antiochus against

Coin of Samos.

Rome: in the Mithridatic it adopted a similar policy. It was united with the province of Asia B.C. 84. Its prosperity gradually decayed under the Roman emperors. Samos was the birth-place of the philosophers Pythagoras[1] and Melissus, and the poets Asius, Chœrilus, and Æschrion.

§ 23. The small island of **Psyra**, *Ipsara*, lies about 6 miles from the N.W. point of Chios, and the Œnussæ between Chios and the mainland. **Icarus** or **Icaria**,[2] *Nikaria*, is 10 miles distant from Samos, and may be regarded as a continuation of the elevated chain which forms that island. It extends from N.E. to S.W., with a length of about 17 miles. Its inhabitants were originally Milesians, but it afterwards belonged to the Samians. It possessed the towns of Isti, Œnoë, and Drepanum, or Dracanum—the latter situated near the promontory of the same name at the E. end of the island. The surrounding sea was named Icarium Mare.

---

[1] Vir fuit hic, ortu Samius : sed fugerat una
Et Samon et dominos, odioque tyrannidis exsul
Sponte erat.   Ov. *Met.* xv. 60.
Samii sunt rata dicta senis.   In. *Trist.* III. 3, 62.

[2] The name is connected in mythology with Icarus, the son of Dædalus.

Transit et Icarium, lapsas ubi perdidit alas
Icarus, et vasto nomina fecit aquæ.—Ov. *Fast.* iv. 283.

Ceratis ope Dædalea
Nititur pennis, vitreo daturus
Nomina ponto.   Hor. *Carm.* iv. 2, 2.

Ruins of Miletus.

CHAPTER VIII.

Asia Minor (*continued*).—Caria, Lycia, Pamphylia, Cilicia.

III. CARIA.

§ 1. **Caria** occupied the south-west angle of Asia Minor, and was bounded on the W. and S. by the sea, on the N. by Messogis dividing it from Lydia, and on the E. by the river Glaucus and Lycia. Though generally a mountainous country, it contains some extensive valleys and a great deal of rich land in the basin of the Mæander. The Peræa, or southern district, is a beautiful country, and contains some fertile tracts. Timber is abundant, and the country produces good grain and fruits, the fig[1] and the olive. The

---

[1] Dried figs were named *Carica*, lit. "Carian figs," by the Latins :—

Ille nux, his mixta est rugosis carica palmis.—Ov. *Met.* viii. 674.

climate varies with the varying altitude: the highest tracts arc
cold and wintry, while it is hot in the lower grounds. The former
supplied pasturage for large flocks of sheep, and even now the green
slopes near Alabanda are covered with flocks. The wool of Miletus
and the wine of Cnidus were the chief exports. The limestone of
the country furnished excellent building-material.

§ 2. The mountain-ranges of Caria are connected with the Taurus
system. The watershed which divides the basin of the Mæander
from the Calbis and the other streams that seek the Mediterranean,
is formed by a range which emanates from Cadmus in the N.E. angle
of the province, and which takes first a southerly and then a
westerly direction, terminating in the peninsula of Halicarnassus :
near the southern coast the ridge was named Lide. From this range
lateral ridges strike off towards the N.W., in the direction of the
Mæander, and form the valleys in which its tributaries flow : the
most westerly of these was named Latmus,[2] terminating in the
subordinate ridge of Grion, near Miletus. The sea-coast is irregular ;
the Latmicus Sinus once extended inland to the roots of the hills, but
has long since been filled up by the alluvium of the Mæander :
between Grion and Lide lies the Iasius Sin., *Gulf of Asynkalessi*,
with a much indented line of coast : between Lide and the high
ground which forms the peninsula of Cnidus, the deep inlet named
Ceramicus Sin., *G. of Budrum*: and on the other side of Cnidus the
irregular gulf in front of the isle of Syme, containing the three
lesser bays named Bubassius, Schœnus, and Thymnias. The peninsulas form the most striking feature in the outline of Caria : that
on which Miletus stood was of a triangular shape, the southern
point forming the promontory of Poseidium; the peninsula of Halicarnassus narrowed at the point where the town stood, and again
expanding ended in the promontories of Zephyrium, Astypalæa, and
Termerium; the peninsula of Cnidus is about 40 miles long, and nowhere more than 10 miles broad, and terminates in the promontory
of Triopium, *Crio :* it is contracted to a narrow neck in two places,
viz. at the point where it connects with the mainland, and midway
at the Bubassius Sinus : there is thus a double peninsula, to which
Herodotus (i. 174) gives the distinctive names of the Triopian and
the Bybassian, and it was at the junction of these two that the
Cnidians cut their canal in the Persian War. The peninsula on the
eastern side of the bay of Schœnus is formed by a ridge named
Phœnix, which terminates in Cynossema, "the Dog's tomb," now
C. Volpo : lastly, another peninsula is formed between the Calbis

---

[2] Latmus was the fabled scene of Diana's interviews with Endymion, to whom
the epithet *Latmius* is hence applied by the Latin poets (Ov. *Trist.* ii. 299 ; Val.
Flacc. viii. 28 ; Stat. *Silv.* iii. 4, 40).

and the Gulf of Glaucus, which terminates in the promontory of **Pedalium** or **Artemisium**. The scenery along the coast is very fine, the rocks in many places rising abruptly from the sea.

§ 3. The chief river of Caria is the **Mæander**, *Meinder*, which rises near Celænæ in Phrygia, having its sources in a lake, whence issues also one of its tributaries, the Marsyas: its course takes a south-western direction, skirting the southern slopes of the range of Messogis, and is remarkable for its extreme tortuousness,[3] whence the term "to mæander"[4] owes its origin. The stream is deep, but not broad: it frequently overflows its banks, and carries down an immense amount of deposit.

The less important streams were, for the most part, tributaries of the Mæander: on its right bank it receives the **Lethæus**, which joins it near Magnesia, and the **Gæsus**, which flowed by Priune; on its left bank, the **Orsinas** or **Mosynus**, *Hagisik*; the **Harpasus**, *Harpa*; and the **Marsyas**, *Tshina*, which rises near Stratouicea, and joins the Mæander opposite to Tralles. We have yet to notice the **Calbis** or **Indus**, *Taros*, which rises in Mount Cadmus, and, flowing to the S., joins the Mediterranean near Caunus.

§ 4. Caria was occupied by the following races—the Carians, who believed themselves to be autochthonous, but, according to the Greeks, were emigrants from Crete—the Caunians, who may have been allied to them, and who were settled on the south coast—and the Hellenic races of the Ionians and Dorians, the former of whom occupied the western coast as far as the Iasian bay, while the latter held the promontories of Halicarnassus and Cnidus. The Carians are represented as a warlike race,[5] serving as mercenaries under any one who was willing to pay them. Their language differed from that of the Greek settlers,[6] although the two people probably became intermixed. The southern coast between these peninsulas and the Calbis was designated Peræa, or more fully Peræa Rhodiorum, as it once belonged to Rhodes.

§ 5. Caria possessed some of the most flourishing and magnificent towns of Asia Minor, especially Milétus, the metropolis of Ionia,

---

[3] Non secus ac liquidus Phrygiis Mæandros in arvis
Ludit, et ambiguo lapsu refluitque fluitque :
Occurrensque sibi venturas aspicit undas :
Et nunc ad fontes, nunc in mare versus apertum,
Incertas exercet aquas.　　　　　　Ov. *Met.* viii. 162.
Mæandros, toties qui terris errat in isdem,
Qui lapsu in se sæpe retorquet aquam.—In. *Herald.* ix. 55.

[4] The following lines supply us with an instance of the metaphorical use of the term :—
Victori chlamydem auratam, quam plurima circum
Purpura *Mæandro duplici* Melibœa cucurrit.—Vira. *Æn.* v. 250.

[5] Theocritus (*Id.* xvii. 89) describes them as φιλοπτολέμους.

[6] Hence Homer characterises them as βαρβαροφώνους (*Il.* ii. 867).

and the first maritime power of Western Asia—Mylasa, the ancient
capital of Caria—Halicarnassus, the greatest of the Dorian colonies
—Tralles and Alabanda, which passed into a proverb for wealth and
luxury—and Cnidus, a seat both of commerce and art. Most of
these towns possessed buildings of celebrity : we may instance the
temple of Branchidæ near Miletus, the Mausoleum at Halicarnassus,
and the temple of Labranda near Mylasa. In addition to these, the
following less important towns had magnificent temples—Magnesia,
Aphrodisias, and Eurômus; while others can show to this day the
remains of fine theatres and other public buildings. These towns
and works of art testify to the extent of Greek influence in this
country : with the exception of Mylasa, indeed, they all claimed a
Greek origin. Three towns belonging to the Ionian confederacy—
Priéne, Myus, and Miletus—were grouped on the shores of the
Latmian bay : they decayed through natural causes, the alluvium
of the Mæander gradually turning the bay into a pestilential marsh :
the two former ceased to exist even in classical times : Miletus
survived until the Middle Ages, but the period of its commercial
greatness terminated with its capture by the Persians, B.C. 494.
The Dorian towns were situated on the southern peninsulas : the
position of Halicarnassus was one of great natural strength, and it
became, during the Persian period, the virtual capital of Caria : it
fell after its capture by Alexander. Cnidus was, from its central
position, the metropolis of the Dorian confederacy, and flourished
down to the period of the Roman empire. A few towns rose under
the Seleucidæ : they were situated in the valley of the Mæander :
Antiochia, Stratonicéa, and probably Aphrodisias, belong to this
period : these towns continued to exist under the later Roman
empire. The great fertility of the soil seems to have been the
foundation of the wealth of the towns of the interior : Tralles,
Alabanda, and Mylasa, were all surrounded by remarkably fertile
districts. We shall describe these towns in their order from N. to S.
taking firstly those which stood on or near the sea-coast, and
secondly, those of the interior.

1. **Magnesia** stood on the Lethæus, a short distance from the right
bank of the Mæander, surrounded by a plain of great fertility. Originally an Æolian town, it was destroyed by the Cimmerians about
B.C. 726, and was re-occupied by Milesian colonists ; it is known as the
residence of Themistocles, and as possessing a splendid temple of
Artemis Leucophryne, the ruins of which are found at Inek-bazar.
Priéne was well situated on the terraced slope of Mycale, and in ancient
times stood immediately on the coast of the Bay of Latmus, from
which, however, it was removed a distance of 40 stades even in
Strabo's time by the alluvial deposits of the Mæander. The two ports,
which it originally possessed, were thus filled up, and the town early
sunk into insignificance. It was the birth-place of Bias. Remains of it
exist near Samsoon, particularly the ruins of the temple of Athena

Polias.[7] **Myus**, the smallest of the Ionian towns, was on the southern bank of the Mæander, about 30 stades from its mouth: it was one of the towns given to Themistocles by the Persian king: it was afterwards connected with Miletus, which finally received its inhabitants. **Milétus** occupied a peninsula at the southern entrance of the Bay of Latmus ; it consisted of an inner and an outer city, with their separate

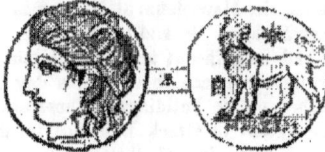

Coin of Miletus.

fortifications, and four harbours, which were protected on the sea-side by Lade and the other islands which formed the Tragasæan group. Down to the period of the Ionian revolt, B.C. 494, Miletus enjoyed the highest commercial prosperity, and planted its colonies along the shores of the Ægean, the Hellespont, the Propontis, and the Euxine: it was exposed to contests with the Lydian kings Ardys, Sadyattes, and Alyattes, and ultimately yielded to Crœsus.

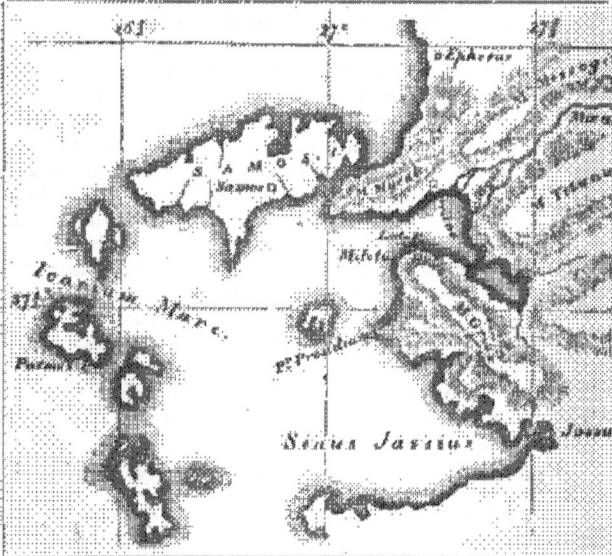

Chart of the Coast about Miletus.

From 494, when the city was plundered and its inhabitants removed by Darius, it was subject to Persia until the battle of Mycale, B.C. 479, when it became independent, and joined Athens, with which it was

---

[7] The ruins of this temple afford a fine specimen of Ionic architecture, of about the same date as the Mausoleum.

connected until nearly the end of the Peloponnesian War. In B.C. 334 it was taken and partly destroyed by Alexander. St. Paul visited it on his return from Macedonia. Miletus holds a conspicuous place in Greek literature, as the birth-place of the philosophers Thales, Anaximander, and Anaximenes, and of the historians Cadmus and Hecatæus.[2] Its manufactures of furniture, woollen cloths, and carpets,[3] were also celebrated. At Branchidæ, or Didyma, 12 miles S. of Miletus, and about 2 miles inland from Prom. Posidium, was the famous temple of Apollo Didymeus, with an oracle, which was consulted alike by Ionians and Æolians, as well as by foreigners : the kings Crœsus of Lydia and Necos of Egypt paid reverence to it. The temple was destroyed by the Persians, B.C. 494, and afterwards rebuilt by the Milesians on an enormous scale. A road called the "sacred way," lined with seated statues led to it from the sea. Only two columns now remain; the rest is a heap of ruins. The length of the temple was 304, and its breadth 165 feet ; in point of size it ranked next to the great temple of Diana at Ephesus. Iassus, Asyn' Kalessi, on a small island close to the north coast of the bay named after it, had a mixed population of Greek settlers, whose chief occupation was fishing. It was taken by the Lacedæmonians in the Peloponnesian War, and was besieged by the last Philip of Macedonia.

Halicarnassus, Budrum, was situated on the Ceramian Gulf, and was regarded as the largest and strongest city in all Caria. Its principal acropolis was named Salmacis after a well near it, whose waters were supposed to have an enervating influence.[1] It possessed two harbours, the entrance to the larger one being guarded by a pier on each side. The most remarkable building was the Mausoleum erected to the memory of Mausolus by his widow Artemisia (B.C. 352): it was situated in the centre of the town. Halicarnassus originally belonged to the Dorian confederacy, but was expelled from it : it became subject to Persia, and, at the same time, the seat of a tyranny founded by Lygdamis, and carried on by Artemisia, who fought at Salamis : this dynasty gradually established its supremacy over the whole of Caria. Halicarnassus was besieged by Alexander, and, with the exception of the acropolis, was taken and destroyed. It was the birth-place of the historians Dionysius and Herodotus. The remains of Halicarnassus consist of the ancient polygonal walls, which are in a good state of preservation, part of a mole on the E. side of the harbour, the foundations of a large Ionic temple, and of a Doric colonnade near the Mausoleum, and some cemeteries outside the walls. The Mausoleum[2] itself

---

[2] The morals of the Milesians were so lax that *Milesius* became a synonym for "wanton."

Junxit Aristides Milesia carmina secum.—Ov. *Trist.* ii. 413.

[3] Quamvis Milesia magno

Vellera mutentur, Tyrios incocta rubores.—Viro. *Georg.* iii. 306.

Eam circum Milesia vellera Nymphæ

Carpebant, hyali saturo fucata colore. In. iv. 334.

[1] Unde sit infamis ; quare male fortibus undis

Salmacis cnervet, tactosque remolliat artus ;

Discite. Causa latet : vis est notissima fontis.—Ov. *Met.* iv. 285.

[2] The name was applied by the Romans, as it is by ourselves, to any fine sepulchral monument :—

Nec mausolei dives fortuna sepulcri

Mortis ab extrema conditione vacat.—Prop. iii. 2, 19.

Nam vicina docent nos vivere mausolea :

Cum doceant ipsos posse perire deos.—Mart. v. 64.

is correctly described by Pliny as having been a quadrangular building surrounded by 36 columns and surmounted by a pyramid, which was crowned with a colossal group of a chariot with four horses. The height of the whole edifice was 140 ft. and its circumference 411. It was decorated with sculptures in relief, executed in Parian marble and of the highest merit. The site of the Mausoleum was explored in 1857 by Mr. C. Newton, who discovered two colossal figures, one of which is supposed to represent Mausolus himself, the halves of two horses forming a portion of the crowning group, some slabs of the frieze, several lions, and other interesting objects. These objects are deposited in the British Museum.

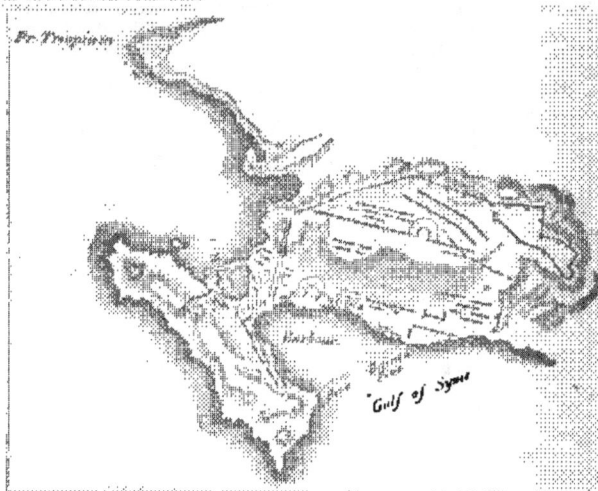

Plan of Cnidus, and Chart of the adjoining Coast.

Cnidus stood at the very extremity of the peninsula already described as terminating in Prom. Triopium. Triopium: a portion of it was built on the mainland, and a portion on an island which was joined to it by a causeway. The island sheltered the two harbours which lay on each side of the causeway, the larger of them, on the south side, being protected by moles of great strength. Cnidus was a

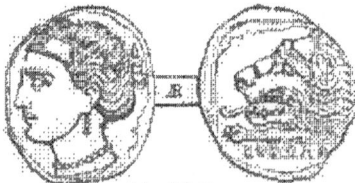

Coin of Cnidus.

member of the Dorian confederacy, the members of which met at the temple of the Triopian Apollo. It surrendered to the Persian general Harpagus, in the time of Cyrus, and was attacked by the Athenians in the Peloponnesian War. Cimon defeated the Lacedæmonian fleet under Pisander near it, B.C. 394. Cnidus had considerable trade, and produced many eminent men—Eudoxus, Ctesias, and Agatharcides—and acquired some remarkable works of art, particularly the statue of Aphrodite by Praxiteles, and others which were set up at Olympia and Delphi. The worship of Venus[1] was prevalent at Cnidus. **Caunus**, in Peræa, stood on the banks of a small stream now called *Koi-gez*, which communicates with a lake about 10 miles from the sea: it is frequently mentioned in history: it was taken by Ptolemy, B.C. 309; it was subsequently given by the Romans to the Rhodians, taken from them, B.C. 167, but again restored to them: it was the birth-place of Protogenes the painter.

2. **Tralles**[2] stood on the slope of Messogis, not far from the Mæander, and was centrally situated at a point where roads from the S., E., and W. converged. Its origin is uncertain, some assigning its foundation to the Argives, others to the Pelasgians. The place was chiefly famous for the wealth of its inhabitants, derived partly from the fertility of the surrounding district, partly from its commercial importance. Extensive ruins of it exist at *Ghiuzel Hissar*. **Alabanda** was situated about 18 miles S. of Tralles, and was also a place of great wealth and luxury: under the Roman empire it became the seat of a Conventus Juridicus, or court-house; its site is supposed to be at *Arab-Hissar* on the Marsyas, where are the remains of a temple and other buildings. **Mylasa** was situated in a fertile plain, not far from the head of the Iassian Bay, and at the foot of a mountain which yielded the beautiful white marble, out of which the town was built: Physcus served as its port. The town boasted a high antiquity, and possessed two splendid temples, one of which stood in the village of **Labranda**, and was connected with the town by a Via Sacra about 9 miles long. Its resistance to Philip, the son of Demetrius, is the only historical event of interest connected with it. The remains at *Melasso* consist of a marble archway, the vestiges of a theatre, and ranges of columns. The temple at Labranda was sacred to Jupiter Stratius, and was of immense size: it was surrounded by a grove of plane trees. It was situated in the mountains between Mylasa and Alabanda, where extensive ruins have been found. **Aphrodisias** stood on the Mosynus, S. of the Mæander, not far distant from the eastern border: it was a very large and fine city, as the ruins at *Ghera* testify, particularly those of the temple of Aphrodite; of its history we know nothing beyond the fact that under the Romans it became a free city. **Antiochia**, surnamed "ad Mæandrum," stood on the Mosynus, and was named after Antiochis, the mother of Antiochus, son of Seleucus. Cn. Manlius encamped here, B.C. 189, on his way to Galatia: the supposed remains, about 5 miles S.E. of *Kuyujo*, are inconsiderable. **Stratonicea**, S.E. of Mylasa, derived its name from Stratonice, the wife of Antiochus Soter, who founded it, probably on the site of the more ancient Idrias. Mithridates resided there: at a later period its resistance to Labienus

---

[1] Nunc, O cœrulco creata ponto,
Quæ sanctum Idalium, Syrosque aperios,
Quæque Ancona, Cnidumque arundinosam,
Colis.   CATULL. XXXVI. 11.

O Venus, regina Cnidi Paphique.   HOR. Carm. L. 30, 1.

[2] Roma was much frequented by the inhabitants of Tralles and Alabanda:—
Ille Andro, ille Samo, hic Trallibus aut Alabandis
Expulsus dictumque petunt a vimine collem
Viscera magnarum domuum dominique futuri.—JUV. III. 70.

attracted the notice of the Romans to it, and Hadrian took it under his special care. The remains at *Eski-Hissar* are very extensive; some columns still stand erect, and the theatre still preserves its seats and a part of the proscenium.

Among the less important towns we may briefly notice—(1.) on the sea-coast, **Heraclea**, whose agnomen "ad Latmum," sufficiently explains its position—**Bargylia**, on the Bay of Iassus, which was sometimes named after it Bargyliaticus, once occupied by a garrison of Philip III. of Macedonia—**Caryanda**, on an island off the north coast of the Halicarnassian peninsula—**Myndus**, a few miles N.W. of Halicarnassus, strongly fortified, and possessing a good harbour, probably at *Gumishlu*—**Pedasa**, probably at the entrance of the Halicarnassian peninsula, where the Persians were defeated in the Ionian revolt—**Physcus**, on the coast of Ceres, with a magnificent harbour, now called *Marmorice*, whence communication with Rhodes was maintained—**Loryma**, near Cape Cynossema, supposed to be at *Port Aplotheca*, where walls and several towers show that a strong place once existed—**Calynda**, near the borders of Lycia, about 7 miles from the sea, and probably on the Calbis, though its site has not been made out. (2.) In the interior—**Nysa**, in the valley of the Mæander, at *Sultan-Hissar*, where are the remains of a theatre with the rows of seats almost entire, an amphitheatre and other buildings; a place of literary distinction—**Alinda**, between Alabanda and Mylasa, one of the strongest places in Caria,—and **Eurōmus**, N.W. of Mylasa, at *Iaklee*, where are the ruins of a magnificent temple.

History.—The Carians do not come prominently forward in history. After they were driven from the sea-coast by the Greek settlers, they lived in villages, and formed a confederation, the members of which met at the temple of Zeus Chrysaoreus, on the site of the later Stratonicea. Caria formed a portion of the Lydian and Persian empires. In the Ionian revolt it joined the Greeks, and after the suppression of the revolt it returned to its former masters, who established a monarchy at Halicarnassus. After the defeat of Antiochus, the Romans divided Caria between the kings of Pergamus and the Rhodians. In the year B.C. 129, the portion assigned to the former was added to the province of Asia.

§ 6. The island of **Cos**[3] lies off the coast of Caria, separated by a narrow channel from the Halicarnassian peninsula, of which it may be deemed a continuation. Its length from N.E. to S.W. is about 23 miles. Its soil was very productive, and its wines and ointment were well known to the Romans:[6] its textile fabrics, consisting of a kind of gauze,[7] were also celebrated. The most fertile portion

---

[5] The modern name, *Stanchio*, is a corruption of ἐς τὰν Κῶ.

[6] Albo non sine Coo.—Hor. *Sat.* ii. 4, 29.
    Lubrica Coa.   Pers. *Sat.* v. 135.

[7] Illa gerat vestes tenues, quas femina Coa
    Texuit, auratas disposuitque vias.—Tibull. ii. 3, 53.
    Quid juvat ornato procedere, vita, capillo,
    Et tenues Coa veste movere sinus.—Prop. i. 2, 1.
    Sive illam Cois fulgentem incedere vidi
    Totum ut Coa veste volumen erit.—Id. ii. 1, 5.
    The term Coa is sometimes used by itself for these robes:—
    Cois tibi pæne videre est.   Hor. *Sat.* i. 2, 101.

of the island was towards the N. and E., where the ground was level: the rest was mountainous. The capital, also named Cos,

was situated at the eastern extremity of the island, and possessed a well sheltered roadstead, much frequented by the numerous vessels which passed through the channel between the island and the main-land; it was thus visited

Coin of Cos.

by St. Paul (Acts xxi. 1). It was also famed for a temple of Æsculapius, to which a school of physicians was attached. Cos was a member of the Dorian Pentapolis: under the Romans it became a free state. The town was fortified by Alcibiades: having been destroyed by an earthquake, it was rebuilt by Antoninus Pius. It was the birth-place of Ptolemy Philadelphus, the painter Apelles, and the physician Hippocrates.

Between Cos and Icaria are the less important islands—Calymna, famed for its excellent honey, but not-meriting the praises bestowed upon its foliage,[a] being a bare island—Lerus, about 30 miles S.W. of Miletus, colonized successively by Dorians and Milesians,[b] with a sanctuary of Artemis, which witnessed, according to mythology, the metamorphosis of Meleager's sisters into guinea-fowls—Patmos, to the N.W., interesting as the spot whither St. John was banished, and where he is believed to have composed the Apocalypse—and the Corassiæ, a group of two larger and several smaller islands. Between Cos and Rhodes are Nisyrus, of volcanic origin, well known for its wine, its millstones, and its hot springs, occupied by a Dorian population, with a town of which the remains of the acropolis still exist—Telos, celebrated for its ointment—Syme, at the entrance of the Sinus Schœnus, high and barren, and hence at times wholly deserted—and Chalcia, off the west coast of Rhodes. These islands retain, with but slight variation, their ancient names.

§ 7. The large island of Rhodus[c] is distant about 0 or 10 miles from the south coast of Caria: its length from N. to S. is about 45 miles, and its width varies from 20 to 25. A range of mountains

---

[a] Fecundaque melle Calymne.—Ov. Met. viii. 222.
    Silvis umbrosa Calymne.   Id. Art. Am. ii. 81.

[b] Its inhabitants enjoyed an unfortunate celebrity for their extreme ill-temper, according to the subjoined verses of Phocylides:—

Λέριοι κακοί, οὐχ ὁ μὲν, ὃς δ' οὔ,
Πάντες, πλὴν Προκλέους καὶ Προκλέης Λέριος.

Even in modern times they are unpopular from their stinginess.

[c] The name was supposed to be derived from ῥόδον, "a rose," which appears as the national emblem on the coins.

traverses the island from N. to S., culminating in Mount Atabýris,
at a height of 4560 feet, the very summit of which was crowned
with a temple of
Zeus. Though gene-
rally mountainous, and
especially so about
the towns of Rhodes
and Lindus, the island
was very fertile, the
soil being rich, and
the climate unrival-
led :[2] its wine,[3] dried

Coin of Rhodes.

raisins, figs, saffron, and oil, were much valued, as also its marble,
sponge, and fish; its inhabitants were skilled In the manufacture of ships, arms, and military engines: hence, even in the
days of Homer, the island obtained fame for great wealth.
The early inhabitants, named Telchines, enjoyed a semi-mythical
fame : the race that succeeded them, the Heliadæ, were of a similar
character: they were followed by settlers from various foreign
countries, among whom the Dorians became dominant, and at length
gave a decidedly Doric character to the island. The three most
ancient towns, Lindus, Ialysus, and Camirus, which were known in
the Homeric age,[4] were members of the Doric Pentapolis, along with
Cos and Cnidus. The later capital, Rhodus, was not founded until
B.C. 408 : its rise proved fatal to the existence of Lindus and Ialysus,
whose inhabitants were removed thither.

Rhodus was at the N.E. end of the island, and was built in the form
of an amphitheatre, on ground gradually rising from the shore, and
with such regularity that it was said to appear like one house. The
acropolis was posted at the S.W. of the town, and there were two
excellent harbours. In addition to many remarkable works of art,
both in sculpture and painting, Rhodes boasted of one of the seven
wonders of the world In the brazen statue of Helios, commonly known
as the Colossus. It was erected, B.C. 280, by Chares, overthrown by
an earthquake, B.C. 224, and appears to have been afterwards restored:
its height was 70 cubits, and it stood at the entrance of one of the
ports. Rhodes produced many men of literary eminence. St. Paul

---

[2] There was a proverb that the sun shone every day at Rhodes :—
    Claramque relinquit
    Sole Rhodon.   Luc. Phars. viii. 247.

Virgil highly praises the Rhodian grape :—
    Non ego te, dis et mensis accepta secundis
    Transferim, Rhodia.   Georg. II. 101.

[4] Τληπόλεμος δ' Ἡρακλείδης, ἠΰς τε μέγας τε
    Ἐκ Ῥόδου ἐννέα νῆας ἄγεν Ῥοδίων ἀγερώχων,
    Οἳ Ῥόδον ἀμφενέμοντο διὰτρίχα κοσμηθέντες.
    Λίνδον, Ἰήλυσόν τε καὶ ἀργινόεντα Κάμειρον.—Hom. Il. ii. 653.

touched there on his voyage from Macedonia to Phœnicia. Lindus stood on the eastern coast, and contained the revered sanctuaries of Minerva and Hercules: it was the birth-place of Cleobulus, one of the seven sages, and of Chares, the maker of the Colossus : the site of the town is marked by the remains of a theatre, and of many highly ornamented tombs. Ialysus stood on the northern coast, about 7 miles from Rhodes. Camīrus was about midway down the western coast ; the Homeric epithet ἀργινόεις had reference to the colour of the soil.

History.—Rhodes did not rise to any political importance until after the erection of its capital in B.C. 408, when the equally balanced state of its parties offered an opening at one time for Sparta, at another time for Athens, according as the oligarchical or democratical faction was uppermost. The naval power of Rhodes rose about the time of Demosthenes, and the town distinguished itself for its resistance to Demetrius Poliorcetes after the death of Alexander. Rhodes sided with Rome in her eastern wars, and received a portion of Caria in reward. In the civil wars it took the part of Cæsar, and, after his death, resisted Cassius, and suffered in consequence most severely. From this period, B.C. 42, Rhodes sunk in power, but retained fame as a seat of learning. In Constantine's division, Rhodes became the metropolis of the Provincia Insularum.

§ 8. S.W. of Rhodes lies Carpăthus, *Skarpanto*, which gave to the surrounding sea the title of Carpathium Mare. It consists for the most part of bare mountains, rising to a central height of 4000 feet, with a steep and inaccessible coast. It was originally a portion of Minos's kingdom ; it was afterwards colonized by Dorians, but seems to have been dependent on Rhodes. It possessed four towns, of which Nisyrus was the chief. The small island of Casus, *Kaso*, lies off its southern extremity.

IV. LYCIA.

§ 9. Lycia was bounded on the N.W. by Caria, on the N. by Phrygia and Pisidia, on the N.E. by Pamphylia, and on the S. by the Mediterranean, which also washes a portion of its eastern and western coasts. It is throughout a mountainous district, being intersected in all directions by the southerly branches of the Taurus range : it was, nevertheless, fertile in wine, corn, and other productions. The scenery is highly picturesque, rich valleys, wooded mountains, and precipitous crags, being beautifully intermingled. Among the products peculiar to Lycia we may notice a particularly soft kind of sponge found at Antiphellus, and a species of chalk possessed of medicinal properties. It also contained springs of naphtha, which attest its volcanic character.

§ 10. The principal mountains in Lycia were named—Dædăla, on the border of Caria—Cragus and Anticragus, two lofty peaks, separated from each other by an elevated plain, and terminating in a cluster of rugged heights on the western coast, Cragus being the most southerly of the two—Massicytus, in the centre of the province, running from N. to S. parallel to the river Xanthus—and Climax, on the eastern coast, the name (meaning "ladder") being originally applied to a mountain which overhung the sea near Phaselis so closely, that at certain times the road at its base was impassable, while the mountain was surmounted only by a difficult pass : the name was afterwards extended to the whole ridge between Lycia and Pamphylia.

Rock-cut Lycian Tomb (Texier's Asia Mineure).

A portion of this mountain is the **Chimæra**, which Ctesias describes as having a perpetual flame issuing from it : this is no doubt a reference to the inflammable gas found in that neighbourhood. The ancient poets[1] frequently refer to this phenomenon, the nature of which they did not understand[2]. To the S. of this range was a volcanic mountain named **Olympus** or **Phœnicus**. Numerous promontories occur on the coast, the most conspicuous being—**Prom. Sacrum**, *Yedy-Booroon*, at the termination of Cragus—and another at the S.E. point, also called **Sacrum**, but sometimes **Chelidonium**, *Chelidonia*, off which lay a group of five rocky islands of the same name : the

[1] Πρῶτον μὲν ῥα Χίμαιραν ἀμαιμακέτην ἐκέλευσε
Πεφνέμεν· ἡ δ᾽ ἄρ᾽ ἔην θεῖον γένος, οὐδ᾽ ἀνθρώπων,
Πρόσθε λέων, ὅπιθεν δὲ δράκων, μέσση δὲ χίμαιρα,
Δεινὸν ἀποπνείουσα πυρὸς μένος αἰθομένοιο.—Hom. Il. vi. 179.

Vix illigatum te triformi
    Pegasus expediet Chimæra.   Hom. Carm. i. 27, 23.

        Flammisque armata Chimæra.   Virg. Æn. vi. 298.

Καὶ Χίμαιραν πῦρ πνέουσαν,
Καὶ Σολύμους ἐπέφνεν.   Pind. Olymp. xiii. 120.

promontory was regarded as the commencement of Taurus. The most important river is the **Xanthus**, *Etchen*, which rises in Taurus, and flows in a S.W. direction through an extensive plain between the ranges of Cragus and Massicytus to the sea : the name, meaning "yellow," has reference to the colour of the water : this river was known to Homer,[e] and was regarded as a favourite stream of Apollo,[f] to whom indeed the whole of Lycia was sacred. In the eastern part of the province a smaller stream was named Limyrus, to which the Arycandus, *Fineka*, is tributary.

§ 11. The most ancient inhabitants of Lycia were a Semitic race, divided into two tribes named Solymi and Termilæ or Tremilæ. The Lycians entered from Crete before that island received its Hellenic character, and subdued the Termilæ on the sea-coast with ease, but had to maintain an arduous struggle with the Solymi, who had retreated into the mountainous district on the border of Pisidia, named Milyas. The Solymi appear to have assumed the name of this district, as they were afterwards known as Milyæ. The Lycians, though "barbarians" in the Greek sense of the term, were an enlightened nation, enjoying a free constitution consisting of a confederacy of 23 towns, cultivating the arts of sculpture and architecture,[g] and probably having a literature of their own.

§ 12. The towns of Lycia were very numerous ; Pliny states that it once contained seventy, though in his day the numbers had sunk to twenty-six ; the higher estimate is justified by the numerous ruins scattered over the face of the country, many of them representing towns, the very names of which are unrecorded. The six largest towns of the confederacy were—Xanthus, Patāra, Pināra, Olympus, Myra, and Tlos. The first of these was the capital of the country, and was situated in the rich plain of the Xanthus : Pinara and Tlos were not far distant from it : the other three were on the coast. Phasēlis, on the eastern coast, though not a member of the Lycian confederacy, rose to great importance as a commercial town. The dates at which these and other towns were built can only be conjectured from the character of the architecture, which, in many cases, indicates a high antiquity. Their flourishing

---

[e] Τηλόθεν ἐκ Λυκίης, Ξάνθου ἀπο δινήεντος.—Hom. *Il*. ii. 877.
'Αλλ' ὅτε δὴ Λυκίην ἷξε, Ξάνθον τε ῥέοντα.
Προφρονέως μιν τίεν ἀναξ Λυκίης εὐρείης.—Id. vi. 172.

[f] Phœbe, qui Xantho lavis amne crines. Hor. *Carm*. iv. 6, 26.
Qui Lyciæ tenet
Dumeta, natalemque silvam,
Delius et Patareus Apollo. Id. *Carm*. iii. 4, 62.

[g] The architecture is partly of a Cyclopean, partly of a Greek character, the latter exhibiting a high state of art. The monumental architecture has a peculiar character, consisting in the use of a pointed arch, not very unlike that used in Gothic architecture.

period appears to have been about the time when the Romans first became connected with the country; it terminated with the fall of Xanthus, and the exactions imposed by Brutus. We shall describe these towns in order from W. to E.

Telmessus stood on the shores of the Bay of Glaucus, and was once a flourishing town, as the remarkable remains at Myra—a theatre, porticoes, and sepulchral chambers in the solid rock—still testify: its inhabitants were highly skilled in augury. Patara, the port of Xanthus, was situated near the mouth of the Xanthus, and possessed a fine harbour, as well as a celebrated temple and oracle of Apollo,[a] hence surnamed Patareus. The harbour was much visited by vessels trading to Phoenicia: St. Paul touched there (Acts xxi. 1). The ruins are very extensive, particularly those of a theatre built in the time of Antoninus Pius; but the harbour has become choked with sand. Xanthus, the capital, was beautifully situated on the left bank of the Xanthus, about 6 miles from its mouth. The city is famous for its determined resistance to Harpagus in the reign of Cyrus, and again to Brutus—on each of which occasions it was destroyed. The ruins near Koonik are magnificent, consisting of temples, tombs, triumphal arches, and a theatre: the sculptures on the tombs are in the best style of art, and very perfect. Tlos stood higher up the valley of the Xanthus: though almost unknown to history, it was a splendid town, and strongly placed, its acropolis being on a precipitous rock. The theatre still remains, with highly worked seats of marble; the side of the acropolis rock is covered with excavated tombs with ornamental entrances. Pinara stood on the declivity of Mount Cragus, and was one of the largest towns in Lycia. A round rocky cliff rises out of the centre of the town, the sides of which are covered with tombs; the rock-tombs, as elsewhere, are highly decorated, and the theatre is in a very perfect state: the ancient name survives in Minara. Antiphellus stood on a small bay on the southern coast; the remains are extensive: it served as the port of Phellus, which was probably more to the N., at Tchookoorbye. Opposite Antiphellus is the island of Megiste, Kastelorizo, which is now the chief place of business along this coast. Myra, Dumbre, stood on a plain about 2½ miles from the sea coast, and at the entrance of a mountain-gorge that leads into the interior: Andriace served as its port, and was much frequented by vessels bound westward from Syria: St. Paul touched there on his voyage to Rome (Acts xxvii. 5). The theatre at Myra is one of the finest in Asia Minor, and the other ruins are very beautiful: the bas-reliefs in some of the tombs still preserve their original colouring. Limyra was more to the E., in the valley of the Limyrus; its site is marked by extensive ruins, some of the inscriptions on the tombs being richly coloured, and the bas-reliefs representing stories from Greek mythology. Olympus was situated at the foot of the mountain of that name, at Deliktash. Lastly, Phaselis, Tekrova, on the eastern coast, with three harbours, formed an entrepôt for the trade between Greece and Phoenicia: it became the haunt of pirates, and was taken by Servilius Isauricus, after which it sunk.[1] The light boats called phaseli were

[a] Hor. Carm. iii. 4, 64. See above, note [2].
Phœbe parens, seu te Lyciæ Pataraea nivosis
Exercent dumeta jugis.   Stat. Theb. I. 696.

[1] Te primum, parva Phaselis
Magnus adit.   Luc. Phars. viii. 251.

said to have been built here, and were the usual devices on the coins of the place.

History.—The Lycians appear as allies of the Trojans in the Homeric poems, but are not again mentioned until the time of Crœsus, who failed in his endeavour to subdue them. Cyrus was more successful, and added Lycia to the Persian empire. Alexander traversed a portion of it, and easily conquered it. It then passed successively to the Ptolemies, the Seleucidæ, and the Romans, who handed it

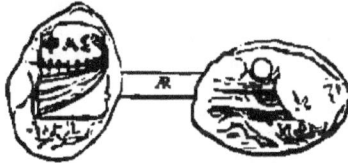

Coin of Phaselis.

over for a time to the Rhodians, but afterwards restored it to independence. The country suffered severely from Brutus on suspicion of its having favoured his opponents, and never recovered its prosperity. Claudius made it a Roman province in the prefecture of Pamphylia, with which it remained united until the time of Theodosius II.

Greek Lycian Tomb (Texier's Asia Minoure).

V. PAMPHYLIA.

§ 13. **Pamphylia** was bounded on the W. by Lycia, on the S. by the Mediterranean, on the E. by the river Melas separating it from Cilicia, and on the N. by Pisidia. It consists of a narrow strip of

land, skirting in a semicircular form the coast of the Pamphylium
Mare. The name was extended by the Romans to Pisidia on the
northern side of Taurus. The country is generally mountainous,
the spurs of Taurus pressing closely on the sea : the most extensive
plain is that which surrounds Attalia.

§ 14. The rivers have a southerly course through the lateral ridges
of Taurus, and fall into the Pamphylian Sea. Following the line of
coast from W. to E., we meet with the **Catarrhactes**, *Duden-su*, deriving its ancient name from the manner in which it precipitates
itself over the cliffs into the sea near Attalia : its lower course across
the plain is continually changing, and hence some difficulty has
arisen in fixing the sites of the towns—the **Cestrus**, *Ak-su*, which
was formerly navigable up to Perga, but has its entrance now closed
by a bar—the **Eurymedon**, *Capri-su*, which has undergone a similar
change : at its mouth Cimon defeated the Persians, B.C. 466 ;
lastly the **Melas**, *Menavgat-su*, in the eastern part of the district.
The coast is regular, the only promontory being Leucotheum, near
Side.

§ 15. The inhabitants of this district were a mixed race of aborigines, Cilicians, and Greeks : hence their name "Pamphýli" (from
πᾶς and φυλή), resembling in its origin the later "Alemanni." Of
their history we know little : they were chiefly devoted to maritime
pursuits, and joined the Cilicians in their piratical proceedings. The
chief towns were either on the sea-coast or on the navigable rivers.
In earlier times the Greek colonies of Side and Aspendus were the
more important ; but at a later period Attalia, which was founded
by Attalus II. of Pergamus, when this province was attached to his
kingdom. Perga was also a considerable town, situated on the road
between Phaselis and Aspendus.

Olbia was the most westerly of the Pamphylian towns, and appears
to have been about 3½ miles W. of *Adalia*, near the coast : it has
been by some geographers incorrectly identified with Attalia. **Attalia**
was situated at the inmost point of the Pamphylian Bay, near the
shifting course of the Catarrhactes : it was founded by Attalus, probably with a view to command the trade of Egypt, and even to this
day it retains its ancient name and importance. **Perga** was beautifully
situated between two hills bordering on the valley of the Cestrus, and
was the seat of a famous temple of Diana : the ruins of a theatre, stadium, aqueduct, and other buildings mark its site. **Aspendus** was on a
hill near the Eurymedon, about 8 miles from the sea ; it was visited by
Alexander in his Asiatic expedition, and appears to have been a populous place. **Syllium** was a fortified place between the Eurymedon and
the Cestrus. **Side**, on the coast, was a colony from Cyme in Æolis : it
possessed a good port, which became the principal resort of the pirates
of this district : it retained its importance under the Roman emperors,
and became the metropolis of Pamphylia Prima : its ruins at *Eski
Adalia* are extensive, the most remarkable being the theatre, on an eminence in the centre of the town : the harbour is choked up with sand.

History.—The Pamphylians never acquired any great political importance. They were subject to Persia, Macedonia, and Syria, in succession. After the defeat of Antiochus they were handed over to the kings of Pergamus. At the death of the last Attalus they were included in the province of Asia, but were afterwards attached to Cilicia. In the reign of Augustus, Pamphylia became a separate province, including a portion of Pisidia, and under Claudius a part of Lycia also.

St. Paul's Travels.—St. Paul visited Pamphylia in his first apostolical journey: having sailed from Cyprus, he disembarked at Perga, and thence crossed the range of Taurus, probably by the course of the Cestrus, into Pisidia. He returned to the same point, but instead of taking ship at Perga, he crossed the plain to Attalia, and thence sailed for Antioch.

VI. CILICIA.

§ 16. **Cilicia** was bounded on the W. by Pamphylia, on the N. by the range of Taurus separating it from Lycaonia and Cappadocia, on the E. by the range of Amanus separating it from Syria, and on the S. by the Mediterranean. Within these limits are included two districts of an entirely different character—the western being mountainous, and hence named Trachĕa, or "rough;" the eastern containing extensive plains, and hence named Pedias, or Campestris, "level:" the river Lamus forms the division between them. The second of these districts is naturally subdivided into two, viz. the plain of Tarsus and Adana, and the plain of Issus. The province is inclosed on the N. and E. by a continuous wall of mountains, and possesses a lengthened line of coast on the S. The length from E. to W. is about 250 miles; the breadth varies from 30 to 50 miles; the length of the coast-line is about 500 miles.

§ 17. The position and physical character of Cilicia bring it into frequent notice in ancient geography. Situated between Syria on the one side, and the rest of Asia Minor on the other, it became the highway between the East and the West, and was of special value to the rulers of Syria. The extent of its seaboard and the supplies of timber which it yielded rendered it a valuable acquisition to Egypt. The beauty of its scenery and its luxurious climate attracted the wealthy Romans thither, and were the indirect means of elevating Tarsus into a seat of learning. Lastly, the fertility of its soil was so great that it was independent of all other countries in regard not only to the necessities but the luxuries of life: in addition to corn, wine, and oil, it was famed for its saffron, and for the goats'-hair cloth named *cilicium*.

§ 18. The chief mountain-ranges of Cilicia are **Taurus** in the N., and **Amanus** in the E. The former fills the western district with lateral ridges extending to the very edge of the sea. Eastward of the Lamus the mountains recede from the coast, and attain such an elevation that their peaks are covered with snow even in June. Between them and the sea-coast intervenes the broad and fertile plain of Tarsus. Amanus consists of a double range, which may be distinguished as the Cilician and Syrian branches: the former descends to the sea in a S.W. direction, between the Pyramus and the Bay of Issus; the latter takes a due southerly direction parallel to the eastern shore of the bay, and terminates abruptly in the promontory of Rhosus at the southern entrance of the bay: these branches unite in the N., and enclose the plain of Issus.

The passes across these mountains deserve special notice. The most frequented pass across Taurus, named **Cilicia Pylæ** or **Portæ**, now *Golek Boghras*, was situated at the head of the valley of the Cydnus, and led across to Tyana: it is a remarkable fissure in the mountain, and easily defensible at several points. It was crossed by the younger Cyrus, and by Alexander the Great, and was selected by Niger as his point of resistance against Septimius Severus. In the western part of the province a pass crosses from Laranda in Lycaonia to one of the lateral valleys of the Calycadnus. The Cilician Amanus had a pass named by Strabo **Amanldes Pylæ** (11), between Mallus and Issus: this is now named *Kara Kapu*. The Syrian Amanus was crossed at two points, to each of which the name of **Amanldes Pylæ** was again applied: one of these, which may be termed the *lower* pass, answers to the *Pass of Beilan* (2), between the Gulf of Issus and Antioch; while the other, or *upper* pass, lies E. of *Bayas* (4): it was by the latter that Darius crossed before the battle of Issus. Lastly, at the point where the mountain approaches the coast most nearly, and where the little stream Cersus, *Merkes* (7), reaches the sea, a double wall with gateways was thrown across, one on each side of the Cersus: these were the "Cilician and Syrian Gates" described by Xenophon (*Anab.* i. 4), through which Cyrus passed, and which Alexander passed and repassed before the battle of Issus.

§ 19. The coast is varied both in outline and character: in Trachea it assumes a convex form, and presents a jagged outline with numerous small bays and promontories: it is here rock-bound and dangerous. The chief promontories are—**Anemurium**, *Anamour*, the most southerly point of Cilicia—**Sarpedon**, *Lissan el Kapeh*, near the Calycadnus—**Zephyrium**, which is perhaps close to the mouth of that river—and **Coryeus**,[1] more to the E., celebrated for its beds of saffron, and for a cave[2] with a remarkable spring. In Campestris two im-

---

[1] Utque solet pariter totis se effundere ignis
Corycii pressura croci, sic omnia membra
Emisere simul rutilum pro sanguine virus.   Luc. ix. 808.

Hoc ubi confusum acetis inferbuit herbis,
Coryeloque croco sparsum sietit.   Hor. Sat. ii. 4, 67.

[2] Deseritur Taurique nemus, Perseaque Tarsos,
Coryciumque patens exesis rupibus antrum,
Mallos, et externo resonant navalibus Ægre.—Luc. iii. 225.

Τὰς γηγενὴ τε Κιλικίων οἰκήτορα
Ἄντρων ἰδὼν ῷκτειρα, δάϊον τέρας
Ἑκατογκάρηνον πρὸς βίαν χειρούμενον
Τυφῶνα θοῦρον, πᾶσιν δς ἀντέστη θεοῖς,
Σμερδναῖσι γαμφηλαῖσι συρίζων φόνον.   Æsch. Prom. 351.

Amanides Pylæ (See pp. 132, 136).

1. *Ras-el-Khanzir*, the promontory at the southern
   extremity of the Gulf of Issus.
2. *Beilan Pass* (Lower Pass of Amanus).
3. *Baghras Pass.*
4. *Pass from Bayas* (Upper Pass of Amanus).
5. *Sinana.*
6. *Alexandria* (*Iskenderun*)

7. River Carsus (*Merkes*).
8. *Bayas.*
9. River Pinarus.
10. Ruins of Issus (I).
11. Pass of the Cilician Amanus, with Gate, now
    Kara Kapu

portant bays penetrate inland, divided from each other by the promontory of **Megarsus**, *Karadash* : the western of the two is wide and open, and received no specific name; the eastern is the **Sinus Issicus**, *G. of Iskunderun*, which runs up in a N.E. direction for 47 miles, with a general width of 25. The coast between the river Lamus and Prom. Megarsus is a low sandy beach : this is followed by a slightly elevated plain in the neighbourhood of Ægæ, and this again by a shelving coast at the head of the bay.

§ 20. The chief rivers are—the **Calycadnus**, *Ghiuk-su*, which rises in the western part of Trachea, and pursues an easterly course through a wide and long valley to the sea near Prom. Sarpedon—the **Cydnus**, *Tersoos Chai*, which rises in Taurus near the Cilician Gates, and in a southerly course traverses the fertile plain of Tarsus to the sea ; its water, like that of the other streams which flow from Taurus, is cold, and nearly proved fatal to Alexander after bathing in it—the **Sarus**, *Sihun*, which in its lower course crosses the rich Aleian plain—and the **Pyramus**, *Jyhun*, which holds a parallel course more to the eastward : the two latter rivers have been already noticed in the introductory section (p. 87).

§ 21. The Cilicians were an Aramaic race, and, according to Greek tradition, derived their name from Cilix, the son of Agenor, a Phœnician. They occupied the whole of the country until the days of Alexander the Great, when the Greeks, who had previously made some few settlements on the coast, gradually drove the Cilicians from the plains into the mountains, where they maintained themselves in independence under the name of "Free Cilicians." The inhabitants of Trachea belonged to neither of these parties, but were connected with the Pisidians and Isaurians, whom they resembled in their freebooting habits.

§ 22. The towns of Cilicia belonged to various historical eras. Tarsus was undoubtedly a Syrian town, and the other towns of Campestris had probably a similar origin, though no evidence can be adduced to that effect. Greek colonies were reputed to have settled at the most favourable points, as Tarsus, Soli, Mallus, Ægæ, and Celendëris. The Seleucidæ founded several new towns, as Seleucia on the Calycadnus, Antiochia ad Cragum, and Arsinoü. Lastly, the Romans revived many of the old towns, and gave them Roman names, such as Cæsarea, Pompeiopolis, Claudiopolis, and Trajanopolis. Six cities are noticed as " free " under the Roman dominion, viz. Tarsus, Anazarbus, Seleucia (which formed the capitals of the three divisions of Cilicia in Constantine's arrangement), Corÿcus, Mopsuestia, and Ægæ. With regard to the position of the Cilician towns, those in Trachea are for the most part on the coast, which offered numerous strong and secure sites on the cliffs : Seleucia on the Calycadnus is the most marked exception. In Campestris, on the other hand, where the coast is low, they are on the rivers: Tarsus on the Cydnus, Adana on the Sarus, Mopsuestia and Anazarbus on the Pyramus.

Commencing with the towns on the coast from W. to E.—Coracesium, *Alaya*, on the border of Pamphylia, was a place of remarkable natural strength, and had a good harbour: it was the only town that held out against Antiochus, and it became the head-quarters of the pirates. **Selinus**[4] was equally strong in position, being placed on a cliff jutting out into the sea: Trajan died there, A.D. 117, after which event the name was changed to Trajanopolis: remains still exist of a mausoleum, agora, theatre, &c., at the mouth of the *Selenti*. Celendëris is also described as a strong fortress on the coast, with a small but sheltered

port, now called *Gulnar:* originally a Phœnician town, it received a Samian colony: Its coins were remarkably fine. **Seleucia,** on the west bank of the Calycadnus, a few miles from its mouth, was founded by Seleucus Nicator, and attained a speedy eminence, rivalling even

Coin of Celenderis.

Tarsus: it was much frequented on account of the annual celebration of the Olympia, and for an oracle of Juno: it was the birth-place of the Peripatetic philosophers Athenæus and Xenarchus: the town still exists under the name of *Selefkieh*, and has remains of an ancient theatre, temples, and porticoes. **Soli** was a highly flourishing maritime town in the western part of Campestris, founded by Argives: it was destroyed by Tigranes, king of Armenia, but restored by Pompey, and thenceforth named **Pompeiopolis:** the philosopher Chrysippus and the poets Philemon and Aratus were born there: the town derives its chief notoriety, however, from the term "solecism," originally descriptive of the corrupt Greek spoken by the Solians: Its remains at *Mezitlu* consist of a beautiful artificial harbour, an avenue of 200 columns, of which 42 still stand, and numerous tombs. **Tarsus,** *Tersoos*, stood on both sides of the Cydnus, about 8 miles from its mouth, where a lagoon served as its port: its situation was most

favourable, being central in regard to the means of communication in Cilicia, and surrounded by a fertile and beautiful plain: originally a Syrian town, it was early colonized by Greeks, and was in the days of Cyrus the Younger the capital of the country: it was visited by Alexander:

Coin of Tarsus.

in the civil wars it sided with Cæsar, and was hence named Juliopolis: Antony received Cleopatra there, and Augustus constituted

---

4 Quas portu mititque rates recipitque Selinus.—Luc. viii. 260.

it a "libera civitas." It was a seat of philosophy, and produced
many eminent men, particularly ·the Apostle St. Paul. **Mallus**
was situated on an eminence near the mouth of the Pyramus, and
was visited by Alexander: its port was named Megarsa. **Ægæ** stood
on the N. coast of the Issicus Sinus at *Kalassy* : in Strabo's time it
was but a small city, with a port. **Issus** stood near the head of the
Issicus Sinus, and is memorable for the great battle fought here between
Alexander and Darius, B.C. 333: the precise position of the town is
uncertain, being by some fixed S. of the river Pinarus (9), but probably
being to the N (See Map, p. 133). **Epiphania** was probably near the
head of the bay; **Baiæ** was at *Bayas* (8), on the eastern shore: **Alexandria
ad Issum** and **Myriandrus** were probably the same place, the latter
being the earlier name ; they stood at or near *Iskenderun* (8). In the
interior, **Mopsucrène**, on the southern slope of Mount Taurus, was the
place where the Emperor Constantius died, A.D. 361. **Adana** was
situated on the military road from Tarsus to Issus, and on the W.
side of the Sarus. **Mopsuestia**, *Messis*, was on the same high road,
at the point where it crossed the Pyramus. **Anazarbus**, or **Cæsarea**, was
higher up the Pyramus, near a mountain of the same name: its site is
now named *Anazarby*.

History.—The early annals of Cilicia are lost to us: we know that
it formed a part of the great Assyrian empire, and that, after the fall
of Nineveh, its king Syennesis was sufficiently powerful to act as
mediator between Crœsus and the Medes. It remained independent
until the rise of the Persian empire, and even under that it enjoyed
its own princes. It was traversed and subdued by Alexander the
Great, and after his death it fell to the Seleucidæ. As the power of
the Syrian monarchy decayed, the Cilicians rose to independence, and
carried on a nefarious system of piracy and slave-hunting over the
whole of the neighbouring coasts. War was prosecuted by the Roman
generals, M. Antonius, B.C. 103, Sulla, 92, Dolabella, 80-79, P. Servilius Isauricus, 78-75, and finally Pompey, 67, with a view to
extirpate these pirates, and under Pompey the eastern part of the
country was organized as a Roman province. The western district
remained independent until the time of Vespasian. In the period after
Constantine, Cilicia was divided into three parts, Prima, the southern
portion of Campestris, Secunda, the northern portion, and Isauria
embracing Trachea.

St. Paul's Travels.—St. Paul visited Cilicia very shortly after his
conversion (Acts ix. 30), entering it probably by way of Antioch (comp.
Gal. i. 21): he went to Tarsus, and is supposed to have founded the
churches in Cilicia. In his second journey he visited these churches,
entering again from Syria, probably following the coast-road by Issus
to Mopsuestia and Tarsus, and thence crossing Taurus by the Cilician
Gates into Lycaonia (Acts xv. 41).

§ 23. The important Island of **Cyprus** lies midway between the
coasts of Cilicia and Phœnicia, nearer to the former in point of
actual distance, but more connected with the latter in regard to
race, history, and the character of its civilization. The length of
the island from W. to E. is about 150 miles: its greatest breadth
about 40 : the principal or S.W. portion has the form of an irregular parallelogram, which terminates in a long narrow peninsula
running in a N.E. direction. The surface of the country is almost

entirely occupied by the elevated range of Mount Olympus, which descends on each side in bold and rugged masses, divided from each other by deep picturesque valleys. The island produced copper (æs Cyprium), as well as gold and silver and precious stones. The lower tracts were eminently fertile, and are described as flowing with wine, oil, and honey, while from the abundance of its flowers it received the epithet of *εὐώδης*. The whole island was regarded as sacred to Venus.[a]

§ 24. The range of Olympus runs from W. to E., and attains an elevation of 7000 feet. Numerous promontories run out into the sea, of which the most important are Acamas, *Haghios Epiphanios*, in the W.; Crommyon, *Cormachiti*, in the N.; Dinaretum, *St. Andre*, in the E., with the small group of islands named Cleides, "the Keys," just off it; Pedalium, *Della Grega*, at the S.E., above which rose a hill named Idalium, with a temple sacred to Venus;[b] and Curias, *Delle Gatte*, at the extreme S. The chief river is the Pedieus, which has an easterly course, and waters the plain of Salamis; the other numerous streams are unimportant. The chief plains were those of Salamis and Citium.

§ 25. The oldest towns of Cyprus (Citium, Amathus, and Paphos) were colonies from Phœnicia: the two former bear Phœnician names, while the latter was the chief sanctuary for the worship of the Phœnician Venus. The Greek colonies hold the next rank in point of age, and a higher rank in point of importance: Salamis, on the S.E. shore, was the most flourishing commercial city in the island; Soloë on the northern coast, was well situated for the Cilician trade; New Paphos became a frequented port, and at one time the seat of govern-

---

a Αἰδοίην χρυσοστέφανον καλὴν Ἀφροδίτην
'Ασομαι, ἣ πάσης Κύπρου ἐρήδεμνα λέλογχεν
Εἰναλίης, ὅθι μιν ζεφύρου μένος ὑγρὸν ἀέντος
'Ήνεικεν κατὰ κῦμα πολυφλοίσβοιο θαλάσσης,
'Αφρῷ ἐνὶ μαλακῷ·   Hom. *Hymn. in Ven.* II.

O, quæ beatam, Diva, tenes Cyprum.—Hor. *Carm.* iii. 26, 9.
O Venus, regina Cnidi Paphique,
Sperne dilectam Cypron.   Id. i. 30, 1.

Tunc Cilicum liquere solum, Cyproque citatæ
Immisere rates, nullas cui prætulit aras
Unda diva memor Paphiæ, si numina nasci
Credimus, aut quenquam fas est complere deorum.
        Luc. viii. 456.

b Δουπου', ἃ Γολγώς τε καὶ Ἰδάλιον ἐφίλησας,
Αἰπεινάν τ' Ἔρυκα, χρυσῷ παίζουσ' Ἀφροδίτη.
        Theocr. *Idyl.* xv. 101.

Hunc ego sopitum somno, super alta Cythera,
Aut super Idallum, sacrata sede recondam.—Virg. *Æn.* i. 680.

Qualis Idalium colens
Venit ad Phrygium Venus
Judicem.   Catull. lxi. 17.

ment. The Egyptian monarchs added some towns, to three of which they gave the name of Arsinoë. Little is known of the history of the towns of Cyprus: they owe their chief celebrity to the worship of Venus. We shall describe them from W. to E. along the northern shore, and from E. to W. along the southern.

Arsinoe stood on the N. coast, near the western promontory Acamas; it was destroyed by Ptolemy Soter. Soli or Soleï was the most important port on the northern coast, and had valuable mines in its neighbourhood; it was said to be an Athenian settlement. Salamis stood at the mouth of the Pediæus on the E. coast; it was an important town in the 6th century B.C., and under an independent dynasty: a famous sea-fight took place off its harbour between Menelaus and Demetrius Poliorcetes, B.C. 300; it was partially destroyed in Trajan's reign and wholly by a subsequent earthquake; it was rebuilt by a Christian emperor, with the name of Constantia. On the S. coast the principal town was Citium, the remains of which are still visible near Larnika, consisting of a theatre, tombs, and the foundations of the walls: the death of Cimon the Athenian, B.C. 449, and the birth of the philosopher Zeno, are the chief events of interest connected with it. Amathus stood more to the W., and was celebrated for the worship of Venus,[7] Adonis, and the Phœnician Hercules or Melkart, as well as for its wheat and mineral[8] productions. Paphos was the name of two towns on the S.W. coast: the older one, named Palæpaphus by geographers, but simply Paphos by the poets, stood on a hill[9] about 1½ miles from the sea, on which it had a roadstead: it was the most celebrated seat of the worship of Venus,[1] whose fane there is mentioned even by Homer. The foundations of the later temple erected by Vespasian are still discernible, and its form is delineated on the coins of some of the Roman emperors. New Paphos, Baffa, was on the coast, about 7½ miles N.W. of the old town, and took a prominent part in the reverence paid to the goddess Venus: it was the residence of the Roman governor in St. Paul's time; the harbour is now almost blocked up. Of the less important towns we may notice—Lapethus, on the northern coast—Golgi, whose position is unknown, also famous for the worship of Venus[2]—Marium, between

[7] Est Amathus, est celsa mihi Paphos, atque Cythera,
Idaliæque domus. Virg. Æn. x. 51.
Calte puer, puerique parens Amathusia culti;
Aurea de campo vellito signa meo.—Ov. Amor. iii. 15, 15.

[8] Fecundam Amathunta metalli. Ov. Met. x. 220.

[9] Celsa Paphos. Virg. Æn. x. 51.

[1] 'Η δ' ἄρα Κύπρον ἵκανε φιλομμειδὴς Ἀφροδίτη,
'Ἐς Πάφον· ἔνθα δέ οἱ τέμενος βωμός τε θυήεις.
Hom. Od. viii. 362.
Ipsa Paphum sublimis adit, sedesque revisit
Læta suas: ubi templum illi, centumque Sabæo
Thure calent aræ, sertisque recentibus halant.—Virg. Æn. i. 415.
Quæ Cnidon
Fulgentesque tenet Cycladas, et Paphon
Junctis visit oloribus. Hor. Carm. iii. 28, 13.

[2] Nunc, o cæruleo creata ponto,
Quæ sanctum Idalium, Syrosque apertos,
Quæque Ancona, Cnidumque arundinosam
Colis, quæque Amathunta, quæque Golgos.—Catull. xxxvi. 11.

Amathus and Citium—and Tamassus, on the northern slope of
Olympus, supposed to be identical with Homer's Temesa.[1]

History.—Cyprus appears to have been subject to the Syrians as
early as the time of Solomon. Under Amasis it became attached to
the Egyptian kingdom. On the invasion of Egypt by Cambyses it
surrendered to the Persians. It took part in the Ionian revolt, but
was subdued by Darius. After the battle of Salamis the Athenians
reduced the greater part of it. The brilliant period of its history
belongs to the times of Evagoras, king of Salamis. It again fell under
the Persians until Alexander's time. In the division of the Macedonian
empire, it was assigned to the Egyptian Ptolemy, and it remained the
most valuable appendage of the Egyptian kingdom until it was annexed
to the Roman empire in B.C. 58.

St. Paul's Travels.—Cyprus was visited by the Apostle in his first
missionary tour. He crossed the sea from Seleucia in Syria to Salamis,
and then probably followed the Roman road to Paphos, whence he set
sail for Pamphylia (Acts xiii. 4-13). In his voyage to Rome he "sailed
under Cyprus," i. e. kept under the N. coast of the island (Acts
xxvii. 4).

---

[1] Νῦν δ᾽ ὧδε ξὺν νηΐ κατήλυθον ἠδ᾽ ἑτάροισι,
    Πλέων ἐπὶ οἴνοπα πόντον ἐπ᾽ ἀλλοθρόους ἀνθρώπους;
    Ἐς Τεμέσην μετὰ χαλκόν, ἄγω δ᾽ αἴθωνα σίδηρον.

                                            HOM. *Od.* l. 182.

    Est ager, indigenæ Tamasæum nomine dicunt;
    Telluris Cypriæ pars optima: quem mihi prisci
    Sacravere senes, templisque accedere dotem
    Hanc jussere meis.   Ov. *Met.* x. 644.

Copper Coin of Cyprus under the Emperor Claudius.

Mount Argæus, Cappadocia (From Texier).

CHAPTER IX.

Asia Minor, *continued*.

VII. Cappadocia.

§ 1. **Cappadocia** was an extensive province in the eastern part of Asia Minor, bounded on the E. by the Euphrates, on the S. by Taurus, on the W. by Lycaonia, and on the N. by Galatia and Pontus, from the latter of which it was separated by the upper part of the range of Antitaurus. These limits include the district named Armenia Minor, but exclude the extensive province of Pontus, which formed a portion of Cappadocia in the time of Herodotus (p. 36). The northern part of the province is mountainous; the central and southern parts consist of extensive plains lying at a high elevation, bare of wood, in some places fertile in wheat and wine, and elsewhere affording fine pastures for cattle and horses. Among the mineral products we may notice a species of crystal, onyxes, a white stone used for sword-handles, and a translucent stone adapted for windows. There are extensive salt-beds near the Halys.

§ 2. The chief mountain-range is Antitaurus, which intersects the country in a north-easterly direction, and attains its highest elevation in the outlying peak of Argæus (p. 86). The chief river is the Halys (p. 67), whose middle course falls within the limits of this province, and which receives the tributary streams of the Melas, Kara-su, flowing by the roots of Argæus; and of the Cappadox, supposed to be the small river of Kir-Shehr, on the border of Galatia. The Carmalas in Cataonia is a tributary of the Cilician Pyramus, while a second Melas, Koramas, in the eastern part of the province, seeks the Euphrates. The great salt lake of Tatta falls partly within the limits of Cappadocia.

§ 3. The inhabitants of this district were regarded by the Greeks as a Syrian race, and were distinctively named " White Syrians." The name "Cappadox" is probably of Persian origin; and some ethnologists regard the Cappadocians as an Arian and not a Semitic race. The Cataonians were deemed a distinct people. The political divisions varied at different eras: the eastern district, between Antitaurus and the Euphrates, was divided into three parts—Armenia Minor, Melitêne, and Cataonia; the western was divided into six portions in the time of the native dynasty. Under the Romans Cataonia was subdivided into four, and Armenia Minor into five districts, the names of which need not be specified. The emperor Valens (about A.D. 371) divided Cappadocia into two provinces named Prima and Secunda, to which Justinian subsequently added Tertia.

§ 4. The towns of Cappadocia offer few topics of interest in connexion with classical literature. The country was so shut out from the great paths of communication that the Greeks were wholly unacquainted with it; and it was only in the century proceding the Christian era that the Romans had occasion to cross its boundaries. The information which we have respecting its towns belongs almost wholly to the period of the Roman empire, when the provincial organization was introduced. We may assume that in most instances the sites of the towns which the Romans built had been previously occupied by the Cappadocians, as we know to have been the case in some instances, where the change of name indicates the change of masters. Thus the old capital, Mazāca, in the valley of the Halys, became Cæsarea ; Mocissus, Justinianopolis ; and Halāla, Faustino-polis. The chief towns were Cæsarea in the N., Tyāna in the S., and Melitēne in the E. The latter was situated on the great mili-tary road which led from Asia Minor to Armenia and Mesopotamia. Many of the towns were of importance as military positions : this was particularly the case with Melitēne, which commanded the passage of the Euphrates ; Cilica and Dascūsa, which were on the same river ; and Satāla, which was the key of Pontus. All these were stations of Roman legions.

Commencing in the western part of Cappadocia Proper, we meet first with Mocissus, on the borders of Galatia, which was enlarged by the emperor Justinian, and made the capital of Cappadocia Tertia, with the name Justinianopolis. Mazāca was situated at the foot of Mount Argœus, and was the residence of the old Cappadocian kings : it was taken by Tigranes, and again by Sapor in the reign of Valerian. The emperor Tiberius enlarged it, constituted it the capital of the province, and changed its name to Cæsarea. The town is still important, and retains its ancient name in the form *Kaisariyeh*. Archelāis was situ-ated on the borders of Lycaonia, probably on the site of the older Garsaura ; and owed its name to its founder, Archelaus, the last king. It was made a Roman colony by the emperor Claudius. The chief town of the southern district was Tyāna, N. of the Cilician Gates, and thus, from its position in reference to that pass, as well as from its natural strength, a place of great importance. It became a Roman colony under Caracalla : afterwards, having been incorporated with the empire of Palmyra, it was conquered by Aurelian, A.D. 272, and was raised by Valens to the position of capital of Cappadocia Secunda. The famous impostor Apollonius was born there. There are considerable ruins of the town at *Kis-hissar*, particularly an aqueduct of granite about 6 miles long. Cybistra, S.W. of Tyana, was once visited by Cicero when he was proconsul of Cilicia. Nora, on the borders of Lycaonia, was a strong fortress in which Eumenes was besieged by Antigonus for a whole winter. Faustinopolis, S. of Tyana, derived its name from Faustina, the wife of the emperor M. Aurelius, who died there, and was deified, a temple being built to her honour. In Catania, the chief town was Comāna Aurea, at the eastern base of Antitaurus, famed for the worship of Enyo, which was traced back to Orestes : it was made a colony by Caracalla : a considerable town, *Al-Bostan*, occupies its site. Melitēne was the most important town in the district of the same name : it stood not far from the junction of the Melas with the Euphrates, at *Malatiyeh* : it owed its first rise to Trajan : it was afterwards embel-lished by the emperors Anastasius and Justinian, and it became the capital of Armenia Secunda : it was the station of the famous Christian *Legio XII. Fulminata:* the Romans defeated Chosroes I. near it, A.D. 577. In Armenia Minor, in addition to the border-fortresses of Cilica, Dascusa, and Satala, already noticed, Nicopolis must be mentioned, as founded on the spot where Pompey conquered Mithridates : its site is probably at *Derriki*. The fortress of Sinoria, built by Mithridates, was somewhere on the frontier between Armenia Major and Minor. Though Cappadocia only receives passing notices in the Bible (Acts ii. 9 ; 1 Pet. I. 1), it is famous in ecclesiastical history from its having given birth to Gregory of Nazianzus, in the western part of the province, of which place he afterwards became bishop, and to Basil, who became bishop of his native town Cæsarea. Nyssa, in the N.W., was equally famous as the see of Gregory.

History.—Cappadocia formed a portion of the Assyrian, Median, and Persian empires. Under the latter it was governed by satraps, who had the title of kings. After the death of Alexander it was annexed to the Syrian empire, but still retained a native dynasty, in which the names of Ariarathes and Ariamnes alone occur, until about B.C. 93, when the royal family became extinct. A new dynasty, in which the name of Ariobarzanes is most frequent, was then seated on the throne under the patronage of the Romans: this terminated with Archelaus, A.D. 17, at whose death Cappadocia was made a Roman province.

Armenia Minor is first noticed as a separate district after the defeat of Antiochus by the Romans. It was then under its own kings, who extended their sway at one time over Pontus. The last of them surrendered to Mithridates; and it afterwards passed into the power of the Romans, who transferred it from one king to another, and finally united it to Cappadocia in the reign of Trajan.

VIII. LYCAONIA AND ISAURIA.

§ 5. **Lycaonia** was bounded on the E. by Cappadocia, on the S. by Cilicia, on the W. by Phrygia and Pisidia, and on the N. by Galatia. Its limits, in reference to the adjacent provinces, were very fluctuating, particularly under the Romans, who handed over portions of Lycaonia sometimes to one, sometimes to another sovereign, and incorporated a large portion of it at one time with Galatia, at another with Cappadocia. Isauria was regarded sometimes as a separate district, sometimes as belonging to Lycaonia: it was the mountainous district on the S.W. border of the latter country, adjacent to Pisidia. Lycaonia is generally a level country, high, and bleak, badly watered, but well adapted for sheep-feeding. The central plain about Iconium is the largest in Asia Minor. The soil is strongly impregnated with salt. Lofty mountains rise both in the northern and southern districts, none of which, however, received specific names in ancient times. The lakes of **Tatta** on the border of Cappadocia, **Coralis** and **Trogitis** in Isauria, are the only physical objects worthy of notice.

§ 6. The Lycaonians were undoubtedly an aboriginal population, and the tradition which connected them with the Arcadian Lycaon is void of all foundation. They were a hardy and warlike race, living by plunder and war. The Isaurians had a similar character, but appear to have been rather connected with the Pisidians in point of race. The towns were both few and small: Derbe was the early, and Iconium the later capital of Lycaonia, as Isaura was of Isauria: Laodicea owed its existence to Seleucus I.

Iconium was situated in the midst of an extensive plain in the western part of the province. Xenophon assigns it to Phrygia: Strabo describes

It as a small place, but it soon rose to importance, and both Pliny and the Acts of the Apostles represent it as very populous: it became the metropolis under the Byzantine emperors, and is still a large place under the name of *Koniyeh*. Laodicea lay to the N.W. of Iconium, and received the surname of *Combusta*, probably from having been burnt down: numerous remains at *Ladik*, consisting of altars, columns, capitals, &c., show that it was a fine and large town. Derbe was a fortified town in the S. of the province, probably at or near *Dirle*, and not far from the base of Taurus: it was the residence of the robber Antipater, and subsequently of Amyntas. Lystra was near Derbe, but its position is quite undecided: it may be at *Bin-bir-Kilisseh*, on the N. of the mountain called *Karadagh*, where there are extensive ruins of churches. Laranda, in the S.W., is known only for its destruction by Perdiccas, and as a subsequent resort of the Isaurian robbers. Isaura was a large town at the foot of Taurus, which was twice ruined, firstly by Perdiccas, and afterwards by Servilius, when it was rebuilt by Amyntas of Galatia: the new town became the residence of the rival emperor Trebellianus.

History.—The Lycaonians never submitted to the Persians, but they yielded to Alexander the Great, and passed successively to the Seleucidæ, Eumenes of Pergamus, and the Romans: the only period when they became at all powerful was under the rule of Amyntas, just before their annexation to Cappadocia. The Isaurians offered a prolonged resistance to the Romans, to whom their marauding habits made them particularly obnoxious. Servilius (B.C. 78) attacked them with success; and subsequently the Romans found it necessary to surround them with a *cordon* of forts, but they repeatedly broke out, and remained the terror of the surrounding countries down to a late period.

St. Paul's Travels.—Lycaonia was visited by St. Paul in his first and second missionary tours. In the former he entered it from Pisidia, and first visited Iconium, then much frequented by Jews; and afterwards Lystra and Derbe, whence he retraced his footsteps to Pisidia (Acts xiv. 1-21). On the second occasion he entered it on the side of Cilicia, and passed through Derbe and Lystra to Iconium, and thence continued his course probably to Antioch in Pisidia (xvi. 1-5). On the latter occasion he took away with him Timothy, whose birth-place was probably Lystra, though it may have been Derbe.

IX. PISIDIA.

§ 7. Pisidia bordered in the E. on Isauria and Cilicia, in the S. on Pamphylia, in the W. on Lycia, Caria, and Phrygia, and in the N. on Phrygia. The limits with regard to these provinces were fluctuating, particularly the northern portion, which was sometimes attached to Phrygia, with the title of Phrygia Pisidica. The country is rough and mountainous, but contains several fertile valleys and plains. The mountain-ranges of Pisidia emanate from Mount Taurus, and generally run from N. to S.: the only one to which a specific name was assigned was Sardemisus in the S.W. The upper courses of the Catarrhactes, Castrus, and Eurymedon, fall within the limits of Pisidia, and flow through the heart of the Taurian range into the Pamphylian plain. These rivers are fed by numerous mountain torrents, which after rain rush down the ravines with extraordinary violence. The districts of Milyas and Caballa, which wo have already noticed in connexion with Lycia, extended over the south-western portions of Pisidia.

§ 8. The Pisidians were a branch of the great Phrygian stock, intermixed with Cilicians and Isaurians, the latter of whom they resembled in their lawless and marauding habits of life. The towns were situated either on or amid inaccessible cliffs, and were so many natural fortresses : such was the position of Termessus, which alarmed even the skilled warriors of Alexander's host; of Selge and Sagalassus, which played a conspicuous part in the Roman wars with Antiochus the Great; and of Cremna, as its name ("the precipice") implies. Antioch, which in accordance with Scriptural notices (Acts xiii. 14) we shall regard as a Pisidian town, though assigned by Strabo to Phrygia Parorios, was situated in the northern plain, and was a Greek rather than a pure Pisidian town, having been founded by Seleucus Nicator. Most of these towns survived to a late period, as the character of their remains proves. Antioch and Cremna became Roman colonies.

Antioch was situated on the S. side of a mountain range on the border of Phrygia : originally it belonged to Syria, but after the battle of Magnesia, B.C. 190, it was annexed to Pergamus : it afterwards became the capital of the Roman province : its remains at *Yalobatch* are numerous, consisting of a temple of Dionysus, a theatre, and a church. Seleucia, surnamed Sidera, probably from ironworks in its vicinity, stood S.W. of Antioch, at *Ejerdir* : it was perhaps founded by Seleucus Nicator. Sagalassus, in the N.W., was situated on a terrace on the side of a lofty mountain, with a fertile plain stretching away below it : Alexander took it by assault ; Manlius reduced it by ravaging the plain : the ruins at *Aglasonn* are very fine, consisting of a theatre, a portico, &c., with innumerable tombs hewn out of the perpendicular face of the cliff. Cremna, S.E. of Sagalassus, occupied the summit of a mountain, three sides of which were terrific precipices : it was taken by the Galatian king Amyntas : there are remains of a theatre, temples, &o., at *Germe*. Selge was situated near the Eurymedon, in the S. of the province, on a lofty projection surrounded by precipices and defiles : it was so populous a place that its soldiers numbered 20,000 ; it was besieged and taken by Achæus : the supposed ruins of Selge, near *Boojak*, are magnificent, and extend for more than 3 miles : about 50 temples, with innumerable tombs and other buildings, have been noticed. Termessus was situated on a precipitous height near the Catarrhactes, at *Karabunar Kiui*, and commanded the ordinary road between Lycia and Pamphylia. Cibyra was the chief town in Caballa, and the head of a tetrapolis, of which Bubon, Balbura, and Œnoanda were the other confederates : it stood on a tributary of the Calbis, and overlooked a wide and fertile plain : it was visited by Manlius, and became subsequently a place of great trade, particularly in wood and iron :[1] the ruins at *Horzoom* consist of a theatre and some temples. The exact positions of Cretopolis and of

[1] Ne Cibyratica, ne Bithyna negotia perdas.—Hor. *Ep.* L 6, 33.

Islands are unknown : they were somewhere in the S.W., on the borders of Pamphylia.

History.—The Pisidians resisted all attempts at permanent subjection. Even the Romans failed : for though they conquered the inhabitants, and handed over the province to Eumenes of Pergamus, and afterwards adjoined it to their province of Pamphylia, yet they never really repressed its lawless inhabitants, nor did they ever introduce a provincial organization.

St. Paul's Travels.—St. Paul visited Pisidia in his first journey, crossing Taurus from Pamphylia to Antioch, where the Jews appear to have been numerous, and returning by the same route after having visited Lycaonia (Acts xiii. 14 ; xiv. 21) : he probably visited Antioch again in his second journey, though the place is not specified (xvi. 4).

Hierapolis in Phrygia (Laborde).

X. PHRYGIA.

§ 9. The important province of **Phrygia**, or, as it was more fully termed, **P. Major**, to distinguish it from P. Minor in Mysia, bordered in the E. on Galatia and Lycaonia, in the S. on Pisidia, in the W. on Caria, Lydia, and Mysia, and in the N. on Galatia. Its boundaries cannot be fixed with any degree of precision, as they varied at different historical eras : it may be described generally as the western part of the central plateau, and as coextensive with the limits of the plateau itself. The country is mountainous and well watered: some portions, particularly the valleys of the Hermus and Mæander, were very fertile and produced the vine:[2] the other parts were adapted to sheep-feeding. The chief productions were wool, which was of a very superior quality, and marble, especially the species found near Synnáda. The western district was much exposed to earthquakes; and the presence of volcanic agency is attested by hot springs.

§ 10. The mountains of Phrygia consist of irregular offsets from the border ranges of **Olympus** in the N., **Taurus** in the S., and **Cadmus** in the S.W. The only name applied specifically to any of the Phrygian hills is **Dindymum**, which appears to have been equally given to a hill about the sources of the Hermus, and to a second near Pessinus.[3] Phrygia contains the upper courses of the **Hermus** and **Mæander**, which seek the Ægæan, and the **Sangarius**, which flows northward to the Euxine: the **Thymbres** and **Alander**, tributaries of the latter, belong wholly to Phrygia, as also do the **Marsyas** and the **Lycus**, tributaries of the Mæander: the Marsyas joined the Mæander almost immediately after its rise:[4] it was connected in mythology with the victory of Apollo over Marsyas.[5] Several large salt lakes occur in the southern part of the province, of which **Anava** has been identified with *Chardak*, and **Ascania** with *Buldur* to the S.E., though not improbably it may be only another name for Anava.

§ 11. The inhabitants of this province came of the same stock as the Thracian tribes, and were in early times the masters of the whole western part of Asia Minor. The affinities that existed between them and the surrounding nations have been already pointed out (p. 89). They were deprived of portions of their territory by the advance of the Semitic races in the S. and W., of the Cappadocians in the E., and finally of the Galatians in the N. From being a warlike race, they became, after the conquest of their country by Persia, purely agricultural, and were regarded with con-

[2]　Πὸη καὶ Φρυγίην εἰσήλυθον ἐμπελόωσαν. Hom. *Il.* III. 184.

[3] The latter of the two was the mountain known to the poets as being sacred to Cybele, who is hence called Dindymene:—
　　O vere Phrygiæ, neque enim Phryges! Ite per alta
　　Dindyma, ubi assuetis biforem dat tibia cantum.
　　　　　　　　　　　　　　　　Virg. *Æn.* ix. 617.

　　Non Dindymene, non adytis quatit
　　Mentem sacerdotum incola Pythius,
　　Non Liber æque.　　　　　Hor. *Carm.* i. 16, 5.
　　Agite, ite ad alta, Gallæ, Cybeles nemora simul;
　　Simul ite, Dindymenæ dominæ vaga pecora.
　　　　　　　　　　　　　　　Catull. lxiii. 12.

[4]　Icarium pelagus Mycaleæque littora juncti
　　Marsya Mæanderque petunt: sed Marsya velox,
　　Dum suus est, flexuque carens, jam flumine mixtus,
　　Mollitur, Mæandre, tuo.　　Claudian. *in Eutrop.* ii. 263.

[5]　Quique colunt Pitanen, et quæ tua munera, Pallas,
　　Lugent damnatæ Phœbo victore Celænæ.　　Luc. iii. 205.

tempt, the Phrygian names of Midas and Manes being given to
slaves. Phrygia was divided into four portions — **Salutaris**, the
central and largest; **Pacatiana**, on the borders of Caria; **Epictetus**
(*i. e.* "acquired") in the N.; and **Parorios**, the mountainous region
in the S. Epictetus was so named as having been transferred from
the Bithynian to the Pergamenian kings about B.C. 190 : the two first
designations did not come into vogue until the 4th century A.D.

§ 12. The foundation of many of the Phrygian towns was carried back to the mythical ages : such was the case with Celenæ,
Hierapolis, and Metropolis. Celenæ appears to have ranked as the
capital in the time of Cyrus the Younger, and Colossæ was then an
important place. These towns waned with the rise of those founded
by the Syrian monarchs, viz. Apamea and Laodicea. Many of
the Phrygian towns were places of extensive trade under the Romans,
particularly the two just mentioned. Some important roads passed
through Phrygia : the great lines of communication between Ephesus
and the East centred at Synnada, whence roads led to Cilicia, to
Cæsarea in Cappadocia, and thence to Armenia, and again northwards
to Dorylæum and Bithynia.

Commencing in the N.E. of the province, **Dorylæum**, *Eski-Shehr*,
was centrally situated on a small stream which flows into the Thymbres, with hot baths in the neighbourhood; Lysimachus made an in-
trenched camp there. **Synnada** stood on a plain in the centre of the
province, and was particularly famous for its marble, which was
streaked with purple veins :[*] ruins of the town exist at *Eski-Kara-Hissar*.
Ipsus lay S.E. of Synnada, and is only famous for the great battle
fought there in B.C. 301, between Antigonus and Demetrius on the one
side, and Cassander, Lysimachus, Ptolemy, and Seleucus on the other.
Philomelium was on the high road between Synnada and Iconium, not
far from the Pisidian Antioch : its ruins are at *Ak-Shehr*. **Celenæ** was
situated at the source of the Mæander, with an acropolis on a hill to
the N.E. : Cyrus the Younger had a palace and park there, and the
sources of the Mæander are said to have been in the palace : the
Catarrhactes, which Xenophon describes as rising in the agora, was the
same as the Marsyas : the inhabitants, and probably the materials, of
Celenæ, were removed to the neighbouring Apamea, and the place disappeared. **Apamea**, surnamed **Cibotus**, was founded by Antiochus
Soter, and named after his mother Apama : it stood a little lower down
the Mæander at *Denair*, where are the ruins of a theatre and other
buildings : the name "Cibotus" (from κιβωτός, "a coffer") may have
referred to its wealth as a commercial emporium, for which its position on the great high road adapted it : it was much damaged by earthquakes, particularly in the reign of Claudius, but it continued a flourishing place to a late period. **Colossæ**, on the Lycus, was an important

---

[*] Sola nitet flavis Nomadum decisa metallis
Purpura, sola cavo Phrygiæ quam Synnados antro
Ipsa cruentavit maculis lucentibus Attys. — STAT. *Silv.* L. 3, 36.

Pretiosaque picto
Marmore, purpureis credit cui Synnada venis.
CLAUD. *in Eutr.* II. 278.

place at the time when Xerxes visited it in B.C. 481, and Cyrus in
B.C. 401; but it fell as the neighbouring city of Laodicea rose, and
was but a small place in Strabo's time; it was finally supplanted by
a town called Chonæ, about 3 miles to the S., which still exists as
Chonos: at Colossæ the Lycus is said by Herodotus to have disappeared
in a chasm for about half a mile : a gorge still exists, which is probably
the chasm referred to, the upper surface having fallen in :[7] Colossæ was
one of the early Churches of Asia, to which St. Paul wrote an Epistle.
Laodicea, lower down the Lycus, was so named after Laodice, the wife
of Antiochus Theos, its reputed founder: it suffered severely in the
Mithridatic war, but soon revived, and became one of the greatest commercial towns of Asia Minor, especially as a mart for wool : it was also
the seat of one of the Seven Churches, to which St. Paul addressed an
Epistle (Col. iv. 16): it was then a very wealthy town, and continued to
flourish down to the middle ages : the ruins of it at *Eski-Hissar* consist
of a stadium, gymnasium, theatres, and aqueduct, erected for the most
part during the Roman period. **Hierapolis** was 5 miles N. of Laodicea
on the road to Sardis: it was famous for its hot springs, and for a cave
whence issued mephitic vapours : a Christian Church was planted there
(Col. iv. 13), and at a later period it claimed to be the metropolis of
Phrygia: it was the birthplace of Epictetus : extensive ruins of it exist
at *Pambuk-kalassi.*

Azani (Texier's ' Asia Minoure').

Among the less important towns we may briefly notice—**Midaïum**,
in the N.E., on the road between Dorylæum and Pessinus, where
Sextus Pompeius was captured by the generals of M. Antony—**Metropolis**, N. of Synnada, at *Piemesh Kalasi*, the capital of the ancient kings

---

f Sic ubi terreno Lycus est epotus hiatu,
   Existit procul hinc, alioque renascitur ore. – Ov. *Met.* xv. 273.

of Phrygia, and the place where Midas was buried—Peltæ, near the source of the Mæander, but of uncertain position, visited by Cyrus the Younger—Cerâmon Agôra, on the borders of Mysia, probably at Ushak—Cayêtri Campus, a place noticed by Xenophon on Cyrus's route, not connected with the well-known river Cayster, but on the E. border of Phrygia, near the lake named Eber Ghieul—Eumenia, N.W. of Apamea, so named by Attalus II. after his brother Eumenes—Blaundus, probably the ancient name of a town the ruins of which are seen at Suleimanli, consisting of an acropolis, theatre, gateway, and a beautiful temple—Ancyra, a small town in the N.W. angle, near the lake of Simaul, near which also stood Synnaus—and Aâni, a place on the Rhyndacus. historically unknown, but from its remains evidently an important place: a beautiful Ionic temple, theatre, and other buildings at Tchavdour-Hissar, mark its site.

History.—Phrygia was the seat of a very ancient dynasty, in which the names of Gordius and Midas are prominent. This was terminated in B.C. 580 by Crœsus, who incorporated Phrygia with his kingdom. Thenceforward its history is merged in that of the surrounding countries, as it never afterwards attained an independent position. The Romans indeed declared it a free country after the death of Mithridates V., in B.C. 120, but soon afterwards they divided it into jurisdictiones, and in B.C. 88 they assigned the districts of Laodicea, Apamea, and Synnada, to Cilicia, from which they were at length permanently transferred to the province of Asia in B.C. 49. In the new division of the empire in the 4th century A.D., Parorica was added to Phrygia, and a district on the Mæander to Caria: the rest was divided into Salutaris and Pacatiana.

St. Paul's Travels.—St. Paul visited Phrygia in his second journey as he passed from Lycaonia into Galatia (Acts xvi. 6): the route he followed is purely conjectural, as no particulars are given in reference to it: he probably followed the course of the Roman road which diverged from Synnada to Cilicia, and passed through the towns of Laodicea in Pisidia and Philomelium, whence perhaps he diverged to Antioch, and struck into the high road again near Synnada: thence he took the high road to Ancyra in Galatia. On his return from Galatia he probably traversed the northern district by Cotyæum and Azani to Mysia. In his third journey he again visited Phrygia (Acts xviii. 23); on this occasion he passed out of the province to Ephesus, probably by the valley of the Hermus.

XI. GALATIA.

§ 13. Galatia, or Gallo-Græcia, bordered in the W. on Phrygia, in the N. on Bithynia and Paphlagonia, in the E. on Pontus, and in the S. on Lycaonia and Cappadocia. The northern portion of the province is rough and mountainous: the southern is also uneven, but has extensive and fertile plains, adapted for sheep-feeding. The eastern district was regarded in ancient times as the most fertile. The chief mountain ranges of Galatia are Olympus in the N. and Dindymus in the W., both of which have been previously noticed. A range named Magâba rises in the central district near Ancyra, and another, named Adorêus, *Elmah Dagh*, on the border of Phrygia. The river Halys in its middle course bisects Galatia from S. to N., and then skirts its northern border for some distance, receiving several unimportant feeders. Galatia also contains the upper course of the **Sangarius**, with its tributaries the **Sibéris**, which rises W. of Ancyra, and joins the main stream near Juliopolis, and the **Scopas**, *Aladan*, which has a parallel course more to the W.

§ 14. The inhabitants of Galatia were a Celtic race, who migrated westward from their settlements in Gaul,[*] and entered Asia Minor under the chieftainship of Leonorius and Lutarius in three bands named Tolistoboii, Tectoságes, and Trocmi. They were engaged by Nicomedes I. king of Bithynia, B.C. 278, to act as mercenaries in his army against his brother Zybœtes. Having succeeded in this war, and having received some land as a reward, they divided into three bands, and ravaged the whole of the surrounding districts. They were resisted and defeated by Antiochus Soter in the first instance, then by Attalus of Pergamum in B.C. 238, afterwards by Prusias I. of Bithynia in 216, again by the Roman consul Manlius in 180, and finally by Eumenes of Pergamum in 167, after which they settled quietly down in the district to which they gave their name. This had been previously occupied by Phrygians, Paphlagonians, and Greeks, of whom the latter were predominant in influence at the time the Gauls entered, as their language was usually spoken, and was adopted even by the invaders for literary purposes. The three tribes of the Gauls divided the country between them, the Tolistoboii occupying the W., the Tectosages the centre, and the Trocmi the E. Each tribe was divided into four parts, named tetrarchies. The twelve tetrarchs formed a senate, and were assisted by a council of 300 deputies, who met at Drynæmētum. The Gauls adopted the Phrygian and Greek superstitions, and became thoroughly Græcised, as their name Gallo-Græci implies : but they appear to have retained their native tongue down, to the 4th century A.D.

§ 15. The only important towns in Galatia were Pessinus the capital of the Tolistoboii, and Ancyra the capital of the Tectosages : these were situated on the great high road of the Romans from Ephesus to the E., and were places of great commercial importance : at Ancyra the road from Ephesus fell in with that leading from Byzantium. Tavium, the capital of the Trocmi, in the E. of the province, was also a considerable place. The only Roman colony was Germe.

Pessinus was situated on the S. side of Mount Dindymus, and owed its chief celebrity to the worship of Agdistis, or Cybele, whose temple was magnificently adorned by the kings of Pergamum, and was visited from all parts of the world : the ruins of a theatre and other buildings,

---

[*] Galatæ and Keltæ are but different forms of the same word : and Galatæ and Galli are respectively the Greek and Latin designations of the same race.

about 10 miles S.E. of *Sevri-Hissar*, show that Pessinus was a remarkably fine town. Ancyra was centrally situated to the N.E. of Pessinus,
and appears in history as the place where Manlius defeated the Tectosages in B.C. 189 : the most famous building was a temple of Augustus.
with an inscription, named Marmor Ancyranum, containing a record
of his achievements : this is still in existence, as also are various
sculptured remains of the citadel : *Angora* is still a very important
place. Tavium was chiefly celebrated for its temple of Jupiter : the
position of the town is probably marked by the ruins of *Boghas Kieui*,
at some distance from the E. bank of the Halys.

Of the less important towns we may notice—Germa, *Yorma*, between
Pessinus and Ancyra, a Roman colony—Bludium, belonging to the
Toliatoboii, the residence of Deiotarus—Corbeus, S.E. of Ancyra—
and Dankla, a town of the Trocmi, where Cn. Pompeius and Lucullus
had an interview. Some places have names of a more or less Celtic
character, as Eccobriga and Drynœmetum.

History.—The history of Galatia commences with the time when one
of the tetrarchs, Deiotarus, was invested by the Romans with the
rights of sovereignty, not only over the Toliatoboii, but also over
Pontus and Armenia Minor. He was succeeded by his son Deiotarus,
Cicero's friend, and he by Amyntas, who received from M. Antony
Pisidia in B.C. 39, and Galatia with other districts in 36. Amyntas died
B.C. 25, and his territories were formed into a province by Augustus.

St. Paul's Travels.—St. Paul visited Galatia in his second missionary
journey : his route through the province is purely conjectural, no town
whatever being specified in the narrative (Acts xvi. 6) : he probably
entered it on the side of Phrygia at Pessinus, and visited Ancyra,
returning by the same route. He again visited Galatia on his third
journey, probably entering it from Cappadocia, and leaving it by way
of Phrygia (xviii. 23). He afterwards addressed an Epistle to the
Galatian Church.

XII. BITHYNIA.

§ 16. **Bithynia** was bounded on the N. by the Euxine, on the
N.W. by the Propontis, on the S.W. by Mysia, on the S. by
Phrygia, on the S.E. by Galatia, and on the N.E. by Paphlagonia :
the limit in the latter direction was generally fixed at the river
Parthenius. It is throughout a mountainous district, but fertile,
particularly the part W. of the Sangarius, which contains some fine
plains : wood was abundant, and extensive forests still exist in the
district E. of the Sangarius. The scenery of the western district
about the shores of the Propontis is magnificent. Among the special
products for which Bithynia was famed, we may notice the cheese
of Salône near Bithynium, aconite (so named from Aconæ, where it
was found), marble, and crystal.

§ 17. The chief mountain range is **Olympus**, of which there are
two great divisions—one on the border of Mysia near Prusa, and
another on the border of Galatia : the former is capped with snow
to the end of March. We have also to notice the lesser ranges of
Arganthonius, between the bays of Astacus and Cius, in the W. ;
and **Orminium**, in the N.E. of the province. The coast of the Pro-

pontis is irregular: two bays penetrate far into the interior, separated from each other by Arganthonius: they were named **Sinus
Cianus**, and **Sin. Astacenus**, after the towns of Cius and Astacus:
the mountain range terminates in **Prom. Posidium**, *C. Bozburun*: a
second promontory named **Acritas**, *C. Akrita*, stands at the northern
entrance of the Bay of Astacus. The northern coast runs nearly
due E. from the mouth of the Bosporus to some distance beyond
the Sangarius, the only marked features being the promontories of
Melæna, *C. Tshili*, near the Bosporus, and **Calpe**, with an adjacent
port, now *Kirpe Liman*, W. of the Sangarius.

§ 18. The chief rivers of Bithynia are—the **Sangarius**, which
bisects the province from S to N., in an extremely devious course—
the **Billæus**, *Filyas*, more to the E., which divides into two branches
in its upper course—and the **Parthenius**, *Bartan-Su*, on the eastern
frontier. Of the smaller streams we may notice—the **Rhebas**, which
joins the Euxine near the Bosporus, commemorated in the story of
the Argonauts[a]—the **Psilis**, more to the E.—the **Hyplus**, E. of the
Sangarius, at the mouth of which the fleet of Mithridates wintered
—and the **Cales** or **Calex**, near Heraclea, the sudden rise of which
destroyed the ships of Lamachus, as they were lying off its mouth.
A large lake named **Ascania**, about 10 miles long by 4 wide, lies E.
of the Bay of Cius.

§ 19. The inhabitants of the western part of Bithynia were an
immigrant race from Thrace, who displaced the previous occupants,
the Mysians, Phrygians, and others. They were divided into two
tribes, named Thyni[1] and Bithyni, the former on the sea-coast, the
latter in the interior. The coast E. of the Sangarius was held by
the Mariandyni. The chief towns in Bithynia were situated either
on or adjacent to the shores of the Propontis. The Greeks occupied
with their colonies the most eligible spots on the coasts: thus the
Megarians settled at Chalcedon and Astacus, and at Heraclea
Pontica on the Euxine; the Milesians at Cius; the Colophonians
at Myrlea. The successors of Alexander founded the flourishing
town of Nicæa, and the Bithynian kings the future capital,

---

[a]
            'Ην δὲ φύγητε
Σύνδρομα πετρέων ἀσσηθέες ἐνδόθι Πόντου,
Αὖτικα Βιθυνῶν ἐπὶ δεξιὰ γαῖαν ἔχοντες
Πλώετε, ῥηγμῖνας πεφυλαγμένοι, εἰσόκεν αὖτε
'Ρήβαν ἀκυρόην ποταμὸν, ἄκρην τε Μέλαιναν
Γνάμψαντες, νήσου Θυνηΐδος ὅρμον ἵκησθε.
                            Apoll. *Argon.* II. 319.

Nec prius obscenum scopulis renpexit ad æquor,
Aut scelis tentata quies, nigrantia quam jam
Littora, longinquaeque exirent flumina Rhebæ.
                            Val. Flacc. iv. 090.

[1] Thyni Thraces arant, quæ nunc Bithynia fertur.
                            Claudian. in *Eutrop.* II. 247.

Nicomedia. The Roman emperors did much for the enlargement
and adornment of these towns, attracted partly by the beauty of the
scenery, and partly by the convenience of the locality in respect to
their Eastern possessions: they also constructed an important road
from Byzantium to Ancyra, where it fell into the grand route from
Ephesus to Armenia. Hadrian particularly favoured this province.
The towns continued to flourish to the latest ages of the empire.

Gate of Nicæa (Texier's 'Asia Minoure').

Prusa, surnamed "ad Olympum," stood at the northern base of
Olympus, and is said to have been named after King Prusias, who
founded it by the advice of Hannibal: it was celebrated for its warm
baths: it is now, under the name of Brusa, one of the most flourishing
towns of Asia Minor. Nicæa was situated at the E. end of Lake
Ascania, on the edge of a wide and fertile plain: it was built by Antigonus on the site of an earlier town, probably after his victory over
Eumenes in B.C. 316, and it received the name of Antigonia, for which
Lysimachus substituted that of Nicæa in honour of his wife: it soon
rose to eminence, and the Bithynian kings often resided there: it vied
with Nicomedia for the title of metropolis: it is chiefly famous for the
Council held there, A.D. 325, in which the Nicene creed was drawn up:
having suffered from earthquakes, it was restored by Valens in A.D. 368:
the remains of its walls are still visible at Isnik. Cius stood at the
end of the inlet named after it, and on a river of the same name,[1]

[1] Τίμος ἀρ' οἳ γ' ἐφύτοντο Κιανίδος ἤθεα γαίης,
    'Αμφ' 'Αργανθώνειον ὄρος· προχοάς τε Κίσιο.
Apoll. Argon. i 1178.

which communicated with Lake Ascania: the town was taken by the Persians, B.C. 499, and again by Philip, son of Demetrius, who destroyed it: it was soon after rebuilt by Prusias, who gave it his own name. **Nicomedia**, on the N. coast of the Bay of Astacus, was founded by Nicomedes I., B.C. 264, and peopled with the inhabitants of Astacus; under the native kings it became the capital of Bithynia: the Roman emperors frequently resided there, especially during their eastern wars: it was a Roman colony, the birthplace of Arrian the historian, and the place where Hannibal died:[1] the modern *Ismid*, which occupies its site, contains many ancient remains. **Chalcedon** stood near the junction of the Bosporus with the Propontis, and nearly opposite to Byzantium: it was founded by Megarians, about B.C. 674, and was a place of considerable trade: it was taken by the Persians after the Scythian expedition of Darius, and in the Peloponnesian War appears to have sided at one time with the Athenians, at another with the Lacedæmonians: in the Mithridatic War it was occupied by the Romans, but was taken by Mithridates; it afterwards became a free city: on its site the village of *Kadi-Kioi* now stands. The Megarian colony of **Heraclea Pontica** was the most important place in the E. of Bithynia, possessing two good harbours, and exercising a supremacy over the whole adjacent coast: it sunk, however, under the kings of Bithynia, and received its deathblow in the Mithridatic War, when it was plundered by the Romans under Cotta. In the interior, to the S. of Heraclea, stood **Bithynium** or **Claudiopolis**, as it was probably named in the time of Tiberius; it was reputed to have been founded by Greeks, and noted for the rich pastures about it: it was the birthplace of Hadrian's favourite, Antinous. Still more to the S. was the ancient town of **Gordium**, the residence of the Phrygian kings, and well known as the place where Alexander severed the "Gordian knot:" it was rebuilt in the time of Augustus, with the name of Juliopolis.

Among the less important towns of Bithynia we may briefly notice —**Dascylium**, on the border of Mysia, where, in the time of Xenophon, the Persian satraps had a residence and park—**Myrlèa**, on the shore of the Bay of Cius, presented by Philip of Macedonia to his ally Prusias, who changed its name to **Apamèa**; it was afterwards a Roman colony —**Drepàne**, on the S. coast of the Bay of Astacus, the birthplace of Helena, the mother of Constantine, by whom it was enlarged and named **Helenopolis**—**Astacus**, at the head of the bay named after it, a Megarian colony, destroyed by Lysimachus in his war with Zipœtes —**Libyssa**, between Nicomedia and Chalcedon, the burial-place of Hannibal—**Chrysopolis**, *Scutari*, opposite to Byzantium, the spot where the Athenians, by the advice of Alcibiades, levied toll on all vessels passing in or out of the Euxine, and the scene of the defeat of Licinius by Constantine the Great, A.D. 323.

History.—The history of Bithynia commences with the accession of Dœdalsus to the sovereignty about B.C. 435, and terminates with

---

[1]
    Post Itala bella
    Assyrio famulus regi, falsusque cup& dt
    Ausonias motus, dubio petet æquora velo;
    Donec, Prusiacas delatus segniter oras,
    Altera servitia imbelli patietur in ævo,
    Et latebram, munus regni. Perstantibus inde
    Ænsadis, reddique sibi poscentibus hostem,
    Pocula furtivo raplet properata veneno,
    Ac tandem terras longa formidine solvet.—Sil. Ital. xiii. 885.

Nicomedes III., who bequeathed his kingdom to the Romans, B.C. 74. Of the eight kings who intervene between these, the most illustrious were Nicomedes I., who founded the capital; Prusias I., who received and betrayed Hannibal; and his son Prusias II., who carried on war with the king of Pergamus. After the death of Nicomedes III. the Romans reduced Bithynia to a province, and, after the death of Mithridates, annexed to it the western part of the Pontic kingdom. Under Augustus Bithynia was assigned to the senate; but Hadrian gave Pamphylia in exchange for it.

In the Bible Bithynia is casually mentioned in two passages (Acts xvi. 7; 1 Pet. l. 1), from the first of which we learn that St. Paul designed to enter it, but failed to do so. It derives an interest from the correspondence of its governor Pliny with Trajan, in relation to the persecution of the Christians, as well as from the great council of Nicæa, to which we have already adverted.

XIII. PAPHLAGONIA.

§ 20. **Paphlagonia** was bounded on the W. by Bithynia, on the N. by the Euxine, on the E. by Pontus, and on the S. by Galatia; it thus occupied the coast-district between the rivers Parthenius and Halys, and extended inland to the range of Olympus. At one time the Paphlagonians appear to have advanced beyond the Halys. Paphlagonia is on the whole a rough and mountainous country, but contains in its northern parts some extensive and fertile plains, on which even the olive flourished. Its hills were well clothed with forests, and the boxwood of Mount Cytorus was particularly celebrated.[4] Paphlagonia was especially noted for its horses, mules, and antelopes. A kind of red chalk was found there in abundance.

§ 21. The chief mountain range, named **Olgassys**, *Ulgaz*, extends from the Halys towards the S.W., and sends its ramifications sometimes to the very shores of the Euxine; of these, **Cytorus** was the one best known to the ancients. The coast protrudes northwards in a curved form, and has two promontories, **Carambis**, *C. Kerempe*, and **Syrias**, *C. Indje*, more to the E. The only important rivers are the border-streams **Halys** and **Parthenius**, which have been already noticed: numerous small rivers intervene, of which we may enumerate, from W. to E., the Sesamus, Amastris, Ochosbanes, Evarchus, and Zalecus. The **Amnias**, a tributary of the Halys, is noted for the engagement that took place on its banks, in which Nicomedes was defeated by the generals of Mithridates, B.C. 88.

§ 22. The Paphlagonians, who are noticed even in the Homeric poems,[1] appear to have been allied in race to the Cappadocians. They are described as a superstitious and coarse people, but brave,

---

4   Et juvat undantem buxo spectare Cytorum,
    Naryclæque picis lucos.   VIRG. *Georg.* ll. 437.
    Sæpe Cytoriaco deducit pectine crines.   OV. *Met.* iv. 311.
    Amastri Pontica et Cytore buxifer.   CATULL. iv. 13.
1   Παφλαγόνων μεγαθύμων ἀστιστάων.   *Il.* v. 577.

and particularly noted for their cavalry.   In addition to the Paphlagonians, the more ancient races of the Henёti and Caucōnes continued to occupy certain districts.   The towns lined the coast, and were for the most part Greek colonies, such as Amastris and Sinôpe, the latter of which was by far the most important in the country, together with the lesser towns Cromna, Cytorus, Abōnitcichos, and Carūsa.   In the interior Gangra and Pompeiopolis were at different eras leading towns.

Amastris, in the W., occupied a peninsula, on each side of which was a harbour: its name was originally Sesamus, which was changed in honour of Amastris, niece of the last Persian king Darius, and which appears to have extended beyond the old town of Sesamus to a tetrapolis of which Tetum, Cytorus, and Cromna were the other members. Amastris was a handsome city, and flourished until the 7th century of our era.   Sinope[*] was situated on a peninsula E. of Prom. Syrias: its foundation was attributed to the Argonauts: it was colonized by the Milesians, seized from them by the Cimmerians, and recovered by the Ephesians, B.C. 632: in the time of Xenophon it possessed a fine fleet, and was mistress of the Euxine: it was unsuccessfully besieged by Mithridates IV. in B.C. 220, but successfully by Pharnaces in 183: thenceforth it was the residence of the kings of Pontus, and gave birth to Mithridates the Great: Lucullus captured it, and restored its independence: it became a Roman colony in the time of Julius Caesar.   It is further known as the birthplace of Diogenes the Cynic: the modern Sinub is still an important place, and contains a few relics of the old town.   Pompeiopolis, on the Amnias, probably owed its name and existence to Pompey the Great.   Gangra was S. of Mount Olgassys, and was the residence of Deiotarus the last king of Paphlagonia: it was made, after the 4th century A.D., the capital of the province, with the name Germanicopolis.   We may further briefly notice—Abonitichos, the birthplace of the impostor Alexander, at whose request the name was changed to Ionopolis—and the small harbours of Cimōlis, Stephâne, Potâmi, Armēne which the 10,000 visited, and Carūsa: all these were trading stations.

History.—Until the time of Croesus, Paphlagonia was under its native princes: it was then annexed to the Lydian empire, and passed with the rest of it to the Persians, under whom the native princes regained their independence.   After Alexander's death Paphlagonia fell to the share of Eumenes, but again reverted to its princes, until it was incorporated with Pontus by Mithridates.   Under the Romans it was united first to Bithynia, and afterwards to Galatia, but in the 4th century A.D. was made a separate province.

XIV. PONTUS.

§ 23. Pontus bordered in the W. on Paphlagonia, in the S. on Cappadocia, in the E. on Armenia and Colchis, and in the N. on

---

[*]   Mox etiam Cromnae juga, pallentemque Cytoron,
Teque cita penitus condunt, Erythea, carina.
Jamque reducebat noctem polus : alta Carambis
Raditur, et magnae pelago tremit umbra Sinopes.
Amyrica complexa sinus stat optima Sinope.
VAL. FLACC. V. 106.

the Euxine: the Halys, the ranges of Antitaurus and Paryadres, and the Phasis, formed its natural boundaries in the three former directions. It derived its name from the "Pontus," *i.e.* the Euxine, on which it bordered. Though this district is surrounded with lofty mountains, which send their ramifications to the very shores of the Euxine, yet the plains on the coast, especially those in the western parts, were extremely fertile, and produced, in addition to grain, excellent fruit. Honey, wax, and iron were among its most valuable productions.

§ 24. The chief mountain ranges are **Paryadres** in the N., and **Scordises** in the E., which have been already noticed. The former sends out two branches, **Lithrus** and **Ophlimus**, to the N., which form the eastern boundary of the fruitful plain of Phanarœa: the position of **Theches** cannot be fixed with certainty; it must have been considerably E. of Trapezus, as no distant view of the Euxine can be obtained from any point due S. of that place. The most important headlands from W. to E. are—**Heracleum**, which bounds the bay of Amisus on the E.; **Jasonium**, near Sido; **Zephyrium**, near Tripolis; **Coralla**, near Cerasus; and **Hieron**, more to the E. Two bays occur on this coast, the **Sinus Amisenus**, *G. of Samsun*, between the mouth of the Halys and Prom. Heracleum; and **Sin. Cotyorus**, between the promontories of Jasonium and Coralla. The most important rivers are—the **Halys**, which both rises and terminates in this province—the **Iris**, *Kassalmak*, which rises in Antitaurus in the S. of Pontus, and flows at first to the N.W. as far as Comana; then to the W. until it receives the Lycus, *Kulei Hissar*, a stream almost as large as itself, from the mountains of Armenia Minor; and finally to the N., in which direction it traverses the plain of Themiscyra to the sea—the **Thermodon**, *Thermeh*, which rises near Phanorœa, and joins the sea near Thorniscym, famed for its connexion with the Amazons[7]—the **Acampsis** or **Apsirus**, *Tchoruk*, which rises in Armenia, and joins the sea at its S.E. point —and the **Phasis**, on the border of Colchis. The less important

---

[7] Qualis Amazonidum nudatis bellica mammis
Thermodontiacis turma vagatur agris.—PROPERT. lll. 14, 15.
Et tu, feminaæ Thermodon cognita turmæ.
Ov. *ex Pont.* iv. 10, 51.

Ἀστραγείζοντας δὲ χρὴ
Κορυφὰς ὑπερβάλλουσαν, ἐς μεσημβρινὴν
Βῆναι εἴλευθον, ἵνθ' Ἀμαζόνων στρατὸν
Ἥξει στυγάνορ', αἱ Θεμίσκυράν ποτε
Κατοικισοῦσιν ἀμφὶ Θερμώδονθ', ἵνα
Τραχεῖα πόντου Σαλμυδησσία γνάθος
Ἐχθρόξενος ναύταιοι, μητρυιά νεῶν.
Æsch. *Prom.* 72.

Quales Thrëïciæ cum flumina Thermodontis
Pulsant, et pictis bellantur Amazones armis.
Virg. *Æn.* xi. 659.

streams from W. to E. are—the Lycastus, near Amisus; the
Chadisius, near Themiscyra; the Sidēnus, near Side; the Tripŏlis,
near the town of the same name; and the Hyssus, more to the E.

§ 26. The population of Pontus consisted of a number of tribes,
whose mutual relations are very obscure. Among the more prominent names appear the Leucosȳri, who were the same as the
Cappadocians; the Tibarēni, identical with the Tubal of Scripture;
the Chalȳbes,[a] who occupied the iron districts of Paryadres; the
Colchi, about Trapezus, allied to the proper Colchians; the Macrŏnes
or Sanni, who lived S.E. of Trapezus; and the Bechīres, on the
sea-coast in the same neighbourhood. The chief towns were of two
classes—the commercial ports on the coast, in most of which the
Greeks settled, such as Amīsus, Trapezus, Cotyŏra, and others of
less importance; and the towns of the interior, which were either
strongholds of the Pontic kings, or entrepôts of trade with Central
Asia: these were in many instances enlarged by the Romans. In
the latter class we have Amasia and Comāna in the valley of the
Iris, Cabīra on the Lycus, and Sebastia in the upper valley of the
Halys. The period at which the coast-towns became known dates
from the return of the 10,000: the interior was first opened to the
world by the Mithridatic wars. The history of the towns is comparatively uninteresting, and they do not appear to have possessed
much architectural beauty.

Amisus stood on the W. side of the bay named after it, on a promontory about 1½ miles N.W. of the still flourishing town of *Samsun:*
its origin is uncertain, but it became, next to Sinope, the most
flourishing of the Greek settlements, and was occasionally the residence
of Mithridates Eupator: it was captured by Lucullus, B.C. 71, and
again by Pharnaces, but restored to freedom by Cæsar after the battle
of Zela: remains of the ancient pier, and of Hellenic walls at *Eski
Samsun,* mark its site. Polemonium was placed at the mouth of the
Sidenus, and probably owed its name and existence to Polemon, king
of those parts, who made it his capital. Pharnacia was founded by
Pharnaces, grandfather of Mithridates VI., and peopled with the
Cotyoræans: it was prosperous from its commerce, and from the
neighbouring iron-works: it is now named *Kerasunt,* from the idea that
it occupied the site of Cerasus. Trapezus, a Sinopian colony, was built
on the slope of a hill near the coast, with a port named Daphnus,
formed by a jutting rock on which the acropolis stood. Even in
Xenophon's time it was an important place, but it reached its highest
prosperity under the emperors Hadrian and Trajan, the latter of whom

<hr>

[a]   Ἀσίας δὲ χειρὸς οἱ σιδηροτέκτονες
        Οἰκοῦσι Χάλυβες, οὓς φυλάξασθαί σε χρή.
        Ἀνήμεροι γὰρ, οὐδὲ πρόσπλαστοι ξένοις.—ÆSCH. *Prom.* 714.

        Stridentque cavernis
        Stricturæ Chalybum, et fornacibus ignis anhelat.
                            VIRG. *Æn.* viii. 420.

        Jupiter! ut Chalybon omne genus pereat,
        Et qui principio sub terra quærere venas
        Institit, ac ferri fingere duritiem!   CATULL. lxvi. 48.

made it the capital of Pontus Cappadocicus; it is still, as *Trebizond*, one of the most flourishing cities of Asia Minor. **Phasis** stood on the S. side of the river of the same name, and thus within the limits of Pontus; it was a Milesian colony, and a place of considerable trade: it possessed a temple of Cybele. In the interior—**Amasia**, once the residence of the kings of Pontus, stood on the river Iris; it gave birth to Mithridates the Great and to the geographer Strabo: it still retains its ancient name, and is a considerable town. **Comana Pontica** stood in the upper valley of the same river, and was a commercial entrepôt for the Armenian trade: it was the chief seat of the worship of Enyo, whose priests exercised an authority second only to that of the kings: a few remains of the place have been discovered at *Gumenek*. **Cabira** was situated on the Lycus, some distance above its junction with the Iris: Mithridates the Great had a palace and treasury there, which Cn. Pompeius succeeded in capturing: **Neocaesarea** was probably a later name for the same place, assigned to it in the reign of Tiberius, a place of ecclesiastical importance as the seat of a council in A.D. 314, and the birthplace of Gregory Thaumaturgus. **Sebastia** was on the N. bank of the Upper Halys, and was enlarged by Pompey, under the name of Megalopolis; the old name, however, returned to it, and still exists under the form *Sivas*: it was a flourishing place under the Byzantine emperors.

Of the less important places we may notice—(1.) on the sea-coast from W. to E. **Amisus**, a small port at the mouth of the Iris—**Themiscyra**, at the mouth of the Thermodon, said to have been built by the Amazons; destroyed by Lucullus—**Cotyora**, a colony from Sinope, with a port whence the 10,000 took ship—**Argyria**, with silver mines—**Cerasus**, a colony from Sinope, visited by the 10,000; the place whence Lucullus introduced the cherry into Italy—and **Apsarus**, a place of some importance at the mouth of the Acampsis, the reputed burial-place of Absyrtus. (2.) In the interior—**Gaziura** on the Iris, the ancient residence of the kings of Pontus—**Phasemon**, N. of Amasia, with hot mineral springs, made a Roman colony by Pompey, with the name Neapolis—and **Zela**, on the left bank of the Iris, rendered illustrious by the victory of Mithridates over the Romans, and still more by that of Cæsar over Pharnaces, reported in the brief despatch, "Veni, Vidi, Vici."

History.—The history of Pontus commences in B.C. 363, with the foundation of a sovereignty over many of the Pontic tribes by Ariobarzanes. His successor, Mithridates II., extended and consolidated his kingdom, and it prospered under the succeeding sovereigns, until it reached its greatest extent under Mithridates VI., who reigned from B.C. 120 to 63. But the wars which he carried on with the Romans proved fatal to his empire: the western portion was annexed by Pompey to Bithynia, B.C. 65; the district between the Iris and Halys was given to the Galatian Deiotarus, and hence named Pontus Galaticus: that between the Iris and Pharnacia was subsequently handed over by M. Antonius to Polemon, and hence named Polemoniacus: and the eastern portion fell shortly after into the hands of Archelaus, king of Cappadocia, and was distinguished as Cappadocicus. Pontus was made a Roman province, A.D. 63: and under Constantine was divided into Helenopontus in the S.W., and Polemoniacus in the centre and E.

Pontus is but seldom noticed in the Bible: Jews from that province were present at Jerusalem on the day of Pentecost (Acts ii. 0); the Jewish Christians were addressed by St. Peter (1 Pet. i. 1); and Aquila was a native of that country (Acts xviii. 2).

Libanus, or Lebanon.

CHAPTER X.

SYRIA — PHOENICIA — ARABIA.

I. SYRIA. § 1. Boundaries and natural divisions. § 2. Mountains. § 3. Rivers. § 4. Political divisions. § 5. Towns; history. II. PHOENICIA. § 6. Boundaries, &c. § 7. Geographical position. § 8. Mountains and rivers. § 9. Inhabitants; towns; history. § 10. Colonies. III. ARABIA. § 11. Boundaries and natural divisions. § 12. Mountains. § 13. Inhabitants. § 14. Divisions; towns; islands; history.

I. SYRIA.

§ 1. Syria, in its widest extent, comprised the whole of the eastern coast of the Mediterranean sea from Cilicia in the N. to the Arabian desert in the S., and extended eastward to the Euphrates From this, however, we must except the southern region of Palestine, and the strip of coast occupied by Phœnicia; its boundaries may then be more accurately defined thus: in the W. the Mediterranean Sea down to near Aradus, and thenceforward the range of Libanus; on the S. an imaginary line, leaving Libanus opposite Sidon, and stretching across the desert somewhat S. of Damascus and Palmyra, to the Euphrates near Thapsacus; on the N.W. the range of Amanus; on the N. Taurus, separating it from Cappadocia; and on the E. the Euphrates, separating it from Mesopotamia. It

is naturally divided into the following three parts—(1.) the coast
district; (2.) the upper valley of the Orontes between the ranges of
Libanus and Antilibanus, to which the name of Cœle-Syria, i. e.
"hollow Syria," was properly applied; and (3.) the extensive desert
which intervenes between these ranges and the Euphrates. These
districts differ widely in climate, character, and productions; thickly-
wooded mountains and well-watered plains characterise the two
former; while the third consists of a series of plateaus rising to
about 1500 feet above the sea, and traversed by undulating hills,
devoid of interest, and, in the absence of artificial irrigation, unproductive. The inhabitants were a Semitic race, allied to the Phœnicians, Hebrews, and Assyrians.

§ 2. The mountain system of Syria is very distinctly marked:
the range of **Amānus**, after skirting the Mediterranean coast closely
in the neighbourhood of Issus, sinks at the spot where the road
leaves the coast and crosses by the Portœ Syriœ, but rises again in
the heights of **Pieria**, which take a westerly direction and form a
considerable promontory. S. of this, the range is broken by the
plain of the Orontes, but is resumed in the maritime range of
Casius (which culminates in a conical peak 5000 ft. high completely
clothed with forest), as well as in the more inland range of **Bargylus**,
Nusairyeih, which is carried on to the border of Phœnicia. Here
the chain is again broken by the valley of the Eleutherus, to the S.
of which the range of **Libanus** rises, and runs in a long unbroken
line to the border of Palestine. The parallel ridge of **Antilibanus**
is separated from it by the river Leontes, and forms the connecting
link with the ranges which traverse the whole length of Palestine.
Of all the Syrian mountains, Libanus, more familiar to us under the
Scriptural name of Lebanon, is the most magnificent. It derives its
name from its *whitened* appearance, arising partly from the snow
which lingers in some spots all the year round, and partly from the
natural colour of the rock. Its greatest elevation is about 10,000
feet. In former times it was clothed with forests of cedar and fir,
which supplied the materials for the Temple at Jerusalem; a single
grove, containing about 400 trees, of which 12 bear marks of great
antiquity, is generally regarded as the representative of the "cedars of
Lebanon:" this grove is situated in the high slopes of the mountain
near *Tripoli*; the tree still exists, however, in other parts. Antilibanus terminates southwards in the well-known peak of Hermon at
an elevation of about 10,000 feet; this will be described in a future
chapter.

§ 3. The most important river in Syria is the **Orontes**,[1] which rises

---

[1] Juvenal uses the name of the Orontes as equivalent to Syria:
In Tiberim defluxit Orontes.—*Sat.* iii. 62.

between the ranges of Libanus and Antilibanus, not far from the
Leontes, and takes a N. course until it reaches the neighbourhood
of Antioch ; there it sweeps round to the W., and again to the S.W.
until it joins the sea ; its modern name, *el-Asy*, "the rebellious,"
may have reference to these sudden alterations in its course. The
scenery of the lower course of the river is not unlike that of our own
Wye. The upper course of the *Litany* also falls within the limits
of Syria. There are numerous coast-streams of but little importance. In the interior the rivers of Damascus—the well-known
"Abana and Pharpar" of the Bible (2 K. v. 12), though small, are
very valuable; the first was named Chrysorrhoas, "golden-flowing,"
by the Greeks, and is now the *Barada*; the second was of less importance, and is now named *Nahr el-Awaj*; the former rises in Antilibanus, the latter in Hermon; they flow in an easterly direction
across the plain of Damascus, communicating to it its extraordinary
fertility and beauty, and fall into two lakes to the E. of the town.

§ 4. Syria was divided into the following 10 districts—**Comma-
gène**, in the extreme N. between Taurus and the Euphrates; **Cyr-
rhestice**, between Amanus and the Euphrates; **Pieria**, about the
mountain of the same name; **Seleucis**, about Antioch; **Chalybonitis**,
thence to the Euphrates; **Chalcidice** to the S.W.; **Apamène**, stretching away from Apamea towards the S.E.; **Palmyrène**, along the
southern frontier about Palmyra; **Laodicène**, westward about Laodicea in Cœle-Syria; and **Casiotis**, on the sea-coast about Mount
Casius. In addition to these we must notice the Biblical **Abilène**,
a district on the eastern slopes of Antilibanus about the town Abila,
which, at the time of our Saviour's birth, belonged partly to Philip,
and partly to Lysanias (Luke iii. 1), and which was handed over to
Herod Agrippa by Caligula.

§ 5. The towns of Syria were of two classes—(1.) the ancient
Biblical towns, which owed their importance partly to military and
partly to commercial considerations, such as Damascus, Tadmor,
Hamath, and the towns commanding the passages of the Euphrates—Samosata and Thapsacus; and (2.) the towns which were
called into existence by the Syrian monarchs, such as Antioch, Seleucia, Apamea, Zeugma. Occasionally the old towns were entirely
rebuilt, at all events highly adorned, either by the Seleucidæ, as was
the case with Epiphania (the ancient Hamath), Berœa (Chalybon),
and Heliopolis (Bambyce), or at a later period by the Roman emperors or governors, as was the case with Heliopolis and Palmyra.
The towns of the first class are situated in the south, those of the
second class for the most part in the north of the country. Damascus was the chief town of the former class; but Antioch was
the capital of the country after it was raised to an independent
position.

Ruins of Palmyra.

Antiochia was situated at the western extremity of a fine alluvial plain on the left bank of the Orontes, near the spot where that river enters the defile that conducts it to the sea. Its position was well chosen for a great capital. It had easy access to the sea by the defile just noticed, to Lower Syria and Egypt by the valley of the Orontes, to Cilicia by the pass commanded by the Portæ Syriæ, and to Mesopotamia by various routes across the desert. It was founded B.C. 300 by Seleucus Nicator, and named after his father Antiochus. It was regularly laid out in streets intersecting each other at right angles, and adorned with temples and public buildings by successive kings, particularly by Antiochus Soter. A new quarter was added by Seleucus Callinicus on an island in the river, which was joined to the shore by five bridges; and another by Antiochus Epiphanes on the side adjacent to the mountain. It was subsequently much adorned by the Roman emperors. Antioch is chiefly interesting from its associations with early Christian history. A church was founded there by fugitive disciples from Jerusalem (Acts xi. 19), and there the honoured name of "Christian" first came into use. It was for some time the head-quarters of St. Paul, whence he started on his two first apostolic journeys. Afterwards it became the seat of a patriarchate which ranked with Constantinople and Alexandria. Its capture by the Persians under Sapor, A.D. 260, is otherwise the most prominent event in its history. Seleucia, *Selefkieh*, surnamed Pieria, was an important maritime city, situated on a plain between Mount Pieria and the sea, about six miles N. of the mouth of the Orontes. It was built by Seleucus Nicator, and served as the port of Antioch. The harbour was excavated out of the plain, and connected with the sea by a canal. St. Paul sailed from here to Cyprus (Acts xiii. 4). An immense tunnel led from the upper part of the city to the sea. Laodicea, *Ladikiyeh*, surnamed ad Mare, stood on the sea-

coast S. of Seleucia, with an excellent harbour, and surrounded by a
rich vine-growing country: it was built by Seleucus Nicator, and furnished with an aqueduct by Herod the Great, of which a fragment still
remains; it was partly destroyed by Cassius, B.C. 43, in his war with
Dolabella. Apamia, in the valley of the Orontes, owed its prosperity
to Seleucus Nicator, who named it after his wife Apama, and established
a commissariat station there; its ruins testify to its former magnificence.
Epiphania was the name given probably by Antiochus Epiphanes to
the ancient Hamath, on the Orontes. Emesa, Hums, was situated near
the Orontes, on a large and fertile plain, and was celebrated for a temple
of the Sun. Heliopolis, Baalbek, in Cœle-Syria, must have been one of
the chief towns of Syria, although unmentioned in early history. It
stood at the neck of the elevated ground whence the Orontes and Litany
flow in different directions; and, as the high road of commerce followed
these rivers, it was undoubtedly an important place of trade. In what
age the worship of the Sun, to which the town owes its name, was first
introduced we know not. The magnificent edifices, so beautiful even
in their ruins, were probably erected in the age of the Antonines, but
the platform on which the great temple stands is of older date, and
probably of Phœnician origin. The chief buildings remaining are three
temples, distinguished as the "Great Temple," the "Temple of
Jupiter," and the "Circular Temple." Julius Cæsar made Heliopolis
a colony, and Trajan consulted its oracle before entering on his Parthian
expedition. Damascus stands on a plain, about a mile and a half from
the lowest ridge of Antilibanus, at an elevation of about 2200 feet above
the sea. This plain, watered by the rivers Abana and Pharpar, is well
clothed with vegetation and foliage. The town now stands on both
banks of the Abana, but it was formerly confined to the south bank.
Damascus is frequently noticed in the Bible, and its history may be
almost said to be the early history of Syria itself. It derives a special
interest, however, from its connexion with St. Paul's life. Near it he
was converted, and in its synagogues he first preached; the "street
called Straight," in which he lodged, is still the principal one in Damascus. Palmyra, "the city of palms," lies about 140 miles N.E. of
Damascus, in the heart of the desert, where it served as an entrepôt for
the caravan trade. Its position is somewhat elevated above the plain,
and the supply of water is comparatively scanty. The history of this
place from the days of Solomon to the Christian era is a blank. Appian
tells us that M. Antony designed an attack upon it; and it is noticed
by Pliny. About A.D. 130 it submitted to Rome, and was made a colony
with the name Adrianopolis by Hadrian, who adorned it with the beautiful buildings the remains of which still strike the traveller with wonder.
Under Odenathus and his widow Zenobia, Palmyra attained an imperial
dignity; but after the defeat of Zenobia and the capture of Palmyra by
Aurelian, A.D. 273, it fell into decay, in spite of the attempt at restoration made by Diocletian. Of the ruins the Temple of the Sun is the
finest; the Great Colonnade is also a striking object, 150 out of the
1500 columns originally erected still remaining. The tombs of this
place are also peculiar—lofty towers divided into stories.
Of the less important towns we may briefly notice—Chalcis, the
capital of Chalcidice, S.E. of Antioch; Chalybon, or Berœa (as it
was named by Seleucus after the Macedonian town), representing
the modern Aleppo, on the road between Hierapolis and Antioch;
Hierapolis, the "Holy City," so named from its being a seat of the
worship of Astarte, an emporium between Antioch and the Euphrates;

Damascus.

its earlier name, Bambўce, was changed to the Greek name by Seleucus
Nicator; Thapsacus, sometimes considered as a Syrian, sometimes as
an Arabian town; as its position attached it rather to the former country we shall notice it here; the most frequented passage of the Euphrates
was opposite Thapsacus, probably near *Dair*; it was here that the
armies of Cyrus the younger, of Darius, and his competitor Alexander
the Great, crossed the river; Zeugma, deriving its name from the
bridge of boats across the Euphrates at this point; the town was founded
by Seleucus Nicator to secure the passage of the river from the capital,
Antioch; it stood opposite Apamea or *Bir*; and lastly, Samosata, in
Commagene, which commanded the most northern passage between
Cappadocia and Mesopotamia.

History.—The history of Syria, as an independent country, commences with the establishment of the dynasty of the Seleucidæ, B.C.
312. Seleucus Nicator, the first of that dynasty, acquired nearly all
the provinces of the old Persian empire. His successors gradually lost
these vast possessions: his son, Antiochus Soter (280–261) lost a great
part of Asia Minor by the establishment of the sovereignties of Bithynia
and Pergamus. Under Antiochus Theos (261–246) Parthia and Bactria
revolted. Seleucus II. (246–226) in vain attempted to recover these
possessions. Antiochus the Great (223–187) was not more successful
against those remote countries, and suffered a further loss of Palestine
and Cœle-Syria: in addition to this he was defeated by the Romans at
Magnesia (B.C. 190), and was obliged to yield up all the provinces

within Taurus to the king of Pergamus. Thenceforward the empire of
Syria rapidly sank, and was gradually reduced to the limits of Syria
Proper and Phœnicia. It became a Roman province in B.C. 65.

II. PHŒNICIA.

§ 6. The limits of Phœnicia are clearly defined on the W. and E.
by the natural boundaries of the Mediterranean Sea and Mount
Lebanon; on the N. and S. they are not so decided; in the latter
direction it intruded, for a considerable distance into Palestine,
terminating below Mount Carmel, about midway between Cæsarea
and Dora; in the former direction the boundary touched the sea
somewhere N. of Aradus. It had a length of 120 and an average
breadth of 12 miles. The country, though not extensive, was fertile
and varied in its productions. While the lowlands yielded corn
and fruit, the sides of Lebanon were an inexhaustible storehouse of
timber for ship-building. The purple shell-fish and the materials
for the manufacture of glass were sources of great wealth.

Name.—The name "Phœnicia" is probably derived from the Greek
word φοῖνιξ—"palm-tree"—which grew abundantly on its soil, and
was the emblem of some of its towns. It has also been connected with
φοῖνιξ—"the red dye"—which formed one of its most important productions.

§ 7. The causes which combined to render this country the earliest
seat of extended commerce are connected partly with its position
relatively to other nations, and partly with the internal capacities of
the country itself. Phœnicia was well adapted to become the entrepôt of European and Asiatic commerce. Centrally situated on
the eastern coast of the Mediterranean Sea, it was the point which
the trade of Palmyra, Babylon, the Persian Gulf and India, Bactria,
and China, would naturally seek. The shores of Europe were easily
accessible from it. Cyprus, Crete, Rhodes, the Cyclades, were so
many stepping-stones to Greece, as were Chios, Lesbos, and Lemnos,
to the Euxine; Sicily and Sardinia were stages on the route to
Spain and the Pillars of Hercules; the open Atlantic thence
invited to the shores of northern Europe. Equally favourable was
its position relatively to Africa. Egypt and the Red Sea were
easily accessible; Cyrene and Carthage answered to the peninsulas
of Greece and Italy; and from the Pillars of Hercules the shores of
Western Africa were open to them. But these advantages in the
position of Phœnicia would probably have been lost if the country
itself had not possessed peculiar advantages for the prosecution of
trade. It may be observed then, that it was protected from intru-
sion at its rear by the lofty barrier of Lebanon intervening between
it and the open plains of Asia, and at its sides by the spurs which
that chain sends forth to the immediate neighbourhood of the sea.

Though easily accessible from the north and south, Phœnicia was still no thoroughfare. The high-road from Egypt to Antioch, which followed the sea-coast as far as Tyre, turned inland from that point, and followed the valleys of the Leontes and Orontes between the ranges of Libanus and Antilibanus. Lastly, the coast is sufficiently broken to supply several harbours amply large enough for the requirements of early commerce.

§ 8. The physical features of Phœnicia are easily described ; the range of **Lebanon** or **Libanus** runs parallel to the coast, throwing out a number of spurs in that direction, which break up the whole country into a succession of valleys. Some of these spurs run into the sea and form promontories, of which the most important are—**Theu-Prosôpon**, *Ras-es-Shekah*, **Prom. Album**, *Ras-el-Abiad*, S. of Tyre, and **Carmêlum**, *Carmel*: the latter will be hereafter described ; Album rises to a height of 300 ft., and intercepts the coast road, which was originally carried over it by a series of steps, hence called Climax Tyriorum, " the Tyrian Staircase ;" a roadway was afterwards cut through the solid rock. Another Climax of a similar character existed in the north, about 25 miles below Theu-Prosopon. The rivers are necessarily short; the principal streams from N. to S. are—the **Eleuthêrus**, *Nahr-el-Kebir*, which drains the plain between Bargylus and Libanus—and the **Leontes**, *Kasimieh* or *Litani*, which rises between the ranges of Libanus and Antilibanus, and flowing for the greater part of its course towards the S.W., turns sharply round towards the W. and gains the sea near Tyre. The small stream **Adônis**, *Nahr el Ibrahim*, which joins the sea near Byblus, derives an interest from its connexion with the legend of the death of Adonis, who is said to have been killed by a wild boar on Libanus. The blood-red hue of the water in time of flood may have given origin to the story.[*]

§ 9. The Phœnicians of historical times were undoubtedly a Semitic nation. Their language bears remarkable affinity to the Hebrew, as evidenced by an inscription discovered at *Marseilles* in 1845, of which 74 words out of 94 are to be found in the Bible. The Mosaic table, however, describes Canaan as the son of Ham (Gen. x. 15), and connects that race with the Egyptians and other Hamitic nations. We must therefore assume, either that there was a later immigration, or that the Phœnicians left their original seats at a time when the difference between the Hamitic and Semitic races were not so strongly marked as they were in later ages. Their

---

[*] Milton alludes to this legend in the lines—

" While smooth Adonis from his native rock
Ran purple to the sea, supposed with blood
Of Thammuz yearly wounded."—*Paradise Lost*, viii. 18.

first settlements were on the shores of the Persian Gulf. Traces of their presence there survive even to the present day in the names *Arad*, *Sidodona*, and *Szur* or *Tur*, the prototypes of Aradus, Sidon, and Tyre. The towns of Phœnicia were situated either on or adjacent to the sea-coast, and owed their importance partly to their manufactures, but still more to the trade which passed through them from Asia to Europe. Sidon appears to have been the original capital, but Tyre subsequently surpassed it both in beauty and celebrity, and had the further advantage of being a strong military position. Arädus and Berytus enjoyed a certain amount of commercial prosperity. Ptolemäis did not acquire in early times the reputation which it now possesses, under the familiar name of *Acre*.

Sidon, *Saida*, was situated on a small promontory about two miles S. of the river Bostrënus. Its harbour was naturally formed by a low ridge of rocks running out from the promontory, parallel to the line of coast. It was famed in early times for its embroidered robes,[2] its metal work,[4] its dyes,[3] and its manufacture of glass; but it was obliged to yield to the growing prosperity of Tyre. It derives an interest to the Christian from St. Paul's visit there. **Tyrus**, *Sûr*, stood more to the S., and consisted of two separate cities—Palæ-Tyrus (" Old Tyre "), which was on the mainland—and New Tyre, subsequently built upon an island about half a mile from the coast, which now rises about twelve feet above the sea, and is three-quarters of a mile long by half a mile broad, but which was probably larger in ancient times. A neck of sand about half a mile broad now connects the rock with the mainland: this, however, has been wholly formed by the sand which has accumulated about the causeway made by Alexander. The harbour was formed at the N.E. end of the island, and there was a double roadstead between the island and the mainland; one (the Sidonian) facing the N., the other (the Egyptian) facing the S. It was famed for its purple dye,[4] which was extracted from shell-fish found on the coast. The origin of Tyre, and the periods in which the New and Old Towns were respectively built, are unknown. Its subsequent history is, in short, the history of Phœnicia itself. The present town contains about 4000 inhabitants, and is in a state of great decay, its commerce giving employment only to a few crazy fishing-boats. For a graphic descrip-

---

Ἔνθ᾽ ἔσαν οἱ πέπλοι παμποίκιλοι, ἔργα γυναικῶν
Σιδονίων, τὰς αὐτὸς Ἀλέξανδρος θεοειδὴς
°Ἤγαγε Σιδονίηθεν,——Hom. *Il.* vi. 289.

'Ἀργύρεον κρητῆρα τετυγμένον· ἓξ δ᾽ ἄρα μέτρα
Χάνδανεν, αὐτὰρ κάλλει ἐνίκα πᾶσαν ἐπ᾽ αἶαν
Πολλόν, ἐπεὶ Σιδόνες πολυδαίδαλοι εὖ ἤσκησαν,
Φοίνικες δ᾽ ἄγον ἄνδρες ἐπ᾽ ἠεροειδέα πόντον—Hom. *Il.* xxiii. 741.

b —————————— pretiosaque murice Sidon.——Luc. iii. 217.
Quare ne tibi sit tanti Sidonia vestis,
Ut timeas, quoties nubilus Auster erit.—Propert. ii. 18, 85
Non qui Sidonio contendere callidus ostro
Nescit Aquinatem potentia vellera fucum.—Hor. *Ep.* i. 10, 26.

c Ille caput flavum lauro Parnasside vinctus
Verrit humum, Tyrio saturata murice palla.—Ov. *Met.* xi. 165.

tion of what Tyre *was* and what it now *is* compare the 27th and 26th
chapters of Ezekiel.

The less important towns were—**Arādus**, in the N., also built on an
island rock, about two miles from the coast, a colony of Sidon, and
still a place of importance under the name of *Ruad*; **Antarādus**, on
the mainland *opposite* Aradus, as its name implies; **Tripōlis**, on a small
promontory, deriving its name from being the metropolis of the *three
confederate towns*, Tyre, Sidon, and Aradus; **Byblus**, the chief seat of
the worship of Adonis, or Thammuz, who was held to have been born
there; the modern name *Jubeil* is derived from the biblical name
Gebal, the residence of the Giblites ; **Berytus**, *Beirut*, the seat of a
famous Greek university from the third to the sixth century of our era,
and now the most important commercial town in Syria; and **Ptolemāis**,
the biblical Accho, whence its modern name *Acre*, at the northern extremity of the bay formed by Prom. Carmel. It was named Ptolemais
after Ptolemy Soter.

History.—The history of Phœnicia is well-nigh a blank, from the loss
of its archives and literature. The few particulars we have are gathered
chiefly from the Bible, Josephus, and the Assyrian Inscriptions. The
country appears to have been parcelled out into several small independent kingdoms, which confederated together as occasion required, and
over which, at such periods, the leading town naturally exercised a
supremacy. Sidon held the post of honour until about B.C. 1200, when
it was attacked by the king of Ascalon (who probably headed the
Pentapolis of the Philistines), and was reduced to the second rank, Tyre
henceforth becoming the metropolis. We know little of Tyre until the
time of Solomon's alliance with Hiram, the mutual advantages of
which were great; Solomon drawing from Phœnicia his supplies of
wood and stone for the erection of the Temple, as well as shipbuilders
and seamen for carrying on his commerce, and Hiram gaining in return
supplies of corn and oil, and a territory in Galilee containing 20 towns
(1 Kings, v: 6-12, ix. 11). After the death of Hiram a series of revolu-
tions and usurpations followed, during which the only names of interest
are Pygmalion (whose sister Elisa, or Dido, founded Carthage) and Itho-
balus, or Eth-baal, the father of Jezebel (1 Kings xvi. 31), a priest of
Astarte, who gained the throne by assassinating Phales. In his reign
the Assyrians, under Sardanapalus I., first invaded the country, and exacted tribute from Tyre, Sidon, Byblus, and Aradus. From the
intimations of the early prophets, Joel and Amos, we infer that the
Phœnicians carried on a vexatious warfare on the borders of Palestine.
Phœnicia was from henceforth subjected to constant invasions from the
Assyrian kings. On the fall of Nineveh Nabopolassar asserted his
authority over Phœnicia, and his son Nebuchadnezzar besieged Tyre for
13 years, after having previously captured Sidon. The result of the
Tyrian siege is uncertain : from Ez. xxix. 17, we may almost infer that it
was unsuccessful—a conclusion which is supported by the fact that the
line of kings was not then disturbed. Shortly after this Cyprus was
seized by Amasis, king of Egypt. Phœnicia seems to have declined from
this time, and to have gradually succumbed to the preponderating influence of the Persian empire without any actual conquest. It formed
along with Palestine and Cyprus the fifth Persian satrapy, and contributed
a contingent to the fleet of Darius in the Greek war. In B.C. 352 a vain
attempt was made to shake off the Persian yoke. Sidon, which was
again the chief city of Phœnicia, was taken, and her population almost
destroyed by Artaxerxes Ochus. At the approach of Alexander the

Great, Aradus, Byblus, and Sidon, received him, but Tyre held out, and was not taken until after a laborious siege of seven months, when its inhabitants were utterly destroyed, and a Carian colony introduced in their place. Alexander formed Phœnicia, with Syria and Cilicia, into a province. In the subsequent arrangement of his dominions Phœnicia fell to the lot of Ptolemy of Egypt, but was shortly after (B.C. 315) seized by Antigonus, and from this time formed a bone of contention between the Egyptian and Syrian kings. In the year B.C. 83, the Phœnicians obtained the aid of Tigranes, king of Armenia, against the latter, and he held it for fourteen years. Ultimately it fell, along with Syria, into the hands of the Romans.

§ 10. The commerce of Phœnicia was prosecuted on a most extensive scale. The chief routes in the continent of Asia have been already described; it remains for us to give a brief account of their maritime colonies on the coasts of Europe and Africa.

Their colonies lined the shores of the Mediterranean to its western extremity. We can trace their progress to Cyprus, where they founded Citium and Paphos; thence to Crete (the scene of the myth of Europa) and the Cyclades, which were chiefly colonised by them; thence to Eubœa, where they once dwelt at Calchis, and to Greece, where Thebes claimed connexion with them. Chios, Samos, and Tenedos, were united to Phœnicia by ancient rites and myths, as also Imbros and Lemnos. The mines of Thasos and of Mount Pangæus, on the opposite coast of Thrace, had been worked by them. They had settled in greater or less force on the southern and western coasts of Asia Minor, and on the coast of Bithynia, where they founded Pronectus and Bithynium, which were doubtless but stations for carrying on trade with the shores of the Euxine Sea. Proceeding yet farther to the west, we find them stretching across to Sicily, Sardinia, Æbusus (Ivica), and Spain (the Tarshish of Scripture), where they founded Gadeira (Cadiz) and numerous other colonies. The northern coast of Africa was thickly sown with their colonies, of which Utica, Hippo, Hadrumetum, Leptis, and more especially Carthage—the centre of an independent system of colonies—were the most important. Outside the Pillars of Hercules, they possessed, according to Strabo (xvii. p. 826), as many as 300 colonies on the western coast of Africa. They are supposed to have traded to the Scilly Isles and the coasts of England for tin, and even beyond this to the shores of Cimbria for amber; and thus, as Humboldt (Kosmos, ii. 132) remarks, "the Tyrian flag waved at the same time in Britain and the Indian Ocean." How far their knowledge of the world extended beyond these limits we have no means of ascertaining. It is stated that they circumnavigated Africa under the direction of Necho, king of Egypt (Herod. iv. 42). The truth of this has been questioned; Herodotus himself disbelieved it, but the reason he gives for his disbelief, viz. that the navigators alleged that the sun was on their right hand, is a strong argument in favour of its truth.

III.—ARABIA.

§ 11. The peninsula of Arabia is bounded on three sides by water, viz. on the N.E. by the Persian Gulf and the Sinus Omana, *Gulf of Oman*; on the S.E. and S. by the Erythræum Mare, or *Indian Ocean*; and on the W. by the Arabicus Sinus. In the N.

its boundary is not well defined. The peninsula itself may be regarded as terminating at a line drawn between the heads of the Persian and Ælanitic gulfs, distant from one another about 800 miles; but it was usual to include in Arabia two outlying districts, viz. the triangular block of desert [7] to the N. of this line, intervening between Palestine and Babylonia, and the peninsula of Sinai, between the arms of the Red Sea. Arabia was, therefore, contiguous to Egypt in the W., Palestine in the N.W., Syria in the N., and Mesopotamia in the N.E. Its physical character is strongly marked: it consists of a plateau of considerable elevation, surrounded by a low belt [8] of coast-land, varying in width according as the mountains which support the plateau approach to or recede from the sea. In modern geography these portions are distinguished as the *Nejd*, "highlands," and the *Tehama*, "lowlands," but no corresponding terms occur among ancient writers. The country, though generally arid and unfit for cultivation, nevertheless abounded in productions of great commercial value,[9] such as spices,[1] myrrh,[2] frankincense,[3] silk,[4] precious stones, and certain kinds of fruits. An extensive trade was carried on between the southern coasts of Arabia and the shores of India and southern Africa, and hence various productions were assigned to it by ancient writers which really belonged to those latter countries.

§ 12. The physical features of Arabia were but little known to the ancients. The ranges of Palestine may be traced down to the head of the Ælanitic arm of the Red Sea, on either side of the remarkable depressed plain named *Akaba*. The high ground on

[7] The name as used in St. Paul's Ep. to the Gal. i. 17 has reference exclusively to this northern district.

[8] This belt appears to have been once covered by the sea, and has been gradually elevated: the process of elevation is still going on, and the increase of land on the W. coast is very observable within historical times. Musa, which Arrian describes as on the sea-coast, is now several miles inland.

[9] Hence the wealth of the Arabs passed into a proverb among the Romans:
Plenas aut Arabum domos. Hor. Carm. ii. 12, 24
    Intactis opulentior
        Thesauris Arabum. Id. iii. 24, 1.

[1]   Sit dives amomo,
Cinnamaque, costumque suam, sudataque ligno
Thura ferat, floresque alios Panchaia tellus;
Dum ferat et Myrrham. Tanti nova non fuit arbos.
                                Ov. *Met* x. 307.

[2] Non Arabo noster rore capillus olet.   Ov. *Her*. xv. 76.
Et gravidæ maduere comæ, quas rore Sabæo
Nutrierat.   Val. Flacc. vi. 709.

[3] Urantur pia tura focis: urantur odores
    Quos tener e terra divite mittit Araba. Tibull. ii. 2, 3.
India mittit ebur, molles sua tura Sabæi. Virg. *Georg.* i. 57.
Totaque thuriferis Panchaïa pinguis arenis. Id. ii. 139.

[4] Nec si qua Arabio lucet bombyce puella. Propert. ii. 3, 15.

the W. side gradually rises towards the S., and terminates in a confused knotty mass of lofty mountains, near the point where the two arms of the Red Sea separate. The general name for these mountains in classical geography was *Nigri Montes*; they are now called *El Tor*, the most conspicuous heights in the group being named *Um Shomer* (8850 feet high), *Jebel Catharine* (8705), *Jebel Nousa*, "Moses' Hill," a little to the E. of *Jebel Catharine*, the reputed scene of the delivery of the law, and *Jebel Serbal* (6759 feet), which stands apart from the central group, near the W. arm of the Red Sea. On the E. side of the *Akaba* are the mountains of Idumæa, or Edom, composed of red sandstone, the most conspicuous height of which is the **Mount Hor** of the Bible, near Petra, the scene of Aaron's death. Of the other chains in Arabia we have notice in Ptolemy of **Zaméthus**, *Jebel Aared*, in the interior; the **Marithi Montes**, near the Persian Gulf; and the **Nigri Montes**, near the Gulf of Oman.

§ 13. The Arabians were mainly a Semitic race, though there appears to have been a Hamitic element mixed with it. The most important tribes known to ancient geographers were, the Scenitæ,[3] "dwellers in tents," the progenitors of the modern Bedouins; the Nabatæi,[4] in Arabia Petræa, about Petra and the Ælanitic Gulf; the Thamydeni, or Thamyditæ, more to the S.; the Minæi, in the S. of *Hedjaz*; the Sabæi[7] and Homeritæ, in the S.W. angle; the Chatramotitæ and Adramitæ, in *Hadramaut*; the Omanitæ on the shores of the *Gulf of Oman*; the Attæi and Gerrhæi, on the Persian Gulf.

§ 14. Arabia was originally divided into two parts : **Deserta**, the northern extension, to which we have already adverted, and **Felix**,[8] which comprised the whole of the proper peninsula. To these a

---

[3] The name Saracéni was afterwards applied to them, though originally restricted to a tribe on the borders of Petræa.

[4] The Nabatæi were well known to the Romans in consequence of their proximity to the Red Sea and their piratical habits : the name is used as equivalent to Arabian.

> Et quos deposuit Nabatæo bellus saltu
> Jam nimios capitique graves.   Juv. *Sat*. xi. 126.
> Euros ad Auroram Nabatæaque regna recessit.   Ov. *Met*. l. 61.

[7] The Sabæans were the chief traders in frankincense :

> Thuris odoratæ cumulis et messe Sabæo
> Pacem conciliant aræ.   Claudian. *de Laud Stil*. l. 36.
> ——— ubi templum illi centumque Sabæo
> Thure calent aræ, sertisque recentibus halant.   Virg. *Æn*. l. 416.

[8] The title of Felix, "happy," though not inappropriate to certain parts of Arabia, and particularly to the N.W. angle, may have originated in a mistaken interpretation of the Semitic *Yemen*, which signifies primarily the *right hand*, and secondarily the *south*, but which the Greeks understood in the secondary sense of *fortunate*, just as the Latins used *dexter*. Certainly the title of Felix is a perfect misnomer for a great portion of the peninsula.

third was subsequently added, of which the earliest notice occurs in Ptolemy, named Petræ, applying to the district surrounding the town of Petra. The towns of ancient Arabia possess few topics of interest. They occupied the sites of the modern towns, and correspond with them in great measure in name: thus, in Macoraba we recognise *Mekka Babba*, "the great Mecca;" in Jambia, *Yembo*; in Mariaba, *Mareb*; in Adana, the modern *Aden*, at present a British possession, and serving the same purpose to which it owed its ancient celebrity, as a station for Indian commerce; in Jathrippa, *Jathret*, the earlier name of *Medina*. The modern *Jeddah* is supposed to be represented by the ancient Thebæ; *Mokka*, however, stands on ground which was not in existence in ancient times, and has supplanted Muza as the chief port of that part. The only towns of which we have any special knowledge were situated in the N. of the country, such as Petra, Ælana, and a few others.

Petra, the capital of the Nabathæi, was by far the most important town in northern Arabia. It was situated between the head of the Ælanitic Gulf and the Dead Sea, and was the central point whence the caravan-routes radiated to Egypt, the Persian gulf, Syria, and southern Arabia. Its position is remarkable: a ravine (*Wady Musa*) of about a mile in length, about 150 feet wide at its entrance, and only 12 feet at its narrowest point, conducts to a plain about a square mile in extent: on this plain stood the town, while the ravine itself served as a necropolis, the tombs being excavated out of the sides of the cliffs, and adorned with sculptured *façades*, which are still in a high state of preservation. The remains of a theatre, hewn out of the rock, are also a remarkable object. These buildings were probably erected during the period that the town was held by the Romans, commencing in the reign of Trajan, in whose reign it was subdued, and lasting for about a couple of centuries. Ælana, which we have already noticed under the Biblical name of Elath, remained a port of commercial importance under the Romans. The names of the other important ports on the Red Sea from N. to S. were—Jambia, *Yembo*, Zabram, Badeo, and Muza: the last was identical with *Moushil*. Sapphar was an important town in the interior, E. of Muza, probably at a spot named *Dhafar*. Saba ranked as the capital of the south, but its position is quite uncertain; it was probably identical with Mariaba in the interior, and is further noticed under the names Sabotha or Sabtha. Mariaba was famed for its enormous reservoir, which received the water of no less than 70 streams for the purpose of irrigation: the bursting of the great dam was regarded as so great a catastrophe that it became an era in Arabic history; it occurred probably about the time of Alexander the Great. The remains of this reservoir have been discovered at *Mareb*. Adana was the chief port on the southern coast, and hence received the name of Arabia Felix; it was the emporium of the trade between Egypt, Arabia, and India. Ælius Gallus destroyed it, but it soon revived. On the Persian Gulf Rhegma and Gherra may be noticed as places of importance in connexion with Indian trade.

Islands.—Off the Arabian coast were the islands Dioscoridis, *Socotra*, and Sarapidis, *Massera*, in the Arabian Sea; and Tylus, or Tyrus, *Bahrein*, and Aridus, *Arad*, in the Persian Gulf. The two latter are of interest in connexion with the history of the Phœnicians. Tylus is also described as abounding in pearls.

History.—The history of Arabia in ancient times is well nigh a blank. No conqueror has ever penetrated the interior to any distance. Antigonus made some unsuccessful attempts to conquer the Nabathæi in the years 312, 311 B.C. The next expedition was undertaken by Ælius Gallus in the reign of Augustus, B.C. 24. Starting from Myus Hormus he landed at Leuce Come, and proceeded by an overland route to a place named Marsyabæ,[9] whence he returned under pressure of the extreme heat and drought. In A.D. 105 the district adjacent to Palestine was formed into a Roman province by A. Cornelius Palma under the name of Arabia.

---

[9] The scene of this expedition was probably quite in the north of the peninsula: as Leuce Come was only two or three days' sail from Myus Hormus, it could not have been S. of *Moilah*: Marsyabæ cannot possibly be identified with the southern Mariaba of the Sabæi, but was perhaps on the site of *Merab*, at the eastern base of the *Nejd* mountains. The following passages relate to this expedition:

> Icci beatis nunc Arabum invides
> Gazis, et acrem militiam paras
> Non ante devictis Sabæm
> Regibus.   Hor. *Carm.* L 29, 1.

> India quin, Auguste, tuo dat colla triumpho,
> Et domus intactæ te tremit Arabiæ. PROPERT. iL 10, 19.

Mount Hor.

Jerusalem.

CHAPTER XI.

PALESTINE.

§ 1. **Palestine** was bounded on the W. by the Mediterranean or "great" sea; on the S. and E. by the desert of Arabia, and on the N. by Syria. Its boundary in the latter direction is not well defined; it ran somewhere N. of Sidon (Judg. i. 31), and along the southern extremity of Hermon (Deut. iii. 8), or Hor (Num. xxxiv. 7, 8): on the S. a range of heights extends from the southern end of the Dead Sea to the Mediterranean: on the E. the limit is again undefined; in the northern part it extends as far as Salcah (Josh. xiii. 11) in nearly the 37° of long., and thence returns to a range of hills skirting the desert, which it follows towards the S. to the junction of the two branches of the Jabbok, and thence to the Arnon. The surface of Palestine is greatly varied. The greater part of the interior is a highland district, diversified in some places with hills, in others with broad undulations. Low plains intervene between this district and the sea, and a remarkable sunken plain, in some parts below the level of the sea, cleaves the highlands from N. to S., along the course of the Jordan. The temperature varies with the varying altitude. While the plains suffer from a tropical heat, the highlands, in which the bulk of the population has in all ages been settled, enjoy a tolerably moderate and equable climate. The productions are consequently equally varied. The palm-tree and the walnut, the balsam and the cedar, find temperatures adapted to their several natures. That the soil, under the most careful cultivation, was pre-eminently fertile,[1] not only the glowing descriptions of the Bible, but the statements of classical writers also inform us. In addition to wheat, barley, and other cereals, a profusion of fruits—the vine, olive, fig, pomegranate, date, almond, &c.—ripened in great perfection. In the highlands, particularly in those on the other side of Jordan, the finest pastures abound.

Names.—Palestine formed a portion of the "land of Canaan," which extended, as we have already shown, beyond the borders of Phœnicia: this, therefore, was its earliest designation in Scripture (Gen. xi. 31). It did not, however, apply to the Trans-Jordanic region, this being styled in contradistinction Gilead (Josh. xxii. 9–11). Before the Exodus it was styled the "land of the Hebrews" (Gen. xl. 15), and after the Exodus the "land of Israel" (Judg. xix. 29), and occasionally the "land of Jehovah" (Hos. ix. 3; compare Lev. xxv. 23; Ps. lxxxv. 1). The expression "Holy Land" which we have adopted occurs but once in Scripture (Zech. ii. 12). Palestine is derived from the Greeks, who described this portion of Syria under the specific title of "Syria Palæstina," *i. e.* "Syria of the Philistines"[2] (Herod. i. 105). After the return from the Babylonian captivity, the name of Judah, which had previously applied only to the tribe of that name and afterwards to the kingdom, was extended over the whole country, and the people were named Judæans or Jews.

§ 2. The geographical position and physical character of Palestine adapted it in many respects for its special office in the world's history. (1.) Its boundaries were well defined: the wilderness encompassed it on the E. and S., while on the N. the mountainous district of Lebanon, and on the W. the Mediterranean Sea closed it

---

[1] The present condition of Palestine presents in this respect a most melancholy contrast. The change may be traced to various causes :—the destruction of the terraces and water-channels—the extirpation of the forests—and the constant wars that have desolated the country.

[2] This was the name by which it was known to the Romans:

   Alba Palæstino sancta columba Syro.—Tibull. *E*. i. 7, 18.

in. Thus the Jews were distinctly separated from all other nations.
(2.) It was well situated with regard to the early scats of empire
and civilization, having Egypt on the one side and Mesopotamia on
the other. All intercourse between these countries was necessarily
conducted through Palestine: in a military point of view especially
Palestine was the gate of Egypt. From these causes both the
Egyptians and Assyrians must have become well acquainted with its
institutions and religion. (3). It possessed no facilities for extended
commerce; the coast-line is regular, and offers no harbourage, except
at the small port of Joppa; the country was not gifted with any
peculiar productions which called forth a spirit of inventive genius.
(4.) The varied character of its soil yielded all the productions
requisite for the necessities and even the luxuries of its inhabitants, and made them comparatively independent of other countries.

§ 3. The mountain system of Palestine is connected with the
great range of Taurus by the intervening chains of Amanus,
Bargylus, and Libanus or Lobanon. From the latter of these a
high mountainous district emanates which runs parallel to, but at
some distance from the Mediterranean coast through the whole
length of the land, broken only at one point by the plain of Esdraelon,
and the valley of the river Kishon. The mountains S. of Esdraelon
are subdivided into two sections by a depression, which occurs in
the neighbourhood of Jerusalem: the southern of these sections
comprised the "hill country of Judæa," the northern the "mountains of Ephraim:" the elevation of this district gradually increases
towards the S., and attains a height of 3250 feet above the level of
the sea in the neighbourhood of Hebron. The regularity of the
coast-line is broken by the protrusion of a lofty spur that bounds
the plain of Esdraelon on the S., terminating in the promontory of
Carmel. The district on the eastern side of Jordan may be regarded
as a prolongation of the range of Antilibanus, which is continued in
the ranges of el-Heish and el-Furus to the head of the Sea of Galilee,
and then subsides into the table-land of Hauran. On the southern
side of the Hieromax the ground rises again, and attains its greatest
elevation in Mount Gilead S. of the Jabbok. The plateau which
succeeds towards the S. rises abruptly from the valley of the
Jordan, and falls off gradually eastward to the desert of Arabia.
The most remarkable height in the whole of Palestine is the northern
peak of Hermon at the extremity of Antilibanus: it received various
names, Sirion, Senir, and occasionally Sion (Deut. iv. 48), the two
former signifying "breastplate," and suggested by the glittering
appearance of the summit under the influence of the sun's rays: it
is now called Jebel-esh-Sheikh, "the old man's mountain," or
"the chief mountain;" its height is about 10,000 feet, and its
summit is streaked with snow even in the middle of summer.

§ 4. Next to the mountains, the plains demand our notice, from the strong contrast which they present in point of elevation and character. These plains extend on each side of the hill-country of Western Palestine: on the W. a rich district stretches from Carmel along the coast of the Mediterranean to the borders of the desert, divided into two portions, **Sharon**, " the smooth," forming the northern division, and **Shephela**, " the low," the southern, while N. of Carmel follows the beautiful plain that surrounds *Acre*. On the E. lies the plain of the Jordan, deeply sunk below the level of the sea, and presenting in almost every respect a remarkable contrast to the hill-country : it was described by the Hebrews as " the desert," by the Greeks as *Aulon*, " the channel," and by the modern Arabs as *el-Ghor*, " the sunken plain." The difference in point of elevation of these closely contiguous districts is best shown by a reference to the accompanying diagram. Jerusalem stands about 3500 feet above the Dead Sea, about the same elevation at which a spectator overlooks the sea at Carnarvon from the top of Snowdon.[1]

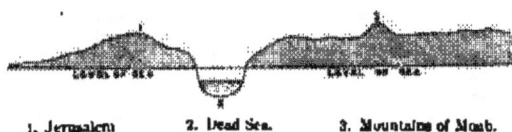

1. Jerusalem   2. Dead Sea.   3. Mountains of Moab.

§ 5. The only river of importance in Palestine is the **Jordan**, which rises at the base of Hermon, and flows with a rapid stream (whence its name, meaning " the swiftly descending ") through the lakes of Merom and Galilee into the Dead Sea, its valley sinking far below the level of the Mediterranean. The Arabs name it *Sheriat-el-Khebir*, " the watering place."

Its early course lies along a level and swampy plain to the Lake of Merom: at this point the depression of its bed commences, and it descends 300 feet to the Sea of Galilee. Emerging from this it descends again 1000 feet by a series of rapids to the Dead Sea, receiving on its left bank the tributary streams of the **Hieromax** and the **Jabbok**. This last stage of its course lies along a deep valley, about eight miles broad, enclosed between two parallel mountain walls. As the river flows in the lowest part of this valley, it is incapable of fructifying it, and hence it was specially termed " *the* desert " (*Ha-arabah*) by the Hebrews. In the midst of this barrenness, the banks of the river are fringed with a prolific growth of trees and grass. It is crossed by fords at four points,

---

[1] This contrast of mountain and plain exercises an influence on the political arrangements, and even on the language of the country. From it arises the broad division of the population into the Amorites, "dwellers in the mountains," and the Canaanites, "dwellers in the plain." Hence also the expressions so frequent in Scripture, "going down," *e.g.* to Jericho, "going up" to Jerusalem. To the same feature we may also attribute the extensive views which are to be obtained from various points of the hill-country.

viz. below the Sea of Galilee, below the confluence of the Jabbok, and at two points opposite Jericho. In the latter part of its course the bed of the river is depressed about 50 to 80 feet below the level of the plain: its breadth varies from 80 to 100 feet, and its depth from 10 to 12 feet. At the time the Israelites crossed it, it was full up to its banks —an occurrence still occasionally witnessed in the beginning of May. The Jordan with its singularly depressed valley formed a natural division of Palestine into two portions, described in Scripture as "this side" and "the other side Jordan."

The Jordan was connected with a system of lakes, which were fed by it; they were named—the first Merom, now Ard-el-Huleh; the second, by the several names of the Sea of Chinnereth or Chinneroth, perhaps from its oval, "harp-like" form, the Sea of Galilee from the province in which it lay, and the Lake of Gennesareth or Tiberias from places on its coast: the third, the "great" or "salt" sea of the Hebrews, the Lacus Asphaltites of the Romans, the Bahr Lut, "Lot's Sea," of the Arabs, and the "Dead Sea" of some classical writers and of modern geography.

Merom is about 4½ miles long by 3½ broad, and is surrounded by an impenetrable mass of jungle: on the plain in its neighbourhood was fought the last battle between Joshua and the Canaanites. The Sea of Galilee is about 13 miles long by 6 wide; it lies in a deeply sunk basin, surrounded by hills of great elevation. On the eastern shore these hills rise almost immediately from the edge of the lake: on the western shore a fertile strip of land intervenes, and at one point, about midway from the ends of the lake, there is a considerable plain about 5 miles wide by 6 broad, formed by the receding mountains. The lake still abounds with fish as in our Saviour's time. The Dead Sea is 40 miles long by 8½ broad, and lies at a depression of above 1300 feet below the level of the sea. The lower part of the sea is narrowed by the projection of a broad promontory: a great alteration in the depth occurs at this point, the northern portion being deep, the southern quite shallow. The whole is enclosed by a double mountain wall, the continuation of that which bounds the Ghor. The saltness of the water is remarkable, the per-centage of salt being 26½, while that of the ocean is only 4. This arises from a barrier of fossil salt at the southern end of the lake, aided by the effects of evaporation. Masses of asphaltum are sometimes thrown up from the bottom. Along the shore are numerous salt marshes, on which pure sulphur is often found, and near the southern end are salt-pits. A number of springs pour into the lake, of which the most famous were En-eglaim, probably the Callirhoë in which Herod bathed, at the N.E. end, and En-gedi on the western coast, surrounded by a small oasis of verdure. The lake receives a further supply from some tributary streams on its eastern shore, of which the Arnon is the most important. Changes have probably occurred in the condition of the lake within historical times: the description of Lot (Gen. xiii. 10) is now inappropriate, and the fact of a Pentapolis, or confederacy of five cities, viz., Sodom, Gomorrah, Admah, Zeboïm, and Lasha, having existed near the southern part of the lake renders it likely that the shallow part of the lake has been recently submerged, and was formerly a rich plain. The opinion formerly entertained, that the Jordan may formerly have found a channel by the Arabah into the Red Sea, has been proved incorrect by the discovery that the ground rises S. of the lake.

§ 6. The population of Palestine was composed of numerous races, which succeeded one another in the possession of the country.

i. Its earliest inhabitants probably belonged to those "Giant" races, of which but a few isolated communities remained in historical times. They were most numerous in the Trans-Jordanic district, where we have notice of the Rephaims in Ashteroth-Karnaim, the Zuzims or Zamzummim in Ham, and the Emim in Shaveh Kiriathaim (Gen. xiv. 5). Og, the King of Bashan, was the last survivor of the race in that district (Deut. iii. 11). They were also found W. of Jordan, viz. the Anakim about Hebron (Num. xiii. 22; Josh. xiv. 15); the Rephaim, who gave name to a valley to the S.W. of Jerusalem (2 Sam. v. 18); and perhaps the Avim in Philistia (Deut. ii. 23). The origin and history of these races is a matter of conjecture.

ii. The Canaanites were, like the Phœnicians, a Semitic race. There is certainly some difficulty in reconciling the Biblical statement (according to which Canaan was a son of Ham, Gen. x. 6) with the conclusions to be derived from language and other ethnological indications. It is clear that when Abraham first entered Canaan the language spoken by the inhabitants was the same as the later Hebrew: not only did Abraham converse with the Hittites without an interpreter (Gen. xxiii.), but the names Melchizedek, Salem, and others, are clearly of a Semitic origin.

iii. The Philistines were a Hamitic race; according to Gen. x. 14, they were connected with the Casluhim, and according to Jer. xlvii. 4, and Am. ix. 7, with the Caphthorim. As these two tribes were closely allied, it is possible that the Caphthorim immigrated into the country of the Casluhim at a later period. The Philistines were intimately connected with Egypt: the name Caphthor survived in Coptos, and Philistine perhaps in Pelusium; the name Philistine is supposed to be of Coptic origin, betokening "strangers" (hence, in the LXX. they are termed ἀλλόφυλοι), indicating their immigration from Upper to Lower Egypt.

iv. The Hebrews were also a Semitic race, who immigrated at a later period from the northern part of Mesopotamia. When they first appear in history they were a nomadic tribe, who merely fed their flocks and herds by the permission of the older occupants. Their growth as a people took place in Egypt, whence they issued as an invading host and took forcible possession of the land of the Canaanites, in many instances exterminating the inhabitants, in others reducing them to the position of bondsmen. It is clear, however, that the Hebrews were at no period possessors of the whole of the country. The Philistines in the S. and the Phœnicians in the N. held their ground permanently; and for a long period the Canaanites occupied strongholds in the midst of the Hebrews (1 Sam. vii. 14; 2 Sam. xxi. 2, xxiv. 7). The population was thus of a mixed character, foreign races holding the extremities of the land, while in the central districts Canaanites were found even to the latest times of the monarchy (Ezr. ix. 12), much in the relative positions of the Spartans and Helots of Laconia (1 Kings, ix. 20, 21).

v. The Samaritans were a mixed race of Hebrews and Babylonians. Their existence, as a people, dates from the period of the Ismelitish captivity, when Shalmaneser introduced colonies of Babylonians into Samaria to supply the place of the inhabitants whom he had carried off. A certain portion of the latter appear to have remained behind, or perhaps they returned gradually from the place of their captivity. Religious teachers were supplied at their own request, and thus both the people and their religion assumed a hybrid character, which led to extreme jealousy on the part of the pure Jews, and ultimately to the estrangement indicated in John iv. 9.

(vi.) We have, lastly, to notice some tribes which were connected with the Israelites by ties of relationship; such as the Moabites and Ammonites, who were descended from Lot, and the Kenites, to whom Hobab, the father-in-law of Moses, belonged.

§ 7. The divisions of Palestine varied in the different periods of its history.

i. The earliest of these periods may be termed the Canaanitish, and lasted from the time when the country is first known to us down to the entrance of the Hebrews. During this it was occupied mainly by the Canaanitish tribes, and partly by the Philistines and the descendants of Abraham and Lot.

The Canaanitish Period.—The Canaanites were divided into the following tribes:—1. Hivites in the northern districts about the roots of Lebanon (Josh. xi. 3), and at one period about Shechem (Gen. xxxiv. 2). 2. Girgashites, whose abode is not specified in the few passages in which the name occurs (Deut. vii. 1; Josh. xxiv. 11; Neh. ix. 8). 3. Jebusites, about Jerusalem (Josh. xv. 8; Judg. i. 21). 4. Hittites, more to the S., in the neighbourhood of Hebron (Gen. xxiii. 3). 5. Amorites, about the western shores of the Dead Sea (Gen. xiv. 7, 13), and across the valley of the Jordan to the opposite highlands, where, at the time of the Exodus, they had two kingdoms, with Heshbon for the southern (Num. xxi. 13, 26) and Ashtaroth for the northern capital (Deut. i. 4; Josh. ix. 10). 6. Canaanites (properly so called), on the sea-shore N. of Philistia and in the plains of the Jordan (Num. xiii. 29), the two branches being described as the "Canaanite on the east and on the west" (Josh. xi. 3). Whether the Perizzites were a Canaanitish tribe or not is uncertain; they are not enumerated in Gen. x. 15-19. It has been surmised, however, that the name is significant, and that the Perizzites were the "agriculturists" in opposition to the Canaanites, "the merchants," and that thus Canaanite and Perizzite formed the two great divisions of the people, according to their occupations (Gen. xiii. 7, xxxiv. 30). Some of the above names are applied in an extended sense to the whole of Palestine, as the Hittites (Josh. i. 4) and the Amorites (Gen. xv. 16; Josh. xxiv. 18).

At the time of the Exodus the Moabites, who had previously occupied the district E. of the Jordan and the Dead Sea, had been expelled from it by the Amorites and were living S. of the Arnon (Num. xxi. 13, 26). The name "field" or "plains of Moab" was, nevertheless, always applied to their former territory (Deut. i. 5; Josh. xiii. 32). The Ammonites lived originally to the N., in the highlands adjacent to the valley of the Jordan, between the Arnon and Jabbok, but had been driven to the borders of the wilderness by the Amorites, eastward of the Jabbok in its upper course (Deut. iii. 16). The Kenites roamed about the country, and are found at one period in the wilderness of Judah (Judg. i. 16), at another in northern Palestine (Judg. iv. 11), and again among the Amalekites (1 Sam. xv. 6).

The Philistines were settled in the southern maritime plain of Judæa, where they had a confederacy of five cities—Ashdod, Gaza. Ekron, Gath, and Ascalon.

ii. The second period may be termed the Israelitish, lasting from the time of Joshua to the Babylonish captivity, when Palestine was divided among the twelve tribes, the earlier nations occupying certain positions. In the latter part of this period the whole country was divided into the two kingdoms of Judah and Israel—the former comprising the southern portion of western Palestine as far as the boundary of Benjamin and Ephraim, and the latter the whole remaining district.

iii. The third period may be termed the Roman, and is contemporaneous with the New Testament history. Western Palestine was then divided into three portions—Judæa, Samaria, and Galilee —while eastern Palestine was divided into several districts, of which Peræa was the most important, extending from the southern frontier to the Sea of Galilee, the northern district being subdivided into Ituræa, Gaulonitis, Auranitis, and Trachonitis. We shall adopt the divisions of this third period in the following detailed description of the country, retaining the tribes as subdivisions.

iv. Finally, at the commencement of the 5th century A.D., Palestine was divided into three provinces; Palæstina Prima, consisting of the northern part of Judæa, Samaria, and Philistia; P. Secunda, Galilee and Northern Peræa; and P. Tertia or Salutaria, the southern parts of Judæa and Peræa, with a part of Arabia Petræa.

I. JUDÆA.

§ 8. Judæa comprised the territories of the tribes of Simeon, Judah, Dan, and Benjamin, together with the maritime district of Philistia. Within these limits were included districts differing widely from each other in physical character, climate, and productions. There was first the "south country," consisting of an undu-

lating plain between the mountains of Judah and the desert of
et-Tih : secondly, the "hill country," the central district, which was
highly elevated and richly cultivated ; thirdly, the "desert," which

Jericho.

intervened between this and the Dead Sea ; and, lastly, the maritime
plain, named Shephela, which was remarkably fertile.

§ 9. The tribe of Simeon occupied the "south country," which was
unfavourably situated, being exposed to the attacks of the Amalok-
ites and other desert tribes : it consequently possessed no towns of
importance, but had several stations about wells, such as Beersheba,
Laharoi, and others.

Beersheba, "the well of the oath," is connected with many Incidents of interest: the well was originally dug by Abraham, and
named after the treaty which he formed with Abimelech : here the
patriarch planted a grove and received his order to slay Isaac ; and
Jacob obtained the blessing from Esau, and offered up sacrifices before
leaving his native country. Samuel here appointed his sons Judges,
and it was visited by Elijah on his journey to Horeb: it was the most
southerly town of Palestine.   There are still at this spot two wells
furnishing pure living water.

§ 10. The "south country" was succeeded by the "hill country," occupied by the tribe of Judah, a broad district of hill and vale overlooking in one direction the Dead Sea, in the other the maritime plain of Philistia. Its fertility was great: it was (and even still is in spots) well covered with corn-fields and vineyards; the ravines were clothed with forests, and the various mountain-tops afforded secure sites for fortified towns. The most elevated part is in the neighbourhood of Hebron, which stands 3000 feet above the level of the sea. The territory of Judah extended on each side of this mountain district into the plain that lies adjacent to it on the W., and over the wide plateau which extends eastward to the precipitous heights overhanging the Dead Sea, and which from its desolate character well deserves the title of the "wilderness" of Judah.

The chief town in the hill country was Hebron,[4] originally Kirjath-arba, situated on a hill overlooking the fertile valley of Eshcol, which is still well clothed with orchards, oliveyards, and vineyards; it is first noticed as the abode of Ephron the Hittite, and afterwards as the place where Abraham settled; Caleb selected it as his portion at the conquest of Canaan, and drove out Arba and his sons; it was the central spot to which the tribe of Judah rallied under David and Absalom. Near it was the cave of Machpelah, where the patriarchs were buried, now marked by a building called the *Haram*; and a little N. of the town is Mamre, *Ramch*, beneath the shelter of whose grove ("plain" in the English translation, Gen. xiv. 13, xxiii. 15) Abraham pitched his tent. Bethlehem, "the house of bread," surnamed of Judah, to distinguish it from another in Zebulun, and also Ephratah, "fruitful," stands a short distance E. of the road leading from Hebron to Jerusalem, on a narrow ridge which protrudes eastward from the central range, and which descends steeply into valleys on all sides but the W. It was here that Jacob buried Rachel—that Ruth gleaned in the fields of Boaz—that David spent his youth—and, above all, that the Saviour of the world was born, and in the adjacent fields the good news was first told from heaven to the shepherds.

Of the other towns in this district we may notice—Maon, on the summit of a conical hill, overlooking the desert of Judah—Carmel, somewhat westward, the scene of the story of Abigail and David—Engedi, a spot on the western shore of the Dead Sea, which gave name to the surrounding wilderness—Lachish, in the maritime plain just at the foot of the hills, an important military post commanding the south country; it was fortified by Rehoboam, and was besieged by Sennacherib—Libnah, to the N.W., also besieged by Sennacherib; it was an old Canaanitish town, and sufficiently strong to revolt from king Jehoram—Etham, *Urtas*, a little S. of Bethlehem, where are certain reservoirs, now named "Solomon's Pools," with which the Temple at Jerusalem was supplied with water. On the heights overlooking the wilderness of Judah were situated the fortresses of Modin, Herodion, and Masada: the site of Herodion is identified with the *Frank Moun-*

---

[4] The modern names of the towns of Palestine are generally identical with the Biblical ones. Hence it is unnecessary to give them, except in cases where there is considerable variation, or for the purpose of identifying the positions.

tain, E. of Bethlehem: Masada was above Engedi: the position of Modin is unknown.

§ 11. The district of **Philistia** comprised the southern portion of the maritime plain of Palestine to Ekron in the N. This district is divided into two belts—one consisting of a sandy strip of coast, and the other of a cultivated district slightly elevated, and with occasional eminences, on which the strongholds of the country were built. This part of the country is remarkably fertile both in corn and in every kind of garden fruit. The five chief towns formed in the early period of Jewish history a confederacy of five cities, viz. Gaza, Ascalon, Ashdod, Ekron, and Gath: the last has not been identified, but the others are still in existence.

Gaza, *Ghuzzeh,* stands near the southern frontier, at present above 3 miles from the sea, but formerly (as some suppose) within 2 miles of it. It ranked as one of the oldest towns of Palestine (Gen. x. 19); though nominally within the borders of Judah, and conquered by them, it was not retained: Samson's death took place there. The position of Gaza, as the "key of Egypt," exposed it to various sieges: it was taken with difficulty by Alexander the Great, and was twice ruined in the 1st century of our era: it now contains about 15,000 inhabitants. **Ascalon,** on the sea coast, was similarly captured but not retained by the tribe of Judah, and was from an early period the seat of the worship of *Derceto,* the Syrian Venus: the site is almost covered with sand, and ere long the words of Zephaniah (ii. 4) will be verified that "Ashkelon shall be a desolation." **Ashdod,** *Esdud,* the Azotus of the New Testament, stands about 4 miles from the sea, and was the scene of the fall of Dagon at the presence of the ark: it was strongly fortified, and was dismantled by Uzziah: Psammetichus of Egypt besieged it for 29 years: here Philip was found after his interview with the eunuch (Acts viii.). **Ekron,** *Akir,* stood more inland, on the borders of Dan: thither the ark was sent from Gath, and thence forwarded to Bethshemesh (1 Sam. v.). **Gath** is supposed to have stood near the frontier of Judah, S.W. of Jerusalem.

§ 12. The tribe of **Dan** occupied a small district between the Mediterranean Sea and the hill country of Benjamin, about the point where the two portions of the maritime plain, Sharon and Shephela, meet.

The chief town was **Joppa,** *Yâfa,* which has in all ages served as the seaport of Jerusalem: its situation is remarkably beautiful, as the name itself, meaning "beauty," implies — the surrounding district being remarkable for its fertility and the brilliancy of its verdure: the materials for the erection of the Temples under Solomon and Ezra were landed here, and it was here that Jonah took ship for Tarshish: it was visited by Peter, who received a remarkable vision there, and raised Tabitha to life. **Lydda,** the later **Diospolis,** was centrally situated at the point where the road from Jerusalem to Joppa crosses that which follows the plain from S. to N.: it was the scene of the healing of Æneas. **Nicopolis** stood between Lydda and Jerusalem; it was a place of military importance under the Maccabees, and the adjacent plain was the scene of the remarkable victory of Judas Maccabæus over the Syrians (1 Mac. iv.): it was regarded by early Christian writers as identical with the Emmaus (Luke xxiv. 13) whither the disciples were returning from Jerusalem, and the place is still named *Amwás*; but as the latter place was only 60 stades, and Nicopolis 160 from Jerusalem, the two places cannot be the same: the site of Emmaus is really unknown.   On the borders of Dan and Benjamin was Upper **Beth-hôron**, *Beit-ur-el-Fóka*, on the summit of a conical hill, commanding the pass leading down to the maritime plain, through which Joshua passed in his pursuit of the Amorites: the Roman road to Cæsarea passed this way, and down the same defile the Jews pursued the Romans under Cestius: a little to the S. was Aijalon, on a spur overlooking a plain—the valley over which Joshua bade the moon to stand still.   The modern *Ramleh*, near Lydda, has been traditionally identified with the Arimathæa of the New Testament, where Joseph lived, as well as with the Ramathaim Zophim of the book of Samuel: the grounds for this are very insufficient: *Ramleh* was probably not in existence before the 8th century A.D.

§ 13. The tribe of **Benjamin** occupied that part of the mountainous district which extends from Jerusalem in the S. to Bethel in the N., and from Bethhoron in the W. to Jordan in the E.   Though this district was insignificant in point of extent, it was important through its central position, commanding the passes that lead down to Jericho in one direction, and to the maritime plain in another, as well as the great high-road that traverses central Palestine from N. to S.   The numerous eminences[s] of this district offered almost impregnable positions for fortresses; and the defiles leading down to the plains were easily defensible.   Hence the tribe of Benjamin acquired a warlike character, "ravening as a wolf" (Gen. xlix. 27) in his mountain fastnesses.

The towns of Benjamin possess much interest from their historical associations.   Jerusalem stood within its boundaries, but deserves a separate notice as the capital of Palestine.   The next in point of importance was Jericho, *Riha*, in the plain of Jordan, and at the entrance of the defile leading to Jerusalem.   The road which connects it with the capital ascends a steep and narrow ravine, and from the head of this pass it traverses a remarkably savage and desolate region, where the traveller is still, as in our Saviour's time, in danger of "falling among thieves."   Jericho itself was the first city which the Israelites took after crossing the Jordan: it was then destroyed, but rebuilt about 500 years afterwards; it then became the seat of a school of prophets, and is illustrious from its connexion with the lives of Elijah and Elisha: the town fell into decay, and was rebuilt on a new site, about 1½ mile S. of the old town, by Herod the Great: this was the town which our Lord visited, and where Zacchæus lived.   The surrounding plain was in early ages remarkable for its fertility—a "divine region" as Josephus terms it; and Jericho was known as the "City of Palm-Trees" (Deut. xxxiv. 3), from the luxuriant palm-groves about it: this plain is now an utter wilderness.   Between Jericho and the Jordan was **Gilgal**, where the Israelites first set up the tabernacle,

---

[s] The names of the towns of Benjamin are frequently significant of this feature; as Gibeah, Geba, Gibeon, "hill;" Mizpeh, "look out;" Ramah, "eminence."

and where in the time of Samuel the people were wont to meet for purposes of public business.

Returning to the hill country, we meet with a number of spots of interest in connexion with the religious and military events of Jewish history. In the N. was Bethel, "the house of God," the Luz of the Canaanites, now Beitin, a short distance off the great northern road; it stood on a low ridge, between two converging valleys; it was the spot where Abraham first pitched his tent, and where Jacob was favoured with his vision: in the time of the Judges it became a place of congress, and was selected by Jeroboam as one of his idolatrous sanctuaries, whence its name was changed into Bethaven, "house of idols" (Hos. x. 5); Josiah purified it by the destruction of the altar and grove: it is now a heap of ruins, as predicted by Amos (v. 5). Gibeon, El-Jib, stood N. W. of Jerusalem on "the way that goeth up to Bethhoron," posted on an isolated hill in the midst of a rich plain: it was originally the chief town of the wily Gibeonites; near it was the "great high place" where the tabernacle was set up after the destruction of Nob: the defeat of Abner and the murder of Amasa occurred here; and here Solomon was favoured with his vision. Gibeah stood about 4 miles N. of Jerusalem at a spot now called Tuleil-el-Fûl: it must not be confounded with the Gibeah, or more properly the Geba, of 1 Sam. xiii. 15: Gibeah was the birth-place, and general abode of Saul, and on its hill the sons of Rizpah were hung.

Places of less importance were—Nob, immediately N. of Jerusalem, the city of the priests whither David fled, and where the priests were in consequence massacred—Anathoth, further N., the birthplace of Jeremiah, and on the road by which Sennacherib advanced to Jerusalem—Geba (also called "Gibeah" in A. V.), Jeba, the scene of Jonathan's adventure against the Philistines—Michmash, on the edge of a ravine leading down to the valley of the Jordan, named "the passage of Michmash;" it was garrisoned by Saul against the Philistines, and the latter people were encamped close to it at the time of Jonathan's exploit: the hosts of Sennacherib selected it as the place to "lay up their baggage" on their advance to Jerusalem—Ai, between Michmash and Bethel, on a ridge overlooking the descent to Jordan, chiefly famous for its capture by Joshua; between it and Bethel was the elevated spot, whence Abraham and Lot surveyed the land and chose their respective quarters; further on towards the N. rise the white peak of Rimmon, where the 600 Benjamites took refuge (Judg. xx. 47), and the dark conical hill of Ophrah, Taiyibeh, whither the Philistines sent out one of their bands (1 Sam. xiii. 17), probably the same place as is afterwards called Ephraim in 2 Chron. xiii. 19 and John xi. 54—Beeroth, S. of Bethel, one of the cities of the Gibeonites, and the place where the caravans from Jerusalem to the N. generally make their first halt; it is thus reputed the place where our Lord was sought by his parents—Ramah "of Benjamin," er-Ram, between Beeroth and Gibeon, to which reference is probably made in Jer. xxxi. 15, the captives being carried this way by the Babylonians: the Ramah at which Samuel lived is a different place, and has not yet been identified—Mizpah, on a hill (now named Neby Samwil, from a tradition that Samuel was buried there), which rises conspicuously above the plain of Gibeon; it was fortified by Asa, and was frequently used as a place of national congress—Kirjath-jearim, W. of Jerusalem, whither the ark was brought from Bethshemesh — lastly, Bethany, now called el-Azariyeh, "the village of Lazarus," situated on the eastern slope of

Olivet—a place consecrated to the mind of the Christian by the residence of our blessed Lord during the last trying scenes of his life.

Jerusalem from the South.

§ 14. The chief town in Palestine was Jerusalem, the Salem, "city
of *peace*," of Ps. lxxvi. 2, and probably of Gen. xiv. 18, the Jebus of
the Canaanites, the Ariel, "Lion of God," of Is. xxix. 1, the Hiero-
solyma of the Greeks, the Ælia Capitolina of the Romans, and the *El-
Kuds*, "Holy Place," of the modern Arabs. Its situation is striking;
it is neither on a hill-top as most of the Jewish strongholds, nor yet
in a valley, but on the edge of a rocky platform in the central ridge
between the Mediterranean and the Dead Sea. On three sides this
platform is severed from the adjacent high land; viz., by the deep
defile of Ge-ben-Hinnom, "the cleft of the son of Hinnom," corrupted into Gehenna by Greek writers, on the W. and S.; and by the
still deeper vale of Jehoshaphat on the E., along which the Kedron
flowed, and which thence continues its course towards the Dead Sea.
On the N. Jerusalem lay open to the country, and in this direction
alone did the city admit of any extension. The elevation of its site
above the sea amounts to 2200 feet, and it stands at the highest
point of the ridge; the ground rises towards the S., but in other directions falls: towards the E., however, the Mount of Olives exceeds
the height of Zion by about 180 feet, and it is to this range, and
perhaps to the yet higher but more distant range of the hills of
Moab on the other side of Jordan, that the Psalmist alludes in the
well-known words, "The hills stand about Jerusalem" (Ps. cxxv.

2). Looking at its position in a political point of view, it will be
observed that it was situated centrally on the borders of the two
most powerful southern tribes, Judah and Benjamin, and equally
accessible to persons traversing the land in its length through the
mountainous district, or in its breadth from the valley of the Jordan
to the maritime plain.

Hills of Jerusalem.—The site of Jerusalem itself was broken by
various elevations: the most conspicuous of these was in the S.W.,
and is now known as **Mount Zion.** On the W. and S.W. it overlooks
the valley of Hinnom at a height of 150 feet, and at the S.E. the
valley of Jehoshaphat at a height of 300 feet above the Kidron: on the
E. and N. it was separated from the rest of the city by a valley called
Tyropœon, which joins those of Hinnom and Jehoshaphat at Enrogel,
gradually deepening as it approaches this point. Whether this hill
was identical with the **Zion** of the Old Testament, must be considered
doubtful. Recent researches have made it probable that the *ancient*
Zion was on Moriah. In this case the modern Zion was the site of the
city of the Jebusites and of the Upper Market-Place of Josephus, while
David's city and sepulchre would be on the opposite height. **Moriah**
was the central portion of the eastern ridge, separated from Zion on
the W. by the Tyropœon, and overlooking the valley of Jehoshaphat on
the E. at an elevation of about 150 feet. This was the spot where
Abraham offered up Isaac, where in David's time Ornan had his
threshing floor, and where Solomon erected the Temple: the fortress
of Antonia was erected at the N.W. angle of the Temple. The site of
the Temple is now covered by the enclosure of the *Mosque of Omar.*
A remarkable rock, now named *Sakrah*, rises in the centre of this space,
and has been supposed to mark the place of the altar. The southern
continuation of this ridge was named **Ophel**, which gradually came
to a point at the junction of the valleys Tyropœon and Jehoshaphat;
and the northern, **Besetha,** "the New City," first noticed by Josephus.
which was separated from Moriah by an artificial ditch, and overlooked
the valley of Kidron on the E.; this hill was enclosed within the
walls of Herod Agrippa. Lastly, **Acra** lay westward of Moriah and
northward of Zion, and formed the "Lower City" in the time of
Josephus. In this portion of the town are the sites which tradition
has connected with the most awful events of our Saviour's life—Golgotha,—and the sepulchre in which his body was laid. These events
may, after all, have really taken place on the eastern hill, or Moriah.

Pools and Fountains.—Among the objects of interest about Jerusalem
the pools hold a conspicuous place. Outside the walls on the W. side
were the Upper and Lower Pools of **Gihon,** the latter close under Zion,
the former more to the N.W. on the Jaffa road. At the junction of the
valleys of Hinnom and Jehoshaphat was **Enrogel,** the *Well of Job,* in the
midst of the king's gardens. Within the walls, immediately N. of Zion,
was the "Pool of Hezekiah." A large pool existing beneath the Temple
(referred to in Ecclus. l. 3), was probably supplied by some subterranean aqueduct. The "King's Pool" was probably identical with
the *Fountain of the Virgin,* at the southern angle of Moriah. It possesses the peculiarity that it rises and falls at irregular periods; it is
supposed to be fed from the cistern below the Temple. From this a
subterranean channel cut through the solid rock leads the water to the
pool of **Silôah,** or **Siloam,** which has also acquired the character of

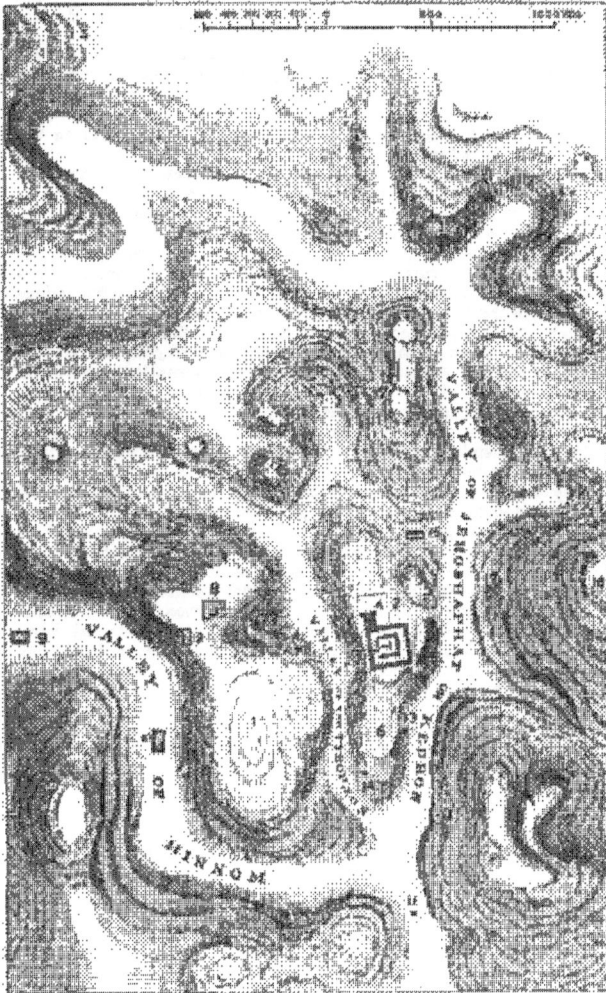

Plan of Jerusalem.

1. Mount Zion. 2. Moriah. 3. The Temple. 4. Antonia. 5. Probable site of Golgotha. 6. Ophel.
7. Bezetha. 8. Church of the Holy Sepulchre. 9, 10. The Upper and Lower Pools of Gihon. 11.
Enrogel. 12. Pool of Hezekiah. 13. Fountain of the Virgin. 14. Siloam. 15. Jotham. 16.
Mount of Olives. 17. Gethsemane.

being an intermittent fountain. The pool to which tradition has
assigned the name of Bethesda is situated on the N. side of Moriah:
it is now named *Birket Israil*, and appears from the character of the
mason-work about it to have been originally designed for a reservoir.

Burial Places.—Burial places were formed in the valleys surrounding
Jerusalem ; in the valley of Hinnom, where is the reputed site of
Aceldàma—" the field of blood ;" in the valley of Jehoshaphat, where
the ancient tombs were excavated out of the rock in tiers ; and on the
Mount of Olives, where were the tombs of the prophets.

History of Jerusalem.—The earliest notice of Jerusalem in the Bible
is as the capital of Melchizedek, the Salem there noticed being now
recognized as identical with it. It next appears as the stronghold of
the Jebusites, who held out against the Israelites for above five centuries. David took it (about B.C. 1049), and established it as his
capital. Solomon further enhanced its importance by erecting the
Temple there. Under the Jewish kings it was taken by the Philistines
and Arabs in the reign of Jehoram ; by the Israelites in the reign of
Amaziah ; by Pharaoh Necho, king of Egypt (B.C. 609); and by Nebuchadnezzar on three occasions, in the years B.C. 607, 597, and 586 ;· in
the last of which it was utterly destroyed. Its restoration commenced
under Cyrus (B.C. 538), and was completed under Artaxerxes I., who
issued commissions for this purpose to Ezra (B.C. 457) and Nehemiah
(B.C. 445). In B.C. 332 it was captured by Alexander the Great.
Under the Ptolemies and the Seleucidæ the town was prosperous, until
Antiochus Epiphanes sacked it (B.C. 170). In consequence of his
tyranny the Jews rose under the Maccabees, and Jerusalem became
again independent, and retained its position until its capture by the
Romans under Pompey (B.C. 63). The Temple was subsequently plundered by Crassus (B.C. 54), and the city by the Parthians (B.C. 40).
Herod took up his residence there as soon as he was appointed sovereign, and restored the Temple with great magnificence. On the death
of Herod it became the residence of the Roman procurators, who occupied the fortress of Antonia. The greatest siege that it sustained, however, was at the hands of the Romans under Titus, when it held out
nearly five months, and when the town was completely destroyed (A.D.
70). Hadrian restored it as a Roman colony (A.D. 135), and among other
buildings erected a temple of Jupiter Capitolinus on the site of the
Temple. The emperor Constantine established its Christian character
by the erection of a church on the supposed site of the holy sepulchre
(A.D. 336), and Justinian added several churches and hospitals 'about
A.D. 532).

II. Samaria.

§ 15. **Samaria** embraced the central district of Palestine from the
borders of Benjamin on the S. to the plain of Esdraelon on the N.,
and from the Mediterranean on the W. to the Jordan on the E.
It was co-extensive with the territories assigned to Ephraim and
the half tribe of Manasseh. Like Judæa it consists of two districts
widely differing in character, the mountain region in the centre,
with the plain of Sharon on the one side and the valley of Jordan
on the other. The mountainous region is more diversified than
that of Judæa, broad plains and valleys frequently intervening.
The maritime plain of Sharon has in all ages supplied abundant

pasture for sheep, but possessed no towns of importance, probably from its exposure to the inroads of the desert tribes of the south.

Cæsarea.   (From a sketch by Wm. Tipping, Esq.)

§ 16. The tribe of Ephraim occupied the greater part of Samaria, and was one of the most powerful of the Jewish confederacy. Its prosperity was due partly to the fertility, and partly to the security of its district. The vales and plains are remarkably rich and well sheltered, and the olive, fig, and vine, still flourish there: Scripture speaks in glowing, yet not exaggerated, terms of the land which fell to the lot of Joseph's younger son (Gen. xlix. 22; Deut. xxxiii. 13-16). Its security also was great: well protected on the N. by the difficult ravines which lead to the plain of Esdraelon, and on the E. by the deep valley of Jordan, it was only on the S. that it was easily assailable; and in this direction its command of the high road through central Palestine gave it an advantage likely to secure peaceful relations with its neighbours. The tribe of Manasseh held a subordinate position to Ephraim, only half the tribe being located on this side of Jordan, in the district adjacent to the plain of Esdraelon.

Towns of Samaria.—**Shechem**, the original capital of Samaria (now *Nablûs*, a corruption of the name Neapolis given to it by Vespasian), stood in a remarkably fertile valley, between the ranges of Gerizim and Ebal, and on the edge of a wide plain. It carries off the palm for beauty of situation from all the towns of Palestine, and is not behind any in historical interest. Abraham first pitched his tent under the terebinths of Moreh, probably at the entrance of the glen. Jacob visited it on his return from Mesopotamia, and settled at Shalem, *Sâlim*, about two miles distant. He bought the "parcel of the field," and sunk the well, which passes by his name to the present day, about a mile and a half from the town—the scene of our Lord's conversation with the woman of Samaria. The adjacent heights of Ebal and Gerizim witnessed the proclamations of the curses and blessings of the Law. It was next the scene of Abimeloch's conspiracy and of the parable delivered by Jotham. At the division of the kingdoms Jeroboam established his government here, and after the return from Babylon it became the head-quarters of the sectarian worship of the Samaritans, who (about B.C. 420) erected a temple on the top of Gerizim. **Samaria**, which succeeded Shechem as capital, was situated six miles N.E. of it, on a steep flat-topped hill, which stands in a basin encircled with hills; the strength of its position was great, and it was well chosen by Omri as the site of his capital. It was besieged, but not taken, by the Syrians under Benhadad (1 Kings, xx.). It was, however, taken by the Assyrians (B.C. 720). Augustus gave it to Herod the Great, who restored it with the name of Sebaste, still preserved in the modern *Sebustieh*. Philip preached there, and it was the abode of Simon the Sorcerer. **Cæsarea**, the capital not only of Samaria but of Palestine under the Romans, stood on a rocky ledge running out into the Mediterranean, at a spot formerly known as Stratônis Turris. It was built by Herod the Great with a view to closer communication with Rome. It was successively visited by Philip, who took up his abode there—by Peter, at the time of Cornelius' baptism—and by Paul, on his journey to Rome. The road to Jerusalem followed the line of the plain through **Antipatris**, *Kefr Saba*—also built by Herod the Great, and noticed in Acts xxiii. 31—to Lydda, where it fell into the road from Joppa. The site of **Tirzah**, which preceded Samaria as a royal residence, is supposed to have been at *Tulluzah*, about seven miles E. of Samaria. The beauty of its situation was proverbial (Cant. vi. 4). **Shiloh**, *Seilûn*, stood on a plain just N. of the border of Benjamin. Its site does not present any natural features of interest, but it is connected with many of the events of Scripture. The tabernacle was first set up there, and Eli died there; it was also the abode of Ahijah the prophet. **Dothan**, or Dothain, "the two wells," near *Kir bâliyeh*, the fertile valley where the sons of Jacob fed their flocks, and the place where Elisha was so wonderfully delivered from the Syrians, was in the northern part of Samaria.

III. GALILEE.

§ 17. **Galilee** extended from the ridge of hills which bounds the plain of Esdraelon on the S. to the extreme N. of Palestine, and from the neighbourhood of the Mediterranean Sea in the W. to the Jordan and the Sea of Galilee in the E.: the sea-coast itself was

held by the Phœnicians. It was divided into two districts—Upper
and Lower Galilee—the former to the N., about Lebanon and Tyre,
distinguished as "Galilee of the Nations," and the latter to the S.
The name originally applied to a "circle" or "circuit" about
Kadesh, in which were the 20 cities presented by Solomon to Hiram :
it was thence extended to the whole district. It included the tribes
of Issachar, Asher, Zebulun, and Naphthali.

Sea of Galilee. (From a Sketch by Wm. Tipping, Esq.)

§ 18. Issachar occupied the fertile plain of Esdraelon, and the
adjacent parts from Carmel on the sea shore to the Jordan : it was
a "pleasant land," for the quiet possession of which Issachar con-
sented to forego political prominence, "bowing his shoulder to bear,
and becoming a servant to tribute" (Gen. xlix. 14, 15). The district abounds in spots of great interest : foremost among these is
Mount Carmel—a series of connected heights bounding for a distance
of 18 miles the plain of Esdraelon on the S., and terminating in a
bold promontory on the Mediterranean coast : its wooded dells and
park-like appearance justify its appellation of Carmel, "a park ;"
the western extremity is now crowned with a famous convent, and
the cliffs abound with caves naturally formed in the limestone,

K 2

which have been frequented by devotees in all ages. The extreme
eastern summit of the hill was the spot selected by Elijah for
the decisive trial between Jehovah and Baal, the memory of which
is preserved in the name of the spot, *el-Maharrakah*, "the burning."
At the foot of Carmel runs the river **Kishon**, *Mukutta*, which in
summer derives its whole supply of water from the sides of the
hill, but at other periods of the year flows throughout the whole
length of the plain, and sometimes with so violent a stream as to be
dangerous to ford; it was in this state when the hosts of Sisera
were swept away by it. The plain of **Esdraelon** runs across Palestine from the Mediterranean Sea to the Jordan in a south-easterly
direction, swelling out to the breadth of about 12 miles in its
central part, but contracting towards either extremity, and terminated towards the E. by the isolated heights of Gilboa, the so-
called Little Hermon, and Tabor: the valley of Jezreel, properly
so called (for the name under the Greek form of Esdraelon extended
over the whole plain), lies between the two former of these ridges,
and leads down to the valley of the Jordan. The plain itself is
remarkable for its fertility and for its adaptation to military movements, particularly those of cavalry and war-chariots; for the latter
reason it was the selected battle-field of the Canaanites under
Sisera against the Israelites—of the Philistines in their victorious
conflict with Saul—and of Josiah in his fatal engagement with
Pharaoh Necho. Its fertility led to frequent incursions from the
Arabian tribes, who sometimes settled there with their flocks and
herds: one such incursion is recorded in Judges vi. vii. in connexion
with the exploits of Gideon. The tribe of Issachar appears from
this cause to have been reduced to a semi-nomadic state, "rejoicing
in their tents" (Deut. xxxiii. 18). **Tabor**, *Tûr*, rises at the N.E.
angle of the plain to a height of 1400 feet above it—an isolated and
picturesque hill, its sides well clothed with herbage and wood, and
its summit crowned with an ancient town, which was in existence
in our Saviour's time—a circumstance subversive of the tradition
which assigns this as the scene of our Lord's transfiguration.
Mount **Gilboa**, *Jebel Fukua*, bounds the plain of Esdraelon on the
S.; it presents a strong contrast to Tabor by being entirely devoid of
wood. Between these two hills is a range, now named *Jebel-ed-Duhy*,
which has been unnecessarily identified with the "little hill of Hermon" in Ps. lxxxix. 12.

The chief town in this district was **Jezreel**, situated on a spur of
Gilboa, and commanding the central passage—"the valley of Jezreel"
—which leads down to Jordan. Jezreel was, under Ahab, the capital
of Samaria. **Bethshan** stood eastward, on the edge of the Jordan
valley, with its acropolis posted on an eminence. The Israelites never
succeeded in wresting it from its Canaanitish occupants, and on its walls
the bodies of Saul and his sons were exposed after the battle of Gilboa.

Its name was changed to **Scythopolis**, perhaps in consequence of the Scythian incursion into Asia, which occurred in the reign of Josiah. This has been again superseded by the old name in the form *Beisân*. On the northern slope of Little Hermon stood the village of **Nain**, where our Saviour raised the young man to life ; and somewhat to the E. was **Endor**, the scene of Saul's interview with the witch. **Megiddo** stood in the western portion of the plain of Esdraelon, and, though within the limits of Issachar, was assigned to Manasseh. It was in this portion of the plain that Josiah was defeated, the place of his death being named Hadad-rimmon in that neighbourhood. The name of Megiddo has been perpetuated in the form of **Armageddon**—"the mountain of Megiddo"—the prophetic scene of the final conflict between the powers of good and evil (Rev. xvi. 16).   •

§ 19. The tribe of **Zebulun** held the district adjacent to the western shore of the Sea of Galilee, and skirting the northern edge of the plain of Esdraelon : thus he is said in Scripture to "suck of the abundance of the seas" in reference to the former, and to "rejoice in his goings out" in reference to the latter (Deut. xxxiii. 18, 19). The hills of this district have a character distinct from the rest of Palestine ; just below their summits they have not unfrequently platforms or basins of size sufficient for the sites of towns ; and in such basins, and not on the very tops of the hills as elsewhere, most of the towns are found. The hills are well clothed with wood, and possess a fertile soil. In addition to this, the Sea or Galilee itself was a valuable possession : its waters afforded an easy means of communication, and at the same time were well supplied with fish. The western shore, well watered and enjoying a tropical heat from the depression of the lake, had a prolific vegetation ; and the "land of Gennesareth," i. e. the plain about the centre of the lake, was the richest spot in Palestine. But these natural features do not form the highest claim to our attention : these shores and waters are hallowed by their association with the ministry of our blessed Lord ; and hence, although the scenery of the lake is uninviting from the monotonous and dreary appearance of the surrounding hills, the Sea of Galilee always has been and will be beautified in the imagination of the Christian.

The chief town of this district in the New Testament period was **Tiberias**, situated at the southern extremity of the plain of Gennesareth, with some famous warm baths in its immediate neighbourhood. It was founded by Herod Antipas (about A.D. 16) and named after the emperor Tiberius : after the destruction of Jerusalem it became the metropolis of the Jewish race. The next important town was **Julias**, situated near the head of the lake on the left bank of the Jordan, and on the site of that **Bethsaida** near which our Lord fed the 5000 : it was built by Philip the tetrarch of Iturea, and named after Julia, the daughter of Augustus. Between these towns were several places of scriptural interest, the sites of which are not satisfactorily ascertained—**Chorazin**, *Tell Hûm*, near the N.E. angle of the lake—**Bethsaida**, *et-Tâbighah*, on a little bay farther down, the home

of the fishermen Peter and Andrew, Philip, James, and John, and the
scene of the miraculous draught of fishes: It must be distinguished
from the Bethsaida before mentioned—Capernaum, perhaps near the
fountain named *Ain et-Tin*, at the northern extremity of the "land
of Gennesareth," the scene of numerous interesting gospel events,
and the town in which our Lord dwelt, and hence called "His own
City;" the identification of its site is more than usually uncertain—
and Magdala, at present the only inhabited spot in the plain of
Gennesareth, the abode of Mary Magdalene. A short distance from the
lake, near Tiberias, is a low ridge, terminating in two points, and hence
named *Kurun Hattin*, "the horns of Hattin." It is the reputed scene
of the delivery of the Sermon on the Mount, and is hence known as
"the Mount of Beatitudes." Nazareth, the early abode of our blessed
Lord, is situated high up on a hill on the northern edge of the plain
of Esdraelon, in one of those basins which we have already described.
It is encircled by a series of rounded hills, one of which, on the N.,
rises to a height of some 400 feet, and is perhaps the hill whence the
inhabitants threatened to precipitate our Saviour. Cana, associated
with our Lord's first miracle, stands considerably to the N. of Nazareth
at *Kana el Jelil*. Sepphoris, to the N.W. of Nazareth, was the strongest
city of Galilee in the Roman age: its name was changed to Diocæsarea
by Antoninus Pius.

§ 20. The tribe of Naphthali occupied the western half of the
valley of the Jordan from the Sea of Galilee to its source, together
with a portion of the central hilly region: their district was remote,
and little frequented, but rich, and remarkably well wooded, con-
firming the prediction that Naphthali should be "full with the
blessing of the Lord" (Deut. xxxiii. 23).

The places of interest in this district are—*Safed*, remarkably situated on an isolated peak, and reputed to be the "city set upon an
hill" to which our Saviour alludes (Matt. v. 14); Kedesh-Naphthali,
W. of Lake Merom, the city of refuge for the northern tribes, and the
birthplace of Barak; Dan, situated in the upper valley of the Jordan,
and the most northerly town of Palestine; it was originally a Phoenician colony named Laish, but was seized by the Danites and its name
changed; and, lastly, Cæsarea Philippi, which, though perhaps not
strictly within the limits of Naphthali, must yet be regarded as a town
of Galilee: it was most beautifully situated at the base of Hermon, near
one of the sources of the Jordan. Herod the Great first erected a
splendid temple here in honour of Cæsar Augustus, and Philip the
tetrarch enlarged the place, and named it, in honour of Tiberius,
Cæsarea, with the addition of Philippi to distinguish it from the
other on the Mediterranean coast.

§ 21. The tribe of Asher received the maritime district parallel to
Naphthali, commencing near Tyre and terminating at Carmel.
The whole of this was fertile, and some portions preeminently so:
Asher "dipped his foot in oil," and his "bread was fat" (Deut.
xxxiii. 24; Gen. xlix. 20). The natural capacities of the region
were thus great: its position, commanding all access to Palestine
from the N., and possessing the only good harbour on the coast,
gave it additional importance; but Asher was unable to expel

the Phœnicians from the eligible sites on the coast, and so fell back into a state of inglorious ease. The history of its towns wholly belongs to Phœnicia.

Rabbath-Ammon (Philadelphia).

IV. PERÆA.

§ 22. Peræa was, as its name implies, the land "on the other side of" Jordan, and sometimes included the whole district, but more properly a portion of it, extending from the river Arnon in the S. to the Hieromax in the N., and from the Jordan to the edge of the Syrian desert. This region presents a striking contrast to western Palestine ; it consists of high undulating downs, which commence with the edge of the lofty ridge bounding the valley of the Jordan, and thence gradually slope off to the desert : in some places trees are but thinly scattered over the country, but in the northern district there are still extensive forests of oak and terebinth. The scenery of the district between Mount Gilead and the Jabbok is described as highly picturesque and park-like. Its extensive pasture-grounds have in all ages sustained a large quantity of sheep and cattle, and on this account Reuben and Gad selected this as their abode. The country is well watered, but the only rivers of

importance are the **Hieromax**, *Sheriat el-Mandhur*, in the N., which
rises in the mountains of *Hauran*, and joins the Jordan a little
below the Sea of Galilee—the Jabbok, *Zurka*, which rises on the
borders of the desert, where it receives the river of Ammon, and
flows in a deeply-sunk channel into the Jordan, forming in ancient
times the boundary between the territories of Sihon and Og, the
two kings of the Amorites, and afterwards between Gad and
Manasseh—and the Arnon, *Mojib*, which separated at one time the
kingdoms of the Moabites and Amorites, and afterwards formed the
southern limit of Palestine in this part; it is a stream of no great
size, discharging itself into the Dead Sea through a deep cleft.

This district was occupied by the tribes of Reuben and Gad, and
partly by the half-tribe of Manasseh. The precise limits of their
various districts cannot be very well defined ; for these tribes led a
pastoral, nomadic life, shifting their quarters from time to time, and
intermixing probably with each other, and with the older inhabitants of the district : their positions may be generally described as
follows :—**Reuben** to the S. from the Arnon to the head of the
Dead Sea : **Gad**, thence to the Jabbok : and half-**Manasseh**, N. of
the Jabbok.

Gadara. (From a Sketch by Wm. Tipping, Esq.)

. The towns in Peræa were neither numerous nor important. **Heshbon** ranked as the capital of Sihon, one of the kings of the Amorites. It stood E. of the head of the Dead Sea, on a slight elevation above the rest of the plateau; it is now an entire ruin. The remains of a reservoir may represent "the fishpools in Heshbon" which Solomon notices (Cant. vii. 4). **Jazer**, where Sihon was defeated, was somewhere to the S.; and in the same direction was **Baal-meon**, "the habitation of Baal," with a high peak near it, whence perhaps Balaam viewed the people of Israel. This may also have been the height whence Moses viewed the promised land. **Rabbath-Ammon**, the capital of the Ammonites, stood on both sides of a small stream tributary to the Jabbok, and is hence described as the "city of the waters," in contradistinction to the citadel, which stood high up on an isolated hill: it was known as **Philadelphia** in the Roman era, having been rebuilt by Ptolemy Philadelphus in the 3rd century B.C.: on its site are remains that testify to its importance, particularly a very large theatre; it is now the haunt of jackals and vultures (comp. Ez. xxv. 5). **Ramoth-Gilead** probably stood on the site of the modern *es-Salt*, on an isolated hill forming one of the heights of Mount Gilead: the modern name represents the ecclesiastical *Salton*, and is also applied to the neighbouring mountain. Ramoth-Gilead was one of the cities of refuge: having been captured by the Syrians, it was unsuccessfully attacked by Ahab and Jehoshaphat, and again by Joram and Ahaziah. **Gerasa** was an important town N. of the Jabbok, situated in a valley leading down to that river. It is first noticed by Josephus as having been taken by Alexander Jannæus, and it afterwards formed the chief town of the Decapolis, or confederacy of ten cities, formed in this district. It was burnt by the Jews at the commencement of the Roman war, and again by Vespasian; but it was afterwards rebuilt with great splendour, and subsequently adorned by the Antonines: the ruins of the theatre, the forum, the temple of the sun, and many other buildings still remain. **Jabesh-Gilead** is supposed to have stood somewhat S.E. of Pella, where there is a valley named *Wady Yábes*. It is noticed in connexion with the war against the Benjamites, and with the threatened cruelty of Nahash. **Pella**, *Fahil*, stood on a small plain or terrace of the mountains of Gilead, overlooking the valley of the Jordan, at an elevation of some 1000 feet; the connexion of its name with the Macedonian Pella is doubtful. The first historical notice is its capture by Antiochus B.C. 218, but it owes its chief interest to its having been the asylum of the Christians at the time of the destruction of Jerusalem. **Gadara**, *Um-Keis*, stood on a spur of Gilead, just S. of the Hieromax, and possessed numerous edifices of the Roman era, among which the remains of two theatres are the most conspicuous: numerous tombs are excavated out of the limestone rock, and in these a troglodyte population still exists, living as the demoniacs of the Gospel age (Matt. viii. 28). **Gergesa**, which is noticed in the passage just quoted, was probably a village in the territory of Gadara. Gadara was taken by Antiochus (B.C. 218) and by Alexander Jannæus (about B.C. 198): it was destroyed in the civil wars, but rebuilt by Pompey, and became under Gabinius the principal town in Peræa. **Mahanaim** is supposed to have stood N. of Gerasa, where there is a place still called *Mahneh*: it derived its name from the "two hosts" of angels who appeared to Jacob, and was the place where Ishbosheth was crowned. In the neighbourhood was fought the battle in which Absalom perished.

§ 23. The territory of **Moab** may be included in our review of this part of Palestine: it lay S. of the Arnon, and eastward of the Dead Sea at its southern extremity—now a bleak and desolate region, but in earlier times very possibly of a more inviting character. The Israelites traversed it in their journey from Egypt, and it is of further interest as the native land of Ruth, and the refuge of David.

The capital of this district was named **Ar Moab**, or **Rabbath Moab**, and at a later period **Areopolis**. It stood some distance S. of the Arnon, on a low hill: under the Romans it was the metropolis of Palæstina Tertia until its destruction by an earthquake, A.D. 315. **Kir-Moab** was more to the S., on the top of a hill about 3000 feet above the Dead Sea, and surrounded by mountains. It was the only town which Joram failed to take. In the ravine that leads hence to the Dead Sea was **Zoar**, the "little city" where Lot took refuge.

§ 24. To the N. of the Hieromax, the plateau of **Bashan** stretches from the valley of the Jordan and the Sea of Galilee far away to the eastward until it meets with a chain of hills, named by classical writers **Alsadámus**. This extensive district formed the ancient kingdom of Bashan, far-famed for its rich pastures and fine forests, whence the expressions proverbial among the Hebrews, "bulls of Bashan," and "oaks of Bashan." It consists of several distinct tracts: (i.) The portion of the country lying to the N.W. of Alsadamus, which is remarkably wild and rocky, abounding with every variety of cliff, gully, and ravine, and hence termed by the Hebrews **Argob**, "rocky," by the Greeks **Trachonitis**, and by the Arabs *Lejah*, "retreat," in reference to its inaccessible character. (ii.) The hills of Bashan themselves, which, though stony, are fertile. (iii.) The wide plain between these and the Jordan, which possesses a remarkably rich soil, and is the district so much praised by the Hebrews. (iv.) The mountainous district about the ridge of Hermon. These formed separate regions in the time of our Saviour, viz. **Batanæa**, in the S.E., about the ranges of Alsadamus, representing the Hebrew name Bashan; **Auranitis**, about the upper valleys of the Hieromax, a name still preserved in the modern *Hauran*; **Trachonitis** to the N.E.; **Ituræa** in the N.W., about the roots of Hermon, named after Jetur, a son of Ishmael, and still called *Jedúr*; and **Gaulonitis**, *Jaulán*, between Hermon and the upper course of the Jordan.

§ 25. The whole of this district was at one time thickly studded with towns: in Argob alone "threescore great cities, besides a great many unwalled towns," are said to have existed (Deut. iii. 4, 5), and the remains everywhere visible render this number not improbable. Many of these remains are in a state of high preservation, being built of large blocks of black basalt, which neither time nor the hand of man have been able to displace. The towns may be classified as belonging to two wholly distinct periods, which we may term the Biblical and the Roman: the remains in many instances show that the Romans adopted the old cities.

Bosrah (Bostra).

(1.) The towns belonging to the Biblical era.—**Edrei**, *Edhra*, strongly situated on the border of Argob, was the scene of the defeat of Og, king of Bashan. It has sometimes been identified with *Dera*, or *Edraha*, a good deal more to the S. **Ashtaroth**, the other of the capitals, named after the patron deity Astarte or Venus, and sometimes hence called Ashtaroth *Carnaim*, " of the two horns " (Gen. xiv. 5), was situated not far from Edrei. Its site has not been satisfactorily made out; it has been identified sometimes with *Ashareh* on one of the branches of the Hieromax. **Kenath**, the Canatha of the early geographers, was situated among the hills of Alsadamus, and is also noticed under the name of Nobah, after its conqueror (Num. xxxii. 42; Judg. viii. 11; 1 Chr. ii. 23). The remains of the town are numerous, consisting of a theatre, a hippodrome, mausoleums, a peripteral temple, and other objects of Greek architecture. **Salcah**, *Sulkhad*, at the S.E. end of the range, and the farthest town in the kingdom of Bashan, possessed a citadel situated on a conical hill. Numerous inscriptions of the Roman period exist, and the remains of vineyards and groves of fig-trees testify to the former prosperity of the place. **Kerioth**, *Kureiyeh*, stood at the S.W. end of *Jebel Haurân*; its remains bear a cyclopean character; inscriptions have been found bearing date A.D.

140, 296: it is noticed by the prophets (Jer. xlviii. 24; Am. ii. 2). **Bozrah** of the Moabites, the **Bostra** of the Romans, now *Busrah*, was on a large and fertile plain S.W. of the range of hills: it is noticed by Jeremiah (xlviii. 24) among the cities of the Moabites, and in 1 Macc. v. 26, as having been taken by Judas. Trajan constituted it the capital of eastern Palestine with the title **Nova Trajana Bostra**, and the year in which this was done (A.D. 106) was the commencement of the Bostrian era observed in these parts. Bostra was raised to the dignity of a colony by Alexander Severus (about A.D. 230): after the introduction of Christianity it became the seat of a primacy, with thirty-three subject bishoprics. The ruins are very extensive and handsome, consisting of a theatre, temple, triumphal arch, and many other monuments.

(2.) The towns belonging exclusively to the Roman era were—**Phæno**, *Musmeih*, the capital of Trachonitis, due S. of Damascus: the beautiful ruins of a temple (bearing date about A.D. 165) and other public buildings remain—**Batanæa**, on the northern declivity of *Jebel Hauran*, noticed by early Arab authors, with numerous Greek remains—**Suæma**, noticed by Ptolemy, in the hill-country, with the ruins of large churches (bearing date A.D. 369, 416) and other buildings—**Neapolis**, to the S., with Greek remains and inscriptions—and **Philippopolis**, *Ormân*, near the S.E. extremity of the range, founded by Philip the Arabian on his election to the empire A.D. 244.

(3.) In addition to these are the remains of numerous towns, of which the modern names alone are known, such as *Hit*, with buildings of about the 2nd century—*Shuhba*, perhaps the same as Dionysias, with a Roman gateway, numerous Greek inscriptions (dates about A.D. 165, 248), and some fine temples—*Suweideh* in *Jebel Hauran*, with most extensive ruins and inscriptions (dates A.D. 103, 135, 196): it is still the chief town in this district; and *Hebrân*, near the southern end of the range, with a temple bearing date A.D. 155.

§ 20. The history of Palestine as an independent state commences with the Exodus from Egypt, and terminates with the Babylonish captivity. It may be divided into three periods, viz. the Judges, the United Kingdom, and the Divided Kingdom.

(1.) *The Judges.*—Under the Judges the Israelites were chiefly engaged in protecting themselves against the attacks of the neighbouring nations—the Philistines, Canaanites of Hazor, Midianites, Amalekites, and Ammonites. The only distant people with whom they came in contact were the Mesopotamians under Chushan-rishathaim. The tribes during this period lived under their own elders, without any bond of political union: in time of war they had their special leaders or judges, who were sometimes elected (Judg. iv. 6, xi. 5), and at other times assumed the office (iii. 9, 15, 31, x. 1, 3). The office of Judge, in the proper sense of the term, originated with Eli, with the exception of Deborah, who also held the office of prophetess (Judg. iv. 4).

(2). *The United Kingdom.*—Under the earliest king, Saul, the border warfare was sustained by the Ammonites, Philistines, and Amalekites, and the boundaries of the empire did not advance; but under his successor David the addition of the territories of Hadadezer king of Zobah, and Hadad king of Damascus, carried the boundary to the Euphrates; while the defeat of the Edomites in the S. by Abishai, one of David's generals, secured the route to the Dead Sea and prepared the way for the commerce afterwards carried on by the Red Sea. His border was effectually secured by the defeat of the Ammonites. The alliance with Hiram king of Tyre, which was commenced by David, was another important step. Under Solomon the Jewish state reached the climax of its greatness; he extended his relations with foreign nations by his alliance with the sovereign of Egypt, and by the commercial intercourse which he carried on with that country: he continued the alliance with Hiram, king of Tyre, and was thus enabled to carry on trade with the distant coasts of Arabia, Africa, and India. The extent of his dominions was from Phœnicia in the N. to the Red Sea in the S., and from the river of Egypt to the Euphrates. Within his own territories the Canaanites were reduced to bondsmen, and on his border the Philistines, Edomites, Moabites, Ammonites, Syrians, and even some of the Arab tribes, yielded a peaceable subjection. Before the termination of his reign, however, the kingdom showed symptoms of decline. Damascus was again raised to an independent position under Rezin. On the other side he was pressed by Hadad, one of the royal family of Edom, who obtained an independent position on his border, while inward disaffection broke out under Jeroboam.

(3.) *The Divided Kingdom.*—On the death of Solomon a disruption of the tribes took place, ten of them combining to form the northern kingdom of Israel, while the remaining two, Judah and Benjamin, formed the southern kingdom of Judah. The latter, though smaller in point of extent, had a counterpoise in the possession of the capital, Jerusalem, and in the compactness of its territory. Israel was, moreover, peculiarly open to the encroachments of the eastern empires, no barrier being interposed between the trans-Jordanic district and the desert, while the heart of the country might be reached from the north by the "entering in of Hamath" between the ranges of Libanus and Antilibanus. Judah, on the other hand, was accessible only on the side of Egypt. Hence, as we might have expected, the former kingdom was the first to succumb beneath the growing influence of Assyria.

The kingdom of Judah lasted from B.C. 975 to B.C. 588, under 20 kings; that of Israel from B.C. 975 to B.C. 721, under 19 kings. The capital of the former was Jerusalem, of the latter Shechem, and after the accession of Omri, Samaria. The history of these kingdoms consists of a constant succession of wars, either among themselves or with the powerful nations on either side of them. Into the details of these wars it is unnecessary for us to enter, as they did not affect the territorial divisions of Palestine until the final extinction of the kingdoms. Israel was incorporated with the Assyrian empire, and at the dissolution of that empire passed, with the remainder of the western provinces, into the hands of the Babylonians. Judah, though occasionally reduced to subjection by the Assyrians, was not totally subdued until after the establishment of the Babylonian empire.

Palestine remained an integral portion, first of the Babylonian, and afterwards of the Persian empire. In the reign of Cyrus the Jews were restored to their native land (B.C. 525), and the Temple was rebuilt; commissions were issued to Ezra under Artaxerxes I. (B.C. 457) and Nehemiah (B.C. 445) for the completion of the works necessary to the re-establishment of the Jewish polity. The conquest of Palestine by Alexander the Great, and the subversion of the Persian empire, led to disastrous results. Palestine was for a lengthened period the debateable ground between the monarchies of Syria and Egypt. Annexed in the first instance to Syria (B.C. 323), it was conquered by

Ptolemy (B.C. 312), and it remained a portion of the Egyptian dominion
from B.C. 301 to B.C. 203. The Jews then sought the assistance of the
Seleucidæ, and a succession of struggles for independence followed,
under the leadership of the Maccabees, terminating in the establishment of an independent dynasty under John Hyrcanus (B.C. 130). The
disputes which disgraced his successors ultimately opened the way for
the interference of Pompey (B.C. 63), and Judæa became henceforth
dependent upon Rome. Antipater, an Idumæan, was appointed procurator by the influence of Julius Cæsar (B.C. 48); and his second son
Herod was elevated to the dignity of king of Judæa (B.C. 38), and afterwards of the whole of Palestine and Idumæa (B.C. 31). On the death
(B.C. 4) of this Herod—distinguished as "the Great"—the kingdom was
divided into three portions, Archelaus receiving Judæa, Samaria, and
Idumæa; Philip, Galilee, with the title of Tetrarch; and Antipas, Trachonitis, Batanæa, and Ituræa. These districts were again consolidated
into one kingdom under Herod Agrippa (A.D. 41) and his son Agrippa
II.; but the Roman authority was really paramount, and the Jews
suffered severely from the rapacity of the governors imposed upon
them. A fierce struggle ensued, terminating in the destruction of
Jerusalem under Titus (A.D. 70), and in the extinction of the national
existence of the Jews.

Roman Remains in the South Wall of Haram at Jerusalem. (From a Sketch by
Wm. Tipping, Esq.)

Temple of Birs-Nimrod at Borsippa

CHAPTER XII.

MESOPOTAMIA, BABYLONIA, ASSYRIA, ARMENIA, &c

I. MESOPOTAMIA.

§ 1. **Mesopotamia** was bounded on the N. by Mons Masius, separating it from Armenia, on the E. by the Tigris, on the W. by the Euphrates, and on the S. by the Median Wall, separating it from Babylonia. It consists for the most part of an immense plain, broken only in one place by the range of **Singaras**, *Sinjar*, which crosses it for a considerable distance towards the S.W. in the latitude of Nineveh. The plain affords excellent pasturage during the spring and early summer months, but afterwards becomes parched up in the absence of artificial irrigation. Hence in modern times it presents, at one period, the most rich and delightful aspect, luxuriant with grass, and enamelled with flowers, at another period the appearance of an arid barren wilderness. In ancient times the remains of cities prove that it was more densely populated, and . better cared for than at present. Timber was both abundant and of fine growth, so much so that the emperors Trajan and Severus built fleets on the banks of the Euphrates. Among its special products may be noticed naphtha, amomum, and *gangitis*, probably a kind of anthracite coal. The remote districts were the haunts of the lion, the wild ass, and the gazelle.

Name.—Mesopotamia is derived from the Greek words μέσος, ποταμός, expressive of its position between the Tigris and Euphrates; it thus closely corresponds with the Hebrew designation *Aram-naharaim*, "Aram of the two rivers," and the modern Arabic *Al-Jezireh*, "the island." The name Mesopotamia is of comparatively recent introduction, not appearing either in Herodotus or Xenophon: this district was probably first recognized by a special name about the time of Alexander the Great.

§ 2. The most important mountain-range is **Masius**, which skirts the N. boundary, and throws out numerous spurs towards the S., imparting a hilly broken character to the northern district: **Singaras** may be regarded as a distant offset of this chain. The chief rivers are the **Tigris** and the **Euphrates**, from which the country derives its name: these have been already noticed, as skirting the borders of the plain. The rivers which traverse the plain are for the most part tributaries of the Euphrates: the most important is the **Chaboras**, *Khabûr*, which rises in Masius, and after a course first towards the S.E. and then towards the S.W. joins the Euphrates at Circesium: at the point where its course changes it receives several tributaries, particularly the **Mygdonius** from Nisibis. The **Bilecha** or **Bilas**, *Belikke*, flows through the N.W. of the district, and joins near Callinicum: on its banks the army of Crassus first encountered the Parthians.

§ 3. Under the Romans the country was divided into two parts —**Osrhoëne** to the W., and **Mygdonia** to the E. of the Chaboras: the former was so named after Osrhoës, an Arabian chief who established himself there in the time of the Seleucidæ. The inhabitants were a Semitic race—a branch of the Aramaic family which extended over Syria. The towns lined the banks of the Euphrates and Tigris, and were thickly strewed over the plain at the foot of the Masian range. We know singularly little of them, and the few particulars recorded belong almost wholly to the period of the Roman empire, when Mesopotamia became a battle-field against the Parthians. The openness of the country and its liability to sweeping invasions may very much account for this: towns rose and fell without any record of their existence. Some, as Corsôte, were in ruins in Xenophon's time; others, as Carmande, were large and prosperous, and yet are never heard of again; while others, like the Cænæ which he notices, are known only by the stupendous mounds under which they are buried.

The most important town in Osrhoëne was **Edessa**, situated on the Scirtus, a tributary of the Bilissus, and otherwise named **Antiochia Callirhoes**, from a fountain of that name: it was probably built by Antigonus, though a much earlier date has been assigned to it, and it has even been identified with the Scriptural Ur: Edessa became in Christian times the seat of a famous theological school). **Nisibis**, the capital of Mygdonia, stood on the Mygdonius, near the base of the Masian range: it was also reputed a town of great antiquity, and probably was so, though not to be identified with any Scriptural town: it is first noticed by Polybius under the name of Antiochia Mygdoniæ; it figures frequently in the wars between the Romans and Parthians, and remained an outpost of the Roman empire to a late date. **Carrhæ**,[1] on a branch of the Belias, was an old town of commercial importance: the same character, though in a higher degree, attached to **Batnæ**, which stood between Carrhæ and the Euphrates, and was the scene of an annual fair of great importance: it was fortified by Justinian. **Apamea**, on the Euphrates, was built by Seleucus opposite Zeugma for the defence of the bridge of boats. **Nicephorium**, lower down the river, was probably founded by Seleucus I., though by some writers attributed to Alexander the Great. **Circesium**, at the junction of the Chaboras, is noticed by Procopius as the φρούριον ἔσχατον of the Romans in his day. **Is**, near the Babylonian frontier, represents the modern *Hit*. **Singara**, near the eastern end of the range of the same name, appears to have been the chief town in the central district: it was the scene of several conflicts in the Eastern wars of the Romans, and particularly of one between Constantius II. and Sapor. **Atra** or **Hatra**, near the Tigris, to the S.E. of Singara, is described as a place of great strength, which held out successfully against Trajan and Septimius Severus: extensive ruins of it still remain under the name of *Al Hathr*.

Of the less important towns we may notice—**Anthemusia**, between the Euphrates and Edessa—**Rhesæna**, *Ras-al-Ain*, near the sources of the Chaboras, afterwards named **Theodosiopolis**, probably as having been rebuilt by Theodosius—**Constantia** between Nisibis and Charræ—Ichnæ, a fortified town or castle on the Bilecha—and **Dura**, near Circesium, the place where a military monument to Gordian was erected.

History.—In early times, Mesopotamia formed a portion of the great Eastern monarchies of Assyria, Media, and Persia. The authority exercised by those powers was of a very lax and indefinite character,

---

[1] Crassus took refuge at Carrhæ after his defeat by the Parthians.
　　　　—— sic, ubi supra
　Arma ducum dirimens miserando funere Crassus
　Assyrias Latio maculavit sanguine Carras,
　Parthica Romanos solverunt damna furores.—Luc. i. 103.

and in all probability the western district, adjacent to the Euphrates, was practically independent. The Assyrian inscriptions make mention of the *Nairi*, as a tribe in that part with which the monarchs were frequently at war. The history of these wars and of the heroes who conducted them is, however, sunk in oblivion: nor do we hear of any conqueror ever issuing from this country, with the exception of Chushan-rishathaim, noticed in the Bible (Judg. iii. 8) as having held Israel in subjection for eight years: his name, "Chushan of the double aggression," seems to bespeak a chieftain versed in the practices of border warfare. The Seleucidæ extended their sway over the northern part of Mesopotamia more particularly, and nominally over the whole of it. Trajan conquered it, but Hadrian relinquished possession of it. It was again conquered under M. Aurelius, but after repeated struggles the greater part was given up to the Persians by Jovian. A.D. 363.

View of Babil from the West.

II. BABYLONIA.

§ 4. **Babylonia** was bounded on the N. by the Median Wall, on the E. by the Tigris, on the S. by the Persian Gulf, and on the W. and S.W. by the Arabian desert. The natural limit on the N. was formed by the approximation of the rivers Tigris and Euphrates to each other. The name was sometimes, however, extended over the whole of Mesopotamia. Babylonia consists of an almost unbroken plain, which in early times under a system of skilful irrigation possessed the very highest fertility, but which at present is for the most part a barren and desolate wilderness. Its soil was well fitted for the growth of cereals, and among the other productions for which the country was famous in ancient times we may notice—the date-palm, sesamum, and asphalt.

§ 5. There are no hills in Babylonia: nor are there any rivers

except the two great border streams of the Euphrates and the Tigris,
which have been already described. Artificial works take the place
of natural features : a network of canals conducted the fructifying
waters of the rivers over the face of the country, and presented,
next to the rivers themselves, the most striking objects in its general
aspect. Of these, four are described by Xenophon (*Anab.* i. 7, § 15)
as crossing from the Tigris to the Euphrates, each sufficiently large
to convey a corn vessel : the longest, named *Nahr-Malcha,* "the
king's canal," entered the Tigris near Seleucia, and was ascribed by
Herodotus (i. 185) to Nitocris. In addition to these, there were two
very important canals on the W. of the Euphrates, designed apparently to regulate the flow of the river, and to prevent it from over-
flowing its banks : the first, named **Maarsares**, left the river above
Babylon, and terminated in a marsh some distance to the S. ; the
second, **Pallacōpas**, commenced about 75 miles S. of Babylon, and
joined the Persian Gulf at Terēdon. Numerous marshes lay along
the courses of these canals W. of the Euphrates, commencing immediately below Babylon. We must also notice the Median Wall
of Xenophon (*Anab.* ii. 4, § 12), which crossed between the rivers
in a north-easterly direction, coming upon the Tigris about 35 miles
above *Baghdad.*

§ 6. The earliest occupants of this country in historical times
were a Cushite or Hamitic race. The name of Cush (which was
more generally restricted to the Ethiopians of Africa). appears in
Asia under the forms Cossæi, Cissia, and Susiana ; Nimrod, the
reputed founder of Babylon, is described in the Mosaic genealogy as
the son of Cush (Gen. x. 8). The indigenous appellation of this
race seems to have been *Akkad,* and its dominant tribe appears
under the familiar name of "Chaldees," or *Kaldaï,* as they are called
in the Assyrian Inscriptions. The wide extension of the name of
Chaldees to the very borders of Armenia seems to imply that at one
period this race had spread over the whole of Mesopotamia. This
original Hamitic race was either superseded by, or, perhaps we
should rather say, was developed into the Semitic race, which
issued hence along the courses of the Tigris and Euphrates northwards, and across the Arabian desert westwards to the shores of the
Mediterranean. Probably, a Scythic or Turanian element was
superadded, representing a still earlier aboriginal population ; this
may be represented by "the nations" noticed in conjunction with
the Hamitic Shinar and the Semitic Elam (Gen. xiv. 1).

§ 7. Babylonia was not parcelled out into any systematic arrangement of provinces or districts, but certain portions of the
plain received special designations, as **Chaldæa**, the position of which
has been described (p. 12) ; **Mesēna**, about the head of the Persian
Gulf, and a second district of the same name in the N., probably at

the point where the Euphrates and Tigris approach each other most
nearly ; **Auranitis**, and **Amordocia**, on the right bank of the Euphrates.

The towns of Babylonia belong to three distinct periods : (i) the
ancient capitals whose history is unknown, except in so far as
the ruins themselves declare it ; (ii) the historical towns erected
during the flourishing period of the Babylonian empire ; and (iii)
those subsequently built by the Seleucidæ for commercial objects,
and which continued to exist under the Roman empire as border
fortresses.   The sites of the first class are marked by those wonderful mounds which rise so conspicuously out of the plain, and of
which the *Birs-i-Nimrúd*, near Babylon, *Akkerkuf*, near *Baghdad*,

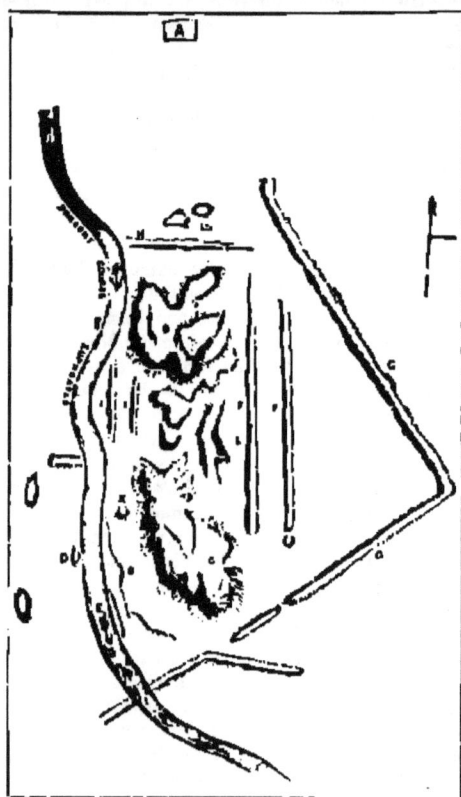

Plan of the Ruins of Babylon.

Niffer, in the central plain, *Warka* and *Senkereh*, about the marshes of the Euphrates, and *Mugheir*, on the western side of that river, besides many others which might be enumerated, are still in existence. Some of these have been identified with the old Biblical capitals of the land of Shinar; of others, even the names are unrecorded in history; but may yet be deciphered from the monograms on the bricks. These cities perished at a very early period, and were in many cases converted into the abodes of the dead, being used as Necropolises by the succeeding towns: this is the case particularly at *Warka* and *Niffer*, where coffins are piled up tier on tier in prodigious numbers. In the second class may be placed the famed capital of Babylon, and its suburb Borsippa. In the third class, Seleucia on the Tigris, Apamêa, Charax Spasinu, and others.

Babylon stood on both sides of the Euphrates, near the modern *Hillah*. Its size was enormous: Herodotus estimates the circuit of the walls at 480 stades, and Ctesias at 360: there appear to have been two walls; and the discrepancy between these writers may be explained on the ground that the former refers to the *outer*, and the latter to the *inner* wall. Even the lowest of these computations would imply an area of above 100 square miles, or nearly five times the size of London. The height of the walls[1] was no less remarkable; according to Herodotus, 200 royal cubits or 337½ feet, nearly the height of the dome of St. Paul's, and their thickness 50 royal cubits or 85 feet. It was entered by a hundred gates of brass, and protected by 250 towers. The more remarkable buildings were—the ancient temple of Belus, represented by the mound of *Babil* (A), an oblong mass about 140 feet high, 600 long, and 420 broad—the palace of Nebuchadnezzar, identified with the mound of the *Kasr* (B), an irregular square of about 700 yards each side—a more ancient palace, contained in the mound of *Amram* (C), more to the S.—and another palace, the "lesser" one of Ctesias, the ruins of which (DD) exist on both sides of the river. There are also remains of an enclosure in two parallel mounds (FF), probably a reservoir. The present remains are almost wholly on the left bank of the river, which has probably changed its course, and formerly ran between the two ridges marked II. The hanging gardens formed one of the greatest ornaments of Babylon. The lines GG are the remains of one of the walls. About six miles to the S.W. of Babylon was Borsippa, represented by *Birs-Nimrúd*, where a mound of a pyramidal form, built up in a series of seven stages to a height of 153 feet, is crowned by the remains of the temple of Nebo: it was

---

[1] The construction of these walls was commonly attributed to Semiramis :—

ὅτη κλατὸ τεῖχος

'Ασφάλτῳ θέσασα Σεμίραμις ἐμβασίλευεν.—THEOCR. Idyl. xvi. 99.

ubi dicitur altam

Cœtilibus muris clauisse Semiramis urbem.—Ov. *Met.* iv. 57.

Perarum statuit Babylona Semiramis urbem,
  Ut solidam coeto tolleret aggere opus;
Et duo in adversum mitti per mœnia currus,
  Ne possent tacto stringere ab axe latus.
Duxit et Euphraten medium, quam condidit, arci,
  Jussit et imperio surgere Bactra caput.—PROPERT. iii. 11, 21.

erected by Nebuchadnezzar, and has been erroneously identified with the ' Tower of Babel ' (Gen. xi. 4).

View of the Kasr, or ancient Palace of Nebuchadnezzar.

The early history of Babylon is involved in much obscurity: it was not the original capital of the country, and its existence is not carried back by historical evidence to a period anterior to the 15th century B.C., when it is noticed in an Egyptian inscription. The earliest notice in the Bible occurs in the reign of Hezekiah, B C. 712. At that time it was ruled by its own king; but generally speaking it was subject to the kings of Nineveh during the period of Assyrian ascendancy. After the fall of Nineveh it rose to be the head of a mighty empire, and was enlarged and adorned by Nebuchadnezzar. It was taken by Cyrus, B.C. 538, who regularly resided there for a certain period of the year: the fortifications were destroyed by Darius Hystaspis, and the temple of Belus by Xerxes. Babylon retained its position until the time of Alexander the Great, but soon afterwards sunk into insignificance through the erection of Seleucia on the banks of the Tigris, B.C. 322.

Seleucia, on the Tigris, near the junction of the Nahr-malcha, was erected by Seleucus Nicator with materials brought from Babylon, and became a place of great commercial importance: it was ruined in the wars between the Romans and Parthians. Not far from it was Coche, a place of military strength in the later days of the Roman empire. Perisabor was a very strong post on the Euphrates, perhaps at Anbar: it is noticed in the history of Julian's wars. Cunaxa, the scene of the battle between Cyrus and Artaxerxes, B.C. 401, was situated in the midst of the canal district, near the Euphrates. Orchoë on the borders of the Arabian desert, W. of the Euphrates, was the principal seat of the Orchêni, a people who obtained celebrity both as an astronomical sect, and for their hydraulic skill.

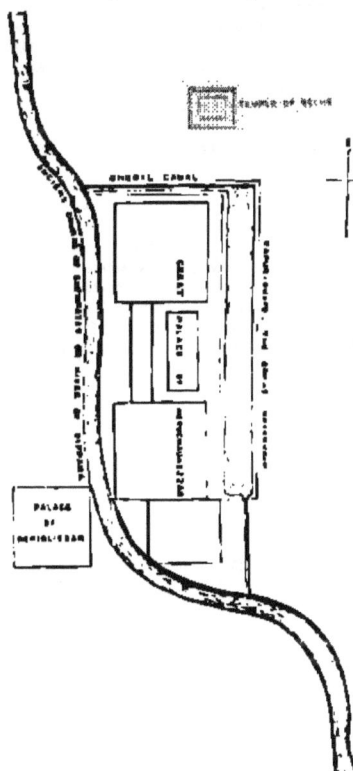

Portions of Ancient Babylon distinguishable in the present Buins.

Apaméa, described as being in Mesene, is of doubtful position. Several towns stood about the shores of the Persian Gulf, whose sites cannot be identified in consequence of the great change that has taken place in the coast: among these we may notice — Ampe, whither the Milesians were transported by Darius, B.C. 494—Apologi Vicus, a considerable place of trade, probably at *Old Bosrah*—Charax Spasinu, near the mouth of the Tigris, founded by Alexander the Great with the name Alexandria, restored by Antiochus Epiphanes with the name of Antiochia, and occupied by Spasines, an Arab chieftain, after whom it received its *agnomen* of Spasini; it was a place of considerable trade —and Teredon, at the mouth of the Pasitigris.

History of the Babylonian Empire.—Babylon remained sunk in comparative insignificance throughout the whole period of Assyrian supremacy. It had nevertheless its own monarchs, with whom the Assyrians frequently carried on war. The era of Nabonassar, B.C. 747, seems to mark a political change, but its nature is uncertain. One of his successors, *Mardoc-empadus*, is undoubtedly the Merodach-baladan of Scripture, who sent ambassadors to Hezekiah: he was expelled from his throne by Sargon, and a second time by Sennacherib, who appointed Belibus as his viceroy from B.C. 702 to B.C. 699, and afterwards *Asshur-nadin* (Assaranadius) from B.C. 699 to B.C. 693. It is uncertain whether the succeeding governors were viceroys or native princes. Esar-haddon, the Assyrian monarch, assumed the crown himself, and held his court there occasionally; but he appears in the later part of his reign to have appointed a viceroy, Saosduchinus, from B.C. 667 to B.C. 647, who was succeeded by Ciniladanus, B.C. 647-625. *Nabopolassar* was the last of these viceroys or subject kings: he aided Cyaxares in the overthrow of Nineveh, and established himself on the throne of Babylon, which he occupied from B.C. 525 to B.C. 604. The Babylonian territory under him consisted of the valley of the Euphrates as high as Carchemish, Syria, Phœnicia, Palestine, and probably a part of Egypt. He carried on war, in conjunction with the Medes, against the Lydians, and afterwards against the Egyptians who had aided the Lydians. His son Nebuchadnezzar gave the Egyptian king Necho a total defeat at Carchemish. *Nebuchadnezzar*, B.C. 604-561, was equally distinguished for his martial achievements and for the gigantic works which he executed in his country, and particularly at Babylon. He reduced Tyre after a siege of thirteen years; sacked Jerusalem, and carried off its inhabitants; and invaded Egypt. There is little to record of his successors, *Evil-Merodach*, B.C. 561-559; *Neriglissar*, B.C. 559-556; and *Laborosoarchod*, B.C. 556-555. *Nabonidus* commenced his reign just as Cyrus was entering upon his Lydian war: he entered into alliance with Crœsus, and fortified his own territory against the Medes. Cyrus commenced his invasion of Babylonia B.C. 540, and, having defeated the enemy in the open field, he laid siege to Babylon, which was then under the care of *Bil-shar-uzur*, the Belshazzar of the Bible, and, entering by the dry bed of the Euphrates, captured the city. Nabonadius had retired to Borsippa, where he was taken prisoner by Cyrus, B.C. 538. Henceforth Babylonia formed a portion of the Persian empire.

Mound of Nimroud. (From Layard's 'Nineveh.')

III. ASSYRIA.

§ 8. Assyria was bounded on the N. by the range of Niphates; on the E. by that of Zaghus; on the S.E. by Susiana; on the W. and S.W. by the Tigris. The northern and eastern portions of Assyria are mountainous, the former being covered with ranges emanating from the Armenian highlands, and the latter with the secondary ridges of the Zagrus chain. The southern and western districts, as high up as Nineveh, on the other hand, partake more of the character of the Mesopotamian plain, though more diversified with heights and river-conrses. The plains of Assyria, as of Mesopotamia, are alternately a garden and a wilderness, the excessive heat of summer completely parching up the vegetation. The hilly district varies in character, the rising ground adjacent to the plain being well watered and productive, the intermediate hills of an arid character, and the higher elevations of Zagros well wooded and offering rich pastures during the summer months.

§ 9. The rivers which water Assyria all flow into the Tigris, and have courses very nearly parallel to each other towards the S.W. Most of them rise in Zagros, but some penetrate through the central chain into the highlands of Media. The chief rivers from N. to S. are—the Zabātus or Lycus, *Great Zab*, which rises in the angle where Niphates and Zagros effect their junction, and, doubling about among the parallel ranges that beset its middle course, reaches the Tigris in 36° of lat.—the Caprus or Zerbis, *Lesser Zab*, which rises in Media, and reaches the Tigris near 35° lat.—the Physcus or Tornadōtus, *Odorneh*, which joins a short distance below the Median Wall—and the Gyndes,[2] *Diala*, which joins a little above Ctesiphon.

§ 10. The inhabitants of Assyria were a Semitic race, Asshur being described in Gen. x. 22 as a son of Shem. There appears to have been, as we have already observed, a close connexion between the population of Babylonia and Assyria; for we are told (Gen. x. 11), that "out of that land (*i. e.* Babylonia) went forth Asshur," or according to another rendering of the words, "out of that land he (*i. e.* Nimrod) went forth to Asshur." Whichever of the two senses we adopt, the general fact indicated remains the same, viz. that there was an affinity between the two races—a view which is supported by indications both of language and history. The political divisions were numerous: few of the names present any feature of interest; we may, however, specify Arrapachitis in the N.E., which is thought to represent the Scriptural Arphaxad; Adiabēne, the district about the course of the *Great Zab*; Aturia, about the metropolis Nineveh; and Sittacēne in the S.

§ 11. The remarks made in reference to the towns of Babylonia apply in great measure to those of Assyria also. The banks of the Tigris are lined with mounds, marking the sites of once flourishing cities, whose histories and even names remain a matter of doubt. It seems tolerably certain that Nineveh itself was not the earliest capital; Scripture notices Resen as surpassing it in size, and places Calah and Rehoboth on a par with it. We have already (p. 12) endeavoured to identify some of these places; we will now add that *Calah Shergat* appears to have been the first capital, and to have been built about B.C. 1273—that the seat of government was thence moved higher up the river to *Nimrûd* by Sardanapalus, B.C. 930—

---

[2] Nec qua vel Nilus, vel regia lympha Choaspes
Profluit, aut rapidus, *Cyri dementia*, Gyndes
Radit Araccaeos haud una per ostia campos.—TIBULL. iv. 1, 140.
—For the allusion to Cyrus see above, p. 32.

and that this remained the capital until the time of Sennacherib,
B.C. 702, who again removed the seat of power to Nineveh. In
addition to these places, there are numerous mounds which undoubtedly mark the sites of large towns, such as *Abu Khameera*
and *Tel Ermah*, on the western bank of the Tigris, *Khorsabad*,
Shereef-khan, and others on the eastern side of it. These towns
were mostly destroyed either before or at the time of the fall of
Nineveh : when Xenophon passed by their sites, he observed the
mounds, but heard little of the famous cities that lay buried beneath
them ; even the name of Nineveh is unnoticed, and the place is described as Mespila, while that of Resen appears under the Græcised
form Larissa. Some few towns of a later date are found in the
southern part of Assyria, of which Ctesiphon is the only one that
attained celebrity.

Vaulted Drain beneath the Palace at Nimroud.   (From Layard's ' Nineveh.')

The capital of Assyria was Ninus or Nineveh; it is described in the
book of Jonah as "a city of three days' journey" (iii. 3), and its
population (judging from the statement in iv. 11) must have amounted

to 600,000. Though it had disappeared before classical times, yet
the memory of its greatness was preserved. Both Strabo (xvi. p. 737)
and Diodorus (ii. p. 7) give striking accounts of its size. The mounds
opposite *Musul*, named *Kouyunjik* and *Nebbi Yunus*, represent the site of
Nineveh, or, at all events, a portion of it. The doubtful point is, how
far Nineveh extended on either side. It has been noticed that the four
mounds* *Kouyunjik, Khorsabad, Karamless,* and *Nimroud,* stand at the

Subterranean Excavations at Kouyunjik. (From Layard's 'Nineveh.')

---

* A brief description of the contents of these mounds will not be out of place
[1.] The mounds of *Kouyunjik* and *Nebbi Yunus* stand in close proximity to each
other. The former contains the magnificent palace of Sennacherib, erected about
B.C. 700, covering an area of 100 acres. The chambers, of which more than
seventy have been explored, were covered with bas-reliefs commemorating the
wars of Sennacherib: many of these are now in the British Museum. On the
northern side of the mound a second palace was erected by Sardanapalus III.,
grandson of Sennacherib: the apartments were decorated with hunting scenes,
executed in the highest style of Assyrian art. Some of these also adorn the
British Museum. The mound of *Nebbi Yunus* derives its name from an unfounded
tradition that Jonah was buried there. The whole enclosure of *Kouyunjik* covers

angles of a quadrangle, the size of which would correspond tolerably
well with the statements of Jonah and Diodorus: hence it has been
conjectured that the whole of the space enclosed between these points
was termed Nineveh, the area being occupied by extensive gardens and
parks surrounding palaces, and temples, and private houses, much as is
the case in modern Oriental towns. This, however, must be regarded
as doubtful, particularly as *Nimroud* probably represents *Resen*.
Nineveh was destroyed, B.C. 625, by the combined armies of the Medes
and Babylonians. **Arbëla**, between the Zabatus and Caprus, has gained
notoriety from the battle between Darius and Alexander the Great,
which was fought, however, at **Gangamëla**, about 20 miles to the N.W.
Apollonia and **Artemita** are supposed to have stood respectively N.
and S. of the Delas in its mid course: more to the E., **Chala** and the
neighbouring **Celönæ**, on the banks of the *Holwan*, commanded the
pass across Zagros. On the banks of the Tigris, in the S. of the
province, were the important towns of **Opis**, probably at the junction of
the Physcus—**Sittace**, further down the stream—and **Ctesiphon**, which
rose into importance after the decay of Seleucia, and became the winter
residence of the Parthian kings; it was strongly fortified: its site is
now named *Al Madain*, "the two cities."

History of the Assyrian Empire.—We pass over the earliest kings, until
we come to *Tiglath-Pileser* I., B.C. 1110, who extended his conquests
over Cappadocia, Syria, and Armenia,[5] and attacked Babylon without

about 1800 acres, and is about 7¼ miles in circumference. (2.) **Khorsabad** stands
about 15 miles N.E. of *Kouyunjik*: it appears to have been named *Sarghun* after
the monarch Sargon, who established it as his capital about B.C. 720. His palace
is covered with a double mound nearly 1000 feet in length. It was richly adorned
with sculptures, representing for the most part processions of tribute-bearers,
sieges of towns, punishments of prisoners, and buildings. The *Louvre* contains a
rich collection of these. (3.) *Nimroud* lies on the left bank of the Tigris, about
17 miles S. of *Kouyunjik*. The great mound is 1800 feet long by 900 broad, and
rises to a conical elevation at the N.W. angle. The buildings here were erected
by a succession of kings—Sardanapalus I., who founded the N.W. palace, B.C. 900,
in which the celebrated black obelisk was found; *Shamas-iva*, B.C. 850, and *Iva-
Lush* (Pul), B.C. 800, who enlarged that palace; *Esar-haddon*, B.C. 680, who built
the S.W. palace with materials plundered from the other palaces; and his son
Sardanapalus III., who built the S.E. palace. (4.) *Kileh-Shergat* is situated
on the right bank of the Tigris, about 60 miles S. of Kouyunjik. The mound is
of a triangular form, 60 feet high, and about 2½ miles round. The most remarkable object discovered here is the cylinder, now in the British Museum, containing
the annals of Tiglath-Pileser I.

[5] The conquered countries are described on the Assyrian monuments by names
which are in themselves instructive, as illustrating both Biblical and classical geography. It may be noticed that many of the nations with whom the Assyrians carried
on the most frequent wars sank into comparative insignificance in after times.
Northwards we can identify the *Mannai* about Lake *Urumiyeh* with the Minni of
Scripture; *Ararat*, or *Khorkhar*, with central Armenia, as described in the Bible;
Muzr with Colchis, whose inhabitants were probably a Hamitic race, as described
by Herodotus, and as indicated by the Assyrian name which answers to the Biblical
Mizraim. Westward of Armenia, the most important tribes were *Tuplai*, the Tubal
of the Bible, the later *Tibareni*; and the *Muskai*, Mesech, the later Moschi, in
Cappadocia; *Khilak*, Cilicia, is also noticed. On the northern and western frontier
of Mesopotamia were the *Nairi*, with whom the Assyrians were constantly engaged.
Along the course of the Euphrates lived the *Tsukhi*, probably the Shuhites of
Scripture; and on the side of Syria the *Khatti*, the Scriptural Hittites, of whom
a tribe named *Patena* evidently represents *Padan-Aram*. The town of Samaria is

success. The celebrated Sardanapalus I., B.C. 930, carried his arms successfully from the shores of the Persian Gulf to the Mediterranean, subduing Tyre, Sidon, Byblus, and Aradus in the latter direction, Babylon and Chaldæa in the former. *Shalmaneser*, B.C. 900, conquered Armenia, Media, Cappadocia, Babylonia, Syria, and Phœnicia. He also received tribute from Jehu, king of Israel, who is named *Yahua*, son of *K'humri*, *i. e.* successor of Omri. *Shamas-iva*, B.C. 850, attacked the Syrians, Medes, and Babylonians, taking two hundred towns either belonging to or confederate with the latter. *Iva-Lush* III., B.C. 800, the Pul of the Bible, received tribute from the Medes, Persians, Armenians, Syrians, Samaritans, Tyre, and Sidon. The name of Menahem, king of Israel, appears in the list of his tributaries, as recorded in 2 Kings xv. 19. *Tiglath-Pileser* II., B.C. 747, carried on wars in Upper Mesopotamia, Armenia, Media, and Syria, where he defeated Rezin, king of Damascus. He is the monarch who invaded the northern districts of Palestine (2 Kings xv. 29). *Shulmaneser*, B.C. 730, is not noticed in Assyrian inscriptions. He carried on war with Hoshea, king of Israel, and besieged Samaria (2 Kings xvii. 3-5). He appears to have died before the city was taken; for "the king of Assyria" (2 Kings xvii. 6) who actually carried off the Israelites was named *Sargon*, who came to the throne B.C. 721, and who is recorded in the inscriptions to have transplanted 27,280 families of the Israelites. Sargon waged war with Merodach-baladan, king of Babylon, and invaded Susiana, Armenia, and Media: he also came into contact with the Egyptian monarchs, one of whom, Sabichus, the second of the Ethiopian dynasty, had formed an alliance with Hoshea . 2 Kings xvii. 4). In this war he took Ashdod (Is. xx. 1) and Gaza; he also extended his expeditions to Cyprus. *Sennacherib*, B.C. 702, subdued and deposed Merodach-baladan, appointing a viceroy over Babylon. In the third year of his reign he defeated the Hittites, the kings of Tyre and Sidon, and, descending southwards, subdued the towns of Philistia, particularly Ascalon. He twice invaded Palestine, on the first sion receiving tribute from Hezekiah (2 Kings xviii. 15), on the second occasion besieging Lachish and Libnah, and shutting up Hezekiah in occa-Jerusalem (2 Kings xviii. 17, xix. 8). The destruction of his army in Egypt has been already referred to. *Esar-haddon*, B.C. 680, renewed the wars with Phœnicia, Syria, Armenia, Susiana, Media, Babylonia, and Asia Minor: he also describes himself as the "conqueror of Egypt and Ethiopia." He is probably the king who carried Manasseh to Babylon (2 Chron. xxxiii. 11). Sardanapalus III., B.C. 660, undertook a campaign against Susiana, but was otherwise unknown for martial deeds. *Asshur-emit-ili*, B.C. 640, was either the last or the last but one of the Assyrian kings, it being doubtful whether he is identical with the *Saraous* of Berosus or not. With the latter monarch the Assyrian empire terminated, Nineveh being destroyed by the conjoined forces of the Medes under Cyaxares, and the Babylonians under Nabo-polassar.

---

described as *Beth Khumri* ("the house of Omri"); Judæa as *Jehuda*; Idumæa as *Hudum*; and Merâe as *Mirukha*. The island of Cyprus is referred to under the name *Yavan* (Javun). Eastward of the Zagrus range were races, some of whose names we cannot identify; the *Hupuska*, who lived eastward of Nineveh; the *Namri*, whose territory extended to the shores of the Persian Gulf; the *Bikni* in Parthia; the *Partsu* in Persia; *Mada* in Media; and *Gimri*, the Sacæ, or Scythians. Southwards, Babylonia is termed *Kan-Duniyas*, Susiana *Nuvaki*, the Karoon being notified under the name *Ula* (Ulai of Daniel, *Eulæus*), and the Shat-el-Arab as the "great salt river." Many of the towns of Phœnicia and Syria are notified under names but slightly varying from the classical or Biblical forms.

The Town and Rock of Wan.

IV. ARMENIA MAJOR.

§ 12. The boundaries of Armenia cannot be very accurately defined : speaking generally, Armenia may be described as the high mountainous country between the Euxine, Caspian, and Mediterranean seas and the Persian Gulf, whence the mountain chains of Western Asia radiate in various directions.   On the S. the limit of this district may be placed at the ranges which overlook the Mesopotamian and Assyrian plains, viz. Masius and Niphates, and more to the E., Caspius Mons, which separated it from Media : the eastern boundary was formed by the converging streams of the Araxes and the Cyrus ; and the latter river may be regarded as its northern boundary also, until it approximates to the Euxine, whence the south-westerly direction of the mountain-chains carried the boundary towards the upper valley of the Euphrates, which formed its limit on the W. Armenia is an elevated plateau, forming the westerly continuation of the great plateau of *Irán*. The general elevation of its central plains above the level of the sea may be stated at about 7000 feet.   Out of this plateau, as from a new base, spring mountain chains of great elevation, the central range culminating in the splendid conical peak of *Aghri Tagh* (17,260 feet), to which the Biblical name of Ararat has been more particularly assigned.   The uplands, though exposed to a long and severe winter, afford most abundant pasture in the summer months, and have been in all ages the resort of the shepherds of the Mesopotamian lowlands during that season. A fine breed of horses roamed over the wide grassy plains, and formed the most valued production of the country.

§ 13. The mountain ranges have been already generally described: we need here only repeat that three lines of mountain chains may be traced through this country; the most northerly consisting of **Paryadres** and its eastern continuations, which separate the upper courses of the Araxes and Cyrus; the central one consisting of the chain, which under the name of **Abus**, first divides the two branches of the Euphrates from each other, and then bounds the upper course of the Araxes on the S., terminating in the twin heights of the *Greater and Less Ararat;* while the southerly one, which is the most continuous and best defined of the three, in the first place separates the upper courses of the Euphrates and Tigris, then under the name of **Niphates** [*] passes southwards of Lake Arsissa, and after parting with Zagrus, proceeds under the name of Caspius Mons to the shores of the Caspian Sea. The yet more southerly range of **Masius**, which bounds the Mesopotamian plain, is an offset from Niphates; it strikes across from the Euphrates in a south-westerly direction to the Tigris, and is continued on the eastern side of that river under the name of **Gordiaei Montes**, which return back in a northerly direction to the central chain. The chief rivers are—the **Euphrates** and **Tigris**, which seek the Persian Gulf—the **Araxes** and the **Cyrus**, which seek the Caspian Sea, uniting, just as the two former, previously to their discharge—and the **Acampsis**, which flows northwards into the Euxine. These rivers are described elsewhere (p. 75, 77). There are, as might be expected in a country where the watershed is so undecided, several lakes. Of these the most important, named **Arsene** or **Thospitis**, *Wan,* is in the S., while **Lychnitis**, *Goutcha,* is in the N.E.

§ 14. The Armenians were an Indo-European race, their country having probably been the very cradle of that branch of the human family. Of the tribes the Carduchi may be specially noticed, the progenitors of the modern *Kurds,* and occupying the same country, viz. the mountain ranges eastward of the Tigris on the borders of Assyria. Armenia was divided into a large number of districts, the titles of which are for the most part devoid of interest: we may notice, however, the following—**Gogarine**, in the extreme N.,

[*] This name is sometimes referred to as equivalent to Armenia itself:
Addam urbes Asiæ domitas pulsumque Niphaten. —Virg. *Georg.* III. 30.
Canteruus Augusti tropæa
Cæsaris et rigidum Niphaten.—Hor. *Carm.* II. 9, 19.

probably the original seat of the people named Gog in Scripture: **Chorsine**, representing the modern name *Kars*: **Sophene**, a considerable district about the sources of the Tigris; and **Gordyene**, about the Gordyei Montes, both of which names contain the elements of the name *Kurdistan*. The towns are unnoticed until the period when the Romans entered into the country. We need not infer that the places which come prominently forward in the history of their wars were the only or the chief towns in existence. We have evidence in the inscriptions[7] found at *Wan* that an ancient capital stood on the impregnable rock which rises on the eastern shores of Lake Arsissa, and it is doubtful whether the Roman historians have mentioned even its name. From the tenour of the inscriptions it may be inferred that the flourishing period of *Wan* lasted from B.C. 850 to B.C. 700: tradition assigns the foundation of the city to Semiramis. It is hardly probable, however, that the towns of Armenia attained any very great importance: the only purpose that they would serve would be as trading stations on the routes which have crossed the highlands from time immemorial. The majority of the population would naturally be scattered over the face of the country in those villages of subterraneous houses, which Xenophon (*Anab.* iv. 5, § 25) describes, and which still exist in precisely the same state.

The capital, **Artaxata**, stood on the banks of the Araxes, below the heights of *Ararat*: it was built under the superintendence of Hannibal, and named after the Armenian ruling sovereign Artaxias: having been destroyed by Corbulo, A.D. 58, it was rebuilt by Tiridates with the name Neronia. **Tigranocerta**, "the city of Tigranes," was situated on the banks of the Nicephorius, a tributary of the Tigris: it was built and strongly fortified by Tigranes, and shortly after dismantled by Lucullus, who defeated Tigranes before its walls: its exact position is unknown. **Amida**, on the Tigris, occupied the site of the modern *Diarbekr:* the only event of interest in its early history is the siege it sustained from the Persian king Sapor, A.D. 359. **Artemita** stood either at or near the ancient town of *Wan*, on the eastern shore of Lake Arsissa: the **Buana** of Ptolemy, and the **Salban**, captured in the reign of Heraclius, were probably in the same neighbourhood. We may briefly notice **Arsamosata**, a fortress in the valley of the Euphrates near the junction of the two branches — **Carcathiocerta**, in the same neighbourhood — **Arsen**, probably at *Erzerum* — **Theodosiopolis**, identified by some writers with Arzen, but by others placed about 35 miles to the E.: it derived its name from Theodosius II., who founded it — **Naxuana**, *Nachdjevan*, in the valley of the Araxes — and **Elegia**, near *Erzerum*, the scene of a battle between Vologaeses III. and the Romans, A.D. 162.

History—The history of Armenia is unimportant; it has been a

---

[7] They are found on the face of the rock, and in excavated chambers, which may have been used as sepulchres: detached stones and slabs also bear inscriptions. Some of these resemble the most ancient Assyrian inscriptions, others are of the time of the Persian empire.

scene of constant warfare, but at no period the seat of an independent empire—exposed to the invasions of the more powerful masters of the surrounding plains, Assyrians, Medes, Greeks, Syrians, and at a later period the battle-field on which the armies of Rome contended for the empire of the East.   Armenian historians record the names of the kings who held rule in the country from the earliest times: the first dynasty was named after Haïg, who is said to have lived B.C. 2107 : there were fifty-nine kings belonging to this, the last of whom, Wahe, fell in a battle with Alexander the Great, B.C. 328.   This dynasty was followed by a succession of seven governors appointed by Alexander, and after his death by the Seleucidæ, from B.C. 328 to B.C. 149.   The independent dynasty of the Arsacidæ established itself, according to the Roman historians, in the year B.C. 188 in the person of Artaxias ; but according to the Armenians, in B.C. 149, in the person of Valarsaces, a brother of Tigranes III.   The Arsacidæ were divided, according to the latter authorities, into two branches, the elder of which reigned from B.C. 149 to A.D. 62, and the younger at Edessa from B.C. 38, and afterwards in Armenia Magna from A.D. 62 until A.D. 428.   The most illustrious of these rulers was Tigranes I., the ally of Mithridates against the Romans.

§ 15.   The countries which we have described in the preceding part of this chapter were the scene of one of the most interesting adventures recorded in ancient literature, viz. the advance and retreat of the 10,000 Greeks, who aided Cyrus the younger in his expedition against his brother Artaxerxes.   As the narrative presents some few geographical difficulties, we shall give a brief account of the route described in Xenophon's Anabasis.

The early part of the course lay across the plateau of Asia Minor, from Ephesus to Dana or Tyana, and thence over the Taurus range into the maritime plain of Cilicia, which was traversed to the eastern extremity of the Bay of Issus ; thus far the route requires no elucidation.   We now approach the border of Syria.   South of Issus the Amanian range approaches close to the sea-shore : the Kersus (*Merkas-su*) discharges itself at this point : and on each bank was a fort, one belonging to Cilicia, the other to Syria, which guarded the pass of the "Cilician and Syrian Gates :" Cyrus passed through these to Myriandrus.   The narrative is then singularly defective in the omission of all notice of the difficult *Pass of Beilan*, and the rivers which must have been crossed before reaching the Chalus (*Kowoik* or river of *Aleppo*).   The river Daradax and the Castle of Belesis must have been met with close to the Euphrates, although no mention is there made of the river : Belesis may be represented by the ruins of *Balis*, and the river Daradax by a canal drawn from the Euphrates to the town.   The Euphrates was crossed at the ancient ford of Thapsacus, the later Sura, *Suriyeh*, and the army entered on the plain of Mesopotamia, which Xenophon (i. 5) calls Syria in this part as far as the river Araxes, better known as the Chaboras, *Khabur*—Araxes being apparently an appellative for any river.   Thenceforward the plain is termed Arabia (i. 5), as being occupied by Scenite Arabs : the Masca was merely a channel of the Euphrates surrounding the site of the town Corsote, *Irzah* : Pylæ was situated about 70 miles N. of Cunaxa, at the point where the plain and the mountains meet: Carmande may have been *Hit*,

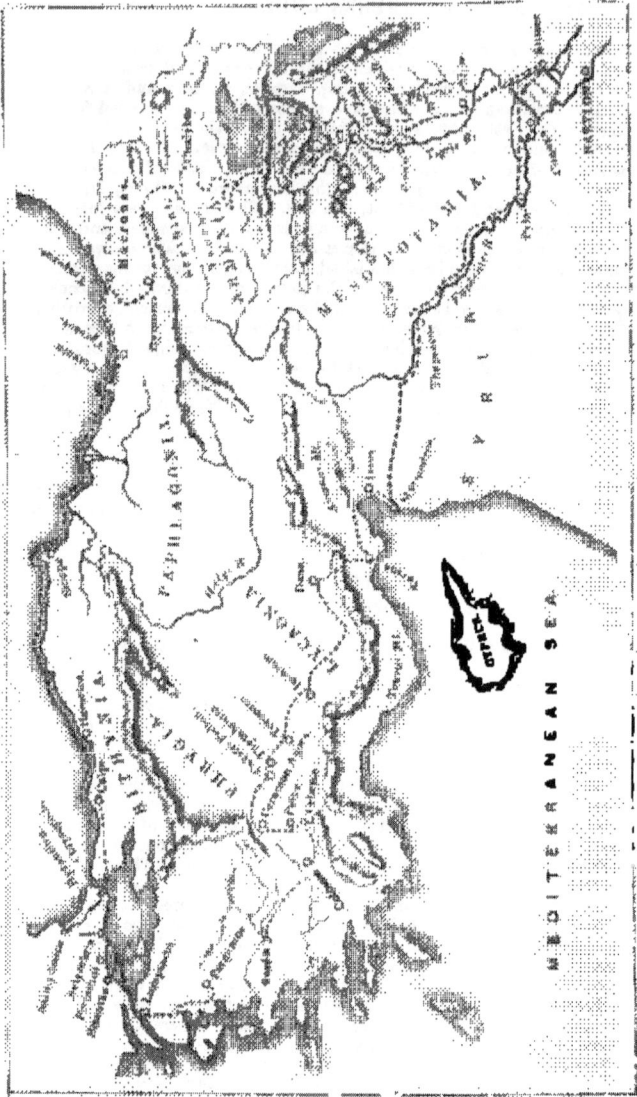

Map of the Route of the Ten Thousand.

Babylonia was now entered: Xenophon describes four canals as crossing the plain from the Tigris to the Euphrates; these may yet be distinguished, the third of them being the *Nahr Malcha* of modern maps. Xenophon does not give the name of the place where the battle was fought; this is supplied by Plutarch, as Cunaxa, the exact position of which cannot be ascertained: Plutarch states that it was 500 stades or nearly sixty miles from Babylon.

After the battle the Greeks retreated northwards over the plains of Babylon, by a somewhat circuitous route, until they reached the Median Wall, the remains of which (named *Sidd Nimrud, i. e.* 'Wall of Nimrod') may still be traced across the plain from the Euphrates to the Tigris, near Opis, in a north-easterly direction. This wall they are said to have passed through (ii. 4), but must have again passed through it in order to reach Sittace (perhaps at *Akbara*), where they crossed the Tigris. The river Physcus and the town Opis cannot be identified with certainty: the former is supposed to be either the *Adhem*, on the banks of which extensive ruins have been found, or the *Nahr-wan*, an artificial channel. in which case Opis would be near *Eski Baghdad*, in about 34° 30' latitude. The Lesser Zabatus (*Zab*) was crossed without being noticed by the historian: Cænæ was probably *Kalah Shergat*. The Zabatus (*Great Zab*) was forded at a point about 25 miles from its confluence with the Tigris: the torrent which they next crossed (iii. 4) was the Bumâdus, *Ghazir*, which joins the right bank of the Zabatus about three miles below the ford: thence they reached Larissa (*Nimrúd*) and on the following day Mespila (*Konyunjik*), the site of ancient Nineveh. They followed the ordinary route towards the north, leaving the Tigris at a considerable distance to their left, by *Balnai*. They forsook this route, however, as they approached the *Khabour*, and instead of fording it near its confluence with the Tigris, deviated to the right, and crossed a range of hills to *Zakko*: the passage of the *Khabour*, and of its confluent the *Hazel*, are not noticed, though the former is a difficult operation. Crossing the triple ridge in the neighbourhood of *Zakko*, they reached, after four days, the mountains of *Kurdistan*, which, in the neighbourhood of *Fynyk*, press close upon the bank of the Tigris. Xenophon resolved to cross Armenia instead of following the other routes which offered themselves: he crossed the mountain range to *Finduk*, which he reached probably at the end of the first day's march, and thence by a series of difficult passes reached the Centrites or eastern Tigris, which receives the waters of the rivers *Bitlis, Sert*, and *Bohtan*. They crossed the Centrites near *Tilleh*; then proceeded northwards, and in six days reached the Teleboas, which Ainsworth identifies with the *Kara-su*, a confluent of the Southern Euphrates, but Layard with the river of *Bitlis*: assuming the latter as the more probable, Xenophon would have passed a little westward of the lake of *Wan*, a range of mountains intervening, and would have reached the Euphrates (*Murad-su*) in six days from the Teleboas. After leaving the Euphrates, the course, as described by Xenophon, is quite uncertain. Ainsworth identifies the Phasis with the *Pasin Chai*, a tributary of the Araxes or *Arus*, and the Harpasus with the *Arpa Chai*, another tributary of the same river, and the town Gymnias with *Erz Rum*: Layard and others identify the Phasis with the Araxes or perhaps the Cyrus, and the Harpasus with the *Tcherouk*, which flows into the Euxine. In the former case the holy mountain Theches would be the range between the sources of the Euphrates (*Kura-su*), and the *Tcherouk*; in the latter, it would be more to the eastward, between *Batoun* and *Trebizond*. Arrived at Trapezum, *Trebizond*, they followed the line of coast, partly by land and partly by sea, back to their native country.

The Caucasus.

V.—COLCHIS, IBERIA, ALBANIA, SARMATIA.

§ 16. **Colchis** lay along the eastern coast of the Euxine, from the Phasis in the S. to the Corax in the N.W.: on the N. it was bounded by Caucasus, on the E. by Iberia, and on the S. by Armenia. It answers to the modern provinces of *Mingrelia* and part of *Abbasia*. The chief mountain range is **Caucasus**, which in this part of its course approaches close to the shores of the Euxine: little was known of this extensive range by the ancients: it was the fabled scene of the sufferings of Prometheus,[a] and supplied the poets with a picture of wild and desolate scenery.[b] The chief river of Colchis was the **Phasis** in the S.; numerous lesser streams pour

---

[a]   Caucasiasque refert volucres, furtumque Promethei.
                                        VIRG. *Ecl.* vi. 42.

[b]   Duris genuit te cautibus horrens
          Caucasus.   IB. *Æn.* iv. 366.
     Sive per Syrtes iter æstuosas,
     Sive facturus per inhospitalem
     Caucasum.   HOR. *Carm.* i. 22, 5.

down from the Caucasus to the Euxine. The inhabitants were subdivided into numerous tribes, of which we may notice the Lazi, who
communicated to this district its later name of Lazica; and the
Abasci, whose name survives in the modern *Abbasia*. The only important towns were **Dioscorias**, on the sea-coast, a Milesian colony,
where Mithridates wintered B.C. 60: on its site the Romans afterwards built Sebastopolis; and **Cutatisium**, the reputed birthplace of
Medea,[1] in the interior. There were numerous lesser towns on the
coast, which carried on an active trade in timber, hemp, flax, pitch,
gold-dust, and especially linen.

History.—Colchis occupies a prominent place in mythology as the
native land of Medea, and the scene of the capture of the golden fleece
by the Argonautic expedition :[2] it was regarded by poets as the native
seat of all sorcery,[3] a credit which it may perhaps have gained from the
abundant growth of the plant iris, whence the medicine called colchicum
is extracted. Colchis was reputed the most northerly portion of the
Persian empire, but was practically independent of it. Mithridates
annexed it to the kingdom of Pontus, and made his son king of it. The
Romans deposed him, and appointed a governor; but Pharnaces regained the territory, and under his son Polemon it was part of the
kingdom of Pontus and Bosporus.

§ 17. **Iberia** was bounded on the N. by Caucasus, on the W. by
Colchis, on the E. by Albania, and on the S. by Armenia: it answers
to the modern *Georgia*. The chief mountain ranges in it are—
Caucasus, which was here traversed by the celebrated pass named
Caucasiae Portae, now the *Pass of Dariel*, in the central range; and
the **Moschici Montes** on the side of Colchis. The only important
river is the **Cyrus**, the upper course of which falls within the limits
of Iberia: it received, on its left bank, the **Aragus**, *Arak*, which
rises near the Caucasian Gates. The inhabitants, named Iberi or
Iberes, were divided into four castes—royal, sacerdotal, military,
and servile: they are described as a peaceful and industrious race.
The modern Georgians, their descendants, are still named *Virb*, probably a form of Iberi, by the Armenians. The chief towns were—
Harmostica, the later capital, S. of the Cyrus, near the borders of
Armenia; and **Mestleta**, the earlier capital, near the confluence of
the Aragus with the Cyrus.

History.—The Iberians were probably nominal subjects of the Persian
empire. They afterwards acknowledged the supremacy of Mithridates.

---

[1] Hence named Cytaeis—
    Tunc ego crediderim vobis, et sidera et amnes
    Posse Cytaeeis ducere carminibus.   PROPERT. l. 1, 24.

[2] Εἰ' ἐφεξ' Ἀργοῖς μὴ διαπτάσθαι σκάφος
    Κόλχων ἐς αἶαν κυανέας Συμπληγάδας.   EURIP. Med. 1.

[3] Sed postquam Colchis arsit nova nupta venenis,
    Flagrantemque domum regis mare vidit utramque.
                         OV. Met. vii. 394.

The Romans penetrated into the country under Lucullus and Pompeius, the latter of whom subdued the inhabitants, B.C. 65. It remained, however, under its own princes, even after it had been nominally attached to the province of Armenia in A.D. 115. The Romans, by the treaty of Jovian, renounced their supremacy in favour of the Persians.

§ 18. **Albania** was bounded on the W. by Iberia, on the N. by Sarmatia, on the E. by the Caspian, and on the S. by Armenia, the river Cyrus forming the line of demarcation in this direction: it answers to the present *Shirwan* and part of *Daghestan*. The mountain ranges in this district consist of the eastern portion of **Caucasus**, which is here traversed by an important pass named **Albaniæ Portæ**, *Pass of Derbend*; an important offset from the central chain, the **Ceraunii Montes**, strikes off towards the N.E. The chief river is the **Cyrus**, which here receives two important tributaries—the **Cambyses**, *Yori*, and the **Alazon**, *Alasan*, which unite shortly before their confluence with the main stream: Pompey followed the course of the Cambyses in his pursuit of Mithridates, B.C. 65. The Albani are a race of doubtful origin, but probably Scythians, and allied to the more famous Alani: they were divided into twelve hordes, the name of one of which, Legæ, is preserved in the modern *Leghistan*: these tribes were in Strabo's time united under one king, but formerly had each its own prince. The only towns of importance were—**Albana**, *Derbend*, which commanded the pass on the shore of the Caspian; and **Chabala**, which ranked as the capital.

§ 19. Under the title of **Sarmatia Asiatica** is included the vast region lying N. of the Caucasus and E. of the Tanais, stretching northwards to an undefined extent, and eastwards as far as the Rha, which separated it from Scythia. The mountain ranges assigned to this region emanated from **Caucasus**, and were named **Coraxici Montes**, on the borders of Colchia, and **Hippici** between the Tanais and Rha. The rivers were—the **Tanais**, *Don*, which formed the limit between Europe and Asia—the **Atticitus**, *Kuban*, which discharged itself partly into the Palus Mæotis and partly into the Euxine—the **Rha**, *Wolga*, flowing into the Caspian—the **Udon**, *Kouma*, and the **Alonta**, *Terek*, falling into the same sea more to the S. The inhabitants of this district were broadly classed together under the name of Sarmatæ or Sauromatæ, and were subdivided into a vast number of tribes, whose names and localities, though interesting in an ethnological point of view, need not be specified here. The only towns known to the ancients were situated on the shores of the Euxine, and were for the most part Greek colonies. We may notice **Pityus**, *Pitsunda*, N. of Dioscurias, described in the reign of Gallienus as a strong fortress with an excellent harbour—**Phanagoria**, on the E. side of the Cimmerian Bosporus, founded by the Teians, a great emporium for the trade of these districts, and the Asiatic capital of the kings of Bosporus, with a remarkable temple of Aphrodite: numerous tombs stand on the site, but the town itself has disappeared, the materials having been carried away to other places—and Tanais, at the mouth of the river of the same name, a colony of the Milesians, and a place of large trade: it was destroyed by Polemon I., but probably restored: ruins of it exist near *Nedrigoska*.

Pass of the Caucasus.

Persepolis.

CHAPTER XIII.

THE PROVINCES OF THE PERSIAN EMPIRE.

§ 1. I. Persis.—Of the provinces of the Persian empire Persis de-
mands the earliest notice, as being the original seat of the race, and
containing the metropolis, Persepolis. It was bounded on the N.
by Media and Parthia, from which it was separated by the range
of Parachoathras; on the W. by Susiana; on the S. by the Persian
Gulf; and on the E. by the desert of Carmania. The name still
survives in the modern *Fars.* It is throughout a mountainous district, with some extended plains and a few valleys of great beauty
and fertility. The mountain chains are continuations of Zagrus,
under the names of Parachoathras, *Elwend,* and Ochus, and run for
the most part in a direction parallel to the coast of the Persian Gulf;

hence the rivers are in many cases confined to the interior, and discharge themselves into lakes. This is the case with the **Araxes**, *Bend-amir*, which rises on the borders of Susiana and flows eastward, receiving the tributary waters of the **Cyrus** or **Medus**, *Pulwan*, and discharging itself into a lake now named *Bakitgan*, about 40 miles E. of Persepolis. The only river of any importance that reaches the sea is the **Arosis** or **Oroätis**, *Tab*, on the border of Susiana. The sea-coast was almost uninhabitable from the extreme heat and unhealthiness of its climate.

§ 2. The Persians were the most important nation of the Ariac branch of the Indo-European race. They were originally called Artæi, a form of Arii and of the Sanscrit *Arya*, "noble." The name Persæ is also of Indian origin. The Persians were divided into three castes, warriors, agriculturists, and nomads; and these were subdivided into ten tribes, which have been already noticed in connexion with the geography of Herodotus.[1] They were reputed both by the Greeks and Romans a most warlike[2] race, good riders, and skilful in the use of the bow, but superstitious[3] and effeminate.[4] Persia was divided into numerous districts, of which Paraetacene was the most important. The name is probably derived from a Persian or Sanscrit root signifying "mountaineers." Of the towns but few are known to us. **Pasargadæ** ranked as the ancient capital of Cyrus, and **Persepolis** as that of the later sovereigns. The former was on the banks of the Cyrus, N.E. of Persepolis, its position having been identified by the discovery of the tomb of Cyrus at *Murghàb*; the latter was finely situated at the opening of an extensive plain, near the junction of the Araxes and Medus, and is represented by the extensive and beautiful ruins now named *Chel-Minar*, "the forty columns." A town named **Ispadāna**, in the N. of the province, occupied the site of *Ispahan*.

---

[1] See p. 37.

[2]   Ταγοὶ Περσῶν,
      Βασιλῆις Βασιλέως ὕποχοι μεγάλου,
      Σούνται, στρατιᾶς πολλῆς ἔφοροι,
      Τοξοδάμαντές τ' ἠδ' ἱπποβάται,
      Φοβεροὶ μὲν ἰδεῖν, δεινοὶ δὲ μάχην
      Ψυχῆς εὐλήμονι δόξη.   Æsch. Pers. 23.
      Quaeque pharetratæ vicinia Persidis urget.
                                    Virg. Georg. iv. 290.

[3]   Discat Persicum haruspicium.
      Nam Magus ex matre et gnato gignatur oportet,
      Si vera est Persarum impia religio,
      Gnatus ut accepto veneretur carmine Divos,
      Omentum in flamma pingue liquefaciens.   Catull. sc. 2.

[4]   Persicos odi, puer, apparatus;
      Displicent nexæ philyra coronæ;
      Mitte sectari, rosa quo locorum
      Sera moretur.   Hor. Carm. i. 38, l.

Some doubt exists as to the date of the edifices which adorned Persepolis. It seems probable that they were subsequent to the time of Cyrus, and were erected by Darius Hystaspis and Xerxes. The city was surrounded, according to Diodorus (xvii. 71), by a triple wall of great strength. Persepolis was burnt by Alexander the Great, and is afterwards only noticed in 2 Mac. ix. 1, as having been attacked by Antiochus Epiphanes. The ruins stand on an immense artificial platform, originally some 40 or 50 ft. in height above the plain, which is approached by a remarkably fine flight of steps. The buildings are adorned with bas-reliefs, and the columns are elaborately chiselled. In the neighbourhood of Persepolis are some places which bear marks of high antiquity, but which are unnoticed by any early writer. About five miles off is the steep conical hill named *Istakr*, crowned with the remains of a fortress, and surrounded by a plain which is thickly covered with fragments of sculpture of all kinds. *Naksh-i-rustam* is another cliff in the same neighbourhood, in the face of which numerous tombs have been excavated. The sculptures with which these are ornamented belong partly to the Persian, but more generally to the Sassanian period.

Tomb of Cyrus at *Murghâb*, the ancient *Pasargadæ*.

§ 3. 11. SUSIANA.—Susiana was bounded on the N. by Media; on the W. by the Tigris and a portion of Assyria; on the S. by the Persian Gulf; and on the E. by Persia, from which it is separated by the ranges of Parachoathras: the name survives in the slightly altered form *Khuzistan*. The country is in its eastern half intersected by the various ramifications of Parachoathras; the western portion is a plain, and suffers from intense heat. In addition to the **Tigris**, which skirts its western border, we may notice the **Choaspes**. *Kerkhah*, which rises in Media, not far from Echatana, penetrates the chain of Zagrus, and, emerging into the plain, passes by the ancient Susa, and falls into the Tigris below its junction with the Euphrates. Its course appears to have undergone considerable change within historical times. It formerly divided above Susa, and sent off two arms, one of which joined the Eulæus, while the other flowed into the Chaldæan lake. **Eulæus**, *Karun*, or river of *Shuster*, which rises in Parachoathras, and pursues a westerly course through the mountains, but on gaining the plain turns southwards. It receives from the N. an important tributary, the **Coprātes**, *Dizful*, which approaches within eight miles of the Choaspes in the neighbourhood of Susa. After the junction of the Eulæus and Coprates the river assumes the name of **Pasitigris**, and formerly discharged itself directly into the Persian Gulf, but now into the *Shat-el-Arab*.

§ 4. Susiana appears to have been originally occupied by a Hamitic race; the name of Cush being preserved not only in Susiana, but, still more evidently, in Cossæi and Cissia, the former being the name of a tribe, perhaps identical with the Cuthæans of the Bible, and the latter being the title by which Herodotus describes the whole of the province. These retired towards the mountains, and a Semitic race, the Elymæi, the Elam of Scripture, occupied the maritime plain. Both of these races, however, gave way before the Ariana, who ultimately formed the dominant race here as in Persia and Media. Susiana was divided into numerous districts, of which we need only notice **Elymāis**, in the N.W., about the upper valleys of the Choaspes; **Cossæa**, the mountainous region in the same district

Mound of Susa.

bordering on Media ; Cissia, in its restricted application, the district
about Susa ; and the Elymaei in the maritime plain. Of the towns
we know but little. The only important one was Susa, the Shushan
of the Bible, centrally situated near the junction of the hills and the
plain on the left bank of the Choaspes.

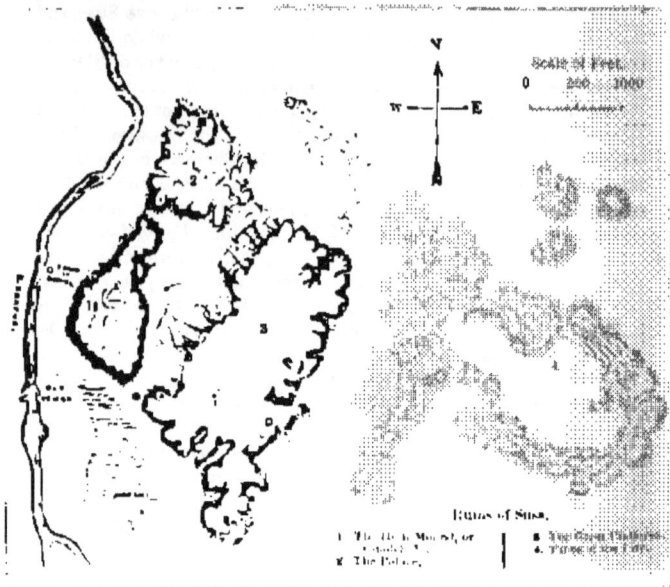

Ruins of Susa.

Susa rose to importance as one of the royal residences[a] of the Persian monarchs. Among the causes which led to this selection may be
noticed its excellent water,[b] the beauty of its scenery, and the retired-

---

[a] Hence the name became familiar to the Greek and Latin poets.

> Οἵτε τὸ Σούσων, ἠδ' Ἐκβατάνων,
> Καὶ τὸ παλαιὸν Κίσσινον ἕρκος
> Προλιπόντες ἔβαν.   Aesch. Pers. 16.

> Nos tot Achæmeniis armantur Susa sagittis,
>   Spicula quot nostro pectore fixit Amor.—Propert. II. 13, 1.

>   Achæmeniis decurrant Medica Susis
>   Agmina.   Lucan. II. 49.

[b] The water of the Choaspes is said to have been specially reserved for the use
of the monarchs. Hence Milton describes it as the

>   "amber stream,
>   The drink of none but kings" (Par. Reg. III. 288),

and Tibullus (iv. 1. 140) as "regia lympha Choaspes."

ness of its situation. The name probably refers to the number of *lilies* (in the Persian language *Shushan*) that grow there. It is sometimes described as on the Eulæus, sometimes on the Choaspes; we have already stated that a branch stream connected these two rivers. The ruins at *Sus* are at present a mile and a half from the latter and six miles from the former stream. The modern *Shuster* has inherited the name but not the site of the old town. The most famous building was the Memnonium, or palace, described in the book of Esther (i. 5, 6), the site of which has been recently explored. It was commenced by Darius and completed by Artaxerxes Mnemon, and consisted of an immense hall, the roof of which was supported by a central group of 36 pillars arranged in a 'square;' this was flanked by three porticoes, each consisting of two rows of six pillars each.

Tomb of Darius.   (From Rawlinson's 'Herodotus.')

§ 5. III. MEDIA.—Media was bounded on the N. by the Caspian Sea; on the W. by the Carduchi Montes and Zagrus, separating it from Armenia and Assyria; on the S. by Susiana and Persis; and on the E. by Parthia and Hyrcania. In the latter direction its limit may be somewhat indefinitely fixed at the line where the mountains subside into the central plain. The province answers to the modern *Azerbiján, Ghílán, Irak Adjem,* and the western part of *Masenderán.* The limits above laid down comprised three districts of very different character:—(i.) the low alluvial strip along the shores of the Caspian; (ii.) the mountainous district of Atropatēnē in the N.W.; and (iii.) Media Magna, the central and southern portion, which abounded in fine plains and fertile valleys, with a climate moderated by their general elevation above the level of the sea. These plains, particularly the Nisæan, produced a breed of horses celebrated far and wide in ancient times. The country was on the whole remarkably fertile. The chief mountains of Media were—Zagrus and Parachoathras in the W.; Caspius Mons, Orontes, Jasonius, and Corönus, in the N., Jasonius representing the lofty peak of Demavend. The western range was crossed by a pass named Portæ Zagricæ or Medicæ, *Kelishin*, on the road leading to Nineveh. A still more important pass, Caspiæ Portæ, formed the main line of communication between Media and Parthia; it was situated E. of Rhagæ at *Dereh*. The only important river[1] is the Amardus, *Kizil Ozien*, which rises in Zagrus and flows northwards into the Caspian. A large lake named Spauta or Martiäna, *Urumiah*, is situated in the N.W., notorious for the extreme saltness of its waters.

§ 6. The Medes were a branch of the Arian stock, and were anciently called Arians, according to Herodotus (vii. 62). They were closely allied to the Persians, as proved by their similarity of dress, by the high official position enjoyed by Medes under the Persian kings, and even by the term "medize" as expressive of deserting to the Persian side. They are first noticed in Assyrian inscriptions under the form *Mada* about B.C. 880. The name has been explained as meaning "middle land," from an idea that Media was centrally situated in regard to the other nations of western Asia. Their name is frequently given by the Roman poets to the Parthians.[2] Their skill in poisoning[3] was noted. Media was divided into two large portions :—(i.) Atropatêne, in the N., named after Atropates, a satrap who rendered himself independent in the time of the last Darius; and (ii.) Media Magna. We have already observed that this division was based on the phy-

---

[1] Virgil (*Georg.* iv. 211) speaks of the Hydaspes as a Median river : he must use the term "Medus" in an extended sense as meaning "eastern:" the Hydaspes is really in India. Horace (*Carm.* ii. 9, 21) similarly describes the Euphrates as "Medum flumen."

[2]   Ille magnos potius triumphos ;
            Ille amen dici pater atque princeps :
            Neu sinas Medos equitare inultos,
            Te duce, Cæsar.   Hor. *Carm.* L 2, 49.
            Triumphatisque possit
            Roma ferox dare jura Medis.   *Id.* iii. 8, 43.
            Horribilique Medo
            Nectis catenas.   *Id.* i. 29, 4.

And so Propertius—
            Vel tibi Medorum pugnaces ire per hastas
            Atque onerare tuam fixa per arma domum.— iii. 9, 25.

[3] Nulla manus illis, fiducia tota veneni est.—Luc. viii. 398.

sical character of the country, and must have been in existence previous to the introduction of the name Atropatene. Of the towns we
know but little. The capital was **Ecbatana**, the Achmeta of Scrip-

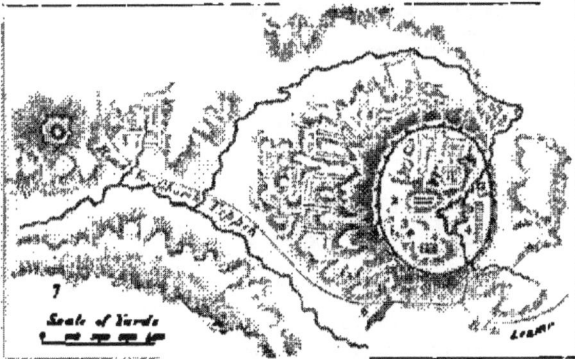

Plan of Ecbatana.

EXPLANATION.

1. Remains of a Fire-Temple.
2. Mound Musper.
3. Ancient Buildings with Shafts and Capitals.
4. Ruins of the Palace of Abdul Khan.
5. Rocky Hill of Zindani-Soleiman.
6. Cemetery.
7. Ridge of Rock called "the Dragon."
8. Hill called "Tawelah," or "the Stable."
9. Ruins of Nakorah.

ture (Ezr. vi. 2), each of these forms being probably a corruption of
Hagmatana as found in the Assyrian inscriptions. The site of this
town has been much discussed. It seems probable that there were
two towns of the name; one in the northern division of Atropatene,
at a place now called *Takht-i-Soleiman*, which was the older capital
of Arbaces, and one in the southern division at *Hamadán*, which
was in existence in Alexander's age.

The city was surrounded, according to Herodotus, by seven concentric walls, increasing in height from the outer to the inner, and
each of a different colour. This story had its origin in the circumstance that the seven colours specified were typical in oriental philosophy of the seven great heavenly bodies. The earlier Ecbatana was
the same place which under the Parthians was described by the various
names of **Phraäta, Praaspa, Vera, Gaza,** and **Gazica.** The later Ecbatana, *Hamadán*, was the residence of the Persian kings, and was more
than once visited by Alexander the Great. It was in existence in the
time of the Seleucidæ, and even later. **Rhagæ,** near the border of Parthia, is first noticed in the book of Tobit (i. 14) under the form
Rhages. It was rebuilt by Seleucus Nicator under the name Europus,
and subsequently by one of the Arsacidæ under the name Arsacia. Its
position near the Caspiæ Portæ made it at all times an important place.
Near the southern border of Media there is a very remarkable hill with
a precipitous cliff, formerly named Bagistanus Mons, now *Behistún*, on

the face of which are a series of sculptures with trilingual inscriptions descriptive of the victories of Darius. They are placed at an elevation of about 300 ft. from the base of the rock, and must have been executed with the aid of scaffolding. Semiramis was reputed to have made a paradise at this spot.

History.—The early history of Media is wrapped in great obscurity. Ctesias furnishes us with a list of kings preceding Cyrus, the first of whom, named Arbaces, would have commenced his reign about B.C. 875: Herodotus, on the other hand, notices only four, of whom the first, Deioces, began his reign B.C. 708, his successors being Phraortes (who is probably identical with the Arphaxad of Tob. i. 2), Cyaxares, and Astyages. The impression derived from the Assyrian annals is, that Media was in a state of semi-subjection to Assyria from the time of the Assyrian king *Shalmanubar*, about B.C. 860, until the accession of Cyaxares, B.C. 644; for the inscriptions record constant invasions, particularly under Tiglath-Pileser, who, about B.C. 740, transplanted the Syrians of Damascus to Kir, supposed to be the *Cyrus* (2 Kings xvi. 9), and under Sargon, about B.C. 710, who attempted a permanent subjection

Caption: Mons Bagistanus, Rock of Behistun.

by planting colonies of captive Israelites in the country (2 Kings xvii. 6). The attempt does not appear to have succeeded; for the inscriptions of Sennacheri and Esar-haddon describe it as a country that had never been subdued by their predecessors. During the whole of this period Media probably retained its own rulers, who acknowledged the supremacy of Assyria by the occasional payment of tribute. The authentic history of Me-

Caption: Sculptures on Rock of Behistun.

dia commences with Cyaxares, B.C. 634. The chief events of his reign were—his struggle with the Scythians, who still held a portion of the country, particularly the line of Zagrus; the capture of Nineveh, B.C. 625; and his war with Alyattes, king of Lydia, which was terminated by the well-known eclipse of Thales, probably B.C. 610 Cyaxares evidently endeavoured to grasp the supremacy which Assyria had exercised over Western Asia, or at all events over the northern portion of it, leaving the southern to Babylon. He is probably the Ahasuerus of the book of Tobit (xiv. 15). Cyaxares was succeeded by Astyages, B.C. 593, who led an uneventful life until the invasion of Cyrus, B.C. 558, when Media was absorbed in the Persian empire.

§ 7. IV. ARIANA.—Under the collective name of Ariāna the provinces in the eastern part of the plateau of Iran were included, viz. Gedrosia, Drangiana, Arachosia, the mountain-district of Paropamisus, Aria, Parthia, and Carmania. The title was originally an ethnological one, expressive of the district occupied by the Arian races, but, like the modern *Irán*, which is undoubtedly derived from it, it has acquired a purely geographical sense. Of the provinces enumerated very little information can be gathered from classical writers. The chief interest that attaches to them is in connexion with the military expedition of Alexander the Great, of which a review will be given after the description of the physical features of the various provinces.

1. **Carmania** was bounded on the S. by the Persian Gulf; on the W. by Persis; on the N. by Parthia; and on the E. by Gedrosia, from which it was separated near the sea-coast by a chain of hills named Persici Montes. It answers in name and position to the modern *Kirman*, but includes beyond that the greater part of *Laristán* and *Moghostán*. It was divided by Ptolemy into Carmania Deserta and Carmania Vera, or "Proper." The former consisted of the interior plain in the N., the latter of the mountainous district in the S., extending from the sea-coast, to a considerable distance inland. As the chains run generally in a direction parallel to the coast, no rivers of any importance reach the sea. The valleys and plains in the latter district are described as fertile, and the mountains themselves yield various mineral productions. The capital was **Carmāna**, in the interior, still existing under the name of *Kirman*. **Harmūsa**, on the sea-coast, was a place of considerable trade.

2. **Parthia** was bounded on the N. by Hyrcania; on the W. by Media; on the S. by Persis and Carmania; and on the E. by Aria and Drangiana. It thus comprehended the southern part of *Khordsan*, nearly the whole of *Kohistan*, and a portion of the great *Salt Desert*. It was inclosed on the N. and S.W. by mountains, viz. Labūtas, Elburz, and Masdorānus in the former direction, and Parachoathras in the latter; and on other sides by a vast desert. The Parthians were undoubtedly an Arian race; the name appears in the Sanscrit language under the form *Párada*. They were particularly celebrated in ancient times for the skill with which they discharged their arrows [1] as they

---

[1] Tergaque Parthorum, Romanaque pectora dicam ;
Telaque, ab averso quae jacit hostis equo.
Qui fugis, ut vincas, quid victo, Parthe, relinquis !
                              Ovid. *de Ar. Am.* i. 209.

Fidentemque fuga Parthum, versisque sagittis.
                              Virg. *Georg.* III. 31.

Navita Bosporum
Poenus perhorrescit, neque ultra
Caeca timet aliunde fata ;
Miles sagittas et celerem fugam
Parthi.   Hor. *Carm.* ii. 13, 14.

retreated. There were few towns of any importance. Hecatompylos, one of the capitals of the Arsacidæ, stood somewhat eastward of the Caspian Gates, probably near Jah Jirm, where an opening northwards exists between Labutas and Masdoranus. It owed its Greek name probably to Seleucus. Apamæa, surnamed Rhagiana, in the western part of the country, was built by the Greeks after the Macedonian conquest. Tagæ stood near the chain of Labutas, probably at Domeghan.

Parthia was the seat of an independent sovereignty from B.C. 250, when Arsaces threw off the supremacy of the Seleucidæ, until A.D. 226, when the Sassanian dynasty rose to power. After the decay of the Syro-Macedonian empire Parthia became the dominant state in western Asia, with Seleucia on the Tigris for its capital, and it offered a stout and protracted resistance to the arms of Rome. The Parthians defeated Crassus, B.C. 53, and were defeated by Cassius, B.C. 51. The surrender of the standards taken on the former occasion by the voluntary act of Phraates, B.C. 20, is referred to by Horace in more than one passage adulatory of Augustus.[2]

3. **Aria** was bounded on the N. by the Sariphi Montes, separating it from Margiana; on the E. by Bagöus Mons, the *Ghor* range; on the S. by Carmania; and on the W. by Parthia. It embraces the eastern portion of *Khordsan* and the western portion of *Afghanistán*. It is watered by the river **Arius**, *Heri Rúd*, which rises in Paropamisus, and runs towards the N.W., where it is absorbed in the sands. The valley of the *Heri Rúd*, as well as many other portions of the province, are very fertile. The chief towns were—**Aria**, the capital, on the river Arius, built, or more probably enlarged, by Alexander the Great, under the name of **Alexandria Arion**, and occupying the site of the present *Herat*. Not improbably the same place is described under the name of **Artacoana**.

4. **Paropamisidæ** is the collective name of a number of tribes occupying the southern spurs of Paropamisus from the upper course of the Etymandrus, *Helmund*, to the Indus, or in other words the provinces of *Cabulistán* with the northern part of *Afghanistán*. Their district was throughout rugged, but well watered, and possessing some fine fertile valleys. The rivers were the **Cophes** or **Cophen**, *Cabul*, which flows eastward into the Indus, receiving in its course the tributary waters of the **Choes**, *Kamah*, otherwise called the **Choaspes** and **Evaspla**; and the **Gurœus**, probably the *Punjkora*, but sometimes regarded as identical with the **Suastus**, which flows into the Choes. The chief town was

---

Nec patitur Scythas,
Et versis animorum equis
    Parthum dicere.   Hor. *Carm.* l. 19, 10.

Tela fugacis equi, et braccati militis arcus.
                     Propert. III. 4, 17.

[2] Et signa nostro restituit Jovi,
    Derepta Parthorum superbis
    Postibus.   Hor. *Carm.* iv. 15, 6.

Ille, seu Parthos Latio imminentes
Egerit justo domitos triumpho. . . . . Hor. *Id.* l. 12, 53.

                   Denique servam
Militiam puer, et Cantabrica bella tulisti
Sub duce, qui templis Parthorum signa refigit.
                   Hor. *Epist.* l., 18, 54.

Carūra or **Ortospāna**, the capital of the Cabolitæ (otherwise called Bolitæ), on the site of the modern *Cabul*. **Nicæa** was probably another name for the same place imposed by Alexander. **Gauzāca** is supposed to represent the modern *Ghizneo*. *Cabul* was the seat of an Indo-Scythian dynasty which established itself after the fall of the Bactrian empire. Its flourishing period appears to have been about A.D. 100.

5. **Arachosia** was bounded on the N. by the Paropamisadæ; on the E. by the Indus; on the S. by Gedrosia; and on the W. by Drangiana. It embraced the modern *Kandahar* with parts of the adjacent provinces. The country derived its name from the river **Arachōtus**, probably the *Arkand-ab*, one of the tributaries of the Etymandrus. The eastern part of this district is covered with the spurs and secondary ranges of the *Soliman* Mountains — the ancient **Paryēti Montes**. The site of the old capital, **Cophen**, also named **Arachōtus**, has not yet been satisfactorily determined ; it may have been at *Ulan Robat*, S.E. of *Kandahar*. A later capital was named **Alexandria** after Alexander the Great, but not founded by him : its position is wholly unknown.

6. **Drangiana** was bounded on the N. by Aria; on the E. by Arachosia; on the S. by Gedrosia; and on the W. by Carmania. It answers to the modern *Seistan*. The eastern part of it is mountainous : the western partakes of the character of the Carmanian plain. It is watered by the **Erymanthus** or **Erymandrus**, *Helmend*, which rises in the lower ranges of Paropamisus and flows towards the S.W. into the **Aria Lacus**, *Zaruh*. A second river, the **Pharnacōtis**, *Perrah-Rūd*, flows from the N. into the same lake. The inhabitants were named either Drangæ, Sarangæ, Darandæ, or Zarangæ. The appellation probably means "ancient," and points to this as the country in which the Arian race first established themselves. The capital, **Prophthasia**, stood N. of Lake Aria, probably at a place where ruins have been discovered between the modern towns of *Dushak* and *Furrah*.

7. **Gedrosia** was bounded on the N. by Drangiana and Arachosia; on the E. by the Indus; on the S. by the Indian Ocean; and on the W. by Carmania. It occupies about the same space as *Beloochistan* and *Mekran*. The northern part is mountainous, a considerable range named **Baetii Montes**, *Washáti*, intersecting the country throughout its whole length: another range, **Arbīti Montes**, *Bala*, skirts the eastern frontier, running parallel to the Indus: the Persici Montes, on the border of Carmania, have been already noticed. The rivers are unimportant, and in many cases are confined to the interior. The largest is the **Arābis**, *Purally*, which joins the Indian Sea at the point where it turns southwards. Gedrosia suffers from excessive heat and drought, and is hence for the most part unfruitful. Its most remarkable productions were myrrh, spikenard, and palms. The inhabitants of the coast appear to have lived very wretchedly, in huts of shells, roofed over with fish-bones, and subsisting wholly on fish. They were an Arian race, and were divided into various tribes. Along the southern coast were two tribes of Indian extraction, the **Arabitæ**, who lived between the Indus and the Arabis, and the **Oritæ**, to the westward of the latter river. The principal towns were **Rhambacia**, not far from the coast, perhaps at *Hawr ;* **Orea**, *Urmara*, founded by Nearchus at the mouth of the Tomerus; **Omāna**, a considerable port on the western part of the coast; and **Pura**, in the interior, perhaps at *Bunpur* the name is an appellative for a "town."

§ 8. V. THE NORTHERN PROVINCES. — It remains for us to

describe the northern provinces of the Persian empire—Hyrcania, Margiana, Bactriana, and Sogdiana:

1. **Hyrcania** lay along the south-eastern shore of the Caspian Sea, bounded on the W. by Media, from which it was separated by Mons Coronus and the river Charindas; on the E. by Margiana; and on the S. by Parthia, the range of Labutas intervening. It comprehended the eastern part of *Mazanderán*, and the district of *Astrabad*. With the exception of a narrow strip of coast, it is throughout mountainous and savage, and infested with wild beasts ;[a] this feature is expressed in its ancient name, Hyrcania, or *Vehrkóna*, "the land of wolves," which is still preserved in the name of the modern town *Gourgan*. The chief river was the **Sarnius**, or *Atrek*, in the eastern part of the country. The Hyrcanians were an Arian race. Their chief town was named Carta or Zadracarta, perhaps the same as Tape, in the W.

2. **Margiana** was an extensive district, lying between the Oxus on the N. and the Sariphi Montes on the S.; on the E. it was contiguous to Bactria, and on the W. to Hyrcania. It includes portions of *Khorasan, Balk*, and *Turcomania*. It contains tracts of great fertility, wherever water is attainable : elsewhere it is barren. The only river is the **Margus**, *Murgh-ab*, which rises in the Sariphi Montes, and flows towards the N.W.; formerly it joined the Oxus, but it now loses itself in the sands. The inhabitants were a Scythian race, the principal tribe being the **Massagetæ**. The capital, **Antiochia Margiána**, occupied the same site as the modern *Merv* on the Margus ; it is said to have been founded by Alexander, and to have been restored by Antiochus Soter.

3. **Bactria**, or **Bactriana**, was bounded on the N. and N.E. by the Oxus, separating it from Sogdiana; on the S.E. and S. by Paropamisus, and on the W. by the desert of Margiana. It answers both in name and position to the modern *Balk*,[b] but included also the eastern provinces of *Badakshan* and *Kundus*. The country is generally mountainous, offsets from Paropamisus covering the eastern and southern portions, and penetrating nearly to the valley of the Oxus. The valleys which intervene are fertile;[c] occasionally steppes and sandy tracts occur. The

---

[a]   Hyrcaniæque admôrunt ubera tigres.—VIRG. Æn. iv. 367.
Its dogs were also famous—
                Canis Hyrcano de semine.   LUCRET. iii. 750.
[b] The Zend form of the name, *Bakhdhi*, supplies the connecting link between the ancient and modern forms.
[c] Its fertility was known to the Romans ; in other respects its remoteness was the most prominent notion.

        Sed neque Medorum, silvæ ditissima, terra,
        Nec pulcher Ganges, atque auro turbidus Hermus,
        Laudibus Italiæ certent ; non Bactra, neque Indi, -
        Totaque thuriferis Panchaia pinguis arenis.— VIRG. *Georg.* ii. 136.
        Illæ ope barbarica variisque Antonius armis
        Victor ab Aurora populis et litore Rubro
        Ægyptum virresque Orientia, et *ultima* secum
        Bactra vehit.   ID. *Æn.* viii. 685.
In the following passage Bactra appears to be used as synonymous with Parthia —
        Urbi sollicitus times,
    Quid Seres et regnato Cyro
        Bactra parent.   HOR. *Carm.* iii. 29, 26.

chief river is the Oxus on its northern border, which has been already described, and which received several tributaries in Bactria—the Bactrus or Dargidus, Dehas, on which the capital stood, with its tributary the Artamis, Dakash,—the Dargomanes, Gorea, higher up—and the Zariaspis, which must be the same as the Bactrus, if the towns Bactra and Zariaspa are to be considered as identical.

The Bactrians were an Arian race, differing but little from the Persians in language, and using very nearly the same equipment as the Medes. The names of some of the tribes are evidently of Indian origin, the Khomari, for instance, representing the modern Kumáras, the Tokhari, the Thakurs, and the Varni, the word Varna, "a caste." The capital, Bactra or Zariaspa, was situated on the river Bactrus, on the site of the present capital Balk: the town lays claim to the very highest antiquity, and is to this day described as "the mother of cities;" it has in all ages been a great commercial entrepôt for the merchandise of eastern Asia; Alexander visited it in the winter of B.C. 328-7. The conqueror erected a city, Alexandria, in this province, probably at Khulm, E. of Bactra. Drepsa or Drapsaka, was probably at Anderáb, in the N.E. of the province.

Bactriana occupies a very conspicuous place both in the mythical and historical annals of the Greeks. It was visited by Bacchus, according to Euripides (Bacch. 15), and conquered by Ninus with the aid of Semiramis, according to Ctesias. The Bactrians aided at the destruction of Nineveh, and for a while resisted the arms of Cyrus. Bactria formed the 12th satrapy of Darius, and remained an integral portion of the Persian empire until its overthrow by Alexander. It was placed under satraps by the conqueror, and after his death fell to the Seleucidæ. In the reign of Antiochus II., Theodotus threw off the Syrian yoke, and established an independent sovereignty (B.C. 250). One of his successors, Eucratides, about B.C. 181, extended his sway over the western part of India, and another, named Menander, advanced his frontier to the Ganges. The power of this dynasty was overthrown by the advance of the Scythian tribes, probably about B.C. 100. It ultimately formed a portion of the Sassanian empire.

4. **Sogdiana** was bounded on the N. by the Jaxartes, and on the S. by the Oxus; eastward it was limited by the lofty chain of mountains, which under the name Comedárum Montes, Muztagh, runs northwards from Paropamisus; westward it stretched away to the Caspian Sea. It embraced Bokhara and the greater part of Turkestan. The eastern part of this province is mountainous, a considerable range of mountains named Oxii Montes, Ak-tagh, penetrating westward between the upper courses of the Oxus and Jaxartes; while another, the Sogdii Montes, Kara-tagh, emanated from the central range more towards the S. The only important rivers are those which have been noticed as forming the northern and southern boundaries: of the tributary streams which joined them we need only notice the Polytimétus, "the very precious" river, as the Greek historians rendered the indigenous name, Sogd, which waters the far-famed valley of Samarcand; the modern name of the stream, Zar-asshan, means "gold-scattering," and contains a similar allusion to the fertility which it spreads about its banks. It flows into the Lake of Karakoul, which probably represents the ancient Oxia Palus.

The Sogdians were allied in race to their neighbours the Bactrians; many of the names of the tribes point to a connexion with India. These are for the most part devoid of interest; we may, however, notice the Chorasmii as representing the modern *Kharism*, or the desert between the Caspian and the Sea of Aral. The towns of importance were — **Maracanda**, *Samarcand*, on the Polytimetus, which has been in all ages a great commercial entrepôt; **Cyreschata** or **Cyropolis**, on the Jaxartes, deriving its name from the tradition that it was the extreme limit of Cyrus's empire; **Alexandria Ultima**, also on the Jaxartes, enters at or near *Khojend*, its name implying that it was the farthest town planted by Alexander in that direction; **Alexandria Oxiana**, probably situated at *Kurshee*, S. of *Samarcand*, where is a fertile oasis; and **Tribactra**, probably representing the modern *Bokhara*.

§ 9. The countries, which we have just described as the northern and eastern provinces of the Persian empire, derive a special interest from the military expedition of Alexander the Great, which gave occasion to the only satisfactory account of them that has reached us. We therefore append a brief review of that expedition in as far as its geographical details are concerned, commencing with the departure of Alexander from Susa.

The Expedition of Alexander the Great.—Alexander started on his Asiatic expedition, in B.C. 334, from his Macedonian capital, Pella. His early course lay along the N. coast of the Ægean Sea by the towns of Amphipolis, Abdera, and Maronea: he reached the shores of the Hellespont at Sestus, and, while his army crossed directly to Abydos, he himself went to Elæus, and crossed to the harbour of the Achæans, the old landing-place of Ilium. Having visited the most interesting spots connected with the history of Troy, he rejoined his army, and advanced along the coast of the Hellespont by Percote and Hermotus to the river Granicus, where his first great victory over the Persians was gained. From the banks of the Granicus he turned southwards through the interior of Mysia and Lydia to Sardis, and thence to Ephesus, both of which surrendered to him without a contest. Miletus was the next important point, and here he met with determined but ineffectual resistance. Thence he advanced to the siege of Halicarnassus, which detained him for a considerable time. Having reached the S. angle of Asia Minor, he turned eastward, and entered Lycia, following the line of coast by Telmissus and Pinara to Patara, and thence crossing to Phaselis. In advancing along the coast N. of Phaselis, he traversed with difficulty the dangerous pass at the foot of Mount Climax, and reached Perge in Pamphylia, whence he advanced to Side on the sea-coast, and to Syllium, a place of uncertain position between Side and Aspendus. He returned to Perge, and struck northwards through the defiles of Taurus by Sagalassus to Celænæ in Phrygia, and thence across the plains of that province to Gordium in Bithynia, which he reached in the early part of the year 333. He halted there for some two or three months, and resumed his course in an E. direction as far as Ancyra, and then S. across Cappadocia to the Cilician Gates of Taurus, which dangerous pass he traversed without molestation, and descended on the S. side of Taurus to the fertile plains of Cilicia. At Tarsus he halted for some time, and made an excursion thence to Anchialus and Soli in the W. of Cilicia. Resuming his course from Tarsus in a S. E. direction, he crossed the Aleïan plain to Mallus at the mouth of the Pyramus, and then followed the line of coast to Issus, and through the gates of Cilicia and Syria to Myriandrus in Syria. Meantime

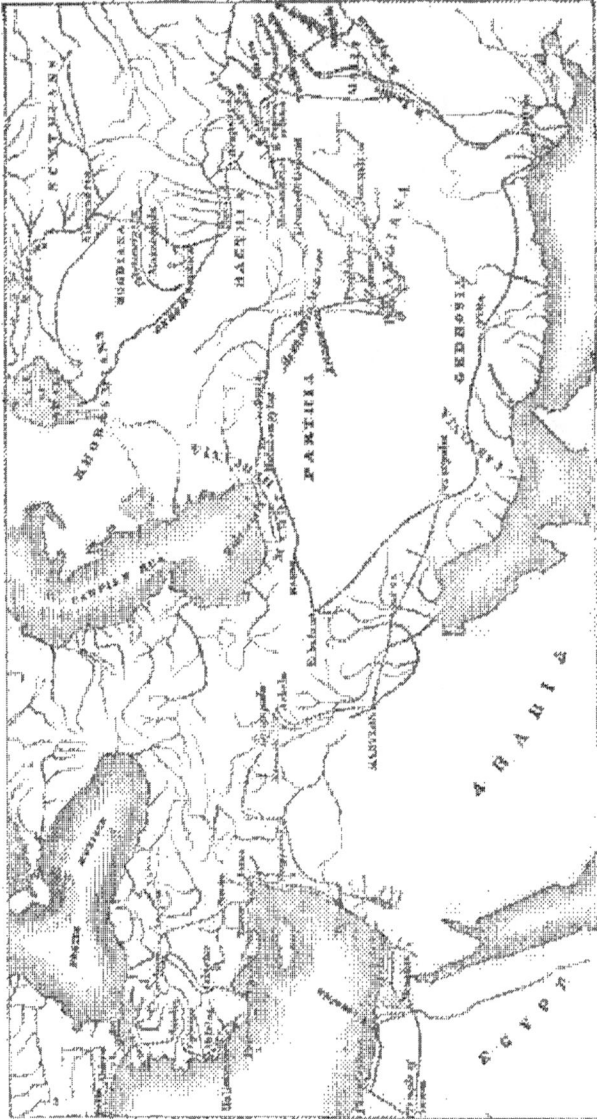

March of Alexander.

Darius was crossing the Amanian range by the northern pass which
descends into Cilicia near Issus. Alexander therefore retraced his
steps, and met the enemy on the banks of the Pinarus, where he again
triumphed in the important battle of Issus. From this point Alexander hastened southwards through Syria to Phœnicia, the chief towns
of which (Marathus, Byblus, Sidon) surrendered, with the exception
of Tyre, which sustained a siege of seven months. Thence (in 352) he
followed the coast southwards, and met with no further obstacle until
he reached Gaza, which held out against him for two months. In
seven days he crossed from Gaza to Pelusium on the frontier of Egypt:
he ascended the eastern branch of the Nile to Memphis, and descended
by the western branch to Canopus. After the foundation of Alexandria, he made his famous expedition to the oracle of Jupiter Ammon,
reaching it by way of Parætonium on the Mediterranean coast, and
returning to Memphis across the desert. In 331 he retraced his steps
to Phœnicia, and struck across from Tyre to Thapsacus on the Euphrates, and having crossed that river took a northerly route under
the roots of Masius to the Tigris at Nineveh, and again succeeding in
the passage of the river, he advanced to meet the hosts of Darius on the
plain of Gaugamela. A decided victory awaited him, the fruits of which
he reaped in the surrender of Babylon and Susa, which he visited in
succession, remaining a short time in each. Leaving Susa, he struck
across the mountainous region that separates Susiana from Persis,
defeating the Uxians at the defile that commands the western, and the
Persians at that which commands the eastern entrance to the "Persian
Gates," and reached Persepolis. In 330 he went in pursuit of Darius
to Ecbatana (Hamadan), and Rhagæ, and passed through the Caspian
Gates to Hecatompylus (near Jah Jirm). The lofty range of Elburz
was surmounted in the invasion of Hyrcania on the borders of the
Caspian Sea, and the forest haunts of the Mardians on the confines of
Ghilan and Mazanderan were scoured: Zadracarta (Sari) witnessed the
triumphal entry of the conqueror. From Hyrcania Alexander proceeded to Parthia, rounding the ridge of Elburz at its eastern extremity, and reached Susia (near Meshed); Aria yielded, and he started
for Bactria; but he was summoned to Artacoana in consequence of a
revolt, and passing through the plain of the Arius (Heri-rūd), decided
on founding the city of Alexandria Ariorum, which still survives under
the name Herat. The next point was Prophthasia (near Furrah), the
capital of Drangiana. In 329 Alexander passed up the valley of the
Etymander into Arachosia, where he founded another Alexandria,
now Candahar. The range of Paropamisus intervened between this
and Bactria: at the southern entrance of the pass of Bamian, about
50 miles north-west of Cabul, another Alexandria, surnamed "ad Cau-
casum," was founded. Surmounting the lofty barrier, he descended
by Drapsaca and Aornus to Bactra, Balk, in the valley of the Oxus.
He crossed the Oxus, probably at Kilif, and traversed the desert north
of that river to the fertile banks of the Polytimetus, Kohik, and the
town of Maracanda, Samarcand; thence on to Jaxartes, the farthest
limits of the known world, where another Alexandria, surnamed
"Ultima," was planted, probably on the site of Khojend. He crossed
the Jaxartes to attack the Scythians, and received homage, not only
from them, but from the distant Sacæ. The disaster of his general,
Pharnuches, recalled him to Maracanda, and led him in pursuit of the
enemy down the valley of the Oxus to the edge of the desert of Khiva.
He returned by the course of the Polytimetus, and passed the winter

of 329 at Bactra. The visit of Pharasmanes, king of the Choras-mians, gave him an opportunity of acquiring some information relative to the extensive steppe about the *Sea of Aral*. In 328 Alexander re-entered Sogdiana, and achieved the capture of a stronghold named the "Sogdian Rock," probably near the pass of *Derbend*, whence he returned to Maracanda. He next visited the district of Xenippa, about 10 miles N. of *Bokhara*, and returned to winter at Nautaca. In 327 Alexander invaded Paretacene, somewhere eastward of Bactria, and took the stronghold of Chorienes. He returned to Bactra, whence he started for his Indian campaign. Having crossed Paropamisus, he descended the course of the Cophen, *Cabul*, by Nicæa, probably the same as Ortospana or Cabura (the modern *Cabul*), to its junction with the Choes, also called Choaspes and Evaspla (the modern *Kamah*), where he turned off into the mountain district intervening between the Cabul and the Indus: the river Gursæus in that district is probably the *Punjhora*, which runs parallel to the Choes; the towns Gorydala and Arigœum stood at the foot of the Indian Caucasus, near the sources of these streams; descending the Gursæus he seized Massaga and the strongholds Ora and Bazira, between the Gursæus and Indus; he returned to the Cophen at Peucela, a place not far westward of the junction of the Cophen and Indus—descended the stream to Embolima—followed up the right bank of the Indus for a short distance to attack the stronghold of Aornus, and having captured it, onwards to Dyrta, probably at the point where the Indus forces its passage through the Hindoo Koosh, whence he returned to the junction of the Cophen. In 326 he crossed the Indus at this point and advanced into the *Punjab* by Taxila (the ruins of which still exist at *Manikyala*) to the banks of the Hydaspes, *Jelum*, one of the five rivers of the district; the spot at which he crossed that river, as well as the sites of the towns Nicæa and Bucephala, which were built to commemorate, the former his victory over Porus, the latter his passage of the river, cannot be identified. Proceeding eastward, he reached the Acesines, *Chenab*, and the Hydraotes, *Ravee*, which he crossed to Sangala, the modern *Lahore*. Proceeding still eastward, he reached the banks of the Hyphasis, *Gharra*, below the junction of the Hesudrus, *Sutledj*. This formed the eastern limit of his discoveries. He returned to the Hydaspes, where a fleet had been prepared for his army, and dropped down that stream to its junction with the Acesines, turning aside to the capture of the city of the Malli, *Mooltan*—then down the Acesines to its junction with the Indus, at which point he built an Alexandria, probably at *Miltun*—and then down the Indus to Patiala at the head of the Delta. In 326 he separated from his fleet, sending Nearchus to explore the coasts of the Indian Ocean to the mouth of the Tigris, while he himself took a land route through Gedrosia and Carmania. His intention had been to follow the line of coast, but finding this impracticable from the excessive heat and sterility of that district, he struck into the interior, and passing by Pura, probably *Bunpur*, he reached the frontier of Carmania, his army having endured terrible sufferings in the passage across the Gedrosian desert. His route through Carmania and Persis was comparatively easy; passing through Pasargadæ and Persepolis in the latter province, he finally gained Susa. The voyage of Nearchus was successful, but presents few topics of interest to us; he followed the coast to the entrance of the Persian Gulf, put in near the mouth of the Anamia, *Ibrahim*, a little eastward of the isle of *Ormus*, and thence resumed his course to the mouth of the Tigris.

§ 10. India was a term used somewhat indefinitely for the country lying eastward of the river Indus. Down to the time of Alexander, it was confined to the districts immediately adjacent to that river; under the Seleucidæ, it was extended to the banks of the Ganges; in Ptolemy's geography, it comprehends all the countries between the Indus and the Eastern Ocean, which were grouped into two great divisions, India intra Gangem, and India extra Gangem. The details of the geography of these vast regions are for the most part devoid of interest to the classical student; but they have their special interest both for those who are acquainted with Indian topography, and in connexion with the history of geography and commerce. It would be out of our province to go into the former subject, and therefore we shall confine ourselves to a general sketch, with a special reference to the latter subject.

(1.) In addition to the more important physical features already noticed,* we may further adduce the following as being known to the ancient geographers: (1.) *Mountains*—Bettigo (the *Ghats*), and Vindius (*Vindhya*). (2.) *Promontories*—Comaria (*Comorin*), Cory or Calligicum (near the S.W. end of the peninsula), Prom. Auræ Chersonesi, the southern termination of the Sinus Sabaricus; Malei Colon, on the W. coast of the Golden Peninsula; and Prom. Magnum, the western side of the Sin. Magnus. (3.) *Gulfs and Bays*—S. Canthi (*G. of Cutch*), S. Barygazenus (*G. of Cambay*), S. Colchicus (*B. of Manaar*), and S. Argaricus, opposite Taprobane (probably *Palk's Bay*). (4.) *Rivers*—Namadus (*Nerbudda*), Namaguna (*Tapty*), along the shores of the Indian Ocean; along the W. side of the Bay of Bengal, Chaberis (*Caveri*), Tyndis (*Kistna*), Mæsolus (*Godavery*), Dosaron (*Mahanadi*), and Adamas (*Brahmini*).

(2.) The principal states on the coast from W. to E. were—Pattalene (*Lower Scinde*), with its capital Pattala (*Tatta*); Syrastrene, W. of the *G. of Cambay*; Larice, along the Indian Ocean from the *Nerbudda* to the *G. of Cambay*, with Ozene (*Oujein*) as its capital; further S., Ariaca, with Hippocura (*Hydrabad*); Dachinabades (*Deccan*); Limyrica, near *Mangalore*, with Carura (*Coimbatore*) for its capital; Cottiara (*Cochin*) and Comaria, at the end of the peninsula; Pandionis Regnum, on the S.E. coast with Modura (*Mathura*) for its capital; then in order up the eastern coast, the Arvarni with Malanga (*Madras*); Mæsolia, in the part of the coast now called *Circars*; the Calingæ; and the Gangaridæ,[7] with Gange (somewhere near *Calcutta*) for their capital. In the interior, commencing from the W., a race of Scythians occupied in the days of Ptolemy an extensive district on the banks of the Indus, comprising the modern *Scinde* and *Punjab*; Caspiria (*Cashmir*), lay more to the N.; the Caspirei between the Hyphasis and the Jomanes: on the course of the Ganges, the Gangani; the

---

* P. 78.

[7] The conquest of this remote people was attributed to Augustus in the most fulsome style of adulation—

In foribus pugnam ex auro solidoque elephanto
Gangaridum faciam, victorisque arma Quirini.

Virg. *Georg.* III. 27.

Mandalm with the town Palimbothra (*Patna*) ; and the Marundæ, thence to *Calcutta*.

(3.) The chief commercial towns, were—along the western coast of Hindostan, Pattala (*Tatta*), Barygaza (*Baroch*), Calliene (*Gallian*), Muziris (*Mangalore*), and Nelkynda (*Neliceram*) ; while there were three principal emporia for merchandize—Ozene (*Oujein*), the chief mart of foreign commerce, and for the transmission of goods to Barygaza, Tagara (probably *Deoghir* in the *Deccan*), and Plithana (*Pultnuah* on the *Godavery*).   Along the Regio Paralia, and on the *Coromandel* coast were several important ports ; in the kingdom of Pandion, were extensive pearl-fisheries.   Further to the N. were—Masolia (*Masulipatam*), famous for its cotton goods ; and Gange, near the mouth of the Ganges, a mart for muslin, betel, pearls, &c.

(4.) The productions of India best known to the Romans were its ivory, its gold and gems,[a] its frankincense,[b] and its ebony.[c]

§ 11. The important island of **Taprobāne**, otherwise called **Salice**, *Ceylon*, has been frequently noticed in connexion with the history of geography.   It was well known to the ancients from its commercial importance.[d]   According to Pliny it contained no less than 500 towns, the chief of which was named **Palæsimundam**, probably the same as is elsewhere called Anurogrammon, which remained the capital from B.C. 267 to A.D. 769.   The island is but seldom alluded to in classical literature.[e]

§ 12. The **Sinæ** occupied a district of undefined limits to the N.E. of India extra Gangem, stretching to Serica in the N.   It probably included the modern districts of *Tonquin*, *Cochin-China*, and the southern portion of *China*.   This district is first described by Ptolemy, who evidently had but a very imperfect knowledge of it.   The towns of most importance were—**Thinæ**, either *Nankin*, or *Thsin* in the province of *Schensi* ; and **Cattigara**, perhaps *Canton*.

§ 13. **Serica** was a district in the E. of Asia, the position of which is variously described by ancient writers, but which is generally supposed to have occupied the N.W. angle of *China*.   The name of Serica as a country was not known before the first century of our era, but the Seres as a people are mentioned by Ctesias and other early writers.   It is uncertain whether the name was an indi-

---

[a] India mittit ebur.   VIRG. *Georg.* l. 57.

Indum sanguineo veluti violaverit ostro
Si quis ebur.   ID. *Æn.* xii. 67.

   Non aurum, aut ebur Indicum.   HOR. *Carm.* l. 31, 6.

            Gemmis et dentibus Indis.   Ov. *Met.* xi. 167.

[b] Et domitas gentes, thurifer Inde, tuas.   Ov. *Fast.* iii. 720.

Thura nec Euphrates, nec miserat India costum.   *Id.* i. 341.

[c]   Sola India nigrum
Fert ebenum.   VIRG. *Georg.* ii. 116.

[d] It consisted of pearls and precious stones, especially the ruby and the emerald.

[e] Aut ubi Taprobanen Indica cingit aqua.   Ov. *ex Pont.* l. 5, 80.

gonous one, or was transferred from the silkworm to the district in
which the insect was found. The country is described as very
fertile, with an excellent climate, its most valuable production being
silk.[4] The method by which commerce was carried on with this
distant people has been already described (p. 80).

§ 14. The vast regions lying between Serica in the E., Sarmatia
Asiatica in the W., and India in the S., were included under the
general name of Scythia, the limits to the N. being wholly unknown.
The modern districts of *Tibet, Tartary*, and a large portion of
Siberia, may be regarded as answering to it. Very little was known
of these remote regions: Herodotus was only acquainted with the
names of the tribes to the N. of the Euxine and Caspian Seas, and
no succeeding writer adds much to his information until we come
down to the age of Ptolemy. By him the country was divided into
two parts, Scythia *intra* and S. *extra* Imaum, in other words
Scythia W. and E. of Imaus, by which he designated the northern
ranges of *Bolor* and its continuations. The mountains and rivers,
which received special names in ancient geography, have been
already noticed (pp. 74, 77).

§ 15. The origin and ethnological affinities of the Scythians are
involved in great obscurity. Into these questions it is unnecessary for us to enter, particularly as we have no reason to
suppose that the name, as applied by Ptolemy, indicated any one
special race, but rather included all the nomad tribes of Central
Asia. It is a matter of more interest for us to know that these
tribes have left traces of their existence amid the gold mines of the
Altai ranges, and in numerous sepulchres and ruined buildings, the
high antiquity of which is undoubted. The conclusion drawn from
these remains is that those nations had attained a higher degree of
civilisation than we should have expected: their skill in metallurgy
is particularly conspicuous. Of the special tribes we may notice—
the **Aorsi**, between the Daix and the Jaxartes, a people who carried
on an extensive trade with India and Babylonia; the **Massagetæ**,
who frequented the steppes of *Independent Tartary* about the *Sea of
Aral*; the **Sacæ**, who occupied the steppes of the *Kirghiz Khasaks*
and the regions both E. and W. of *Bolor*, through whom the trade
was carried on between *China* and the west, as already described;
the **Arsippæi**, the progenitors of the *Calmucks*, who lived in the
Altai; and the **Issedones**, in the steppes of *Kirghiz of Ichim*.

---

[4] Quid, quod libelli Stoici inter Sericos
Jacere pulvillos amant?   Hor. *Epod.* viii. 15.

It was supposed at one time that the Seres obtained the substance from the
leaves of trees. Virgil alludes to this in the line—

"Velleraque ut foliis depectant tenuia Seres l" — *Georg.* ii. 121.

The Nile during the inundation, with the two Colossi of Thebes (Wilkinson).

BOOK III.

AFRICA.

—o—

CHAPTER XIV.

AFRICA.

§ 1. THE continent of Africa, as known to the ancients, was bounded by the Mare Internum on the N.; the Oceanus Atlanticus on the W.; and the Isthmus of Arsinoë, the Arabicus Sinus, and the Mare Erythræum on the E. Its southern limit was unknown: Herodotus indeed correctly describes it as surrounded by water, but the progress of geographical knowledge tended to weaken rather than confirm this belief, and the latest opinion was, that below the equator the coast of Africa trended eastward, and formed a junction with the coast of Asia, converting the Indian Ocean into an inland sea. How far the continent may have extended to the S. does not appear to have been even surmised; the actual knowledge of the interior was limited to the basin of the *Niger*, while the E. coast had been partly explored to about 10° S. lat., and the W. coast to about 8° N. lat., or the neighbourhood of *Sierra Leone*. But even

the greater part of the continent within these limits was, and still
is, a *terra incognita.* The portion of the continent of which the
ancients possessed any adequate knowledge was restricted to the
districts contiguous to the N. coast and the valley of the Nile.

Names.—The history of the names "Libya" and "Africa" is strikingly analogous to that of "Asia." When we first hear of this continent in the Homeric poems no general name is given to it. "Libya" is
the name only of a *district* contiguous to Egypt on the W. The Greeks
early became acquainted with the use of this name through their intercourse with Egypt, and thus gradually extended it to the whole of the
continent, in the first instance exclusive, and finally inclusive, of Egypt
itself. The origin of the name is doubtful. It was referred by the
Greeks to a mythological personage who was either a daughter of
Oceanus or a hero. In later times it has been variously connected with
the Biblical "Lubim"—who were not, however, a maritime, but probably an inland people—and with the Greek λίψ (from λείβω), "the
south-west wind," which blew to Greece from that quarter, and derived
its name from its moist character. The name "Africa" originated
with the Romans in the district adjacent to Carthage, which constituted
their first province on this continent. It was probably the name of a
native tribe, but its origin is still a matter of great uncertainty. Josephus connects it with Epher, a grandson of Abraham and Keturah. It
may perhaps have a Phœnician origin, and mean "Nomads," in which
case it would be equivalent to the Greek Numidia.

§ 2. The seas that surround the continent of Africa are singularly
deficient in bays and estuaries, and hence the coast-line bears a
very small proportion to the area, as compared with either of the
other continents. The uniformity of the Mediterranean coast is
indeed broken by the deep indentations named **Syrtes Major** and
Minor, answering to the *Gulfs of Sidra* and *Khubs.* These are
really the innermost angles of an extensive sea which penetrates
between the highlands of Cyrene on the E. and the Atlas range on
the W. The special names for the parts of the sea adjacent to Africa
were, **Mare Ægyptium,** off the coast of Egypt, and **Libycum Mare,**
more to the W. The shores of the **Oceanus Atlanticus** were explored
by the Carthaginians, but the records contain no topics of interest
connected with it. Of the Southern Ocean the ancients knew still
less. The portion adjacent to the coast was named generally **Mare
Æthiopicum,** and a portion of it S. of *Cape Guardafui* **Mare Bar-
baricum.**

§ 3. Libya, or Northern Africa W. of Egypt, was divided by Herodotus into three parallel belts or districts—the cultivated, the
wild-beast district, and the sandy desert. The first and third of
these denominations answer respectively to the *Tell* of the Arabs
and the *Sahara.* The second is a misrepresentation, and the true
intermediate district is better described by the modern Arabic name
Beled-el-Jerid—"the date-district"—the chain of oases, in which

that fruit is found most abundantly, lying between the cultivated district of the coast and the great sandy desert of the interior. It is a mistake, however, to suppose that the three belts are marked off from each other by any well-defined lines of demarcation; on the contrary the limits are shifting; the *Tell* and *Sahara* are often intermixed, even in the W., where the range of Atlas would seem to form a barrier between the two. The true distinction is one of *production*, and not of position, and the remarks of Herodotus must be accepted as only generally true.

§ 4. The mountains of Africa do not present the same uniformity as those of Asia. In the W. there is an extensive but isolated system, to which the ancients transferred the mythological name of Atlas,[1] occupying that division of the continent which lies between the Syrtes and the Atlantic Ocean. The extreme points of this range may be regarded as *C. Ghir* in the W. and *C. Bon* in the E., and the general direction would therefore be from W.S.W. to E.N.E. It is divided into two portions by the valley of the Moloenth. The W. division, or *High Atlas*, strikes northwards along the course of that river, and in the neighbourhood of the sea sends out lateral ridges parallel to the coast towards the W., to which the ancients gave the specific name of Atlas Minor. The eastern division consists of the range of *Jebel Amer* and a series of subordinate parallel ridges, which gradually approach the Mediterranean coast and decline into the desert in the neighbourhood of the Syrtes.

§ 5. The only river in Africa that holds an important place in ancient geography is the Nile, which was at once the great fertilizer of Egypt and the high-road of commerce and civilization.

The Nile, more than any other river in the world, attracted the attention of writers of all classes. The discovery of its sources was regarded

---

[1] We have already noticed the Homeric sense of the term Atlas (p. 30). The same idea was sustained by the later poets, as when Æschylus speaks of the giant Atlas:—

ὃς πρὸς ἑσπέρους τόπους
ἕστηκε κίον' οὐρανοῦ τε καὶ χθονὸς
ὤμοιν ἐρείδων, ἄχθος οὐκ εὐάγκαλον.   *Prom. Vinct.* 348.

Ubi coelifer Atlas
Axem humero torquet stellis ardentibus aptum.—*Virg. Æn.* vi. 797.

Atlas en ipse laborat
Vixque suis humeris candentem sustinet axem.—*Ov. Met.* ii. 297.

Quantus erat, mons factus Atlas. Jam barba comæque
In silvas abeunt; juga sunt humerique manusque;
Quod caput ante fuit, summo est in monte cacumen,
Ossa lapis fiunt. Tum partes auctus in omnes
Crevit in immensum (sic Di statuistis), et omne
Cum tot sideribus cœlum requievit in illo.—*Ov. Met.* iv. 658.

as the greatest geographical problem of antiquity.[1] It was indeed
believed that it issued from marshes at the foot of the Lunæ Montes,
but the true position [2] of the *Mountains of the Moon* was unknown, and
the description will apply to other Abyssinian rivers, which generally
rise in lagoons. It appears moreover probable that the ancients regarded the Astapus, or *Blue Nile*, to be the true river, and that their
observations applied to that rather than to the *White Nile*, which
moderns generally regard as " the true Nile," as being the larger stream.
At the same time it should be observed that the "blue," or rather
the "black," Nile—for that is the meaning of the Arabic *Azrek*—has
the true characteristics of the Nile. These two branches form a junction S. of Meroë, and for some miles flow together without mixing their
waters. N. of Meroë the united stream receives the **Astabōras**, *Tacazze*;
between that point and the border of Egypt is the region of the "Cataracts," as they are called, which are in reality nothing more than rapids
formed by ridges of granite, which rise through the sandstone, and, by
dividing its stream, increase its rapidity. The fall is, after all, not so
considerable as the imagination of the poets pictured it, the Great
Cataract having a descent only of 80 feet in a space of five miles.
Below the junction of the Astaboras the river flows N. for 120 miles,
then makes a great bend to the S.W.—skirting in this part of its course
the desert of *Bahiouda*—and finally resumes its northerly direction to
the head of the Delta, where it is divided into seven channels,[3] which
were named from E. to W.,—the Pelusian, now dry; the Tanitic, probably the canal of *Moueys*; the Mendesian, now lost in Lake *Men-
saleh*; the Phatnitic, or Bucolic, the lower portion of the *Damietta*
branch; the Sebennytic, coinciding with the upper part of the *Damietta*
branch, and having its outlet covered by the lake of *Bourlos*; the Bol-

---

[1] Nile pater, quænam possum te dicere causa,
    Aut quibus in terris occuluisse caput.—Tibull. i. 7, 23.

    Te, fontium qui celat origines
    Nilus.   Hor. Carm. iv. 14, 45.

    Ille fluens dives septena per ostia Nilus,
    Qui patriam tantæ tam bene celat aquæ.—Ov. Amor. iii. 6, 39.

    Qui rapido tractu mediis elatus ab antris,
    Flaminigeræ patiens zonæ Cancrique calentis,
    Fluctibus ignotis nostrum procurrit in orbem,
    Secreto de fonte cadens, qui semper inani
    Quærendus ratione latet; nec contigit ulli
    Hoc vidisse caput: fertur sine teste creatus,
    Flumina profundens alieni conscia cœli.— Claud. Idyl. iv. 8.

    Aut septemgemini caput haud penetrabile Nile.— Stat. Silv. iii. 5, 21.

Cæsar is represented as willing to relinquish all his schemes of grandeur for
the solution of the problem—
                         spes sit mihi certa videndi
    Niliacos fontes, bellum civile relinquam.—Luc. x. 191.

[2] The source of the ' White Nile ' has been recently discovered in lake Victoria
Nyanza, lying between 0° 20' N. and 3° 50' S. lat.

[3] Et septemgemini turbant trepida ostia Nili.— Virg. Æn. vi. 801
    Et septem digestum in cornua Nilum.   Ov. Met. ix. 773.
    Sive qua septemgeminus colorat
    Æquora Nilus.   Catull. xi. 7.

bitic, the lower part of the *Rosetta* branch ; and the Canopic, or Nau-
cratic, coinciding in its upper part with the *Rosetta* branch, from which,
however, it diverged at 31° lat., and ran more to the W., discharging
itself at the Lake of *Madieh*, near *Aboukir*.

§ 6. The Oases form a peculiar and a very important feature in
the continent of Africa. The word *Oasis* is derived from the Coptic
ouah, "a resting-place." It was a general appellation for spots of
cultivated land in the midst of sandy deserts, but was more especially applied to those verdant spots in the Libyan desert which connect eastern with western and southern Africa. The ancients describe these as islands rising out of the ocean of the wilderness, and
by their elevation escaping the waves of sand which overspread the
surrounding districts. They are, however, *depressions* rather than
elevations—basins which retain the water through the circumstance
of a stratum of clay or marble overlying the sand. The moisture
thus secured produces in the centre of the basin a prolific vegetation, which presents the most striking contrast to the surrounding
desert, and justifies the appellation of the "Island of the Blessed,"
which the ancients [a] applied to one of them. Their commercial importance was very great. They served as stations to connect Egypt
and Ethiopia with Carthage in one direction, and with central Africa
in another. Their full advantage indeed was not realised until the
camel was introduced from Asia by the Persians. After that time
they were permanently occupied and garrisoned by the Greeks and
Romans. Herodotus describes a chain of oases [b] as crossing Africa
from E. to W. at intervals of ten days' journey. With the exception
of the two most westerly—the Atarantes and Atlantes—the localities admit of easy identification, but the distances require a little
adjustment, for Ammonium is twice ten days from Thebes, and a
similar interval exists between Augila and Phazania. In the first
instance he probably computes the distance from the Oasis Magna,
which is midway between Thebes and Ammonium ; in the second,
he omits the intervening oasis of *Zala*.

§ 7. The commerce of Africa was known to classical writers
chiefly through the two nations in whose hands the foreign trade
rested, viz. the Egyptians in the E. and the Carthaginians in the
W. These regulated the trade of the interior, whence they obtained
certain articles of luxury, and ornament highly prized by the
wealthy of Greece and Rome, and received in exchange the oil and
wine of which they themselves stood in need. But though Egypt

---

[a] Herod. III. 26.
[b] They are Ammonium, *el-Siwah* ; Augila, *Aujileh* ; the Garamantes, *Fezzan*,
the Atarantes, who may represent a place on the outskirts of *Fezzan*, and the
Atlantes, whose name bears reference to the range of Atlas.

The Little Oasis.

and Carthage were thus the great marts of African commerce, the trade with the interior was actually carried on by certain tribes who were fitted by birth and habit to endure the privations and dangers incident to the long journeys across the desert. The Nubians were the carriers of Egypt; the Nasamonians and other tribes that lived about the Syrtes were the carriers of Carthage. These tribes conducted their business very much in the same manner, and by the same routes, as the Africans of the present day, the physical character of the continent necessitating the adoption of the caravan as the only secure mode of travelling, and fixing the routes with undeviating certainty by the occasional supplies of water.

§ 8. The most valued productions of the Interior were gold, precious stones, ivory, ebony, and slaves.

(1.) Gold was abundant both in the Æthiopian mountains and in the very heart of the continent S. of the Niger.

(2.) Precious stones were procured from the mountains of Central Africa. The most common species was the carbuncle, which derived its classical name, "calcedonius," from the Greek name of Carthage, whence it was exported to Italy.

(3.) Ivory was found in all parts. The Ptolemies had their stations on the shores of the Red Sea for the express purpose of hunting elephants. In the interior of Æthiopia and the adjacent districts of *Kordofan* and *Darfur*, it was the staple commodity, while even on the western coast of the Atlantic the Carthaginians found it abundant.

(4.) Slaves were perhaps the largest article of African commerce. Not only did the Egyptians and Carthaginians require them for their own domestic use, but the latter people exported them, particularly females, in immense numbers to Italy and the Mediterranean islands. The supply was obtained from the interior of the continent, particularly the districts about the *Niger*. Herodotus tells us that the Garamantes had regular slave-hunts, and his statement is verified by the modern practice of the chieftains of *Fezzan*, who hunt down the *Tibboos*.

As the trade was chiefly carried on by means of barter, it becomes an interesting question what productions were given in exchange by the merchants. The same articles appear to have formed the *media* of exchange in ancient as in modern times. The northern part of the desert is abundant in salt; Central Africa is deficient in it; and a scarcity of this necessary article operates as a famine in the districts S. of the great desert; this, therefore, forms the great staple of trade in exchange for gold and slaves. Dates are another valuable commodity. The region of dates lies between 26° and 29° N. lat., and from this district it is exported largely in all directions—southwards as far as the Niger, and northwards to the shores of the Mediterranean, whence the agricultural tribes, in the time of Herodotus as at the present day, made periodical journeys to obtain their supply. With regard to the Carthaginian trade on the shores of the Atlantic we are told that trinkets, harness, cups, wine, and linen, were given to the natives.

§ 9. We are acquainted with several of the main routes by which the traffic was carried on. In Africa, as in Asia, there were certain

spots which were the focusses of the caravan-trade. Thebes in
Egypt was the chief emporium in the lower valley of the Nile;
Meroë in Æthiopia was the chief one on the Upper Nile; Phazania,
Fezzan, was the chief one in the interior. These were connected
by chains of posts, forming the great lines of communication, and
each post, in its measure, becoming a commercial mart. Lastly,
Coptos was the chief emporium for the Indian trade, which passed
through the ports of Myos Hormos and Berenice.

(1.) From Thebes a route led westward through the oases of Ammonium and Augila to Phazania, whence it branched off either southwards
to the Niger or northwards to Leptis and Carthage. Two routes led
northwards from Thebes to Meroë; one by the course of the Nile
throughout, another by the course of the Nile until the point where
it makes its great bend, and thence across the Nubian desert.

(2.) From Meroë a route led westward to the shores of the Red Sea,
whence ports, such as Adule, were found, communicating either with
Lower Egypt or with the opposite coast of Arabia. Another route undoubtedly led from Meroë southwards to the districts of *Senaar* and
Abyssinia.

(3.) From Phazania routes led northwards to the coast of the Mediterranean, where Leptis formed the great emporium, and southwards to
the districts of Central Africa.

(4.) From Coptos, roads, with caravanserais, were constructed by the
Ptolemies to Myos Hormos and Berenice, and a vast amount of traffic
passed by this "overland route" between India and Europe. Pliny
estimated the annual value of the imports from the East at about
1,500,000 pounds sterling.

§ 10. The ethnology of ancient Africa is not a subject of much
interest. The nations with whom the Greeks and Romans came in
contact were almost wholly of Asiatic origin. The north Africans,
though darker than Europeans, and hence occasionally described in
terms which seem only applicable to negroes, were really allied to
the races of Europe and Asia, as the Mosaic genealogy indicates
when it represents the sons of Ham, the brother of Shem and
Japheth, as occupying Æthiopia, Egypt, Libya, and Canaan. This
opinion prevailed even in ancient times. Juba, according to Pliny,
pronounced the Egyptians to be Arabs; while far away to the W.,
in Mauritania, a tradition of the Asiatic origin of the people was
perpetuated. The Æthiopians were perhaps the nearest approach
to the negro; but the ancient monuments prove that there was a
wide distinction, even in their case, and that they were no more
true negroes than their modern representatives, the *Bisharies* and
Shangallas. The other great divisions of the family of Noah were
represented in the colonies on the coast of the Mediterranean—the
Semitic in the Phœnicians, the Japhetic in the Greeks and
Romans.

Memphis.

CHAPTER XV.

§ 1. The boundaries of *Ægyptus*, or *Egypt*, were—on the N., the Mediterranean Sea; on the E., the Arabicus Sinus, and that portion of Arabia which intervenes between the head of the Sinus Heroöpolites and the Mediterranean, now called the *Isthmus of Suez*; on the S. Æthiopia, from which it was divided at Syene; and on the W. the Libyan desert. Its length is estimated at 526 miles, and the total area at about 9070 square miles, the upper valley amounting to 2255, the Delta to 1975, and the outlying districts to 4840. In shape it resembles an inverted Greek *upsilon* (*ʌ*), as it consists of a single long valley, spreading out on either side at its base. It was naturally divided into two parts—Lower and Upper Egypt: the former the wide alluvial plain of the Delta, the latter the narrow valley of the Nile with its primitive formations of granite, red sandstone, and limestone. Each of these had its characteristic productions—the papyrus being the symbol of the Delta; the lotus, that of Upper Egypt: and each had its own peculiar deities.

The Name.—The name "Ægyptus" first appears as the designation of the Nile (Hom. *Od.* iv. 477), and was thence transferred to the country in which that river forms so prominent an object. The name appears to have been specially applied to the Thebaid, where it was perpetuated in that of the town Coptos. It may perhaps be connected with the Biblical Caphthor; the modern name "*Cupts*" is evidently a relic of it.

§ 2. The position and physical character of Egypt account to a great degree for its importance in the ancient world. Situated midway between the continents of Asia and Africa, it was the gate, as it were, through which all intercourse between those two continents was carried on. With the Mediterranean on one side, and the Red Sea on the other, it held easy communication with the southern peninsulas of Europe, and with the coasts of India; and was, even in early times, the link to connect the west with the east. Surrounded by deserts, the valley of the Nile formed a large oasis, isolated from the adjacent countries, yet easily accessible on all sides by means of routes which nature has formed. The wonderful fertility of its soil admitted of the maintenance of an immense population, and supplied the material wealth and comfort which are essential to the early advance of civilization. The climate has been at all times famed for its salubrity, and the natural productions were not only varied, but in some instances had a direct tendency to encourage art and manufacture. Among the more important articles we may notice—grain of all kinds (wheat, barley, oats, and maize), vegetables in great profusion (onions, beans, cucumbers, melons, garlic, &c.), flax, cotton, papyrus (a most valuable fibrous plant, used for making boats, baskets, rope, paper, sails, sandals, as well as an article of food), the lotus, olives, figs, almonds, and dates. Stone of the finest quality for building abounded in Upper Egypt, while various ornamental species, such as porphyry, were also found.

§ 3. The chief physical feature of Europe is the river and valley of the Nile. The valley is enclosed between two parallel ranges [1] of limestone hills, the eastern shutting it off from the Red Sea, the western from the Libyan desert. The average breadth of this valley

[1] Hinc montes natura vagis circumdedit undis,
Qui Libyæ te, Nile, negant: quos inter in alta
It convalle tacerns jam moribus unda receptis.
Prima tibi campos permisit, apertaque Memphis
Rura, modumque vetat crescendi ponere ripas. —Luc. x. 327.

as far as 30° N. lat. is about 7 miles. Between this point and 25°
its width varies from 11 miles at the widest to 2 at the narrowest
point : S. of 25° to Syene, the valley contracts so much•that in
some places the hills rise almost immediately from the river's banks.
The plain is generally more extensive on the W. than on the E.
side of the river, and hence the towns are situated almost invariably
on the left bank. The length of the river from the sea to Syene is
732 miles, and its fall throughout this distance is estimated at 865
feet, or about ½ a foot per mile. We have already described the
general course of the river, but there are a few topics connected
with it that deserve further notice in this place.

Name.—The name "Nile" appears to have been of Indian origin,
and to signify the "blue river." The indigenous name was "Hapi."
Homer names it the "Ægyptus." [2]

Its Inundation.—The Nile begins to rise about the beginning of July.
About the middle of August it is high enough for purposes of irrigation,
and between the 20th and 30th of September it reaches its maximum
height : it remains stationary for a fortnight, and then gradually
recedes. An elevation of 30 feet is ruinous from excess of moisture,
but one of 24 is necessary to insure a good harvest ; below 18 is again
ruinous from deficiency of moisture. Various theories were pro-
pounded by the ancients as to the cause of the inundation: Agathar-
chides of Cnidus correctly attributed it to the rains of Abyssinia,
which thoroughly saturate that country.

Its Importance.—Egypt was in truth the "gift of the Nile." Its soil
was due to the action of the river: each succeeding inundation deposited a rich stratum, which is now known to exist to a depth of
above 60 feet below the present level of the land. Its fertility was
wholly dependent upon the periodical inundations. [3] Its commerce

---

[2] Οὐ γάρ τοι πρὶν μοῖρα φίλους τ' ἰδέειν, καὶ ἱκέσθαι
Οἶκον εὐκτίμενον, καὶ σὴν ἐς πατρίδα γαῖαν,
Πρίν γ' ὅτ' ἂν Αἰγύπτοιο διιπετέος ποταμοῖο
Αὖτις ὕδωρ ἔλθῃς.   *Od* Iv. 475

[3] The references to this subject in the classics are very numerous.
Aut pingui flumine Nilus,
Cum refluit campis, et jam se condidit alveo.—VIRG. Æn. ix. 31.
Qualis et, arentes cum findit Sirius agros,
Fertilis aestiva Nilus abundet aqua ?
 .   .   .   .
Te propter nullos tellus tua postulat imbres
Arida nec Pluvio supplicat herba Jovi.—TIBULL. L. 7, 21-23, 25, 26.
Sic ubi deserult madidos septemfluus agros
Nilus, et antiquo sua flumina reddidit alveo,
Ætherioque recens exaruit aethere limus ;
Plurima cultores versis animalia glebis
Inveniunt, et in his quaedam modo coepta, sub ipsam
Nascendi spatium : quaedam, imperfecta, suisque
Trunca vident numeris : et eodem in corpore saepe
Altera pars vivit ; rudis est pars altera tellus.—OV. *Met.* L. 122.

Virgil specially refers to the contrast of the black subsoil and the brilliant verdure of the fields :—
Et viridem Ægyptum nigra fœcundat arena."—*Georg.* iv. 291.

passed up and down the broad stream as on a high road. Add to this, that the water was deemed so pure that the Persian kings imported it, and that the supply of fish and fowl formed one of the staples of food, while the reeds which grew on its banks served for sails, material for paper, and other useful purposes. We can hardly then be surprised that the Egyptians paid divine honours to this river, and worshipped it under the form of a bull.

§ 4. The hills of Egypt are of secondary importance. The ranges that bound the valley of the Nile were named **Arabici Montes**, *Jebel Mokattem* on the E., and **Libyci Mts.**, *Jebel Silsili* on the W. In addition to these we may notice—**Casius**, *El Katieh*, on the borders of Arabia Petræa, near the Mediterranean, its summits once crowned with a temple of Zeus Ammon,—**Troïcus Mons,** *Gebel Musarah*, whence the stone for the casing of the Pyramids was taken : the name was probably the corruption of some Egyptian word—**Alabastrites**, S. E. of the town of Alabastra—**Porphyrites**, E. of Antæopolis—and **Smaragdus**, N. of Berenice : these three last hills were so named after the geological character of the rocks.

§ 5. Numerous canals intersected the country, and conveyed the waters of the Nile to the distant parts of the valley. The maintenance of these canals was essential to the well-being of the country, and accordingly Augustus (B.C. 24) ordered a general repair of them as one of his first measures for the improvement of the province. In addition to the agricultural canals, there were two constructed for commercial purposes. The most important one joined the Nile and the Red Sea, and was named at different periods "Ptolemy's River" and "Trajan's River." It was commenced by Pharaoh Necho, B.C. 610, continued by Darius Hystaspis about 520, completed by Ptolemy Philadelphus in 274, and restored by Trajan in A.D. 106 : it originally began in the Pelusiac branch of the Nile near Bubastus, and terminated at Arsinoë on the Sinus Heroöpolites ; Trajan's began higher up the river at Babylon opposite Memphis, and entered the Red Sea 20 miles S. of Arsinoë at Klysmon : this existed for 700 years. The other, named the Canôpic Canal, connected the city of Canopus with Alexandria and Lake Mareotis.

§ 6. There were several important lakes in the N. of Egypt. **Mœris**, near Arsinoë, is described by ancient writers as an artificial lake of wonderful construction. At present there is a natural lake, named *Birket-el-Kerun*, 30 miles long from S.W. to N.E., and 7 broad ; it is connected with the Nile by the canal named *Bahr-Jusuf*, "Joseph's Canal" and until recently it was supposed that the canal was the artificial work to which the ancients referred ; traces of a large reservoir have, however, been discovered, which was probably part of Lake Mœris. The object of the lake was to irrigate the fertile nome of Arsinoë, the water being conveyed in different directions by subordinate channels. The **Amâri Lacus** were a cluster of salt lagoons E. of the Delta near Heroöpolis. **Sirbönis**, *Sebaket Bardoil*, was a vast morass, E. of the Delta, and near the Mediterranean Sea, with which it was once connected by a channel. The Persian army under Darius Ochus was partly [4] destroyed here in B.C. 350. **Nitris**, the *Natron Lakes*, were a group of six, situated in a valley S.W. of the Delta: the sands about these lakes were formerly the bed of the sea; they are all salt, and some few contain natron, or sub-carbonate of soda, which was extensively used by the bleachers and glass-makers of Egypt. **Mareötis**, *Birket-el-Mariout* lay S.W. of the Canopic arm, and ran parallel to the Mediterranean, from which it was separated by a ridge of sand; its breadth was 22 miles, and its length 42, and it was originally connected by canals with the Canopic arm, and with the harbour of Alexandria. These canals became gradually choked, and the lake had almost disappeared, until in 1801 the English army made a new channel, and let in the waters of the sea. The shores of Mareotis were formerly laid out in olive-yards and vineyards: [5] a very fine kind of papyrus also grew there.

§ 7. The Egyptians believed themselves to be autochthonous, and the Greeks considered them to belong to the same stock as the Indians and Ethiopians. They were, however, a distinct branch of the great Hamitic family, intermixed indeed, in certain parts of the country, with the Arabian, Libyan, and Ethiopian races, but essentially separate from them. The population was undoubtedly much larger in ancient, than in modern times, but the estimates that have come down to us are not trustworthy; Diodorus gives it as seven millions, while from the statement in Tacitus (*Ann*. ii. 60), we may estimate it at six millions: it is now put at less than two millions. The inhabitants were divided into castes, the number of which is variously given: it appears that the possession of the land was vested in the king, the priests, and the soldiers; these, therefore, were the three great estates of the realm: the husbandmen were included under the soldiers.

§ 8. The earliest division of Egypt was the twofold one, based on

---

[4] Diodorus (L 30) incorrectly represents the whole of the army as having been swallowed up in it, and he is followed by Milton, who speaks of

that Serbonian bog
Betwixt Damiata and Mount Casius old
Where armies whole have sunk.— *Par. Lost*, ii. 293.

[5] Sunt Thasiæ vites, sunt et Mareotides albæ.—Virg. *Georg.* ii. 91.
Mentemque lymphatam Mareotico.—Hor. *Carm.* i. 37, 14.

Mareoticus is frequently used for Egyptian generally, as in the following reference to the Pyramids:—

Par quota Parrhasiæ labor est Mareoticus aulæ.—Mart. viii. 36.

the natural features of the country, of Upper and Lower Egypt, the latter being co-extensive with the Delta. Subsequently, Upper Egypt was divided into two parts—Thebäis, to which the title of Upper Egypt was henceforward restricted, and Heptanômis or Middle Egypt. This triple division is still retained by the Arabs, who denominate the three districts from N. to S. *El-Rif, Wustani,* and *Said.* Egypt was further subdivided into *nomes,* or cantons, the number of which varied at different eras: Herodotus mentions only 18 ; under the Ptolemies the total number was 36 ; under the later Roman emperors as many as 58. The nomes were subdivided by the Romans into *toparchies,* and the toparchies into *aroura.* Under the later Roman emperors the Delta was divided into 4 provinces—Augustamnica Prima and Secunda; and Ægyptus Prima and Secunda ; and the Thebaid into two parts—Upper and Lower.

§ 9. The towns of Egypt were exceedingly numerous : Herodotus states their number at 20,000, Diodorus at 18,000 : in this estimate, however, must be included walled villages, as well as proper towns. Each town was specially devoted to the religious worship of some deity or animal, and they appear to have been generally named after their tutelary god. The Greeks, who identified the Egyptian gods with their own, translated these names into the corresponding terms in their own language, and hence the original names have been for the most part lost to us. Occasionally, however, both are recorded ; thus we have the Egyptian Chemmis, and the Greek *Panopolis ;* Busiris, "the burial-place of Osiris," and *Taposiris ;* Atarbechis and *Aphroditopolis.* Occasionally the Bible gives the original name, as in the case of On for *Heliopolis,* though even in this case we have also the name translated into the Hebrew *Beth-shemesh ;* Ammon for *Thebes ;* Sin for *Pelusium.* In cases where the significance of the name was not so clear, the old Egyptian form has been retained with but slight variation, as in the case of Thebes for *Tape,* "the capital ;" Memphis for *Menofre,* "the place of good ;" Canôpus for *Kahi-noub,* "the golden soil." In some instances the indigenous name still adheres to the site of the place, as in the case of *Sin* for Pelusium. We shall describe the towns under their respective districts : it will be only necessary to remark here that there were two ancient capitals—Thebes and Memphis ; and one comparatively modern one—Alexandria. Of the two former, Memphis appears to have the best claim to be regarded as the prior capital, but at certain periods of history they were contemporaneously capitals of the two kingdoms of Upper and Lower Egypt. It may further be remarked that the Egyptians were not a sea-faring people, and that hence their capitals were high up the valley of the Nile ; the position of the later capital, Alexandria, was due to the commercial genius of the Greeks, to whom the other maritime emporia—Naucratis, Berenice, and Myus Hormos—also owed their existence.

§ 10. The Delta was the most northerly of the three divisions of Egypt; It derived its name from the similarity of its shape to the Greek letter Δ, the two sides of the triangle being formed by the outer arms of the Nile, and the base by the Mediterranean Sea. The Delta, as a political division, extended beyond the Canopic and Pelusiac arms, as far as the alluvial soil extended.[*] The true boundaries of the Delta were thus the Libyan and Arabian deserts: the apex of the Delta was formerly more to the S. than it is at present. The soil is not nearly so fertile as that of Upper Egypt; hence much is devoted to such crops as flax, cotton, and other plants that succeed on second-rate soils. The nitre which is abundant in many parts, produces positive barrenness. The Delta contained, according to Strabo, 10, and according to Ptolemy, 24 nomes.

§ 11. The towns of the Delta are invested with associations of a varied character, extending over a vast number of centuries. The proximity of this district to the borders of Asia brought it into early communication with Syria and Mesopotamia. The Bible introduces us to various towns in connection partly with the early sojourn of the Israelites in Goshen, and partly with the later alliance between Judæa and Egypt during the era of Assyrian supremacy. From this source we first hear of Heliopolis, the seat of the most famous college of learned priests in Egypt—of Pelusium, the most important border-fortress—of Tanis, the seat of royalty under some of the early dynasties—of Bubastus, also occasionally the residence of the kings of Lower Egypt, and of other less important places. These were all first-rate towns in the days of Egyptian greatness, and were highly favoured by the most renowned monarchs. We may add to the list Sais, the royal residence of Psammitichus and Amasis, as well as of other earlier sovereigns—Mendes, the chief seat of the worship of Pan—and Canôpus, the early port of Egypt. At a later date, Naucratis became the most busy place as the emporium of Greek commerce. But this was in turn superseded by Alexandria, which became the capital of the whole of Egypt under the Ptolemies: its rise proved fatal to the prosperity of many of the towns of the Delta. The Ptolemies restored or adorned many of the towns, as the character of their remains still testifies. Their final ruin was in some cases produced by the changes of the river's course; but the majority

---

[*] The term Delta was not peculiar to the lower course of the Nile, but was used in all cases where rivers have formed an alluvial deposit, and have hence divided before entering the sea, as in the cases of the Rhone, the Indus, and the Achelous.

probably survived until the latest period of the Roman Empire.
We shall describe the towns in order from N. to S., commencing
with those which lay W. of the Delta proper.

Alexandria stood on a tongue of land between Lake Mareotis and
the Mediterranean Sea. It was founded by Alexander the Great, B.C.
332, on the site of a small town called Rhacôtis. Its position was
good : the Isle of Pharos[7] shielded it on the N., and the headland of
Lochias on the E., while Lake Mareotis served as a general harbour both
for the town and for the whole of Egypt. The town was of an oblong
shape, about 4 miles in length from E. to W., and about a mile in
breadth. Two grand thoroughfares bisected the city in opposite directions, communicating at their extremities with the four principal gates.
A mole 7 stadia long, and hence named Heptastadium, connected the
Isle of Pharos[8] with the mainland. On the E. side of the mole was
the " Greater Harbour," extending as far as the headland of Lochias,
the portion at the innermost angle, which was reserved for the royal
galleys, being separated from the rest, and named the " Closed Port."
On the W. side of the mole was the haven of Eunostus, " Happy
Return." The Isle of Pharos contained at its E. extremity the
celebrated lighthouse,[9] said to have been 400 feet high : it was built by
Sostratos of Cnidus under Ptolemy[1] Soter and his successor. The city
itself was divided into three districts—the Jews' quarter in the N.E.
angle ; the Brachium or Pyruchium, the royal or Greek quarter, in the
E. and centre ; and the Rhacôtis, or Egyptian quarter, in the W. The
second contained the most remarkable edifices, including the Library
with its Museum and Theatre, connected together by marble colonnades, the Palace, the Stadium &c. The Library is said to have
contained 700,000 volumes, some of which were deposited in the
Serapeum in the quarter Rhacotis. The collection was begun by
Ptolemy Soter, and was carried on by succeeding sovereigns, especially

---

[7] Νῆσος ἔπειτά τις ἐστὶ πολυκλύστῳ ἐνὶ πόντῳ,
Αἰγύπτου προπάροιθε (Φάρον δέ ἑ κικλήσκουσι),
Τόσσον ἄνευθ', ὅσσον τε πανημερίη γλαφυρὴ νηῦς
Ἤνυσεν, ᾗ λιγὺς οὖρος ἐπιπνείῃσιν ὄπισθεν.
Ἐν δὲ λιμὴν εὔορμος, ὅθεν τ' ἀπὸ νῆας ἐΐσας
Ἐς πόντον βάλλουσιν, ἀφυσσάμενοι μέλαν ὕδωρ.—Hom. Od. iv. 354.

[8] Tunc claustrum pelagi cepit Pharon. Insula quondam
In medio stetit illa mari, sub tempore vatis
Protëos : at nunc est Pellæis proxima muris.—Luc. x. 509.

[9] Septima nox, Zephyro nunquam laxante rudentes,
Ostendit Phariis Ægyptia littora flammis.—Lcc. ix. 1004.

Claramque serena
Arce Pharon.   VAL. FLACC. vii. 84.

, Teleboumque domos, trepidis ubi dulcia nautis
Lumina noctivagæ tollit Pharus æmula Lunæ.—STAT. Silv. iii. 1, 100.

From the celebrity of this lighthouse Pharos became a synonym for Egypt
itself, as in STAT. Silv. iii. 2, 102, " regina Phari ;" Lcc. viii. 443, "petImus
Pharon arvaque Lagi." So also Pharius for Ægyptius in numerous places.

[1] Hence the allusion in the following lines :—
Et Ptolemæw littora capta Phari.—PROPERT. ii. 1, 30.
Nupta Senatori comitata est Hippia Ludium
Ad Pharon et Nilum famosaque mœnia Lagi.—JUV. Sat. vi. 83.

by Energetes. The library of the Museum was destroyed during the blockade of Julius Cæsar; that of the Serapium, though frequently injured, existed until A.D. 640, when it was destroyed by the Khalif Omar. Alexandria was the seat of a university, and produced a long roll of illustrious names, among which we may notice Euclid, Ctesibius, Callimachus, and Ptolemy. The modern town occupies the Hepta-stadium, the site of the old town being partly covered with modern villas. The most interesting remains of the ancient town are the two obelisks, commonly called "Cleopatra's Needles," which bear the distinctive sign of Thothmes III., and were brought from Heliopolis by one of the Cæsars—Pompey's Pillar, erected by the eparch Publius in honour of Diocletian, and named "Pompey's" according to one ex-planation from the Greek word πομπαῖος "conducting," inasmuch as it served as a landmark—and, lastly, the Catacombs, or remains of the ancient Necropolis. Alexandria prospered during the reigns of Ptolemy Soter and Philadelphus, and began to decline under Philo-pator. In B.C. 80 it was bequeathed to Rome by Ptolemy Alexander; and from 55 to 30 it occupies a prominent place in the civil wars of the Roman leaders. Under the emperors it was generally prosperous: the erection of Nicopolis as a rival town by Augustus—serious com-motions under Diocletian—and a general massacre by Caracalla, were the chief adverse events. In A.D. 270 it was subject to Zenobia, and in 297 it was taken by Diocletian after it had joined the side of Achilleus. It was taken by the Arabs in 640. Alexandria holds a prominent place in the history of the Christian religion. From the time of the Babylonish Captivity the Jews resorted to Egypt in great numbers, and under the Ptolemies they occupied, as we have seen, one of the quarters of Alexandria, where they lived under their own ethnarch, and sanhedrim. Here they became versed in the Greek language, and for the use of the Alexandrian Jews the Greek translation of the Old Testament, named the Septuagint, was made under the auspices of the Ptolemies. Violent disputes frequently occurred between the Jews and Greeks, partly on religious, partly on political matters. Alexan-dria received the Christian faith at a very early period, and became the seat of a patriarchate. A violent persecution occurred here in Diocle-tian's reign, in which the bishop Peter perished. Nicopolis, which Augustus founded, in B.C. 24, as a rival to Alexandria, stood on the banks of the canal which connected Canopus with the capital, and about 3½ miles from its eastern gate. It was named in commemoration of the victory gained on the spot over M. Antonius. The town soon fell into decay. Canôbus or Canôpus was situated about 15 miles E. of Alexandria, near Aboukir, at the mouth of the Canopic branch of the Nile. Before the rise of the later capital it was the chief port of the Delta;[2] it was also celebrated for the worship of Zeus-Canobus, under the form of a pitcher with a human head: the numerous festivals made it notorious for the profligacy[3] of its inhabitants: a scarlet dye for

---

[2] Hence the early acquaintance which the Greeks had with it :—

Καὶ μὴν Κάνωβον πᾶσι Μέμφιν ἐστιν.—ÆSCH. Suppl. 311.

Ἔστιν πόλις Κάνωβος ἐσχάτη χθονὸς.

Νείλου πρὸς αὐτῷ στόματι καὶ προσχώματι.—In. Prom. Vinct. 846.

[3] Ut strepit assiduo Phrygiam ad Niloticus loton
Memphis Amyclæo passim lascivu Canopo.—Sil. ITAL. xl. 432.

staining the nails was prepared here. **Hermopolis Parva,** *Damanhur,* stood 44 miles S.E. of Alexandria, on a canal connecting Lake Mareotis with the Canopic arm. **Andropolis,** *Chebur,* more to the S.E., is supposed to have been so called from the worship of the Shades of the Dead : it was probably the same as **Anthylla,** which was assigned to the Egyptian queens for pin-money. **Letopolis,** named after the deity Leto or Athor, stood near the apex of the Delta, a few miles S.W. of Cercasorum. **Cercasorum,** *El-Arbas,* stood at the apex of the Delta, on the Canopic branch, and from its position was a town of great military and commercial importance. The Delta now commences about 7 miles N. of it.

Towns of the Delta proper.—**Sais,** at one time the capital of the Delta, stood on the right bank of the Canopic branch, on an artificially elevated site, now partly occupied by *Sa-el-Hadjar.* It was famous for the worship both of Neith (Minerva), and of Isis: the great annual festival, entitled "the Mysteries of Isis," was celebrated on a lake near the town; it was also one of the supposed burial-places of Osiris. Sais was a royal city under the 17th, 24th, 26th and 28th dynasties, and attained its highest prosperity under the 26th, from B.C. 697 to 524 ; Psammitichus and Amasis were its most illustrious kings. It was still more famous as a seat of learning, and was visited by Pythagoras and Solon. The ruins of Sais consist of a boundary wall 70 feet thick, enclosing a large area, vast heaps of bricks, and traces of the lake. **Naucratis** stood on the E. bank of the Canopic arm, about 30 miles from the sea, and was originally an emporium founded by Milesian colonists at the invitation of Amasis, B.C. 550, and endowed by him with various privileges. It possessed a monopoly of the Mediterranean trade probably down to the foundation of Alexandria, after which it sunk. Its chief manufactures were porcelain and flower-wreaths. It was visited by Solon, and probably by Herodotus. The exact site is uncertain, but is supposed to have been at *Salhadschar.* **Mendes** was situated at the point where the Mendesian arm flows into the lake of Tanis. Under the Pharaohs it was a place of importance ; but it declined early, probably through an encroachment of the river. It was famed for the worship of Mendes, or Pan, and for a species of ointment. **Tanis** was seated on the Tanitic arm, and was one of the chief cities of the Delta, and even the capital under various kings from the 13th to the 24th dynasties. It is the Scriptural Zoan, said to have been founded only 7 years after Hebron, and was regarded as the capital of Lower Egypt in Isaiah's time. Its position near the coast and near the E. frontier made it an important military post, and the marshes which surrounded it rendered it inaccessible to an enemy. It was the stronghold of the Memphite kings during their struggle with the Shepherds. The vestiges of the old town at *San* consist of an enclosure, 1000 feet long, and 700 wide, with a gateway on the N. side, numerous obelisks and sculptures belonging to the temple of Pthah, two granite columns, and lofty mounds. The name of Rameses the Great occurs frequently on the sculptures. **Thmuis** stood on a canal between the Tanitic and

---

Prodigia et mores Urbis damnante Canopo.—Juv. *Sat.* vi. 31.
Sed luxuria, quantum ipse notavi,
Barbara famoso non cedit turba Canopo.—Ib. xv. 45.

Canopus is used by Lucan as a synonym for Egypt—
Et Romana petit imbelli signa Canopo.— x. 64.

Mendesian branches, at *Tel-ebmni*. It was, like its neighbour Mendes, devoted to the service of Thmu, or Pan. It retained its importance down to a late period, and was an episcopal see. **Sebennytus**,[a] *Semennhood*, was favourably situated between a lake and the Sebennytic arm, and was a place of commercial importance.   About 6 miles above Sebennytus, on the course of the river, was **Busiris**, considerable remains of which exist at *Abousir*. It possessed a very celebrated temple of Isis, which stood at *Bebayt*, and of which there are most extensive ruins of the Ptolemaic era.   The temple of Isis stood on a platform 1500 ft. by 1000, surrounded by an enclosure, and was itself 600 ft. by 200, built of the finest granite, and adorned profusely with sculptures.   It was erected by Ptolemy Philadelphus.   **Xois** stood nearly in the centre of the Delta, and was the residence of the 14th dynasty, who probably held out against the Hyksos here.   It is supposed to be identical with the **Paprêmis** of Herodotus.   **Leontopolis** stood S.E. of Xois, and appears to have been a comparatively modern town.   In the reign of Ptolemy Philometor the Jews built a temple here similar to that of Jerusalem, which remained the head-quarters of a large Jewish community until the time of Vespasian.   Its site is supposed to be at *El-Mengaleh*.   **Bubastus**, the Scriptural Pi-beseth, was situated on the E. side of the Pelusiac arm, S.W. of Tanis.   It was sacred to Pasht,[b] who was worshipped under the form of a cat, and hence it became a depository for the mummies of that animal.   Some monarchs of the 22nd dynasty reigned here.   The great canal left the Nile just N. of the town.   Bubastus was captured by the Persians B.C. 352, and thenceforth declined.   Its ruins at *Tel-Basta* are very extensive, and consist of an enclosure three miles in circumference, large mounds intended to restrain the Nile, and heaps of granite blocks.   **Athribis** stood on the E. bank of the Tanitic branch, and was sacred to the goddess Thriphis.   Extensive mounds and the basement of a temple are found on its site at *Atrieb*, and the character of the ruins indicates their erection in the Macedonian era.   The town had been embellished by the old Egyptian kings, and a granite lion still exists bearing the name of Rameses the Great.

Towns E. of the Delta proper.—**Pelusium**, the Sin of the Bible, stood E. of the Pelusiac arm about 2½ miles from the sea, and was the key of Egypt on this side.   It is connected with several events in the history of Egypt—particularly the advance of Sennacherib, king of Assyria; the defeat of the Egyptians by Cambyses, in B.C. 525 ; the advance of Pharnabazus of Phrygia and Iphicrates the Athenian in 373 ; and the several captures of it by Alexander the Great in 333, by Antiochus Epiphanes in 173, by Marcus Antonius in 55, and by Augustus in 31. The surrounding district produced lentils[c] and flax.[d]   The Pelusiac mouth, which was shallow even in classical[e] times, was choked by

---

[a] The name in Egyptian form is *Gemnouti* "Gem the God."

[b] Sanctaque Bubastis, variusque coloribus Apis.— Ov. *Met.* ix. 690.

[c] Nec *Pelusiacæ* curam aspernabere lentis.— Virg. *Georg.* I. 228.
Accipe Niliacum, *Pelusia* munera, lentem :
    Tillor est alien, carior illa faba.—Mart. xiii. 9.

[d] Et Pelusiaco filum componere lino.— Sil. Ital. III. 375.

[e]   Qua dividui pars maxima Nili
    In cada decurrit Pelusia septimus amnis. — Luc. viii. 465.
'Αμφ προστομίων λεπτοψαμάθων
    Neιλου.   Æsch. *Suppl.* 3.

and as early as the first cent. A.D., and the coast-line is now far removed from the site of Pelusium, the modern *Tineh*. **Magdolum**, the scriptural Migdol, stood about 12 miles S. of Pelusium, on the coast-road to Syria. Here Pharaoh Necho is said to have defeated the Syrians, about 608 B.C. **Heroöpolis** was near the mouth of the Royal Canal, and gave name to the W. arm of the Red Sea, though it did not stand immediately on the coast. Its ruins are at *Abu-Keyscheid*. It must have been a place of commercial importance. **Heliopolis**, the Scriptural On and Beth-shemesh, stood on the verge of the eastern desert, N.E. of Cercasorum, and near the right bank of Trajan's Canal. It was a town of the highest antiquity, and the seat of a famous university, which is said to have been visited by Solon, Thales, Plato, and Eudoxus, and to have possessed the archives from which Manetho constructed his history of the Egyptian dynasties. It was also visited by Alexander the Great, and it has acquired a special interest in connection with sacred history, as the place where Moses was probably instructed in Egyptian science, and where Jeremiah wrote his Lamentations. The place was especially devoted to the worship of the Sun, and the bull Mnevis was also honoured there. The remains at *Matarieh* consist of a remarkable obelisk of the age of Osirtasen I., some fragments of sphinxes, a statue belonging to the temple of the Sun, and the boundary-walls of brick, 3750 ft. long, by 2370. **Babylon**, *Baboul*, stood on the right bank of the Nile, near the entrance of the Great Canal, and probably owed its name and foundation to some Babylonian followers of Cambyses in B.C. 525. Under Augustus it was a place of some importance, and the head-quarters of three legions. **Arsinoë** stood at the N. extremity of the W. gulf of the Red Sea, and was one of the principal harbours of Egypt. It was named after the sister of Ptolemy Philadelphus, and its revenues belonged to her and the succeeding queens. Its position near the entrance of the canal, and on the shore of a fine bay, insured it a share of the Indian trade; but its exposure to the S. wind, and the dangerous reefs in approaching it, were serious checks to its prosperity. Its site is at *Ardscherud* near *Suez*.

Of the less important towns in the Delta we may notice from N. to S.—**Menaläus**, named after a brother of Ptolemy Lagus, between Alexandria and Hermopolis, on the Canopic arm—**Momemphis**, "Lower Memphis," on the E. shore of Lake Marœtis, a place of some strength from the nature of the approaches—**Marea**, S. of Lake Marœtis, one of the chief fortresses on the side of Libya, where Amasis defeated Pharaoh Apries—**Bolbitine**, *Rosetta*, on the Bolbitic branch of the Nile, the site of the famous Rosetta stone, in which the beneficent acts of Ptolemy Epiphanes are recorded—**Buto**, *Kem-Kasir*, on the Sebennytic arm, celebrated for its monolithic temple and oracle of the goddess Buto—and **Tamiathis**, at the mouth of the Phatnitic arm; its modern representative *Damietta* occupies a site about 5 miles higher up the river.

§ 12. **Heptanomis** was the central district of Egypt, and contained, as its name implies, 7 nomes;[*] it extended from Cercasorum in the N. to Hermopolis in the S. Under the emperor Arcadius it

---

[*] More than seven nomes were occasionally assigned to Middle Egypt; Strabo so divides them, and Ptolemy adds an eighth, the Arsinoïte.

received the name of Arcadia. The width of the valley fluctuates: near Hermopolis it is contracted on the E. side of the river, and tolerably broad on the W. Lower down, the hills diverge still more to the W., and embrace the district of Arsinoë, returning to the river on the N. side of it. Below this it again expands until it attains, near Cercasorum, almost the breadth of the Delta. This district comprised the greatest works of Egyptian art—the Pyramids, the Labyrinth, and the artificial district formed by the canal of *Bahr-Jusuf*. It is also remarkable for its quarries and rock-grottoes; of the first we may notice the Alabastrites E. of Hermopolis; the quarries of veined alabaster 8 miles to the N., chiefly used for sarcophagi; and the quarries E. of Memphis, whence they obtained the stone for casing the Pyramids. The most remarkable grottoes were those of Speos Artemidos, *Beni-Hussan*, and of *Koum-el-Ahmar* more to the N. The towns were numerous and important: Memphis, the earliest metropolis of Egypt, and the capital of one of the nomes, stood near the N. boundary; while the following towns from N. to S. represented the capitals of the other six nomes—Arsinoë, Heracleopolis, Aphroditopolis, Oxyrynchus, and Hermopolis.

Memphis,[1] the Noph of Scripture, stood on the W. bank of the Nile, 15 miles S. of Cercasorum. Its origin was ascribed to Menes, and it was the first capital of the whole of Egypt. The site of the town was originally a marsh, formed by a southerly bifurcation of the Nile. Menes diverted the branch into the main stream, by means of an embankment. The town was some 15 miles in circumference, much of the area being, however, occupied by gardens, and by the soldiers' quarters, named the "White Castle." The soil was extremely productive, and ancient writers dilate upon its green meadows, its canals covered with lotus-flowers, its vast trees, its roses, and its wine. Its position was highly favourable. The Arabian and Libyan hills converge here for the last time, and it could thus command the trade of the valley of the Nile. It was centrally placed as regards Upper Egypt and the Delta, and sufficiently near the border to have communication with Syria and Greece. It was quite the Pantheon of Egypt, and possessed temples of Isis,[2] Proteus, Apis, Serapis, the Sun, the Cabeiri, and particularly of Pthah, or Hephæstus. It was visited by Solon, Hecatæus, Thales, Herodotus, Strabo, and Diodorus Siculus. Its site is at *Mitranieh*, and its remains consist of blocks of

---

[1] The Egyptian name signified "the place of good."

[2] Te canit, atque suum pubes miratur Osirim
    Barbara, Memphiten plangere docta bovem.—TIBULL. I. 7, 27.
    Neu fuge linigeræ Memphitica templa juvencæ.—OV. *Art. Am.* I. 77.
    His quoque deceptus Memphitica templa frequentat,
    Assidet et cathedris mœsta juvenca tuis.—MART. li. 14.
    Barbara Pyramidum sileat miracula Memphis.—MART. *de Spect.* i. 1.
    Regia pyramidum, Cæsar, miracula ride;
    Jam tacet Eoum barbara Memphis opus.—ID. *Epig.* viii. 30.

granite, a large colossus of Rameses II., broken obelisks, columns, and statues, spread over many hundred acres of ground. Memphis was the seat of the 3rd, 4th, 6th, 7th, and 8th dynasties. The Shepherd Kings retained it as the seat of civil government. The house of Rameses, the 18th dynasty, though they made Thebes their capital, paid great attention to Memphis. Under the 25th dynasty it again became the seat of a native government. It suffered severely from the Persians under Cambyses. In the reign of Artaxerxes I. the Persians took refuge here after their defeat by Inarus, and were besieged for a year. After the expulsion of Nectanebus II. it sunk to the position of a provincial city, and in Strabo's time a large portion was in ruins. Near Memphis at a place now called *Geezeh*, are the three celebrated Pyramids ; the largest, attributed by Herodotus to Cheops, was originally 750 ft. square at its base, and 480 ft. high ; it covered about the same space as Lincoln's Inn Fields ; its dimensions are now reduced to 732 ft. square, and 460 ft. high. The second, attributed to Chephren, was formerly 707 ft. square, and 454 ft. high ; its dimensions now being 690 and 446. The third, attributed to Mycerinus, whose coffin has been found there, was 354 ft. square, and 218 high ; these are now reduced to 333 and 203. On the S. of this are three small pyramids, one of which has the name of Mencheres (Mycerinus) inscribed upon it. Another cluster of three also stands E. of the great pyramid. The object for which they were built is uncertain : they probably served for tombs, and their uniform position, facing the cardinal points, makes it probable that they were used for astronomical purposes. About 200 ft. N. of the second pyramid is the Sphinx, cut out of the solid rock ; it bears the name of Thothmes IV. of the 18th dynasty, and appears to have been an object of divine worship. **Arsinoë,** otherwise called **Crocodilopolis,** from the divine honours here paid to the crocodile, stood S.W. of Memphis, between the river and Lake Mœris. The surrounding region was the most fertile in Egypt, and produced, in addition to grain of all sorts, dates, figs, roses, and olives. Near it were the necropolis of crocodiles, and the celebrated Labyrinth.[*] Its ruins are at *Medinet-el-Fyoum.* **Heracleopolis Magna,** *Anasieh,* was situated at the entrance of the valley of the *Fyoum,* and was the royal residence of the 9th and 10th dynasties. The ichneumon was worshipped there. **Oxyrynchus** derived its name from the worship of a fish of the sturgeon species. A Roman mint existed there in the age of Hadrian and Antoninus Pius. Some broken columns and cornices at *Hehnewh* mark the site of the town. **Hermopolis Magna,** *Eshmoon,* stood on the borders of Upper Egypt, and was a place of resort and opulence. A little S. was the castle, at which the river boats paid toll. On the opposite side of the river was the necropolis, at the well-known grottoes of *Beni Hassan.* The god Thoth, or Mercury, was worshipped at Hermopolis. The portico of his temple still exists, and consists of a double row of pillars, six in each. **Antinoöpolis,** nearly opposite Hermopolis, was built by the emperor Hadrian, A.D. 122, in memory of Antinous, to whom divine honours were paid. The ruins at *Enseneh* attest its former magnificence.

We may further notice briefly—**Acanthus,** *Dashour,* about 14 miles S.

---

[*] The Labyrinth was a stadium in length, and had twelve courts, six facing the N., and six the S. The chambers in it contained the monuments of the kings who built it, and the mummies of the crocodiles.

of Memphis, the seat of a temple of Osiris, enclosed with a hedge of acanthuses—Cynopolis, *Samallut*, S. of Oxyrynchus, and so named from the worship of the dog-headed deity Anubis—Nilopolis, near Heracleopolis Magna, built on an island in the Nile—and Aphrodite-polis, *Atfych*, a considerable town, a short distance from the E. bank of the river.

§ 13. **Thebais** was the most southerly division of Egypt, extending from Hermopolis Magna in the N. to Syene in the S., and at certain periods beyond the latter town to Hiera Sycamina. It was divided into 10 nomes, though occasionally a greater number is given. The cultivable soil between Syene and Latopolis is a narrow strip of alluvial deposit, skirting the banks of the Nile, and bounded by steep walls of sandstone. These are succeeded below Latopolis by limestone rocks, which continue to the head of the Delta. The valley expands into plains at Latopolis and Thebes, but below these points it contracts to a narrow gorge. The soil was remarkably fertile, though the ordinary fall of rain was very small. The population was probably of a purer Egyptian stamp than that of the Delta. The towns were very numerous, and attained the highest importance in early times. Among them Thebes stands foremost as the metropolis of Upper Egypt, and the seat of the most magnificent temples and palaces of Egypt. Coptos held high rank under the Ptolemies as the entrepôt of Indian commerce. Among the more remarkable objects of art we may notice the temples of Apollinopolis Magna, the temples of Athor and Isis at Tentyra, the canal of *Jusuf* commencing at Diospolis Parva, the necropolis of Abydos, the sepulchral chambers at Lycopolis, and the superb portico of Hermopolis Magna. The chief supply of stone was obtained from the sandstone quarries of Silsilis, below Ombos.

Pavilion of Rameses III. at Thebes.   (From Wilkinson.)

Thebes,[1] the No-Ammon of the Bible, and the Diospolis Magna of the Greeks and Romans, stood on both sides of the Nile, at a point where the hills on each side recede from the river, leaving a plain some 12 miles wide from E. to W., and about the same in length from N. to S. The population chiefly lived on the E. bank; on the W. were the temples,[2] with their avenues of sphinxes, and the necropolis. The site is now partly occupied by four villages—Luxor and Karnak on the E. bank, Gournah and Medinet Aboo on the W. The western portion, which was named Pathyris, as being under the protection of Athor, and was the "Libyan Suburb" of the Ptolemaic age, contained the following buildings :—the Menophthium, or temple and palace of Setei-Menephthah ; the Memnonium,[3] or Ramesium, occupying a succession of terraces at the base of the hills, containing the colossal statue of Rameses,[4] and numerous chambers adorned with hieroglyphics ; the Amenophium, or temple of Amunoph III., the Memnon of the Greeks, and near it the colossal statues Tama and Chama, rising to a height of 60 ft. above the plain, the most easterly of which was the celebrated vocal Memnon[5] ; the Thothmesium, a temple erected by several sovereigns of the name of Thothmes ; and the southern Ramesium, adorned with sculptures relating to Rameses IV. The necropolis extends for 5 miles along the Libyan hills, the most interesting portion being that which contains the Royal Sepulchres. On the E. side of the river the most conspicuous objects are :—at Luxor, the obelisk of Rameses III., the fellow to which stands in the Place de la Concorde at Paris ; two monolithal statues of the same monarch ; a court, with a double portal and colonnades attached ; and at Karnak the palace of the kings, containing the great court, the great hall, 329 ft. long, by 173 broad, and 80 high, and other chambers, one of which has the great Karnak Tablet sculptured upon it. The quarters of Karnak and Luxor were connected by an avenue of andro-sphinxes. These various buildings were erected at vastly different periods, commencing with Sesortasen I., and descending through the Amunophs, Rameses, and Thothmes, down to the time of the Ptolemies, and even the Roman emperors. The period of the eminence of Thebes commenced with

---

[1] The name is derived from the Coptic Ap, "head," which with the article became Tape : the more correct form of the name is therefore Thebe, as given by Pliny.

[2]

αἴθ' ὅσα Θήβας
Αἰγυπτίας, ὅθι πλεῖστα δόμοις ἐν κτήματα κεῖται,
Αἵ θ' ἑκατόμπυλοί εἰσι, διηκόσιοι δ' ἀν' ἑκάστην
Ἀνέρες ἐξοιχνεῦσι, σὺν ἵπποισι καὶ ὄχεσφιν —Hom. Il. ix. 381.

The "one hundred gates" of the poet were not (as we should naturally suppose) entrances through the walls of the town, but the propylæa of temples. Thebes does not appear even to have been surrounded by a wall.

[3] The word Memnonium appears to be a Greek corruption of Miamun, attached to the name of Rameses II., and hence applied to the buildings erected by that monarch at Thebes and Abydos.

[4] The weight of this gigantic statue has been estimated at 887 tons 5¼ cwt.

[5] The statue of Memnon was fractured by an earthquake before Strabo's time : Juvenal refers to its condition :—

Dimidio magicæ resonant ubi Memnone chordæ.—Sat. xv. 5.

The statue was said to utter a metallic sound a little after sunrise; this was no doubt produced by a deception of the priests : in the lap of the statue is a stone which, when struck, emits a metallic sound.

the 18th dynasty, when the Hyksos were expelled from Lower Egypt, and continued for nearly 8 centuries, from 1600 to 800. Its decline may be attributed to the rise of Memphis, and to the gradual increase of communication with the Greeks and other foreigners. In the Persian era it ceased to hold rank as a metropolis. Its chief buildings were destroyed by Cambyses. It suffered severely after its capture by Ptolemy Lathyrus in B.C. 86; but it continued to exist until the irruption of the Saracens, and was a considerable place in the 4th cent. A.D. **Lycopolis**, *E'Syout*, was S.E. of Hermopolis, and was so named from the worship of Osiris under the form of a wolf: in the adjacent rocks are chambers containing mummies of wolves. This, or **Abydus**, on the *Bahr Yusuf* about 7½ miles W. of the Nile, was the birth-place of Menes, and the burial-place of Osiris, and ranked next to Thebes itself in point of importance. It had sunk before Strabo's time. The ruins at *Arabat-el-Matfoon* consist of a large pile called the "Palace of Memnon," erected by Rameses II. of the 18th dynasty; and a temple of Osiris, built by Rameses the Great; the celebrated *Tablet of Abydos*, now in the British Museum, was discovered here in 1818; it contains a list of Egyptian kings prior to Rameses the Great. **Tentyra** stood about 38 miles N. of Thebes, and probably derived its name from the goddess Athor, or Venus, *Thyn-Athor*, meaning the "abode of Athor." Its inhabitants abhorred the crocodile, and hence arose sanguinary conflicts with the inhabitants of Ombos, one of which Juvenal seems to have witnessed.[1] The remains of the town at *Denderah* are striking, though of a late period of Egyptian art. The chief buildings are—the temple of Athor, the portico of which has on its ceiling the so-called "Zodiac," which, however, is probably a mythological subject, executed in A.D. 35; the chapel of Isis; and the Typhonium, so named from the representations of the Typhon on its walls. The inscriptions range from the time of the later Ptolemies to Antoninus, the names of the Cæsars from Tiberius to Antoninus being most frequent. **Hermonthis**, *Erment*, stood 8 miles S.W. of Thebes, and was celebrated for the worship of Isis, Osiris, and their son Horus. Its ruins show its former magnificence: the chief building, the Iseum, was erected by Cleopatra (B.C. 51–29), to commemorate the birth of her son Cæsarion. **Latopolis**, *Esneh*, derived its name from the large fish *lato*, under which form the goddess Neith was worshipped. Its temple was magnificent; but the jamb of a gateway is the only relic of the original structure; the other remains belong to the Macedonian and Roman eras, the names of Ptolemy Euergetes and Epiphanes, of Vespasian, and Geta, appearing in the sculptures. **Apollinopolis Magna** stood about 13 miles below the Lesser Cataract, and became under the Romans the seat of a bishop's see, and the head-quarters of the Legio II. Trajana. The remains at *Edfoo* consist of two magnificent temples; the larger one founded by Ptolemy Philometor, and dedicated to Noum, 424 ft. long, by 145 wide, and having a gateway 60 ft. high; the lesser one founded by

---

[1] Inter finitimos vetus atque antiqua simultas,
Immortale odium, et nunquam sanabile vulnus
Ardet adhuc Coptos et Tentyra. Summus utrimque
Inde furor vulgo, quod numina vicinorum
Odit uterque locus, cum solos credat habendos
Esse deos, quos ipse colit.—JUV. XV. 35.

Terga fugæ celeri præstantibus omnibus instant
Qui vicina colunt umbrosæ Tentyra palmæ.—Ib. XV. 70

Ptolemy Physcon. **Antæopolis**, on the E. or right bank of the river, was so named from the worship of Antæus, introduced from Libya. The plain adjacent to it was the traditional scene of the combat between Isis and Typhon. Under the Christian emperors it was an episcopal see. **Chemmis**, or as it was later called, Panopolis (the Greek Pan representing the Egyptian Chem) was celebrated for the worship of Pan, and also of Perseus, who was said occasionally to visit the place. The modern name *Ekhmim* is a corrupted form of Chemmis. **Coptos**, *Kouft*, stood about a mile from the river, and was the spot where the route for Berenice on the Red Sea left the valley of the Nile. Subsequently to B.C. 266, when Berenice was built, it was a prosperous and busy place, and remained so down to the latest period of the Roman Empire. **Ombi** was about 30 miles N. of Syene, and was devoted to the worship of the crocodile-headed god Sevak. The remains of two fine temples still exist, mainly of the Ptolemaic age, with a few specimens of an earlier date : the larger one was a kind of Pantheon, the smaller was sacred to Isis : they stand on a hill, and present an imposing appearance. **Syene**,[2] *Assouan*, was the most southerly town of Egypt, and stood on a peninsula immediately below the Great Falls. The granite quarries about it produced the fine stones out of which the colossal statues and obelisks of Egypt were cut. Syene[3] was important both as a military and commercial post. Opposite Syene is the small island of **Elephantine**, which commanded the navigation of the river from the S. : it was thus regarded as the key of the Thebaid, and hence was garrisoned by the successive owners of Egypt, whether Egyptians, Persians, Macedonians, or Romans. Its fertility and verdure present a strong contrast to the sterility that surrounds it. The most striking remains on it are a temple of Kneph built by Amenoph III., and the Nilometer. About 6½ miles above Syene were the two small islands of **Philæ** ;[4] the lesser one, to which the name was more particularly applied, was reputed the burial-place of Osiris, and hence regarded as specially sacred. Both islands abound in temples and monuments, erected for the most part by the Ptolemies. The chief temple, dedicated to Ammon Osiris, was at the S. end of the small island, and was approached from the river through a double colonnade ; the walls are covered with sculptures representing the history of Osiris. The Pharoahs kept a strong garrison on the island. Philæ was also the seat of a Christian Church.

On the coast of the Red Sea there were two ports of consequence— **Myos-Hormos** and **Berenice**, founded by Ptolemy Philadelphus for the purposes of the Indian and South African trade. The first was probably so named from the pearl-mussel found there (" Harbour of

---

[2] Its position, very nearly under the tropic of Cancer, is frequently noticed by Lucan.

Calida medius mihi cognitus axis
Ægypto, atque umbras nusquam flectente Syene. —II. 587.
Nam quis ad exustam Cancro torrente Syenen.
Ibit,——   viii. 851.
Cancroque suam torrente Syenen,
Imploratus adest.   x. 234.

[3] It was the place to which Juvenal was banished.

Quos dirimunt Arabam populis Ægyptia rura
Regni claustra Philæ.   Luc. x. 312.

the Mussel"), the second after the mother of Philadelphus. They
stood respectively at 27° and 23° 56' N. lat. The more southerly
position of Berenice rendered it ultimately the most prosperous of the
two places. It stood on a small bay at the extremity of a deep gulf,
named Sinus Immundus. Myos-Hormos seems to have declined in the
reigns of Vespasian and Trajan.

The Memnonium at Thebes during the inundation. (From Wilkinson.)

§ 14. Three of the Oases were closely connected with Egypt.
Oasis Magna, *El-Khargeh*, or as it was sometimes simply termed
"Oasis," lies in the latitude of Thebes. It is 80 miles long, by
about 9 broad; and is bounded by a high calcareous ridge. None
of the monuments on it reach back to the Pharaonic era, the
principal buildings bespeaking the Macedonian or even the Roman
period. It was a place of exile for political offenders, and for
Christian fugitives. It was visited by Cambyses on his expedition
against the Ammonians. The great temple, 142 ft. by 63, and
about 30 in height, was dedicated to Ammon; the other remains
are a remarkable necropolis, and a palace of the Roman era. **Oasis
Parva**, *El-Dakkel*, lies N. of Oasis Magna, from which it was
separated by a high ridge, and contains several warm springs. It
has a temple and tombs of the Ptolemaic era. Under the Romans
it was celebrated for its wheat; now its chief productions are dates,
and other fruits. **Ammonium**, *El-Siwah*, was about 20 days'
journey distant from Thebes, from which point it was most easily
accessible, though it was also approached from Parætonium. This
Oasis is about 6 miles long, by 3 broad, well irrigated by water
springs (one of which "the Fountain of the Sun," was particularly
celebrated for the apparent coldness of its water), and remarkably
fertile in dates, pomegranates, and other fruits, which were largely
exported. The oasis derived, however, its chief celebrity from the

temple' and oracle of Jupiter Ammon, which ranked with those of
Delphi and Dodona, and was visited by Alexander the Great. The
ruins of the temple exist at *Ummebeda*, and probably belong to the
Persian era of Egyptian history. The walls were covered with
hieroglyphics, and the colours still remain in some places. The
soil of the oasis is strongly impregnated with salt.

History of Egypt —The history of Egypt may be divided into four
periods, viz.—the Pharaonic, down to B.C. 525; the Persian, from 525
to 332 ; the Macedonian or Hellenic, from 332 to 30 ; and the Roman
from B C. 30 to A.D. 640.

I. The first of these, the Pharaonic, may be divided into three
portions :—the old monarchy, extending from the foundation of the
kingdom to the invasion of the Hyksos ; the middle, from the entrance
to the expulsion of the Hyksos ; and the new, from the re establishment of the native monarchy by Amosis to the Persian conquest.

(1.) *The Old Monarchy.*—Memphis was the most ancient capital, the
foundation of which is ascribed to Menes, the first mortal king of
Egypt. The names of the kings, divided into dynasties, are handed
down in the lists of Manetho,' and are also known from the works
which they executed. The most memorable epoch in the history of
the Old Monarchy is that of the Pyramid kings, placed in Manetho's
fourth dynasty. Their names are found upon these monuments : the
builder of the great pyramid is called Suphis by Manetho, Cheops by
Herodotus, and *K'hufu*, or *Shufu*, in an inscription upon the pyramid.
The erection of the second pyramid is attributed by Herodotus and
Diodorus to Chephren; and upon the neighbouring tombs has been
read the name of *K'hafra*, or *Shafre*. The builder of the third pyramid
is named Mycerinus by Herodotus and Diodorus; and in this very
pyramid a coffin has been found bearing the name *Menkura*. The most
powerful kings of the Old Monarchy were those of Manetho's 12th
dynasty : to this period are assigned the construction of the Lake of
Mœris and the Labyrinth.

(2.) *The Middle Monarchy.*—Of this period we only know that a
nomadic horde for several centuries occupied and made Egypt tributary; that their capital was Memphis; that in the Sethroïte nome
they constructed an immense earth-camp, which they called Abaris;
that at a certain period of their occupation two independent kingdoms

---

' Ventum erat ad templum, Libycis quod gentibus aram
Inculti Garamantes habent : stat corniger illic
Jupiter, at memorant, sed non ant fulmina vibrans,
Ant similis nostro, sed tortis cornibus Ammon.
Non illic Libycæ posuerunt ditia gentes
Templa, nec Eois splendent donaria gemmis.
Quamvis Æthiopum populis, Arabumque beatis
Gentibus, atque Indis unus sit Jupiter Ammon,
Pauper adhuc Deus est, nullis violata per ævum
Divitiis delubra tenens : moremque priorum
Numen Romano templum defendit ab auro.—Luc. ix. 511.

' Manetho was an Egyptian priest who lived under the Ptolemies in the 3rd
century B.C., and wrote in Greek a history of Egypt, in which he divided the kings
into thirty dynasties. The work itself is lost, but the lists of dynasties have been
preserved by the Christian writers.

were formed in Egypt, one in the Thebaid, which held intimate relations with Ethiopia; another at Xois, among the marshes of the Nile;
and that, finally, the Egyptians regained their independence, and expelled the Hyksos, who thereupon retired into Palestine.

(3.) *The New Monarchy* extends from the commencement of the 18th
to the end of the 30th dynasty. The kingdom was consolidated by
Amosis, who succeeded in expelling the Hyksos, and thus prepared the
way for the foreign expeditions [4] which his successors carried on in Asia
and Africa, extending from Mesopotamia in the former to Ethiopia in
the latter continent. The glorious era of Egyptian history was under
the 19th dynasty, when Sethi I., B.C. 1322, and his grandson, Rameses
the Great, B.C. 1311, both of whom represent the Sesostris of the Greek
historians, carried their arms over the whole of Western Asia and
southwards into *Soudân*, and amassed vast treasures, which were expended on public works. Rameses originated the project of connecting
the Red Sea with the Nile. He is further known as the builder of the
rock temples of *Aboo-simbel*, as well as of temples at Napata, Tanis,
Thebes, Memphis, and other places. Under the later kings of the
19th dynasty the power of Egypt faded: the 20th and 21st dynasties
achieved nothing worthy of record; but with the 22nd we enter upon a
period that is interesting from its associations with Biblical history,
the first of this dynasty, Shishonk I. (Sesonchis), B.C. 990, being the
Shishak who invaded Judæa in Rehoboam's reign and pillaged the
temple (1 Kings xiv. 25): the extent of his rule is marked by the forces
he commanded, consisting of Libyans, Sukkiims (who are supposed to
be the Troglodytes from the western shores of the Red Sea), and Ethiopians (2 Chron. xii. 3). In the reign of Osorkon I. the expedition of
Zerah, the Ethiopian, took place (2 Chron. xiv. 9); this expedition is
nowhere else noticed, and it appears almost unavoidable that we
should identify Zerah with Osorkon. The 25th dynasty consisted of
Ethiopians, the two first of whom, Sabaco and Sebichus, ruled over
the whole of Egypt, while the third, Taracus, was restricted to Upper
Egypt. The second of these monarchs is the So with whom Hoshea,

---

[4] We find in inscriptions the names of foreign nations subdued by the Egyptian
monarchs. Of these the most important are: *Naski*, undoubtedly the negroes;
the name survives in *Nammones = Naski Amen*, "negroes of Ammon;" *Cush*,
as in Scripture, the Greek Ethiopia: *Shaso*, the general name of the Arabs:
Pulishto, the Philistines, who were connected with the Egyptians by descent, as
is implied in the name Caphthor, mentioned in the Bible as the primitive seat of
the Philistines (Jer. xlvii. 4; Am. ix. 7): *Khita*, or *Sheta*, Hittites, to whom
belonged the fortress of *Ateah*, or Kadesh, perhaps Ashteroth-Karnaim: *Shaire-
tana*, supposed to be the Sharatinians who lived near Antioch: *Tokkari*, a people
whose residence is unknown, represented as wearing helmets similar to those in
the sculptures of Persepolis: *Rebo*, a nation probably from the northern part of
Assyria: *Pount*, probably dwelling on the borders of Arabia: *Shari* (compare
Scriptural Shur), a tribe of Northern Arabia: *Rot-n-no*, probably in Northern
Syria; the name may be connected with Aradus: *Nahrayn*, undoubtedly the
Naharaim of Scripture (Mesopotamia), with the town *Ninivu* (Nineveh): *Shinar*,
the Scriptural Shinar, Babylonia: *Turrsha*, *Mashoash* (Moschi?), and *Kufa*, Asiatic
races whose residences have not been identified: *Armoori* (Samaria?): *Lemanon*,
a Syrian tribe about Lebanon: *Kanana*, the Canaanites: lastly, *Hyksos*, with
regard to whom great doubt exists; the name is of Arabian origin, and may
signify either "Shepherd kings" or "Arab kings;" but whether they were
Canaanites, Arabians, or Philistines, is not agreed.

king of Israel, made a treaty (2 Kings xvii. 4), in whose reign Egypt
came into collision with Assyria. Taracus, the Tirhakah of Scripture,
succeeded So in the rule of the Thebaid, while native princes governed
Lower Egypt. The Assyrian war was continued in his reign, and the
sieges of Libnah and Lachish by Sennacherib, which took place in each
of the two expeditions noticed in Scripture (2 Kings xviii. 13, 17), had
reference to the Egyptian rather than the Jewish campaign. It was probably during the reign of Tirhakah that the dodecarchy prevailed in
Lower Egypt: these twelve contemporaneous rulers were probably the
heads of the nomes. The Æthiopian dynasty in Upper and the dodecarchy in Lower Egypt were followed by the re-establishment of a
native dynasty in the person of Psammetichus I., B.C. 671. He introduced Greek auxiliaries into his army, to the great dissatisfaction of
the native troops, who seceded in a body, and settled to the south of
Meröe. The long siege of Azotus, stated at twenty-nine years (Her. ii.
157), and the threatened invasion of the Scythians, were two chief
events of his reign. His son Neco, or Necho, B.C. 617, made a vain
attempt to regain the supremacy which Egypt had once enjoyed over
Western Asia: he defeated Josiah at Megiddo (2 Kings xxiii. 29), but was
himself utterly defeated by Nebuchadnezzar at Carchemish (Jer. xlvi. 2).
Psammetichus II., or Psammis, B.C. 601, passed an uneventful reign of six
years, and was succeeded by Apries, the Pharaoh-Hophra of the Bible,
B.C. 595, the king with whom Zedekiah, king of Judah, entered into
alliance. He was successful in the early part of his reign, capturing
Gaza and Sidon, and obliging the Chaldæan army to retire from Jerusalem; but his attempt on Cyrene was a failure, and terminated in the
revolt of his troops, and his own deposition and death: it would appear
from some passages in the Bible (Is. xix. 2; Jer. xliii. 10, xliv. 1, 30)
that Nebuchadnezzar undertook an expedition into Egypt. Amasis,
B.C. 570, who deposed and succeeded Necho, cultivated friendly relations with the Greeks, and gave them Naucratis as an emporium: his
works of art, particularly the monuments at Sais, were numerous and
splendid. Psammenitus came to the throne just as Cambyses reached
the frontier of Egypt, B.C. 525. He was defeated at Pelusium, and
afterwards besieged and captured at Memphis; and from this time
Egypt formed an integral part of the Persian empire.

II. *The Persian Era.*—The 27th dynasty consisted of eight Persian
kings, who were satraps of the Persian emperor. The chief events
during this period were the two revolts in 488 and 456, the first of
which delayed the second invasion of Greece. The 28th dynasty contains only one name, Amyrtæus the Saite, who reigned over the whole
land, and whose sarcophagus is preserved in the British Museum.
The 29th contained four, and the 30th three kings, the last of whom,
Nectanebus II., was dethroned by the generals of Darius Ochus.

III. *The Hellenic Era.*—This commences with the conquest of Egypt
by Alexander the Great (B.C. 332). On the dissolution of the Macedonian empire in 323, Egypt fell into the hands of Ptolemy Soter, the
founder of the dynasty of the Lagidæ. The early kings of this dynasty
were engaged in frequent contests with the kings of Syria. Soter himself (323-283) conquered Phœnicia and Cœle-Syria; Philadelphus (283-
247) secured peace by giving these provinces as the marriage-portion of
Berenice, the wife of Antiochus Theus; Euergetes (247-222) took up
arms to revenge the death of Berenice, and reduced the Syrian provinces to the confines of Bactria and India; Philopator (222-205, de-

fealed Antiochus the Great at Raphia, and thus regained the disputed
possessions which had previously been conquered by the Syrians; but
under Epiphanes (205-181) they were finally lost, and the attempt to
regain them under Philometor (181-146) ended in the total defeat of
the Egyptians at Pelusium in 170. The succeeding reigns of Euergetes II. (146-117), Lathyrus (117-107, and again 89-81), Alexander I.
and Cleopatra (107-90), and Auletes (80-51), are chiefly notorious for
the profligacy of the successive sovereigns and the frequent insurrec-
tions of the Alexandrians. The disputes that prevailed opened the
door for the interference of the Romans, and the last of these kings
was restored to his throne by A. Gabinius, proconsul of Syria. In the
reign of his successors, Ptolemy and Cleopatra, the Alexandrian war
arose, in which Cæsar took the part of Cleopatra, and Ptolemy perished
in 47. Cleopatra thenceforward reigned in conjunction with another
brother: her eventful life was terminated by her own hand in 30, and
the dynasty of the two Ptolemies ended. As to the internal state of
Egypt under the Hellenic monarchs, it was on the whole prosperous.
Commerce was fostered not only by the foundation of Alexandria, but
subsequently by the opening of the Indian trade through the Red Sea
by Philadelphus; literature flourished greatly at Alexandria; even the
old Egyptian edifices came in for a share of royal patronage, and many
of the temples were either restored or enlarged.

IV. *The Roman Era.*—For a long period Egypt enjoyed peace and
prosperity under the Roman emperors, who treated it generally with
consideration, and aided in the maintenance of the religious edifices.
In the reign of Aurelius a serious rebellion occurred (A.D. 171-175);
in 269 the country was for a few months occupied by Zenobia, queen
of Palmyra; and thenceforward troublous times set in through the
resistance offered to Aurelian in 272, Probus in 276, and Diocletian
in 295. The religious disputes of the Arians and Athanasians form
prominent topics in the history of this period; and the extent to which
monasticism prevailed on the banks of the Nile exercised a prejudicial
influence on the country. In A.D. 379 Paganism was denounced by an
imperial edict, and all the temples were overthrown. The only subsequent events were the subjugation of Egypt by Persia in A.D. 618; and
its conquest by Amron, the general of the Khaliph Omar, in 640.

The Ruins and Vicinity of Philæ. (From Wilkinson.)

II.—ÆTHIOPIA.

§ 15. Æthiopia, in its strictly territorial sense,[1] was bounded on
the N. by Egypt, on the W. by the Libyan Desert, on the S.[2] by the
Abyssinian highlands, and on the E. by the Indian Ocean and the
Red Sea, from Prom. Prasum in the S. to Prom. Bazium in the N.
It embraces *Nubia, Sennaar, Kordofan,* and northern *Abyssinia.*
It is for the most part a mountainous country, rising gradually towards the S. Water is abundant there, and the country seems to
have been famed for its fertility in ancient times. In addition to
various kinds of agricultural produce, it possessed some articles of
great commercial value, particularly gold, ebony, and ivory.

Name.—The Greeks derived " Æthiopia" from αἴθω, and ὄψ, according to which it would betoken the land of the *dark-complexioned.* It
is probable, however, that it was a Græcized form of *Ethosh,* the name
by which the Egyptians described it.

§ 16. The mountain-ranges of this vast district were but imperfectly known. A lofty chain skirts the sea-coast, and shuts out the
interior from easy access to the sea. On the W. a range, named
Æthiopici Montes, forms the natural limit on the side of the desert.
Far away to the S. were the **Lunæ Montes,** reputed to contain the
sources of the Nile. The sea-coast was tolerably well known from
the visits of merchants. The *Straits of Bab-el-Mandeb* are not
noticed under any specific name. Two bays only are described, viz. :
Adulicus Sinus, *Annesley Bay,* in the Red Sea ; and **Avalites Sin.**
somewhat S. of the Straits. Of the promontories we may notice—
Bazium, *Ras-el-Naschef,* nearly in the parallel of Syene ; **Aromata,**
C. Guardafui, the most easterly point of Africa ; and **Prasum,** *C.
Delgado,* in the extreme S. The positions of others that are
noticed on the shores of the Indian Ocean, such as **Zingis, Noti
Cornu,** and **Rhaptum,** are not well ascertained. The chief river is
the Nile, which has been already described as dividing into two
branches in this part of its course, to one of which (probably the
Blue Nile) the name of **Astapus** was given, and which also receives,
near Meroë, an important tributary, now named the *Tacazze,* and
probably formerly the **Astaboras.** The lakes, in which the Nile was

---

[1] The name Æthiopia was sometimes used in a broader sense to signify all the
inhabitants of interior Africa, and in this case the inhabitants of Æthiopia proper
were distinguished as the Æthiopians beyond *Egypt.* We have already (p. 19)
referred to the mythical Æthiopians.

[2] Æthiopia was the most southerly land known to the ancients ; hence Lucan
describes it as—

Æthiopumque solum quod non premeret ur ab ulla
Signiferi regione poli, nisi poplite lapso
Ultima curvati procederet ungula Tauri.—iii. 253.

reputed to have its sources, fell within the limits of Æthiopia : in addition to these we have to notice the lake Coloe, or Psebôa. *Dembea*, through which the Astapus flows.

§ 17. The inhabitants of this vast region were a mixture of Arabian and Libya nces with the genuine Æthiopians. They were divided into a number of tribes, designated according to their diet or employment, such as the Rhizophagi, "root-eaters," Acridophagi, "locust-eaters," &c. The residences of these tribes are uncertain, with the exception of the following four :—The Blemmyes and Megabari, between the Red Sea and the Astaboras ; the Icthyophagi, "fish-eaters," on the coast of the Red Sea, N. of the Bay of Adule ; and the Troglodytæ, "cave-dwellers," in the mountains skirting the Red Sea, S. of Egypt. The Macrobii, "long-lived," had a settled residence, but its locality cannot be considered as known. The Sembritæ are deserving of notice, as being in all probability the descendants of the Automoli, noticed by Herodotus (ii. 30) as the war-caste of Egypt, who deserted in the reign of Psammetichus, B.C. 658. The Sembritæ appear to have lived on the Astapus, not far from Auxume, which has been derived by some from the Egyptian name of the caste "Asmach." The Nubæ [a] originally lived on the western bank of the Nile, S. of Meroë, in *Kordofan:* they were the water-carriers and caravan-guides engaged in the trade between Egypt and Inner Africa, and derived their name from the *gold* ("noub" in Egyptian) imported from *Kordofan.* Originally they were isolated tribes, but in the 3rd cent. A.D. they were consolidated, and in the reign of Diocletian (about A.D. 300) were transferred by the Romans to the Nile, as a barrier against the Blemmyes : they thus gave to that district the name of *Nubia,* which it still retains. The country may be considered as divided into the following districts ;—Dodecaschœnus, in the N., extending for 12 *schœni* (as its name implies) from Philæ to Pœcelcis : by the Romans it was annexed to Egypt ; *Æthiopia Proper,* or the kingdom of Meroë, which extended southwards from Pœcelcis to the junction of the Blue and White Niles ; *Regio Auxomitārum,* between the upper course of the Blue Nile and the Red Sea, nearly coextensive with *Abyssinia ;* and *Barbaria* or *Azania, Ajan,* the coast-district from the promontory of Aromata to that of Rhaptum : the latter name, according to Ptolemy, applied more particularly to the interior. The southern portion of Meroë was named the "Isle," as being bounded on three

[a] Illa simul, insmitam testantes corpore solem,
Exusti venere Nubæ. Non ærea cassis,
Nec lorica riget ferro, nec tenditur arcus ;
Tempora multiplici mos est defendere lino,
Et lino munire latus, sceleratæque socris
Spicula dirigere, et ferrum infamare veneno.—SIL. ITAL. III. 266.

of its sides by rivers, viz.: the Nile on the W., the Astapus or
Blue Nile on the S., and the Astaboras on the N.E. It was
bounded on the E. by the Abyssinian highlands, and on the W. of
the Nile was the desert of *Bahiouda*. This district was rich[4] in
productions of every kind—minerals, animals, and vegetables; and
its fertility, combined with its central position, led to the high
prosperity which it attained.

§ 18. The towns of Æthiopia, with which we are acquainted
through the Greek historians and geographers, may be distinguished
into two classes: the genuine Æthiopian towns, which were chiefly
situated in the valley of the Nile; and the Greek *emporia* on the
shores of the Red Sea. The latter belong to the period of the Ptolemies, and include Ptolemais-Theron, Adûle, Arsinoë, and Berenice
Epidelres. From these an active trade was carried on, not only with
the interior, but with Arabia, Western India, and Ceylon. These
towns flourished until the Saracen invasion in the 7th cent. A.D. Of
the Æthiopian towns, the southern capital Meroë was undoubtedly
the first in importance. The remains of temples and pyramids prove
the existence of numerous towns in the same district. Napâta[5]
comes next, and as the northern capital of Æthiopia was even more
important in relation to Egypt. Numerous important towns were
erected by the Pharoahs between Napâta and the Egyptian frontier,
the history of which is lost, but the ruins remain, testifying to the
former grandeur of the temples:[6] these are found at *Dendoor*, a short
distance S. of Talmis; at *Derr*; at *Aboosimbal* or *Ipsambol* (perhaps
the ancient Aboccis), about two days' journey below the Second
Cataract; at *Semneh*, above the Great Cataract, a place probably intended to guard the Nile; at *Soleb*, below the Third Cataract; and at
numerous other places. Subsequently to the fall of Meroë, Auxûme
rose to importance as a seat both of art and of commerce. Most of
the towns of the interior were entrepôts for the Central African
trade: to this circumstance Meroë, Auxûme, and Napâta owed their
wealth. Some of the towns in Dodecaschœnus were border-fortresses,
and are hence noticed in connexion with the campaigns of Petronius.

---

[4]   Late tibi gurgite rapto
Ambitur nigris Meroe fecunda colonis,
Laeta comis ebeni : quæ, quamvis arbore multa
Frondeat, æstatem nulla sibi mitigat umbra :
Linea tam rectum mundi ferit illa Leonem.—Luc. x. 302.

[5] The pyramids and temples near *Gebel-el-Birkel* are supposed to mark its site ;
while the thirty-five pyramids of *Nouri* stand eight miles higher up.

[6] These temples were chiefly built by the Egyptian monarchs ; the temple of
the Sun at *Derr*, and the richly sculptured temples at *Aboosimbal* are of the date
of Rameses the Great. At *Hassaia* is a temple bearing the sign-manual of
Thothmes III. These buildings probably survived to a late age, and were
beautified or enlarged at various eras : at *Dendoor*, for instance, there are remains of the Augustan age.

(1.) *In Dodecaschœnus.*—**Talmis** stood on the left bank of the Nile, about five days' journey S. of Philœ. The ruins of it at *Kalabschh* are highly interesting, consisting of a rock-temple dedicated to Mandulis, with bas-reliefs and beautiful sculptures. This temple was originally built by Amunoph II., was rebuilt by one of the Ptolemies, and repaired in the reigns of Augustus, Caligula, and Trajan. A fac-simile of these sculptures stands in the British Museum. A curious Greek inscription of Silco, probably one of the kings of the Nubæ who protected the Roman frontier, has been found there. Another temple of great interest belongs to the Pharaonic era. **Pselcis**, on the left bank of the Nile at *Dakkeh*, was one of the strongholds which Petronius took from the Æthiopians, and constituted a Roman fortress (B.C. 23). There is a temple of Hermes Trismegistus at *Dakkeh*, founded by Ergamenes, a contemporary of Ptolemy Philadelphus. **Hiera Sycaminus** was an extensive mart on the southern frontier, probably at *Wady Maharrakah*. The lesser towns in this district were—**Parembole**, *Debot*, a fortress on the Egyptian border, with a temple of Isis founded by Ashar-Amun, and adorned by Augustus and Tiberius, of which there are considerable remains; **Taphis**, *Teffa*, with large stone-quarries near it; **Tutsis**, the ruins of which are at *Gerf Hossayn*, consisting of a rock-temple of the reign of Rameses the Great, with numerous figures; **Tachompso**, on an island opposite Pselcis, and hence named **Contra-Pselcis**, when the latter place rose to importance: its position cannot be ascertained, as no island exists opposite the site of Pselcis: the lake noticed by Herodotus (ii. 29) was merely a reach of the Nile.

(2.) *In Æthiopia Proper.*—**Napāta**, the northern capital, was situated probably at the E. extremity of the great bend which the Nile makes in about 19° N. lat., and near *Gebel-el-Birkel*, where are found, on the left bank of the Nile, two temples dedicated to Osiris and Ammon, richly decorated with sculptures, and some pyramids. The two Egyptian lions which now adorn the British Museum were brought from this spot. Judging from its ruins, Napata must have been a very wealthy place, in consequence of its being the terminus of the routes from Gagaudes in the N.W., and Meroë in the S.E. It was the capital of Æthiopia under the Sabacos and Tirhaka, who extended their sway over Upper Egypt; and it was the most southerly point that the Romans reached. It sunk after its capture by Petronius, B.C. 22. The town of **Meroë** stood about 90 miles S. of the junction of the Astaboras with the Nile, at *Damkalah*, where its site is marked by some pyramids. In addition to this, ruins of cities, whose names have perished, extend for a considerable distance near the Nile between 16° and 17° N. lat., consisting of numerous temples, colonnades, and mounds of bricks. The architecture bespeaks a late age of Egyptian art. Meroë was the seat of a powerful state, in which the priesthood exercised great influence, while the sceptre was often held by females, with the official name of Candace. When the Egyptian monarchs extended their sway over Northern Æthiopia, Meroë remained independent. In the time of the Romans, however, it was an unimportant place. In the same district were two towns, named **Primis Parva** and **P. Magna**, the former of which, also named Premnis, is placed near the northern frontier at *Ibrim*, and was a fortress captured by Petronius, and afterwards retained by the Romans as an advanced post; the other was to the S. of Napata, not far from Meroë. **Auxime** stood E. of the Astaboras, in about 14° 7' N. lat., and is represented by *Axum*, the capital of *Tigre:* it was a place of considerable trade, and attained a high degree of prosperity after the fall of Meroë in the 1st or 2nd cent. of our era. From the fact of Greek being spoken there, it was not improbably a colony of Adulo. The most interesting relics of the old town are an obelisk 60 feet high, and a square enclosure with a seat, reputed to be the throne of the old kings. Auxume was the seat of a bishoprick, as we learn from a rescript of Constantius Nicophorus about A.D. 350.

(3.) *On the Coast.*—**Adûle**, *Thulla*, on the bay of the Red Sea named after it, is said to have been founded by fugitive slaves from Egypt. Under the Romans it served as the port of Auxume, and it was then a place of extensive trade. It possessed a famous inscription, named *Monumentum Adulitanum*, copied by Cosmas in the 6th cent. A.D., in which the proceedings of Ptolemy Euergetes are recorded. **Ptolemais Theron**, originally a town of the Troglodytæ on the Red Sea, was selected by Ptolemy Philadelphus (B.C. 282-246) as the spot whence elephant-hunting should be prosecuted: it hence became a place of large trade, both in elephants and in ivory. Its position is uncertain, but it was probably not far from Adule. Equally uncertain is the position of **Saba** in the same neighbourhood, one of the places at which the Sabæans of the Bible dwelt, while another place of the same name stood on the opposite coast of Arabia.

Of the other towns on the coast we may briefly notice—**Arsinoë**, a port in the country of the Troglodytæ, once called Olbia; **Berenice Panchrysus**, in the Troglodyte country, named the "All-golden," from the mines of *Jebel Ollaki* near it; a second **Arsinoë**, near the entrance of the Red Sea; and **Berenice Epidaires**, deriving its surname from its position "on a neck" of land at the Straits of *Bab-el-Mondeb*: it was also called **Deire**; Ptolemy Philadelphus favoured it, and named it after his sister Berenice.

(4.) *On the Indian Ocean.*—**Malao**, probably at *Berbera*, was a mart for gum, cattle, slaves, and ivory. **Rhapta** was the collective name of several villages (probably opposite the isle of *Pata*), so called from the "sewed" boats, *i.e.* fastened by fibres instead of nails, which were used there: it was the most distant trading station known on this coast.

History.—Æthiopia was intimately connected with Egypt, and not unfrequently was under the same sovereign. Among the predecessors of Sesortasen were eighteen Æthiopian kings. Sesortasen himself is said to have conquered Æthiopia. The 13th dynasty took refuge there during the occupation of the Hyksos. The 16th and 18th dynasties also conquered it; and the monuments of Thothmes I., II., III., and IV., prove the extent of their sway to have reached as far as Napâta. In the 8th cent. B.C. an Æthiopian dynasty extended their sway over Lower Egypt, under the kings Sabaco, Sebichus (the So of Scripture), and Taracus (Tirhakah). In the reign of Psammetichus (B.C. 630) the whole of the war-caste of Egypt migrated to Æthiopia, and settled probably in the district we have assigned to them. Cambyses endeavoured to conquer Æthiopia, but failed: nevertheless the Persian occupation of the Nile-valley opened the country considerably; and subsequently, under the Ptolemies, the arts and commerce of the Greeks were fully introduced. In the reign of Augustus an Æthiopian army advanced to the borders of Egypt: they were repulsed by Petronius, and pursued as far as Napâta. The Roman supremacy was acknowledged from that time (B.C. 23) until Diocletian's reign (A.D. 284-305). The frequent notices of Æthiopia in the Old Testament have been already referred to. In the New Testament, the only occasion on which the name occurs is in connexion with the conversion of the eunuch of Queen Candace.

Ruins of Cyrene. (From Hamilton.)

CHAPTER XVI.

Marmarica, Cyrenaica, Syrtica, Africa Propria Numidia, Mauretania, Libya Interior.

I.—Marmarica.

§ 1. **Marmarica** was a barren and sandy strip skirting the Mediterranean from the valley of the Nile in the E. to Cyrenaica in the W. : it answers to the modern *Desert of Barkah.* It was divided by

Ptolemy into two parts, Libycus Nomos in the E., and Marmaricus
Nomos in the W., the point of separation being at the Catabathmus
Magnus. The chief physical features in this district are the two
singular " descents " (καταβαθμοι, *Akabah*), where the land
slopes off from a considerable elevation on the shore down to the
interior : they were named **Catabathmus Magnus**, which rises to
900 feet, and which extends towards the Oasis of Ammonium in the
S.E. ; and **C. Minor** 500 feet high, more to the E. near Paraetonium.
The only river is the **Paliurus**, *Teminah*, on the W. border. The
Marmaridæ, after whom the district was named, are not noticed by
Herodotus,[7] but appear as the principal tribe in these parts between
the age of Philip of Macedon and the third cent. of our era: the
limits assigned to their abode by the ancient geographers vary
considerably. The chief towns were **Taposiris**, " the tomb of Osiris,"
about 25 miles from Alexandria, where Justinian constructed a
town-hall and baths; **Apis**, about 12 miles W. of Paraetonium ;
and **Paraetonium** or **Ammonia**, *Baretoun*, possessing a fine harbour.
Alexander started from this point to visit the oracle at Ammon,
B.C. 332 ; and Antony stopped here after the battle of Actium: it
was fortified by Justinian. There were numerous lesser ports, one
of which, **Plynus**, was probably the same as Panormus ; another
owed its name, **Menelai Portus**, to the tradition that Menelaus
landed there ; while **Chersonesus Magna** stood near the promontory
of the same name on the border of Cyrenaica, and was named
" Magna " in contradistinction to " C. Parva " near Alexandria.

II.—CYRENAICA.

§ 2. The district generally called **Cyrenaica** after its chief town
Cyrene, and occasionally **Pentapolis** after the five confederate towns

[7] It is not improbable that the Giligammæ of Herodotus are the same people
as the Marmaridæ of later writers : no subsequent writer notices the Giligammæ.
The Marmaridæ are frequently noticed by the later Latin poets :

    Gens unica terras
Incolit a sævo serpentum innoxia morsu,
Marmaridæ Psylli : par lingua potentibus herbis :
Ipse cruor tutus, nullumque admittere virus,
Vel cantu comante, potest.   Luc. ix. 891.

    Misti Garamante perusto
Marmaridæ volucres.   Luc. iv. 679.

Marmaridæ, medicum vulgus, strepuere catervis :
Ad quorum cantus serpens oblita veneni,
Ad quorum tactum miles jacuere corastæ.—Sil. Ital. iii. 300.

The Adyrmachidæ of Herodotus, whom we have already noticed (p. 38) as living
on the coast, appear to have retired into the interior : they are noticed by Silius
Italicus—

  Versicolor contra castra, et faleatus ab arto
  Ensis Adyrmachidis ac lævo tegmine crure.—iii. 278.

on it, extended along the coast of the Mediterranean from Chersonesus Magnus in the E., where it touched Marmarica to Aræ Philenôrum at the bottom of the Greater Syrtis in the W. The portion of this territory which was actually occupied by the Greeks consisted of the table-land and the adjacent coast, which here projects in a curved form into the sea to the N.E. of the Syrtis. The position and physical character of this region were highly favourable. It lies directly opposite Peloponnesus at a distance of 200 miles. Its centre is occupied by a moderately elevated table-land, which sinks down to the coast in a succession of terraces, and is throughout clothed with verdure and intersected by mountain streams running through ravines filled with the richest vegetation. Rain is abundant; and the climate is tempered by the sea breezes from the N., and by ranges of mountains, which shut out the heat of the *Sahara* from the S. It produced corn, oil, wine, dates, figs, almonds, and other fruits, and especially the plant *silphium* or *laserpitium*, whence the medical gum called *laser* was extracted, and which was the emblem of the country. Its honey and horses were also famed.

§ 3. The most striking physical features in this district are the promontories, of which we may notice from E. to W., Chersonesus Magna, *Ras el-Tin*; Zephyrium, C. Derne; Phycus, *Ras Sem*, the most northern headland in this part of Africa; and Borium, *Ras Teyonas*, on the E. coast of the Syrtis. The range of hills, which runs parallel to the coast of the Syrtis, was named Herculis Arênæ, "the sands of Hercules;" S.W. of these were the Velpi Mts., and more to the E., on the S. frontier, the Bacceilous Mt. The only river was the small stream Lathon, which joins the sea N. of Boreum. Near it was the little lake called Triton or Lacus Hesperidum, which some of the ancients confounded with that at the bottom of the Lesser Syrtis.

§ 4. The inhabitants of this district in the age of Herodotus were the Libyan tribes of the Giligammæ in the E., the Asbystæ in the centre, and the Auschisæ in the W. These were driven from the coast by Greek settlers who first entered under Battus, the founder of Cyrene, B.C. 631, and who gradually gained possession of the whole coast, erecting, in addition to Cyrene, Apollonia which served as its port, Teuchira and Hesperides on the coast of the Syrtis, and Barca about 12 miles from the N. coast. These five formed the original Pentapolis. Under the Ptolemies, various changes took place: the name of Hesperides was supplanted by that of Berenice, and Teuchira by Arsinoë. Barca sank and its port assumed its position under the name of Ptolemäis: Cyrene also waned before the growing prosperity of its port Apollonia. Henceforward the Pentapolis consisted of the cities of Cyrene, Apollonia, Ptolemais,

Arsinoë, and Berenice. The country continued to flourish under
the Romans until the time of Trajan, when the Jews who had
settled there in large numbers under the Ptolemies, rose and massacred the Romans and Cyrenæans. From this time it declined,
and the ruin of the Greek towns was completed by the Persian
Chosroes in A.D. 616.

Ruins of Ptolemais, the port of Barca.   (From Hamilton.)

Taking the towns in order from E. to W., we first meet with Apollonia, originally only the port of Cyrene, but afterwards the more
important town of the two: it was the birthplace of Eratosthenes, the
geographer. Its site at *Marsa Sousah* is marked by the splendid ruins
of several temples, the citadel, a theatre, and an aqueduct. **Cyrene**,
founded by colonists from Thera,[a] stood on the edge of the upper of

---

[a] The foundation of Cyrene is described in the following lines, Callista being
the poetical designation of Thera: the city is dignified with the title "divine,"
and its tutulary goddess represented as seated on a golden throne :—

Καὶ, Λατοίδαι·
μανίων μιχθέντες ἀνδρῶν
Πθεσιν, ἐν ποτι Καλ-
λίσταν ἀπώκησαν χρόνῳ
Νᾶσον· ἔνθεν δ' ὕμμι Λατοί-
δας ἔπορεν Λιβύας πεδίον
Σὺν θεῶν τιμαῖς ὀφέλ-
λειν, ἄστυ χρυσοθρόνου
Διανέμειν θεῖον Κυράνας
'Ορθόβουλον μῆτιν ἐφευρομένοις.—PIND. Pyth. iv. 457.

In

two terraces some 1800 feet above the sea, from which it was 10 miles distant ; the spot was selected in consequence of a beautiful fountain, named Cyre,[1] which bursts forth there and which the Greeks dedicated to Apollo. Its commerce was considerable, particularly in *silphium*,[1] and it held a distinguished place in literature, as the birthplace of Aristippus, the founder of the Cyrenæan school ; of Carneades, the founder of the New Academy at Athens ; and of the poet Callimachus. Its ruins at *Grennah* are very

Coin of Cyrene.

extensive, and contain remains of streets, aqueducts, temples, theatres, and tombs. In the face of the terrace, on which the city stands, is a vast subterraneous necropolis. Cyrene was governed by a dynasty, named the Battiadæ,[2] in which the kings bore alternately the names of Battus and Arcesilaus, from B.C. 630 to about 430, after which it became a republic. It was made a Roman colony with the name of Flavia. Ptolemais was erected by the Ptolemies, and was peopled with the inhabitants of Barca on

Coin of Barca.

the former site of the port of that town. Its ruins are in part covered by the sea. Barca stood on the summit of the terraces which overlook the W. coast of the Syrtis, in the midst of a well-watered[3] and fertile plain.

In anoth. passage of the same poet we have other characteristics of the place noticed—its fertility, the white colour of its chalk cliffs, and the celebrity of its horses :—

> Χρῆσεν οἰκιστῆρα Βάττον
> Καρποφόρου Λιβύας, ἱερὰν
> Νᾶσον ὡς ἤδη λιπὼν
> Κτίσσειεν εὐάρματον
> Πόλιν ἐν ἀργινόεντι μαστῷ.—ID. *Pyth.* iv. 10.

[1] Οἱ δ᾽ οὔτω σφῆς Κυρῆς ἐθήκαντο κελάδουσι
    Δωριέων.   CALLIM. *Hymn. in Apoll.* 88.

[1] Quam magnus numerus Libyssæ arenæ
    Laserpiciferis jacet Cyrenis,
    Oraculum Jovis inter æstuosi,
    Et Batti veteris sacrum sepulcrum.—CATULL. vii. 3.

[2]   Et iniquo e Sole calentes
    Battiadas lato imperio acceptisque regebat.—SIL. ITAL. ii. 60.

    Nec non Cyrene Pelopei stirpe nepotis
    Battiadas pravos fidei stimulavit in arma.—SIL. ITAL. iii. 252.

[3] The epithet *arida* in the following passages must be held to refer, not to the actual site of the town, but to the neighbouring desert table-land :—

Adfult

It was founded about B.C. 554, by some disaffected citizens of Cyrene joined by some Libyans, and it soon became so powerful as to deprive Cyrene of her supremacy over the western district. In B.C. 510 it was besieged by the Persians at the instigation of Pheretima, mother of Arcesilaus III., and after a siege of nine months was taken and its inhabitants transplanted to Bactria. The name however survived, and is somewhat vaguely applied by Virgil[4] to a Libyan tribe in the neighbourhood. *Barca* still forms one of the divisions of *Tripoli*. *Teuchira* or *Tauchira*, afterwards *Arsinoë*, was particularly noted for the worship of Cybele. It was founded by Cyrene, and its site is still called *Tochira*. *Hesperides*, afterwards *Berenice*, derived its first name from the notion that the fabled gardens of the Hesperides[5] were found in the fertile districts of Cyrene,[6] and its second from the wife of Ptolemy Euergetes, who raised it to a state of commercial prosperity. Off the northern coast is the small island of *Platea*, on which the Theræans first settled.

History.—The early history of Cyrenaica has been already given: it was subjected to Egypt by Ptolemy son of Lagus, B.C. 321. The last of the Cyrenæan kings, Apion, bequeathed it to the Romans B.C. 95, who gave the cities their freedom, but, in consequence of their dissensions, reduced it to a province (probably in B.C. 75), and united it with Crete, B.C. 67. In Constantine's division it was constituted a distinct province. Its connexion with Biblical history is briefly told. We have already mentioned that vast numbers of Jews were settled there : these visited Jerusalem periodically, as on the day of Pentecost (Acts ii. 10). One of them, named Simon, was selected to carry our Saviour's cross to Calvary (Luke xxiii. 26).

§ 5. In the interior, S. of Cyrene, dwelt the important tribe of the **Nasamōnes**, who extended their territory as far as the shores of the Syrtis westward, and inland to the Oasis of Augila : they had

---

Adfuit undosa cretus Berenicide milea
Nec, teretí dextras in pugnam armata dolore,
Destituit Barce altlentibus arida venis.— Sil. Ital. lil. 249.

Æternumqne arida Barce.—Id. ll. 62.

[4] Illne deserta siti regis lateqne furentes
Barcæl.   Æn. iv. 42.

[5]   Fuit aurea silva,
Divitiisque graves et fulvo germine rami,
Virgineusque chorus, nitidi custodia luci,
Et numquam somno damnatus lumina serpens,
Robora complexus rutilo curvata metallo.
Abstulit arboribus pretium, nemorique laborem
Alcides : pomisque inopes sine pondere ramos,
Retulit Argolico fulgentia poma tyranno. - Luc. ix. 360.

[6] The following extract from a modern writer justifies the selection as a matter of taste : "The rest of the journey (to Grennah) was over a range of low undulating hills, offering perhaps the most lovely sylvan scenery in the world. The country is like a most beautifully arranged *Jardin Anglais*, covered with pyramidal clumps of evergreens, variously disposed, as if by the hand of the most refined taste ; while *bosquets* of junipers and cedars, relieved by the pale olive and the bright green of the tall arbutus tree, afford a most grateful shade from the midday sun.'—Hamilton's *Wanderings in Africa*, p. 81.

a bad reputation among the Romans as wreckers.[7] The Oasis of Augila lies due S. of Cyrene between the 29° and 30° of N. lat., and was in ancient times the source whence the Nasamonians obtained their annual supply of dates, which they carried northwards to their head-quarters near the sea. It consists in reality of three oases, the largest of which retains the name of *Aujilah*, and is still famous for its dates. Each of the oases is a small hill rising out of an unbroken plain of red sand.

III.—SYRTICA REGIO.

§ 8. **Syrtica** was a narrow strip of coast land extending along the Mediterranean Sea for about 100 miles between the Greater and Lesser Syrtes. Its character is sufficiently attested by its name Syrtis (from the Arabic *sert* "desert"): it is so overwhelmed with sand that men and even vessels are sometimes buried beneath the accumulations carried by storms. The **Syrtes** are the two large bays which form the angles of the Syrtic sea, as already described. The dangers connected with the navigation of this sea existed chiefly in the imaginations of poets.[8] The most important promontories were **Cephalæ** or **Trišron**, *Cefalo*, at the W. extremity of the Greater Syrtis, and **Zeitha**, at the E. extremity of the Lesser.

---

[7] Hoc tam segne solum raras tamen exserit herbas,
Quas Nasamon gens dura legit, qui proxima ponto
Nudos rura tenet, quem mundi barbara damnis
Syrtis alit.   Nam littoreis populator arenis
Imminet, et nulla portus tangente carina
Novit opes.   Sic eum toto commercia mundo
Naufragiis Nasamones habent.   Luc. ix. 438.

Hoc colt æquoreus Nasamon, invadere fluctu
Audax naufragia, et prædas avellere ponto.—Sil. Ital. III. 320.

[8] Syrtes vel primam mundo Natura figuram
Cum daret, in dubio pelagi terræque reliquit :
(Nam neque subsedit penitus, quo stagna profundi
Acciperet, nec se defendit ab æquore tellus
Ambigua sed lege loci jacet invia sedes :
Æquora fracta vadis, abruptaque terra profundo,
Et post multa sonant projecti littora fluctus.
Sic male deseruit, nullosque exegit in usus
Hanc partem Natura sui :) vel plenior alto
Olim Syrtis erat pelago, præltusque natabat ;
Sed rapidus Titan ponto sua lumina pascens
Æquora subduxit zonæ vicina perustæ :
Et nunc pontus adhuc Phœbo siccante repugnat.
Mox ubi damnosam radios admoverit ævum,
Tellus Syrtis erit : nam jam brevis unda superne
Innatat, et late periturum deficit æquor.—Luc. ix. 303.

Tres Euros ab alto
In brevia et Syrtes urget, miserabile visu ;
Illiditque vadis, atque aggere cingit arena.—Æn. i. 110.

There are two small rivers—the Cinyps[9] in the E., which has not been identified; and the Triton,[1] el-Hammah, in the W., which formerly flowed through a series of lakes, Libya palus, Pallas, and Tritonitis: it now gains the sea by a direct course, and the three lakes are merged in one named *Shibk-el-Lowdeah*. The most valued productions of this country were the lotus, and a species of precious stone known as *Syrtides gemmæ*.

§ 7. The native tribes occupying this district in the time of Herodotus were the Lotophāgi about the Syrtis Minor, and the Gindānes more to the W. The former were so named from the custom, which still prevails there, of eating the fruit and drinking a wine extracted from the juice of the *Zizyphus Lotus* or jujube tree,

---

[9] The Cinyps was famed for the fine goats' hair produced about it :

> Nec minus interea barbas incanaque menta
> Cinyphii tondent hirci, setasque comantes.—Virg. *Georg.* iii. 311.

> Rigetque barba,
> Qualem forficibus metit supinis
> Tonsor Cinyphio Cilix marito.—Mart. vii. 95.

Its banks were also proverbially fertile :—

> Cinyphiæ segetis citius numerabis arista.—Ov. *ex Pont.* ii. 7, 25.

It was frequently used as a synonym for African generally, *e.g.* :—

> Cinyphias inter pestes tibi palma nocendi est.—Luc. ix. 787.
> Cinyphiumque Jubam.   Ov. *Afri.* xv. 755.

[1] The Triton and its lakes were connected with some of the Greek legends : It was there that the Argonaut Euphemus, the ancestor of Battus, received the promise of a settlement in Africa :—

> τόν ποτε
> Τριτωνίδος ἐν προχοαῖς
> λίμνας θεῷ ἀνέρι εἰδομένῳ
> Γαῖαν διδόντι ξείνια
> Προπράων Εὔφαμος καταβάς
> δέξατ'.   Pind. *Pyth.* iv. 33.

It is doubtful whether the term Triton-born, applied to Pallas, originally referred to this lake : it is more probable that in Homer and Hesiod the Bœotian stream is meant. The later poets, however, undoubtedly connected Pallas with the African river, which Euripides hence describes as—

> Λίμνης τ' ἐνύδρου Τριτωνίδος
> Πόρναν ἀστόν.   *Ion.* 871

So also the Latin poets—

> Huc, qui stagna colunt Tritonidos alta paludis,
> Qua virgo, ut fama est, bellatrix edita lympha
> Invento primam Libyen perfudit olivo.—Sil. Ital. iii. 322.

> Torpentem Tritona adit limosa paludem.
> Hanc, ut fama, Deus, quem toto littore pontus
> Audit ventosa perflantem marmora concha,
> Hanc et Pallas amat: patrio quæ vertice nata
> Terrarum primam Libyen (nam proxima cælo est,
> Ut probat ipse calor) tetigit : stagnique quieta
> Vultus vidit aqua, posuitque in margine plantas,
> Et se dilecta Tritonida dixit ab unda.—Luc. ix. 347.

which according to the Homeric legend[2] produced a state of dreamy forgetfulness. In addition to these, the Nasamones, Psylli, and Macæ roamed over portions of the district. Egyptian, Phœbician, and Cyrenæan colonists settled on the coast and intermixed with these Libyan tribes. Ptolemy mentions, in place of these, numerous tribes whose names are not noticed by any other writer. The chief towns were the Phœnician[3] colonies of Leptis Magna, Œa, and Sabrata, which having received Roman colonists became important places, and gave to the whole region the name of Tripolitana, which still survives in the modern *Tripoli*.

Leptis Magna was favourably situated on a part of the coast where the central table-land descends to the sea in a succession of terraces, as at Cyrene. It possessed a roadstead, well sheltered by the promontory of Hermæum.[4] The old Phœnician city was situated similarly to Carthage, upon an elevated tongue of land at the point where a small river discharges itself into the sea; the remains of sea walls, quays, fortifications on the land side, and moles are to be seen on its site, which is still called *Lebda*. At a later period a new city, named Neapolis, grew up on the W. side of the old town, which henceforth served as the citadel alone. This became the great emporium for the trade with the eastern part of Interior Africa, and under the Roman emperors, particularly Septimius Severus who was a native of the place, it was adorned with magnificent buildings, and flourished until the 4th cent. A.D., when it was much injured

Coin of Leptis.

by a native tribe named Ausuriani. Though partly restored by Justinian, it never recovered this blow. Its ruins are deeply buried in the sand, and a small village, *Legatah*, occupies its site. Œa became a Roman colony about A.D. 50 and flourished for 300 years, when it was ruined by the Ausuriani. On its site stands the modern capital *Tripoli*: a very perfect marble arch, dedicated to M. Aurelius Antoninus and L. Aurelius Verus, is the principal relic of the old town. Sabrata, or Abrotonum, was a

---

[2] Οἵ δ᾽ ἄρα Λωτοφάγοι μήδοντ᾽ ἑτάροισιν ὄλεθρον
  'Ημετέροις, ἀλλά σφι δόσαν λωτοῖο πάσασθαι.
  Τῶν δ᾽ ὅστις λωτοῖο φάγοι μελιηδέα καρπόν,
  Οὐκ ἔτ᾽ ἀπαγγεῖλαι πάλιν ἤθελεν, οὐδὲ νέεσθαι·
  'Αλλ᾽ αὐτοῦ βούλοντο μετ᾽ ἀνδράσι Λωτοφάγοισι
  Λωτὸν ἐρεπτόμενοι μενέμεν, νόστου τε λαθέσθαι. – Hom. *Od.* ix. 92.

[3] The Phœnician origin of the first and last of these towns is implied in the following lines :—

  Sabrata tum *Tyrium* vulgus, *Sarranaque* Leptis,
  Œaque Trinacrios Afris permixta colonos.—SIL. ITAL. iii. 256.

[4] Proxima Leptis erat, cujus statione quieta
  Exegere hiemem, nimbis flammisque carentem.—LUC. ix. 948.

considerable mart for the trade of the interior. In the Roman period, it was chiefly famed as the birthplace of Flavia Domitilla, wife of Vespasian: extensive ruins of it remain at *Tripoli Vecchio*. Of the less important towns we may notice *Tacape*, *Khabs*, at the innermost point of the Lesser Syrtis, noted for its hot sulphur-baths, in a fertile district, but with a bad harbour; *Zuchis*, in the same neighbourhood, noted for its purple dyes; and **Automāla**, on the borders of Cyrenaica. Off the coast were the islands of **Meninx**, *Jerbah*, S.E. of the Lesser Syrtis, occupied by the Lotophagi, and hence named **Lotophagitis**; and **Cercina**, *Karkenah*, and **Cercinitis**, *Jerbah*, at the N.W. extremity of the same gulf, which lay so close together that they were joined by a mole.

IV.—AFRICA PROPRIA.

§ 8. The Roman province of **Africa**, in its restricted sense,[*] embraced that portion of the continent which lies between the Lesser Syrtis in the E., the desert of *Sahara* in the S., the river Tusca in the W., and the Mediterranean in the N. It answers nearly to the modern *Tunis*. The name was used in a broader sense to include Syrtica in the E., and Numidia in the W., and sometimes even some portions of Mauretania beyond the Ampsaga, which formed the western limit of Numidia.

§ 9. The position and physical character of this country deserve particular notice. It occupies that great angle on the northern coast of Africa, of which Mercurii Prom., *C. Bon*, is the apex, and which is formed by the southerly deviation of the coast, at right angles to its general course, in the neighbourhood of the Lesser Syrtis. It thus approaches very near the continent of Europe, standing directly opposite the southern peninsula of Italy and the island of Sicily, from which it is about 90 miles distant, and in easy communication with the coasts of Spain. As regards the Mediterranean, it stands just at the junction of the two great basins, eastern and western, into which that sea is divided, and thus commanded the navigation of each, forming as it were a new starting point for the commerce of the Phoenicians, without which they perchance might have been confined, as the Greeks generally were, to the eastern alone. As regards Africa, this district is shut off from

---

[*] The limits of the Roman province varied at different periods: as originally constituted in B.C. 146, it consisted of the possessions of Carthage of *that time*, i.e. the districts of Zeugitana and Byzacium: the rest of the old Carthaginian possessions were handed over to the Numidian kings. In the Jugurthine war the Romans gained Leptis Magna and some other towns in Syrtica. In the civil war Cæsar added Numidia, as far as the Ampsaga, under the title of New Africa. In B.C. 30 Augustus restored this to Juba, but resumed it again in B.C. 25, and fixed the western boundary at Saldæ, thus including a portion of Mauretania also in Africa. Finally, Caligula gave up this latter portion, and refixed the boundary at the Ampsaga. In the 3rd cent. (probably in Diocletian's reign) the whole was re-arranged into four provinces—Numidia, Africa Propria or Zeugitana, Byzacium, and Tripolis. The term Africa was occasionally applied to all of these.

the general body of the continent by the range of Atlas in the S., and the desert regions of Syrtica in the E. The country was also highly favoured in regard to climate and soil. The great range of Atlas forms a barrier between it and the sands of the Sahara, and provides an adequate amount of moisture. On the N. side it descends in a series of terraces towards the sea, and offers a most fertile soil to the agriculturists. In the southern district only does the desert approach the sea, and the soil become unfruitful. The grain produced a hundredfold,[*] the vine a double vintage, and fruit of every kind grew in the greatest profusion.

§ 10. The mountains were offsets from the great chain of Atlas, some few of which only received special names, as **M. Jovis** S. of Carthage; **Cirna**, which runs parallel to the northern coast; and **Mampsarus** in the S.W. The promontories are—**Brachōdes** at the N.W. point of the Lesser Syrtis; **Prom. Mercurii**, C. Bon, the N.E. point; **Prom. Apollinis** or **Pulchrum**, C. Farina, at the W., as Mercurii is at the E. of the bay of Carthage; and **Prom. Candidum**, C. Blanc, N. of Hippo. Two bays must be noticed—**Sinus Neapolitānus**, G. of Hammamet, on the E. coast; and the **S. Carthaginiensis** between the promontories of Mercury and Apollo on the N. coast. The chief river is the **Bagrādas**, Mejerdah, which rises in Mount Mampsarus and flows in a N.E. course into the bay of Carthage: its lower course[†] has been much altered through the soil it has brought down, and its mouth has been removed some 10 miles northward.

§ 11. The inhabitants of this district in the time of Herodotus were the native Libyan tribes named the **Maxyes** and **Zauēces** in the S.; the **Gyzantes**, undoubtedly the same as the later **Byzantes** and **Byzacii**, on the W. coast of the Syrtis; and the **Machlyes** in the S.E. near the Triton, perhaps the same as the Maxyes already mentioned. In addition to these the Phœnicians were settled at various spots on the coast. In the Roman period the Phœnicians and Libyans had intermixed, and their descendants formed a distinct race, named **Libyphœnices**, whose settlements were chiefly about the river Bagradas. The towns of this district were in almost every instance

---

[*]   Byzacia cordi
Rura magis, centum Cereri fretimentia culmis,
Electos optare dabo inter præmia campos.—SIL. ITAL. ix. 204.

[†] The character of this river is well described in the following passages:—
Primaque castra locat cano procul æquore, qua se
Bagrada lentus agit, siccos sulcator arenæ.—LUC. iv. 587.

Turbidus arentes lento pede sulcat arenas
Bagrada, non ullo Libyæis in finibus amne
Victus limosas extendere latius undas,
Et stagnante rado patulos involvere campos.—SIL. ITAL. vi. 140.

founded by the Phœnicians. The names alone sometimes indicate
this: as in the case of Carthage, from *carth*, "a town;" Leptis,
"fishing station;" and Utica, "ancient." Others, as Neapolis and
Hadrumētum, are known on other grounds to have belonged to
them. Aspis alone is doubtful, as its existence cannot be traced
earlier than the time of Agathocles. Under the Carthaginians, the
metropolis was Carthage. After its destruction Utica succeeded to
that position; and after the separation of Byzacium, Hadrumētum
became the capital of the latter division. The towns appear to have
enjoyed a large degree of prosperity under the Romans, which they
retained until the entrance of the Vandals. The history of Carthage is in reality prior to the existence of the Roman province of
Africa, and therefore deserves a special notice.

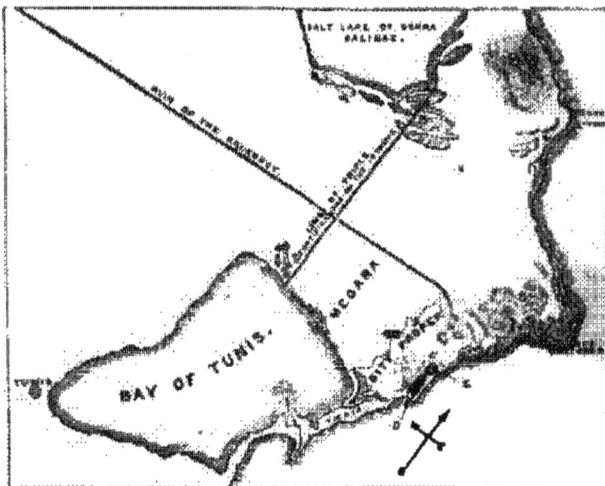

Map of the site of Carthage

§ 12. The city of **Carthage** stood on a peninsula on the W. side
of the Sinus Carthaginiensis between two bays, that on the S. being
the present *G. of Tunis*, and that on the N. a lagoon, now called the
Salt Lake of *Sokra*. The peninsula is formed by a line of elevated
ground attaining the height of 300 ft. at its western, and 400 ft.
at its eastern extremity, the two points being named *C. Camart*
and *C. Carthage*. Inland it slopes down and was contracted to
an isthmus between the two bays. The circuit of the peninsula

was about 30 miles. Great changes have been effected on its site
through the deposits of the river Bagradas : the northern bay has
become partly a lagoon, and partly firm land ; the southern bay,
once a deep and open harbour, is now a lagoon about 6 ft. deep, and
with a very narrow entrance. The isthmus which connected the
peninsula with the mainland has been enlarged from 25 stadia,
which was its width in Strabo's time, to 40. On the S. side, on the
other hand, the sea has somewhat encroached, and has covered a
portion of the ancient site ; the coast-line has receded considerably
inland to the N. of the town. Finally the river Bagradas itself,
which formerly joined the sea about 10 miles to the N., is now 20
miles distant.

The original city of the Phœnicians probably stood on the S.E. of
the peninsula, near *C. Carthage*. From this point a tongue of land
(the Tænia of Arrian) stretched to the S. The port was on the S. side
of the peninsula, and consisted of an outer[a] and inner harbour, connected together by a channel and with an entrance from the sea 70 feet
wide. The outer one (*b*) was for merchant vessels, and the inner,
named Cothon (*a*), from an island in it, for ships of war, of which 220
could be put up in separate docks.[b] The latter was probably entirely
excavated. Adjacent to the port on the W. stood the Forum, containing the senate-house, the tribunal, and the temple of Apollo ; and
to the N. of the port was the Byrsa, or citadel, containing the temple of
Æsculapius on the highest point.[i] The whole town was surrounded
with walls to the extent of 360 stadia, the strongest defences being on
the land side, where there was a triple line, each 30 cubits high, with
strong towers at intervals. Water was conveyed to the city by an
aqueduct 50 miles long, and was stored in vaulted reservoirs. The
suburb of Megara, or Magalia, stood W. of the City Proper.

Name.—Carthage derived its name from the Phœnician word *Carth*,
"a city ;" it appears to have been fully called *Carth-Hadeshoth,* " new
city," in contradistinction perhaps to Utica " the old city." This
name the Greeks converted into Καρχηδών, and the Romans into *Carthago:* the inhabitants were named sometimes after the city, but more
usually after the mother country ; the Greeks calling them Φοίνικες, and
the Latins *Pœni*. At a late period the epithet *Vetus* was added, in
order to distinguish it from its colony Carthago *Nova* in Spain.

---

[a] According to Mannert the outer port was a portion of the *Lake of Tunis,* and
the entrance to it was *inside* the Tænia. The recent researches of Dr. Davis have
led him to the conclusion that the ports were more to the N., and that the outlet
from the outer port was by a channel communicating directly with the open sea.
He states that the remains of Scipio's mole are still visible at the entrance of this
channel (*Carthage,* p. 128).

[b] In the final siege of the city, Scipio constructed an embankment across the
entrance of the harbour (D), whereupon the Carthaginians opened a new entrance
(E) to the inner harbour.

[i] Dr. Davis has transferred the site of the Byrsa from the *Hill of St. Louis,* on
which Mannert places it, to a height near the sea, more to the N.E., where he
has discovered ruins which he identifies with the temple of Æsculapius, consisting
of massive walls arranged in the form of a temple, together with a staircase leading up to it.

History.—Carthage was a colony of Tyre, established probably about 100 years before the foundation of Rome as an emporium jointly by the merchants of the mother city and of Utica. Tradition assigned its origin to Dido,[1] who on the death of her husband fled from Tyre and purchased of the natives as much ground as she could enclose with a bull's hide:[2] the latter part of the legend originated in the Phœnician word *Bosrah* "fortress," which the Greeks confounded with *βύρσα* "a hide." Carthage soon rose to a supremacy over the older Phœnician colonies, and herself planted numerous colonies on the coasts of Africa, from the Greater Syrtis in the E. to the most southerly parts of Mauretania in the W., as well as in Sardinia, Corsica, Sicily, and on the coasts of Gaul and Spain. The district which formed the proper territory of Carthage extended over Zeugitana and the strip of coast along which lay Byzacium and the Emporia. Her wealth was derived partly from agriculture and partly from commerce, and her population is said to have been 700,000 at the time of the Third

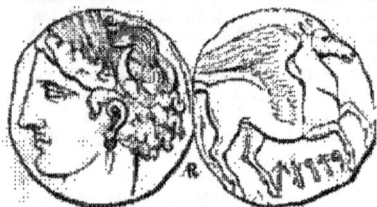

Coin of Carthage.

Punic War. Carthage became the great rival of Rome, and was engaged in a series of wars with that power. In the first (B.C. 264—241) she lost Sicily and the Liparian islands; in the second (B.C. 218—201) she lost the whole of her foreign supremacy; and in the third (B.C. 150—146) she was taken and utterly destroyed. After an interval of 24 years an abortive attempt was made by C. Gracchus to colonise the place from Rome under the name of Junonia. Julius Cæsar renewed the attempt in 46; and it was successfully accomplished by Augustus in 19, who sent 3000 colonists there. The new town which probably occupied the site of the old one, though placed by some at Megara, became one of the most flourishing towns of Africa, and the seat of a Christian church which could boast of Cyprian and Tertullian as its bishops. In A.D. 439 it was made the Vandal capital. It was retaken by Belisarius in 533, and finally destroyed by the Arabs in 647.

§ 13. The Romans divided Africa into two portions—**Byzacium**

---

[1] Urbs antiqua fuit, Tyrii tenuere coloni,
Carthago, Italiam contra, Tiberinaque longe
Ostia; dives opum, studiisque asperrima belli.—*Æn.* i. 12.

[2] Condebat primæ Dido Carthaginis arces,
Instabatque operi subducta classe juventus.
Molibus hi claudunt portus: his tecta domusque
Partiris, justæ Bitia venerande senectæ.—*Sil. Ital.* ii. 108.

[3] Devenere locos, ubi nunc ingentia cernes
Mœnia, surgentemque novæ Carthaginis arcem :
Mercatique solum facti de nomine Byrsam,
Taurino quantum possent circumdare tergo.—*Æn.* i. 365.

Fatali Dido Libyes adpellitur oræ :
Tum pretio mercata locos, nova mœnia ponit,
Cingere quæ arcto permissam littora tauro.—*Sil. Ital.* i. 23.

or **Byzacena** in the S. (named probably after the Byzantes or
Gyzantes, a native tribe of that district), and **Zeugitana** in the N.
(said to be named after a mountain called Zeugis, whose position is
unknown). The line of division between the two was coincident
with the parallel of 36° N. lat. The division was not authori-
tatively recognized until the time of Diocletian, nor does the name
of Zeugitana occur in any writer earlier than Pliny. We adopt the
division more for the purpose of convenience, than for any importance attaching to it in connexion with classical literature. We
shall describe the towns of Byzacium in the first instance.

I. *Towns in Byzacium.* (i.) *On the Coast from S. to N.* **Thenae** was
opposite to Cercina, and became a Roman colony with the name of
Ælia Augusta Mercurialis. **Thapsus** stood on the edge of a salt lake;
it was strongly fortified, and celebrated for Cæsar's victory over the
Pompeians in B.C. 46:[4] its ruins are at *Demass.* **Leptis** surnamed
Minor, in order to distinguish it from Leptis in Syrtica, was a flourishing Phœnician colony in the district of Emporia, just within the S.E.
headland of the Bay of Neapolis. Under the Romans it became a
libera civitas and perhaps a colony. **Hadrumetum**, the capital of Byzacium, stood just at the S. entrance of the Bay of Neapolis. It was
a Phœnician colony, and under the Romans a *libera civitas* and a
colony. It was surrounded by a fertile district and became one of
the chief ports for the export of corn, and is further known as the
birthplace of Cæsar Clodius Albinus. Having been destroyed by the
Vandals, it was restored by Justinian with the name of **Justiniana.**
The remains at *Susa* consist of a mole, several reservoirs, and fragments
of pillars. (ii.) *In the interior.* **Thysdrus**, between Thenæ and Thapsus,
a Roman colony, is known as the place where the Emperor Gordianus
set up the standard of rebellion against Maximin. Extensive ruins,
especially a fine theatre, exist at *Jemme.* **Capsa**, *Cafsa*, in the S.,
stood on an oasis surrounded by an arid desert: it was the treasury of
Jugurtha and was destroyed by Marius, but was afterwards rebuilt and
made a colony. **Thala** or **Telepte** lay N.W. of Capsa, and had a
treasury and arsenal in the Roman period. **Suffetula** was centrally
situated, N.E. of Thala, at a spot where several roads met. The magnificent ruins at *Sfaitla* prove its importance.

II. *In Zeugitana.* (i.) *On the Sea-Coast.* **Neapolis** stood on the bay
named after it, and was the nearest point to Sicily. It was a Phœnician factory and afterwards a Roman colony: some remains exist at
Nabel. **Aspis** or **Clypea** was so named from the "shield-like" form of
the hill[5] on which it was built, and which stood S. of Prom. Mercurii.
It possessed a sheltered harbour, and, being backed by a large plain, it
was the most convenient landing-place on this part of the coast: whether
a Phœnician town existed on the spot is uncertain, but the later town
was built by Agathoclea, B.C. 310. In the First Punic War the troops
of Manlius and Regulus landed here in 256, and took ship again in 255.
In the second, it was the scene of a naval skirmish in 208, and of

---

[4] Et Zama et uberior Rutulo nunc sanguine Thapsus.—Sil. ITAL. iii. 261.

[5] Tum, quae Sicanio praecinxit littora muro,
In clypei speciem curvatis turribus, Aspis.—Id. iii. 243.

Masinissa's narrow escape in 204. In the third, it was besieged to no
purpose by Piso both by land and sea in 148. Tunes was a strongly
fortified town about 15 miles S.W. of Carthage at the head of the bay,
which is now named after its great representative *Tunis*. Utica was
situated at the mouth of the western branch of the Bagradas, near the
promontory of Apollo, and 27 miles N.W. of Carthage. It possessed a
good artificial harbour, and was strongly defended both on the land
and the sea side. It was founded by the Tyrians 287 years before
Carthage,[6] but soon became independent of the mother country. It
appears as the ally or dependent of Carthage in the Roman treaties of
B.C. 509 and 348, as well as in that formed between Hannibal and
Philip of Macedon in 215. In the two first Punic wars it generally,
though not consistently, aided Carthage; but in the third it seceded,
and hence rose high in favour with the Romans,[7] who made it their
chief emporium and the seat of government. The name is associated
with numerous events in the African wars of the Romans, but especially
with the death of the younger Cato. It was made a free city and,
under Hadrian, a colony; and was endowed with the *Jus Italicum* by
Septimius Severus. It was also the seat of a Christian bishoprick.
It was destroyed by the Saracens. The remains of temples and castles
at *Duar* mark the site of the town. The most interesting relic is an
aqueduct, carried over a ravine on a treble row of arches near the
town. **Hippo**, surnamed Diarrhytus, *Biceria*, stood on the W. side of
the outlet of a large lake, and derived its second name, according to
the Greek version, from the inundations to which it was liable, though
not improbably it had in reality a Phœnician origin. The town was
fortified by Agathocles, and was made a free city and colony by the
Romans. (ii.) *In the interior*. **Zama, Jama**, stood five days' journey
S.W. of Carthage, and is renowned as the scene of Scipio's victory over
Hannibal in B.C. 202. It was a very strong place, and was selected as a
residence by Juba. It was probably made a colony by Hadrian. **Vacca**
or **Vaga** was an important town S.W. of Utica at *Bayjah*: it was
destroyed by Metellus, but afterwards restored by the Romans.
Justinian fortified it and named it Theodoris.

History.—After the fall of Carthage and the constitution of the
Roman province, the country was the scene of important events in the
civil war of Pompey and Cæsar, particularly of the battle of Thapsus,
and again in the wars of the second triumvirate. Subsequently to this
the province remained quiet and prosperous, the most serious dis-
turbance being the insurrection under the two Gordians, A.D. 238.
The struggles of Constantine and his competitors extended to this
region, and were followed by fresh commotions under his successors.
The African provinces were united to the western empire in A.D. 395,
and were disjoined in the reign of Valentinian III. The introduction of
the Vandals by Boniface in 429 in support of the Donatist schism
proved fatal to the prosperity of the province: they held it for about
100 years, when they were exterminated by Belisarius under Justinian,
in 534. That emperor expended immense sums on the towns, but

---

[6] Proxima Sidoniis Utica est effusa maniplis
   Prisca alta veterisque ante arces condita Byrsæ.—SIL. ITAL. III. 241.

[7] We may conclude from the following line (which Horace addresses to his
book) that Roman literature was cultivated there:—

   Aut fugies Uticam, aut vinctus mitteris Ilerdam.—HOR. *Ep.* i. 20, 13.

the incursions of the Arabians rendered the tenure of the African provinces difficult, and a series of struggles ensued commencing in 647 and terminating with the final withdrawal of the Romans in 709.

VI.—NUMIDIA.

§ 14. The boundaries of **Numidia** were the river Tusca in the E., the Ampsaga in the W., the Mediterranean in the N., and the range of Atlas in the S. It lay between the Roman province of Africa on the E. and Mauretania on the W., and corresponds to the modern *Algeria*. The maritime district is remarkably fertile, and produced besides the usual grain crops, every kind of fruit. Its marble was particularly celebrated, being of a golden yellow hue with reddish veins. The interior consists of a series of elevated plains, separated from each other by spurs of the Atlas range, and adapted only to a nomad population, partly from the severity of the climate in winter, and partly from the nature of the soil which yields a luxuriant herbage only in the early spring.

§ 15. The mountain ranges emanate from Mount **Atlas**, and occasionally were known by special names, as **Thambes**, which contained the sources of the Rubricatus, and **Aurasius** in the S.W. The coast line is broken by numerous promontories of which we may notice from E. to W.—**Hippi Prom.**, *Ras el Hamlah*; **Stoborrum**, *C. Ferro*; and **Tretum**, *Seba Rus*. The most important bays are the **Sinus Olchacites**, *G. of Estorah*; and the deep and extensive **Numidious Sin.**, which has no specific name in modern times. The chief rivers were the **Tusca**, on the eastern boundary; the **Rubricatus** or **Ubus**, *Seibouse*, which flows E. of Hippo Regius; and the **Ampsaga**, *Wad-el-Kibbir*, on the borders of Mauretania.

§ 16. The general name for the inhabitants of this district was **Numidæ**, a Latinized form of the Greek νομάδες, "nomads." This describes generically their character as known to the Romans. They are described as living[*] (very much as their modern representatives the *Kabyles*) in *Mapalia*, i.e. huts made of branches overspread with clay, and as excelling in the management of the horse.[*]

---

[*] Virgil gives a most graphic description, applicable alike to the ancient Numidian and the modern *Kabyle*:

Quid tibi pastores Libyæ, quid pascua versu
Prosequar, et raris habitata mapalia tecta?
Sæpe diem noctemque et totum ex ordine mensem
Pascitur, itque pecus longa in deserta sine ullis
Hospitiis: tantam campi jacet. Omnia secum
Armentarius Afer agit, tectumque, Laremque,
Armaque, Amyclæumque canem, Cressamque pharetram.

Georg. III. 339.

[*] Et Numidæ infreni cingunt, et inhospita Syrtis.—*Æn.* iv. 41.

Ille passim exsultant Nomades, gens inscia freni;
Queis inter geminas per ludum mobilis aures
Quadrupedem flectit non cedens virga lupatis.—SIL. ITAL. I. 215.

They were sometimes more specifically called **Maurali Numidæ**, while later writers used the general name of **Mauri**. They were divided into numerous tribes, of which the most important were the **Massyli**[1] who lived between the river Ampsaga and Prom. Tretum; and the **Massæsyli** who, though living W. of the Ampsaga, were of Numidian origin. The towns of Numidia first came into notice in the period of the Roman wars in Africa. The names of several of them furnish indications of a Phœnician origin, as in the case of the capital Cirta, which we have already noticed as a Phœnician word, and again in those where the worship of Venus was carried on, as Aphrodisium and Sicca Veneria. Hippo and Collops were their principal stations on the coast. When Numidia fell into the hands of the Romans, the chief towns were endowed with various privileges as free cities and colonies; and some were very much enlarged and adorned with magnificent buildings, as we know from the ruins of Constantia, Lambèse, Theveste, and others. The ruin of the Numidian towns was caused by the Vandals in the middle of the 5th cent. of our era.

(i.) *On the Coast from E. to W.*—The first town of importance was **Hippo**, surnamed **Regius**, as being the residence of the Numidian kings;[2] it stood W. of the Ubus on a bay to which it communicated its name. It was originally a Tyrian, and in later times a Roman colony; but it owes its chief interest to St. Augustine who was bishop of it, and who died shortly before its destruction by the Vandals in A.D. 430. Its ruins are S. of Bonah. **Rusicade**, which served as the harbour of Cirta, was at the mouth of the small river Thapsus and at the head of the Sinus Olchachites. Its site is at *Stora*. Out of its materials *Philippeville* was partly built. **Collops Magnus** or **Culla**, *Collo*, stood on the W. side of the Sin. Olchachites, and was celebrated for its purple-dyeing establishments.

(ii.) *In the Interior.*—**Bulla Regia**, near the E. frontier, probably derived its surname from being a residence of the Numidian kings. Under the Romans it was a *liberum oppidum;* the name *Boul* still attaches to its ruins. **Cirta** was beautifully situated on a steep rock, round the base of which flowed a tributary of the Ampsaga. It was the residence of the kings of the Massyli, who possessed a splendid palace there: it was the strongest fortress in the country, and the point where the lines of communication centred. Hence it is frequently mentioned in the history of the Punic, Jugurthine, and Civil wars. Under the Romans it was a colony with the surname of Julia. It was also called Colonia Sittianorum from Sittius, to whom it was given. Having fallen into decay, it was restored by Constantine with the name **Constantina**, which its site still retains in the slightly altered form of *Constantineh*. The finest relic is a triumphal arch, now in Paris. **Lambèse** lay near the confines of Maurelania, and was the station of

---

[1] Massylique ruunt equites.—Æn. iv. 132.
Et gens, quæ nudo residens Massylis dorso
Ora levi flectit frenorum nescia virga.—Luc. iv. 682.

[2] Antiquis dilectus regibus Hippo.—Sil. Ital. iii. 259.

an entire legion : its ruins at *Lemba* are magnificent, consisting of the remains of an amphitheatre, a temple of Æsculapius, a triumphal arch, &c. *Thaveste* was situated not far from the frontier of Byzacium. It was a Roman colony, and a centre of communication for the interior districts. Its history is unknown, but the extensive ruins of it at *Tebessa* prove it to have been an important town. **Sicca Veneria** stood on the river *Bagradas*, and derived its surname from the worship of Venus. It was built on a hill, and was a Roman colony; its site is supposed to be at *Kaff.*

Of the less important towns we may briefly notice: on the coast, **Tabrica**[1] at the mouth of the *Tusca*, the scene of the death of Gildo; and **Aphrodisium**, a port and Roman colony near Hippo Regius. In the interior: *Tibilis*, 54 miles E. of Cirta, with hot baths in its neighbourhood ; **Tagaste**, the birthplace of St. Augustine, S.E. of Hippo Regius; and **Naraggara**, W. of Sicca, the spot where Scipio had an interview with Hannibal before the battle of Zama. The positions of **Thirmida**, where Jugurtha murdered Hiempsal, and **Suthul**, where the former had a treasury, are wholly unknown.

History.—The Romans became acquainted with the Numidians in the First Punic War, when they served with great effect in the Carthaginian ranks. In the Second Punic War they joined Rome, in reward for which their prince Masinissa was made king of a territory extending from the Mulucha in the W. to Cyrenaica in the E., the proper territory of Carthage excepted. Masinissa was succeeded by Micipsa, who associated with himself his sons Adherbal and Hiempsal, and his brother's illegitimate son Jugurtha. The latter murdered Hiempsal, and declared war against Adherbal, who sought the aid of Rome. The dispute was settled for a time, but broke out again. Adherbal was murdered, and Jugurtha in turn was put to death by the Romans, B.C. 106. After the reigns of Hiempsal II. and Juba I., Numidia was made a province by Julius Cæsar in B.C. 46. Numidia holds a conspicuous place in ecclesiastical history as the head-quarters of the Donatist heresy ; violent disputes followed, and the entrance of the Vandals completed the ruin of the country.

VI.—MAURETANIA.

§ 17. **Mauretania** was bounded by the river Ampsaga on the E., the Mediterranean on the N., the Atlantic on the W., and the range of Atlas on the S. It corresponds to the western part of *Algeria* and the empire of *Morocco.* Under the Romans it was divided into two large portions—**Cæsariensis** and **Tingitana**, named after their respective capitals, Cæsarea and Tingis, and separated from each other by the river Mulucha. It may be described generally as the highlands of N. Africa, the level of the land rising from the Mediterranean to Mt. Atlas in three great steps, each of which stretches out into extensive plains. These plains, though deficient in wood, possessed a soil of extraordinary fertility, which, aided by the cultivation bestowed on them in ancient times, rendered Mauretania the

---

[1] Quales, umbriferos ubi pandit Tabraca saltus,
   In vetula scalpit jam mater simia bucca.—JUV. x. 194.

"granary of the world." The productions specially noticed by ancient writers were—elephants, now no longer found there; crocodiles, which could hardly have existed in such a country; scorpions; and copper, which is still found there.

§ 18. The mountain-chains of this province are all connected with the great range of **Atlas**, and have a general direction from N.E. to S.W. The special names attached to them are devoid of interest, with the exception of **Atlas Minor**, which is inappropriately given by Ptolemy to a range parallel to the Mediterranean Sea. The most important of the ranges is that which, striking northwards from the main chain of Atlas, forms the watershed between the rivers which seek the Mediterranean, such as the Molochath, and those which, like the Subur, seek the Atlantic. S. of the Subur, this range sends out numerous ramifications towards the Atlantic, which formed a natural division between the N. and S. portions of ancient Mauretania, as it still does of *Morocco*. The promontories from E. to W. are—**Iomnium**, *Ras-al-Katanir*; **Apollinis**, near Cæsarea; **Metagonium**, *Ras-al-Harshah*, forming the W. point of the bay into which the Mulucha falls; **Rusadir**, *C. Tres Forcas*, the most marked projection along this coast; **Abyla**, *Jebel-el-Mina*, the southern of the Pillars of Hercules, opposite to Calpe in Spain; **Cotes** or **Ampalusia**, *C. Spartd*, the extreme W. point of Mauretania; **Solois**, *C. Cantin*, more to the S.W.; **Herculis Prom.**, *C. Mogador*; and **Usadium**, *Osem*. The chief rivers on the N. coast are—the **Ampsaga**, on the E. border; the **Usar** or **Blaar**, probably the *Ajebby*; the **Chinalaph**, *Shellif*, the most important of all, joining the sea, after a north-westerly course, near Prom. Apollinis; the **Mulucha**, probably the same as the **Molocath**, and the **Malva**, now the *Muluwi*, which joins the sea near Metagonium Prom.; and on the W. coast, flowing into the Atlantic, the **Subur**, *Subu*, joining the sea 50 miles S. of Lixus; the **Sala**, *Bu-Regrab*, still more to the S.; the **Phuth**, *Wady Tensift*; and the **Lixus**, *Al-Haratch*.

§ 19. The inhabitants were known generally as the **Maurusii** or **Mauri**,[4] whence the modern *Moors*. Tradition assigned to them an

---

[4] The notices of this people among the Latin poets are frequent : the chief points that attracted attention were their dark colour and their skill in archery :—

<blockquote>
Maurus concolor Indo.—Luc. iv. 678.
Nigri manos cæca Mauri.—Juv. v. 53.
Mauro obscurior Indus.—Id. xi. 135.
</blockquote>

<blockquote>
Integer vitæ, sceleriæque purus
Non eget Mauri jaculis neque arcu,
Nec venenatis gravida sagittis,
    Fusce, pharetra.   Hor. Carm. i. 22, 1.
        Et hærens
Lorica interdum Maurusia pendet arundo.—Sil. Ital. x. 401.
</blockquote>

Asiatic origin; and, according to Procopius, an inscription on two pillars at Tipasa pronounced them to be Canaanites who had fled from Joshua. They were divided into a vast number of tribes, of which we need only notice the powerful **Massæsyli** on the borders of Numidia. The towns were exceedingly numerous, partly perhaps on account of the insecurity of the country, which necessitated defences even for the villages. No fewer than one hundred and seventy-nine episcopal towns are enumerated, the majority of them being probably insignificant places. The Romans instituted a vast number of commercial colonies even before they took possession of the country.[b] Augustus founded three in Tingitana, namely, Julia Constantia, Julia Campestris, and Banasa Valentia; and eight in Cæsariensis. Claudius added two in the former, and two in the latter; and there were subsequently added two and eleven in the respective provinces: thus making a total of twenty-eight. The capitals were Cæsarea and Tingis, and, after the subdivision of Cæsariensis, Sitifis, while Salda served as the chief port of this district. In addition to the Roman towns, the Carthaginians planted a number of colonies on the W. coast, which fell into decay with the power of Carthage itself.

(1). *Towns in Cæsariensis.*—Igilgili, *Jijeli*, stood on a headland on the coast of the Numidicus Sinus. It possessed a good roadstead, and was probably the emporium for the surrounding country. **Salda** possessed a spacious harbour, and was a Roman colony. It was an important point on this coast, having formed the boundary at one time of the kingdom of Juba, and at another of Sitifensis. A flourishing city, *Bujeijah*, occupied its site in the Middle Ages. **Icosium**, the ancient representative of *Algiers*, ranked as a Roman colony, and was endowed by Vespasian with the *Jus Italicum*. **Jol** or **Cæsarea**, as it was named in honour of Augustus, was originally a Phœnician colony, and afterwards the capital of Bocchus and Juba II., the latter of whom beautified it, and gave it its new name. Under the Romans it became the capital of Cæsariensis and a colony. It was burnt by the Moors in the reign of Valens, but was again restored. The magnificent ruins at *Zershell*, in 2° E. long., mark its site. **Cartenna**, *Tenes*, was a Roman colony, and the station of a legion. **Siga** was a commercial town at the mouth of a river of the same name. Neither the river nor town have been identified. It was destroyed in Strabo's time, but was afterwards restored. In the interior, **Sitifis** was the most important town in the eastern district, and became the capital of Sitifensis. It stood near the frontier of Numidia at *Setif*. **Tubusuptus** stood about 18 miles S.E. of Saldæ, and was a Roman colony under Augustus. **Auzia**, *Hamzah*, was near the Gariphi Mts., and was a considerable town under the Romans.

Horace uses the term Maurus as tantamount to African:—

Barbaras Syrtes, ubi Maura semper

Æstuat unda.—*Carm.* ii. 6, 3.

[b] The colonies in Tingitana were connected with the trade of Spain: so close was the connexion between the two countries that in the later division of the empire by Theodosius Tingitana was attached to Bœtica.

(2). *In Tingitana.*—On the coast we meet with **Rusádir**, a Roman colony near Metagonium Prom. **Tingis**, *Tangier*, W. of Abyla, ranked as the capital of the province, and a Roman colony. Its origin is carried back to the mythic age. **Zilia**, *Azzila*, 24 miles from Tingis, was originally a Phœnician town, afterwards a Roman colony with the name of *Julia Constantia*. **Lixus**, at the mouth of the river of the same name, was a great trading station on this coast, and a Roman colony. Lastly, **Thymiaterium**, probably at *Mamora*, was the first Carthaginian colony planted by Hanno. The position of **Banása** on the Subur is uncertain, some authorities representing it as a maritime, others as an inland town : in the former case its site corresponds to *Mehediah*, in the latter to *Mamora*. It was a Roman colony, with the name of Valentia. **Volubilis** was a town of considerable importance on the Subur, 25 miles from Banasa. Near its site are the splendid ruins of *Kasr Faraun*, "Pharaoh's Castle," with Roman inscriptions. **Babba**, which Augustus constituted a colony with the title of **Julia Campestris**, has been variously placed on the *Guarga*, one of the tributaries of the Subur, and on the more northerly *Wadi al Khous*.

History.—The Romans first became acquainted with Mauretania in the Punic and Jugurthine wars. In the latter, Bocchus is noticed as king : he was succeeded by his two sons, Bogudes and Bocchoris, who took different sides in the wars of the Triumvirate. Their territory was handed over to Juba II. in B.C. 25, in exchange for Numidia. His son Ptolemy succeeded to the throne, and was put to death by Caligula in A.D. 41. In the following year Claudius divided the country into the two provinces of Cæsariensis and Tingitana. Twenty-one colonies were planted in these provinces, besides several *Municipia* and *Oppida Latina*. About A.D. 400 we find Tingitana forming a portion of the diocese of Spain ; and Cæsariensis, which was still attached to the diocese of Africa, subdivided into Mauretania Prima, or Sitifensis, and Mauretania Secunda, or Cæsariensis. The Vandals seized these provinces in 429 ; Belisarius recovered them for the Eastern Empire. Incursions of the Moors followed ; and the Arab conquest in 698-700 finally dissevered the connexion between Mauretania and Rome.

VII.—LIBYA INTERIOR.

§ 20. Under the somewhat indefinite term **Libya Interior** is included the vast region lying S. of the countries we have hitherto been describing, from the Atlantic in the W. to Æthiopia in the E. The limit southwards was fixed at no definite point : it advanced with the advance of commerce and navigation, until in the age of Ptolemy it reached the 11° N. lat. on the western coast. The information that we have in reference to it is unimportant, being restricted merely to the names of the various physical features. We shall therefore confine ourselves to a very brief notice of them.

(1.) *Mountain Chains.*—**Mons Ater**, *Harusch*, running from E. to W., and separating Phazania from the Roman province of Africa ; **Usargāla**, more to the W., a continuation of Atlas, S. of Numidia and Mauretania; **Girgiri**, *Tibesti*, running N. to the confines of Numidia ; **Sagapōla**, running parallel to the coast of the Atlantic, and containing the sources of the Subur ; **Mandrus**, more to the S., reaching to the parallel of the Fortunatæ Insulæ ; **Caphas**, containing the sources of the Daradus.

and its westerly prolongation Ryssadium, terminating in a headland of
the same name, *C. Blanco*; and Theon Ochema, *Sierra Leone*. Numerous ranges in the interior highlands, as far S. as the latitude of Sierra
Leone, are noticed by name in Ptolemy's writings; these, however,
have not been identified.

(2.) *Promontories.* on the W. coast from N. to S.—Gannaria, *C. Non*;
Soloentia, *C. Bojador*; Arsinarium, *C. Corveiro*, the most westerly
point of the continent; Ryssadium, *C. Blanco*; Catharon, *C. Darca*;
Hesperion Ceras, *C. Verde*; and Notium, *C. Rozo*.

(3.) *Rivers.*—The Subur, *Sus* (probably the same as the Chretes of
Hanno and the Xion of Scylax), which enters the sea just below the
most western projection of Atlas; the Darkdus, *Rio de Ouro*, discharging itself into the Sinus Magnus, and said to have crocodiles in
it; the Stachir, probably the *St. Antonio*; the Nia or Pambotus, *Senegal*,
frequented both by the hippopotamus and crocodile; and the Masitholus,
Gambia. Some few rivers of the interior are noticed, which were said
to discharge themselves into vast inland lakes: of these the Gir* and
the Nigir are probably branches of the great river *Niger*, of which some
reports had certainly reached the ancients. The Gir is described as
having a course of above 300 miles, with a further curvature to the N.
of 100. The lakes connected with the Nigir were designated Libya
Palus, and Nigritis, probably the modern *Dibbeh*; and with the Gir,
Nuba, *Lake Tchad*, and Chelonides, perhaps *Fittre*.

§ 21. The inhabitants of the interior were but very imperfectly
known to the ancients. The races that come most prominently forward are—the Gætuli, who lived in the W. between the Atlas range
and the basin of the Nigir; the Garamantes, whose district lay S. of
the Syrtes; and the Nigritæ, about the rivers Gir and Nigir, and
their lakes.

The first of these races, the Gætulians, followed a nomad life, and
were reputed a warlike and savage race. They first came under the
notice of the Romans in the Jugurthine war, when they were serving as
cavalry under Jugurtha. Some of them remained in Numidia under
the Roman government; but they became so troublesome that an expedition was sent against them under Lentulus, surnamed Gætulicus, in
the year A.D. 6. Thenceforward they are described as living in the
desert S. of Mauretania. They were not themselves negroes, but some
of the tribe intermixed with negroes, and were hence named Melano-
gætuli. The Gætulians seem to be the progenitors of the great aboriginal people of modern Africa, named *Amazergh*, of which the Berbers
and *Tuaricks* are the branches most generally known. Garamantes
was a name applied generally to all the tribes inhabiting that part of
the Great Desert which lay E. of the sources of the Bagradas and Mount
Usargala, and S. as far as the river Gir. The name was, however, more
specifically applied to the people of Phazania, *Fezzan*, a very large
oasis lying S. of the great Syrtis. This oasis and its inhabitants are
described by Herodotus, and most of his statements are borne out by
modern investigation. It is surrounded by hills of stone and sand,

---

*　　　Gir notissimus amnis
　Æthiopum simili mentitus gurgite Nilum
　　　　　　　　　　CLAUDIAN. *Laud. Stil.* i. 252.

attaining a height of 1200 feet, and intersected by ridges from 300 to
600 feet high. It is deficient in water, and hence not above one-tenth
of it is cultivable. Its chief produce is dates. Salt is abundant, and
is applied as manure to the date-trees. White clay is used for arable
land, and this is probably what Herodotus' informants mistook for
salt. The story of the oxen with the long forward horns has a foundation in the practice which still prevails of giving artificial forms to the
horns. The Troglodyte Æthiopians, whom the Garamantes hunted,
have their representatives in the Tibboos, who are still hunted by the
chieftains of Fezzan. The Romans, from whom our next notice of these
people is derived, found them troublesome neighbours, and sent an
expedition against them under Cornelius Balbus Gaditanus, B.C. 19.
Ethnologically they were allied to the Gætulians. Their chief town
was Garama, *Gherma*, whence a considerable trade was carried on.
The Nigritæ lived on the banks of the Nigir in the modern *Soudan*.
Very little was known of them. Their chief town was Nigeira, perhaps
Gona.

§ 22. Off the W. coast of Africa lie the Insulæ Fortunatæ,
Canaries, and *Madeira*, to which the name, originally connected
with the mythic idea of the "isles of the blessed," was not unnaturally transferred, when the ancients became acquainted with the
existence of islands in the fancied position of Elysium, and blest
with so delicious a climate. These islands became known to the
Romans about B.C. 82, through the reports which Sertorius received
at Gades from some sailors. The geographers describe only six instead of seven islands, viz.: Junonia or Autolala, *Madeira*; Junonia
Minor or Aprositus, *Lanzarote*; Canaria or Planaria, *Gran Canaria*;
Nivaria or Convallis, *Teneriffe*; Capraria or Caspiria, *Gomera*; and
Pluitalia or Pluvialia, *Ferro*. Ptolemy selected this group as the
point through which he drew his first meridian: one of the islands
(*Ferro*) was used for the same purpose by geographers down to a
late period. The Purpurariæ Insulæ, described by Pliny, were probably the above-noticed *Lanzarote*, with the smaller ones of *Graciosa*
and *Alegranza*.

The isle of Cerne, off the W. coast, has been variously identified
with *Fedellah* in 33° 40′ N. lat., with *Agadir* in 30° 20′, and with
Arguin in 20° S.: the latter is the most probable view. Off the E.
coast an island named Menuthias has been variously identified with one
of the islands of *Zanzibar*, and with *Madagascar*. The probability is
that the island has been incorporated with the coast at *Shamba*, about
80 miles S. of the river *Govind*.

Europa. (From an ancient Gem.)

BOOK IV.

EUROPE.

CHAPTER XVII.

EUROPE.

§1. Boundaries; Name. § 2. General Features. § 3. Internum Mare.
§ 4. Externum Mare. § 5. Mountains. § 6. Rivers. § 7. Climate
and Productions. § 8. Commerce. § 9. Inhabitants.

§ 1. The boundaries of Europe, though better known than those
of the two other continents, were nevertheless not accurately fixed
until a late period of ancient geography : in the extreme N. indeed
the true boundary remained a problem even in the days of Ptolemy,
and the vast regions of Northern Russia were a *terra incognita*.
It was, however, generally believed that the continent was bounded
on that side by an ocean, the exact position of which was unknown,
but which was supposed to extend eastward from the northern
point of the Baltic Sea.   In the N.W. the British Channel formed
the limit; in the W. the Atlantic Ocean ; in the S. the Mediterranean Sea ; in the S.E. the chain of seas connecting the Mediterranean with the Euxine, viz. the Hellespontus, Propontis, and
Thracian Bosporus ; and in the E. the Pontus Euxinus, the Palus

ANC. GEOG.   P

Mæotis, and the river Tanais.[1] The boundary on this side was very fluctuating in the early days of ancient geography, as we have already had occasion to observe. The modern boundary is more to the E., and is fixed at the river Ural and the Caspian Sea.

Name.—The name " Europa " (Εὐρώπη) may be derived either from a Semitic word *Oreb*, "the sunset," or from the Greek words *εὐρὺς ὤψ*, the "*broad-looking*" land. The first accords best with the westward progress of the human race, and the probability that the Phœnicians were the first civilized nation of Asia who had communication with the coasts of Europe: It is also supported by the analogy of the classical Hesperia, the "western land" of Europe, and by the probable origin of Arabia, "the western land" of Asia. The second accords best with the early use of the term in the Homeric Hymn to Apollo,[2] where it seems applied to the broad open land of Northern Greece as distinct from the Peloponnesus and the islands of the Ægean Sea. The mythological account[3] that it was derived from Europa, the daughter of the Phœnician king Agenor, was probably based on the early intercourse established by the Phœnicians with the shores of Greece.

§. 2. The general configuration of the continent of Europe is remarkable for its extreme irregularity.[4] In these respects it presents a strong contrast to the other continents. If we compare the African with the European coast-line, we find the former straight and unbroken, the latter varied by the projection of three important peninsulas as well as by a vast number of lesser sinuosities. Or, if we compare the interior of Asia with that of Europe, we find the former spreading out into extensive plains and abounding in elevated plateaus, while the latter is intersected in all directions by rivers and mountains, and broken up into valleys. Contrasted with Africa, we may describe Europe as the continent of *peninsulas*; contrasted with Asia, as the continent of *valleys*. Hence in a great measure arose the social and political characteristics of the continent. Easily accessible by sea, it was well adapted for commerce and colonization; inaccessible by land, it gained security

---

Hence Lucan describes the Tanais as—

　　　Asiæque et terminus idem
Europæ, mediæ dirimens confinia terræ.—III. 274.

'Ημὲν ὅσοι Πελοπόννησον πίειραν ἔχουσιν,
'Ηδ' ὅσοι Εὐρώπην τε καὶ ἀμφιρύτας κατὰ νήσους
Χρησόμενοι.　　　　　　　　　　　　Hom. *Hymn. in Apoll.* 290.

[2] According to this, Europa was carried off by Zeus under the form of a bull from Phœnicia to Crete. The story is told at length by Ovid (*Met.* ii. 838, *seq.*), and is alluded to by Horace:—

　　　Sic et Europe niveum doloso
　　　Credidit tauro latus, et scatentem
　　　Belluis pontum mediasque fraudes
Palluit audax.　　　　　　　　　　　Carm. iii. 27, 25.

[4] Hence Strabo (ii. 126) describes Europe as πολυσχημονεστάτη the " most variously figured " of the earth's divisions.

for the growth and consolidation of its institutions. These natural
advantages, combined with its admirable geographical position, its
climate, and its productiveness, rendered it the central seat of power
to the whole civilized world.

§ 3. In describing the seas which wash the shores of Europe, we
shall commence with that one with which the ancients were most
familiar and which they designated **Mare Nostrum** from its proximity to them, or **Mare Internum**, in contradistinction to the sea
outside the Pillars of Hercules. The importance of this sea in the
early ages of history cannot be over-estimated; it lay in the centre
of the civilized world, touching the three continents of Europe,
Asia, and Africa, which it united rather than separated, furnishing
a high-road for the interchange of commerce and the arts of social
life. Its size was unduly magnified by the geographers: its real
length is about 2000 miles, its breadth from 80 to 500 miles, and
its line of shore, including the Euxine, is 4500 leagues. It is
divided physically into three basins—the Tyrrhenian or western,
the Syrtic or eastern, and the Ægæan or northern. The line of
demarcation between the two first is formed by a submarine ledge
connecting C. Bon in Africa with Sicily, and between the second
and third by a curved line connecting the S. points of the peninsulas
of Greece and Asia Minor, the course of which is marked by the
islands of Cythēra, Crete, and Rhodes.

The subdivisions of this sea in ancient geography are numerous, the
waters about each particular country being generally named after
it. We have already noticed those connected with the continents
of Asia and Africa. Adjacent to the coasts of Europe were the following : (I.) In the Tyrrhenian basin, **Mare Hispānum, Ibericum,** or **Palea-
ricum,** between the coast of Spain and the Balearic Isles ; **M. Gallicum,** *G. of Lyons,* along the S. coast of Gaul ; **M. Sardôum** or
Sardonicum, about Sardinia ; **M. Ligustícum,** *G. of Genoa,* in the N.W.
of Italy ; and **M. Tyrrhēnum,**[1] along the W. coast of Italy, sometimes
named also **M. Infĕrum,**[2] "the *lower* sea," in contradistinction to the
Adriatic, which was designated **M. Supĕrum,** "the *upper* sea." (II.) In
the Syrtic basin, **M. Sicŭlum**[3] or **Ausonium,** about the E. coast of
Sicily, its limits eastward not being clearly defined ; **M. Ionium,**[4]

---

[1] Oens inimica nihil Tyrrhenum navigat æquor.—Virg. *Æn.* I. 67.
Cæmentis licet occupes
    Tyrrhenum omne tuis et mare Apulicum.—Hor. *Carm.* III. 24, 3.

[2] An mare, quod *supra*, memorem, quodque *alluit infra* ?
    Virg. *Georg.* II. 158.

[3] The term Siculum Mare is somewhat indefinitely used : Horace extends it to
the sea W. of Sicily, and even over the Tyrrhenian Sea :
    Nec Siculum mare
    Pœno purpureum sanguine.—*Carm.* II. 12, 2.
    Nec Siculâ Palinurus undâ.—*Id.* III. 4, 28.

[4] The name "Ionian" is derived by Æschylus from Io ; the extent of the sea

between Southern Italy and Greece as far N. as Hydruntum in the former, and Acroceraunia in the latter; and **M. Adriaticum**, or, as the poets named it **Hadria**,⁰ the limits of which were gradually extended from the upper portion of the *Adriatic* over the whole of that sea and sometimes even over the Ionian Sea. In the Ægean basin, now the *Archipelago*, **M. Creticum**, to the N. of Crete; **M. Myrtōum**,¹ named after the small island of Myrtus and extending along the eastern coast of Peloponnesus; and **M. Thracium**, along the coast of Thrace.

§ 4. The Mare Internum was connected at its western extremity with the **Mare Externum** by a narrow channel formerly named **Fretum Gaditānum**,² now the *Straits of Gibraltar*, at the neck of which stood the projecting rocks of Calpe on the European, and Abyla on the African coast, generally regarded by the ancients as the **Herculis Columnæ**,³ " Pillars of Hercules." The names by which

---

was not well defined, the passages quoted below from Euripides and Pindar showing that it was extended by the Greeks as far W. as Sicily.

Χρόνον δὲ τὸν μέλλοντα πόντιος μυχὸς
σαφῶς ἐπίστασ', Ἰόνιος κεκλήσεται
τῆς σῆς πορείας μνῆμα τοῖς πᾶσιν βροτοῖς.—ÆSCH. *Prom.* 839.

Καί κεν ἐν ναυσὶν μόλον Ἰ-
ονίαν τάμνων θάλασσαν,
Ἀρέθουσαν ἐπὶ
Κράναν——.   PIND. *Pyth.* III. 120.

Ἰόνιον κατὰ πόντον ἐλάτα
πλεύσασα, περιρρύτων
ὑπὲρ ἀκαρπίστων πεδίων
Σικελίαν——;   ÆSCH. *Phœn.* 208.

The Latin poets altered the quantity of the first syllable for scansional convenience, e. g.—

Namæ quot Ionii veniant ad littora fluctus.—VIRG. *Georg.* II. 108.

Jactari quos cernis in Ionio immenso.—OV. *Met.* IV. 534.

⁰ The Adriatic had but an ill fame among the mariners of Italy on account of the violent gusts which swept over it; Horace repeatedly alludes to this:—

Quo (i. e. noto) non arbiter Hadriæ
Major, tollere seu ponere vult freta.—*Carm.* I. 3, 15.
                          Auster
Dux inquieti turbidus Hadriæ.   *Id.* III. 3, 4.
                          Improbo
Iracundior Hadriâ.   *Id.* III. 9, 22.

¹ Nunquam dimoveas, ut trabe Cypria
Myrtoum pavidus nauta secet mare.—*Id.* I. 1, 13.

² These straits are referred to by Horace:—

Horrenda late nomen in ultimas
Extendat oras, qua *medius liquor*
Secernit Europen ab Afro.   *Id.* III. 3, 45.

The violence of the current is characterized by an old poet quoted by Cicero:
Europam Libyamque rapax ubi dividit unda.—*De Nat. Deor.* III. 10.

³ Much doubt existed in ancient times both as to the nature and position of the " Pillars of Hercules." It was usual to erect columns or pillars at the extreme point reached by any traveller; and hence the pillars of Hercules denoted the

the ancients described the Atlantic Ocean were numerous. The Greeks described it as ἡ ἔξω θάλασσα, "the outer sea," with special reference to the sea *within* the Pillars of Hercules ; also as ἡ Ἀτλαντίς, "the Atlantic," in reference to the mountain Atlas in the W. of the world ; and again as Ὠκεανὸς Ἑσπέριος, "the western ocean"; and lastly as ἡ μεγάλη θάλασσα, "the great sea." The Latins not unfrequently described it simply as *Oceanus*, and sometimes *Oceani mare*.[4] The Northern Ocean was described by various names indicating either its position as ὁ βόρειος Ὠκεανός, Oceanus Septentrionalis, &c. ; or its character as a frozen sea, as ἡ πεπηγυῖα θάλασσα, Mare Concretum, M. Pigrum, &c.

The subdivisions of these oceans were as follows. In the Atlantic, Oceanus Gaditanus, just outside the pillars of Hercules ; O. Cantaber, B. of Biscay ; O. Gallicus, off the N.W. coast of Gaul, at the mouth of the *English Channel;* and Mare Britannicum, the E. part of the channel as far as the *Straits of Dover*. In the Northern Ocean, M. Germanicum or Cimbricum, *German Ocean*, united by the Fretum Gallicum, *Straits of Dover*, with the M. Britannicum; and M. Sarmaticum, or Suevicum, *Baltic Sea*, united with the German Ocean by the Sinus Lagnus, *Little Belt*, and the Sinus Codanus, *Kattegat*, and subdivided into the Sinus Venedicus, *Gulf of Dantzic*, and M. Cronium, *Kurisches Haff* near *Memel*.

§ 5. The mountain system of Europe is clearly defined. A series of ranges traverses the continent from E. to W., dividing it into two unequal portions, of which the northern is by far the most extensive, but the southern the most important in ancient geography. There is thus far a general similarity between the continents of Asia and Europe ; so much so indeed that we may regard the

---

farthest limit to which the achievements of the god were carried : but whether these pillars were artificial or natural, and, if the latter, whether they were rocks or islands, seems to have been involved in much doubt. The earliest notice of them in Greek poetry is by Pindar, who regarded them as the *ultima Thule* of his day, beyond which the fame of his heroes could not advance.

Νῦν γε πρὸς ἐσχατιὰν Θή-
ρων ἀρεταῖσιν ἱκάνων ἅπτεται
Οἴκοθεν Ἡρακλέος σταλᾶν. τὸ πόρσω
Δ' ἐστι σοφοῖς ἄβατον
Κἀσόφοις. οὐ μὴν διώξω. κεινὸς εἴην.—PIND. Olymp. iii. 77.

Οὐκέτι πρόσω
Ἀβάταν ἅλα κιόνων
Ὑπὲρ Ἡρακλέος περᾶν εὔμαρὲς,
Ἥρως θεὸς ἃς ἔθηκεν
Ναυτιλίας ἐσχάτας
Μάρτυρας κλυτάς.   ID. Nem. iii. 33.

[4]   Simul ipsa precatur
Oceanumque patrem rerum Nymphasque sorores.—VIRG. Georg. iv. 391.
Usque ad Hyperboreos et mare ad Oceanum.—CATULL. cxv. 6.
Et quas Oceani refluum mare lavit arenas.—Ov. Met. vii. 267.

mountain systems of the two continents as but parts of a single grand system, the point of union between them being at the Thracian Bosporus. There is, however, this marked distinction between the two continents: in Asia the central mountain range is remote from the sea; in Europe it is closely contiguous to it. The most important links in the European range from E. to W. are—Hæmus, and its continuations between the Euxine and the Adriatic Seas; the Alps, between the Adriatic and Tyrrhenian Seas; and the Pyrenees, between the Tyrrhenian Sea and the Atlantic Ocean.

Hæmus,[5] properly so called, rises on the shores of the Euxine near Mesembria, and runs in a westerly direction to the valley of the Strymon, where it divides into the diverging ranges of Scomius and Scardus. A lateral range, which leaves it not far from the Euxine, and which runs parallel to the coast of that sea, terminates at the entrance of the Thracian Bosporus. The name seems to be connected with the Greek χεῖμα and the Sanscrit *hima*, in which case it betokens the rough and stormy character of the range.[6] From its westerly extremity a series of ranges connects Hæmus with the Alps; occasionally all of these were included under the general name of Hæmus, but they were more properly known by the specific names of Scardus between Macedonia and Mœsia, Bebii Montes between Illyria and Mœsia, Adrius and Albanus in Northern Illyria. The great range of the Alps connects with the Illyrian ranges at the head of the Adriatic Sea, and curves round in the form of a bow to the Ligurian shore near Genoa. The name is probably derived from a Celtic word *Alb* or *Alp* "a height." This range was but imperfectly known until the time of the Roman empire;[7] it was then thoroughly explored and crossed by

---

[5] The height of Hæmus was over-estimated by the ancients: it does not exceed 3000 ft.

[6] Homer refers to the cold of Hæmus in the following line:

> Ζαλαγ' ἐφ' ἱπποπόλων Θρῃκῶν ὄρεα νιφόεντα.—*Il.* xiv. 227.

So also Virgil:

> O qui me gelidis in vallibus Hæmi
> Sistat, et ingenti ramorum proteget umbra.—*Georg.* ii. 488.

Hæmus, as the chief mountain in Thrace, was regarded as the original seat of music:

> Unde vocalem temere insecutæ
>   Orphea silvæ,
> Arte maternâ rapidos morantem
> Fluminum lapsus, celeresque ventos,
> Blandum et auritas fidibus canoris
>   Ducere quercus.   Hor. *Carm.* i. 12. 7.

[7] The Alps are described at length in the two following passages:—

> Sed jam præteritos ultra meminisse labores
> Conspecto propius demserre paventibus Alpes,
> Cuncta gelu canaque æternum grandine tecta,
> Atque æri glaciem cohibent: riget ardua montis
> Ætherii facies, surgentique obvia l'herbæ
> Duratas nescit flammis mollire pruinas.

Quantum

various frequented routes. The description of these and of the various
subdivisions of the range will fall most appropriately under the
head of Italy. The Pyrenæi Montes[a] rise on the shores of the
Mediterranean, and run in a westerly direction to the *Bay of Biscay*,
forming the boundary between Gaul and Spain. The chain is thence
continued in a direction parallel to the S. coast of the *Bay of Biscay*
to the shores of the Atlantic ; the western prolongations were known
as **Saltus Vasconum** and **Mons Vinnius** or **Vindius**. The name is
probably derived from the Celtic word *bryn* "a mountain."

From the central range already described emanate subordinate
ranges towards the S. which, extending deeply into the Mediterranean,
form three extensive peninsulas. The most westerly of these is Spain,
which owes its existence to the various ramifications of the Pyrenæan
range, taking for the most part a south-westerly direction, and so
communicating a quadrangular form to that peninsula. The central
one is Italy, which is supported by a single range, the **Apennini
Montes**, an offset from the Alps, which forms the back-bone of the
country, passing through its whole extent, and giving it a direction
towards the S.E. The third or most easterly springs similarly from
Hæmus, and may be said to have its base extending from the Adriatic
to the mouth of the Danube, but as it proceeds southwards narrows
into the peninsula of Greece ; the central range of this peninsula may

---

Quantum Tartareus regni pallentis hiatus
Ad manes imos atque atra stagna paludis
A supera tellure patet ; tam longa per auras
Erigitur tellus, et cœlum intercipit umbra.
Nullum ver usquam, nullique æstatis honores.
Sola jugis habitat diris, sedesque tuetur
Perpetuus deformis Hiems : illa undique nubes
Huc atras agit, et mixtos cum grandine nimbos.
Jam cuncti flatus ventique furentia regna
Alpina posuere domo. Caligat in altis
Obtutus malis, obruuntque in nubila montes.—SIL. ITAL. iii. 477.

Sed latus, Hesperiæ quo Rhætia jungitur oræ,
Præruptis ferit astra jugis, paruitque terendam
Vix æstate viam. Multi seu Gorgone visa
Obriguere gelu : multos hausere profundæ
Vasta mole nives, cumque ipsis sæpe juvencis
Naufraga caudenti merguntur plaustra barathro.
Interdum glacie subitam labente ruinam
Mons dedit, et tepidis fundamina subruit Austris
Pendenti malefida solo.   CLAUD. *de Bell. Get.* 340.

The earlier poets refer to the great height of the range, and the consequent
severity of the climate, in general terms :

Tum sciat, *atrias* Alpes et Norica si quis.—VIRG. *Georg.* iii. 474.

Furius *hibernas* cana nive conspuet Alpes.—HOR. *Sat.* ii. 5, 41.

Fontis, et Alpino modo quæ certare rigori.—OV. *Met.* xiv. 794.

Occasionally, the term was extended to the Pyrenees :

Nunc *geminas* Alpes, Apenninumque minatur.—SIL. ITAL. ii. 333.

[a] At Pyrenæi *frondosa* cacumina montis.—SIL. ITAL. iii. 415.

Bimaris juga *ninguida* Pyrenæi.—AUSON. *Epist.* xxiv. 69.

Jamque Pyrenææ, quas nunquam solvere Titan
Evaluit, fluxere nives.   LUC. iv. 83.

be observed to leave Hæmus in about 42° N. lat. and 21° E. long.,
and may be traced through Pindus and the other Greek ranges down
to the island of Cythēra.

The northern projections from the main range are not in themselves
unimportant, but fall into districts that were little known to the
ancients. The ranges of Germany are the most prominent of these,
consisting of the Hercynia Silva, under which name most of the western
ranges of Germany were at one time included, but which was afterwards restricted to the range connecting the Sudētes with the Carpathians; the Sudētes, in the N.W. of Bohemia, where the name is still
retained; and Carpātes, the range which encloses Hungary on the N.
and E., and which is still known as the Carpathians. It may be
observed generally of these northern ranges that they run parallel to
the main chain, thus contrasting strongly with the southern ranges
which are nearly at right angles with it.

§ 6. The rivers of Europe are numerous and important in comparison with the size of the continent. They fall, however, for the
most part into the northern districts, with which the ancients did
not become acquainted until a late period : those of the peninsulas of
Greece and Italy have necessarily (with the exception of the Po) short
courses. The description of the rivers will fall more appropriately
under the heads of the countries through which they flowed, with
the exception of some few which come prominently forward as
boundaries of countries, and which hold an important place in the
history and political geography of the continent. These rivers have,
with but slight variation, retained their ancient names to the present
day : they are the Danube, the Rhine, the Vistula, the Tyras or
Dnieper, and the Tanais or Don.

The Ister or Danubius[*] rises in Mons Abnoba,[1] the Black Forest,
and flows with a general easterly direction into the Euxine Sea. In
its upper course it formed the boundary between Germany on the N.,
and Rhætia, Noricum, and Pannonia on the S. It then skirted the

---

[*] The former of these names more properly belonged to the Greeks, the latter
to the Romans. The Latin poets, however, frequently used the Greek form, e. g.

Arsit Orontes

Thermodonque altus, Gangesque et Phasis et Ister.—Ov. Met. II. 248.

Quaque Istros Tanaisque Getas rigat atque Magynos.

Tibull. iv. 1, 146.

The name Danubius contains the root dan "water," which also appears in
Rho-dan-us, Eri-dan-us, Tan-ais.

[1] The early Greeks had very indefinite notions as to its sources. Pindar represents it as flowing through the country of the Hyperboreans :

Τὰς ποτε
'Ιστρου ἀπὸ σκιαρᾶν παγᾶν ἔνεικεν
'Αμφιτρυωνιάδας,
Μνᾶμα τῶν Ὀλυμπίᾳ κάλλιστον ἄθλων
Δᾶμον Ὑπερβορέων πείσαις.   Olymp. III. 24.

Hesiod knew of it simply as a large river :

Στρυμονα Μαίανδρόν τε, καὶ Ἴστρον καλλιρέεθρον.—Theog. 338.

E. frontier of the last-mentioned country in a southerly direction, dividing it from Dacia, and then, reverting to its easterly course, separated Dacia from Mœsia. For a long period it formed the boundary of the Roman empire.[a] The Rhenus rises in the Alps and flows with a general northerly direction into the German Ocean. In its upper course it deviates to the W. between the Lacus Brigantinus and the town of Basilia, *Bâle;* and in its lower course it again inclines towards the W., and traverses a low country, where its channels have shifted at various times. A description of this part of its course will be given hereafter. The Rhine formed the boundary between Gaul and Germany, and was the great frontier of the Roman empire against the German tribes.[b] The Vistula is noticed as the boundary of Germany on the side of Sarmatia. Little was known of its course: it is described as rising in the Hercynia Silva and discharging itself into the Baltic Sea. The Tyras[c] formed the southern boundary of Scythia in the time of Herodotus, and the division between Dacia and Sarmatia in the time of the Roman empire. It is described as rising in the Carpathian ranges and flowing into the Euxine. Little was known of its course.[d] The Tanais derived its importance from being regarded as the boundary between Europe and Asia.[e] Its source, unknown to the ancients,[f] is in a lake in the province of *Toula;* it flows first in a S.E. and then in a S.W. direction, and discharges itself into the Palus Mæotis.

§ 7. The climate of Europe, particularly of the southern portion of the continent, with which the ancients were best acquainted, presents a favourable contrast to that of the other continents. Surrounded by water, it is equally free from the extremes both of heat

---

[a] Hence we read in Horace:

Non, qui profundum Danubium bibunt,
Edicta rumpent Julia.   *Carm.* iv. 15, 21.

[b] The name is sometimes applied to the tribes living on its E. bank:—

Alter enim de te, Rhene, triumphus adest.—Ov. *ex Pont.* iii. 4, 88.

Non vacat Aretoas acies, Rhenumque rebellem
Pandere.   STAT. *Silv.* i. 4, 88.

[c] The modern name *Dniestr* appears under the form Danastris in the later writers of the Roman empire. The ancient name is still in use among the Turks under the form *Tural.*

[d] Ovid refers to the rapidity of its stream:

Nullo tardior amne Tyras.—*Ex Pont.* iv. 10, 50.

[e] See note 1 (page 314). Hence, also, the epithet in Horace:

Extremum Tanaim si biberes, Lyce.—*Carm.* iii. 10, 1.

[f] Lucan places it in the Rhipæan mountains:

          Qua vertice lapsus
Rhipæo Tanais diversi nomina mundi
Imposuit ripis.   Luc. iii. 272.

Virgil assigns to it a similar locality:

Solus Hyperboreas glacies Tanaimque nivalem
Arvaque Rhipæis nunquam viduata pruinis
Lustrabat.   *Georg.* iv. 517.

and cold, and is adapted to mature all the most valued productions of the vegetable world. The southern peninsulas* produced corn, wine, and oil, and admitted of the introduction of many foreign plants, such as the cherry, the orange, peach, fig, and mulberry. The northern districts, being covered with extensive forests and morasses, were not so favoured in point of climate, and to this circumstance we may partly attribute the unwillingness of the Greeks and Romans to penetrate them. There can be no question that a vast improvement has taken place in this respect through the progress of cultivation.

§ 8. The commerce of Europe, though prosecuted on a most extensive scale, does not present many topics of interest in connexion with ancient geography. Being carried on chiefly by sea, it did not conduce to throw open the interior of the continent to the same extent as we have witnessed in the cases of Asia and Africa. There were, however, two exceptions to this general assertion: viz. the tin and the amber trade, which both led to the formation of commercial routes. In regard to the first of these productions, Diodorus Siculus tells us (v. 22) that the merchants conveyed the tin from Britain to the coast of Gaul, and that it was thence carried on pack-horses to Marseilles (probably by the valleys of the *Seine, Saône,* and *Rhone*). Amber was found on the shores of the Baltic, and was conveyed thence by an overland route to the head of the Adriatic, where it was shipped for various parts: the extent of country traversed by this route will appear from a glance at the map, and it is a matter of regret that we are not in possession of the details relating to the course followed.

---

* Virgil thus eloquently contrasts the superior climate of southern Europe with that of Asia:

Sed neque Medorum silvæ, ditissima terra,
Nec pulcher Ganges, atque auro turbidus Hermus,
Laudibus Italiæ certent; non Bactra, neque Indi,
Totaque thuriferis Panchaia pinguis arenis.
Hæc loca non tauri spirantes naribus ignem
Invertere, satis immanis dentibus hydri;
Nec galeis densisque virûm seges horruit hastis;
Sed gravidæ fruges, et Bacchi Massicus humor
Implevere; tenent oleæ, armentaque læta.
Hinc bellator equus campo sese arduus infert;
Hinc albi, Clitumne, greges, et maxima taurus
Victima, sæpe tuo perfusi flumine sacro,
Romanos ad templa deûm duxere triumphos.
Hic ver assiduum, atque alienis mensibus æstas;
Bis gravidæ pecudes, bis pomis utilis arbor.
At rabidæ tigres absunt, et sæva leonum
Semina; nec miseros fallunt aconita legentes;
Nec rapit immensos orbes per humum, neque tanto
Squameus in spiram tractu se colligit anguis.—*Georg.* ii. 136.

§ 9. The population of Europe belonged in the main to the Japhetic or Indo-European branch of the human race. The divisions of this great family and their mutual relations present many unsolved problems. Without going into these questions, we may point out the following races as among the most important : (i.) the Celts and Cimmerians, who entered this continent from the steppes of Caucasus, and, passing round the head of the Black Sea, spread themselves over the whole of Europe and permanently settled in the West. The countries occupied by them in classical times were Gaul, the British Isles, portions of Spain, Rhætia, parts of Pannonia, and Noricum. (ii.) The Sclavonians, or, as the ancients denominated them, Scythians and Sarmatians, who occupied the east of Europe as far as the *Oder* westward. (iii.) The Teutons, who arrived at different epochs : (1) as Low Germans, from the regions between the Oxus and Jaxartes, and established themselves in the N.W. of Europe, and (2) as High Germans, who, displacing the Celts and Sclavonians, occupied the middle highlands of Germany, and are found in classical times E. of the Rhine and N. of the Danube. (iv.) The Græco-Latin stock, which probably crossed from Asia Minor by way of Thrace and the Ægæan Isles. In Greece it was known by the name of Pelasgian : the Phrygians, early Thracians, and Macedonians, belonged to this race. The element which Italy had in common with Greece, also belonged to it. (v.) The Iberians, who formed the basis of the population in Spain and in the S.W. angle of Gaul, were of the same races as the modern Basques, and therefore did not belong to the Indo-European family. (vi.) The Illyrians, or progenitors of the modern *Shipe-tares.* Of the two but little is known.

Mount Athos.

CHAPTER XVIII.

THRACIA AND MACEDONIA.

I. THRACIA. § 1. Boundaries and general description. § 2. Mountains. § 3. Rivers. § 4. Inhabitants. § 5. Towns; Roads; History; Islands. II. MACEDONIA. § 6. Boundaries; Name. § 7. Mountains. § 8. Rivers. § 9. Inhabitants. § 10. Towns; Roads; St. Paul's Travels; History.

I. THRACIA.

§ 1. THE boundaries of Thracia[1] in the Roman era were—on the E. the Euxine and the Bosporus; on the S. the Propontis, Hellespont, and Ægæan; on the W. the river Nestus, dividing it from Macedonia; and on the N. Mount Hæmus, dividing it from Mœsia. At an earlier period the district N. of Hæmus to the Ister was included within the limits of Thrace; and in the earliest times the name was still more broadly applied to all Europe N. of Greece. The surface of Thrace is generally mountainous, and the coast of the Ægæan is

---

[1] The poetical form of the name is Thraca :

Gemit ultima pulsa

Thraca pedum.   -   VIRO. Æn. xii. 434.

Thracane vos, Hebrusque nivali compede vinctus.—HOR. Ep. i. 3, 3.

extremely irregular. The soil was fertile,[2] particularly in corn (which was exported to Athens and Rome) and in millet. The climate is described as very severe:[3] nevertheless the grape ripened there, and we cannot but suppose that the accounts of the ancients as to the climate are somewhat exaggerated. Horace were abundant, and a breed of a white colour was famous.[4] Cattle and sheep formed the chief wealth of the inhabitants of the interior, while large amounts of gold, existing between the Strymon and Nestus, enriched the inhabitants of the coast, as well as foreign settlers, particularly the Phœnicians and Athenians. Certain kinds of precious stones were also found, particularly one named *Thracia gemma*.

Name.—The most probable derivation of the name is from the adjective τραχεῖα, "rugged," indicative of the character of the country. The transfer of the aspirate from the middle to the beginning of the word gives us the form Θρῃϊκίη.

§ 2. The chief mountain-range in Thrace is Hæmus, which skirts the northern frontier and sends out three lateral ridges towards the

---

[2] Homer characterises it by the epithet *ἐριβῶλαξ*.

'Ρῆγμον, δι ἐκ Θρῄκης ἐριβώλακος εἰληλούθει.—*Il.* XX. 485.

He also represents cargoes of wine as coming from Thrace :

Πλεῖαί τοι οἴνου κλισίαι, τὸν νῆες 'Αχαιῶν
'Ἠμάτιαι Θρῄκηθεν ἐπ' εὐρέα πόντον ἄγουσιν—*Il.* IX. 71.

[3] There is some ground for this belief : several historians (Xen. *Anab.* vii. 4, 3; Floros, iii. 4 ; Tac. *Ann.* iv. 51) relate events which imply an unusual degree of cold. But the exaggerated descriptions of the ancients were doubtless connected with the poetic fiction of Hæmus being the residence of the north wind. To the north of that chain the climate was supposed to be particularly mild. As an instance of exaggeration we refer to the passage commencing with the following lines, in which the country about the Thracian Rhodope is introduced

At non, quæ Scythiæ gentes, Mæotiaque unda
Turbidus et torquens flaventes Ister arenas,
Quaque redit medium Rhodope porrecta sub axem.
Illic clausa tenent stabulis armenta ; neque ullæ
Aut herbæ campo apparent, aut arbore frondes :
Sed jacet aggeribus niveis informis et alto
Terra gelu late, septemque assurgit in ulnas.
Semper hiems, semper spirantes frigora Cauri.

VIRG. *Georg.* iii. 349.

Compare also the expressions quoted in note [1], and the epigram attributed by some to Cæsar.

Thrax puer adstricto glacie dum ludit in Hebro.

[4] Τοῦ δὴ καλλίστους ἵππους ἴδον ἠδὲ μεγίστους·
Λευκότεροι χιόνος, θείειν δ'ἀνέμοισιν ὁμοῖοι.—HOM. *Il.* X. 436.

Quem Thracius albis
Portat equus bicolor maculis, vestigia pripri
Alba pedis frontemque ostentans arduus albam.—VIRG. *Æn.* v. 565

From their skill in horsemanship the Thracians are described by Homer as *ἱππόπολοι* :

Νόσφιν ἐφ' ἱπποπόλων Θρῃκῶν καθορώμενος αἶαν.—*Il.* XIII. 4.

So also *Il.* XIV. 227.

S.E. The most easterly of these three separates the basin of the
Hebrus from the Euxine, and is continued in a line parallel to the
shore of the Propontis and the Hellespont to the extremity of the
Thracian Chersonese. The most westerly, named **Rhodōpe**,[5] *Despoto*,
divides the basins of the Hebrus and the Nestus. Between these a
third range of less importance separates the upper valley of the Hebrus
from that of the Tonzus. In addition to these we have to notice the
isolated height of **Ismārus**, near the S. coast, surrounded by a district
famed for its fine wine.[6] In the S.E. a rocky ridge protrudes far into
the sea, between the Hellespont and the Ægæan Sea, and forms a
long peninsula, the ancient **Chersonēsus Thracica**,[7] now the *Peninsula of Gallipoli*. A wall, crossing the ridge near Agora, severed
the peninsula from the mainland: the breadth at this point is only
36 stadia, and the length from the wall to the extreme point is 420
stadia. The most important promontories on the Euxine are **Thynias**,
N. of Salmydessus, and **Philia**, S. of it; and on the Ægæan, **Mastusia**,
C. *Greco*, the termination of the Thracian Chersonese;[8] **Sarpēdonium**,
C. *Paxi*, N. of Imbros; and **Serrium**, opposite Samothrace.

---

[5] The poetical allusions to Rhodope refer to its height, and to its being the
abode of Orpheus and Rhesus :

    Aut Atho aut Rhodopen aut alta Ceraunia telo
    Dejicit.   VIRG. *Georg.* i. 332.
               In altam
    Se recipit Rhodopen, pulsumque Aquilonibus Hæmon.--Ov. *Met.* x. 76.
    Quam satis ad superas postquam *Rhodopeïus* auras
    Deflevit vates.   Id. x. 11.
    Nec tantum Rhodope mirantur et Ismarus Orphea.   VIRG. *Ecl.* vi. 30.
               Fervunt Rhodopeïæ arces
    Altaque Pangæa et Rhesi Mavortia tellus.   *Georg.* iv. 461.
Sometimes the name is used generally for Thrace ; *e.g.*
    Spicula deposito *Rhodopeïa* pectine torsit.   SIL. ITAL. xii. 400.

[6] . . . . . ἀτὰρ αἴγεον ἀσκὸν ἔχον μέλανος οἴνοιο,
    Ἡδέος, ὅν μοι ἔδωκε Μάρων, Εὐάνθεος υἱός,
    Ἱρεὺς Ἀπόλλωνος, ὃς Ἴσμαρον ἀμφιβεβήκει.—HOM. *Od.* ix. 196.
               Juvat Ismara Baccho
    Conserere.   VIRG. *Georg.* ii. 37.
    Fertur in Ismariis Bacchus amasse jugis.   Ov. *Fast.* iii. 410.
    Tu quoque, O Eurytion, vino, Centaure peristi,
    Nec non Ismario tu, Polypheme, mero.   PROPERT. ii. 33, 82.
    Ismariæ celebrant repetita triennia Bacchæ.—Ov. *Met.* ix. 641.
The plural form *Ismara* is to be observed in the second of these passages : it
occurs also in Lucret. v. 30.

[7] It was here that Polymnestor lived, to whom Priam entrusted his son Polydorus :
    Ὃς τὴν ἀρίστην Χερσονησίαν πλάκα
    Σπείραι, φίλιππον λαὸν εὐθύνων δορί.—EURIP. *Hec.* 8.

[8] Αὐτὴ Δολόγκων σύμμαχος συμπηδότι
    Μαζουσία προὔχουσα, χερσαίου κέρως.—LICOPHR. 533.

§ 3. The chief river of Thrace was 'the **Hebrus**,' *Maritza*, which rises in the N.W., and flows first towards the S.E. as far as Adrianopolis, and then towards the S.W. to the Ægæan, receiving in its course numerous tributaries, of which the **Tonsus**, or **Artiscus**, and the **Agriānes**, on its left bank, were the most considerable. The **Nestus**, on the W. border, rises not far from the Hebrus, and in a S.E. course joins the sea near Abdera. Numerous small streams flow into the Hellespont and Propontis: one of these, named **Ægospotāmi**, "Goat River," in the Chersonesus, was famed for the naval engagement between the Athenians and Spartans in B.C. 405, which took place at its mouth. Two large lakes occur on the coast— **Bistōnis**, *L. Buru*, E. of Abdera, the water of which was brackish; and **Stentōris**, formed by an arm of the Hebrus. An extensive bay, named **Melas Sinus**, *G. of Saros*, penetrates inland W. of the Chersonesus.

§ 4. The earliest inhabitants of Thrace appear to have been of the Pelasgian race;[1] these were supplanted, at a time subsequent to the Trojan War, by an immigrant race from the north, allied to the Getæ and Mysi. These latter are the historical Thracians whom Herodotus and other later writers describe. They were reputed a savage and barbarous race,[2] faithless and sensual, and particularly addicted to drinking. They were brave soldiers, and from the time of the Peloponnesian War were much employed as mercenaries in the armies of

---

* The poetical allusions to the Hebrus refer to its northerly position ἐγγύθεν ἄρκτου—its coldness—and its connexion with the history of Orpheus, the musician's head having been carried down the stream to the sea:

Εἶη δ' Ἡβρώην μὲν ἐν ὥρεσι χείματι μέσσῳ,
Ἔβρον γάρ ποταμὸν, τετραμμένος ἐγγύθεν ἄρκτου.—THEOCR. *Idyl.* vii. 110.

Qualis apud *gelidi* cum flumina conditus Hebri
Sanguineus Mavors clipeo increpat, atque furentes
Bella movens immittit equos : illi æquore aperto
Ante Notos Zephyrumque volant ; gemit ultima pulsu
Thraca pedum.   VIRG. *Æn.* xii. 331.

          ut nec
Frigidior Thracam, nec purior ambiat Hebrus.
        HOR. *Ep.* i. 16, 13.

Tum quoque marmorea caput a cervice revulsum
Gurgite cum medio portans Œagrius Hebrus
Volveret, Eurydicen vox ipsa et frigida lingua,
Ah miseram Eurydicen ! anima fugiente vocabat.
Eurydicen toto referebant flumine ripæ.—VIRG. *Georg.* iv. 523.

[1] The Thracian tribes of the Cicones (*Il.* ii. 846) and the Caucones (*Il.* x. 429) were in close alliance with Priam in the Trojan War.

[2] It is hardly in accordance with the character of the Thracians that they should have been the inventors of music ; yet their country was the reputed abode of Orpheus, Eumolpus, Musæus, and Thamyris, and was regarded by the later poets as the cradle of music. The probability is that the term Thracian was originally of wider use, and was applied to certain districts in Central Greece, from which the associations were in course of time transferred to the northerly country.

more civilized nations. As a people they had no political cohesion : they were divided into a number of tribes, which were engaged in constant feuds with each other. Of these tribes we may notice the Odrysæ, about the upper valleys of the Hebrus; the Bessi, in the mountains near the source of that river; the Bistōnes,[3] on the coast E. of the Nestus; and the Cicōnes,[4] in the same neighbourhood. Their country was divided by the Romans into fourteen districts, the names of which are of no special interest.

§ 5. The towns in Thrace of historical importance were of foreign and not of native origin. They may be divided into two classes—the Greek colonies, which were exclusively on the coast; and the Roman towns of the interior, which were built on the sites of old Thracian towns. The coast presented many sites most admirably adapted to settlement, partly for commercial and partly for warlike purposes. The position of the Thracian Chersonese was most important, as it commanded not only the passage across the Hellespont into Asia, but also that leading up the strait into the Euxine : it was one of the two keys that locked that sea, the other being the Thracian Bosporus commanded by Byzantium. The influence of this district on the corn-trade of Greece was therefore very great. From an early period the Greeks occupied the most favourable spots : the Megarians settled at Selymbria on the Propontis and at Byzantium, and the latter town in turn colonized Mesembria on the shore of the Euxine; the Milesians founded Cardia on the Chersonese, Salmydessus and Apollonia on the Euxine; the Samians occupied Perinthus on the Propontis; while on the N. shore of the Ægean, Ænus was attributed to the Æolians, Maronea to the Chians, Abdēra to the Teians, Mesembria and Stryme to the adjacent islands of Samothrace and Thasos. These towns reached their highest prosperity in the flourishing period of Greek history. The foundation by Lysimachus of Lysimachia, in B.C. 300, as his capital, is significant of the importance attached to the Chersonese in a strategetical point of view. The interior of Thrace was thrown open by the Romans; and several important towns, such as Trajanopolis, Hadrianopolis, and Philippopolis,[5]

---

[3] The name of this tribe is not unfrequently used for the Thracians generally :

Βιστονίη φόρμιγγι λιγείῃ ἔρχεν ἀοιδῆς.—APOLL. RHOD. II. 704.

Sanguineum veluti quatiens Bellona flagellum,
Bistonas aut Mavors agitans——　　　　LUC. vii. 568.
　　　　Phrygiæ contraria tellus,
Ristoniis habitata viris.　　　　OV. Met. xiii. 439.
Nodo coerces viperino
Bistonidum sine fraude crines.　　　　HOR. Carm. ii. 19, 19.

[4] 'Υλάδεν με φίλων ἄκρας Κικόνεσσι πέλασσεν
Ισμάρῳ· ἔνθα δ' ἐγὼ πόλιν ἔπραθον, ὤλεσα δ' αὐτούς.—HOM. Od. ix. 39.

[5] Philippolis is classed as a Roman town, inasmuch as the Macedonians, by whom it was originally occupied, were unable to keep possession of it.

were founded on the most central spots. The selection of the ancient Byzantium as the capital of the Eastern Empire secured to Thrace a large amount of prosperity in the later period of Roman history. We shall describe these towns in the following order :—(1.) Those on the sea-coast from W. to E. ; and (2.) those of the interior.

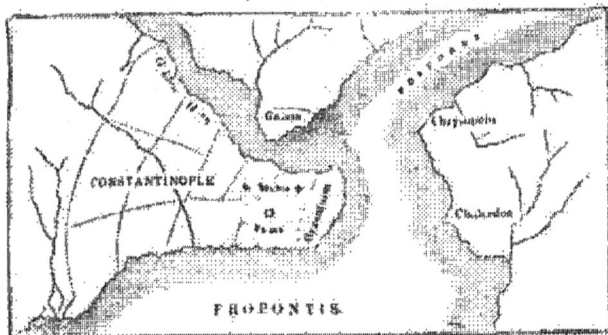

Map of Constantinople.

(1.) *Towns on the Sea-Coast.*—**Abdera** was situated some distance E. of the Nestus. It was originally occupied by a colony from Clazomenæ in B.C. 650, and afterwards by Teians in 541. At the time of the expedition of Xerxes it was a highly flourishing place. It was taken by the Athenians in 408, and appears to have fallen to decay after B.C. 370, when it suffered from a war with the Triballi. It was the

Coin of Abdera.

birth-place of the historian Hecatæus, and of the philosophers Protagoras, Democritus, and Anaxarchus: its inhabitants were nevertheless proverbial for their stupidity.[6] **Maronea,** *Marogna,* was not far from Lake Ismaris, in a district famed for its superior wine.[7] It was taken by Philip V. of Macedon in B.C. 200 ; and, on his being compelled to relinquish his conquests, its inhabitants were cruelly massacred by him. Under the Romans it became a free city. **Ænus,** *Enos,* on a promontory S.E. of Lake Stentoris, was a very ancient town, though

---

[6] Hence the uncomplimentary allusions in the following lines :
Fervecum in patria crassoque sub aere nasci.—Juv. x. 50.
Si patiens fortisque tibi durusque videtur,
Abderitanæ pectora plebis habes. Mart. x. 25.
[7] Cesit et Ænum Neptunius incola rupis,
Victa Maronea fœdatos lumina Baccho.—Tibull. iv. 1, 56.

its origin is uncertain.[a] In the Peloponnesian War it appears as an ally of Athens, and subsequently came into the possession successively

Coin of Ænus.

of Ptolemy Philopator in B.C. 222, Philip of Macedon in 200, and Antiochus the Great: under the Romans it was made a free town. Cardia, Caridia, at the head of the Gulf of Melas, was founded by a colony of Milesians and Clazomenians, and in the time of Miltiades was replenished with Athenian settlers. It was destroyed by Lysimachus; and, though rebuilt, never regained any importance. It was the birth-place of King Eumenes.

Coin of Cardia.

Sestus,[b] Jalowa, was the principal town of the Chersonesus, and stood on the Hellespont nearly opposite to Abydus. It owed its importance wholly to its position, as the point at which the straits were crossed, and consequently it sunk when the Romans transferred the station to Callipolis. The bridge of boats constructed by Xerxes terminated a little S. of the town. It was taken by the Athenians, B.C. 478, and was termed by them the "corn-chest of the Piraeus," as giving them command of the Euxine. It was taken by the Spartans, B.C. 404; was blockaded by Conon without effect in B.C. 394; and again by Cotys, a Thracian king, with a similar result, in 362, at which time it had fallen into the power of the Persians. It was besieged by the Athenians in 353, when its inhabitants were massacred;

---

[a] Ænus is noticed by Homer; it could not therefore have been founded by Æneas, as Virgil asserts:

βῆλε δὲ Θρηκῶν ἀγὸν ἀνδρῶν,
Πείρως 'Ιμβρασίδης, ὃς ἀρ' Αἰνόθεν εἰληλούθει.—Il. iv. 512.

Terra procul vastis colitur Mavortia campis,
Thraces arant, acri quondam regnata Lycurgo
Hospitium antiquum Trojæ, sociique Penates,
Dum fortuna fuit. Feror huc, et littore curvo
Moenia prima loco, fatis ingressus iniquis;
Æneadasque meo nomen de nomine fingo.—Æn. iii. 13.

[b] Sestus has been already noticed in the passages quoted under the head of Abydus. We may add the following, which contain references to the lives of Hero and Leander:

Sestiacos nunc Fama sinus pelasgasque natarum
Jactos. 'Stat. Silv. i. 3, 27.

Mittit Abydenus, quam mallet ferre, salutem,
Si cadat ira maris, Sesti puella, tibi.—Ov. Heroid. xviii. 1.

and lastly it surrendered to the Romans in 190. Gallipolis, *Gallipoli*, stood higher up the coast, opposite Lampsacus, and became a flourishing place under the Romans. Lysimachia, at the N.E. extremity of the Chersonese, owed its name and existence to Lysimachus, who constituted it his capital, and peopled it with the inhabitants of Cardia. After the death of its founder, it passed successively into the hands of the Syrians and Egyptians. Is was destroyed by the Thracians during the war of the Romans against Philip of Macedon; and, though restored by Antiochus the Great, never recovered its prosperity. Perinthus, *Eski Eregli*, was built like an amphitheatre on a small peninsula jutting out into the Propontis. It was originally a Samian colony, founded about B.C. 599. It was famed for its obstinate defence against Philip of Macedon, at which time it was a flourishing commercial town. Its name was changed to Heracles about the 4th cent. of our era. Salymbria, *Silivri*, a colony of the Megarians, was about 22 miles E. of Perinthus, and just inside the wall of Anastatius. It is noticed by Xenophon as the place where he met Medosades, and as being taken by Alcibiades. The Emperor Eudoxius changed its name to Eudoxiupolis.

Byzantium was situated at the extreme point of the promontory which divides the Propontis from the Bosporus, an inlet of the latter, the modern "Golden Horn," bounding the site of the town on the N. Its position was magnificent, commanding the opposite shores of Europe

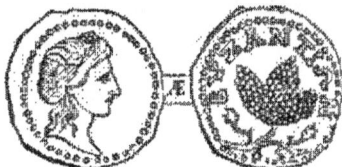

Coin of Byzantium.

and Asia, at the same time secure and well adapted for trade, and surrounded by beautiful scenery. Its foundation is ascribed to the Megarians,[1] who sent thither two colonies in the years B.C. 667 and 629. The chief events in its history are—its capture by Alcibiades in 408, when it was in the hands of the Spartans; its recapture by Lysander in 405; the unsuccessful siege of it by Philip of Macedon in 340, when aid was given to it by Athens; the heavy imposts exacted by the Gauls in 279; its capture by Severus after a three years' siege, in the civil war with Pescennius Niger, A.D. 196, after which the walls were levelled, and the inhabitants treated with great severity; and its final capture by Constantine, when Licinius had retired thither after the battle of Adrianople. That emperor selected the promontory on which Byzantium stood as the site of his new capital; and on May 12, A.D. 330, founded Constantinopolis, or, as it was originally styled, "New Rome."[2] The new town, like old Rome, stood on 7 hills, 5 of which were enclosed within the fortifications that extended from the "Horn," which served as the port, to the Propontis. It was divided into 14 regions, and was adorned by its founder with a similar number of churches and

[1] It is said to have been built on the site of an older town named Lygus; hence in Ausonius—

in cum
Byzantina Lygos, tu Punica Byrsa fuisti.—*Nob. Urb.* 2.

[2] The modern *Stambul* is a corruption of the Greek εἰς τὴν πόλιν.

palaces, as well as with several triumphal arches and 8 public baths. Subsequent emperors added to its edifices: Theodosius the Great built the "Golden Gate;" Theodosius II. added hot baths; Justinian, the "second founder" of the city, built the temple of the Eternal Wisdom, *St. Sophia*, and 25 churches, and restored the palace. The chief events in the history of the town are—its almost total destruction in the reign of Justinian by the factions of the Circus, A.D. 532; the blockade of Chosroes, from 616 to 626; the two unsuccessful sieges of the Arabs in 668 and 673, and 713-718; its capture by the Latins in 1204; and its capture by the Turks in 1453. **Salmydessus** stood on the coast of the Euxine, about 60 miles N.W. of the Bosporus, near *Midjeh*. The coast was extremely dangerous, and the people had the character of being unscrupulous wreckers.[3] The name was applied to the district as well as the town. **Apollonia**, or, as it was later called, **Sozopolis**, whence the modern *Sizeboli*, was a Milesian colony more to the N., with two large harbours. It possessed a temple with a colossal statue of Apollo, which M. Lucullus transported to Rome. **Mesembria**,[4] at the foot of Hæmus, was founded originally by Megarians, and afterwards received colonists from Byzantium and Chalcedon, about 500 B.C. It was a member of the Greek Pentapolis on the Euxine.

Of the less important towns we may notice—**Dicæa**, a Greek town on Lake Bistonis, identified either with *Cerna* or *Bosrón*; **Ismárus**, an old town of the Ciconès, at the foot of the mountain of the same name; **Stryme**, a Thasian colony, near the river Lissus; **Mesembria**, a colony from Samothrace, N. of that island; **Dóriscus**, at the mouth of the Hebrus, where Xerxes reviewed his army; **Aphrodisias**, probably the same as **Agóra**, at the neck of the Chersonese; **Alopeconnésus**, *Alexi*, an Æolian colony on the W. coast of the Chersonese; **Elæus**, a Teian colony on the Hellespont, near Prom. Mastusia, celebrated for its temple and tomb of Protesilaus; it was frequently visited by fleets either entering or leaving the Hellespont; **Madytus**, *Maïto*, opposite Abydos; near it was the promontory of **Cynosséma**, "Dog's tomb," so named as being the burial-place of Hecuba, who was metamorphosed into a dog;[5]

---

[3] Τραχεῖα πόντου Σαλμυδησσία γνάθος
'Εχθρόξενος ναύτῃσι, μητρυιὰ νεῶν—Æsch. Prom. 726.

[4] Hæc precor evincat, propulsaque fortibus Austris
Transeat instabiles strenua Cyaneas :
Thynniacaeque sinus, et ab his per Apollinis urbem
Alta sub Anchiali mœnia tendat iter :
Inde Mesembriacos portus, et Odessum, et arces
Præteream dictas nomine, Bacche, tuo.—Ov. *Trist.* L. 10, 33.

[5] Θανοῦσα δ', ἢ ζῶσ'. ἐτᾶδ' ἐκκλήσω βίον ;
Θανοῦσα· τύμβῳ δ' ὄνομα σῷ κεκλήσεται.
Μορφὴ ἐντεῦθεν, ἢ τί τῆς ἐμῆς ἐρεῖς;
Κυνὸς ταλαίνης σῆμα, ναυτίλοις τέκμαρ.—Eurip. *Hecub.* 1270.

Clade sui Thracum gens irritata tyranni
Troada telorum lapidumque incessere jactu
Corpit. At hæc missum ranco cum murmure saxum
Morsibus insequitur : rictuque in verba parato
Latravit, conata loqui. *Locus exstat, et ex re*
Nomen habet : veterumque diu memor illa malorum,
Tum quoque Sithonios ululavit mœsta per agros.
Ov. *Met.* xiii. 565

Pactye, whither Alcibiades was exiled; and Anchialus, on the Euxine, N. of Apollonia, of which it was a colony.

(2.) *In the Interior.*—Philippopolis, founded by Philip of Macedon, was built on three hills (whence its other name of Trimontium) S.E. of the Hebrus, on the site of a previously existing Thracian town. It was a very populous place, and is still, as *Philippopoli*, one of the most important towns of Thrace. Hadrianopolis, at the junction of the Tonzus with the Hebrus, was founded by the Emperor Hadrian on the site of the older Uscudama. The fertility of the surrounding country and the centrality of its position rendered it a very flourishing place. It carried on several manufactures, especially one of arms. It was besieged by the Goths in A.D. 378. *Adrianople* is still a large place. Trajanopolis was founded either by or in honour of Trajan. It stood in the lower valley of the Hebrus, but its position is uncertain: by some it is placed at *Orikhova*, about 40 miles from the mouth of the river; by others on the Egnatia Via, some distance W. of the Hebrus.

Of the less important towns we may notice—Develtus, *Zagora*, W. of Apollonia; Berœa, or Irenopolis, as it was afterwards named after the Empress Irene, E. of Philippopolis; Nicæ, near *Adrianople*, the scene of the defeat and death of the Emperor Valens in A.D. 378; Isurülum, N.W. of Perinthus, and in the neighbourhood of the Campus Serenus, on which Licinius defeated Maximinus; Cænophrurium, more to the E., where Aurelian was murdered in A.D. 275; Plotinopolis, S. of Hadrianopolis, but of uncertain position, named after Plotina, the wife of Trajan; Tempyra, on the Egnatia Via, near Trajanopolis, situated in a defile (probably the Κορσίλων στενά of Arrian), in which Cn. Manlius was attacked on his return from Asia Minor in B.C. 188; and Nicopolis, near the mouth of the Nessus, probably founded by Trajan.

Roads.—Thrace possessed two high roads, both starting from Byzantium: one of these (called the "King's Road," as having been in part followed by Xerxes) ran parallel to the Ægean coast into Macedonia; the other followed the valley of the Hebrus, through Adrianople and Philippopolis into Mœsia. The former was the route selected by the Romans for their great eastern road; it formed a portion of the Egnatia Via; the time of its construction through Thrace seems quite uncertain.

History.—The earliest historical event of consequence was connected with the expedition of Darius in 513 B.C. against the Scythians. The course which he pursued through Thrace has been already referred to (cap. iii. § 7). On his return he left Megabazus to subdue the country: this was effected, but the Persian occupation was only of short duration. Miltiades was tyrant of the Chersonese at this period. The next events are connected with the expeditions against Greece under Mardonius in 492, and under Xerxes in 480, both of which passed through the country. The Thracians joined the invaders and fought at the battle of Platæa. The Athenians subsequently expelled the Persians from the Thracian towns in the years 478-476. The kingdom of the Odrysæ was the most powerful at this time. In 431 the Athenians entered into alliance with Sitalces, who undertook a campaign against Macedonia. The command of the Bosporus and Hellespont were of the greatest importance to the Athenians, and various engagements took place between them and the Spartans, terminating with the battle of Ægospotami in 405. Subsequently to this the influence of Sparta predominated until the accession of Philip II. to the throne of Macedonia in 359, who succeeded in getting possession of that part of Thrace which lay W. of the Nestus, as well as the remainder of the coast. On the death of Alexander the Great in 323, Thrace fell to the share of Lysimachus; and, after his death in 281, was for a short time subject to Seleucus and Ptolemy Ceraunus. A long period of anarchy and uncertainty followed. In 247 the coast-towns were conquered by Ptolemy Euergetes, and remained subject to Egypt for about 50 years. Philip V. of Macedonia invaded Thrace in the years 211, 205, and 200; but was compelled by the Romans to resign his conquests in 196. In 190 Manlius traversed Thrace on his advance against Antiochus. Philip renewed his invasions in 184 and the following years with no permanent results. After the annexation of Macedonia to the Roman Empire in 148, frequent wars with the Thracians occurred. The country, however, preserved a show of independence until the reign of Vespasian (A.D. 69-79), when it was made a Roman province.

Islands.—The following islands lie off the coast of Thrace: Imbros, Lemnos, Samothrace, and Thasos. Imbros, *Embro,* which may be regarded as a continuation of the Thracian Chersonese, is mountainous [a] and well-wooded, and possessed a town of the same name on its N. coast. It was occupied by Pelasgians, and colonized by Athenians, who retained possession of it to a late period. It was visited by Ovid on his voyage to the place of his exile.[b] The Cabiri were worshipped there. Lemnos, now *Stalimene,* a corruption of εἰς τὰν Λῆμνον, lay S.W. of Imbros about midway between Mount Athos and the Hellespont. It is of an irregular quadrilateral shape, being nearly divided into two peninsulas by two deep bays. It is covered with barren and rocky hills of no great height, which in many places indicate the presence of volcanic agency. Hence the island was connected with

Coin of Imbros.

Hephaestus,[c] and hence also its ancient name of Æthalea "the burning

---

[a] Hence the epithet by which Homer characterizes it:
Μεσσηγὺς δὲ Σάμου τε καὶ Ἴμβρου παιπαλοέσσης.—*Il.* xxiv. 78.

[b] Venimus ad portus, Imbria terra, tuos.—Ov. *Trist.* I. 10, 19.

[c] Ægeo premitur circumflua Nereo
Lemnos, ubi ignifera fessus respirat ab Ætna
Mulciber: ingenti tellurem proximus umbra
Vestit Athos, nemorumque obscurat imagine pontum.
STAT. *Theb.* v. 49.
Vulcanum tellus Hypsipylæa colit. Ov. *Fast.* III. 82.

Ἤδη γάρ με καὶ ἄλλοτ' ἀλεξέμεναι μεμαῶτα
Ῥῖψε, ποδὸς τεταγών, ἀπὸ βηλοῦ θεσπεσίοιο·
Πᾶν δ' ἦμαρ φερόμην, ἅμα δ' ἠελίῳ καταδύντι
Κάππεσον ἐν Λήμνῳ, ὀλίγος δ' ἔτι θυμὸς ἐνῆεν·
Ἔνθα με Σίντιες ἄνδρες ἄφαρ κομίσαντο πεσόντα.—HOM. *Il.* I. 590.

Hence "Lemnius" was an epithet of Vulcan:
Lemnius extemplo valvas patefecit eburnas.—Ov. *Met.* IV. 185.
Hæc pater Æoliis properat dum Lemnius oris.—VIRG. *Æn.* VIII. 454.

isle." On the E. coast is the Hermæan rock to which Æschylus refers.[1] The earliest inhabitants were the Thracian Sinties : these were succeeded by the Minyæ,[1] and these in turn by the Pelasgians.[2] Lemnos belonged generally to the Athenians. It possessed originally only one town of the same name but afterwards two, **Myrina, Kastro**, on the W. coast, and **Hephæstia** on the N. Pliny states that there was a remarkable labyrinth on the island. **Samothracia**, "the Thracian Samos,"[3] *Samothraki*, lies N. of Imbros, opposite the mouth of the Hebrus. It is of an oval shape, and about 8 miles long and 6 broad, and contains a mountain of remarkable height[4] (5240 feet), which renders the island a very conspicuous object from the coasts both of Asia and Europe : the name *σάμος* has reference to this elevation. Samothrace was the chief seat of the worship of the Cabiri. **Thasos,**

Thaso, lies about 3½ miles off the plain of the river Nestus. It is covered with mountains, some of which are bare, others wooded, the highest of them attaining an elevation of 3428 feet :[5] only a few cultivated spots occur near the sea shore. It produced marble,[6] wine,[7] and more especially gold, the mines of which were worked originally by

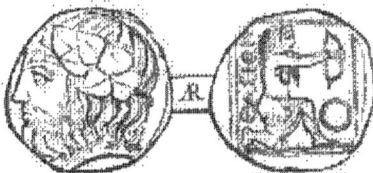

Coin of Thasos.

---

[1] Ἔσχμητ· Ἴδη μεν, πρὸς Ἑρμαῖον λέπας
Δήμετρο.　　　　　　　　*Agam.* 283.

So also Sophocles :
πολλὰ δὶ φωνῆς τῆς ἡμετέρας
Ἑρμαῖον ὄρος παρέπεμψεν ἐμοὶ
στόνον ἀντίτυπον χειμαζομένῳ.—*Philoct.* 1459.

[1] The Minyæ were said to be the offspring of the Argonauts and the Lemnian women, who had all murdered their husbands, and were living under the rule of Hypsipyle, the daughter of Thoas, to whom Ovid refers in the expression "tellus Hypsipylea :" see above, note [6].

[2] The Pelasgians were also guilty of an act of gross cruelty in the murder of their offspring by the Athenian women whom they had carried off. "Lemnian deeds" hence became a proverbial expression for any atrocity.

[3] Threïciamque Samum, quæ nunc Samothracia fertur.—VIRG. *Æn.* vii. 208.
Θρηϊκίῃ τε Σάμῳ, Ἴδης τ' ὄρεα σκιόεντα.　　HOM. *Hymn in Apoll.* 34.

[4] From the top of this rock Homer describes Neptune as surveying the plain of Troy :
Καὶ γὰρ ὁ θαυμάζων ἧστο πτόλεμόν τε μάχην τε
Ὑψοῦ ἐπ' ἀκροτάτης κορυφῆς Σάμου ἐληφόσης,
Θρηϊκίης· ἔνθεν γὰρ ἐφαίνετο πᾶσα μὲν Ἴδη,
Φαίνετο δὲ Πριάμοιο πόλις, καὶ νῆες Ἀχαιῶν.—*Il.* xiii. 11.

[5] Archilochus most truly compares Thasos to an "ass's backbone overspread with wild wood."—(*Frag.* 17, 18.)

[6] Non huc admissæ Thasos aut undem Carystos.—STAT. *Silv.* i. 5, 34.
Hic Nomadum lucent flaventia saxa Thasosque.—*Id.* ii. 2, 92.

[7] Sunt Thasiæ vites.　　　　　　　VIRG. *Georg.* ii. 91.
Hence the head of Dionysus appears on the coins of Thasos.

the Phœnicians, and afterwards by the Greeks of Paros, who settled here under Telesicles, the father of Archilochus, about 720 B.C. These Thasian Greeks also worked the mines on the coast of Thrace. Thasos thus became very wealthy, and was obliged to contribute liberally to the support of the Persian army under Xerxes. The chief town was on the N. coast, and possessed two ports. It was taken by the Athenians in B.C. 462, to whom the island remained generally subject. It was made free by the Romans after the battle of Cynoscephalæ in 197. We have yet to notice the two small islands at the N. entrance of the Thracian Bosporus, named **Cyaneæ Insulæ**, from the greenish *coppery* colour of the rocks, and **Symplegides** from their apparently *clashing together* as vessels approached them. They were an object of dread to mariners.[*]

Philippi.

II.—MACEDONIA.

§ 6. The boundaries of **Macedonia**, in the extent it attained subsequent to the reign of Philip, were—in the S. the Ægean and the Cambunian range, separating it from Thessaly; in the W. Mount Lingon and a southerly offset of Scardus, which formed the limits on the side of Epirus and Illyria respectively; in the N. Scardus, between it and Mœsia; and in the E. the river Nestus and Thracia.

---

[*]   Εἶθ᾽ ὥφελ᾽ Ἀργοῦς μὴ διαπτάσθαι σκάφος
Κυάνων ἐν αἶαν, συανέας Συμπληγάδας.—EURIP. Med. I

The surface of the country is mountainous, but there are several extensive and very fertile plains enclosed between the ridges, and well watered by the rivers which traverse them. The sea-coast is remarkably irregular. Among the special sources of wealth of this country we may notice the gold and silver mines on the S. coast.

Name.—The country derived its name from the Macedónes, whose original territory lay in the S.W. of Macedonia between the hills on the W. border and the neighbourhood of Pella. The extension of the power and name of this tribe over the whole of the country was a gradual process, the more marked stages being the advance of the frontier to the Strymon by Perdiccas (454-413 B.C.) and to the Nestus by Philip (359-336).

§ 7. The mountain ranges of Macedonia are connected with Scordus or Scardus, a continuation of Hæmus, which skirts the northern frontier. Three offsets from this range penetrate southwards through the country. The most westerly divides the Strymon from the Nestus under the name of Orbelus, and is prolonged in an offset named Pangæus,[9] *Pirnari*, famed for its mines of gold and silver. A second divides the basins of the Axius and Strymon and was known by the name of Cercine, *Karadagh*, between Pæonia and Mygdonia, and Dysórum, more to the S. near Lake Prasias. The third in the W. was known by the names of Barnus and Bermius, lower down, near the town of Berœa. The central range gives the most prominent feature to the line of the coast by forming the peninsula of Chalcidice, which is enclosed by the Sinus Thermaicus, *B. of Saloniki*, in the W., and the Sinus Strymonicus, *G. of Rendina* in the E., and which terminates towards the S. in the three lesser peninsulas

---

[9] The following are the classical allusions to this mountain : the deity to whose Euripides refers may be either Bacchus or Lycurgus, king of the Edonians, who is said to have been torn to pieces by horses in this mountain :—

παχέως
Δ' ἀμφὶ Παγγαίου θέμεθλα
Ναιετάοντες ἴδαν.   PIND. *Pyth.* iv. 319.

Βόλβην δ' ἔλιπον ὄσσακα, Παγγαῖόν τ' ὄρος
'Ηδωνίδ' αἶαν.   ÆSCH. *Pers.* 494.

Βάκχου προφήτης, ὅς τε Παγγαίου πέτραν
'Θίκησε σεμνὸς τοῖσιν εἰδόσιν θεός.—EURIP. *Rhes.* 969.

Altaque Pangæa, et Rhesi Mavortia tellus.—VIRG. *Georg.* iv. 462.

Video Pangæa nivoris
Cana jugis.   LUC. I. 680.

Πέμπτα γὰρ δὴ ποταμίους διαβρόχς,
Λάσγμετ ἐκλάθην Στρυμάνος φυταλμίοις,
'Οι' ἤλθομεν γῆς χρυσόβωλον εἰς λίπας
Πάγγαιον—   EURIP. *Rhes.* 916.

of **Acte**, **Sithonia**,[1] and **Pallene**,[2] with the intervening bays named **Sin. Singiticus**, and **Sin. Toronaicus**; the extreme points of the peninsulas were named respectively **Nymphæum**, *Hagio Ghiorgki* : **Derrhis**, *Dhrepano*, and **Canastræum**, *Paliuri*. In addition to these we may notice the promontories of **Ampélus**, *Kartali*, in Sithonia ; **Posidium** or **Posidonium**, *Posidhi*, in the S.W. of Pallene ; **Gigonis**, *Apanomi*; and **Ænus**, *Kara-burnu*, on the W. coast of Chalcidice.

Of all the Macedonian mountains, **Athos**, at the extremity of the peninsula of Acte, possesses the highest interest : the whole of the peninsula is rugged and mountainous, and at its southern extremity Athos rises conspicuously to the height of 6350 feet—an insulated cone of white limestone.[3] Off the adjacent promontory the fleet of Macedonia was wrecked in B.C. 492 : to avoid a similar disaster Xerxes cut a canal[4] across the isthmus about 1½ miles S. of Acanthus : the breadth of the isthmus is 2500 yards, and the traces of the canal are still perceptible, though its existence was disbelieved by the ancients.[5] The mountain and peninsula are now named *Monte Santo* from the number of monasteries and chapels on it.

§ 8. The largest river in Macedonia is the **Axius**,[6] *Vardar*, which

---

[1] Sithonia is used by Virgil as a synonym for any northern country with a severe climate ; by Ovid and Horace for Thrace ; their allusions to Bacchus imply the Thracian tendency to drunkenness :

Sithoniasque nives hiemis subramus æquore.—Virg. *Ecl.* i. 66.

Tempus erat, quo sacra silent Trieterica Bacchi
Sithonias celebrare nurus.   Ov. *Met.* vi. 587.

Monet Sithoniis non levis Evius.   Hor. *Carm.* i. 18, 9.

[2] Pallene, or Phlegra, as it was otherwise called, was the fabled scene of the conflict between the gods and the Titans, as well as of that between Hercules and the giant Alcyoneus, which was sometimes placed at the isthmus of Corinth :

ὅταν θεοὶ ἐν πεδίῳ Φλέ-
γρας Γιγάντεσσιν μάχαν
Ἀντιάζωσιν—   Pind. *Nem.* i. 100.

καὶ τὸν βουβόταν οὔρεἴ ἴσον,
Φλέγραισιν εὑρὼν, Ἀλκυονῆ,
Σφετέρας δ᾽ οὐ φείσατο
Χερσὶν βαρυφθόγγοιο νευρᾶς
Ἡρακλέης.   Pind. *Isth.* vi. 47.

[3] Juno is represented as alighting upon it in her journey from Olympus to Lemnos :

Πιερίην δ᾽ ἐπιβᾶσα καὶ Ἠμαθίην ἐρατεινὴν,
Σεύατ᾽ ἐφ᾽ ἱπποπόλων Θρηκῶν ὄρεα νιφόεντα,
Ἀκροτάτας κορυφάς, οὐδὲ χθόνα μάρπτε ποδοῖιν.
Ἐξ Ἄθω δ᾽ ἐπὶ πόντον ἐβήσατο κυμαίνοντα.—Hom. *Il.* xiv. 226.

[4] Cum Medi peperere novum mare, cuinque juventus
Per medium classi barbara navit Athon.—Catull. lxvi. 45.

[5] Velificatus Athos, et quicquid Græcia mendax
Audet in historia.   Juv. x. 174.

[6] The importance of the Axius is well depicted in the following passage :

Αὐτὰρ Πυραίχμης ἄγε Παίονας ἀγκυλοτόξους,
Τηλόθεν ἐξ Ἀμυδῶνος, ἀπ᾽ Ἀξιοῦ εὐρυρέοντος,
Ἀξιοῦ, οὗ κάλλιστον ὕδωρ ἐπικίδναται αἶαν.—Hom. *Il.* ii. 848.

rises in Mount Scardus, and flows towards the S.E. into the Thermaic Gulf, receiving in its course the **Erigon,** *Tzerna,* from the W. The lower course of the Axius has undergone considerable changes. The **Strymon,**[7] *Struma,* is the next in point of importance: it rises in the N.E. and flowing towards the S. and S.E., passes through the Lake of Prasias, and falls into the Strymonic Gulf near the town of Amphipolis: its banks were much frequented by cranes. The **Haliacmon,** *Vistritza,* in the S., is a considerable stream, rising on the border of Epirus, and after a circuitous course to the S.E. and N.E. flowing into the Thermaic Gulf. In early times it received the Lydias[8] from the Lake of Pella as a tributary; but this stream now joins the Axius. There are several large lakes in Macedonia, one of which, **Prasias** or **Cercinitis,** *Tak-hyno,* has been already noticed as being formed by the river Strymon: Herodotus (v. 16) gives an interesting account of its amphibious inhabitants. **Bolbe,**[9] *Besikia,* lies near the Strymonic Gulf, with which it is connected by a channel flowing through the pass of Aulon or Arethusa; it is about 12 miles long, and 7 broad. **Begorritis** was a small lake in Eordæa, probably *Kitrini.*

Μάκαιρ ὦ Πιερία, σέβεταί σ' Εὔιος,
'Ήξει τε χορεύσων
Ἀμα Βακχεύμασι·
Τόν τ' ὠκυρόαν διαβὰς Ἀξιον
Εἱλισσομένας Μαινάδας ἄξει,
Λυδίαν τε τὸν τὰς εὐδαιμονίας
Βροτοῖς ὀλβοδόταν, πατέρα τε
Τὸν ἵκλυον εὔιππον χώραν ὕδασιν
Καλλίστοισι λιπαίνειν.   EURIP. Bacch. 567.

[7] The poetical allusions to the Strymon have reference to its northerly position and the abundance of cranes on its banks.

σήγγουσιν δὲ τὰν
'Ρίθρον ἁγνοῦ Στρυμόνος.   Æsch. Pers. 496.

Τείχεα μὲν καὶ λᾶες ὑπαὶ ῥιπῆς σε πέσωσιν
Στρυμονίου Βορέαο.   CALLIM. Hymn. in Del. 25.

Πνοαὶ δ' ἀπὸ Στρυμόνος μολοῦσαι
κακόσχολοι—   Æsch. Agam. 192.

Quales sub nubibus atris
Strymoniæ dant signa grues, atque æthera tranant
Cum sonitu, fugiuntque Notos clamore secundo.—VIRG. Æn. x. 264.

Nec quæ Strymonio de grege ripa sonat.—MART. ix. 30.

Deseritur Strymon, tepido committere Nilo
Bistonias consuetus aves   Luc. iii. 199.

[8] This river is referred to in the passage quoted above (note [7]).

[9]

Μακεδόνων
Χώραν ἀφικόμεσθ' ἐπ' Ἀξίου πόρον,
Βάλβῃ θ' ἕλειον δόνακα.   Æsch. Pers. 492

Q 2

§ 9. The Macedonians[1] were allied to the Hellenic race, but were not regarded as pure Hellenes[2]: they formed but one element in the population of Macedonia: the rest were either Thracians, as the Pæonians, Pierians, Bottiæans, Edonians, &c., or Illyrians, as the Lyncestians and Eordæans. Greek colonies were planted along the coasts. The Macedonians were regarded by the Greeks as a semi-barbarous people, but it is tolerably certain that they had attained a considerable advance in the arts: their coinage, which is of a remarkably fine character, is evidence of this.[3] The original Macedonia

Coin of Macedonia.

was divided into two parts, Upper and Lower: the former consisting of the western district adjacent to the hills, the latter of the districts about the tributaries as far as Pella. In addition to this, the country was parcelled out into districts named after the various tribes, of which the most important were as follows: **Edonis**[4] between the Strymon and Nestus, occupied by a Thracian tribe; **Bisaltia** between the Strymonic Gulf and Lake Bolbe; **Sintica**, W. of Lake Prasias; **Mygdonia**,[5] between the Axius in the W. and Lake Bolbe in the E., in the peninsula of **Chalcidice**; **Emathia**[6] between the mid-courses

---

[1] The late Latin poets adopted the form Macetæ in lieu of Macedonia, e. g.

Rursus bella volet Macetûm instaurare sub armis.—Sil. Ital. xiii. 878.

Nec te regnator Macetûm nec barbarus unquam.—Stat. Silv. iv. 6, 106.

[2] The language of the Macedonians bore some affinity in its structure to the Æolian dialect, and contained several words that are found in Latin.
[3] The coin represented above exhibits the head of Artemis Tauropolos, and on the reverse the club of Hercules encircled with a garland of oak.
[4]   Non ego sanius
Bacchabor Edonis.   Hor. Carm. ii. 7, 26.

Utque suum Bacchis non sentit saucia vulnus
Dum stupet Edonis exululata jugis.   Ov. Trist. iv. 1, 41.

Nec minus assiduis Edonis fessa chorea.—Propert. i. 3, 5.

Some of the Latin poets altered the quantity of the penultimate:
Edonis ut Pangæa super tricteride mota
Li juga, et inclusum suspirat pectore Bacchum.—Sil. Ital. iv. 778.

[5] The Mygdonians were a Thracian race. The classical allusions to Mygdonia refer not to this country, but to a district in Asia Minor.
[6] In the Homeric age Emathia was restricted to the southern district near the Haliacmon—a country which well deserves the epithet of "lovely;"

Πιερίην δ' ἐπιβᾶσα καὶ Ἠμαθίην ἐρατεινήν.—Il. xiv. 226.

of the Axius and Haliacmon containing the capital, Pella ; **Bottiæa,** a maritime district between the lower courses of the rivers just mentioned ; **Pieria,** a narrow strip of plain between the mouths of the Peneus and Haliacmon, the reputed birth-place of Orpheus and of the Muses, whence the name of Pierides was transferred into Bœotia; **Elimiôtis** in the upper valley of the Haliacmon ; **Orestis** on the borders of Epirus, and occupied by an Epirot tribe ; **Eordæa,** a secluded district between the basins of the Axius and the Haliacmon to the W. of Mount Bermius ; **Lyncestis** [7] in the W. in the southern half of the basin of the Upper Erigon, where the valley of the Bevus lies ; **Pæonia,** in the N. and N.E., whither the Pæônes, who once occupied the whole valley of the Axius [8] withdrew after the Argolic colonization of Emathia ; the principal tribes to the E. were the Odomanti, Astræi, and Agrianes. The Romans at first divided the whole country into four parts in the following manner :—(1) from the Nestus to the Strymon, with Amphipolis as its capital ; (2) from the Strymon to the Axius, with Thessalonica as its capital ; (3) from the Axius to the Peneus, with Pella as its capital ; (4) the mountain district, with Pelagonia as its capital. They afterwards, however, united it with Illyria and Thessaly as one province. Under Constantine it was divided into Prima and Secunda or Salutaris, the former being the coast-district, the latter the interior.

§ 10. The towns of historical importance in Macedonia were, with the exception of the capitals Edessa and Pella, situated either on or adjacent to the sea-coast. Many of them received colonies from

---

[7] It is sometimes called Lyncus by Livy and Thucydides ; the Egnatian Road traversed it, and it was the scene of operations in Sulpicius's campaign against Philip in B.C. 200. Ovid describes a mineral spring in this district, which has been discovered at a place called *Ecclisso Verbeni :*

> Huic fluit effectu dispar Lyncestius amnis,
> Quem quicunque parum moderato gutture traxit,
> Haud aliter titubat quam si mera vina bibisset.—*Met.* xv. 329.

Perseus traversed this district in his march from Citium to Elymis (Liv. xlii. 53).

[8] In the Homeric age they were near the sea coast :

> Αὐτὰρ Πυραίχμης ἄγε Παίονας ἀγκυλοτόξους
> Τηλόθεν ἐξ ᾿Αμυδῶνος, ἀπ᾿ ᾿Αξιοῦ εὐρυρέοντος.—*Il.* ii. 816.

Emathius is frequently used by the Latin poets as an epithet of Alexander ; as in the expressions *Emathii manes* (Stat. *Silv.* iii. 2, 117), *Emathius dux* (Ov. *Trist.* iii. 5, 39), *Emathia acies* (Luc. viii. 531). Elsewhere it is used as a general term for Macedonia, e. g. :—

> Vel aos Emathiis ad Pæonas usque nivosas
> Cedamus campis.   Ov. *Met.* v. 313.

> Bella per Emathios plusquam civilia campos
> Jusque datum sceleri canimus.   Luc. i. 1.

> Nec fuit indignum superis, bis sanguine nostro
> Emathiam et latos Hæmi pinguescere campos.—Virg. *Georg.* i. 491.

Greece : Potidæa, for instance, from Corinth, Mende and Methône from Eretria, Acanthus from Andros, Torône from Eubœa, Amphipolis

Amphipolis.

and Neapolis from Athens, and Olynthus from the Greeks of Chalcidice itself. Therma, the old name of Thessalonica, bespeaks a Greek origin : so also does Crenides, the former name of Philippi ; and Apollonia, which belonged to two towns, one in Mygdonia, the other in Chalcidice. Some of these towns come prominently forward in the Peloponnesian War—particularly Potidæa, Amphipolis, and Acanthus. The coast district of Macedonia was, down to this period, entirely independent of the Macedonian kings, whose seat of power was fixed in the valley of the Axius. After the conclusion of the Peloponnesian War the Chalcidian Greek towns were formed into a confederacy under the presidency of Olynthus, which lasted until B.C. 379. About the middle of the 4th century B.C., Philip succeeded in reducing them to submission. The towns which underwent a change at this period were Potidæa and Therma, which were respectively named Cassandria and Thessalonica. Several of the Macedonian towns flourished under the Romans, particularly those that stood on the Egnatia Via.

I.—*On the Coast from E. to W.* **Philippi** stood near the eastern frontier about ten miles from the sea, and was named after Philip the father of Alexander, by whom the town, formerly called Crenides,

had been enlarged as a border fortress on the side of Thrace.   A stream
named the Gangitas flowed by
it. The town is chiefly famous
for the two great battles *
between Brutus and Cassius
on the one side, Antony and
Octavian on the other, which
were fought on the plain
S. of the town, B.C. 42.
The republican leaders held
a strong position on a couple

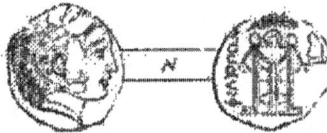

Coin of Philippi.

of hills about 2 miles from the town, with a pass between them: the
triumvirs attacked them from the maritime plains.   Augustus made it
a colony, with the name Col.
Jul. Aug. Philip. Neapolis,
Kavallo, which served as
the port of Philippi, was
probably the same place as
the earlier Datum, which
was originally a colony of
Thasos, and afterwards occupied by Athenian settlers,
who gave it the name of
Neapolis: a range of hills
intervenes between it and
Philippi. Amphipolis stood
on an eminence on the E.
bank of the Strymon about
3 miles from the sea, where
Eïon served as its port: it
derived its name from being
almost surrounded by the
river.   Its position was an
important one, as commanding the only easy communication between Greece and
Thrace: several roads met
here, whence its name of

Plan of the Neighbourhood of Amphipolis.

1. Site of Amphipolis.   polis; the three marks
2. Site of Eïon.   across indicate the gate.
3. Long Wall of Amphi-   4. Lake Cercinitis.

Ennea Hodoi "nine ways": attempts were made to colonize it by Aris-

---

* Many Roman writers describe this battle as fought on the same ground as
Pharsalia:—

                    Pharsalia sentiet illum
        Emathiaque iterum madefacti caede Philippi.—Ov. *Met.* xv. 823.
        Ante novæ venient acies, scelerique secundo
        Præstabis nondum siccos hoc sanguine campos.—Luc. vii. 858.
        Thessaliæ campis Octavius abstulit udo
        Cædibus assiduis gladio.   Juv. viii. 242.
The mistake may have originated in the ambiguity of Virgil's lines:—
        Ergo inter sese paribus concurrere telis
        Romanas acies iterum videre Philippi.   *Georg.* i. 489.
The poet Horace was present at this battle, as he himself tells us:—
        Tecum Philippos et celerem fugam
        Sensi, relicta non bene parmula.   *Carm.* ii. 7, 9.
Lucan takes considerable licence when he describes Philippi as close to Hæmus:—
        Latosque Hæmi sub rupe Philippos.   i. 680.

tagoras of Miletus in B.C. 497, and by the Athenians in 465 ; these faile l,
but a second trial by the Athenians in 437 was successful. It soon

Coin of Amphipolis.

became an important
town : it was captured
by Brasidas in 424, and,
in spite of the attempt
to recover it by the
Athenians under Cleon
in 422, it remained inde-
pendent of them. Philip
of Macedon took it in
358, and it remained at-
tached to Macedonia un-
til 168, when the Romans
made it a free city. A few remains still exist at *Neokhorio*. Olynthus
was favourably situated in a fertile plain at the head of the Toronaic
Gulf, between the peninsulas of Pellene and Sithonia. Originally a
Bottiæan town, it passed at the time of the Persian invasion into the
hands of the Chalcidian Greeks. From its maritime position it became
an important place, and, under the early Macedonian kings, the head
of a powerful confederacy, which was, after a long contest, dissolved by
Sparta in B.C. 379. The growing power of the Macedonian kings
brought Olynthus into alliance with Athens in 352, but the town fell
through treachery into the hands of Philip, and was utterly destroyed
in 347. A few vestiges mark its site at *Aio Mamas*. Potidæa, *Pinaka*,
originally a Dorian city colonized from Corinth, stood on the isthmus
of the peninsula of Pellene. It yielded to the Persians on their march
into Greece, but after the battle of Salamis resisted them, and was unsuccessfully besieged by them. It then attached itself to Athens, and,
having afterwards revolted, was taken after a two years' siege in B.C. 429.
Having passed into the hands of the Olynthians in 384, of the Athenians
in 364, and of Philip[1] who gave the land back again to the Olynthians
but destroyed the town, it was at length rebuilt by Cassander with the
name of Cassandria, and peopled with the Olynthians and others : it
then became one of the most important towns of Macedonia. Its
occupation by the tyrant Apollodorus about 279, and its unsuccessful
siege by the Romans in 169, are the chief events of its later history.
Thessalonica stood at the head of the Thermaic Gulf, partly on the

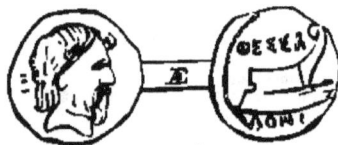

Coin of Thessalonica.

level shore, partly on the
slope of a hill. From
its admirable position in
relation to the valley of
the Axius in the W. and
that of the Strymon in
the E., and also from
its possessing a good
port, it was and still is
(as *Saloniki*) the most
important commercial
town of this district.
Its original name was Thorma, from the hot springs about it : this was
changed to Thessalonica, probably by Cassander, who rebuilt it in B.C.
315, and named it after his wife or daughter. Its early history is unim-

---

[1] Callidus emptor Olynthi.—Juv. xii. 47.

portant. Xerxes rested here in his invasion of Greece : the Athenians
occupied it in B.C. 421, but resigned it to Perdiccas in 419. Under
the Romans it became the metropolis of Macedonia, and from its
central position, "*posita in gremio imperii nostri*," as Cicero says, it
was the chief town between the Adriatic and Euxine seas. Cicero
visited it several times: it was made a free town after the second Civil
War, and was governed by six supreme magistrates. The Via Egnatia
intersected the town from E. to W., and two arches still exist at each
entrance, the western supposed to commemorate the battle of Philippi,
the eastern the victory of Constantine either over Licinius or over the
Sarmatians. **Methône** was a Greek colony of Eretria, situated about
2 miles from the W. coast of the Thermaic Gulf: it was occupied by
the Athenians in their war with Perdiccas, and remained in their hands
until B.C. 353, when it was taken and destroyed by Philip. **Pydna**
was originally built on the coast of the Thermaic Gulf, but having
been taken in B.C. 411 by Archelaus, it was removed to a distance of
about 2½ miles from the sea. It afterwards fell into the power of
Athens, but was betrayed to Philip in 356. The place is chiefly famous
for the great battle between Perseus and Æmilius Paullus in 168,
which sealed the fate of the Macedonian monarchy: two *tumuli* near
Ayan probably mark the scene of the engagement. **Dium**, though not
a large town, was valuable from its position near the W. coast of the
Thermaic Gulf, commanding the coast-road into Thessaly. In the
Social War it was almost destroyed by the Ætolians, but it recovered,
and was occupied by Perseus in B.C. 169 : it afterwards became a
Roman colony. The remains of a stadium and theatre still exist near
Malathria : the town was adorned with numerous works of art, particularly Lysippus's group of the 25 chieftains who fell at the Granicus,
which was placed here by Alexander, and was afterwards transferred
to Rome.

Of the less important towns we may notice :—**Œsyme**, a colony from
Thasos in Pieria, on the coast of the Strymonic Bay. **Phagres**, *Orfana*,
a fortress on the same coast S.E. of Amphipolis. **Eïon**, the port of
Amphipolis at the mouth of the Strymon, the spot where Xerxes
sailed for Asia; it was taken by Cimon in the Persian War, and besieged
by Brasidas in the Peloponnesian War. **Myrcinus**, on Lake Prasias,
N. of Amphipolis; it was selected by Histiæus of Miletus for his settlement, and was the place whither Aristagoras retired. **Siris** or **Serrhæ**
in Odomantice, in the widest part of the great Strymonic plain, visited
by Xerxes in his retreat from Greece, and by P. Æmilius Paulus after
his victory at Pydna. **Argilus**, in Bisaltia, W. of Amphipolis. **Heraclea
Sintica**, *Zervokhori*, somewhat W. of Lake Prasias, the place where
Demetrius, son of Philip
V. was murdered. **Apol-
lonia**, *Pollina*, in Mygdonia, S. of Lake Bolbe.
Stagira, the birth-place
of Aristotle, on the
shore of the Strymonic
Gulf. **Acanthus**, lower
down the coast, cap-
tured by Brasidas in
B.C. 424, and by the Ro-
mans in 200. **Apollonia**,

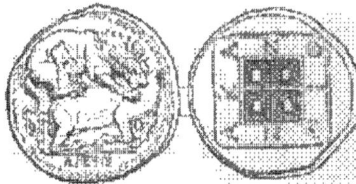

Coin of Acanthus.

Polighero, the chief town of Chalcidice, N. of Olynthus. **Olophyxus**

Charadria, and Acrothol, on the E. coast of the peninsula of Acte; and lastly, Petra, a fortress among the mountains of the S. frontier, commanding a pass which led to Pythium in Thessaly by the back of Olympus; Scipio Nasica here defeated the forces of Perseus, and opened the way for L. Æmilius Paulus.

11. *In the Interior.* Pella, the later capital of Macedonia, stood on a hill, surrounded by marshes, named Borbaros, through which there was communication with the sea by means of the river Lydias. As the metropolis of Philip, and the birth-place of Alexander the Great,[3] it rose from an insignificant town of the Bottiæans to be a place of worldwide renown. Having been the royal residence of all the Macedonian kings except Cassander, it became under the Romans a colony and station on the Egnatian Road. There are remains at *Neokhori*, where a fountain still retains the name of *Pel*. Ægæ or Edessa, the earlier capital of Macedonia, stood N.W. of Pella, at the entrance of a pass, which connected Upper and Lower Macedonia. Philip was murdered here in B.C. 336. After the seat of power was removed, it still remained the hearth of the Macedonian race, and the burial-place of their kings; the tombs were rifled by the Gallic mercenaries in the employ of Perseus. The remains at *Vodhena* are but trifling. Berœa, *Verria*, stood on a branch of the Haliacmon, S.W. of Pella: it was unsuccessfully attacked by the Athenians under Callias in B.C. 432, on their march from Pydna to Therma; it surrendered to the Romans after the battle of Pydna. A portion of the old walls and other remains still exist. Heraclea, the chief town of Upper Macedonia, was surnamed Lyncestis from the district in which it stood: it was on the Egnatian Road, and at the base of the Candavian mountains. Stobi in Pæonia stood on the Erigon, and was a place of some importance under the Macedonian kings: the Romans made the place a depôt of salt. It was the later capital of Macedonia Salutaris. Scupi was the frontier town on the border of Illyricum, in the N.W. of Pæonia.

Of the lesser towns we may notice—Petra, a fortress of the Mædi; Dobērus, at the S. foot of Cercine, in a lateral valley of the Axius; Eurōpus, in Emathia, between Idomene and the plains of Cyrrhus and Pella, on the right bank of the Axius; Physeus, Begorra, and Galadræ in Eordæa, the first alone possessing any historical interest; Celetrum, *Kastoria*, in Orestis, on a peninsula surrounded by the waters of a small lake; it was taken by Sulpicius in B.C. 200; Astræum, in Pæonia, on a tributary of the Strymon; Stymbāra on

---

[3] Pelloeus is a frequent epithet of Alexander :

    Unus Pellæo juveni non sufficit orbis.—Juv. x. 168.

    Hos habuit numen Pellæi mensa tyranni.—Mart. ix. 44.

Sometimes it is used as an equivalent for Macedonian :

    Ergo in Thessalicis Pellæo fecimus arvis
    Jus gladio?   Luc. ix. 1073.

Sometimes it refers to Alexandria in Egypt, or to Egypt generally :

    Non ego Pellæas arces, adytisque retectum
    Corpus Alexandri pigra Marcotide mergam?—Luc. ix. 153.

    Nam qua Pellæi gens fortunata Canopi.—Virg. *Georg.* iv. 287.

Hence the title is transferred even to the Ptolemies :

    Pellæusque puer gladio tibi colla recidit,
    Magne, tuo.   Luc. viii. 607.

the upper course of the Erigon where Sulpicius encamped in B.C. 400); Bylazora, the greatest city of Pæonia, near the passes leading into Mœsia.

Roads.—Macedonia was traversed by the **Via Egnatia**, which entered it on the side of Illyricum at Heraclea, and thence passed by Edessa and Pella to Thessalonica, and across Chalcidice by Apollonia to Amphipolis. This road appears to have been constructed shortly after the reduction of Macedonia by the Romans in B.C. 168. From this, roads diverged in different directions, leading—(1) from Thessalonica along the coast to Tempe in Thessaly ; (2) from Pella through Berœa to the same spot, falling into the coast-road at Dium ; (3) from Heraclea Lyncestis to Stobi ; (4) from Thessalonica to Stobi ; (5) from Stobi to Scopi in the N.W., and (6) from Stobi to Serdica in the N.E.

St. Paul's Travels.—Macedonia was first visited by St. Paul in his second apostolical journey. Starting from Troas he crossed the Ægean by Samothrace to Neapolis, and thence to Philippi "the first city" of that part of Macedonia on the side of Thrace. From Philippi he followed the Egnatian Road through Amphipolis and Apollonia to Thessalonica, where at the suit of Jason he was brought before the " politarchs," as the governors of that free city were styled. From Thessalonica he journeyed to Berœa, where he remained a short time; thence he descended to the sea-coast probably at Dium, and took ship for Athens (Acts xvi. 11, xvii. 15). In his third journey he again visited Macedonia (Acts xx. 1-2), approaching it from Troas (2 Cor. ii. 12), and staying at Philippi, where he was joined by Titus (2 Cor. vii. 5). From Philippi he went "round about unto Illyricum " (Rom. xv. 19) ; but whether by that expression we are to infer that he actually crossed the mountains into that country, is uncertain. His route is quite unknown, and we only know that he next visited Greece. He shortly after returned by the same route, crossing from Neapolis to Troas (Acts xx. 3-6). He addressed two epistles to the church at Thessalonica, and one to the church at Philippi.

History—The earliest Macedonian dynasty claimed a descent from the Temenidæ of Argos and called themselves Heracleids. The first kings of whom we have any special notice were Amyntas (about 520-500 B.C.) and Alexander (about 480), who was contemporary with Xerxes. The capital at this period was Edessa : Alexander and Perdiccas extended their territory to the Strymon, and the latter became the active enemy of Athens. After the death of Archelaus, the son of Perdiccas, in 399, a long period of anarchy succeeded until the accession of Philip in 359, who reduced Olynthus, and advanced his frontier to the Nestus. Under his son, Alexander the Great, Macedonia became the seat of an empire which extended over the whole eastern world. After the death of Alexander, the throne of Macedonia was for a long time an object of constant contention. Cassander first had the title of king; his sons were displaced by Demetrius, son of Antigonus, in 294. Pyrrhus, of Epirus, followed in 287, and after 7 months Lysimachus of Thrace gained the power. After his death in 281 a period of anarchy followed, during which the Gauls invaded the country from 280 to 278. At length, in 278, Antigonus Gonatas obtained a firm seat on the throne, and founded a dynasty which lasted until the conquest of Macedonia by the Romans in B.C. 168. Of this dynasty the kings Demetrius II. and Antigonus II. are known for the part they took in the affairs of Greece. Philip V. first came into contact with the Romans; he was defeated at Cynoscephalæ ; and Perseus the last king, at Pydna

Mounts Olympus and Ossa

CHAPTER XIX.

NORTHERN GREECE—THESSALY AND EPIRUS.

§ 1. The peninsula of Greece, the most easterly of the southern
projections of the continent of Europe, was bounded on the N. by
Macedonia and Illyria, and in all other directions by seas, viz.: by the
Ægæan and Cretan on the E., the Libyan on the S., and the Ionian
on the W. The northern boundary was clearly defined by a chain
of mountains extending from the Ægæan to the Ionian Sea; the
most important links in this chain were Olympus and Cambunii
Montes in the E., Lacmon in the centre, and the Ceraunian range
in the W. The extreme length of the country was about 250
miles, and its extreme breadth from the coast of Acarnania to that
of Attica about 180 miles. Its area was considerably less than that
of *Portugal.*

Names.—The Greeks themselves possessed no general geographical
designation for the land in which they lived. The term **Hellas,** which
approaches most nearly to such a designation, was of an *ethnological*

rather than of a geographical character. It described the abode of
the Hellenic race, wherever that might be, and thus while in the
Homeric age it was restricted to a small district in the south of Thessaly, Herodotus (ii. 182, iii. 136, vii. 157) and Thucydides (i. 12)
extend it beyond the limits of Greece proper to Cyrene in Africa,
Syracuse in Sicily, and Tarentum in Italy, as being Hellenic colonies.
Within the limits of Greece, Hellas proper was restricted to that
portion which lay between the Corinthian Gulf on the S. and the
Ambracian Gulf and the Peneus on the N. Epirus was excluded from
it as not being occupied by Hellenes, and Peloponnesus as having
its own distinctive title. The latter was, however, sometimes included in Hellas, as it had an Hellenic population. Sometimes the
Greek islands were included on a similar ground; and after the spread
of the Hellenic language consequent upon the Macedonian conquest
of Hellas, even Macedonia and Illyria were included. The Romans,
and ourselves in imitation of them, gave the name of Græcia to the
country. The origin of this is uncertain: the Græci are only once
noticed by a Greek writer (Aristot. *Meteor.* i. 14) as a tribe living
about Dodona in Epirus. It has been surmised that the name was
extensively applied to the tribes on the W. coast of Epirus, and thence
spread to the E. coast of Italy, where the Romans first came in contact with the Hellenic race. The name of Græcia was superseded by
that of Achaia as the official title of the country after its conquest by
the Romans.

§ 2. The position and physical characteristics of the peninsula of
Greece were highly favourable to the promotion of early settlement.
As the tide of population flowed westward from Asia, it was guided
to the shores of Greece by the islands which stud the Ægæan Sea.
There it met with a country singularly adapted to its requirements
—an extensive line of coast, broken up into innumerable bays and
inlets, and well furnished with natural harbours; a land protected
by its insular character from sweeping invasion, and subdivided
into a number of separate and sequestered districts, which nature
protected by her mountain barriers; a climate reputed in ancient
times the most healthy and temperate in the world; a bright clear
air; a soil fertile and varied in its productions, producing wheat,
barley, flax, wine, and oil; mountains, whose sides were clothed
with forests, whose uplands supplied rich pasturage for cattle, and
from whose bowels abundance of excellent limestone might be
obtained for building purposes. And when, under these fostering
influences, the population of Greece outgrew the narrow limits of
the land, there was no difficulty in finding settlements, which, under
equally favourable circumstances, gave back power and wealth to
the mother country: in one direction Sicily and Southern Italy, in
another the northern coast of Africa, were near at hand and open to
colonization, while in a third the tide flowed back to the coast of
Asia Minor, and thence ramified to the distant shores of the
Euxine.

§ 3. The mountain chains of Greece are marked with great distinctness. We have already had occasion to notice the series of mountains which divide Greece from Macedonia. **Lacmon** is the connecting link between the **Cambunii Montes** on the E., **Pindus** in the S., **Tymphe** in the W., and the mountains of Macedonia in the N. The Cambunii Montes form the northern limits of Thessaly, and terminate in the far-famed heights of **Olympus**, near the Ægæan Sea.

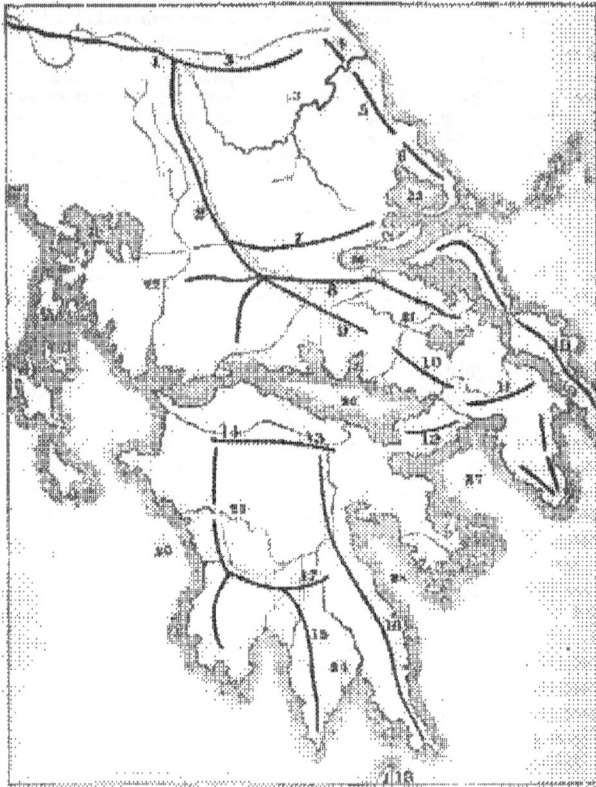

Map of Greece, showing the direction of the Mountain Ranges.

1. Lacmon.	9. Parnassus.	17. Parthenius.	25. Sinus Pagasæus.
2. Pindus.	10. Helicon.	18. Cythera.	26. Sinus Maliacus.
3. Cambunii Mts.	11. Cithæron.	19. Eubœa.	27. Sinus Saronicus.
4. Olympus.	12. Geranea.	20. River Peneus.	28. Sinus Argolicus.
5. Ossa.	13. Cyllene.	21. River Cephissus.	29. Sinus Cyparissus.
6. Pelion.	14. Erymanthus.	22. River Achelous.	30. Sinus Corinthiacus.
7. Othrys.	15. Taygetus.	23. River Alpheus.	31. Sinus Ambracius.
8. Œta.	16. Parnon.	24. River Eurotas.	

Tymphe is continued westward in the ranges which bound Epirus
on the N., and which terminate in the striking promontory of
Acro-ceraunia on the shores of the Ionian Sea. Pindus may be
termed the *backbone* of Greece: it emanates from the northern
range just mid-way between the Ægæan and Ionian Seas in about
40° N. lat., and descends in an unbroken course towards the S.E.
for sixty miles, to about 89°, where it terminates in Tymphrestus.
From this point the central chain divides into five branches, one of
which, named Othrys, takes a due E. direction, skirting the shores
of the Maliac Gulf; a second, Œta, goes off towards the S.E., in a
line parallel to the coast of the Eubœan Sea, assuming, in different
parts, the names of Cnemis. Ptoon, and Teumessus; a third retains
the direction of the parent chain, and assumes the well-known names
of Parnassus, Helicon, Cithæron, and Parnes; a fourth strikes off
towards the S.W., under the name of Corax and Taphiassus, and terminates in the promontory of Antirrhium, on the shores of the
Corinthian Gulf; lastly, a fifth diverges more to the N., and under
the name of Agræi Montes, penetrates to the shores of the Ambracian Gulf. We have yet to notice in Northern Greece a chain
which forms the E. boundary of Thessaly, connecting Olympus and
Othrys, and which contains the well-known heights of Ossa and
Pelion, and terminates in the promontory of Sepias. Southwards
the central range may be traced between the Corinthian and Saronic
Gulfs in the heights of Geranea and Onea, which join Northern
Greece and Peloponnesus. The mountain system of Peloponnesus
presents some interesting points of contrast to that of Northern
Greece. Instead of having a backbone-ridge (like Pindus), Peloponnesus consists of a central region of a quadrangular form, bounded
on all sides by lofty chains. The northern barrier of this rocky
heart is formed by the lofty mountains of Cyllēne in the E.,
and Erymanthus in the W., the Aroanii Montes filling up the interval. The eastern boundary is formed by Artemisium and Parthenium.
The southern and western walls are not so distinctly marked, but
the angle at which they meet is marked by the lofty chain of
Lycæus. The eastern and western walls are continued towards the
S. in the ranges of Parnon and Taygetus, which may be traced down
to the promontories of Malea and Tænarium.

§ 4. The river system of Northern Greece is regulated by that of
the mountains. It may be observed that there are two well-defined
basins in Northern Greece, one of which, Thessaly, is enclosed
between the ranges of Pindus on the W., Olympus on the N., Ossa
and Pelion on the E., and Othrys on the S.; the other is the triangular space enclosed between Œta, Parnassus, and Helicon, and
containing the provinces of Doris, Phocis, and Bœotia. The northern
basin is drained by the Penëus, which escapes through the only

outlet afforded through the mountain wall, viz. the Vale of Tempe : in the southern basin no such outlet exists, and the waters of the Cephissus collect in the lake Copäis, whence they were carried off by subterraneous channels, partly of natural, partly of artificial formation. The western district was drained by the Achelöus, which, rising not far from the Peneus, in the northern extremity of Pindus, flows southwards into the Ionian Sea, after a course of 130 miles, receiving numerous tributaries from either side. The other rivers of Northern Greece will be noticed in the account of the provinces through which they flow. Between the northern and southern basins the Sperchēus receives the waters that collect between Othrys and Œta, and after a course of sixty miles through a beautiful and fertile valley, falls into the Lamiac Gulf. The only rivers of importance in Peloponnesus are—the Alphēus, which drains the central mountain district in a westerly course ; and the Eurōtas, which drains the broad valley lying between Parnon and Taygetus.

§ 5. The coast-line of Greece is singularly extensive, compared with the area of the country. While the latter is less than *Portugal*, the length of its coast exceeds that of *Spain* and *Portugal* together. This is, of course, owing to its extreme irregularity. Commencing our review in the N.E., we find the line regular and unbroken down to the promontory of Sepias. Westward of that point the sea makes an incursion into the Thessalian plain, finding a narrow entrance between the ranges of Othrys and Pelion, and then opening into an extensive sheet of water, known as the Pagasæus Sinus. *G. of Volo.* From the entrance of this gulf it proceeds westward, in the opening afforded by the divergence of Othrys and Œta, and terminates in the Maliacus Sin., *G. of Zeitun.* Thenceforward it resumes its original direction, and with numerous sinuosities follows the line of Œta and its continuation as far as Parnes, from which point it takes a due southerly direction to Sunium. The Saronicus Sin., *G. of Egina,* intervenes between the peninsulas of Attica and Argolis, and the Argolicus Sin., *G. of Napoli di Romania,* between Argolis and Laconia. The southern coast is broken by the bold projections of Malea and Tænarium, bounding the Laconicus Sin., *G. of Kolocythia,* and by the lesser promonotory of Acritas, in the W., enclosing with Tænarium the Messeniacus Sin. These bays give the resemblance to the leaf of the plane-tree, or vine, which was noticed by the ancients. The western coast of Peloponnesus is varied by a large but not deep indenture, named Cyparissius Sin. The Corinthiacus Sin., *G. of Lepanto,* shortly after follows, at first broad, then narrowed by the promontories of Rhium and Antirrhium to a strait, and then expanding to a landlocked sheet, which resembles a lake rather than an arm of the sea: its N. coast is broken by the bays of Crissa and Anticyra ; the S. coast is more regular, until it approaches the E.

extremity, where it is divided by the projections of the Geranean range. The Corinthian Gulf on the W. coast of Greece is met by the Saronic on the E., and the two are separated by a very narrow isthmus of low land to the S. of the Geranean range. The W. coast of Northern Greece is regular, the only interruption in the line of coast being the Ambracius Sin., G. of Arta, a landlocked sheet of water, approached by a narrow passage guarded by the promontory of Actium. The promontory of Acro-ceraunia, on the frontier of Illyricum, completes our review of the coast.

§ 6. The original population of Greece belonged to a stock which we have named Graeco-Latin, as being found equally in the peninsulas of Greece and Italy. In Greece this common element was described under the name of Pelasgi—a name which had almost passed away in the historical age, and which was supposed by the Greeks themselves to indicate an aboriginal population of great antiquity.[1] The later inhabitants of Greece were named Hellenes, and some doubt still exists as to the relation that existed between them and the Pelasgi. Most probably they belonged to the same stock, though of a superior character and standing. In this case we may regard the names as indicating different eras of civilization. The foreign settlements were unimportant: doubt exists as to the Egyptian colonies said to have been planted in Greece under Cecrops in Attica and under Danaus in Argolis, but there can be little question that the Phœnicians settled at Thebes in Bœotia. The abodes of the Pelasgi and Hellenes varied at different periods, and deserve special notice in consequence of their importance in the political divisions of Greece.

(1.) *The Pelasgi.*—The Pelasgi were an agricultural race, and selected the fertile plains for their original abodes. On these they erected walled towns for their protection. They left indications of their presence in the names Argos (= "plain") and Larissa (= "a fortified town"), and in the massive masonry with which they surrounded their towns. Hence we may assume that the Pelasgians lived in the following districts:—Thessaly, which Homer calls "Pelasgic Argos;"[2] the district of Argolis, which he calls "Achaean Argos," or simply "Argos;"[3] and in Peloponnesus generally, which he calls "Mid-Argos,"[4] meaning *the whole breadth* of Argos—particularly the western part, which he terms "Iasian Argos."[5] In the Homeric age branches of the Pelasgian race were known by special names, such as the Arcadians in central Peloponnesus, the Caucones in Elis, the Dolopians

---

[1] Τοῦ γηγενοῦς γάρ εἰμ' ἐγὼ Παλαίχθονος
'Ίνις Πελασγὸν, τῆσδε γῆς ἀρχηγέτης.   Æsch. Suppl. 251

[2] Νῦν δ' αὖ τοὺς, ὅσσοι τὸ Πελασγικὸν Ἄργος ἔναιον.   Il. II. 681.

[3] Ἡμετέρῳ ἐνὶ οἴκῳ, ἐν Ἄργεϊ, τηλόθι πάτρης.   Il. I. 30.

[4] Ἀνέρας, τοῦ κλέος εὐρὺ καθ' Ἑλλάδα καὶ μέσον Ἄργος.—Od. I. 344.

[5] Εἰ πάντες σε ἴδοιεν ἀν' Ἴασον Ἄργος Ἀχαιοί.   Od. xviii. 246.

on the southern borders of Thessaly and Epirus, and the Perrhæbi in northern Thessaly.

(2.) *The Hellenes.*—The Hellenes are noticed by Homer as the Selli,[6] who took care of the oracle of Dodona, as Hellenes[7] In conjunction with the Myrmidones and Achæans, and as Panhellenes[8] in conjunction with Achæans—the latter implying that there were several tribes of Hellenes. Hellas, the residence of the Hellenes, is variously applied by Homer to a district of some size adjacent to Phthia, in a wider sense as including the whole district south of Thessaly to the Corinthian Gulf, and in a wider sense still as descriptive of the whole of Northern Greece in opposition to Mid-Argos or Peloponnesus.[9] The Hellenic race was divided by the Greeks into four large clans—the Dorians, Æolians, Ionians, and Achæans. These migrated from their original seat in the S. of Thessaly, and were dispersed in the following manner in the Heroic or Homeric age:—the Achæans in the original Hellas and in the S. and E. parts of Peloponnesus; the Ionians along the S. shore of the Corinthian Gulf and in Attica; the Dorians in a small mountain district between Thessaly and Phocis; and the Æolians in the centre of Thessaly, in Locris, in Ætolia, and on the W. side of the Peloponnesus, where they were named Epeans. The Minyans were a powerful race, scattered over the peninsula, whose origin is uncertain. By some they are regarded as a branch of the Æolians: their settlements were about the head of the Pagasæan Gulf in Thessaly, in the centre of Bœotia, and about Pylos in western Peloponnesus.

(3.) The first change that took place in this disposition of the Hellenic race occurred in northern Greece through the irruption of the Thessalians, who, crossing over from Epirus into the rich plain of the Peneus, dispossessed the Ætolian Bœotians. These, retiring southwards, settled in the fertile province named after them, where they in turn dispossessed the Minyans and other occupants. The date assigned to these occurrences by the Greeks was B.C. 1124.

(4.) The second and more important change was supposed to have occurred B.C. 1104, but appears really to have happened much later. We refer to the immigration of the Doric race into Peloponnesus under the Heracleids. They crossed the mouth of the Corinthian Gulf in conjunction with the Ætolians, and ejected the Achæans from the southern and eastern districts of Argolis, Laconia, and Messenia. The Achæans retired to the shore of the Corinthian Gulf and permanently occupied the province named after them; the Ionians were obliged to withdraw from this district to Attica; while the Ætolians seized the territory of the Epeans, and occupied it under the name of Elis. Corinth is said to have held out for about thirty years against the Dorian arms. The Æolians were then expelled from it, and took refuge among their emigrant compatriots.

---

[6] Δωδώνης μεδέων δυσχειμέρου· ἀμφὶ δὲ Σελλοι
Σοὶ ναίουσ' ὑποφῆται, ἀνιπτόποδες, χαμαιεῦναι.   *Il.* xvi. 231.

[7] Ἐγχείῃ δ' ἐκέκαστο Πανέλληνας καὶ Ἀχαιούς.   *Il.* ii. 530.

[8] Φεύγον ἔπειτ' ἀπάνευθε δι' Ἑλλάδος εὐρυχόροιο,
Φθίην δ' ἐξικόμην ἐριβώλακα.   *Il.* ix. 474

[9] See above, note [6].

§ 7. The political divisions of Greece were regulated almost entirely by the natural features of the country. The northern basin was named Thessaly, which included also the vale of the Sperchous and the mountainous region to the E. of the basin. Epirus was the corresponding district on the other side of Pindus, extending southwards to the Ambracian Gulf. The southern basin included Bœotia, the greater part of Phocis, and the little state of Doris, which lay at the head of the valley of the Cephissus. Between Œta and the Eubœan Sea lived the Locri Epicnemidii and Opuntii. Locris occupied the triangular district between Parnassus and Corax and the Corinthian Gulf. Then followed Ætolia and Acarnania, divided from each other by the Achelous. Attica was the triangular peninsula S. of Bœotia, and Megaris occupied the isthmus. In Peloponnesus the central mountain district was named Arcadia; N. of this was Achaia and the adjacent territories of Sicyonia, Phliasia, and Corinthia; S. of it Laconia and Messenia, divided from each other by Taygetus; W. of it Elis; and E. of it Argolis, occupying the eastern peninsula.

I.—THESSALIA.

§ 8. The boundaries of Thessalia, in its widest extent, were—the Cambunii Montes and Olympus on the N., Pindus on the W., the Ægæan on the E., and the Malian Gulf and Œta on the S. Within these limits were included *Thessaly Proper* (*i. e.* the plain enclosed between the mountain ranges of Pindus, Olympus, and Othrys) and the outlying districts of Magnesia in the E., Malis in the S.E., and Dolopia and Œtæa in the S.W. The most striking feature in the general aspect of Thessaly is the great central plain which spreads out between the lofty mountain barriers surrounding it, justifying by its appearance the opinion of the ancients that it had once been a vast lake, whose waters at length forced for themselves an outlet by the narrow vale of Tempe. This plain is divided into two parts by a range of inferior heights running parallel to the left bank of the Enipeus; these were named the "Upper" and "Lower" plains, the first being the one nearest Pindus. The rich alluvial soil of this plain produced a large quantity of corn and cattle, which supplied wealth to a powerful and luxurious aristocracy. The horses were reputed the finest in Greece,[1] and hence the cavalry of Thessaly was very efficient.

§ 9. The mountains of Thessaly rank among the most famous, not only of Greece but of the whole ancient world. Olympus towers to the height of nearly 10,000 feet in the N.E. angle of the province, and presents a magnificent appearance from all sides. Its lower sides

---

[1] Hence the horse is the usual device on the coins of Thessaly.

are well wooded, but the summit is a mass of bare light-coloured rock, and is covered with snow for the greater part of the year. Below its summit is a belt of broken ridges and precipices. Olympus was the reputed abode of Zeus and the other gods.[1] A road crossed its southern slopes between Heracleum and Gonnus, by means of which the narrow pass of Tempe might be avoided. Xerxes followed this mountain road, as also did the Romans under App. Claudius in B.C. 191. The Cambunii Mts., which form the barrier between Macedonia and Greece, were surmounted by a route following the course of the Titaresius from the S. This route bifurcated before crossing the mountain, and led either by the Volustana Pass to Phylace, or by a more easterly route to Petra and the sea-coast. To the S. of Olympus, and separated from it by the narrow vale of Tempe, rises Ossa, with a conical peak about 5000 feet high. The ancients supposed that Ossa and Olympus were once united, but were severed either by an earthquake or by the arm of Hercules.[2] This mountain figures, along with Olympus and Pelion, in the description of the war of the giants against the gods.[3] Pelion is a long ridge extending from Ossa southwards to the promontory of Sepias. On its eastern side it rises almost precipitously from the sea, and allows no harbours along this part of the coast.[4] It is still covered with exten-

---

[1] The epithets which Homer applies to this mountain refer to its *height* (αἰπύς, and more commonly μακρός), its *size* (μέγας), its *many ridges* (πολυδειράς), its *declivitous* (πολύπτυχος), its *snowy* top (ἀγάννιφος and νιφόεις), and its *brilliancy*, as the abode of the gods (αἰγλήεις). The passages in which the name occurs are too numerous for quotation. The wooded sides of the mountain are referred to by Virgil, in the epithet *frondosum* (see below, note [4]), and by Euripides in the following passage, where he speaks of the "leafy retreats" in which Orpheus played :—

Τάχα δ' ἐν τοῖς πολυδένδρεσ-
σιν Ὀλύμπου θαλάμοις, ἔν-
θα ποτ' Ὀρφεὺς κιθαρίζων
ξύναγεν δένδρεα Μούσαις,
ξύναγεν θῆρας ἀγρότας.—*Bacch.* 560

Postquam diversit Olympo
Hercules gravis Ossa manu, subitæque ruinam
Sensit aquæ Nereus.   Lec. vi. 347.

Dissiluit gelido vertex Ossæus Olympo ;
Carceribus laxantur aquæ, fractoque meatu
Redduntur fluviusque mari, tellusque coloniæ.
                              Claud. *Rapt. Proserp.* ii. 183.

[3] Οἵ ῥα καὶ ἀθανάτοισιν ἀπειλήτην ἐν Ὀλύμπῳ
Φυλόπιδα στήσειν πολυάϊκος πολέμοιο·
Ὄσσαν ἐπ' Οὐλύμπῳ μέμασαν θέμεν, αὐτὰρ ἐπ' Ὄσσῃ
Πήλιον εἰνοσίφυλλον, ἵν' οὐρανὸς ἀμβατὸς εἴη.—*Od.* xi. 312.

Ter sunt conati imponere Pelio Ossam
Scilicet, atque Ossæ frondosum involvere Olympum.—*Georg.* i. 281.

[4] ἀπὸ ἀλίμενου Πηλίου.—Eurip. *Alc.* 596.

sive forests.[6] Othrys, in the S., is again a lofty and well-wooded range, but not invested with so many interesting associations as the mountains already described.[7] Two routes led across it to Lamia; the most westerly starting from Thaumaci and Pharsalus, the other from Thebæ on the Pagasæan Gulf. Pindus, in the W., is an extensive range,[8] forming the watershed between the basins of the Peneus and the Achelous. The southern part of the range was named Cercetium. It was crossed at two points—by a northern road which followed up the valley of the Peneus, and descended on the W. side by that of the Arachthus to Dodona and Passaron; and by a southern road which led from Gomphi in Thessaly to Argithea, and thence to Ambracia; this pass, now called *Portes*, is of a very difficult character: Philip suffered severely there in B.C. 189, and it was probably the route followed by Q. Marcius Philippus in B.C. 169. The most southerly range of Thessaly, named Œta,[9] divides it from Locris,

---

[6] Hence Homer gives it the epithet εἰνοσίφυλλον (see above, note [4]).

Pelion Hæmoniæ mons est obversus in Austros:
Summa virent pinu : cætera quercus habet.—Ov. *Fast.* v. 381.

Pelion was the original residence of the Centaurs, and more especially of Chiron, the instructor of Achilles ; they were expelled thence by the Lapithæ :

Ἦματι τῷ ὅτε Θῆρας ἐτίσατο λαχνήεντας·
Τοὺς δ᾽ ἐκ Πηλίου ὦσε καὶ Αἰθίκεσσι πέλασσεν.—*Il.* ii. 743.

Talis et ipse jubam cervice effudit equina
Conjugis adventu pernix Saturnus, et altum
Pelion hinnitu fugiens implevit acuto.—*Georg.* iii. 92.

Quorum post abitum princeps e vertice Pelli
Advenit Chiron, portans silvestria dona.—CATULL. lxiv. 279.

The number of medicinal plants growing on the mountain made it a fitting abode for Chiron.

[7] The allusions in the following passages refer to its *woods*, whence "the tawny troop of lions" issued at the sound of Apollo's lyre; and to its *snowy summit*:

Ἔβα δὲ Λατοΐο᾽ Οἴφρο-
οι νάναν Λεόντων
ἁ δαφοινὸς ἴλα.—EURIP. *Alcest.* 594.

At medicæ ignes cœli, rapidique Leonis
Solstitiale caput nemorosus submovet Othrys.—LUC. vi. 337.

Ceu, duo nubigenæ cum vertice montis ab alto
Descendunt Centauri, Homolen Othrymque nivalem
Linquentes cursu rapido.　　　*Æn.* vii. 674.

[8] The poetical allusions to Pindus are of a general character, as one of the most important mountains of Greece :—

Nam neque Parnassi vobis juga, nam neque Pindi
Ulla moram fecere, neque Aonie Aganippe.—VIRG. *Ecl.* x. 11.

Caucasus ardet,
Ossaque cum Pindo, majorque ambobus Olympus.—Ov. *Met.* ii. 234.

[9] Œta is associated with the death of Hercules, which took place on its summit, the hero being there burnt on a funeral pile :

Vixdum clara dies summa lustrabat in Œta
Herculei monumenta rogi.　　　SIL. ITAL. vi. 45.

Doris, and Ætolia. The only practicable route by which this range could be surmounted led through the famous pass of Thermopylæ, and after following the sea-coast for a certain distance, crossed Cnemis into Bœotia. Thermopylæ was thus, in the S. of Thessaly, very much what the vale of Tempe was in the N.—an almost impregnable post against an invading army.

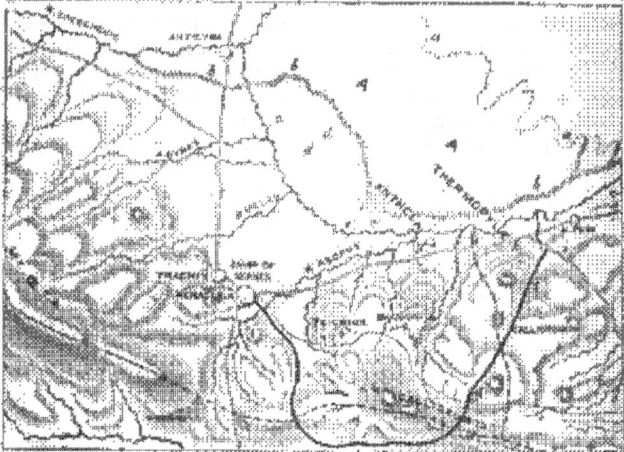

Map of Thermopylæ and the surrounding Country.

The "Gates" or pass of Thermopylæ were formed by a spur of Œta, which protruded to the immediate vicinity of the coast (c c), the interval between the two being for the most part occupied by a morass. Great changes have taken place in this locality: the sea-coast is now removed to a considerable distance (a a) by the alluvial deposits (A A) brought down by the Spercheus, and a broad swampy plain spreads away from the foot of Œta, removing all appearance of a pass. The Spercheus, which formerly fell into the Maliac Bay near Anticyra, now deviates to the S. (b b) by Thermopylæ; while the Asopus, which crossed the plain immediately W. of the pass, now falls into the Spercheus by a course (e e) considerably removed from it. The Dyras has been altered

---

Hence Œtæus became a favourite epithet of Hercules, e.g. :

> Troja, bis Œtæi numine capta dei.   Propert. iii. 1, 32.

> Qualis ubi implicitum Tirynthius ossibus ignem
> Sensit et Œtæus membris accedere vestes.— Stat. Theb. xi. 234.

The allusion in the following line appears to be borrowed from some Greek writer who lived in the vicinity of Œta, and saw the evening star rise over its brow :—

> Sparge, marite, nuces; tibi deserit Hesperus Œtam.
>   Virg. Ecl. viii. 30.

in the opposite direction (*d d*). The springs, whence the first part of
the name " *Hot* Gates" [1] is derived, remain: some are at the W. entrance
(*g*) of the pass, others at its E. entrance (*h*): the latter mark the true
site of Thermopylæ. At each of these points Œta throws out a projection, and between the two was a small plain, about half a mile broad
and more than a mile long, across which the Phocians built a wall (*i*)
for the defence of the pass. As Tempe could be avoided by a circuitous route over the lower limbs of Olympus, so could Thermopylæ
by a mountain-track called Anopæa (*f f*), which surmounted Callidrō-
mus at the back of the pass. Thermopylæ was the scene of many
struggles famous in the history of Greece. In B.C. 480 Leonidas held
it with a small band of Spartans against the hosts of Xerxes until his
position was turned by the path Anopæa; in 279 the Greeks held it
against Brennus with a similar result; in 207 the Ætolians attempted
to make a stand against Philip of Macedonia here; and in 191 Antiochus not only fortified the pass but also the mountain-path against the
Romans, who nevertheless succeeded in forcing their way through
both.

§ 10. The rivers of Thessaly Proper are without exception tributaries of the Peneus. This circumstance results from the peculiar
conformation of the country, the western district being a single basin,
whence but one outlet is afforded to the sea. The various streams
converge with singular uniformity, like the folds of a fan, to a central
point, and thence proceed, in a single sluggish stream, across the
plain to the vale of Tempe. The most important of these rivers is
the Peneus, *Salambria*, which rises in the N.W. angle of the province,
in the central height of Lacmon, and descends with a S.E. course to
Æginium, where it enters on the plain; near Tricca it turns to the
E., and descends to a point where it receives its most important tributaries; then passing through the hills which divide the upper and
lower plains of Thessaly, it slowly traverses the lower plain to Larissa,
where it turns to the N. and flows through the vale of Tempe to the
sea.[2] Its chief tributaries are the Lethæus from the N., the Enipeus,

---

[1] The following lines contain references to the topography of Thermopylæ, and
also to its being the place of congress of the Amphictyonic council :—

> Ὁ ναύλοχα καὶ πετραία
> Θερμὰ λουτρά, καὶ ναγοὺς
> Οἵτε παραναιετάουσιν,
> Οἵ τε μέσσαν
> Μηλίδα πὰρ λίμναν,
> Χρυσαλακάτου τ' ἀκτὰν κόρας,
> Ἔνθ' Ἑλλάνων ἀγοραὶ
> Πυλάτιδες καλέονται.—SOPH. *Trach.* 633.

[2] In its lower course the Peneus is more rapid and is full of small vortices;
hence the Homeric epithet of διερήν and ἀργυροδίνῃ, though the waters are
rather turbid than " silvery " (see below, note [4]).

> Φοῖβε, σὺ μὲν καὶ κύκνος ὑπὸ πτερύγων λίγ' ἀείδει,
> Ὄχθῳ ἐπιθρώσκων ποταμὸν πάρα δινήεντα,
> Πηνειόν.   Hom. *Hymn.* 20 in Apoll

Fersaliti, with its tributaries, the **Apidânus**[3] and **Onarius**, from the S., and the **Pamisus** and **Phœnix** from the W. Near the western entrance of the Pass of Tempe it receives an important tributary from the Cambunian range, named **Titaresius**, *Elassonitiko*.[4] The Vale of **Tempe**, through which the lower course of the Peneus flows, is a narrow ravine between the lower ridges of Olympus[5] and Ossa, about 4½ miles long, and in some places not more than 100 yards broad. The scenery is grand, but has not the sylvan softness which the Latin poets ascribed to it.[6] As a military post the Vale of Tempe was important, commanding as it did the only easy approach from the sea-coast to the interior. A route already described (p. 356) avoided the

---

[3] The Apidânus is sometimes represented as the larger of the two streams. It was the only river in Greece which, according to Herodotus, was not exhausted by Xerxes' army. The Enipeus is rapid (*irrequietus*) throughout the whole of its course, and not, as Lucan suggests, only after its junction with the Apidânus.

> Ἡ Δωρίδος δρόμον εἴαι,
> Ἡ Φθιάδος, ἴνθα τὸν καλλί-
> στων ὑδάτων πατέρα
> Φασὶν Ἀπιδανὸν γύας λιπαίνειν ;—Eurip. Hec. 450.

Irrequietus Enipeus
Apidanusque senex.   Ov. *Met.* I. 579.

Apidanos : *nunquamque celer, nisi mixtus*, Enipeus.—Luc. vi. 373.

[4] The waters of the Titaresius were said to float "like oil" on those of the Peneus :—

> Οἵ τ᾿ ἀμφ᾿ ἱμερτὸν Τιταρήσιον ἔργ᾿ ἐνέμοντο,
> Ὅς ῥ᾿ ἐς Πηνειὸν προΐει καλλίρροον ὕδωρ,
> Οὐδ᾿ ὅγε Πηνειῷ συμμίσγεται ἀργυροδίνῃ,
> Ἀλλά τέ μιν καθύπερθεν ἐπιρρέει, ἠΰτ᾿ ἔλαιον·
> Ὅρκου γὰρ δεινοῦ Στυγὸς ὕδατός ἐστιν ἀπορρώξ.—Il. II. 751.

See also Luc. vi. 375.

[5] Hence Euripides terms it "the most beautiful base" of Olympus :—

> Τὰν Πηνειοῦ σεμνὰν χώραν,
> Κρηνίδ᾿ Οὐλύμπου καλλίσταν,
> Ὄλβῳ βρίθειν φάμαν ἤκουσ᾿
> Εὐθαλεῖ τ᾿ εὐκαρπείᾳ.— *Trand.* 216.

[6] Confestim Peneos adest, viridantia Tempe,
Tempe, quæ silvæ cingunt superincumbentes.—Catull. lxiv. 286.

Speluncæ, vivique lacus; at frigida Tempe,
Mugitusque boum, mollesque sub arbore somni.—Virg. *Georg.* ii. 469

Est nemus Hæmoniæ, prærupta quod undique claudit
Silva : vocant Tempe. Per quæ Peneus, ab imo
Effusus Pindo, spumosis volvitur undis :
Dejectuque gravi tenues agitantia fumos
Nubila conducit, summasque aspergine silvas
Impluit, et sonitu plus quam vicina fatigat.
Hæc domus, hæc sedes, hæc sunt penetralia magni
Amnis : in hoc residens facto de cautibus antro,
Undis jura dabat, Nymphisque colentibus undas.—Ov. *Met.* I. 568.

pass. An important lake, **Bœbéis**,[7] *Karla*, occupies the hollow between the range of Pelion and the plain of Thessaly. It is fed by several small streams, and occasionally by the overflow of the Peneus. A small stream, named **Amphrysus**, flowing into the Pagasæan Gulf, is famed in mythology as the river on whose banks Apollo fed the flocks of Admetus.[8] On the S. the **Sperchéus**, *Elladha*, drains the valley formed by the divergent ranges of Othrys and Œta. It rises in Tymphrestus, and falls into the Maliac Gulf near Anticyra, traversing in its lower course a broad and very fertile plain.[9] The changes that have taken place about its mouth have been already referred to.

§ 11. The original inhabitants of Thessaly were Æolian Pelasgi, after whom the country was named Æolis. These were either expelled or conquered by the Thessalians, an immigrant race from Thesprotia in Epirus, who also drove out the Bœotians from their quarters in the neighbourhood of Arne. The population was divided into three classes :—(i.) The Thessalians Proper, the rich landed proprietors of the plain ; (ii.) the descendants of the original inhabitants, whose position was similar to that of the Laconian *periœci* ; and (iii.) the Penestæ or serfs, who were probably descendants of the original inhabitants reduced to slavery on some account: their position resembled that of the Laconian *helots*. Of the second class we may notice—the Perrhæbi, between Olympus and the Peneus; the Magnētes in Magnesia ; the Acharans in Phthiotis ; the Dolōpes in Dolopia ; and the Malians in Malis. Thessaly Proper was subdivided into four districts :—**Hestiæōtis**, including **Perrhæbia**, in the N., from Pindus in the W. to Olympus in the E., and bounded on the S. generally by the Peneus; **Pelasgiōtis**, S. of the Peneus, and along the W. side of

---

[7] The waters of Bœbeis were reputed "sacred," perhaps because Athena bathed her feet in them :—

> Τοιγάρ πολυμηλοτάταν
> Ἑστίαν οἰκεῖ παρὰ καλλίναον
> Βοιβίαν λίμναν.   EURIP. *Alcest.* 587.

Mercurio et sanctis fertur Bœbeïdos undis
Virgineum primo composuisse latus.—PROPERT. II. 2, 11.

[8] Te quoque, magna Pales, et te memorande canemus
Pastor ab Amphryso.   *Georg.* III. 1.
Et flumine puro
Irrigat Amphrysos famulantis pascua Phœbi.—LUC. vi. 367.

[9]   Μηλιᾶ τε κόλπον, οὗ
Σπερχειὸς ἄρδει πεδίον εὐμενεῖ ποτῷ.—ÆSCH. *Pers.* 488.
Rura mihi et rigui placeant in vallibus amnes ;
Flumina amem silvasque inglorius.   O, ubi campi,
Spercheosque, et virginibus bacchata Lacænis
Taygeta !   *Georg.* II. 485.
Ferit amne citato
Mallaeas Spercheos aquas.   LUC. vi. 366.

Pelion and Ossa; Thessaliôtis, the central plain of Thessaly and the upper conrse of the Peneus; and Phthiôtis, in the S., from the Maliac Gulf on the E. to Dolopia on the W. In addition to these we have to notice the four outlying districts named Magnesia, a long, narrow strip between Lake Ikebeis and the sea, including the ranges of Ossa and Pelion; Dolopia, a mountainous district in the S.W., occupying both sides of Tymphrestus; Œtæa, in the upper valley of the Spercheus, between Othrys and Œta; and Malis, on the southern side of the Spercheus, between it and Œta.

§ 12. The towns of Thessaly conld boast in many cases of a very

Coin of Thessalia.

high antiquity. The name of Larissa bespeaks a Pelasgic origin; Iolcus, on the Pagasæan Gulf, was at a very early period a seat of commercial onterprise; while Ithôme and Tricca in the W., Cmnnon (probably the same as Ephyre), Phoræ, and Gyrton, and many other towns, were of importance in the Homeric age. In the later periods of Greek history the towns owed their celebrity to two very distinct canses: —(i.) as residences of the powerful families,—Larissa, for instance, of the Aleuadæ, Crannon of the Scopadæ, Pharsalus of the Creondæ, and Pheræ of Jason and his successors; (ii.) as military posts commanding the approaches to Southern Greece, such as Gomphi on the side of Epirus, Gonnus near Tempe, and Demetrias on the shores of the Pagasæan Gulf: Pharsalus was also well situated in regard to the passes across Othrys. Many of the mountain forts are noticed in the history of the Roman wars with the Macedonian kings Philip and Perseus, and with Antiochus. Though Thessaly was in possession of a considerable stretch of coast, it was not well provided with harbours. The only sheltered spots were situated in the Pagasæan Gulf, such as Demetrias, Iolcus, and Aphetæ; and their remote position rendered them ill adapted for commercial operations. The Thessalian towns were not, in as far as we know, embellished by the arts of the sculptor or the architect, and consequently the remains now existing possess but little else than topographical interest.

In Hestiæotis.—**Tricca,**[1] *Trikkala,* stood near the left bank of the Peneus, at the spot where the N. route from Epirus entered the plain of Thessaly. It was the first town at which Philip V. arrived after

---

[1] Homer gives it the epithet " horse-feeding " :—

Λαὸν οἳ οἱ ἔχοντε Τρίκην ἐξ ἱπποβότοιο.—*Il.* iv. 202.

his defeat on the Aous. It possessed a famous temple of Asclepius. Pelinnæum,[2] *Old Gardhiki*, was an important place to the E. of Tricca. Alexander the Great passed through it in his march from Illyria to Bœotia; it is also noticed in the war between Antiochus and the Romans. Gomphi, *Episkopi*, in the S.E., was a most important position, as having command of the passes into Athamania and Dolopia. It was taken by Amynander, in B.C. 198, in the Roman war against Philip, and again by Cæsar, in B.C. 48, in his war with Pompey.[3] Gonnus or Gonni, *Lykostomo*, stood on the left bank of the Peneus at the W. entrance of the Vale of Tempe—"in ipsis faucibus saltus quæ Tempe appellatur" (Liv. xxxvi. 10). Philip passed this way after the battle of Cynoscephalæ in B.C. 197, as also did the Roman army under Claudius in 191. It was strongly fortified by Perseus in 171.

In Pelasgiotis.—Gyrton, *Tatari*, was situated on a fertile plain between the Titaresius and the Peneus. It was reputed the original abode of the Phlegyæ, and continued to be a place of importance to a late period, though seldom noticed in history. Larissa was situated in a fertile plain upon gently rising ground on the right bank of the Peneus. It is probably identical with the Homeric Argissa.[4] Democracy prevailed at Larissa, and hence the place sided with Athens in the Peloponnesian war. It was the head-quarters of Philip the son of

Coin of Larissa.

Demetrius before the battle of Cynoscephalæ, in B.C. 197, after which it fell into the hands of the Romans. It is still a very important place and retains its ancient name. Crannon, or Cranon, was situated S.W. of Larissa, and is supposed to be identical with the Homeric Ephyra.[5] It was the residence of the wealthy family of the Scopadæ, whose flocks grazed in the fertile plain surrounding the town. In B.C. 431 Crannon aided the Athenians and in 394 the Bœotians. In 191 it was taken by Antiochus.[6] Some ruins at a place called *Palea Larissa* mark its site. Pheræ, famed in mythology as the residence of Admetus and in history as that of Jason, was situated S.W. of lake Bœbeis and not far from the Pagasæan Gulf, on which Pagasæ served as its port. During the period of the supremacy of Jason and his family (B.C. 374—362) it may be regarded as the capital of Thessaly. It was taken by Antiochus

---

[2] 'Αλλά με Πυθώ τε καὶ τὸ
    Πελιννäιον ἀνίνι.—PIND. *Pyth.* x. 6.

[3] The positions of the towns Phæca, Argenta, Pherinum, Thimirum, Lisiæ, Stimo, and Lampsus, which are noticed by Livy (xxxii. 14, 15) as near Gomphi, are quite uncertain.

[4] "Οἳ δ' Ἄργισσαν ἔχον, καὶ Γυρτώνην ἐνέμοντο.—*Il.* II. 738.

[5] Τὸ μὲν ἄρ' ἐκ Θρῄκης Ἐφύρους μέτα θωρήσσεσθον,
    'Ηὲ μετὰ Φλεγύας μεγαλήτορας.   *Il.* xiii. 301.

[6] It appears to have been a declining place in the time of Catullus:—

    Deseritur Scyros: linquunt Phthiotica Tempe,
    Cranonisque domos, ac mœnia Larissæa.—lxiv. 35.

in B.C. 191. In the centre of the town was the celebrated fountain of Hyperia.[7] Messeïs was also in or near the town. The remains of Pheræ are at *Velestino*. Scotussa, *Supli*, lay W. of Pheræ, near the frontiers of Phthiotis: it was a very ancient town, and reputed to have been the original seat of the oracle of Dodona. In B.C. 367 it was taken by Alexander of Pheræ, and in 191 by Antiochus. In its territory were the hills named Cynoscephalæ, memorable for the battles fought there between the Thebans and Alexander of Pheræ, in 364, and between the Romans under Flaminius, and Philip of Macedon in 197.

In Thessaliotis.—Metropolis was situated on the road from Gomphi to Thaumaci. It derived its name from its having been founded by several towns, whose population coalesced there. Cæsar marched through it on his way to Pharsalus, and it was taken by Flaminius in B.C. 198. Traces of this town exist at *Paleokastro*. Pharsalus, *Fersala*, lay about two

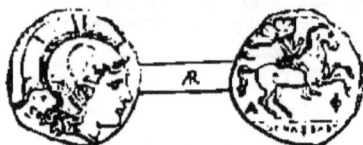

Coin of Pharsalus.

miles and a half from the left bank of the Enipeus, admirably situated for the command of the pass that conducts to southern Greece. It was built on a hill some 600 feet above the plain, which descended precipitously on three of its sides, and contained on its summit a level space for the acropolis. It was besieged without success by Myronides in B.C. 455, and was taken by the Romans under Acilius Glabrio in 191. It is chiefly famous for the battle fought between Cæsar and Pompey, in 48, on the plain just N. of the city.

In Phthiotis.—Thebæ, surnamed Phthiotides, was situated in the N.E. corner of the district, near the Pagasæan Gulf. Previous to the foundation of Demetrias, it was the most important maritime city in Thessaly. It was one of the strongholds of Cassander in his war with Demetrius Poliorcetes in B.C. 302. The Ætolians made it their head-quarters in Northern Greece, until it was taken from them by Philip, son of Demetrius, who changed its name to Philippopolis. It was attacked without success by Flaminius in 197. Its ruins, consisting of the circuit of the walls and towers, and a part of the theatre, stand upon a height near *Ak-Ketjel*. Eretria, near Pharsalus, is noticeable as the spot where Q. Flaminius halted in his march from Pheræ to Scotussa in B.C. 197. Halus[8] was situated on a projecting spur of Othrys, near the sea, and overlooking the Crocian plain. Pteleum,[9]

---

[7] Καὶ μὲν ὕδωρ φορέουσι Μεσσηΐδος, ἢ Ὑπερείης.—*Il.* vi. 457.

Ἐγγὺς μὲν Φέρης, ἀράσων Ὑπερηΐδα λίμνην.　　PIND. *Pyth.* iv. 221.

Ὢ γῆ Φεραία, χαῖρε· σύγγονόν θ' ὕδωρ
Ὑπερεία κρήνη, νᾶμα θεοφιλέστατον.　　SOPH. *Fragm.* 758.

Flurit Amymone, fluunt Messeides undæ,
Fluvit et effusis revocans Hyperia lacertis.—VAL. FLAC. iv. 374.

[8] It is notised by Homer (*Il.* ii. 682).

[9] The Homeric epithet of ἀγχίαλον was possibly more appropriate in early than in late times: a large marsh near the site of the town may once have been a fertile meadow:—

Ἀγχίαλόν τ' Ἀντρῶν', ἠδὲ Πτελεὸν λεχεποίην.—*Il.* ii. 697.

near the entrance of the Pagasæan Gulf, is mentioned by Homer among the possessions of Protesilaus. Antiochus landed here in B.C 192, and the town, having been deserted by its inhabitants, was destroyed in 171. **Larissa Cremaste** received its surname from its position "*Αααφίνι*" upon the side of Othrys : it was occupied by Demetrius Poliorcetes in his war with Cassander in B.C. 302, and was taken by the Romans in their wars with Philip in 200, and with Perseus in 171. **Malitæa** was situated on a lofty hill on the left bank of the Enipeus, a day's march from Pharsalus : it was visited by Brasidas and by the allies in the Lamian War; Philip failed to take it. **Lamia**, originally belonging to the Maliensea, was situated on a height about 8 miles from the sea and 3½ from the Sperchcus. It is well known from the war named after it, carried on in B.C. 323 by the Athenians and their allies against Antipater, who was besieged there. In 192 Lamia submitted to Antiochus, and was consequently attacked by the Romans and taken in 190; its site is fixed at *Zituni*.

In Magnesia.—**Demetrias**, the most important town in this district, was founded about B.C. 290 by Demetrius Poliorcetes, who peopled it with the inhabitants of the surrounding towns. It stood on a declivity overhanging the Pagasæan Gulf on its eastern side. It was favourably situated for the command of the interior of Thessaly as well as of the surrounding seas ; and was hence termed one of the three "fetters" of Greece, the other two being Chalcis and Corinth. In 196 it was taken by the Romans and in 192 by the Ætolians : in 191 it surrendered to Philip, and it was retained by himself and his successor until 169. **Iolchus**[1] was situated on a height a little N. of Demetrias : it was famed in the heroic ages as the place where Jason lived, and where the Argonauts assembled. **Meliboea** was situated on the sea-coast[2] between the roots of Ossa and Pelion : it was plundered by the Romans under Cn. Octavius in B.C. 168.

In Malis.—The chief and only important town in this district was **Trachis**, or **Trachin**, situated in a plain at the foot of Œta, a little W. of Thermopylæ. It derived its name from the "rugged" rocks surrounding the plain. It commanded the approach to Thermopylæ, and hence was valuable as a military position. It is celebrated in mythology as the scene of the death of Hercules, to which Sophocles refers in his Trachiniæ. Historically it is famous for its connexion with **Heraclea**, which the Lacedæmonians erected in its territory in B.C. 426, and which became, after the Peloponnesian War, the head-quarters of the Spartans in Northern Greece, until its capture by the Thebans in 305. It was afterwards a valuable acquisition to the Ætolians, who held out against the Romans under Acilius Glabrio for nearly a month in 191.

Of the less important towns we may notice : In *Hestiæotis*—**Æginium, Stagus**, near the Peneus, an almost impregnable fortress, frequently

---

[1] Homer gives Iolchus the epithets—" roomy," " well built :"—

Πελίης μέν ἐν εὐρυχόρῳ Ἰαωλκῷ

Ναῖε πολύφρονν.   *Od.* xi. 255.

Βοίβην, καὶ Γλαφυράς, καὶ ἐϋκτιμένην Ἰαωλκόν.   *Il.* ii. 712.

[2] The purple shell-fish was found on this coast :—

Jam tibi barbaricæ vestes, Meliboeaque fulgens

Purpura.   LUCRET. ii. 499.

Purpura Mæandro duplici Meliboea cucurrit.—*Æn.* v. 251.

noticed in the accounts of the Roman wars; **Ithōme**, an Homeric town,[2] somewhat E. of Oompbi; **Phacium**, on the left bank of the Peneus, visited by Brasidas in B.C. 424, laid waste by Philip in 198, and occupied by the Roman prætor Bæbius in 191; **Mylæ**, a strong post on the Titaresius at *Dhamasi*, taken by Perseus in B.C. 171; **Mallæa, Cyretiæ**, and **Eritium**, on tributaries of the Titaresius, mentioned in connexion with the Roman wars in Greece—Cyretiæ was plundered by the Ætolians in B.C. 200, taken by Antiochus, but recovered by Bæbius and Philip in 191, and occupied by Perseus in 171; **Oloosson**, an Homeric town,[4] situated on the edge of a plain near Tempe, and now called *Elassona*; **Asōrus, Pythium**, and **Doliche**, three towns in the upper valley of the Titaresius, which formed a tripolis or confederacy; and the Homeric **Orthe**, sometimes identified with Phalanna. In *Pelasgiotis*—**Atrax**, on the left bank of the Peneus, about ten miles above Larissa; **Metropolis**, near Atrax, taken by Antiochus in 191; and **Laceria**, on the W. side of lake Bœbeis, situated on a very remarkable hill with two summits,[5] which rises like an island out of the plain. In *Thessaliotis*—**Asterium**, or **Peiresiæ**, an Homeric town, situated on a hill, with white, calcareous cliffs,[6] near the junction of the rivers Apidanus and Enipeus; **Phyllus**,[7] situated on a hill of the same name on the opposite side of the Apidanus, with a famed temple of Apollo; and **Arne**, afterwards **Cierium**, near the Cuarius, the chief town of the Æolian Bœotians. In *Phthiotis*—**Phyllus**, between Pharsalus and Thebes, an old Homeric town belonging to Protesilaus, and possessing a temple in his honour; **Iton**, or **Itōnus**,[8] on the Cuarius, with a celebrated temple of Athena;[9] **Antron**, at the entrance of the Maliac Gulf, existing in Homer's time,[1] and noticed in the Roman wars as having been purchased by Philip, but taken from him by the Romans; **Proerna**, near the sources of the Apidanus, a place captured by Antiochus, but recovered by Acilius in B.C. 191; **Narthacium**, on a hill of the same name in the valley of the Enipeus, near which Agesilaus conquered the Thessalian cavalry in B.C 301; **Thaumaci**, *Dhomoko*, S. of Proerna, strikingly situated on a precipitous rock, whence the traveller,

---

[2] Homer (*Il.* ll. 729) characterises it as "rocky," Ἰθώμην κλωμακόεσσαν.

[4] Homer gives it the epithet "white," from the argillaceous soil about it :—

Ὄρθην, Ἠλώνην τε, πόλιν τ' Ὀλοοσσόνα λευκήν.—*Il.* ll. 739.

[5] There are the "twin hills in the Dotian plain," of which Hesiod (*ap* Strab. ix. p. 442) speaks :—

ἢ οἵη Διδύμους ἱεροὺς ναίουσα κολωνοὺς
Δωτίῳ ἐν πεδίῳ, πολυβότρυος ἀντ' Ἀμύροιο.

[6] Οἳ δ' ἔχον Ἀστέριον, Τιτάνοιό τε λευκὰ κάρηνα —*Il.* ll. 735.

[7] Aptior armentis Midee, precoramque Phyllos.—STAT. *Theb.* iv. 45.

[8] The Homeric epithet "mother of flocks," was applied to it probably from its possessing a portion of the uplands of Othrys :—

Οἳ δ' εἶχον Φυλάκην, καὶ Πύρρασον ἀνθεμόεντα
Δήμητρος τέμενος, Ἴτωνά τε, μητέρα μήλων.—*Il.* ll. 695.

[9] Πηλιάδες κορυφῆσιν ἐθαμβεον εἰσορόωσαι
Ἔργον Ἀθηναίης Ἰτωνίδος. APOLL. *Argon.* ix. 551.

Ἰλθον Ἰτωνιάδος μιν Ἀθηναίης ἐς ἄεθλα
Ὁρμενίδαι καλέοντες. CALLIM. *Hymn. in Cer.* 74.

[1] The epithet "rocky" is highly appropriate; some of the best millstones in Greece come from the rocks of Antron :—

Καὶ Πέρσον ἀμφιρύτην, Ἀντρωνά τε πετρήεντα.—HOM. *Hymn. in Cer.* 491.

emerging from the defiles of Othrys, gains his first view of the Thessalian plain: it was unsuccessfully besieged by Philip in B.C. 199, and taken by the Romans under Acilius in 191 ; **Xyniæ**, near the district of the Ænianes, and on the borders of a lake of a similar name, now called *Taukli*: it was plundered by the Ætolians in B.C. 198 ; **Phalāra**, the port of Lamia, on the Malian Gulf ; and **Echinus**,[2] between Lamia and Larissa, in a fertile district, at one time held by the Ætolians, and taken from them by Philip, after a long siege. *In Magnesia*—**Bœbe**, on the E. shore of the lake named after it ; **Pagasæ**, at the head of the Pagasæan Bay, celebrated in mythology as the port where Jason built the ship Argo[3]; **Aphētæ**, a port at the neck of the same gulf, whence the Argonauts are said to have sailed, and where the Persian fleet assembled before the battle of Artemisium ; **Homōle**, on a hill of the same name[4] connected with Ossa, near the outlet of the Peneus ; **Eurymēnæ**, on the sea-coast, more to the S. ; **Thaumacia**, still further down the coast, an Homeric town, to be distinguished from Thaumaci in Phthiotis ; **Casthanæa**, at the E. foot of Pelion, noticeable as the place whence the chesnut-tree derived its name ; and **Olizon**,[5] opposite Artemisium in Eubœa, on the neck of land which runs into the Pagasæan Gulf. *In Dolopia*—**Ctimēne**, probably near the sources of the Cuarius. *In Œtæa*—**Hypāta**, at the foot of Œta, S. of the Spercheus, a town whose inhabitants were famed for their skill in sorcery : it belonged to the Ætolian League in the time of the Roman wars in Greece. *In Malis*—**Antioȳra**, at the mouth of the Spercheus.

History.—The history of Thessaly is comparatively devoid of interest. The various tribes and districts were very rarely united in any course of action: rather was it the rule that feuds raged between the leading cities of Larissa, Pharsalus, and Pheræ, and that the power of this wealthy province was frittered away in petty squabbles. In the Persian War the Thessalians designed resistance to the invader, but on the refusal of the allied Greeks to make a stand at Tempe, they *medized* and aided Xerxes. After the battle of Œnophyta the Athenians invaded Thessaly under Myronides, in B.C. 454, without any effect. In the Peloponnesian War the Thessalians took little part, but their sympathies were with Athens ; and although Brasidas succeeded in crossing the country with the aid of the nobles, the people would not suffer

[2] It is noticed by Aristophanes :—

　　Πρώτιστα τὸν Ἐχινοῦντα καὶ τὸν Μηλιᾶ
　　Κόλπον.　　　　　　　　　　　*Lysistrat.* 1169.

[3] Namque ferunt olim Pagasæ navalibus Argo
　　　Egressam longe Phasidos isse viam.—PROPERT. i. 20, 17.

　　Jamque fretum Minyæ Pagasæa puppe secabant.—OV. *Met.* vii. 1.

　　Ut Pagasæa ratis peteret cum Phasidos undas.—LUC. ii. 715.

[4] The hill was regarded as a favourite haunt of Pan, and of the Centaurs and the Lapithæ :—

　　　　Σύγχορτα δ' Ὁμόλας ἔναι-
　　　　ον, εσίκαισιν ὅπερ χέρας
　　　　Πληρούντες, χθόνα Θεσσαλῶν
　　　　　Ἱππείαις ἐθώμαζον.　　　EURIP. *Herc. Fur.* 371.

　　Descendunt Centauri, Homolem Othrymque nivalem
　　Linquentes cursu rapido.　　　VIRG. *Æn.* vii. 675.

[5] Homer characterises it as the "rough" or "craggy" Olizon :—

　　Καὶ Μελίβοιαν ἔχον, καὶ Ὀλιζῶνα τρηχεῖαν.—*Il.* ii. 717.

reinforcements to be sent to him. In 395 the Thessalians joined the Bœotian league against Sparta. Thessaly was afterwards the scene of internal discord through the rise of Pheræ under Lycophron, who defeated the Larissæans and their allies in 404, and introduced the Lacedæmonians into the country. Jason, the successor of Lycophron, succeeded in obtaining the supremacy over Thessaly, with the title of Tagus, in 374, and exercised an important influence in the affairs of Greece, particularly after the battle of Leuctra. The tyranny exercised by the successors of Jason, Polyphron, Polydorus, and Alexander, led to the interference of Alexander of Macedon, and, after his withdrawal, of the Thebans, who invaded Thessaly under Pelopidas in the years 369 and 368, and again in 364; by which the power of Pheræ was checked, but not crushed. It remained for the Macedonians under Philip to effect this in 352, when the last of the tyrants, Lycophron, was defeated and expelled. Thessaly henceforth formed a part of the Macedonian empire, to which they remained attached, in spite of an attempt to throw off the yoke after Alexander's death, until the Romans established their supremacy (B.C. 197).

Islands.—Off the coast of Thessaly lie the following islands :—Sciathus, *Skiatho*, opposite the promontory of Sepias, originally occupied by Pelasgians, afterwards by Chalcidians of Eubœa, with a town of the same name, which was destroyed by the last Philip of Macedonia in B.C. 200: the island produced a good wine. Halonnesus, *Skopelo*, more to the E., now one of the most flourishing isles of the Ægean, in consequence of its excellent wine : it was the cause of a dispute between Philip and the Athenians in B.C. 343. Peparethus,[6] *Kilidhromia*, still more to the E., said to have been colonized by Cretans, famed for its wine and oil,[7] and possessing three towns, the chief one of which was destroyed by Philip in B.C. 200. Scandila, *Skandole*, a small island between Peparethus and Scyrus. And, lastly, Scyros, *Skyro*, so called from its ruggedness, E. of Eubœa, divided into two parts by a narrow isthmus. The town stood on the sides of a high rocky peak[8] on the E. coast, and contained a temple of Athena, who was the patron deity of the island. Scyros is frequently noticed in mythical legends : Thetis concealed Achilles, and Pyrrhus was nurtured there ; Theseus retired thither from Athens, and was treacherously slain there ;[9] his bones were conveyed to Athens in B.C. 469. The island thenceforth belonged to Athens. Its soil was unproductive, but it possessed a famous breed of goats, and quarries of variegated marble.

II. EPIRUS.

§ 13. **Epirus** was the name given to an extensive district in the N.W. of Greece, lying between the Ionian Sea in the W. and Pindus

---

[6] Αἰγαί τ', Εἰρεσίαι τε καὶ ἀγχιάλη Πεπάρηθον.—HOM. *Hymn. in Apoll.* 32.

[7] Nitidæque ferax Peparethos olivæ.—Ov. *Met.* vii. 470.

[8] Αὐτὸς γάρ μιν ἐγὼ κοίλης ἐπὶ νηὸς ἐΐσης
Ἤγαγον ἐκ Σκύρου μετ' ἐϋκνήμιδας Ἀχαιούς.—Od. xi. 507.

Σκῦρον ἑλὼν αἰπεῖαν, Ἐνυῆος πτολίεθρον.   *Il.* ix. 668.

[9] Ὁ Θησέως παῖς, Σκύρου δὲ λυγαίος τάφους
Κρημνῶν ὕπερθεν αἰγιλὴψ βεβηκότων
Πάλαι δακτεῖ τὰς ἀταρχύτους μύχας.   LYCOPHR. 1324

in the E., and extending from the Acroceraunian Promontory in the
N. to the Ambracian Gulf in the S. It is for the most part a wild
and mountainous country; the valleys are numerous, but not extensive, and have at no period supplied sufficient corn for the support
of the inhabitants.
There is but a single
extensive plain, in which
Dodona was situated.
Epirus has always
been a pastoral country.
Among its most valued
productions were oxen [1]
(which supplied the na-
tional emblem), horses,[2]
and dogs.[3]

Coin of Epirus

Name.—The name is derived from ἤπειρος, "mainland," and was
originally applied to the whole W. coast of Greece as far S. as the Corinthian Gulf, in contradistinction to the islands that skirt the coast.[4]
This use prevailed as late as the time of the Peloponnesian War.

§ 14. The mountains that traverse Epirus emanate from the
central range of Pindus. The only one that received a specific
designation was the **Ceraunii Montes** in the extreme N.W., which
attains a great height as it approaches the Ionian Sea, and terminates
in the promontory of **Acroceraunia**, *Linguetta*, the dread of ancient
mariners.[5] This range marks the limit between the valleys which
fall towards the N.W. and those which fall towards the S.W., the
latter being to the S. of the Ceraunian range. The rivers (with the

----

[1] Hence Pindar alludes to the "lofty ox-feeding hills" of Epirus:—

> Θάρσει δὲ κρατεῖ
> Θθίᾳ· Νεοπτόλεμος δ' Ἀ-
> πείρῳ διαπρυσίᾳ,
> Βουβόται τόθι πρῶνες ἔξ-
> οχοι κατάκεινται.   PIND. *Nem.* IV. 81.

[2] Eliadum palmas Epiros equarum.   *Georg.* I. 59.

[3] Veloces Spartæ catulos, acremque Molossum
Pasce sero pingui.   *Georg.* III. 405.

Simul domus alta Molossis
Personuit canibus.   HOR. *Sat.* II. 6, 114.

[4] Ὁι τε Δωδώνην ἔχον, ἠδ' οἳ Σελλοὺς ἀμφενέμοντο,
Ὁι τ' Ἤπειρον ἔχον, ἠδ' ἀντιπέραια' ἐνέμοντο.—*Il.* II. 634.

[5] Quem mortis timuit gradum
Qui siccis oculis monstra natantis,
Qui vidit mare turbidum, et
Infames scopulos, Acroceraunia?—HOR. *Carm.* I. 3, 17.

Et magno late distantia ponto
Terruerunt pavidos accensa Ceraunia nautas.—SIL. ITAL. VIII. 632.

exception of the **Aöus**, the head waters of which, fall within the
limits of Epirus) seek the sea in nearly parallel courses in a south-
westerly direction. The most important of them is the **Achelöus**,
Aspropotamo, which traverses the eastern part of the province.
The others, in order from E. to W., are—the **Arachthus**, *Arta*, which
falls into the Ambracian Gulf, and which was regarded as the
boundary between Hellas Proper and Epirus; the **Acheron**,[*] *Gurla*,
a stream of no great size, which falls into a small bay named
Glycys Limen, "Sweet Harbour," *Port Fanari;* the **Thyamis**,
which joins the sea opposite the island of Corcyra; and the
Calydnus, N. of the Ceraunian range, which formed the N. limit
of Epirus. In the eastern part of Epirus was a lake named **Pambötis**,
now *Joannina*. The line of coast is irregular and forms numerous
inlets: in the S. the **Ambracius Sinus** penetrates into the interior to
a distance of 25 miles, and attains a width of about 10 miles; the
entrance to it is by a narrow and tortuous channel, which we shall
have occasion to describe more minutely hereafter.

§ 15. The inhabitants of Epirus were not considered by the Greeks
as an Hellenic race: the southern tribes were, nevertheless, closely
allied to it, while the northern bore affinity to the Illyrians and
Macedonians. They were divided into numerous clans, of which
three gained a pre-eminence—the **Chaönes**, **Thesprôti**, and **Molossi**.
Epirus was hence divided into three districts—**Chaonia**, upon the
W. coast from the Acroceraunian promontory to the Thyamis;
Thesprotia from the Thyamis to the Ambracian Gulf, including the
district of the Cassopæi in the S.; and **Molossis**, in the interior from
the Aöus to the Ambracian Gulf. In the latter division are included
two districts which were politically distinct from Epirus, viz.:
Ambracia the district about the Hellenic town of the same name on
the N. of the Ambracian Gulf; and **Athamania**, an extensive district
in the valley of the Achelous and on the slopes of Pindus. The
towns of Epirus Proper are few and unimportant; shut off as this
country was from the rest of Greece, and adapted to pastoral pursuits
alone, it can be no matter of surprise that the people lived (as we
are expressly informed that they did) in villages. It was not until
the Molossian kings introduced habits of Greek civilization that any
advance was made in this respect. The only place in Epirus Proper
which gained any fame in early times was Dodöna, the seat of a
famous oracle; and even this must have been unimportant in point
of size, otherwise its site would not have remained doubtful. The
Corinthians planted a colony, Ambracia, on the shores of the Ambra-

[*] This river was invested with many dread associations, as being under the
rule of Aïdoneus the king of Hades. In one part of its course it flowed through
a lake named after it, Acherusia, and it received a tributary, the Cocytus, *two*

cian Gulf, which became historically famous. When the Romans gained possession of Greece, Epirus became a little more " in the world," as several of the ports were favourable for communication with Italy. A large town, Nicopolis, was founded in B.C. 31 by Augustus, at the entrance of the Ambracian Gulf, which became the chief city of Western Greece, and survived to the Middle Ages. We shall notice the towns in their order from N. to S.

Phœnice, in Chaonia, was situated upon the banks of a river at some distance from the sea-coast. It is described in B.C. 230 as the strongest and richest of the cities of Epirus: it was taken in that year by the Illyrians. Peace was concluded there between Philip and the Romans in 204. The hill on which it stood retains the name of *Finiki*. **Buthrōtum** was situated at the head of a salt-water lake,[7] named Pelōdes, which was connected by a river with the sea. It is said to have been founded by Helenus, son of Priam, after the death of Pyrrhus. Cæsar captured it after he had taken Oricum, and it became a Roman colony. **Nicopolis** was founded by Augustus in commemoration of the victory gained at Actium: it was situated on a low isthmus separating the Ionian Sea from the Ambracian Gulf, about 3 miles N. of *Preveza*, the spot on which the town was built being the place where Augustus encamped before the battle. The scene of the engagement is illustrated by the accompanying plan, which shows a double entrance to the Ambracian Bay—the one in the W. guarded by a promontory named *La Punta* (3), the other by *C. Madonna* (4), between which lies the *Bay of Preveza* (P), about 4 miles broad. Actium is to be identified with the former of the two promontories. The battle was fought

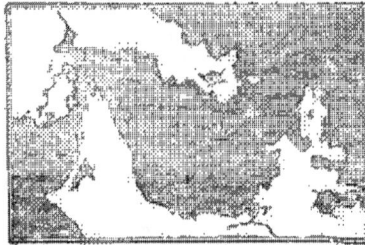

Plan of Actium.

outside the straits, the fleet of Antony having been previously in the Bay of *Preveza*. The position of the temple of Apollo, where Antony's camp was pitched, was at 5; while the ruins of *Preveza* are at 1. Augustus established a quinquennial festival at Nicopolis in commemoration of his victory, and made the place a Roman colony. A church appears to have been planted there, as it is probably the place noticed by St. Paul in his Epistle to Titus. **Dodōna** was probably situated at the S. extremity of Lake Pambotis, where is a ridge, *Mitzikeli*, corresponding to the ancient Tomārus, and a fertile plain surrounding the end of the lake. The oracle of Dodona ranked with those of Delphi and Ammon,

_____

[7] The epithet "celsam," which Virgil gives it, is misplaced, as the town lies low :—

Protinus aërias Phæacum abscondimus arces,
Littoraque Epiri legimus, portuque sublimus
Chaonio, et celsam Buthroti ascendimus urbem.—*Æn.* iii. 291.

and was visited from all parts of the world.[6] The responses were delivered from an oak—in the hollow of which the image of the god was
placed—by means of the rustling of the leaves, which were interpreted
by the priests.[9] The temple was destroyed by the Ætolians in B.C. 219,
and afterwards restored. The ruins at *Kastritza* are supposed to represent the site of the town.[1] *Passaron*, the old capital of the Molossi, is
of uncertain position. It was taken by the Roman praetor Anicius
Gallus in B.C. 167. **Argithea**, the capital of Athamania, was situated
on the road between Ambracia and Gomphi, E. of the Achelous. **Ambracia**, *Arta*, stood on the left bank of the Arachthus, about 7 miles
from the shores of the Ambracian Gulf. Originally a Thesprotian town,
it was occupied by a Corinthian colony about B.C. 635, and became a
most flourishing place. The Ambraciots sided with Sparta in the
Peloponnesian War, and for a time they got possession of Amphilochia
in 432. Their attempts to conquer Acarnania in 429, and to retake
Amphilochian Argos in 426, both failed, and their power was thenceforth checked. Under Pyrrhus, Ambracia became the capital of Epirus.
In 189 it sustained a memorable siege by the Romans, and thenceforth
it declined in power.

Places of less importance were—**Palaeste**, upon the coast of Chaonia,
where Caesar landed from Brundusium in his war against Pompey;[2]
Onchesmus, which served as the port of Phoenice, and was apparently
used as a point of transit to Italy, the wind favourable for crossing
being termed Onchesmites; **Cestria**, on the Thyamis, famed for its
breed of oxen; it appears to have been also called Troy; **Sybota**, a
small harbour opposite the S. point of Corcyra, with two small islands
of the same name before it (the Corinthians erected their trophy, after
their Corcyraean engagement in B.C. 432, at the "continental," the Corcyraeans at the "insular" Sybota); **Chimarium**, more to the S., used
by the Corinthians as a naval station in the war just referred to;

Cassŏpe, the capital of the Cassopæi, near the coast, a city of great size, as its ruins testify ; Pandosia, on the river Acheron, an ancient colony of Elis ; and Ephyra, an old Homeric town,[3] afterwards called Cichyrus, situated near the mouth of the Acheron.

History.—The history of Epirus is almost a blank until the rise of the Molossian dynasty after the Peloponnesian War. Alexander, the brother-in-law of Philip of Macedon, extended his sway over the whole of Epirus. He died in B.C. 326, and was succeeded by Æacides, and Æacides by Alcetas, after whom the celebrated Pyrrhus became king, and raised the kingdom to its greatest splendour. Pyrrhus was succeeded in 272 by his son, Alexander II., who was followed in succession by his two sons Pyrrhus II. and Ptolemy, with whom the family of Pyrrhus became extinct, about 233. A republican form of government then prevailed. After the conquest of Macedonia in 168, the Romans inflicted a most savage revenge on the towns of Epirus on suspicion of their having favoured Perseus : 70 towns were destroyed by Æmilius Paulus, and 150,000 inhabitants reduced to slavery. The country thenceforth became a scene of desolation, and prosperity was confined to the few sea-coast towns which the Romans favoured.

§ 16. Off the coast of Epirus lies the important island of Corcyra, *Corfu*,[4] also named Drepane from its resemblance in shape to a scythe, and probably the same as Homer's Scheria.[5] Its length from N. to S. is about 38 miles ; its breadth varies from 20 miles in the N. to some 3 or 4 in the S. ; its nearest approach to the mainland is in the N., where the passage is only 2 miles wide. It is generally mountainous, and was deservedly celebrated for its fertility in ancient times. The chief town, also named Corcyra, was on the E. coast, a little S. of the modern capital. The only other town of importance was Cassiŏpe in the N.E.

The loftiest mountains are in the N., where *San Salvatore* rises to nearly 4000 feet. From these a ridge runs southwards, forming the backbone of the island. The height named Istŏne was probably near the capital. The promontories were named—Cassiŏpe, *Catharina*, in

---

[3] 'Εξ 'Εφύρης ἀνιόντα παρ' Ἴλου Μερμερίδαο·
'Ὤιχετο γὰρ κάκεισε θοῆς ἐπὶ νηὸς 'Οδυσσεὺς,
Φάρμακον ἀνδροφόνον διζήμενος.   Od. I. 259.

'Ἠὲ καὶ εἰς 'Εφύρην ἐθέλει, πίειραν ἄρουραν,
'Ελθεῖν, ὄφρ' ἔνθεν θυμοφθόρα φάρμακ' ἐνείκη. Od. II. 328.

[4] *Corfu* is a corruption of the mediæval name κορυφώ, applied to the two lofty peaks of the rock on which the modern citadel stands. These were the

                Aeriae Phæacum arces

commemorated by Virgil (Æn. III. 291).

[5] 'Εστι δέ τις πορθμοῖο παροιτέρη 'Ιασοιο
'Αμφιλαφὴς πίειρα Κεραυνίη εἰν ἁλὶ νῆσος,
        . . . . . Δρεπάνη τόθεν ἐκαληίσται
Οὔνομα Φαιήκων ἱερὴ τροφός.   APOLLOS. Argon. iv 9-2.

Ὃς ἄρα φωνήσασ' ἀπέβη Γλαυκῶπις 'Αθήνη
Πόντον ἐπ' ἀτρύγετον· λίπε δὲ Σχερίην ἐρατεινήν.—Od. vii 79.

the N.E.; Phalacrum, *C. Drasti*, in the N.W.; Leucymna, *Lefkimo*, on the E. coast; and Amphipagus, *C. Bianco*, in the S. The town of Corcyra stood on a peninsula formed on one side by the lagoon of *Peschiera*, and on the other by a bay. It possessed two ports—the Hyllaic in the *Peschiera*, and the other in the bay. The acropolis was near the former, on the long undulating promontory S. of *Corfu*. A little N. of the town was the isle of Ptychia, *Vido*. Corcyra was colonized by the Corinthians about B.C. 700. It rapidly rose to a state of high prosperity, and entered into rivalry with the mother country. War broke out about B.C. 664, and the island was reduced by Periander (625-585), but soon regained its independence. The quarrel with Corinth respecting Epidamnus led to the outbreak of the Peloponnesian War in 431, in which Corcyra sided with Athens. The subsequent events of importance are the sieges of Corcyra by the Spartans under Mnasippus in 373, by Cleonymus in 312, by Cassander in 300, and its capture by the Romans in 229.

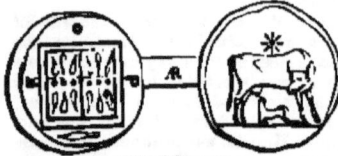

S. of Corcyra are two small islands, anciently named Paxi, now *Paxo* and *Antipaxo*.

Delphi from the West.　(From a sketch by Sir Gardner Wilkinson).

Mount Parnassus and the Hill above Delphi, with the Village of *Chrysó* and the port (*Scala*) below. (From a Sketch by Sir Gardner Wilkinson.)

CHAPTER XX.

CENTRAL GREECE.—ACARNANIA, ÆTOLIA, WESTERN LOCRIS, DORIS, PHOCIS, EASTERN LOCRIS, BŒOTIA.

I. ACARNANIA.

§ 1. **Acarnania** was a maritime province in the S.W. of Northern Greece, bounded on the N. by the Ambracian Gulf and Epirus; on the E. by the Achelous, separating it from Ætolia; and on the W.

and S.W. by the Ionian Sea. In form it resembles a triangle, the apex pointing to the S. The sea-coast is irregular and lined with islands, which render navigation dangerous. The interior is traversed by mountain ranges of moderate height, having a general south-easterly direction, and covered with forests. The soil is fertile, especially the plains about the lower course of the Achelous which sustained large quantities of sheep and cattle; its resources were not, however, much improved by its inhabitants.

Coin of Acarnania.

§ 2. Its physical features were but imperfectly known to the ancients. None of the mountains received special names, and only two of the promontories, viz. Actium. *La Punta*, at the entrance of the Ambracian Gulf, which we have already noticed in connexion with Nicopolis, and Crithôte on the W. coast. The chief river is the Achelôus. *Aspropotamo*, which attains a width of about ⅔ of a mile

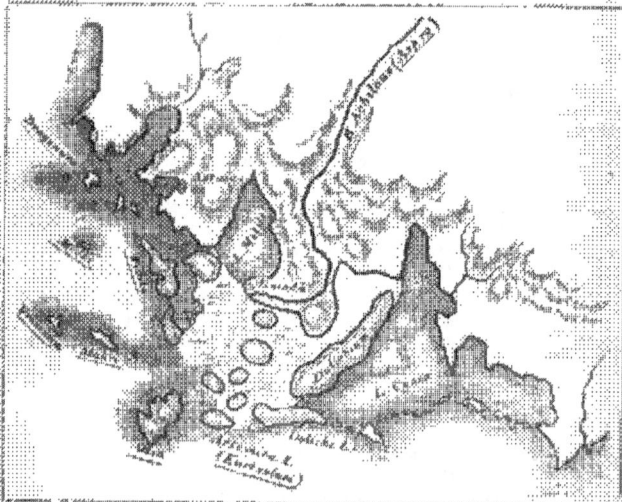
Mouth of the Achelous.

near Stratus, and, as it approaches the sea,[1] crosses over an alluvial
plain of remarkable fertility, named Paracheloïtis, with an exceedingly
tortuous course.[2] It brings down an immense amount of deposit,[3]
which has formed a considerable district near its mouth. There are
several lakes in the interior; the most important of which, named
Melite, lay near the mouth of the Achelous.

§ 3. The early inhabitants of Acarnania were (with the exception
of the Amphilochians) considered to belong to the Hellenic race,
though they were intimately connected with the Epirot tribes.
They were at an early period driven into the interior by the Greek
settlers on the coast: they are described as a rude and barbarous
people, engaged in constant wars with their neighbours, living by
rapine, and famed for their skill in slinging. They lived for the
most part in villages, and had no well-defined form of government.
In times of danger they formed a league, which held its meetings
either at Stratus or at Thyreum, under the presidency of a *stratēgus*
or general. The proper Acarnanian towns were few and unimportant ;
Stratus, on the Achelous, appears to have ranked as the capital.
Colonies were planted by the Corinthians about the middle of the
7th century B.C. at several points on the sea-coast, such as Anactorium
and Sollium. Several of the towns are mentioned in connexion
with the Athenian campaign in 426, and again in the history of
the Ætolian wars. The foundation of Nicopolis proved fatal to
Argos, Anactorium, Sollium, and other places in the N.W., which
were depopulated in order to supply the new town with inhabitants.
We shall describe these towns in order from N. to S.

On the Sea-Coast.—Argos,
surnamed Amphilochicum,
stood on the E. shore of the
Ambracian Gulf, on the
small river Inachus. Its
site has been identified with
Neokhori, now at some short
distance from the shore, but
near a lagoon which formerly
may have extended further
inland. Argos became prominent in the Peloponnesian War : its

Coin of Argos Amphilochicum.

---

[1] It was with this lower portion of the Achelous that the Greeks were best
acquainted. Homer dignifies it with the title of "king" :—

ἀλλ᾽ οὐκ ἔστι Διὶ Κρονίωνι μάχεσθαι·
Τῷ οὐδὲ κρείων Ἀχελώϊος ἰσοφαρίζει,
Οὐδὲ βαθυῤῥείταο μέγα σθένος Ὠκεανοῖο.   *Il.* xxi. 193.

[2] The legend of the contest between Hercules and Achelous for the hand of
Deïaneira, the daughter of Œneus, may have been based upon the efforts made
by the inhabitants to restrain the river within due bounds by dykes and dams;
several of the coins of the country represent the god Achelous as a bull with
the head of an old man.

Et tuus, (Eneu,
Pene gener *crassis oblimat* Echinadas *undis.*—Lvc. vi. 363.

original inhabitants, who were a non-Hellenic race, were expelled by the Ambraciot Greeks, but were afterwards restored by the Athenians in B.C. 432. The Ambraciots invaded the Argive territory in 430 and 426, but were utterly defeated on the latter occasion by Demosthenes.[4] At a subsequent period of history, we hear of Argos as in the hands of the Ætolians, and it was here that the Roman general, M. Fulvius, concluded a treaty with that people. **Anactorium** was on the S. coast

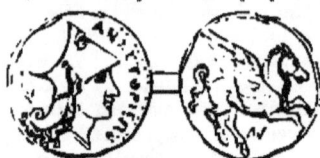

Coin of Anactorium.

of the Ambracian Gulf, at the W. entrance of the promontory, now named *C. Madonna*. It was colonized by Corinthians and Corcyræans jointly, but, in the war between the two states in B.C. 432, it fell into the hands of the former, with whom it remained until 425, when the Athenians restored it to the Corcyræans. **Thyreum** was situated either on or near the Ionian Sea, a short distance S. of the canal which separated Leucas from the mainland. It is first noticed in B.C. 373, when Iphicrates invaded its territory. At the time of the Roman wars in Greece, the meetings of the Acarnanian League were held there. **Œniadæ** was an important place on the right bank of the

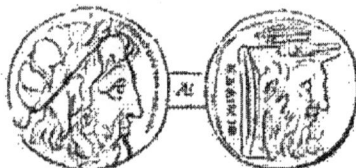

Coin of Œniadæ.

Achelous, about 10 miles from its mouth. It commanded the access to the interior, and was fortified both by art and nature, being surrounded by extensive marshes. The Messenians took it in B.C. 455, but did not retain it. The Athenians under Pericles besieged it without success in 454, and with a different result under Demosthenes in 424. The Ætolians occupied it until 219, when it was taken by Philip, who in turn was deprived of it by the Romans in 211. Its ruins are found at *Trikardho* and consist of remains of a theatre, arched posterns, and a larger arched gateway.

In the Interior.—**Stratus** stood on the right bank of the Achelous and was a military post of importance, as commanding the passes towards the N. In B.C. 429 it was vainly attacked by the Ambraciots. It afterwards fell into the hands of the Ætolians, nor could Philip V. or Perseus wrest it from them. It is frequently noticed in the Macedonian and Roman wars. Extensive remains of it exist at *Surovigli*.

Of the less important towns we may notice. *On the Sea Coast*—**Limnæa**, in Amphilochia, at the S.E. corner of the Ambracian Gulf,

---

[4] The following places are noticed in connexion with this campaign—Olpæ, a fortified hill which the Ambraciots captured, about three miles N.W. of Argos on the shore of the gulf; Crenæ, where the Acarnanians took up their position, somewhat S.W. of Argos; Metropolis, where the Spartan general Eurylochus was posted, a little E. of Olpæ; and the pass which was closed by the Greater and Lesser Idomene, now the *Pass of Makrinoro*, near the coast on the road to Ambracia.

between Argos and Stratus; Palærus, on the sea-coast between Leucas and Alyzia, noticed as an ally of Athens in B.C. 431; Sollium, on the coast near Palærus, but of uncertain position, a Corinthian colony, captured by the Athenians in B.C. 431; Alyzia, about 2 miles from the
sea-coast, with a sanctuary
of Hercules adorned with
works of art by Lysippus ; a
naval battle was fought
near it in B.C. 374, between
the Athenians and Lacedæ-
monians ; and Astacus, near
Prom. Crithote, a colony of
Cephallenia.   *In the Interior*
—Medeon, S. of Limnæa, a

Coin of Alyzia.

strong post unsuccessfully besieged by the Ætolians in B.C. 231, and occupied by Antiochus in 191; Phytia, on a height S. of Medeon, strongly fortified, but nevertheless taken by the Ætolians after the time of Alexander the Great, and by Philip in B.C. 219; and lastly, Metropolis, S. of Stratus, captured by the Ætolians, and taken by Philip in B.C. 219.

History.—The Acarnanians are not noticed in history until the time of the Peloponnesian War, when they appear as allies of the Athenians, and were great supporters of their influence in Western Greece. The Acarnanians particularly distinguished themselves in the battle of Olpæ in B.C. 426. We next hear of them as at war with the Achæans in 391, when the Lacedæmonians, as allies of the latter people, invaded their country. They were afterwards subjected to the Ætolian League; hence they were naturally thrown into alliance with the Macedonian kings, to whom they adhered with great fidelity until the conquest of Greece by the Romans. It is uncertain whether Acarnania was attached to the province of Achæa or of Epirus.

§ 4. Off the coast of Acarnania lie several islands, of which the most important are—Leucadia, *Santa Maura* ; Cephallenia ; *Cepha-lonia* ; and Ithaca, *Thiaki* : and the less important—the Taleboides, consisting of Taphus, *Meganisi*, Carnus and others, between Leucas and the coast ; the Echinades, "sea-urchins" (so named from their jagged outlines), a cluster opposite the mouth of the Achelous, some of which, as Dulichium, have been incorporated with the mainland (see Map, p. 370); and Egilippa E. of Ithaca. To the former class we may add Zacynthus, *Zante*, which, though off the coast of Elis, is evidently a member of the same group.

Leucadia was originally a peninsula of the mainland and is so de scribed by Homer ;[5] it was formed into an island by the Corinthians, who dug a canal across the isthmus.[6] The island is 20 miles in length from N. to S., and from 5 to 8 miles in breadth ; in shape and size it

---

[5] Οἵη Νήσιον ἔλλαν, ἀνεγμένον στολίεθραν,
'Αστὴρ Ἱππέμας, Κεφαλλήνεσσιν ἀνάσσων.   *Od.* xxiv. 376.

[6] The canal was originally dug about B.C. 664 ; it was, however, filled up by sand from the time of the Peloponnesian War until about 200, when it was re-opened by the Romans.

resembles the *Isle of Man*. A range of limestone mountains traverses it from N. to S., terminating in the white cliffs of **Leucata**,[7] *C. Ducato*, which rise out of the sea to a height of above 2000 feet, and were crowned with a temple of Apollo. The chief town, also named **Leucas**,

Coin of Leucas.

a Corinthian colony, was situated on the Dioryctus or canal at *Kaligoni*, about 1½ miles S. of the modern capital; in the Macedonian period it was the chief town of Acarnania: in the Roman wars it sided with Philip, and was taken by the Romans in B.C. 197. In addition to this we have notices of **Hellomenum** and **Phara** in the S.

Cephallenia, the Samos or Same of Homer,[*] lies about 5 miles S. of Leucas, and is the largest island in the Ionian Sea, being in length from N. to S. 31 miles, and varying in width from about 6 to 10 miles. It is mountainous,[*] the most lofty range in the S.E. being formerly named **Ænus** and now *Elato*, from the fir-trees which cover it. From the character of the soil, as well as the want of water, it appears to have been rather unproductive. There were four towns—**Same**, the capital, on the E. coast; **Proni** in the S.E.; **Cranii** in the S.W.; and **Pale** in the W. The chief historical event connected with them is the siege of Same by M. Fulvius in B.C. 189.

Ithaca lies off the E coast of Cephallenia[1] at a distance of 3 or 4

---

[7] This was the scene of the famed lover's leap :—

> Phœbus ab excelso, quantum patet, aspicit æquor :
> Actiacum populi Leucadiumque vocant.
> Hinc se Deucalion, Pyrrhæ succensus amore,
> Misit, et illæso corpore pressit aquas.
> Nec mora : versus amor tetigit lentissima Pyrrhæ
> Pectora, Deucalion igne levatus erat.
> Hanc legem locus ille tenet : pete protinus altam
> Leucada, nec saxo desiluisse time.—Ov. *Heroid.* xv. 165.

The cape was an object of dread to mariners :—

> Mox et Leucatæ nimbosa cacumina montis,
> Et formidatus nautis aperitur Apollo.—Virg. *Æn.* iii. 274.

> Totamque instructo marte videres
> Fervere Leucaten, auroque effulgere fluctus.— *Id.* viii. 676.

> Nec nubifer Actia texit

Litora Leucatea.   Claud. *de Bell. Get.* 183.

[*] 'Εν νορθμῷ 'Ιθάκης τε Σάμοιό τε παιπαλοέσσης. *Od.* iv. 671.
> Οἵ τε Ζάκυνθον ἔχον, ἠδ' οἳ Σάμον ἀμφενέμοντο. *Il.* ii. 634.
> "Οσσοι γὰρ νήσοισιν ἐπικρατέουσιν ἄριστοι,
> Δουλιχίῳ τε, Σάμῃ τε, καὶ ὑλήεντι Ζακύνθῳ. *Od.* xvi. 122

[*] Hence the Homeric epithet παιπαλοέσση. See previous note.
[1] Its position is thus described by Homer :—

> Αὐτὴ δὲ χθαμαλὴ πανυπερτάτη εἰν ἁλὶ κεῖται
> Πρὸς ζόφον, αἱ δέ τ' ἄνευθε πρὸς ἠῶ τ', ἠέλιόν τε.—*Od.* ix. 25.

where χθαμαλή probably refers to the position of the island, *lying under the* mountains of Acarnania, and πανυπερτάτη to its being at the *extremity* of the group of islands formed by Zacynthus, Cephallenia, and the Echinades.

miles: its length from N. to S. is about 17 miles, and its greatest breadth about 4. It consists of a ridge of limestone rock, divided by a deep and wide gulf, *G. of Molo*, into two nearly equal parts, which are connected by an isthmus about ⅓ a mile across. The chief mountain is in the N. and was named **Neritus**;[2] the forests which formerly clothed it have now disappeared. The island is generally rugged and sterile, abounding with bold cliffs and indented by numerous creeks. The localities derive an especial interest from the frequent references to them in the Homeric poems. The capital was probably in the N.W. at *Polis*, in which case **Mt. Neïum**[3] will answer to *Exoge*, the Isle of **Asteris**[4] perhaps to *Dascaglio* and the harbour of **Rheithrum** to the bay of *Afales*. The fountain of Arethusa[5] gushes out of a cliff, still named Corax, at the S.E. extremity of the island. The port of **Phorcys**[6] may be either *Dexia* on the N. side of the *G. of Molo*, or *Skhinos* on the S. side. The Grotto of the Nymphs is a cave on the side of *Mt. Stephanos*, and on the summit of the hill of *Aetos* which forms the isthmus are the ruins of the so-called "Castle of Ulysses." The island appears to have been divided in ancient as in modern times into four parts, of which three were named Neïum, Crocyleium, and Ægirous (the Ægilips of Homer[7]), the two latter probably answering to *Ba ky* and *Anoge*.

Zacynthus lies S. of Cephallenia and about 8 miles from the coast of Peloponnesus: its length is about 23 miles, and its circumference 50. It was celebrated for its fertility, an attribute which has obtained for it in modern times the title of "the flower of the Levant." The most important hill was named **Elatus**, *M. Skopo*,

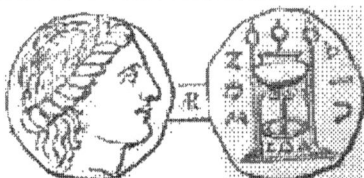

Coin of Zacynthus.

and the most remarkable natural object are the pitch-wells which

[2] Ναιετάω δ' Ἰθάκην εὐδείελον· ἐν δ' ὅρος αὐτῇ
Νήριτον, εἰνοσίφυλλον, ἀριπρεπές.				Od. ix. 21.

[3] Οἳ ῥ' Ἰθάκην εἶχον καὶ Νήριτον εἰνοσίφυλλον.		Il. ii. 632.
Jam medio apparet fluctu nemorosa Zacynthos,
Dulichiumque, Sameque, et Neritos ardua saxis.
Effugimus scopulos Ithacae, Laërtia regna,
Et terram altricem saevi exsecramur Ulixi.— VIRG. Æn. iii. 270.

[4] Νῆσός δέ μοι ἠδ' ἕστηκεν ἐν· ἀγρῷ νόσφι πόληος,
Ἐν λιμένι Ῥείθρῳ, ὑπὸ Νηίῳ ὑλήεντι·			Od. i. 185.

[5] Ἔστι δέ τις νῆσος μέσσῃ ἁλὶ πετρήεσσα,
Μεσσηγὺς Ἰθάκης τε Σάμοιό τε παιπαλοέσσης,
Ἀστερίς, οὐ μεγάλη· λιμένες δ' ἔνι ναύλοχοι αὐτῇ
Ἀμφίδυμοι· τῇ τόν γε μένον λοχόωντες Ἀχαιοί.		Od. iv. 844.

[6]					αἱ δὲ νέμονται
Πὰρ Κόρακος πέτρῃ, ἐπί τε κρήνῃ Ἀρεθούσῃ.		Od. xiii. 407.

[7] Φόρκυνος δέ τίς ἐστι λιμήν, ἁλίοιο γέροντος,
Ἐν δήμῳ Ἰθάκης· δύο δὲ προβλῆτες ἐν αὐτῷ
Ἀκταὶ ἀπορρῶγες, λιμένος ποτιπεπτηυῖαι.		Od. xiii. 96.

[7] Οἳ ῥ' Ἰθάκην εἶχον καὶ Νήριτον εἰνοσίφυλλον,
Καὶ Κροκύλει' ἐνέμοντο, καὶ Αἰγίλιπα τρηχεῖαν.		Il. ii. 632.

are found near the shore of the *Bay of Chieri* on the S.W. coast. The island no longer deserves the epithet of "woody" given to it by Homer and Virgil.[*] The chief town, **Zacynthus**, on the E. coast, was founded by Acharnans, and was hence hostile to the Spartans in the Peloponnesian War. It was taken by the Roman general Valerius Lævinus in B.C. 211, and was finally surrendered to the Romans in 191.

II. ÆTOLIA.

§ 5. **Ætolia** was bounded on the W. by the Achelous; on the N. by the ranges of Tymphrestus and Œta; on the E. by Locris; and on the S. by the Corinthian Gulf. Within these limits are included two districts—Ætolia Proper, along the coast between the Achelous and the Evénus, and Ætolia Epictētus (*i.e.* "acquired") the mountainous district in the N. and E.; these formed in reality independent divisions, and the name *Epictētus* seems merely to indicate the extension of the geographical title to the mountainous region, which otherwise would not have been included in any of the provinces. These districts differed widely in character. The southern consisted of an extensive plain, or rather a double plain, one skirting the sea-coast, the other in the interior, the range of Aracynthus forming the line of demarcation. The soil was very fertile, producing excellent corn, and affording rich pasture grounds, which fed a fine breed of horses. On the slopes of the hills the vine and olive flourished. The interior was a wild unproductive region, infested with wild beasts to a late period.

§ 6. The chief mountains were—**Tymphrestus**, a continuation of Pindus in the N.E.; **Bomi**, containing the sources of the Evénus, the most westerly part of Œta; **Corax**, a S.W. offset from Œta, a lofty mountain crossed by a difficult pass into Doris; **Myénus**, to the S.W. between the Evénus and Hylæthus; **Taphiassus**, running down to the sea a little to the westward of Antirrhium, and terminating in a precipitous cliff, on the face of which the road is carried, whence the modern name *Kaki-Skala* "bad ladder"; **Chalcis**, an offset of Taphiassus to the W.; **Aracynthus**, the range referred to as separating the two plains, running in a S.E. direction between the Achelous and Evénus; and, lastly, **Panætolium**, *Viena*, near Thermum, deriving its name from its being the spot where the Ætolian confederacy assembled. The only important rivers in Ætolia were the **Achelous**, which has been already noticed, and the **Evénus**, *Fidhari*, which takes its rise on the western slopes of Œta and flows with a violent[*] stream in a

[*] Δουλίχιόν τε, Σάμη τε, καὶ ὑλήεσσα Ζάκυνθος.   Od. I. 24.
Jam medio apparet fluctu nemorosa Zacynthos.—Æn. III. 270.
[†] Venerat Eveni *rapidas* Jove natus ad undas.—Or. *Met.* Ix. 104.

south-westerly course to the Corinthian Gulf.[1] In the interior plain
there are two large lakes named **Hyria**,[2] *Zygos*, and **Trichonis**, *Apo-
kuro*, communicating with each other, and also with the Achelous
into which their surplus waters were discharged by the river **Cyäthus**.

§ 7. The original occupants of Ætolia were the Pelasgic tribes of
the Curêtes, Lelêges, and Hyantes, the first being the most important.
These were expelled by the Hellenic tribes of the Epeans under
Ætolis, who crossed over from Elis. Ætolians also settled about
Pleuron. The tribes occupying the interior were—the **Apodoti** above
Naupactus; the **Ophionenses** in the upper valley of the **Evēnus**
with the subordinate divisions of the **Bomienses** and **Callienses** about
the sources of the river; the **Eurytânes** more to the N.W., and the
Agræi in the valley of the Achelous. The towns were more important in the heroic than in the later historical age. Homer notices
five cities as taking part in the Trojan War, viz. Pleuron, Calydon,
Olēnus, Pylēne, and Chalcis : the two first of these were rivals and
were engaged in constant feuds. They were (according to Strabo)[3]
the "ornament" of ancient Greece. Thermum, in the interior, appears
to have been the later capital in the days of the Ætolian confederacy.
The names Arsinoë (applied to the earlier Conōpe) and Lysimachia
originated with the wife of Ptolemy Philadelphus, the founder of those
towns. The final decay of the Ætolian towns was due to the same
cause that ruined those of Acarnania, viz. the foundation of
Nicopolis. We shall describe them from W. to E.

Thermum, *Vlokho*, was strongly placed on a spur of Panætolium, N.
of Lake Trichonis. It was the spot where the meetings of the Ætolian
League were held, and from its impregnable position was regarded as the
acropolis of all Ætolia. It was, nevertheless, surprised by Philip V.
in B C. 218, and in 206. Some remains of its walls and of a public
edifice are still existing. **Pleuron**[4] originally stood on a plain between

[1] It was the fabled scene of the death of Nessus by the hands of Hercules :—

'Ὃς τὸν βαθύρρουν, ποταμὸν Εὔηνον βροτοὺς
Μισθοῦ 'πόρευε χεροῖν, οὔτε πομπίμοις
Κώπαις ἐρέσσων, οὔτε λαίφεσιν ναός.   SOPH. Trach. 559

Et Meleagream maculatus sanguine Nessi
Evenos Calydona secat.   LUC. vi. 365.

[2] Near this lake was a vale where Cycnus was said to have been metamorphosed
into a swan by Apollo : hence the expression *Cycneïa Tempe* :—

At genetrix Hyrie, servati nescia, flendo
Deliacat : stagnumque suo de nomine fecit.—OV. Met. vii. 380.

Inde lacus Hyries videt, et Cycneïa Tempe.—Id. vii. 371.

[3] Τὸ δὲ παλαιὸν πρόσχημα τῆς Ἑλλάδος ἦν ταῦτα τὰ κτίσματα.—Iv. p. 450.

[4] In the following passage Homer represents Pleuron and Calydon as united
under one king :—

Εἰσόμενος φθογγὴν 'Ανδραίμονος υἷι Θόαντι
'Ὃς πάσῃ Πλευρῶνι καὶ αἰπεινῇ Καλυδῶνι
Αἰτωλοῖσιν ἄνασσε, θεὸς δ' ὣς τίετο δήμῳ.   Il. xiii. 216.

the Achelous and the Evenus, at the foot of Mt. Curium. This site
was forsaken about B.C. 230 in consequence of the place having been
ravaged by Demetrius II., of Macedonia; and a new Pleuron was
erected at the foot of Mt. Aracynthus, which was a member of the
Achaean League in B.C. 146. The ruins of this town are near Messo-
longhi, and consist of remains of the walls and of a theatre. Calydon
stood on a fertile plain[5] near the Evenus at some distance from the
Corinthian Gulf. It was a place of great fame in the Heroic age as
the residence of Œneus, the father of Tydeus and Meleager, and grandfather of Diomedes.[6] In B.C. 391 it fell into the hands of the Achaeans,
who retained it until the battle of Leuctra in 371, when it was restored
to the Ætolians. In the civil war between Pompey and Caesar it
appears to have been a considerable town: its inhabitants were shortly
after removed to Nicopolis. Calydon was famed for the worship of
Diana Laphria.

Of the less important towns we may notice—Conope, near the E.
bank of the Achelous, afterwards called Arsinoe after the wife of
Ptolemy Philadelphus who enlarged it; Ithoria, S. of Conope, at the
entrance of a pass and strongly fortified, taken and destroyed by
Philip V, in B.C. 219; Paeanium, yet more to the S., destroyed at the
same time; Lysimachia, on the S. shore of Lake Hyria, probably
founded by Arsinoe and named after her first husband Lysimachus;
Proschium, near the Achelous, said to have been founded by Æolians
from Pylene,[7] which latter stood in the Corinthian Gulf, though its
position is uncertain; Olenus,[8] an old Homeric town at the foot of
Mt. Aracynthus, said to have been destroyed by the Æolians; Elaeus,
belonging to Calydon, a place which was fortified by the aid of

---

Sophocles represents Œneus as king of Pleuron; others make him king of
Calydon: all the legends about Pleuron vary considerably :—

  Ἥτις πατρὸς μὲν ἐν δόμοισιν Οἰνέως
  Ναίουσ' ἐπὶ Πλευρῶνι, νυμφείων ὄτλον
  Ἄλγιστον ἔσχον, εἴ τις Αἰτωλὶς γυνή.   Soph. Trach. 6.

The Curetes noticed in the Iliad (ix. 525) as attacking Calydon, were inhabitants of Pleuron.

  [5] Hence the Homeric epithet of "lovely :"—

   Ὀρτύθ' κάστατον πεδίον Καλυδῶνος ἐραννῆς.   Il. ix. 577.

The epithets "rocky" and "lofty" are supposed to apply to the neighbourhood
rather than the town :—

   Χαλκίδα τ' ἀγχίαλον, Καλυδῶνά τε πετρήεσσαν.   Il. II. 640.

See also Il. xiii. 217, quoted above, note [4].

  [6] References to Calydon are frequent in Ovid : thus we have Calydonis, applied
to Deianeira, daughter of Œneus (Met. ix. 112); Calydonius heros, to Meleager
(Id. viii. 824); Calydonius amnis, to the Achelous, inasmuch as Calydon was the
capital of Ætolia (Id. viii. 727); and Calydonia regna to Apulia, as being the
territory of Diomedes, the grandson of Œneus (Id. xiv. 512).

  [7] Οἳ Πλευρῶν' ἐνέμοντο, καὶ Ὤλενον ἠδὲ Πυλήνην.   Il. II. 639.
    Sensit scopulosa Pylene.—Stat. Theb. iv. 102.

  [8] The Roman poets use Olenius as equivalent to Ætolian :—

   Olenius Tydeus (fraterni sanguinis illum
   Conscius honor agit) eadem sub nocte sopora.—Stat. Theb. i. 402.
   Et praeceps Calydon et quam Jove provocat Idam
   Olenos.   Id. iv. 104.

Attalus, hut was taken hy Philip in 219; and **Chalcis**, also called Chalcla arid Hypocalchis, an old Homeric town E. of the Evenus and at the foot of a mountain of the same name.

In *Epictetus*, on the sea-coast, **Macynia**, at the foot of Mt. Taphiassus, described by the poet Archytas as "the grape-clad, perfume-bearing, lovely Macyna;" **Molycrium**, near Prom. Antirrhium, colonised hy the Corinthians, hut suhject to the Athenians in the early part of the Peloponnesian War, and taken hy the Spartan general Eurylochus, in B.C. 426; **Potidania** and **Crocylium**, on the borders of Locris, S. of the Hylæthus; **Ægitium**, in the mountains hordering the valley of the Hylæthus, the place where Demosthenes was defeated by the Ætolians in B.C. 426; **Callium**, the chief town of the Callienses, on a spur of Mt. Œta, and on the road crossing that mountain to the valley of the Spercheus; it was surprised hy the Gauls in 279; **Aperantia**, in the district of the same name near the Achelous, taken by Philip V. hut recovered hy the Ætolians in 189; and **Agrinium**, also near the Achelous, hut of uncertain position, noticed as in alliance with the Acarnanians in 314.

History.—The Ætolians first come under our notice in the history of the Peloponnesian War, when their country was unsuccessfully invaded by the Athenians under Demosthenes in B.C. 426. They next appear as joining the confederate Greeks in the Lamian War, when their country was again invaded, without any results, in 322.

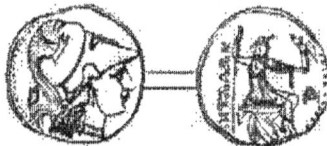

Coin of Ætolia.

They took a prominent part in the expulsion of the Gauls in 279, and particularly in the contest at their own town of Callium. Thenceforward they hecame an important people, and extended their sway over the whole of western Acarnania, the south of Epirus and Thessaly, Locris, Phocis, and Bœotia. They became involved in the Social War, in 220–217, when their country was invaded and Thermum captured by Philip. A second war with Philip followed, in 211–205, in consequence of their alliance with the Romans, and Thermum was again taken. They joined the Romans at Cynoscephalæ in 197, but heing afterwards dissatisfied, they went to war with them in conjunction with Antiochus in 192. They were unfortunate in that war, and were obliged to yield to Rome. The league was dissolved ahout 187, and Ætolia afterwards added to the province of Achaia.

III. WESTERN LOCRIS.

§ 8. Western **Locris** (by which we mean the district of the Locri Ozòlæ,[*] in contradistinction to that of the Epicnemidian and Opuntian Locrians on the shores of the Euboean Sea) was bounded on the W. by Ætolia, on the N. by Ætolia and Doris, on the E. by

---

[*] The name Ozolæ was variously derived from ὄζειν, "to smell," either from a mephitic spring, or from the abundance of asphodel which scented the air; or from ὄζοι, "the branches" of a vine which grew luxuriantly in that country.

Phocis, and on the S. by the Corinthian Gulf. This district is
mountainous, and for the most part unproductive. It was but little
known. The mountains, which emanate either from **Parnassus** in
the N.E. or from **Corax** in the N.W., received no specific names;
and the only river worthy of notice is the **Hylaethus**, *Morno*, which
rises on the slopes of Parnassus, and runs with a S.W. course into
the Corinthian Gulf, near Naupactus. The line of coast extends
from **Prom. Antirrhium** in the W., at the entrance of the Corinthian
Gulf, to the **Sinus Crissaeus** in the E. The towns were unimportant,
with the exception of Amphissa, the capital, in the interior on the
E. frontier; and Naupactus on the coast, for a long period the residence of the exiled Messenians.

Naupactus, *Lepanto*, was situated just within the entrance of the
Corinthian Gulf, a little E. of Prom. Antirrhium, and possessed the
best harbour on the whole of the N. coast of that gulf. The Messenians
were settled there by the Athenians in B.C. 455, and in the Peloponnesian
War it became the head-quarters of the latter power in Western
Greece. It was regained by the Locrians after the battle of Ægospotami. The Achaeans held it before the time of the Theban supremacy, and the Ætolians from the time of Philip II. of Macedonia
until its capture by the Romans in 191. **Amphissa**, *Salona*, was
situated in a pass at the head of the Crissaean plain, and about seven
miles N.W. of Delphi. The Locrians took refuge here at the time of
Xerxes' invasion. The town was destroyed by Philip in B.C. 338 by
order of the Amphictyonic Council, but was soon rebuilt and was able
to withstand a siege from the Romans in 190. On the foundation of
Nicopolis many of the Ætolians betook themselves to Amphissa, which
thus remained a populous place.

Of the less important towns we may notice—**Œnaon**, E. of Naupactus,
where Hesiod was said to have been killed and whence Demosthenes
started on his Ætolian expedition in B.C. 426; **Anticyra**, more to the
E., noticed by Livy (xxvi. 26), and to be distinguished from the Phocian
town of the same name; **Eupalium**, a short distance from the coast,
the place where Demosthenes deposited his plunder in 426, and which
was afterwards taken by Eurylochus; **Erythrae**, the port of Eupalium,
where Philip landed in 207; and **Œanthe**, a port at the W. entrance of
the Crissaean Bay at *Galaxidhi*, the spot whence the Locri Epizephyrii
are said to have embarked.

History.—The Locri Ozolae are first noticed in the time of the Peloponnesian War, when they appear as a semi-barbarous nation along
with the Ætolians and Acarnanians. In B.C. 426 the Locrians promised to aid Demosthenes; but, after his retreat, they yielded to the
Spartan Eurylochus. At a later period they belonged to the Ætolian
League.

IV. DORIS.

§ 9. The small state of **Doris**[1] lay nestled between the ranges of

---

[1] Doris was regarded by the Greeks as the mother country (μητρόπολις, Herod.
viii. 31) of the whole Dorian race. It is, however, very unlikely that so small a
district could supply a military force sufficient for the conquest of the Peloponnesus, and other statements are at variance with the view.

Œta and Parnassus, and bounded by Ætolia on the W., Locris on the S., Thessaly on the N., and Phocis on the E.   It consisted of a single valley watered by the Pindus, *Apostolin*, a tributary of the Cephissus.   It thus opened eastwards into the plain of Phocis, but in other directions was surrounded by mountains.   An important route crossed this district, leading from Heraclea in Malis to Amphissa in Locris.   The Dorian state consisted of a tetrapolis, or confederacy of four towns, named Erineus, Boium, Cytinium, and Pindus, of which the first ranked as capital, while Cytinium commanded the route just referred to, and is hence noticed in the military operations of Demostheues and Eurylochus in B.C. 426, and of Philip in 338.

History.—Doris is seldom noticed in history.   In the invasion of Xerxes it submitted to the Persians.   Subsequently the Dorians received assistance from the Lacedæmonians against the Phocians and others.   The towns suffered much in the Phocian, Ætolian, and Macedonian wars.

V. Phocis.

§ 10. **Phocis** lay between Doris on the N.W., Eastern Locris on the N.E., Bœotia on the S.E., the Corinthian Gulf on the S., and Western Locris on the W.   The only direction in which the boundary was well defined with regard to the contiguous provinces was on the side of Eastern Locris, where the Cnemidian range intervened.   On the side of Doris and Bœotia it lay quite open, the valley of the Pindus connecting it with the former, and that of the Cephissus with the latter.   The country is divided physically into two distinct regions by the range of Parnassus—the northern consisting of the valley of the Cephissus, which opens into a wide plain in the neighbourhood of Elatea ; the southern, of a rugged, broken district, extending from Parnassus to the coast of the Corinthian Gulf.   The line of the coast itself is broken by the bays of Crissa and Anticyra.

§ 11. The chief mountain range in Phocis is **Parnassus**,[2] which attains an elevation of 8000 feet, and terminates in a double peak ; the northern and eastern sides of the summit are covered with perpetual snow.   The highest peak was named Lycorea.   Between the central mass and the precipitous cliffs which overlook Delphi, an

---

[2] The poetical references to Parnassus are numerous, partly from its proximity to Delphi, and partly as the supposed residence of Apollo and the Muses ; we select the following :—

Nec tantum Phœbo gaudet Parnassia rupes.—Virg. *Ecl.* vi. 29.

Hesperio tantum, quantum semotus Eoo
Cardine Parnassus gemino petit æthera colle,
Mons Phœbo, Bromioque sacer.   Luc. v. 71.

Themis hanc dederat Parnassia sortem.—Ov. *Met.* iv. 642.

Vox mihi mentitas tulerit Parnassia sortes.—Val. Flacc. III. 618.

extensive upland district intervenes, partly cultivated, and elsewhere covered with forests. A subordinate range, named Cirphis, runs parallel to Parnassus, on the S. side of the Pleistus. The only important river is the Cephissus, which rises near Lilæa,[2] where it was said to burst forth from the ground with a thundering noise. It first flows towards the N.E., and then to the S.E., through the plains of Elatea : near tne Bœotian border it receives a small tributary, named the Assus, from the slopes of the Cnemis. In the S., the small river Pleistus derived some celebrity from its proximity to Delphi.[4]

§ 12. The Phocians are said to have derived their name partly from Phocus, a grandson of Sisyphus of Corinth, and partly from Phocus, a son of Æacus. They thus seem to have been regarded as a mixed Æolic and Achæan race. Their seats were in the valley of the Cephissus, where they had a confederacy of towns, which held their meetings at Phocium, near Daulis. The Delphians were a distinct people, probably of the Dorian race, who were said to have come from Lycorea in the first instance. They were always bitterly opposed to the Phocians. Among the towns of Phocis, Delphi stands pre-eminent in point of interest and importance, as the seat of the most celebrated fane of antiquity. It brought other places about it into notice, such as Crissa, and its port Cirrha, Daulis, and Panopeus, which lay on the road to Bœotia. The towns in the plain of the Cephissus were important in a strategetical point of view, as they 'commanded the passes across Œta into Northern Greece. Elatéa was one of the keys of Greece, and Hyampolis was hardly less important. Many of the Phocian towns suffered from the position which the country thus occupied. Xerxes destroyed twelve of them in his march southwards. Most of these were rebuilt; but they suffered a more sweeping destruction at the end of the Sacred War, when all the towns, with the exception of Abæ, were destroyed by Philip. They were a second time rebuilt, and are in several instances noticed in the Roman wars in Greece. These towns are described in order, commencing from the N.W., and taking the circuit of the province.

---

[1] Οἵ τ' ἄρα πάρ ποταμὸν Κηφισσὸν δῖον ἔναιον,
Οἵ τε Λίλαιαν ἔχον, πηγῆς ἔπι Κηφισσοῖα.   Il. ii. 522.

Propellentemqne Lilæam
Cephisi glaciale caput.   Stat. Theb. vii. 348.

Κηφισὸν δ' ἂρ ἔπειτα κιχήσαο καλλιρέεθρον,
Ὅστε Λιλαίηθεν προχέει καλλίρροον ὕδωρ.   Hom. Hymn. in Apoll. 240.

[4] Πλειστοῦ τε πηγαὶ καὶ Ποσειδῶνος κράτος
Καλέουσα, καὶ τέλειον ὕψιστον Δία.   Æsch. Eumen. 27.

Οὐδέ τι σε τίθημεν ὄφις μέγας· ἀλλ' ἔτι κεῖνο
Θήραω αἰνογένειαν ἀπὸ Πλειστοῖο καθήρω
Παρνησὸν νιφόεντα περιστέφει ἐννέα κύκλοις.—Callim. Hymn. in Del. 91.

Lilæa was situated at the foot of Parnassus, and at the sources of the Cephissus. It was destroyed at the end of the Sacred War, but was soon afterwards restored. It was taken by Demetrius, but subsequently threw off the Macedonian yoke. Its ruins, at *Paleokastro*, consist of the circuit of the walls, and some of the towers. Delphi was

Map of Delphi.

situated S. of Parnassus, in the narrow valley of the Pleistus. Its position is very remarkable; the uplands of Parnassus terminate towards the S. in a precipitous cliff, 2000 feet high, rising to a double peak,[a] named the Phædriädes (B B), from their "glittering" appearance[b] as they faced the rays of the sun. Below the cliffs the ground slopes off in a double ridge toward the maritime plain, and in a semicircular recess on this slope the town was placed. Between the peaks, the southern of which was sometimes called Hyampēa (R), there is a deep fissure, down which a torrent pours in rainy weather, receiving near

---

[a] These peaks were sometimes supposed to be the summits of Parnassus itself :—

    Mons ibi verticibus petit ardua castra duobus,
    Nomine Parnassus, superatque cacumine nubes.—Ov. *Met.* l. 315.

[b] Σὺ δ' ὑπὲρ διλόφου πέτρας
    στέροψ ὅπωπα λιγνύς, ἵν-
    θα Κωρύκιαι Νύμφαι
    στείχουσι Βακχίδες,
    Κασταλίας τε νᾶμα.   Soph. *Ant.* 1126.

the base of the cliff the waters of the celebrated fountain of Castalia [7] (L), in which visitors to Delphi purified themselves, and whose waters were in a later age supposed to communicate poetic inspiration. [8] On the uplands between the Phædriades and the central mass of Parnassus, about seven miles from Delphi, was the Corycian cave, [9] in which the

Mouth of the Corycian Cave. (From a Sketch by Sir Gardner Wilkinson.)

Delphians took refuge in the Persian War: the main chamber is 200 feet long, and 40 high. The greater portion of Delphi stood W. of the stream, though the walls of Philomelus (A A) enclosed a certain amount of ground on the E. of it. In the former direction was the sacred

---

[7] 'Αλλ', ὦ Φοίβου Δελφοί θέραπες,
Τᾶς Κασταλίας ἀργυροειδεῖς
Βαίνετε δίνας· καθαραῖς δὲ δρόσοις
Ἀφυδρανάμενοι, στείχετε ναούς.   EUR. *Ion.* 94.

Qui rore puro Castaliæ lavit
Crines solutos.   HOR. *Carm.* iii. 4, 61.

Inde ubi libatos irroravere liquores
Vestibus et capiti, flectunt vestigia sanctæ
Ad delubra deæ.   OV. *Met* L 371.

[8]   Mihi flavus Apollo
Pocula Castalia plena ministrat aquæ.—OV. *Am.* i. 15, 35.

Me miserum! (neque enim verbis sollennibus ullis
Incipiam nunc Castaliæ roenilibus undis
Invisus, Phœboque gratis).   STAT. *Silv.* v. 5, 1.

[9] Σέβω δὲ νύμφας, ἔνθα Κωρυκὶς πέτρα
Κοίλη, φίλορνις, δαιμόνων ἀναστροφή.   ÆSCH. *Eumen.* 22.

Πᾶσι Νύμφαι ἄρα τᾶς θη-
ροτρόφου θηροσφαγίε
Φιάσουε, ὦ Διόννυ', ἡ
Κωρυκὶς Κωρυκίαις;   EUR. *Bacch.* 556.

enclosure (τέμενος, πέθα) containing the following buildings: the Temple (1), divided into three parts—the Pronaus, Naus and Adytum: the second containing the hearth with the perpetual fire and the stone

Interior of the Corycian Cave. (From a Sketch by Sir Gardner Wilkinson.)

which was supposed to mark the centre of the earth,[1] and the third the subterranean chamber whence came the oracular responses[2]; the

---

[1] From the numerous references to this stone, we select the following:—

> 'Ορῶ δ' ἐπ' ὀμφαλῷ μὲν ἄνδρα θεομυσῇ
> Ἕδραν ἔχοντα προστρόπαιον.   ÆSCH. Eumen. 40.

> Κἀν τῇδε θαλλῷ καὶ στέφει προσίξομαι
> Μεσόμφαλόν θ' ἵδρυμα Λοξίου πέδον,
> Πυρός τε φέγγος ἄφθιτον κεκλημένον.   ÆSCH. Choeph. 1035.

> Μέλεος μελέῳ ποδὶ χηρεύων,
> Τὰ μεσόμφαλα γᾶς ἀπονοσφίζων
> Μαντεῖα.   SOPH. Œd. Tyr. 479.

[2] In the inmost part of the chamber stood a tripod over a deep chasm in the earth, whence mephitic vapours arose. The priestess sat upon the tripod, when she uttered the oracles:—

> ὁ Φοῖβε, μαντεῖον δ' ἐπι-
> βὰς ζάθεον, τρίποδί τ' ἐν χρυσέῳ
> θάσσεις, ἐν ἀψευδεῖ θρόνῳ,
> Μαντείας βροτοῖς ἀναφαίνων,
> Θεσφάτων ἐμὰν ἀδύτων
> Ὕπερ Κασταλίας μετθμον
> Γείτων, μέσον γᾶς ἔχων μέλαθρον.   EUR. Iph. Taur. 1252.

Great Altar (2) on which sacrifices were daily offered; the Thesauri, or treasuries (3), several detached buildings, in which the most valuable treasures were preserved; the Bouleuterion, or senate-house (4); the Stoa, built by the Athenians (5), which also served as a treasury; the grave of Neoptolemus, son of Achilles (6); the fountain of Cassotis, *Hellenion* (7); the Lesche, a public room where people could meet for conversation (8); and the Theatre (9). The temple was erected by the Alcmæonidæ, and was one of the largest in Greece; the exterior, which was faced with Parian marble, was of the Doric order, and the interior of the Ionic. Outside of the sacred enclosure were the following objects: the Stadium, of which there are still considerable remains; the fountain of Delphusa,² *Kerna* (M), between the Stadium and the enclosure; the Synedrion (N), in a suburb named Pylæ, on the road to Crissa; and, on the E. side of the stream, the Gymnasium (o); the Sanctuaries of Autonous (n) and of Phylacus (F); the temple of Athena Pronœa (E ; and three temples (D). Outside the walls was the ancient cemetery (c), of which there are still considerable remains. The ruins of Delphi are now called *Kastri*. The antiquity of the oracle was very great: even in Homer's age Pytho, as it was then called,⁴ was famed for its treasures;⁵ it was even believed that other deities had owned the place before Apollo. The selection of this spot by the latter deity, on account of its seclusion and beauty, is recorded in the Homeric hymn to Apollo; the first priests were said to have been brought from Crete, and were settled at Crissa. As Cirrha rose to importance, Crissa declined, and was finally merged in Delphi; jealousy arose between Delphi and Cirrha, on account of the exactions practised on pilgrims landing at the latter place, and the Sacred War followed in B.C. 595-585, terminating in the destruction of Cirrha, and in the institution of the Pythian games. Henceforward Delphi became the seat of an independent state, the government of which was of a theocratic character. The temple was destroyed by fire in 548, and a new one of great magnificence erected by the Alcmæonidæ. The Persians approached the place for the purpose of plunder in 480, but were deterred by divine interposition. In 357 the Phocians seized the temple, in revenge for the fine imposed upon them by the Amphictyonic Council: hence the second Sacred War, which terminated with the restoration of the temple to its former possessors, and the punishment of the Phocians. The Gauls visited it in 279, but again heaven (it is said) interfered. The temple was less fortunate in this respect as far as the Romans were

¹ This fountain is referred to, though not by name, in the following passages:
Ἀγχοῦ δὲ κρήνη καλλίρροος, ἔνθα δράκαιναν
Κτεῖνεν ἄναξ, Διὸς υἱὸς, ἀπὸ κρατεροῖο βιοῖο.—HOM. *Hymn. in Apoll.* 300.

Ἀγ᾿ δ᾿ νεφθαλὲς ὣ
Καλλίστας προσέλκνυμα δάφνας,
Ἀ τὰς Φοίβου θυμέλας
Σαίρεις ὑπὸ ναοῖς
Κήπων ἐξ ἀθανάτων,
Ἵνα δρόσοι πήγονσ᾿ ἱεραί
Τὰς ἀέννων παγὰν
Ἐκπρουείσαι   EUR. *Ion.* 112.

² Οἱ Κυπάρισσον ἔχον, Πυθῶνά τε πετρήεσσαν.   *Il.* II. 519.

³ Οὐδ᾿ ὅσα λάϊνος οὐδὸς ἀφήτορος ἐντὸς ἐέργει
Φοίβου Ἀπόλλωνος Πυθοῖ ἐνὶ πετρηέσσῃ.   *Il.* IX. 404.

concerned; Sylla and Nero plundered it; it was restored by Hadrian, and rifled by Constantine: the oracle was silenced by Theodosius. Crissa lay S.W. of Delphi, at the southern end of a projecting spur of Parnassus. It gave name to the bay near which it stood, and on the shore of which Cirrha was subsequently built as its port. Between the two towns was a fertile plain, named indifferently the Cirrhæan or Crissæan, though the terms are more properly applied to two separate portions of the plain, the Crissæan inland and the Cirrhæan on the coast, which were divided from each other by two projecting rocks. Crissa was one of the most ancient cities in Greece, and is described in one of the Homeric hymns[7] as possessing the sanctuary of Delphi; its name is even used by Pindar as synonymous with Delphi. It sunk with the rise of Cirrha, and seems to have become an insignificant place by B.C. 600. Cirrha was destroyed in B.C. 585 by the Amphictyons, on account of the toll which was levied there on pilgrims going to Delphi: it was, however, afterwards rebuilt as the port of Delphi. Antícyra was situated on a bay of the Corinthian Gulf, which was named after it, and where it possessed an excellent harbour. It was supposed to represent the Homeric Cyparissus. Though destroyed at the close of of the Sacred War, it recovered, and was taken by the Roman Consul Flaminius in B.C. 198. It was particularly famed for its hellebore, which was regarded as a cure for madness.[8] Panópeus, or Panope,[9] was near the frontier of Bœotia, between Daulis and Chæronea.[1] It was a very ancient town, originally inhabited by the Phlegyæ. It was destroyed by Xerxes, and again by Philip; was taken by the Romans in B.C. 198, and was a third time destroyed in the war between Sulla

---

[8] Homer gives it the epithets—"divine," "conspicuous," "vine-bearing:"—

Κρίσσαν τε ζαθέην, καὶ Δαυλίδα, καὶ Πανοπῆα.   Il. II. 520.

Ἵξον δ᾽ ἐς Κρίσσην κιθαίελον, ἀμπελόεσσαν.   Hymn. in Apoll. 438.

The Pythian games were celebrated on this plain:—

Ἐν Κρίσᾳ δ᾽ εὐρυσθενὴς εἶ-
δ᾽ Ἀπόλλων μιν, πόρε τ᾽ ἀγλαίαν.   Pind. Isthm. ii. 28.

πᾶν δ᾽ ἐνίμπλατο
Ναυηγῶν Κρισσαίον ἱππικῶν πέδον.   Soph. El. 729.

[7] Ἷξεν δ᾽ ἐς Κρίσσην ὑπὸ Παρνησσὸν νιφόεντα,
Κνημὸν πρὸς Ζέφυρον τετραμμένον, αὐτὰρ ὕπερθεν
Πέτρη ἐπικρέμαται, κοίλη δ᾽ ὑποδέδρομε βῆσσα,
Τρηχεῖ᾽· ἔνθα ἄναξ τεκμήρατο Φοῖβος Ἀπόλλων
Νηὸν ποιήσασθαι ἐπήρατον, εἶπέ τε μῦθον.—Hom. Hymn. in Apoll. 282.

[8] Nescio an Anticyram ratio illis destinet omnem.—Hor. Sat. ii. 3, 83.
Naviget Anticyram.   Id. 166.

Ne dubitet Ladas, si non eget Anticyra, nec
Archigene.   Juv. xiii. 97.

I, bibe, dixissem, purgantes pectora succos,
Quicquid et in tota nascitur Anticyra.—Ov. e Pont. iv. 3, 53

[9] Jam vada Cephisi, Panopesque evaserat arva.—Ov. Met. iii. 19.

Quis tibi Phœbeas acies, veteremque revolvat
Phocida? qui Panopen, qui Daulida, qui Cyparisson.
STAT. Theb. vii. 343.

[1] Αὐτὰρ γὰρ ὄλησσε, Διὸς αὐδρὴν παρεόντιν
Πυθώδ᾽ ἐρχομένην, διὰ καλλιχόρον Πανοπῆσι.   Od. xi. 580.

s 3

and Archelaus. **Daulis**[2] stood W. of Panopeus, on the high road to Delphi. It was a place of importance in the heroic age. It shared the fate of the other Phocian towns in the Persian and Sacred wars. It was subsequently rebuilt, and was reputed impregnable, from its position on a spur of Parnassus. **Hyampolis** stood on a height[3] at the entrance of a valley, which formed a natural route across Cnemis into Locris. It was consequently the scene of several engagements: the Phocians here defeated the Thessalians; Xerxes destroyed it; Jason, in 371, took its suburb, named Cleonae; the Boeotians and Phocians fought near it in 347; and Philip destroyed it. It was rebuilt, and is mentioned in the Roman wars in Greece. The circuit of its walls may be seen at *Vogdhani*. **Abae**, near Hyampolis, derived its fame from its possessing a temple and oracle of Apollo,[4] which was consulted from all quarters, and particularly by Croesus and Mardonius. It was destroyed by fire in B.C. 480 by the Persians, and in 346 by the Boeotians. Hadrian erected a small temple near the site of the old one. **Elatea** stood in the plain of the Cephissus, in command of the most important pass across Mount Oeta, and hence a place of the greatest importance in a military point of view. It was burnt by Xerxes, but afterwards restored and occupied by Philip in B.C. 338, much to the alarm of the Athenians. It successfully resisted Cassander, but was taken by Philip, son of Demetrius, and again by the Romans in 198. The name survives in *Leftu*, where are some few remains of the old town.

Of the less important towns we may notice—**Drymaea**, a frontier town on the side of Doris, taken by Xerxes; **Neon**, at the foot of Tithorea, rebuilt after its destruction by the Persians, and finally destroyed at the end of the Sacred War; **Tithorea**, regarded by Pausanias as occupying the site of Neon, but probably a different place, distant 3½ miles, the former being at *Velitza* and the latter at *Palea Fiva*; at Tithorea the Phocians took refuge from Xerxes, probably in a spacious cavern, which exists behind *Velitza*; **Ambrysus**, N.E. of Anticyra, at the foot of Mount Cirphis, very strongly fortified by the Thebans against Philip, taken by the Romans in B.C. 198; **Stiris**, near the Boeotian frontier, strongly posted on a height, defended by precipitous rocks, destroyed by Philip, but afterwards rebuilt; **Phocicum**, near Daulis, where the meetings of the Phocian confederacy were held; **Parapotamii**, on the left bank of the Cephissus (whence its name), near the border of Boeotia, never rebuilt after its destruction by Philip in the Sacred War; **Anemoria**, an Homeric town (*Il.* ii. 521), said to be named from its exposure to the blasts that descended on it from Parnassus; **Cleonae**, near Hyampolis, on the pass crossing to Locris; and

---

[2] Daulis is famed in mythology as the spot where Procne was turned into a swallow and Philomela into a nightingale: the latter bird is still found there in great numbers. West of Daulis was the spot called Schiste Odos, where the road from Ambrysus fell into the main road leading to Delphi:—

Φωκὶς μὲν ἡ γῆ κλήζεται· σχιστὴ δ' ὁδὸς
Ἐς ταὐτὸ Δελφῶν κἀπὸ Δαυλίας ἄγει.   Soph. Œd. Tyr. 733.

[3] Et vallem Lebadea tuas | et Hyampolin acri
Submigam scopulo?   Stat. Theb. vii. 345

[4] Οὐκ ἔτι τὸν ἄθικτον εἶμι
Γᾶς ἐπ' ὀμφαλὸν σέβων,
Οὐδ' ἐς τὸν Ἀβαῖσι ναόν.   Soph. Œd. Tyr. 807.

Tritæa, somewhere in the valley of the Cephissus, but of uncertain position.

History.—The history of Phocis, apart from Delphi, presents few features of interest. In the Peloponnesian War, the Phocians sided with Athens: after the battle of Leuctra (B.C. 371) they became subject to the Thebans, and their separation from the Thebans led ultimately to the Sacred War. At the battle of Chæronea, and in the Lamiac War, they fought on the side of Grecian Independence.

VI. EASTERN LOCRIS.

§ 13. The territory of the Eastern Locrians consisted of a narrow strip of coast land between the continuations of Œta and the Eubœan Sea, extending from the pass of Thermopylæ in the N.W. to the mouth of the Cephissus in the S.E. This district was divided between two tribes, surnamed **Epicnemidii** and **Opuntii**, the former so styled from the adjacent hill of Cnemis, the latter from their capital, Opus. The range of **Cnemis**, *Talanda*, attains a considerable elevation in the N.; the portion of the range adjacent to Opus was of less height,

Coin of the Locri Opuntii.

and received no special designation. Spurs project in various parts to the vicinity of the coast, and in one instance form a considerable promontory, named **Cnemides**. The rivers necessarily have very short courses: the most important are the **Bogarius** and **Manes**. The valleys were in many cases fertile, as was also the whole of the coast district. Routes cross the mountains between Alpenus and Tithronium in Phocis, between Thronium and Elatea, and between Opus and Hyampolis.

§ 14. The eastern Locrians are noticed by Homer, as taking part in the Trojan War. The distinction into Epicnemidians and Opuntians was not recognized by classical writers, but originated with the geographers, Strabo and others. In classical times Opus was regarded as the capital of the whole district: at a later period Thronium became the chief town of the Epicnemidians.. These were the only towns of importance in the whole district.

Thronium[1] was situated on the Boagrius, about 2½ miles from the coast. It is but seldom noticed: in B.C. 431 it was taken by the

---

[1] Δοκρών δὲ τοῖσδ' ἴσας ἔχων
Ναῦς ἦλθ' Ὀιλεὼς τόκος, κλυτὰν
Θρονιάδ' ἐκλιπὼν πόλιν.   Eur. *Iph. Aul.* 261

Athenians, and in the Sacred War by Onomarchus. Opus stood at the
head of the Opuntian Gulf, a little removed from the coast : it was
reputed one of the most ancient cities of Greece, and was, according to
Homer, the native city of Patroclus.⁴ In the war between Antigonus
and Cassander, Opus was besieged by Ptolemy for its antagonism to the
former. Of the less important places we may notice: **Alpēnus**, at the
southern entrance of the pass of Thermopylæ; **Nicæa**, a fortress close
to the sea commanding that pass, and hence a very important acquisition to Philip in his wars in B.C. 346 and 340 ; **Scarphé**, on the road
to **Elatea**, and hence noticed in the narrative of Flaminius's march by
Livy (xxxiii. 3) ; **Daphnus**, on the sea-coast, originally belonging to
Phocis ; **Alōpe**, on an insulated hill farther down the coast; **Cynus**, the
principal port of the Opuntians, about seven miles N. of Opus ; and,
lastly, **Naryx**, between Opus and Hyampolis, the reputed birth place of
Ajax,⁷ and the scene of an engagement between the Bœotians and
Phocians in B.C. 352.

- *History.*—The history of the Eastern Locrians is unimportant : the
Opuntians are noticed as taking part with the Spartans in the Persian
and Peloponnesian wars.

VII. Bœotia.

§ 15. Bœotia was bounded by the Eubœan Sea on the E., Phocis
on the W., the Corinthian Gulf and Attica on the S., and the
district of the Opuntian Locrians on the N. It thus stretched from
sea to sea, and may be said to close the mouth of the Peloponnesus.
On the S. it possessed a well-defined boundary in Mount Cithæron ;
but towards the N.E. it lay open along the vale of the Cephissus,
though in this direction it was partly closed by the ridge of Hyphantium, an offset from the Opuntian range. Within the limits above
specified were two districts, of a widely different character : (i.)
Northern Bœotia, a large basin of an oval form, completely surrounded by hills, and subdivided by subordinate ranges into two main
portions—one containing the plain of Orchomenus and Lake Copais,
the other the plain of Thebes and Lake Hylica; (ii.) Southern
Bœotia, a long and in some parts wide valley, drained by the
Asōpus. The sea-coast on either side is irregular, but does not offer
good harbours. The climate of Bœotia was much influenced by the
presence of so much stagnant water, which rendered the air heavy
and the winters severe. The soil possessed remarkable fertility, that
about Copais being of a deep alluvial character, equally well suited
to the growth of corn and to the purposes of pasture: the Bœotian
horses were amongst the best in Greece. The vine and other fruits

---

⁶ Deucalion and Pyrrha are also said to have resided near Opus.

⁷ Hence the epithet *Narycius* applied to him, Ov. *Met.* xiv. 468.
The same epithet is applied to Bruttium in Italy, under the idea that Locri was
colonised from Naryx :—
　　Naryciæque picis lucos.　　　　　　　Virg. *Georg.* II. 438.
　　Hic et Narycii posuerunt mœnia Locri.—*Æn.* III. 399.

flourished remarkably well. The mountains yielded iron ore, and black marble. The plain of Thebes abounded with moles, whose skins were made an article of commerce. Lake Copais produced abundance of fish, particularly eels, and water-fowl were numerous; while the reeds that fringed its shore supplied the country with flutes.

§ 16. Bœotia is skirted by mountain ranges in all directions. In the western part of the province rises the long range of Helicon, the soft and sylvan character of whose scenery rendered it, in the eyes of the Greeks, a fitting residence for the Muses;[a] Aganippe and Hippocrēne were two of the numerous rills which coursed down its sides amid groves of myrtle and oleander,—the former rising near Ascra and joining the Termessus, the latter flowing into the Olmeus: the Grove of the Muses was near Aganippe. One of the heights of Helicon was named Leibethrium, *Zagora*; another more to the N., Laphystium, *Granitza*; while between the two was Tilphossium, extending almost to the edge of Lake Copais, and separating the plains of Coronea and Haliartus. On the southern frontier, Cithaeron separated Bœotia from Attica, bounding the plain of Asōpus on the S.: it was a well-wooded, wild chain, and hence was aptly selected as the scene of various mythological events, such as the metamorphosis of Actæon, the death of Pentheus, and the exposure of Œdipus.[b] It was also regarded as the scene of the revels of Bacchus.[c] On the N.E. the range of Cnemis is continued in a line parallel to the sea-coast, rising into the heights of Ptoum, E. of Lake Copais, Messapium, near Anthedon, and Hypaton, more to the S., while in the N.W. a

[a] Μουσάων Ἑλικωνιάδων ἀρχώμεθ' ἀείδειν,
Αἵθ' Ἑλικῶνος ἔχουσιν ὄρος μέγα τε ζάθεόν τε,
Καί τε περὶ κρήνην ἰοειδέα πόσσ' ἀπαλοῖσιν
Ὀρχεῦνται, καὶ βωμὸν ἐρισθενέος Κρονίωνος·
Καί τε λοεσσάμεναι τέρενα χρόα Περμησσοῖο,
Ἢ Ἱππουκρήνης, ἢ Ὀλμειοῦ ζαθέοιο,
Ἀκροτάτῳ Ἑλικῶνι χοροὺς ἐνεποιήσαντο
Καλούς, ἱμερόεντας· ἐπιῤῥώσαντο δὲ ποσσίν.   Hxs. *Theog.* L.

Pandite nunc Helicona, Deæ, cantusque movete.—Virg. *Æn.* vii. 641.

Hence the Muses were named Heliconiades :—

Adde Heliconiadum comites, quorum unus Homerus.
                                            Lucret. iii. 1050.

[b] Ἀλλ' ἵα με ναίειν ὄρεσιν, ἵνα κλῄζεται
Οὑμὸς Κιθαιρὼν οὗτος, ὃν μήτηρ τέ μοι
Πατήρ τ' ἐθέσθην ζῶντι κύριον τάφον.   Soph. *Œd. Tyr.* 1451.

Ὁ ζαθέων πετάλων πολυθηρότα-
τον πέδον, Ἀρτέμιδος χιονότροφον ὄμμα Κιθαιρών,
Μήποτε τὸν θανάτῳ προστεθέντα, λόχευμ' Ἰοκάστης
Ὤφελες Οἰδίποδαν θρέψαι βρέφος ἐκβαλον οἴκων.—Eur. *Phœn.* 801.

Qualis commotis excita sacris
Thyas, ubi audito stimulant trieterica Baccho
Orgia, nocturnusque vocat clamore Cithæron.—*Æn.* iv. 301.

projecting spur of the Cnemidian range, named **Hyphantium**, penetrates close to the banks of the Cephissus, and separates the plains of Bœotia and Phocis. In addition to these, we have to notice a series of elevations which separate the basin of Lake Copais from that of Lake Hylica, the most prominent height being **Phœnicium**, *Fagà*: and again another series between the Theban plain and the valley of the Asopus, of which **Teumessus** is the most conspicuous. The approaches to Bœotia from the N. were (i.) by the valley of the Cephissus, which was commanded by a defile near Chæronea, and (ii.) by a track across Hyphantium.

§ 17. The only river of importance in Northern Bœotia is the **Cephissus**, which enters it from Phocis in the N.W., and, after a short course across the plain of Chæronea, discharges itself into Lake Copais. This lake forms one of the most striking features in Bœotia. So completely do the mountains shut in the basin, that no opening existed for the escape of the waters; these, therefore, collected in the deepest part of the basin, and formed a considerable lake, originally named **Cephissis**, from the chief river flowing into it, afterwards **Copais**, from the town of Copæ, and now *Topolias*, whence the surplus waters escaped by subterranean channels (called *katavothra*) to the Eubœan Sea, distant between four and five miles. These *katavothra* are four in number, three communicating with the sea, and one with Lake Hylica; the central, or main stream, emerges at Upper Larymna, and the two others on either side of it. These natural outlets being found occasionally insufficient, two artificial tunnels were constructed in the heroic age, probably by the Minyæ of Orchomenus. As long as these channels were kept clear, the greater part of the bed of Lake Copais was under cultivation. The size of the lake has varied at different times. Strabo states its circumference at forty miles; it is now sixty, in consequence of the channels becoming choked. Numerous lesser streams poured into Lake Copais from all directions. In the plain of Thebes is a large lake named **Hyllea**, *Livadhi*, filling a deep crater surrounded by mountains: it lies at a lower level than Copais, and received some of its surplus waters by a tunnel. Another lake, now called *Moritzi*, more to the eastward, forms a connecting link between Hylica and the sea. Southern Bœotia is watered by the **Asopus**, which rises in Mount Cithæron, and flows in an easterly course with a sluggish stream[2] to the

---

[2] Homer characterizes the Asopus as "rushy" and "abounding in grass;"—

Ἀσωπὸν δ' ἵκοντο βαθύσχοινον, λεχεποίην   *Il.* iv. 383.

Euripides also speaks of the "low spreading plains" about its banks:—

Πεδίον ὑποτάνυσι, αἳ παρ' Ἀσωποῦ ῥοαῖς
Εὔκαρπον ἐκβάλλουσι Θηβαίων στάχυν,
Γαίας τ' Ἐρεχθεῖδ' τ', αἳ Κιθαιρῶνος λέπας
Νιφῶν καταρχήκασιν.   *Bacch.* 748.

Euboean Sea: its valley (in length about forty miles) is divided into three parts by spurs of Teumessus—the plain of Parasopia along its upper-course, the plain of Tanagra, and the plain of Oröpus.

§ 18. The original inhabitants of Bœotia were a Pelasgic race, and were known by various tribal names. The later inhabitants were an Æolian race, who immigrated into this province from Thessaly. A Phœnician colony also settled at Thebes under the name of Cadmeans. The Bœotian character was supposed to be influenced by the climate, which was dull and heavy: it may, however, have been equally affected by the sensuality of the people. To whatever cause it was due, the stupidity of the Bœotians passed into a proverb.[1] It should at the same time be stated in their favour, that they cultivated a taste for music and poetry, and that they reckoned among their countrymen Hesiod, Pindar, and Plutarch. The Bœotian towns occupy a prominent place in Greek history. This is due to a variety of causes: (i.) their wealth was great, in consequence of the extreme fertility of the soil; (ii.) their situations were secure, the spurs of the ranges surrounding the plain, offering remarkably fine sites; (iii.) the position of Bœotia between northern and southern Greece rendered it the passage of every invading host; and (iv.) the plains of Bœotia offered the very best ground in Greece for military evolutions. Bœotia was what the Low Countries were at one time to Europe, the "cock-pit" of Greece. Orchomenus, at the N.W. extremity of the Copaic Lake, originally took the lead of all the Bœotian towns. After the immigration of the Bœotians, Thebes gained the supremacy, and Orchomenus took the second place, remaining however, for a long period, a powerful rival, and retaining its position as capital of its own plain. The chief towns were formed into a confederacy, under the presidency of Thebes: of these there were originally fourteen, of which we can certainly name ten, viz.: Thebes, Orchomenus, Lebadēa, Coronēa, Copæ, Haliartus, Thespiæ, Tanagra, Anthēdon, and Platæa, while the remaining four are supposed to have been Ocalea, Chalia, Onchestus, and Eleutheræ. Oröpus was probably once a member, but afterwards became subject to Athens; and Platæa withdrew from the confederacy as early as B.C. 519. The towns of Bœotia flourished until the extinction of independence, consequent upon the battle of Chæronea in 338 and the capture of Thebes in 335. They then sunk so fast that in the Roman age Tanagra and Thespiæ were the only ones remaining: the rest were a heap of ruins. We shall

[1] The expressions were Βοιωτία ὗς and Βοιώτιον οὖς:—
Τιμωαί τ᾽ ἔπειτ᾽, ἀρχαῖον ὄνειδος ἀλαθέσιν λόγοις εἰ φεύγομεν, Βοιωτίαν
Ὕν.   Pind. OL. vl. 181.

describe these towns in order, commencing from the N.W., and proceeding round by the W. to the S.

Chæronea, the key of Bœotia on its northern frontier, was situated at the edge of the valley of the Cephissus, with its citadel posted on a steep granite rock. It was the scene of engagements between the Athenians and Bœotians in B.C. 447, between the Macedonians under Philip and the Bœotians in 338, and between the Romans under Sulla and the forces of Mithridates in 86. Orchomĕnus was strongly

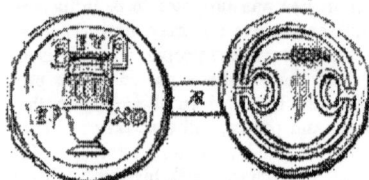

Coin of Orchomenus.

posted on a hill overlooking the marshes of the Copaic Lake, the Cephissus "winding like a serpent"[4] about the base on the S. and E., while the small river Melas washes its northern side. The walls extended to a distance of two miles in circumference : the most remarkable object in the town was the Treasury of Atreus, the ruins of which still remain.

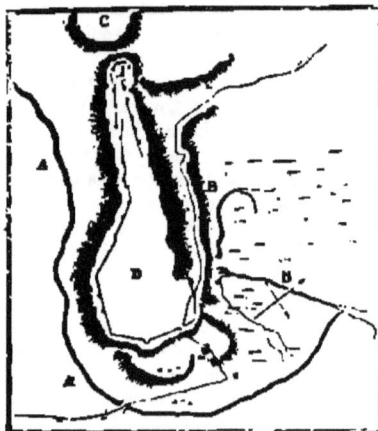

Plan of Orchomenus.

A A. The Cephissus.
C. Mount Acontium.
I. Acropolis.

B B. The Melas.
D. Orchomenus.
5. Treasury of Minyas.

Orchomenus was at one period the first, and after the rise of Thebes continued to be the second city in Bœotia, owing its wealth to the rich alluvial plain on which it stood. It was, in the Homeric age, famed for its treasures,[5] and was the seat of the powerful races of the Minyæ[6] and the Phlegyæ.[7] It took the patriotic side in the Persian War, was on friendly terms with Thebes during the Peloponnesian War, but afterwards joined the Spartans, and suffered utter destruction at the hands of the Thebans, B.C. 368. It was afterwards rebuilt, again destroyed by the Thebans in 346, and restored by the Macedonians;

---

[4] Καί τε δι' Ὀρχομενοῦ εἰλιγμένος εἶσι δράκων ὥς.—Nexton, ap Strab. ix. p. 424.

[5] Οὐδ' ὅσ' ἐς Ὀρχομενὸν ποτινίσσεται, οὐδ' ὅσα Θήβας
Αἰγυπτίας, ὅθι πλεῖστα δόμοις ἐν κτήματα κεῖται.   Il. ix. 381.

[6] Ὅς ποτ' ἐν Ὀρχομενῷ Μινυηΐῳ ἶφι ἄνασσεν.   Od. xi. 283.

[7] Ἵξεν δ' ἐς Φλεγύων ἀνδρῶν πόλιν ὑβριστάων,
Οἳ Διὸς οὐκ ἀλέγοντες ἐπὶ χθονὶ ναιετάασκον
Ἐν καλῇ βήσσῃ, Κηφισίδος ἐγγύθι λίμνης.—Hom. Hymn. in Apoll. 278.

but it never afterwards flourished. **Lebadēa**, *Livadhia*, was situated near the western border, with its acropolis on a spur of Helicon, by whose base the Hercyna flowed. It owed its importance to the possession of the oracle of Trophonius which was delivered from a cave in the rock. Lebadea was taken and plundered both by Lysander and by Archelaus the general of Mithridates. **Coronēa** was situated on a height overlooking the Copaic plain; at this point the roads from Orchomenus and Lebadea in the N. joined those from Thebes and Platæa in the S. It was thus the scene of several important military events—of Tolmides's defeat and death in B.C. 447, of Agesilaus'e victory over the Argives and Thebans in 391, and of a double siege in the Sacred War. **Haliartus** stood on the southern side of Lake Copais, amid well-watered meadows,[*] on the road between Coronea and Thebes. It is chiefly memorable for the engagement in which Lysander perished, B.C. 395. It was twice destroyed—by the Persians in 480, and by the Romans in 171. **Thebes** was situated in the southern plain of Bœotia, on a spur of Mt. Teumessus, which rises about 150 feet above the plain: at the base of the hill, on either side, run the streams Ismēnus and Dirce,[*] which unite in the plain below the city: a third stream of less importance, named Strophia, runs through the city.[*] The

Coin of Thebes.

Cadmeia, or citadel, is supposed to have stood at the southern end of the town, and the temple of Ismenian Apollo a little to the E. of it, while the Agora and other buildings stretched out towards the N. Of the seven gates[*] for which Thebes was so celebrated, three opened

---

[*] Hence the Homeric epithet, "grassy," applied to it :—

Οἵ τε Κορώνειαν, καὶ ποιήενθ᾽ Ἁλίαρτον.   *Il.* II. 503.

Εἶθεν ἄρ᾽ εἰς Ἁλίαρτον ἀφίκετο ποιήεντα.—*Hymn. in Apoll.* 243.

[*] The streams of Dirce and Ismenus are frequently commemorated by the Greek poets, particularly by Euripides, who speaks of them as the "twin streams," and applies to the water of Dirce the epithet "white," or "limpid," and "fair-flowing;" and by Pindar, who applies similar epithets to it.

Διδύμων ποταμῶν, πόρον ἀμφὶ μέσον
Δίρκας, χλοεροτρόφον ἁ πεδίον   /
Πρόπαρ Ἰσμηνοῦ καταδεύει.   EUR. *Phœn.* 823.

Νεαρὸν ἅπαντ᾽ Ἰσμηνὸν ἐμπλήσω φόνου,
Δίρκης τε νᾶμα λευκὸν αἱμαχθήσεται.   *Bacc. Fur.* 571.

Δίρκα δ᾽ ἁ καλλιρρόεθρος.   *Id.* 780.

                              πίσω σφε Δίρκας
Ἀγνὸν ὕδωρ, τὸ βαθύζωνοι κόραι
Χρυσοπέπλου Μναμοσύνας ἀνὰ-
ταιλαν παρ᾽ εὐτειχέσιν Κάδμου πύλαις.—PIND. *Isthm.* vi. 105.

[*] From the two more important streams, Thebes is described as "the two-rivered city :"—

Διπόταμον ἵνα πόλιν μόλω.   EUR. *Suppl.* 621.

[*] The erection of these walls was attributed to Amphion and Zethus :—

Καί ῥ᾽ ἔτεκεν δύο παῖδ᾽, Ἀμφίονά τε, Ζῆθόν τε,
Οἳ πρῶτοι Θήβης ἕδος ἔκτισαν ἑπταπύλοιο,
Πύργωσάν τ᾽ ἐπεὶ οὐ μὲν ἀπύργωτόν γ᾽ ἐδύναντο
Ναιέμεν εὐρύχορον Θήβην, κρατερώ περ ἐόντε.—HOM. *Od.* xi. 762.

towards the S. and one towards the W.; the position of the northern gate is self-evident, that of the two others is doubtful. Thebes was believed to have been founded by a Phœnician colony under Cadmus, whence the title of the citadel, Cadmeia, and the old Homeric name of the people, Cadmeans. The town holds an important place both in mythology, as the birth-place of Dionysus and Hercules, and in the early annals as the scene of the wars of the "Seven against Thebes" and of the Epigoni. Its subsequent history is involved in that of Bœotia, and indeed in that of Greece generally. Its fall dates from its capture by Alexander in 335, when it was utterly destroyed. It was rebuilt, in 316, by Cassander, and again destroyed by Mummius, in 146. Thespiæ was

Coin of Thespiæ.

situated at the foot of Helicon, W. of Thebes. It was generally hostile to Thebes, and took a prominent part in the Persian War on the patriotic side. It was several times dismantled and depopulated by the early Thebans, but it survived to the Roman era and became then one of the chief towns of Bœotia. It derived celebrity both as a seat of fine arts—possessing statues cut by Praxiteles—and as the place where the Erotidia (games in honour of Love) were celebrated. It had a port named Crèusis on the Corinthian Gulf. Platæa stood about 6½ miles S. of Thebes, at the foot of Cithæron, and commanding the pass across that ridge into Attica. It was the scene of the remarkable victory over the Persians in B.C. 479, and of the no less famous siege in the Peloponnesian War in 429–427. After the destruction of the town by the Thebans, Platæa remained in ruins until 387, when it was partly restored, but again destroyed by the Thebans in 374, and permanently restored after the battle of Chæronea in 338. Tanagra

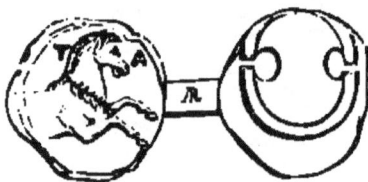

Coin of Tanagra.

was on a circular hill close to the left bank of the Asōpus, and from its proximity to Attica it became the scene of engagements between the Athenians and Lacedæmonians in B.C. 457, between the Athenians under Myronides and the Bœotians, the latter being defeated at Œnophyta in 456, and between the Athenians and Bœotians in 426. Larymna was the name of two towns on the Cephissus, one of which, named Upper Larymna, was at the spot where the river emerged from its subterranean channel; the other, Lower Larymna, at the mouth of the river. The former originally belonged to Locris, the latter was a member of the Bœotian confederacy. The Romans removed the inhabitants of the Upper to the Lower Larymna, which became a considerable town: its ruins are named Kastri, and consist of the circuit of the walls and other vestiges.

Of the less important towns we may notice—**Alalcomēnæ**, at the foot of Mt. Tilphossium, celebrated for the worship of Athena ;[3] **Onchestus**, S.E. of Haliartus, and belonging to it, famed for a temple and grave of Poseidon ;[4] **Ascra**, on Mt. Helicon, W. of Thespiæ, the residence of Hesiod ;[5] **Thisbe**,[6] in the S.W., near the sea, and possessing a low enclosed plain which was liable to be flooded, but was rendered in parts cultivable by means of a causeway made to divert the waters ; **Creusis**,[7] at the head of a small bay of the Corinthian Gulf, serving as the port of Thespiæ but difficult of access in consequence of the storms and headlands ; **Eutrēsis**, an Homeric town between Creusis and Thespiæ, possessing a temple and oracle of Apollo ; **Leuctra**, a little S.E. of Thespiæ, the scene of the celebrated battle between the Thebans and Spartans in B.C. 371, the battle-field being marked by a tumulus in which the Spartans were probably buried; **Hysiæ**, at the N. foot of Cithæron, on the high road from Thebes to Athens, and at one time belonging to Athens; **Erythræ**, a little S. of the Asopus, at the foot of Cithæron, the extreme E. point to which the camp of Mardonius reached; **Scolus**, between Tanagra and Platæa, and hence visited by Mardonius and selected by the Thebans as a spot to throw up an intrenchment against the Spartans in B.C. 377; **Etēonus**, afterwards named **Scarphe**, to the right of the Asopus, under Cithæron;[8] **Delium**, on the sea-coast, close to the border of Attica, with a celebrated temple of Apollo, the scene of the Athenian defeat in B.C. 424, and also of the defeat of a Roman detachment by the troops of Antiochus, in 192; **Aulis**, on the Euripus,[9] the place where the Grecian fleet assembled before they started for Troy,[1] identified with the

---

[3] "Ήρη τ' Ἀργείη, καὶ Ἀλαλκομενηΐς Ἀθήνη. *Il.* iv. 8.

[4] Onchestus was famed for a grove of Neptune near it :—
'Ογχηστὸν δ' ἱερὸν, Ποσιδήιον ἀγλαὸν ἄλσος. *Il.* ii. 506.
'Ογχηστον δ' ἷξε, Ποσιδήιον κυλαὸν ἄλσος·
Ένθα καθμὴν πῶλος ἀναπνέει ἀχθόμενος κῆρ.—*Hymn. in Apoll.* 230.

[5] Hesiod thus describes his native place :—
Νάσσατο δ' ἀγχ' Ἑλικῶνος διζυρῇ ἐνὶ κώμῃ.
Ἀσκρη, χείμα κακῇ, θέρει ἀργαλέη, οὐδέ ποτ' ἐσθλῇ.—*Op. et Di.* 639.

[6] The rocks on the sea-coast have in all ages been the resort of vast numbers of wild pigeons :—
Πολυτρήρωνά τε Θίσβην. *Il.* ii. 502.
Quæ nunc Thisbæas agitat mutata columbas.—Ov. *Met.* xi. 600.
Nysa, Dionæisque avibus circumsona Thisbe.—Stat. *Theb.* vii. 261.

[7] A very difficult route led from this place to Megaris, along the heights of Cithæron. The Spartans passed this way under Cleombrotus in B.C. 378, and after the battle of Leuctra in 371.

[8] Hence the terms which Statius applies to it :—
Qui Scolon densumque jugis Eteonon iniquis.—*Theb.* vii. 266.

[9] Ἔμαλον ἀμφὶ παρακτίαν
ψάμαθον Αὐλίδος ἐναλίας,
Εὐρίπου διὰ χευμάτων
Κέλσασα, στενόπορθμον
Χαλκίδα. *Iph. in Aul.* 164.

[1] It is characterised by Homer as the "rocky," by Euripides as the "tranquil" Aulis :—
Οἳ δ' Ὕρίην ἐνέμοντο, καὶ Αὐλίδα πετρήεσσαν. *Il.* ii. 496.
Αὔλιν ἀκλύστεαν. *Iph. in Aul.* 121.
Ὅτ' ἐς Αὐλίδα νῆες Ἀχαιῶν
Ἡγερέθοντο, κακὰ Πριάμῳ καὶ Τρωσὶ φέρουσαι. *Il.* ii. 303.

modern *Vathy*, a name evidently representing the βαθὺς λιμήν of Strabo
(ix. p. 403); **Mycalessus**, an Homeric town (*Il.* ii. 498) somewhere near the
Euripus, chiefly famous for the massacre of its inhabitants by the Thracians
in B.C. 413; **Salganeus**, on the coast N. of Chalcis, commanding the N.
entrance to the Euripus; **Anthedon**, on the coast, celebrated for its
wine, and occupied by a non-Bœotian race; **Schœnus**, on a small river
of the same name which flows into Lake Hylica, the birth-place of
Atalanta; **Hyle**, on Lake Hylica, erroneously described by Moschus
as the birth-place of Pindar; **Teumessus**, N.E. of Thebes, on a low
rocky hill of the same name, chiefly known from the legend of the
Teumessian fox which ravaged the territory of Thebes; **Acrœphium**,
on the E. of Copais on the slope of Ptoum, with a celebrated oracle of
Apollo near it, which was consulted by Mardonius; **Copæ**, on the N.
extremity of the lake and the site of *Topolia*, a place which, though
a member of the Bœotian confederacy, was of small importance; and,
lastly, **Tegyra**, very near Orchomenus, with a celebrated temple and
oracle of Apollo.

History.—The withdrawal of Platæa from the confederacy was the
first event that involved the Bœotians in a foreign war: Athens sided
with the seceding town and war followed, in which Platæa was rendered
independent, probably in B.C. 519. In the Persian War the Thebans
sided with the invader, much to the dissatisfaction of the other towns ;
and they retained their supremacy only through the aid of the Spartans.
The Athenians invaded Bœotia in 457 and 456, meeting with a defeat
at Tanagra, but succeeding at Œnophyta, and for a while establishing
democracy. The invasion of Tolmides in 447 was unsuccessful, and
oligarchy was reinstated. The attack on Platæa in 431 was the first
act of the Peloponnesian War, throughout which the Thebans steadily
opposed Athens. Jealousy of the Spartans produced an opposite
policy after the conclusion of the war: Thebes and Sparta became
hostile, and the Bœotian War at length broke out in 395, signalized by
the death of Lysander at Haliartus and the victory of Agesilaus at
Coronea in 394. The peace of Antalcidas in 387 and the seizure of
the Cadmea in 382, by which Sparta endeavoured to humble Thebes,
were followed by the expulsion of the Spartans in 379, and the increase
of Theban power. The peace of Callias in 371 permitted the con-
centration of the Spartan efforts against Thebes ; but these were foiled
on the plain of Leuctra in 371, and, under Epaminondas, Thebes became
the leading military power in Greece until the battle of Mantinea in
362. Throughout all this period Orchomenus and Thespiæ had sided
with the enemies of Thebes : the former was burnt in 366, and the
latter deprived of its inhabitants about the same period. War with
Athens ensued in connexion with Eubœa in 358, and this was followed
by the Sacred War in 357, which, through the intervention of Philip,
terminated in the recovery of the cities which Thebes had lost in the
early part of the war. The alliance with Athens was renewed in 339 in
opposition to Philip, who defeated the joint army at Chæronea in 338,
deprived Thebes of its supremacy, and held possession of the Cadmeia.
The attempt to expel the Macedonian garrison led to the total destruction of the city by Alexander in 335. It was rebuilt in 316 ; was twice
taken by Demetrius in 293 and 290 ; its walls were destroyed by
Mummius in 146 ; and it was finally reduced to insignificance by Sulla
in the Mithridatic War.

The Parthenon in its present state.

CHAPTER XXI.

CENTRAL GREECE—*continued.* ATTICA, MEGARIS.

VIII. ATTICA.

§ 1. Attica is a peninsula (as its name, derived from ἀκτή,[1] probably implies) of a triangular form, having two of its sides washed by the sea, viz. by the Ægean on the E. and the Saronic Gulf on the W., and its base united to the land, being contiguous on the N. to Bœotia. In the N.W. it was bounded by Megaris, which naturally belongs to the peninsula, and was originally united to Attica, but was afterwards separated from it. The area of Attica is about 700 square miles; its greatest length is 50, and its breadth 30 miles. The position and physical character of this country destined it for commercial and political supremacy. Standing at the entrance of the Peloponnesus, it commanded the line of communication between Northern and Southern Greece; and yet, being actually off the high road, it was itself tolerably secure from the passage of invading

---

[1] The name would thus have been originally Ἀκτική: this etymology has been questioned of late, and the name referred to the root *Att*, or *Ath*, which we see in Ath-enæ.

armies. On the N. it is shut off from Bœotia by a line of lofty and in most places inaccessible mountains, while on the S. the passes of Megaris were easily defensible. The E. coast was guarded by the isle of Euboea, and by the narrow intervening strait of Euripus, and the W. by the adjacent islands of Salamis and Ægina. As the most easterly part of Greece, it was the nearest point to Asia, with which it held easy communication by the intervening chain of islands. It was also practically the nearest point to Egypt. The soil is light and dry, and little adapted to the growth of corn. The primitive limestone, which is the geological formation of the country, protrudes on the mountain-sides, and even on the plains. The country was too hilly, and the soil too poor, for the breeding of horses or cattle. On the other hand, Attica was rich in mineral productions. The silver mines of Laurium and the marble quarries of Pentelicus were sources of national wealth. Hence, though agriculture was held in honour, maritime commerce was the natural occupation of the population; and this, combined with the centrality of its position, secured that ascendency which rendered Athens so conspicuous in ancient history.

§ 2. The mountain-chain which separates Attica from Bœotia in the W. part of the province, where the line of communication between Northern Greece and Peloponnesus ran, was named Cithæron. This was continued towards the E. in the range of Parnes.[1] *Nozía*; and towards the S. in the Onéan mountains of Megaris. The northern ranges were crossed at three points: viz. in the W. by the Pass of Dryoscephalæ, "Oak-heads," between Platæa and Eleusis; in the centre by the wild and rugged Pass of Phyle, through which ran the direct road between Thebes and Athens; and in the E. by the Pass of Deceléa, leading from Athens to Oropus and Delium. From the N.W. angle of Attica a range runs towards the S., terminating on the W. of the Bay of Eleusis in two summits named Kerāta, "the Horns," now *Kandili*. Another range descends from Parnes, under the name of Ægaleus, to the E. of the Bay of Eleusis. Another, also emanating from Parnes, runs in a parallel direction more to the E., and was named, in its N. portion, Brilessus, or Pentelicus, *Mendeli*, and in its S. portion Hymettus.[2]

---

[1] Parnes was favourable to the growth of the vine :—

Dives et Ægaleos nemorum, Parnesque benignus
Vitibus.   STAT. *Theb.* xii, 620.

[2] Hymettus was famed for its honey; it was also formerly well clothed with wood : the passage quoted from Ovid describes the source of the Ilissus on this mountain :—

Est prope purpureos colles florentis Hymetti
Fons sacer, et viridi cespite mollis humus.
Silva nemus non alta facit; tegit arbutus herbam :
. Ros maris et laurus, nigraque myrtus olent.

The latter is subdivided into two parts by a remarkable break,—the northern or Greater Hymettus, now named *Telo-Vuni*; and the southern or Lesser, which was formerly called Anhydrus, "Waterless," and now *Mauro-Vuni*. Between the ranges specified, plains intervene: viz. the Eleusinian or Thriasian Plain, between Kerata and Ægaleus; and the Athenian Plain, or, as it was frequently termed, "*the* Plain" (τὸ Πέδιον), between Ægaleus and Pentelicus. The mountainous district at the head of the latter, between Parnes, Pentelicus, and the sea, was named Diacria, "the Highlands." S.E. of Hymettus is an undulating district named Mesogæa, "the Midland;" and this is followed by the Paralia, "the Sea-coast," a hilly and barren district, including the whole southern division from Prom. Zoster on the W., and Brauron on the E., down to Sunium. In the S. of this lies the ridge of Laurium,[4] *Legrana*, probably so named from the shafts (λαύρα, "a street" or "lane") sunk for obtaining the silver-ore, some of which still remain, as do also the heaps of scoria. The chief promontories are Zoster, the extreme point of Hymettus; Sunium, at the extreme S. of Attica, rising almost perpendicularly from the sea to a great height, and crowned with a temple of Minerva, to the ruins of which the promontory owes its name of *C. Kolonnes*; and Cynosura, "Dog's Tail," a long rocky projection, bounding the Bay of Marathon on the N.

§ 3. The rivers of Attica are little better than mountain torrents, almost dry in summer, and only full in winter or after heavy rains.

Nec densum follis buxum, fragilesque myricæ
Nec tenues erylæi, cultaque pinus abest.
Lenibus impulsæ Zephyris, auræque salubri,
Tot generum frondes, herbaque summa tremunt.—Ov. *Art. Am.* III. 687.

Hoc tibi Thesei populatrix misit Hymetti
Pallados a silvis nobile nectar apis.—MARTIAL. xIII. 104.

Ingenium, dulcique senex vicinus Hymetto.—Juv. xIII. 185.

The marble of Hymettus was also famed:—

Non trabes Hymettiæ
Premunt columnas ultima recisas
Africa.   HOR. *Carm.* II. 18, 3.

[4] Homer gives it the epithet "sacred;" the epithet "silvery" in Euripides has reference to the mines of Laurium:—

᾽ΑΛΛ' ὅτε Σούνιον ἱρὸν ἀφικόμεθ', ἄκρον ᾽Αθηνῶν   Od. III. 278.

Περαίμας
᾽Ἡν' ὑλᾶεν ἔπεστι πόντου
Πρόβλημ' ἁλίκλυστον, ἄκραν
Τὴν ὑλᾶεα Σουνίου,
Τὰς ἱερὰς ὅπως προστεί-
ποιμ' ἂν ᾽Αθάνας.   SOPH. *Aj.* 1217.

ἧ τε Σουνίῳ
Διαι ᾽Αθάναι σὺν ἐπάργυρος πέτρα.   EUR. *Cycl.* 293.

The Athenian plain is watered by two rivers: the Cephissus,[5] a perennial stream which rises in Parnes and flows on the W. side of Athens into the Phaleric Bay; and the Ilissus, a less important stream rising in Hymettus, and, after receiving the Eridanus, flowing through the S. of Athens towards the Phaleric Bay. These rivers still retain their ancient names. The former is now subdivided into several streams for the purpose of irrigating the olive-groves and gardens; the latter is generally exhausted before it reaches the sea. The Cycloborus[6] was a torrent descending from Parnes, probably the *Megalo Potamo*. The Eleusinian plain is watered by a second Cephissus, *Saraudaforo*, which rises in Cithæron, and by another stream now named the *Janula*.

§ 4. The population of Attica belonged to the Ionian branch of the Hellenic race, and made it their particular boast that they were *autochthonous*,[7] a circumstance which Thucydides (i. 2) attributes to the poverty of the soil. The Athenians were originally named Cranai, and afterwards Cecropidæ, and did not assume their later name until the reign of Erechtheus. The earliest political division of Attica was attributed to Cecrops, who parcelled out the country into twelve independent communities, which were afterwards consolidated into one state by Theseus. Another ancient division, attributed to the sons of Pandion, was based upon the natural features of the country, Ægeus receiving the coast-land (ἀκτή), with the plain of Athens (πεδιάς); another brother the highlands (διακρία); and another the southern coast (παραλία). These districts supplied the basis of the three political parties in the time of Solon and Pisistratus. Another division was into four tribes (φυλαί), the names of which varied at different times, the most important designations being those which prevailed in the time of Cleisthenes into Geleontes, Hoplites, Argades, and Ægicores. This division was superseded by that of Cleisthenes into ten tribes, named after Attic heroes; two more were added in B.C. 307, named after Antigonus and his son Demetrius; and a third in the reign of Hadrian, after whom it was named. There was a further division into townships or cantons (δῆμοι), of which there were 174 in the third century B.C.[8] The

---

[5] οὐδ' Ἰλισσος
Κρῆναι μινύθουσι
Κηφισοῦ νομάδες ῥεέθρων,
'Αλλ' αἰὲν ἐπ' ἤματι
'Ωκυτόκος πεδίων ἐπινίσσεται.   SOPH. Œd. Col. 686.

[6] Aristophanes refers to the roaring sound of its waters:—
'Αρπαξ, ἀκράχτης, Κυκλοβόρου φωνὴν ἔχων.   Equit. 137.

[7] εἶναί φασι τὰς αὐτόχθονας
Κλεινὰς 'Αθήνας οὐκ ἐπείσακτον γένει.   EUR. Ion. 592.

[8] Herodotus (v. 69) appears to give 100 as the original number of the *demi*; there is, however, some little doubt about the meaning of the passage.

tribes and the *demi* were to a certain extent a cross division, the latter being originally a *local*, the former a purely *political* arrangement; and thus adjacent townships belonged in many cases to different tribes. Even the *demus* lost its local character by degrees, as change of abode did not affect the original arrangement, the descendants of a man always remaining members of the *demus* in which their ancestor was enrolled in the time of Cleisthenes. The larger *demi* contained a town or village, the smaller ones only a temple or place of assembly. The names of most of them are preserved, but their positions are very often unknown.

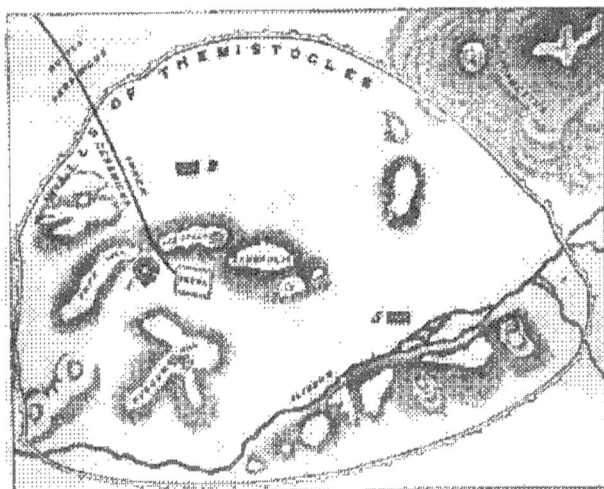

Plan of Athens.

1. Pnyx Ecclesia.   2. Theseum.   3. Theatre of Dionysus.   4. Odeum of Pericles.
3. Temple of the Olympian Jove.

§ 5. **Athens**,[*] the capital of Attica, was situated in the central plain already described, at a distance of about 4½ miles from the sea-coast. The site of the city was diversified by several elevations, the most conspicuous of which was the **Acropolis**, an oblong, craggy rock rising abruptly about 150 feet, with a flat summit 1000 feet long from E. to W., and 500 broad; while grouped around it were the lesser heights of the **Areopagus** and the **Pnyx** on the W., and the **Museum** on the S.W. The river **Ilissus** traversed the southern

---

[*] The name is said to have been derived from the worship of Athena, which was introduced by Erechtheus.

quarter of the city, near the base of the Museum; the Cephissus ran
outside the walls on the W. side of the town, about 1½ mile distant.
In addition to the hills already enumerated, we must notice **Lyca-
bettus**,[1] *Mount St. George*, a lofty conical peak to the E. of the
Acropolis, not included within the limits of the city. The walls of
Themistocles passed along the W. base of the Pnyx, and crossed the
Ilissus near the W. extremity of the Museum; thence they turned
E., and included some heights to the S. of the Ilissus; on the E. side
of the town they passed below Mt. Lycabettus, and returned with a
broad sweep towards the N. to the neighbourhood of the Pnyx.
The town within these limits consisted of two parts—the Acropolis
or Polis, and the *Asty* or "City"—the former consisting of the
central rock already described, on which the original city[2] of Cecrops
stood, and which subsequently formed the citadel of Athens; the
latter consisting of the town, which lay beneath and around it, and
which was divided into the following districts:—Inner Ceramicus,
extending from the gate of Eleusis to the Agora; Melite, comprising

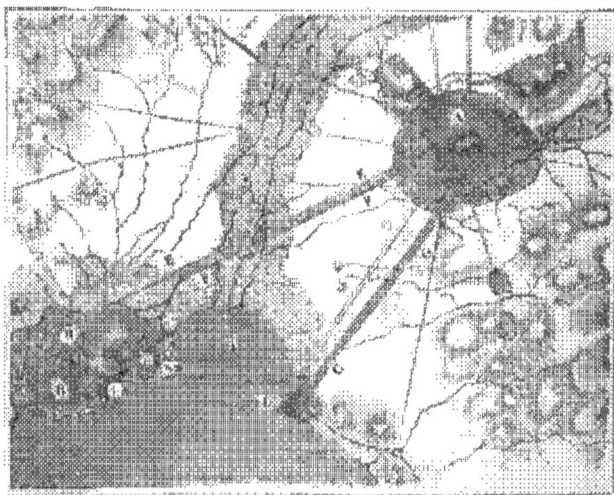

Athens and its Port-Towns.

[1] Aristophanes alludes to Lycabettus as a mountain of some celebrity :—

      Ἦν οὖν σὺ λέγῃς Λυκαβηττοὺς
      Καὶ Παρνασῶν ἡμῖν μεγέθη.   *Ran.* 1064.

[2] This was the "ancient Cecropia :"—

      Αὐτός τ' ἄνασσα, παῖδα ἀλευὸν Αἰγέως
      Καὶ τοὺς σὺν αὐτῷ, δεξιὸν τεταγμένους
      Κέρας, παλαιᾶς Κεκροπίας οἰκήτορας.   *Eurip. Suppl.* 658.

the hills of the Pnyx and Museum; Scambonidæ and Colyttus, in
the same quarter, and sometimes included under Melite; Cœle, between the Museum and the Ilissus; Cydathenæum, on the S. of the
Acropolis; Diomêa, including the whole eastern district; and Agræ,
in the S.E., beyond the Ilissus. The appearance of the town was
striking from the number of fine public buildings in it, and particularly those on the summit of the Acropolis. The streets and private
houses, on the other hand, were of very inferior character. The
port of Athens was on the Saronic Gulf, at a distance of about
4½ miles from the city. The original port was at Phalērum, on the
E. side of the Phaleric Bay (I), at a spot now named *Treis Pyrgoi* (D).
Subsequently to the Persian War this was abandoned for a more
westerly situation, where the Peiraïc peninsula afforded three natural
basins,—the largest being Piræus (H) on the W. side, now named
Drako or *Port Leone;* and the two smaller ones on the E. side,
Munychia, *Fanari* (K), and **Zea,** *Stratiotiki* (L), the former being
the most inland of the two. Gradually the peninsula was covered
with buildings, and important suburbs grew up at the extremity
and on the W. side of it, named respectively Piræus (D) and Munychia (c). The port-towns were connected with the city (A) by three
walls, two of which ran in a S.W. direction to Piræus, in parallel
lines 350 feet apart, and were together named the " Long Walls," or
separately the Northern or Outer (E E), and the Southern or Intermediate (F F), while the third, called the Phaleric (G G), connected
Athens with Phalerum.   The general aspect of Athens thus resembled two circular cities connected by a long street. The port-
town was described as the *Lower* City, in contradistinction to the
Asty or *Upper* City: occasionally, however, the latter term, as
already observed, was applied to the Asty itself, in contradistinction
to the Acropolis, which towered above it. The population of the
whole city is variously estimated at from 120,000 to 192,000 souls.
We proceed to a more minute description of the town and its most
remarkable public buildings.

(1.) *The Acropolis.*—The rock of the Acropolis stood in the centre of
Athens, and was the very heart of the city, its fortress and its sanctuary.[3] On three sides it is inaccessible: towards the W. it is ascended
by a gentle slope. The summit was enclosed with walls, said to have
been originally erected by the Pelasgians, but certainly rebuilt after
the Persian War: the northern, which retained the name of the Pelasgic
Wall, was probably restored by Themistocles, and the southern by
Cimon, after whom it was named. The name of Pelasgicum extended
to a space of ground below the wall, probably at the N.W. angle of the

Acropolis. The rocks on the N. side were named the Long Rocks, a title equally applicable to those on the S. side, but restricted in use to the former, probably as being the more conspicuous from the Athenian plain. The western entrance was guarded by the Propylæa (Plan, 2, 3), erected by the architect Mnesicles in B.C. 437-431, under the direction of Pericles, consisting of a double central portico, through which a magnificent flight of steps led up from the town, and two projecting wings, 26 feet in front of the western portico—the northern one containing a chamber named Pinacothéca, from its walls being covered with paintings, while the southern had no chamber. Opposite the latter stood the small temple of Niké Apteros (Plan, 4), " Wingless Victory," built to commemorate the victory of Cimon at the Eurymedon : the whole was of Pentelic marble, and extended along the whole W. end of the Acropolis, a distance of 168 feet. Of these buildings the inner portico still remains, together with the northern wing. The temple of Niké Apteros has been rebuilt in modern times with the original materials, which were found on the spot. Just in front of the northern wing is the so-called Pedestal of Agrippa, formerly surmounted by the equestrian statues of the two sons of Xenophon (Plan, 5). The chief building within the Propylæa was the **Parthénon** (Plan, 1), which stood on the highest part of the Acro-

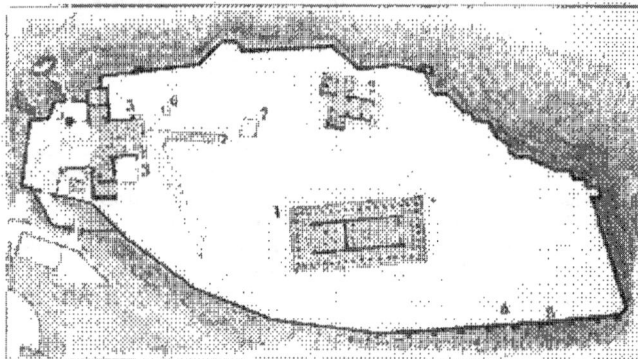

Plan of the Acropolis.

polis : it was built by the architects Ictinus and Callistratus, under the direction of Pericles, and was dedicated to Athena the "virgin," so named as being the invincible goddess of war. It was built entirely of

[1] Ἔστιν γὰρ οὐκ ἄσημοι Ἑλλήνων πόλει,
Τῆι χρυσολόγχου Παλλάδος κεκλημένη,
Οὗ παῖΔ Ἐρεχθέως Βοῦβοι ἐξεφῦσεν γόνους
Βίᾳ Κρέουσαν, ἔνθα προσβλέψας νέφραι
Παλλάδος ὑπ᾽ ἐχθὺ τῆς Ἀθηναίων χθονὸς
Μακρὰς καλοῦσι γῆς ἄπαντες Ἀντίδοι.   EUR. Ion. 1.

[2] From the position of this temple at the entrance of the Propylæa, the goddess was invoked by persons quitting or entering the Acropolis :—

Νίκη τ᾽ Ἀθάνα Πολιὰς, ἢ σώζει μ᾽ ἀεί.   SOPH. Philoct. 134.
Δέσποινα Νίκη ξυγγενοῦ, τῶν τ᾽ ἐν πόλει γυναικῶν.
     ARISTOPH. Lysistrat. 317.

Pentelic marble in the purest Doric style, its dimensions being 228 feet in length, 101 in breadth, and 66 in height to the top of the pediment. It consisted of a *cella*, surrounded by a peristyle, having eight columns in each front, and seventeen at each side—in all forty-six.   Before each end of the cella there was an interior range of six columns.   The cella itself was divided into two chambers, the eastern of which was the *Naos*, or shrine, and specially named the Hecatompědon, being ninety-eight feet long, and the western, named the Opisthodōmus and the Parthenon, in its special sense, forty-three feet long.   The former con-

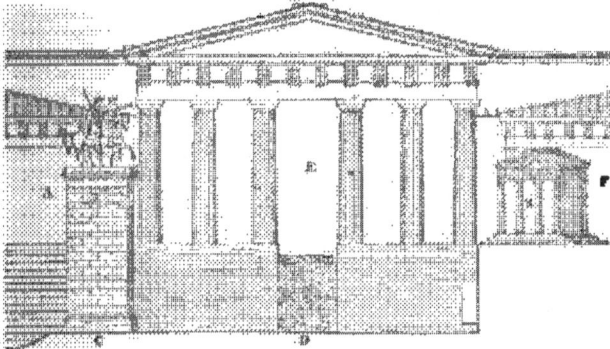

The Propylæa restored.

A. Pinacotheca.
B. Temple of Niké Apteros.
C. Pedestal of Agrippa.
D. Road leading to the Central Entrance.
E. Central Entrance.
F. Hall corresponding to the Pinacotheca.

tained the colossal statue of Athena of ivory and gold, the work of Phidias, while the latter was used as the Treasury of Athens.   Round the summit of the outer walls of the cella was a frieze in low relief, 520 feet in length, representing the Panathenaic procession : the slabs of which it was formed are the well-known Elgin Marbles in the British Museum.   The Parthenon remained almost entire until A.D. 1687, when it was accidentally blown up during the siege by the Venetians ; it was again injured in 1827.   The Erechthëum (Plan, 2) stood N. of the Parthenon, and was the most revered of all the sanctuaries of Athens, being connected with the most ancient legends of Attica.   The original temple was attributed to Erechtheus, and contained the statue of Athena Pollas, of olive-wood, which fell down from heaven, the sacred tree, and the well of salt water—the former evoked by Athena, and the latter by Poseidon in their contest—and the tombs of Cecrops and Erechtheus. The building contained two separate sanctuaries, dedicated to Athena and Pandrosus.   This temple was destroyed by the Persians, and a new one founded on its site about the commencement of the Peloponnesian War, but not completed until about B.C. 393: its form was peculiar, consisting of an oblong *cella*, seventy-three feet long, and thirty-seven broad, with a portico at the E., and two porticoes at the western end, not facing the W., but the N. and S., and thus resembling the transepts of a church. The E. portico consisted of six Ionic columns, of which five are now standing; the N. portico had four columns in front, and two at the sides, all of which remain; the S. portico had its roof supported by six caryatides, instead of columns, and was low : five of these are standing, and the other is in the British Museum. The building contained two principal chambers—the eastern, or larger one, sacred to Minerva, the lesser to Pandrosus: the former contained the olive-wood statue covered with a *peplos*,[a] and the latter the olive-tree,. These compartments were on different levels, the eastern being eight feet higher than the western. The N. portico, which gave entrance to the Pandrosium, contained the sacred well; and the S. portico was the Cecropium, or sepulchre of Cecrops, accessible only from within. The whole was surrounded by a Temenos, or sacred enclosure, within which were numerous statues. The Acropolis further contained the colossal statue of Athena Promachus (Plan, 5), seventy feet high, facing the Propylæa, and so lofty that the point of the spear and crest of the helmet were visible from Sunium ; a brazen quadriga on the left hand as you entered the Acropolis ; the Gigantomachia, a piece of sculpture on the Cimonian wall ; and a temple of Artemis Brauronia, between the Propylæa and Parthenon.

(2.) *The Asty.*—The first object that meets one descending from the Acropolis is the Areopagus, "the hill of Ares, or Mars,"[b] memorable as the place of meeting of the Upper Council, which held its sittings on the S.E. summit of the rock in the open air: a bench of stone excavated in the rock, forming three sides of a quadrangle facing the S., served as their chamber. Here it was that St. Paul addressed the men of Athens (Acts xvii. 22). At the N.E. angle of the hill was a dark chasm, which formed the sanctuary of the Eumenides.[c] About a quarter of a mile from the centre of the Areopagus is the Pnyx, or place of assembly of the people, an area of nearly semicircular form, gently sloping towards the agora, artificially formed out of the side of a rocky hill by excavating at the back, and embanking in front: the *bema*, whence the orators spoke, faced the N.E. in the direction of the agora ; it is a large stone, twenty feet high and eleven broad, and commanded a view of the Acropolis and city. The area of the Pnyx contained 12,000 square yards, and was unencumbered with seats. Behind the *bema*, on the summit of the rock, is an artificial terrace, whence a view of the sea could be obtained: this has been supposed by some to have been the original Pnyx, but it was more probably an appendage of the other. The Agora, or market-

---

[a] This is the image referred to by Æschylus : —

ἴζου παλαιὸν ἄγκαθεν λαβὼν βρέτας.　　　　*Eum.* 80.

[b] Ἔσται δὲ καὶ τὸ λοιπὸν Αἰγέως στρατῷ
Ἀεὶ δικαστῶν τοῦτο βουλευτήριον
Πάγον δ' Ἄρειον τόνδ' Ἀμαζόνων ἕδραν
Σκηνάς θ', ὅτ' ἦλθον Θησέως κατὰ φθόνον
Στρατηλατοῦσαι, καὶ πόλιν νεόπτολιν,
Τήνδ' ὑψίπυργον ἀντεπύργωσαν τότε·
Ἄρει δ' ἔθυον, ἔνθεν ἔσθ' ἐπώνυμος
Πέτρα, πάγος τ' Ἄρειος.　　　　*Æsch. Eum.* 683.

[c] The position of this sanctuary is frequently alluded to by the tragic poets :—

Πάγον παρ' αὐτὸν χάσμα δύεσθαι χθονός.　　　*Eur. Electr.* 1240.

Ἵνε καὶ σφαγίων τῶνδ' ὑπὸ σεμνῶν
Κατὰ γῆς σύμεναι.　　　　*Æsch. Eum.* 1006.

place, was in the depression between the Acropolis, the Areopagus, the Pnyx, and the Museum: it contained several stoæ, or colonnades—the Stoa Eleutherios dedicated to Jupiter; the Stoa Basileios, where the Archon Basileus held his court; and the Stoa Pœcile (so named from the frescoes with which it was adorned), from which the school of the Stoics derived their name. The other public buildings and objects in the Agora were—the Metrônm, where the public records were kept; the Tholus, where the Prytanes took their meals; the Bouleuterion, or council-chamber of the 500; the statues of the ten Eponymi, or heroes of Athens; the Prytanéum; and the central altar of the Twelve Gods. On the hill of the Museum was the monument of Philopappus, who lived in the age of Vespasian; portions of it still remain. Beneath the S. wall of the Acropolis, near its E. end, was the stone Theatre of Dionysus, commenced in B.C. 500 and completed in 340; the middle of it was excavated out of the rock, and its extremities supported by strong masonry. The area was large enough to hold all the population of the city, which here viewed all the grand productions of the Greek drama. The seats were arranged in curved rows hewn out of the rock, and, as the area was unenclosed, the spectators commanded a view of Salamis and the sea, while behind them were the Parthenon and the other buildings of the Acropolis.* Adjacent to the theatre on the S. was the Lenæum, containing within its enclosure two temples of Dionysus; and immediately E. of the theatre was the Odéum of Pericles, the roof of which is said to have been an imitation of the tent of Xerxes. On a height N. of the Areopagus stands the Theséum, founded in B.C. 469 and completed in 465, containing the bones of Theseus, which Cimon brought from the isle of Scyros: it was built of Pentelic marble, and in the Doric style of architecture, 104 feet long by 45 broad, with six columns at each end, and thirteen on each side, thirty-four in all, and divided in the interior into a central cella 40 feet long, with a pronaos, facing the E., 33 feet long, and an opisthodomus facing the W., 27 feet long; the porticoes being included in each case. The pediments of the porticoes and the metopes of the E. front were filled with sculptures, representing the exploits of Theseus and Hercules. The building is nearly perfect at the present time, having been formerly used as a Christian church dedicated to St. George, and now as a national museum. The great temple of Zeus, named the Olympiéum, stood S.E. of the Acropolis, near the right bank of the Ilissus: its erection was spread over nearly 700 years, having been commenced by Pisistratus and his sons, carried on by Antiochus Epiphanes B.C. 174, and again in the reign of Augustus by a society of princes; and finally completed by Hadrian. Its remains consist of 16 gigantic Corinthian columns of white marble, 6½ feet in diameter, and above 60 feet high. The temple was 354 feet long, and 171 broad.

Among the less important objects we may notice—the Odéum of Herodes, near the S.W. angle of the Acropolis, built in the time of the Antonines by Herodes Atticus, and capable of holding about 6,000 persons; the Cave of Apollo and Pan, at the N.W. angle of the

* Allusion is made to its position in the following lines:—
Χαίρετ᾽ ἀστικὸς λεὼς,
Ἱσταρ ἤμεναι Διός,
Παρθένου φίλαι φίλοι
Σωφρονοῦντες ἐν χρόνῳ.
Παλλάδος δ᾽ ὑπὸ πτεροῖς
Ὄντας ἄζεται πατήρ.   ÆSCH. EUM. 997.

Acropolis, 18 feet long, 30 high, and 15 deep, frequently noticed in the Ion of Euripides; the Clepsydra, a fountain so named from its being supposed to have a subterraneous communication with the harbour of Phalerum; the Aglaurium, a cave in the Long Rocks, whence a flight of steps led up to the Acropolis;[1] it was the sanctuary of Aglaurus, a daughter of Cecrops; the Gymnasium of Hadrian, to the N. of the Acropolis; the Horologium of Andronicus Cyrrhestes, commonly called the "Temple of the Winds," which served as the weather-cock and public clock of Athens, supposed to have been erected about B.C. 100; the Street of Tripods, along the E. side of the Acropolis, so named from the tripods which the victorious choragi dedicated to Dionysus in the small temples in this street: one of these temples, erroneously called the "Lantern of Demosthenes," was erected by Lysicrates in B.C. 335, and still exists; Callirhoë, a spring situated S. of the Olympieum, yielding the only good water in Athens; the Pisistratidæ erected over it a building with 9 pipes, hence called Enneacrūnus;[2] the Arch of Hadrian, a poor structure still existing opposite the N.W. angle of the Olympieum, and erected probably, not by, but in honour of Hadrian; and the Panathenaïc Stadium, situated between two parallel heights on the S. side of the Ilissus.

(3.) *Suburbs of the City.*—The most beautiful and interesting of the suburbs was the Outer Ceramicus,[3] outside the Dipylon, through which ran the road to the Academia, some 6 or 8 stadia distant from the gate. The Academy is said to have belonged to the hero Academus: it was converted into a gymnasium, and was adorned with walks, groves,[4] and fountains, as well as with numerous altars and a temple of Athena. Here Plato taught, and hence his school was called the Academic. Sylla had its groves destroyed, but they were afterwards restored. It still retains the name of *Akadhimia*. A short distance beyond it was the hill of Colōnus, immortalized by the tragedy of

---

[1] The position of the Aglaurium and its flight of steps are alluded to by Euripides:—

    Ὦ Πανὸς θακήματα καὶ
    Παραυλίζουσα πέτρα
    Μυχώδεσι μακραῖς,
    Ἵνα χορῶν στείβουσι ποδοῖν
    Ἀγραύλου κόραι τρίγονοι
    Στάδια χλοερὰ πρὸ Παλλάδος ναῶν.   *Ion,* 501.

[2] Et quos Callirhoë novies errantibus undis
    Implicat.   STAT. *Theb.* xii. 629.

[3] The Ceramicus was the burial place for those who were honoured with a public funeral; hence Aristophanes says:—

    Ὁ Κεραμεικὸς δέξεται νώ,
    Δημοσίᾳ γὰρ ἵνα ταφῶμεν.   *Av.* 395.

[4] The olive-trees in the Academy were particularly fine:—

    τῇδε θάλλει μέγιστα χώρᾳ,
    Γλαυκᾶς παιδοτρόφου φύλλον ἐλαίας·
    Τὸ μέν τις οὔτε νεαρὸς οὔτε γήρᾳ
    Σημαίνων ἁλώσει χερὶ πέρσας.—SOPH. *Œd. Col.* 700.
    Ἐν εὐσαίαις δρόμοισιν Ἀκαδήμου θεοῦ.—EUPOL. *Fragm.*

Atque inter silvas Academi quærere verum.—HOR. *Ep.* ii. 2, 45.

Sophocles.[5] On the E. of the city was Cynosarges, where the Cynic School was established by Antisthenes : a grove, which surrounded it, was destroyed by Philip in B.C. 200. A little S. was the Lycéum, the chief of the Athenian gymnasia, where Aristotle and his successors in the Peripatetic School taught : it was sacred to Apollo Lycius.

History.—The foundation of Athens was attributed to Cecrops, the first king of Attica, in whose reign Poseidon and Athena contended for the possession of that country. The greatness of the town, however, dates from the reign of Theseus, who consolidated the 12 states of Attica into one kingdom, of which Athens became the capital. The first attempt to embellish the town was made by Pisistratus and his sons, B.C. 500–514. Xerxes reduced it to a heap of ashes in 480, but it was afterwards rebuilt with great splendour under the direction of Themistocles, Cimon, and Pericles, the former of whom secured the town by walls. On the capture of the city in 404, the Long Walls and fortifications of Piraeus were destroyed by the Lacedæmonians, but were again restored by Conon. After the battle of Chæronea in 338, Athens became a dependency of Macedonia, but it retained nominal independence down to the time of the Roman dominion. Having sided with Rome, it was attacked by the last Philip of Macedonia in 200, when all the suburbs were destroyed. A greater calamity befell it in 86, when Sulla took the town by assault, and destroyed the Long Walls and fortifications of the city and Piraeus. Though the commerce of Athens thenceforward decayed, the town enjoyed a high degree of prosperity as a school of art and literature. The Roman emperors, particularly Hadrian, added new buildings, and the town was never more splendid than in the time of the Antonines. The walls were restored by Valerian in A.D. 258, and it was thus secured against the attacks of the barbarians.

Coin of Athens.

In the sixth century, the schools of philosophy were abolished by Justinian, and the temples converted into churches.

The other Towns of Attica.—Acharnæ, the largest demus of Attica, was situated near the foot of Parnes about 7 miles N. of Athens : its soil was fertile, but the chief occupation of its inhabitants was the manufacture of charcoal for the supply of the capital : its exact site is

[5] An altar of equestrian Neptune stood there, to which reference is made by Sophocles :—

Εὔιππον, ξένε, τᾶσδε χώρας
῀Ικου τὰ κράτιστα γᾶς ἔπαυλα
Τὸν ἀργῆτα Κολωνόν.		Soph. Œd. Col. 668.

οἱ δὲ πλήσιοι γύαι
Τόνδ᾽ ἱππότην Κολωνὸν εὔχονται σφίσιν
Ἀρχηγὸν εἶναι.		Ib. 58.

not known.[6] **Eleusis**, *Lepsina*, stood upon a height near the sea,
opposite the island of Salamis; the fertile Thriasian plain spread inland
from it, and the road from Athens to the isthmus passed through it.
Eleusis owed its celebrity to the worship of Demeter, whose coming
(ἔλευσις) appears to be implied in the name: the road which connected
the place with Athens was named the "Sacred Way,"[7] from the solemn
procession which travelled along it annually at the time of the Eleusinian festival. The temple of Demeter was burnt by the Persians in
B.C. 484; its restoration was commenced by Pericles, who employed
Ictinus as architect, but it was not completed until B.C. 318; it was
the largest in Greece, and regarded as one of the four finest specimens
of Grecian architecture in marble. The only noteworthy remains at
Eleusis are the fragments of the Propylæa, the platform of the temple,
and traces of wharfs. The plain of Eleusis was exposed to periodical
inundations of the Cephissus; to check these Hadrian raised some
embankments. **Orōpus**, *Skala*, was situated on the shore of the maritime plain, which lies about the mouth of the Asopus on the border
of Bœotia: from its position it was a frequent cause of dispute between
the Athenians and Bœotians. In B.C. 412 the latter people gained
possession of it, and in 402 they removed the town 7 or more probably 17 stadia from the sea, to the site now named *Oropo*, whence
it was shortly removed back to its old site. It changed hands frequently; after the battle of Chæronea Philip gave it to the Athenians. In 318 it became independent,[8] but in 312 it was taken by
Cassander, and, after the expulsion of his troops, handed over to the
Bœotians. It possessed a temple of Amphiaraus. **Rhamnus**, *Ovrio-
Kastro*, stood on a rocky peninsula on the E. coast, between Oropus
and Marathon, and was chiefly celebrated for the worship of Nemesis;
the temple stood near the town, and contained a colossal statue of the
goddess by Phidias: traces of two temples have been discovered,—a
smaller one which is supposed to have been destroyed by the Persians,
and a larger one subsequently erected on a contiguous site; the latter
was a peripteral hexatyle, 71 ft. by 33 ft., while the former was only
31 ft. by 21 ft. **Marathon** was the name both of a place and of a
plain[9] about 26 miles N.E. of Athens, the latter of which has obtained
an undying celebrity from the victory which the Athenians here
obtained over the Persians in B.C. 490. The plain skirts a small bay
formed by the promontory of Cynosura on the N. and a projection of
Pentelicus on the S.; inland it is backed by the heights of Drilesus

---

[6] It gives title to a well-known play of Aristophanes, in which the sufferings
of the agriculturists during the Peloponnesian War are depicted, the position and
occupation of the Acharnians exposing them to serious losses.

[7] The Sacred Way left Athens by the Sacred Gate, though it might also be
reached by a branch road passing through the Dipylum. It traversed the Outer
Ceramicus, where it was lined with tombs and statues; it then crossed the
Cephissus and surmounted the range of Ægaleus by the pass of *Dhafni*; the
temples of Apollo and Venus were in this part of its course: it then descended to
the sea, near where the Rheiti or salt-springs gush out from the base of Ægaleus,
and thence followed the line of the shore to Eleusis.

[8] It was noted in mythology as the place where Theseus destroyed the
Cretan bull:—

Te, maxime Theseu,
Mirata est Marathon Cretæi sanguine tauri.—Ov. *Met.* vii. 433.

and Diacria, and on either side it is closed in by marshes.* It is about
6 miles long, and 3 miles at its greatest breadth, and of a crescent
form. A small stream, the *Marathona*, flows through the centre of it.
On this plain stood a Tetrapolis, or confederacy of four towns. viz.:
Marathon, which occupied the site of *Vrana*, on a height fortified by
the ravine of a torrent; Probalinthus, probably at the S.W. of the
plain; Tricorythus at the other extremity, near *Suli*; and Œnoë, at
Inoi, near the head of the valley of *Marathona*. The village which
now bears the name of *Marathona* is on the left bank of the river
below Œnoë. In the plain, about ½ a mile from the sea, is the *Soro* or
artificial mound which covers the bodies of the Athenians slain in the
battle: it is about 30 ft. high and 200 yds. round. Near *Vrana* are
the traces of a temple, probably that of Hercules noticed by Herodotus,
while 1000 yds. to the N. is the *Pyrgos*, or remains of the tower, which
may be the site of the trophy of Miltiades. Brauron, near the E. coast
on the river Erasinus was chiefly celebrated for the worship of Artemis,
who had a temple both here[1] and at its port, named Halæ Araphenides;[2]
the latter contained the statue brought from Tauris by Orestes and
Iphigeneia.

Of the less important places we may notice—Eleutheræ and Œnoë,
which commanded the Pass of Dryoscephalæ over Cithæron; their
positions are uncertain,—the latter is probably represented by the ruins
of *Ghyfto-castro* at the entrance of the pass, and the former by *Myupoli*
about 4 miles to the S.E.; Phyle, *Fili*, a strong fortress on a steep
rock, about 10 miles from Athens, commanding the pass across Parnes,
and memorable as the point selected by Thrasybulus in B.C. 404 as the
base of operations against the Thirty Tyrants; Decelēa, on a circular
and isolated spur of Parnes, which commanded the pass across Parnes
to Oropus, now named the Pass of *Tatoy*, through which the Athenians
drew their supplies of corn from Eubœa; the Lacedæmonians under
Agis seized it in B.C. 413, and thence carried on a guerilla warfare
against the Athenians; Aphidna, between Decelea and Rhamnus,
probably on the hill of *Kotroni*, the birthplace of Tyrtæus the poet,
and of Harmodius and Aristogeiton, and celebrated in mythology as
the place where Theseus deposited Helen; Pallēna, on the road from
Athens to Marathon, between Hymettus and Pentelicus, possessing a
celebrated temple of Athena;[3] Stiria, on the E. coast, S.E. of Brauron,

---

* Large quantities of water-fowl frequented the marshes, as well as the "pleasant mead of Marathon" itself:—

ὅσα τ᾽ εὐδρόσους τε
Τῆς τόπους ἴχετε, καὶ λειμῶ-
να τὸν ἐρόεντα Μαραθῶνος·
Ὄρνις τε πτεροπτείλος
Ἀτταγᾶς, ἀτταγᾶς.   ARISTOPH. Av. 245.

[1] Σὺ δ᾽ ἀμφὶ σεμνάς, Ἰφιγένεια, κλίμακας
Βραυρωνίας δεῖ τῇδε κληδουχεῖν θεᾶς.—EUR. Iph. Taur. 1474.

[2] Χῶρός τίς ἐστιν Ἀτθίδος πρὸς ἐσχάτοις
Ὅρασι, γείτων δειράδος Καρυστίας,
Ἱερός, Ἁλάς τιν οὑμὸς ὀνομάζει λεώς·
Ἐνταῦθα τεύξας ναόν, ἱδρύσαι βρέτας.
Ἐπώνυμον τῆς Ταυρικῆς.   EUR. Iph. Taur. 1462.

[3] Παλληνίδος γὰρ σεμνὸν ἐκ τυμβῶν πάγον
Δῖας Ἀθάνας.   *   EURIP. Heracl. 849.

connected with Athens by a road named the "Stirian Way;" Prasiæ, on the E. coast with an excellent harbour, *Porto Rafti*, whence the Theoria, or sacred procession, used to sail, and with a temple of Apollo; Prasia, the birth-place of Demosthenes, E. of Hymettus; Thoricus, *Theriko*, on the E. coast, about 7½ miles N. of Sunium, celebrated in mythology as the residence of Cephalus, whom Eos carried off to the gods, and a place of importance, as testified both by its ruins and by its occupation by the Athenians in the Peloponnesian War; Sunium on the promontory of the same name, fortified by the Athenians in B.C. 413, and regarded as one of the most important fortresses in Attica: the temple of Athena which crowned the heights was a Doric hexastyle, the only remains of which are 9 columns of the S. flank and 3 of the N., together with 2 columns and 1 of the antæ of the pronaus; Anaphlystus, *Anavyso*, N.W. of Sunium, near the mines of Laurium; Sphettus in the same neighbourhood, connected with Athens by the "Sphettian Way" which entered the city by the N. end of Hymettus; a manufactory of vinegar appears to have existed there;[4] and Halæ Æxonides, nearer Athens, where were some salt-works.

History.—The history of Attica and of its capital Athens is almost synonymous with the history of Greece itself: so prominent is the position which it holds in all ages. Our limits will not permit us to do more than point out the chief periods into which its history may be divided, and which are—

(1.) The early period down to the time of Solon's legislation B.C. 594, during the first portion of which Athens was governed by kings; the historical events during the whole of this period are few and unimportant.

(2.) The growth of the Athenian state from the time of Solon, B.C. 594, to the attainment of its supremacy in 478. This period is signalized by the Persian Wars (490–479), in which Athens took a conspicuous part, and by the gradual extension of the political influence of Athens through its maritime power.

(3.) The period of Athenian ascendency, which lasted from 478 to 413, when the army and fleet were destroyed at Sicily. Under the administration of Pericles Athens arrived at the height of its glory. The Peloponnesian War broke out in 431, and proved destructive of Athenian supremacy.

(4.) From the decline of the ascendency of Athens to the Roman conquest of Greece in 146. The battle of Ægospotami in 405 and the capture of Athens by Lysander in 404 completed the humiliation of Athens. In 378 Athens joined Thebes, and again became the head of an important maritime ascendency, which lasted until 355, when the Social Wars terminated in the independence of her allies. A subsequent alliance with Thebes against Philip was brought to a close by the battle of Chæronea in 338, where the Athenians were totally defeated. On the death of Alexander the Athenians endeavoured to shake off the Macedonian yoke, but the Lamian War ended disastrously in 322, and Athens surrendered to Antipater. The Macedonian governor was expelled in the reign of Cassander by Demetrius Poliorcetes in 307, and Athens was captured by him in 295. Antigonus Gonatas, king of Macedonia, the son of Poliorcetes again reduced Athens in 262. On the death of his successor Demetrius, in 229, Athens joined

---

4   εἶθ' ἅξει διήμενος Σφηττόν,
Κατέκλασεν αὐτοῦ τὰ βλέφαρα.   ARISTOPH. *Plut.* 720.

the Ætolian League. In 200 Philip V. besieged Athens, and she was only relieved by the Roman fleet : she afterwards joined Rome against Philip. Attica was finally added to the other dependencies of Rome in 146.

Islands off the Coast of Attica.—**Salamis**, *Kuluri*, lies between the coasts of Attica and Megaris, closing the bay of Eleusis on the S. Its shape resembles an irregular semicircle facing the W. ; its length from N. to S. is about 10 miles, and its greatest width from E. to W. about the same. It had in early times the names of Pityusa, Sciras, and Cychria,⁴ the former from the pine-trees on it, the two latter after the heroes Scirus and Cychreus. The island is mountainous, and the shore much indented : the most salient points are the promontories of **Silenia** or **Tropæa**, *C. St. Barbara*, at the S.E., off which lies the small isle of **Psyttalia**, *Lipsokutali*, a mile long and from 200 to 300 yards across ; **Sciradium**, probably at the S.W., where stood the temple of Athena Sciras; and **Budorum** at the W. The old city of **Salamis** stood on the S. shore ; the new one on the N. shore. The island is chiefly memorable for the defeat of the Persian fleet by the Greeks in B.C. 480, which took place in the channel⁵ between the island and Attica, and was witnessed by Xerxes from his seat on Mount Ægaleus. Salamis was colonised at an early period by the Æacidæ of Ægina, and was the residence of Telamon and his son Ajax at the time of the Trojan War. It was independent until about B.C. 620, when a dispute arose for its possession between the Athenians and Megarians. The question was ultimately referred to the Spartans, who decided in favour of Athens ; and to this power it belonged until the establishment of the Macedonian supremacy in 318. In 232 the Athenians purchased it of the Macedonians, and expelled the inhabitants in favour of Athenian settlers : thenceforward it was attached to Athens. **Ægina**, *Eghina*, lies in the centre of the Saronic Gulf nearly equidistant from the shores of Attica, Megaris, and Epidaurus. In shape it is an irregular triangle. The S. portion of the island is occupied by the magnificent conical hill named *St. Elias :* the W. side is a well cultivated plain. The ori-

Coin of Ægina.

ginal inhabitants were Achæans,⁷ but these were superseded by Dorians from Epidaurus. The chief town, also called **Ægina**, stood on the N.W. coast, and possessed two harbours and numerous public buildings, particularly the shrine of Æacus. The moles of the ports and walls of the city can still be traced. On a hill in the N.E. of

---

⁴ This name occurs in Æschylos ;—

'Ακτάς ἀμφὶ Κυχρείας.—*Pers.* 570.

⁵ Πλήθουσι νεκρῶν δυστπότμῳ ἐφθαρμένων
Σαλαμῖνος ἀκταὶ πᾶς τε πρόσχωρος τόπος.—*Æsch. Pers.* 272.

⁷ The mythical account of its original population is, that Zeus changed the ants (μύρμηκες) of the island into Myrmidons, over whom Æacus ruled. See Ov. *Met.* vii. 624, *seq.*

the island are the remains of a magnificent temple of the Doric order,
which has been variously regarded as that of Zeus Panhellenius, and
that of Athena noticed by Herodotus (iii. 59). The sculptures which
adorned it, and which were discovered in 1811, represent events connected with the Trojan War. The temple was erected early in the
6th century. Another town named Œa was in the interior of the
island. Ægina, as a dependency of Epidaurus, became subject to
Pheidon, tyrant of Argos, about B.C. 748. It soon became a place of
great commercial activity: as early as 563 it had entered into relations
with Egypt, and about 500 it held the empire of the seas, and planted
colonies in Crete and Italy. The authority of Epidaurus was renounced,
and Ægina became an independent state. As such it entered into a
league with Thebes against Athens in 505, and ravaged the coasts of
Attica. The Æginetans did good service to the Greek cause at the
battle of Salamis. The Athenians, to whom Ægina had become, in
the expressive language of Pericles, the "eye-sore of the Piræus,"
defeated them in 460, took their town in 456, and expelled the whole
population in 431: the refugees were settled at Thyrea by the Spartans,
and were restored by Lysander in 404.

In addition to these we have to notice—**Helēna** or **Macris,** *Makronisi*,
off the E. coast, a long, narrow island, uninhabited in ancient as in
modern times; **Patrocli Insula,** off the S. point, so named after a
general of Ptolemy Philadelphus, who built a fort on it; and **Belbína,**
St. George, at the entrance of the Saronic Gulf, described by Themistocles as one of the most insignificant spots in Hellas.

§ 6. The important island of **Euboea,** *Negropont*,[a] lies opposite to

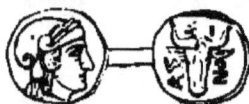

Coin of Euboea.

the coasts of Attica, Bœotia, and Locris. Politically it was closely connected with the first of these countries, to
which we therefore append it. Geo-
graphically it lay in closer contiguity
to Bœotia, the strait separating them,
named **Euripus**, being only 40 yards
across at Chalcis. The length of the island from N. to S. is about 90
miles; its breadth varies from 30 miles to 4. The mountain-range
which traverses it throughout its whole length may be regarded as a
continuation of Pelion and Ossa; on the E. coast it rises to the
height of 7266 feet; it terminates in the promontories of **Cenæum,**[b]
Lithadha, in the N.W.; **Artemisium** in the N., opposite the Thessalian Magnesia, the scene of the Persian defeat in B.C. 480; **Caphē-**

---

[a] The modern name is compounded of *Egripo*, a corruption of Euripus, and
ponte, "a bridge."

[b] It was crowned with a temple of Zeus, surnamed Cenæus:—

Ἀκτή τις ἔστ' Εὐβοίις, ἔνθ' ὁρίζεται
Βωμοὺς τέλη τ' ἔγκαρπα Κηναίῳ Διί.　　　SOPH. Trach. 237.

Ἀκτή τις ἀμφίκλυστος Εὐβοίας ἄκρον
Κήναιόν ἐστιν, ἔνθα πατρῴῳ Διὶ
Βωμοὺς ὁρίζει, τεμενίαν τε φυλλάδα.　　　Id. 752.

rus,[1] *Kavo Doro*, in the S.E.; and Geræstus,[2] *Mandili*, in the S.E. Though generally mountainous, the island contains some rich plains, particularly those about the towns of Histiæa and Chalcis—the latter being named Lelantum.[3] The E. coast is remarkably rocky, and both the prevalent winds and the currents render it extremely dangerous. The part called the "Hollows" was somewhat N. of Geræstus. The streams are of trifling size. The island was fertile: the plains produced corn, and the hills fed sheep. The marble quarries of Carystus were far-famed. The original inhabitants were the Abantes,[4] after whom the island was sometimes named Abantis; but in historical times these gave place to Ionian Greeks, who founded the most important towns; viz. Chalcis, Eretria, Orcus or Histiæa, and Carystus.

Chalcis, *Egripo*, stood on the shore of the Euripus, just where the strait is divided into two channels by a rock, which now forms a central pier for the bridge that connects the island with the mainland. The extraordinary flux and reflux of the currents[5] at this point were noticed by the ancients. Chalcis rose to great commercial importance, and planted colonies in

Coin of Chalcis in Eubœa.

Sicily, Italy,[6] and Macedonia. The chief events in history are—its capture by the Athenians in B.C. 506; its revolt from that state in 445, and its subsequent reconquest by Pericles; its second revolt

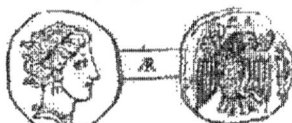

---

[1] On this promontory the Greek fleet was wrecked on its return from Troy :—

Τέρμα πέλαγος Αἰγαίας ἁλός,
Ἄκται δὲ Μυκόνου, Δήλιοί τε χοιράδες,
Γύρας τε Λιμνός θ' αἱ Καφηρεαί τ' ἄκραι
Πολλῶν θανόντων σώμαθ' ἕξουσιν νεκρῶν.—EUR. Troad. 88.

Sett triste Minervæ
Sidus, et Eubolcæ cautes, ultorque Caphereus.—Æn. xi. 260.

[2] Ὦρτο δ' ἐπὶ λιγὺς οὖρος αὔμεναι· αἱ δὲ μάλ' ὦκα
Ἰχθυόεντα κέλευθα διέδραμον· ἐς δὲ Γεραιστὸν
Ἐννύχιαι κατάγοντο. HOM. Od. III. 176.

[3] Κηραίου δ' ἐπέβης ναυσικλείτης Εὐβοίης.
Στῆς δ' ἐπὶ Ληλάντῳ πεδίῳ. HOM. Hymn. in Apoll. 219.

[4] Οἳ δ' Εὔβοιαν ἔχον μένεα πνείοντες Ἄβαντες,
Τῶν δ' αὖθ' ἡγεμόνευ' Ἐλεφήνωρ, ὄζος Ἄρηος,
Χαλκωδοντιάδης, μεγαθύμων ἀρχὸς Ἀβάντων. Il. 540.

[5] Aretatus rapido ferret qua gurgite pontus,
Euripusque trahit, cursum mutantibus undis,
Chalcidicas puppes ad iniquam classibus Aulim.—LUC. v. 234.

[6] The most famous of these colonies was Cumæ, which consequently received the epithet "Chalcidian :"—

Chalcidicæque levis tandem super adstitit arce.—VIRG. Æn. vi. 17.

Hæc ego Chalcidicis ad te, Marcelle, sonabam
Littoribus, fractas ubi Vesbius egerit iras.—STAT. Silv. iv. 4, 78.

in 411, which was again unsuccessful; the attacks upon it by the Romans in 207 and 192; and its destruction by Mummius. From its position in command of the Euripus it was termed by Philip of Macedon one of the "fetters of Greece." **Eretria**, stood S. of

Coin of Eretria.

Chalcis at the S.W. extremity of the plain of Lelantum, which was a bone of contention between the two cities. The original town, near *Vathy*, was destroyed by the Persians in B.C. 490 for the part it had taken in the Ionian revolt, but was again rebuilt more to the S. at *Kastri*. The defeat of the Athenians off its harbour, in 411, led to its revolt from that power. It was governed by tyrants from about 400 to 341; and was taken by the Romans and Rhodians in the war with Philip V. It was the seat of a philosophical school, founded by Menedemus, and the birth-place of the tragic poet Achæus. The remains of the acropolis and of a theatre still exist at *Kastri*. Oreus stood on the N. coast, and was originally named Histiæa;[f] it was occupied by the Persians after the battle of Artemisium, and afterwards became subject to Athens, from which it revolted in B.C. 445, and was in consequence taken by Pericles, its inhabitants banished, and Athenian settlers placed in their stead. After the Peloponnesian War, Oreus became subject to Sparta, and remained so until the battle of Leuctra. In the wars between Philip and the Romans it was taken by the

Coin of Carystus.

latter in the years 207 and 200. **Carystus** was situated on the S. coast, and is chiefly known in history as the place where the Persians landed in B.C. 490. The marble quarries were on the slopes of the neighbouring hill of Ocha: the marble was of a green colour with white bands, and was much prized at Rome.[g]

Of the less important towns we may notice—**Dium**,[g] near Prom. Cenæum, the mother-city of Canæ in Æolis; **Ædepsus**, on the N.W. coast, with some warm baths; **Orobiæ**, opposite Cynus in Bœotia, with an oracle of Apollo Selinuntius: the town was partly destroyed by an earthquake in B.C. 426; **Ægæ**, opposite Anthedon, possessing a famous temple of Poseidon;[h] **Amarynthus**, about a mile from Eretria, with a

---

[f] It is noticed under this name by Homer, as abounding in grapes:—
　　　　πολυσταφύλον θ' Ἱστίαιαν.　　　　　Il. II. 537.

[g] Quidve domus prodest Phrygiis innixa columnis,
　　Tænare, sive tuis, sive, Caryste, tuis?—Tibull. III. 3, 13.
　　Idem beatas lautus exstruit thermas
　　De marmore omni, quod Carystus invenit.—Mart. Ix. 76.

[h] Κάρυστόν τ' ἤμαλον, Δίον τ' αἰπύ πτολίεθρον.—Hom. Il. II. 538.

temple of Artemis Amarynthia; Porthmus, a harbour on the narrowest part of the Eubœan channel opposite to Rhamnus, and hence a place of importance as a point for attacking the coast of Attica; Styra, N. of Carystus, occupied originally by a Dryopian population, a place noticed in the Persian War and subsequently subject to Athens; Geræstus, on the promontory of the same name, with a celebrated temple of Posei-don; and, lastly, Cerinthus,[1] on the N.E. coast.

History.—As Eubœa never formed a single political state, its history resolves itself into that of its separate towns. We have already seen that Chalcis and Eretria were powerful cities in early times: they continued so until the time of the Pisistratidæ, when Chalcis engaged in war with Athens, and lost its territory in consequence in B.C. 506. After the Persian War, the whole of Eubœa became dependent on Athens: it revolted in 445 and again in 411, but was reconquered on each occa-sion. With the decline of Athenian supremacy, tyrants established themselves in the towns; these submitted to Macedonia without a struggle, and the island remained a part of the Macedonian dominions until 194, when the Romans took it from Philip V.

§ 7. Not far distant from the coast of Attica lies an important group of islands, to which the name of Cyclades[2] was given, because they lay in a circle (ἐν κύκλῳ) around Delos, which, though the smallest, was the most important of them. These islands appear to be physically connected with Eubœa, and to be a continuation of the same elevation, rising from the sea at intervals. The numbers and names of them are variously given; but, according to the best authorities, the following twelve constituted the group :—Ceos, Cythnos, Seriphos, Siphnos, Paros, Naxos, Delos, Rhenéa, Mycônos, Syros, Tenos, and Andros. The order in which they are enume-rated is in a circle commencing at the N.W. These islands were for the most part occupied by Ionian colonists.

Ceos or **Cea**, *Zea*, is about 13 miles S.E. of the promontory of Sunium, and is 14 miles in length by 10 in breadth. It was said to have been originally occupied by nymphs who were driven from it by a lion. The Ionians colonised it and built four towns; of which Iúlis, the capital, in the N., was the most celebrated as being the birth-place of the lyric poets Simonides[3] and Bacchylides, and of the philosopher

---

[1] Αἴγας· ἔνθα δέ οἱ κλυτὰ δώματα βένθεσι λίμνης,
    Χρύσεα, μαρμαίροντα, τετεύχαται, ἄφθιτα αἰεί.—*Il.* xiii. 21.

[2] The general appearance of these islands hardly justifies the epithet of "glittering" applied to them by Horace; they are for the most part bare and brown :—

      Interfusa *nitentes*
      Vitæ æquora Cycladas.   *Carm.* l. 14, 19.
      Fulgentesque tenet Cycladas.   *Id.* iii. 28, 14.

[3] Horace alludes to him in the lines :—

      Non, si priores Mæonius tenet
      Sedes Homerus, Pindaricæ latent,
        Ceæque, et Alcæi minaces,
        Stesichorique graves Camœnæ.   *Carm.* iv. 9, 3.

Ariston: its laws were so excellent as to pass into a proverb. The other towns were—Coressia, which served as the port of Iulis; Carthæa,[4] in the S.E.; and Poïessa in the S.W. Cythnos, *Thermia*, is seldom mentioned: its chief celebrity in ancient times was derived from its excellent cheeses, and in modern from some hot springs to which it owes its present name. It possessed a town of the same name on the W. coast at

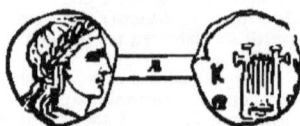

Coin of Cythnos.

Hebræo-kastron, of which some remains still exist: this town was occupied by Philip's troops in B.C. 200, and was unsuccessfully besieged by Attalus and the Rhodians. Seriphos, *Serpho*, was chiefly famed for its poverty and insignificance, and was hence used by the Romans as a place of banishment.[5] It possessed, however, iron and copper mines. It was the fabled scene of the education and exploits of Perseus.[6] Siphnos, *Siphno*, attained a high degree of prosperity from its gold and silver mines, and possessed a treasury at Delphi. These mines, however, were at length worked out, and the

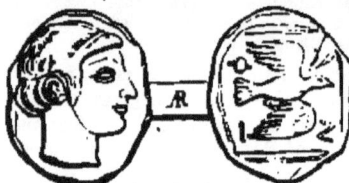

Coin of Siphnos.

inhabitants became poor even to a proverb. They manufactured a superior kind of pottery. The chief town lay on the E. side of the island on the site of the modern *Kastro*. Paros, *Paro*, is one of the largest of the Cyclades: it consists of a single round mountain, sloping evenly to a maritime plain which surrounds it on all sides. It was celebrated for its fine

Coin of Paros.

marble, dug out of the sides of Mt. Marpessa,[7] and for its figs. The

---

[4] Transit et antiquæ Carthea mœnia Cres.—Ov. *Met.* vii. 368.

[5] Æstuat infelix angusto limite mundi,
Ut Gyaræ clausus scopulis, parvaque Seripho.—Juv. x. 169.

[6] Περσέως ἑσπέσι τρίτος ἀ-
ρωστι μακογυητὰν μέρος,
Εἰνολίᾳ Σερίφῳ
Δαοίσι τε μεῖρας ἔγας.—Pind. *Pyth.* xii. 19.

[7] Nec magis incepto vultum sermone movetur,
Quam si dura silex aut stet Marpesia cautes.—Virg. *Æn.* vi. 470.

Σπέλαν θέμεν, Παρίαν
Λίθου λευκοτέραν.   Pind. *Nem.* iv. 131.

Urit me Glyceræ nitor
Splendentis Pario marmore purius.—Hor. *Carm.* i. 19, 5.

capital was on the W. coast: remains of it exist at *Paroikhia*. Its chief historical event is the unsuccessful attempt of Miltiades to subdue it after the battle of Marathon. The poet Archilochus was born there. **Naxos**, *Naxia*, was the largest of the Cyclades, being 19 miles in length by 15 in breadth; it was also eminently fertile, producing corn, wine, oil, and fruit of the finest description. In the centre of the island a mountain, named Drius, rises to the height of 3000 feet. Its capital stood on the N.W. coast, on the site of the modern town. The ruins of a temple

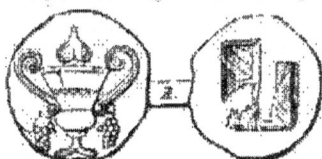

Coin of Naxos.

still exist there. Naxos was the seat of a tyranny before the Persian War. The failure of the Persian expedition against it in B.C. 501 was indirectly the cause of the Ionian revolt. The island was cruelly ravaged by the Persians in 490. After the Persian War it was subject to Athens, from which it revolted in 471 to no good effect. **Delos**,[a] *Dhíles*, lies in the centre of the Cyclades, between Rhenea and Myconos. It is little more than a rock, being only five miles in circumference, but it was regarded as one of the holiest spots in all Hellas, having been called into existence (as

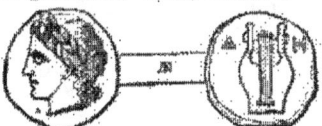

Coin of Delos.

was believed) by the trident of Poseidon, and fixed in its place by Zeus[b] that it might become the birth-place of Apollo and Artemis. It enjoyed a singular immunity from earthquakes—which was attributed to its miraculous origin. The worship of Apollo was celebrated by a great periodical festival, in which the Athenians and other nations took part. The sanctity of the isle is attested by the regard shown to it by Datis and Artaphernes, as well as by its being selected as the treasury of Greece in B.C. 477, and by the purification of it by the Athenians in 426. After the fall of Corinth, in 146, it became the centre of an extensive commerce, and was particularly celebrated for its bronze. It was ravaged by the generals of Mithridates, and thenceforth sank into insignificance. The town stood on the W. side of the island, just under Mount Cynthus,[c] a bare granite rock, about 400 feet

---

[a] Delos had a variety of poetical names, of which the most important was Ortygia, connected with the legend, that Latona was changed by Jupiter into a quail (ὄρτυξ). The name Ortygia occurs in Homer, *Od.* v. 123; xv. 403; but in the latter passage it is described in terms (πέρι Ορτυγίης καθύπερθεν) which make it doubtful whether it can be applied to Delos. See note[a], p. 428.

[b] Sacra mari colitur medio gratissima tellus
　　Nereidum matri, et Neptuno Ægæo:
　　Quam pius Arcitenens, oras et littora circum
　　Errantem, Gyaro celsa Myconoque revinxit,
　　Immotamque coli dedit, et contemnere ventos.—VIRG., Æn. III. 73.

　　Ἣ ὣς σε πρῶτον Λητὼ τέκε, χάρμα βροτοῖσιν,
　　Κλινθεῖσα πρὸς Κύνθου ὄρος κραναῇ ἐνὶ νήσῳ
　　Δήλῳ ἐν ἀμφιρύτῃ.　　HOM. *Hymn. in Apoll.* 25.

[c] Ipse jugis Cynthi graditur, mollique fluentem
　　Fronde premit crinem fingens, atque implicat auro.—VIRG., Æn. iv. 147

high, which served as its acropolis. A small stream, named Inôpus, and an oval lake are noticed by the ancients. The foundations of the theatre, of a stoa, and of a few houses, are all the remains of the once splendid town: the rest of the materials were transported to Venice and Constantinople. **Rhenia**[2] is separated from Delos by a strait about half a mile wide: it is about ten miles in circumference, and is divided into two parts by inlets. It served as the burial-place of Delos. **Myconos**,[3] *Mykono*, is little else than a barren granite rock, ten miles in length and six in its greatest breadth, with two towns on it: its inhabitants were famed for their avarice. **Syros**, *Syra*, was a more fertile island, but hardly deserves the praises bestowed upon it by Homer,[4] though it still produces good wine. It possessed two cities, one on the E. the other on the W. coast. The philosopher Pherecydes was a native of Syros. **Tenos**, *Tíno*, lies about fifteen miles from Delos, and is about fifteen miles long. It is one of the most fertile of the Cyclades. The inhabitants were wealthy, and paid a yearly tribute of 3600 drachmæ to Athens. The capital stood on the S.W. coast, and possessed a celebrated temple of Poseidon. The island was famed for its fine garlic. **Andros**, *Andro*, the most northerly of the group, is twenty-one miles long and eight broad: it was fertile, and particularly famed for its wine. The town lay in the middle of the W. coast: it was besieged by Themistocles after the Persian War, and by the Romans in their war with Philip. S.W. of Andros is the small island of **Gyaros**, *Jura*, a barren rock, about six miles in circumference, which the Roman emperors used as a place of banishment:[5] a purple fishery was carried on there.

IX. MEGARIS.

§ 8. The small district of **Megaris** occupied the northern portion of the Isthmus of Corinth, extending from the confines of Bœotia on the N. to Corinthia on the S.; the limit in the latter direction having been originally at Crommyon on the Saronic, and Thermæ on the Corinthian Gulf, but afterwards more to the N., at the Sci-

---

[2] Νάξον τ', ἠδὲ Πάρον, 'Ρηναίά τε πετρήεσσα.—Hom. Hymn. in Apoll. 44.

[3] The epithet *humilem*, applied to this island by Ovid, is incorrect : it was one of the islands to which Delos was anchored (see note [5], above).

Illine humilem Myconon, cretosaque rura Cimoli.—Ov. Met. vii. 463.

Ipsa tua Mycono Gyaroque revelli,

Delo, times. STAT. Theb. iii. 438.

[4] Νῆσός τις Συρίη κικλήσκεται, εἴ που ἀκούεις,
'Ορτυγίης καθύπερθεν, ὅθι τροπαὶ ἠελίοιο,
Οὔτι περιπληθὴς λίην τόσον ἀλλ' ἀγαθὴ μὲν,
Εὔβοτος, εὔμηλος, οἰνοπληθὴς, πολύπυρος.—Od. xv. 402.

There is room for doubt whether Homer's Syria is identical with Syros, or whether it is not rather a poetic fiction. The question turns partly on the further question whether Homer's Ortygia represents Delos.

[5] Aude aliquid brevibus Gyaris, et carcere dignum.—Juv. i. 73.

Ut Gyaræ clausus scopulis, parvaque Seripho. —Ip. x. 170.

It is noticed by Virgil as one of the rocks to which Delos was anchored (see note [5], p. 427), though it is not particularly near that island. The epithet *celsa* is misplaced, whether it be applied to Gyarus or (as in some copies) to Myconus.

ronian rocks. In the N.E. Megaris was contiguous to Attica; elsewhere it was bounded by the sea, viz. by the Coriuthian Gulf on the W., and the Saronic on the E. It thus lay open on the side of Attica alone, and was naturally connected with that country rather than with any other. It is a rugged and mountainous country, and contains only a single plain about 6 or 7 miles long, and about the same in breadth, which opens towards the Saronic Gulf on the S., and was named Leucon, "the White Plain." The chief mountain-range was named Geranea. Makriplayi, a southerly extension of Cithæron, which stretches across the isthmus like a vast wall, and forms the natural boundary between Northern Greece and the Peloponnesus. It was crossed at three points: on the W. by a road near the sea-coast, little frequented from its distance; in the centre by the pass now named Dervenia, which was probably the main line of communication in early times; and on the E. by a coast-road, which afterwards became the main line of communication, and which is celebrated for its difficulty, being carried for several miles along a narrow ledge cut in the face of the cliff some 600 or 700 feet above the sea. This pass is the Scironia Saxa* of antiquity, the Kakescala, "Bad Ladder," of modern times.[7] On the border of Attica were the heights of Kerata, before noticed. The promontory of Ægiplanctus[8] is on the W. coast.

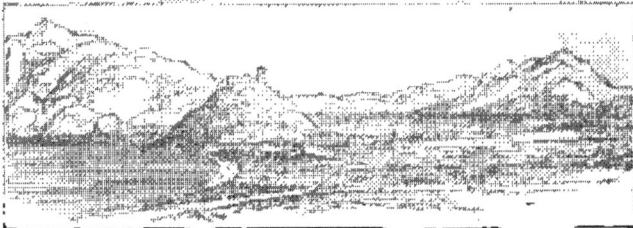

Minoa.   Nisaea.   Megara.

§ θ. The capital, Megara, stood on a low hill with a double summit, in the plain already noticed, about a mile and a half from the

---

* They were said to have been so named after Sciron, a robber whom Theseus destroyed:—

    Tutus ad Alcathoen, Lelegeia mœnia, limes
    Composito Scirone patet: sparsique latronis
    Terra negat sedem, sedem negat ossibus unda:
    Quæ jactata diu fertur durasse vetustas
    In scopulos: scopulis nomen Scironis inhæret.—Ov. Met. vii. 443.

[7] Hadrian rendered this road passable for carriages.

    [8] Ἀθηνῶν δ' ὑπὲρ Γοργῶπιν ἐσκήψεν φάος·
    Ὄρος τ' ἐπ' Αἰγίπλαγκτον ἐξικνούμενον,
    Ὤτρυνε θεσμὸν μὴ χατίζεσθαι πυρός.— Æsch. Ag. 302.

Saronic Gulf. The summits were named Caria and Alcathoë, Caria being probably the highest, and were each the site of an acropolis. Below the city was a port-town named Nisæa, connected with Megara by long walls, which have now wholly disappeared. The port itself was formed by a small island named Minôa, which was united to Nisæa by a bridge over a morass. This island is now, in all probability, incorporated with the mainland, and is a rocky hill on the margin of the sea. It has been otherwise identified with a small island still existing off the coast, but at too great a distance (200 yards) to be connected by a bridge, and with the promontory of Tikho more to the E., which is too distant to accord with the length of the walls. Megara possessed a second port on the Corinthian Gulf, named Pagæ or Pegæ, Psátho.

The town of Megara is said to have been founded by Nisus son of Pandion, and to have been subsequently restored by Alcathous [*] son of Pelops. The Megarians themselves attributed its origin to Car, son of Phoroneus. Its situation was highly favourable for commerce, as all the roads between Northern Greece and Peloponnesus passed through its territory, while its ports gave it communication with the E. and W. It was beautified with numerous edifices, particularly the Olympiæum or inclosure of Zeus Olympius, the Bouleuterion, the Prytaneum, numerous temples and tombs, and a magnificent aqueduct built by Theagenes. The whole of these buildings have disappeared, and modern Megara is a poor place, occupying the western summit.

History.— Megaris was originally a part of Attica, and thus an Ionian state. It was afterwards conquered by the Dorians, and was for a long time subject to Corinth. The Dorians were expelled in Solon's time, and Megara rose to great commercial prosperity, not only attaining its independence, but becoming the mother-city of numerous colonies in Sicily and Thrace. Its power was weakened partly by its internal dissensions and partly by its contests with the neighbouring states of Athens and Corinth. In B.C. 455 the Megarians formed an alliance with Athens which lasted for ten years. In the early part of the Peloponnesian War they suffered severely from Athenian inroads: in 427 Nicias blockaded Nisæa, and in 424 they got possession both of it and of the Long Walls, but did not succeed in taking Megara. The Megarians themselves levelled the Long Walls shortly after. Thenceforward Megara is seldom noticed. It became the seat of a philosophical school, founded by Eucleides, and it obtained under the Romans an ill fame for licentiousness.

---

[*] Apollo is said to have aided Alcathous: the stone on which he deposited his lyre, when struck, returned a musical sound: the stone was preserved in the Prodomeis:—

Φοῖβε ἄναξ, αὐτὸς μὲν ἐπύργωσας πόλιν ἄκρην,
᾽Αλκαθόῳ Πέλοπος παιδὶ χαριζόμενος.—Theogn. 771.

Regia turris erat vocalibus addita muris :
In quibus auratam proles Letoia fertur
Deposuisse lyram : saxo sonus ejus inhæsit.—Ov. Met. viii. 14.

Corinth.

CHAPTER XXII.

§ 1. THE physical features of the Peloponnesus have been already
noticed in the general description of Greece. It only remains for us
here to account for the name, and to enumerate the provinces into
which it was divided. The name of Peloponnesus, "the Isle of
Pelops," came into vogue subsequently to the Dorian immigration,
and embodied the belief of the later Greeks as to the wealth and influence of Pelops, the hero of Olympia. The earlier names, as given
in the Iliad, were Apia [1] (from ἀπό, "the distant land"), and
Argos. Its area is computed at 1779 square miles; and its popula-

---

[1] Καὶ μὲν τοῖσιν ἐγὼ μεθομίλεον ἐκ Πύλου ἐλθών,
Τηλόθεν ἐξ Ἀπίης γαίης· καλέσαντο γὰρ αὐτοί.—Il. l. 269.

γυναῖκ᾽ εὐειδέ᾽ ἀνάγει
Ἐξ Ἀπίης γαίης, τοὺς ἀνδρῶν αἰχμητάων;   Il. lii. 49.

tion, during the flourishing period of Greek history, at upwards of a
million. It was subdivided into numerous states of various sizes,
of which the following six were the most important :—Achaia, Elis,
Messenia, Laconia, Argolis, and Arcadia ; while Corinthia, Sicyonia,
Phliasia, and Cleonæ, were of small size.

I. CORINTHIA, SICYONIA, PHLIASIA, AND CLEONÆ.

§ 2. The territory of Corinth, described by the Greeks under the
name of Corinthia (ἡ Κορινθία), occupied the isthmus which connects
Northern Greece with Peloponnesus, together with a certain amount
of district on either side of it. Towards the N. it extended to the
border of Megaria, from which it was separated by the Geranean
range; towards the S. it bordered on Argolis, and was bounded by
the Onēan range. The Saronic and Corinthian Gulfs approach
one another between these ranges, and are divided by a low ridge
about 3½ miles across, the highest point of which is only 246 feet
above the level of the sea. A glance at the map will show how
favourably this district was situated both for military and commercial purposes. It was the gate* of the Peloponnesus. N. and S.
it was shut off from the adjacent countries by mountain ranges
which were difficult to cross; E. and W. it held easy intercourse
with the shores of the Ægean and of the Ionian seas,* by means of
the Saronic Gulf in the former direction, and the Corinthian in the
latter. The intervening land served to connect as well as separate these seas, and rendered Corinth the entrepot of commerce between Asia and Europe. In addition to these natural advantages,
nature provided an admirable acropolis in the celebrated Acrocorinthus, an outlying member of the Onēan range, which rises in
an isolated mass to the height of 1900 feet,[4] at a short distance from
the Corinthian Gulf. The soil of Corinthia was by no means fertile,
the coast-plain in the direction of Sicyon being the only arable land
in the whole district.

---

* When Agesilaus captured Corinth, he is described as—
ἀναστήσας τὴν Πελοποννήσου τὰς πύλας.—XEN. Ages. 2.
It has been termed in modern times the "Gibraltar of Greece."

* Hence Corinth is described as the "city of the two seas:"—
δίπορον κορυφὴν Ἰσθμου.   EURIP. Troad. 1087.
Laudabunt alii claram Rhodon, aut Mitylenen,
   Aut Ephesum, bimarisve Corinthi
Mœnia.   HOR. Carm. i. 7, 1.

[4] The description of Statius is hardly exaggerated; modern travellers have
remarked on the conical shadow of the rock stretching midway across the
isthmus :—
   Qua summas caput Acrocorinthus in auras
Tollit, et alterna geminum mare protegit umbra.—Theb. vii. 106.

The Isthmus was the most important part of the Corinthian territory, both as the spot where the merchandise was conveyed from sea to sea, and as the scene of the Isthmian games. The name probably comes from the same root as the Greek *i-évac*, and the Latin *i-re* "to go," and thus meant a "passage."[a] The traffic was originally carried on by means of the Diolcus, a level road, on which small vessels could be transported bodily by means of rollers, and the merchandise of the larger ones conveyed in carts. A canal was frequently projected, and actually commenced by Nero, but the scheme was not carried out: it may be traced near the Corinthian Gulf for 1200 yards. A short distance S. of the Diolcus the Isthmus was crossed by a wall, which may still be traced in its whole extent: it was fortified with square towers. The date of this work is uncertain; it probably was re-erected on various occasions. Temporary defences were thrown up at the time of the Persian invasion, and again in B.C. 369 by the Spartans. The Isthmian games were celebrated at a spot immediately S. of the wall. The sanctuary was a level spot of an irregular quadrangular form, enclosed by strong walls, and containing the temple of Poseidon and other sanctuaries. The stadium lay to the S. and the theatre to the W. of the sanctuary. The games were celebrated every two years in honour of Poseidon,[b] under the presidency of the Corinthians, and, during the ruin of Corinth, of the Sicyonians.

§ 3. The mountain ranges have been already noticed. Onia was so named from its resemblance to an ass's back. It closes the entrance of the Isthmus on the S., and was passable at two points— by a ravine between its W. extremity and the Acrocorinthus, and by a road that skirted the Saronic Gulf at its E. extremity. Gerania, in the N., terminates in the promontories of Olmiæ and Hæræum, on the shores of the Corinthian Gulf. The latter, now *C. St. Nikolaos*, was the most westerly point of the Isthmus, and was crowned with a temple of Juno, which did service as a fortress. The only stream of importance is the Nemea, which rises in Apesas and flows north-wards through a deep vale into the Corinthian Gulf, forming the boundary between the territories of Corinth and Sicyon. The inhabitants were mainly Æolians, but the dominant race in historical times were Dorians. The capital, Corinthus, was the only important town in the district. It lay at the northern foot of the Acrocorinthus, with its acropolis on the summits of the rock, and possessed two ports—Lechæum on the Corinthian, and Cenchreæ on the Saronic Gulf.

The site of Corinth was not strictly on the plain, but upon a broad level rock some 200 feet above the plain. It was surrounded with

---

[a] Pindar expressly terms it the " bridge of the sea :"—

> Καὶ γέφυραν ποντιάδα
> Πρὸ Κορίνθου τειχέων.   *Isthm.* iv, 34.

[b] 'Ισθμίαν ἱππικοῖ νίκαν,
Τὰν Ἑπτακρήνει Ποσειδάων ἐνδίκεσις,
ἀμφίων αὐτῷ συνέφραψα τόμαν
Πέλοπν ἀναδείσθαι σελίνων.   PIND. *Isthm.* II. 20.

walls, extending (those of the Acrocorinthus included) to 85 stadia : and it was connected with Lechæum by two walls (Plan, 10, 10), each 12 stadia long. The population has been estimated at from 70,000 to 80,000. The buildings of the old town were almost wholly destroyed

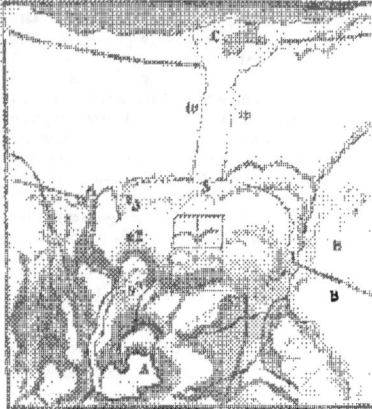

Plan of Corinth.

by Mummius in B.C. 146, and the only account we have of the place refers to the new town, which was visited by Pausanias. The Agora (1) stood in the centre of the town, and was adorned with a vast number of temples and statues: from it four main arteries ran at right angles to each other, leading to Acrocorinthus, and to the gates of Cenchreæ (4), Lechæum (5), Sicyon (6), and Tenea (7). Below Acrocorinthus was an edifice named Sisyphbeium (9). The Propylæa, Odeum, Gymnasium, and other public buildings, were grouped about these streets.

Very few remains now exist of the old Greek town ; we have in the W. seven Doric columns, conjectured, but on insufficient grounds, to belong to the temple of Athena Chalinitis (2), and in the N. foundations, supposed to be of the Temple of Apollo (3): of the Roman town in the E., an amphitheatre, and the ruins apparently of some baths. The Acrocorinthus (A) was partly enclosed with walls : in the greater part of

Fountain of Peirene at Corinth.

its circuit it was inaccessible from its cliffs ; the summit is not perfectly level, but rises into crests ; it was once covered with buildings now in ruins; the ancient temple of Venus stood on the E. crest, but all traces of it have vanished. The celebrated fountain of Peirene[7] (8) still remains: the chief spring is on the summit of the Acrocorinthus: two other springs in the city were supposed to be connected with it, and were also known

---

[7] So celebrated was this fountain, that Pindar describes Corinth as the "city of Peirene :"—

Τοῖσι μὲν ἐξεύχετ᾽ ἐν ἄστει Πειράνας σφετέρου Μὲν πατρὸς ἀρχὰν καὶ βαθὺν Κλᾶρον ἔμμεν καὶ μέγαρον.   Olymp. xiii. 65.

Euripides

by the name of Peirene—one being at the foot of the Acrocorinthus, and now named *Mustapha*; the other, *Paliko*, on the road to Lechæum. Outside the walls, on the E., was the suburb of Cranæum (H), the favourite residence of the wealthy citizens. Lechæum (C), was the chief station of the ships of war, and the emporium of the traffic with the W. coasts of Greece and Italy; the site of the port, which was artificial, is now a lagoon. Cenchreæ, distant about 8½ miles, was the emporium of the trade with Asia, and was a natural port improved by moles: the name of *Kekhries* is still attached to the site, but no town exists there. Corinth was one of the earliest seats of Greek art : painting is said to have been invented there ; the

Coin of Corinth.

most ornate style of Greek architecture still bears the name of Corinthian: statuary also flourished, and the finest bronze[8] for this purpose was known as *Æs Corinthiacum*, while its pottery was hardly less celebrated. Ship-building was carried on, and the first trireme was built there. Though Corinth produced Arion, the second inventor of the dithyramb, and the Cyclic poets Æson, Eumelus, and Eumolpus, yet literature was not much patronized there. The wealth[9] and licentiousness[1] of the place were

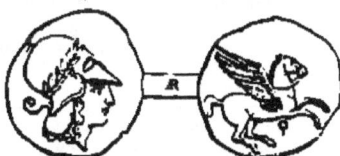

Roman Coin of Corinth.

On the obverse, the head of the Emperor Antoninus Pius.
On the reverse, the port of Cenchreæ.
The letters C. L. I. COR. stand for Colonia Laus Julia Corinthus.

Euripides also speaks of it as the "revered water," and describes it as the resort of the Corinthian elders who played at draughts there ; the fountain to which he refers is the northern one :—

Πισσυνὸν προσελθὼν, ἵνθα δὴ παλαίτεροι
Θάσσουσι, σεμνὸν ἀμφὶ Πειρήνης ὕδωρ.			*Med.* 87.

	Ἡ Πειρήνης ὑδρευομένα
	Πρόσπολος οἰκτρὰ σεμνῶν ὑδάτων.			*Troad.* 205.

The fountain whence Pegasus was caught up by Bellerophon was probably the one on the Acrocorinthus.

	[8] Illaeaque auro vestes, *Ephyreiaque* ære.—Virg. *Georg.* ii. 464

	[9] Even in the Homeric age it was emphatically the "wealthy" Corinth :—
		Ἀφνειός τε Κόρινθον, εὐκτιμένας τε Κλεωνάς.—*Il.* II. 570.

	[1] Hence the well known expression οὐ παντὸς ἀνδρὸς εἰς Κόρινθον ἐστὶν ὁ πλοῦς.
		Non cuivis homini contingit adire Corinthum.—Hor. *Ep.* i. 17, 36.

proverbial ; it was favourably known for its hospitality towards strangers.[1]

Of the other places in Corinthia, we must notice—Schœnus, *Kalamaki*, which stood on the Saronic Gulf at the narrowest part of the Isthmus ; Solygea, on a hill of the same name, S. of Cenchreæ, the scene of an engagement between the Athenians and Corinthians in B.C. 425; Piræus, *Porto Franco*, a harbour on the confines of Epidaurus, where the Athenians blockaded the Peloponnesian fleet in 412 ; Tenea, in the valley that runs S. of Corinth, probably at *Chilimodi*, the town where Œdipus was said to have passed his childhood, and whence Archias drew most of his colonists for Syracuse : its inhabitants claimed a Trojan origin, and were on this account spared by Mummius ; Piræum, *Perachora*, near the Corinthian Gulf, between the promontories Heræum and Olmiæ, and Œnoë, more to the E., each possessing a strong fortress for the defence of this district ; and Crommyon, on the Saronic Gulf, once the property of Megaris : its ruins are near the chapel of *St. Theodorus*.

History.—The foundation of Corinth was carried back by its inhabitants to the mythical ages. In the Homeric poems it is noticed under the two-fold appellation of Ephyra[2] and Corinthus—the first said to have been derived from a daughter of Oceanus and Tethys, the second from a son of Zeus. A Phœnician colony settled on the Acrocorinthus at an early period, and introduced the worship of Aphrodite, for which the town was ever celebrated. The original population was of the Æolian race, but the place was conquered by the Dorians, who thenceforth became the dominant class. The earliest dynasty was that of the Heraclcids, commencing with Aletes and continuing for twelve generations, from B.C. 1074 to 747. This was followed by an oligarchy, under the presidency of the Bacchiadæ, which lasted until 657, and under which the foundations of the commercial greatness of Corinth were laid, and the colonies of Syracuse and Corcyra planted. A tyranny succeeded under Cypselus, 657-627, Periander, 6275-83, and Psammetichus, 585-580, when an aristocracy was catablished under the auspices of Sparta. The Corinthians aided with Sparta in the Peloponnesian War, but after the conclusion of it opposed her, and was engaged in war with her from 395 until the peace of Antalcidas in 387, when the alliance was resumed. After the battle of Chæronea, Corinth was held by the Macedonian kings, and continued in their hands until the battle of Cynoscephalæ, when the Romans declared it free, but retained possession of Acrocorinthus. Corinth afterwards became the head-quarters of the Achæan League, and was consequently taken and utterly destroyed by Mummius in 146; and thus the "light of all Greece," as Cicero termed it, was quenched. It remained in ruins until 46, when Julius Cæsar planted a colony of veterans and freedmen there, and it again became a flourishing town, with the title of Colonia Julia Corinthus.

St. Paul's Travels.—Corinth was visited by St. Paul on his second apostolical journey. A large community of Jews was settled there, and

---

[1] Τρισολυμπιονίκαν ἐπαινέων
Οἶκον, ἄμερον ἀστοῖς,
Ξένοισι δὲ θεράποντα, γνώσομαι
Τὰν ὀλβίαν Κόρινθον, Ἰσθμίου
Πρόθυρον Ποσειδᾶνος, ἀγλαόκουρον.—PIND. *Olymp.* xiii. 1.

[2] Ἔστι πόλις Ἐφύρη μυχῷ, Ἄργεος ἱπποβότοιο,
Ἔνθα δὲ Σίσυφος ἔσκεν, ὃ κέρδιστος γένετ' ἀνδρῶν.—*Il.* vi. 152.

was temporarily increased by the decree of Claudius, which expelled all
Jews from Rome. He remained there eighteen months, and founded a
church, to which he afterwards addressed two epistles. Thence he
went to Cenchreæ, and sailed for Syria (Acts xviii. 1-18). He probably visited it again from Ephesus during his three years' abode at
that place, and certainly at a later period of his third journey (Acts
xx. 3).

§ 4. The territory of **Sicyon** lay along the coast of the Corinthian
Gulf, contiguous to Corinthia on the E., Achaia on the W., and
Phliasia and Cleonæ on the S. It consisted of little beyond the
valley of the **Asōpus**, *St. George*, which, as it approaches the sea,
opens out into a wide and remarkably fertile plain, on which the
olive[4] more particularly flourished. In addition to the Asopus, the
Nemea ran along its E., and the **Sythas** along its W. border: these
were but small streams. The inhabitants of this district were
Ionians, with a dominant race of Dorians. They were divided into
four tribes, of which the Dorians formed three—Hylleis, Pamphyli,
and Dymanatæ; and the old Sicyonians the remaining one—
Ægialeis.

The capital, **Sicyon**, occupied a strong position on a flat hill, about
two miles from the gulf
where the village of *Vasilika* now stands. The
height is defended on
every side by a natural
wall of precipices, and is
accessible only by one or
two narrow passages: the
Asopus flows along its E
side, and the Hellason
along the W. The town
in its greatest extent
consisted of three parts
—the Acropolis, on the
hill; the lower town at
the northern foot of the
hill; and the port-town,
which was fortified, and
connected with the acro-
polis by means of long
walls. The town . pos-
sessed numerous fine
temples and public

Site of Sicyon.
A. *Vasilika*.   b b b. Remains of ancient Walls.

buildings : of these, the remains of the theatre, cut out of the rock;
of the stadium, adjacent to it; and of the temple of Tyche and Dioscuri,
may still be seen.[5] The only other place of importance in Sicyonia was
Titāne, which stood more to the S., on the right bank of the Asopus, and
possessed a temple of Asclepius: the ruins of it are called *Palæokastron*.

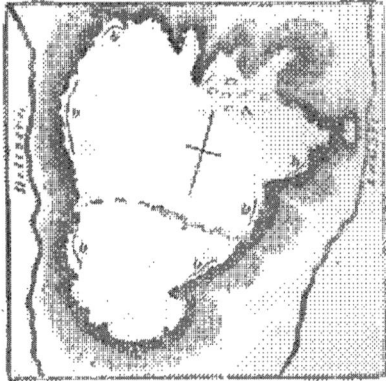

---

[4] Quot Sicyon bacas, quot parit Hybla favos.—Ov. *ex Pont.* iv. 15, 10.
     Venit hiems : teritur Sicyonia bacca trapetis.—Virg. *Georg.* ii. 519.
[5] The modern name *Vasilikà* (τὰ βασιλικά) has reference to these ruins.

History.—Sicyon was one of the oldest cities of Greece, and was in the earliest ages known by the names of Ægialea, Mecōne, which was

Coin of Sicyon.

its sacerdotal designation, and Telchinia, as being one of the earliest seats of workers in metal. In the heroic age it was the abode of the Argive Adrastus.[*] It was at first dependent upon Argos; it then became the seat of the tyranny of the Orthagoridæ from B.C.

676 to 560 : subsequently the Sicyonians were staunch allies of Sparta, and took an active part against Athens in the Megarian and Peloponnesian Wars, as well as against Corinth in 394, and Thebes in 371; the latter power gained possession of the place in 368, but did not retain it. In 323 Sicyon joined the other Greeks in the Lamian War. A series of rulers succeeded one another, and the place had no settled master until its decline about the commencement of the Christian era; the chief events were its capture by Demetrius Poliorcetes in B.C. 303, when its name was changed for a while to Demetrias, and the devastation of its territory by Cleomenes in 233, and by the Ætolians in 221. Sicyon was famed as the earliest school of painting and statuary, and also for the skill of its inhabitants in articles of dress. The painters Eupompus, Pamphilus, and Apelles, and the sculptors Canachus and Lysippus lived here. Its finest paintings were removed to Rome by M. Scaurus.

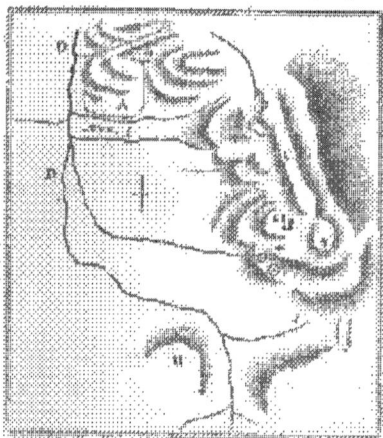

Map of the Neighbourhood of Phlius.

A. Phlius.
B. Arenthyrea or Arcadia.
C. Mount Tricarānon.
D D. The Asopus

1. Ruins, perhaps of Alea.
2. The gate leading to Corinth.
3. Pass-district on Mount Tricarānon.
4. The way to Nemea.

§ 5. The territory of Phlius was bounded by Sicyonia on the N., Arcadia on the W., Cleonæ on the E., and Argolis on the S. : it consisted of a small valley about 900 feet above the level of the sea, surrounded by mountains, from which tributary streams pour down to the river Asopus, in the middle of the plain. The chief heights were named Carneātes, or Arantīnus, *Polyfengo*, in the S., in which the Asopus rises ; and Tricarānon, in the N.E., which rises to three summits. The

* Καὶ Σικυῶν', ὅθ᾽ ἄρ᾽ Ἄδρηστος πρῶτ᾽ ἐμβασίλευεν—*Il.* II. 572.

ancient capital was on Arantinus, and was named Arantia and
Aræthyrea. The later capital, Phlius, stood on one of the spurs of
Tricaranon, above the right bank of the Asopus, near the village of
St. George, where its foundations may still be traced. The town
was commanded by the height of Tricaranon, on which the Argives
built a fortress about B.C. 370, probably represented by the ruins at
Paleokastron.

History.—Phlius was a Dorian state subsequently to the return of the
Heracleids, and was generally in alliance with Sparta. In B.C. 393
internal dissensions occurred, and the Spartan faction was exiled: they
were restored in 383, but the disputes continued, and led to the forcible
entry of Agesilaus in 370, after a siege of twenty months. The opposite faction appears to have been now exiled, and the town was nearly
captured by them, aided by Arcadians and Eleans, in 368. A formidable attack was made in 367 by the Theban commander at Sicyon.
After the death of Alexander, Phlius was subject to tyrants. It is
noted as the birth-place of Pratinas, the inventor of the Satyric drama.

§ 6. The territory of Cleonæ lay between Corinthia on the N.,
Argolis in the S. and E., and Phliasia in the W.: it contained the
upper valleys of the rivers Nemea and Langia, *Longo*, which flow
into the Corinthian Gulf. The road from Corinth to Argos passed
through it, and was commanded by a remarkable pass on the S.
border, named Tretus, "bored," either from the numerous caverns
in the adjacent mountains, or because the path itself appears to be
"bored"; it is now called *Dervenaki*:[7] it might be avoided by a
footpath across the mountains, named Contoporia. In the N. is a
conspicuous mountain, named Apesas,[8] *Fuka*, 3000 feet high, connected with Acrocorinthus by a rugged range of hills.

The town of Cleonæ was small, but well situated on an insulated hill,
and strongly fortified;[9] its site, marked by the traces of its walls, near
Kurtesi, retains the name of *Klenes*. Its history is uneventful: it was

---

[7] This pass was the scene of Hercules's conflict with the Nemean lion, which
occupied one of the caverns :—

> Νεμεαίόν τε λέοντα,
> Τὸν ῥ᾽ Ἥρη θρέψασα, Διὸς κυδρὴ παράκοιτις,
> Γουνοίσιν κατένασσε Νεμείης, πῆμ᾽ ἀνθρώποις.
> Ἔνθ᾽ ἄρ᾽ ὅγ᾽ οἰκείων ἐλεφαίρετο φῦλ᾽ ἀνθρώπων,
> Κοιρανέων Τρητοίο, Νεμείης, ἠδ᾽ Ἀπέσαντος.
> Ἀλλά ἑ ἲς ἐδάμασσε βίης Ἡρακληείης.   HES. *Theog.* 327.

> To cressia mactas

Prodigia, et rursum Nemea sub rupe leonem.—VIRG. Æn. viii. 291.

[8] The appearance of the mountain justifies the description of Statius :—

> Mons erat audaci adsurtus in aethera dorso
> (NomineLernaei memorant Apesanta coloni)
> Gentibus Argolicis olim sacer; inde ferebant
> Nubila suspenso celerem cernerasse volatu
> Perseа.   *Theb.* III. 460.

[9] Ἀφνειόν τε Κόρινθον, ἐϋκτιμένας τε Κλεωνάς.—Il. ii. 570.

Neris et ingenti turrita mole Cleonæ.—STAT. *Theb.* iv. 47.

generally allied to Argos. It owed its chief importance to the public games which were celebrated at Nemea, in its territory, on the road to Phlius. The grove,[1] which was the place of meeting, lay in a deep, well-watered vale,[2] about two or three miles long, and half a mile broad, at the head of the river Nemea. It contained a temple of Zeus, of which three columns, of the Doric order, still remain, a stadium, and other monuments. Near it was the village of Bembina, the site of which is not known.

II. ACHAIA.

§ 7. The province of **Achaia** extended along the Corinthian Gulf from the river Sythas, which separated it from Sicyonia, to the Larissus, on the borders of Elis: on the S. it was contiguous to Arcadia. Its greatest length is about 65 miles, and its breadth from 12 to 20 miles: it was thus a narrow strip of coast-land, as its old name of **Ægialus**[3] implies, skirting the mountain ranges of Arcadia, which form a massive wall, broken only by a few deep gorges, and which send forth numerous spurs to the very edge of the coast. Between these lower ridges are plains and valleys of great fertility, watered by numerous unimportant streams. The coast is generally low and deficient in good harbours. The only important mountain in Achaia itself was named **Panachaïcus**, *Voidhia;* it is in the W., near Patræ, and rises to the height of 6322 feet. There are three conspicuous promontories—**Drepanum**, *Dhrepano*, the most northerly point of Peloponnesus, a low sandy point about four miles E. of Rhium; **Rhium**, *Castle of Morea*, at the entrance of the Corinthian Gulf; and **Araxus**, *Kalogria*, W. of Dyme, and at one time the boundary between Achaia and Elis. Of the streams we need only notice the **Crathis**, *Akrata*, a perennial stream which joins the sea near Ægæ, and which receives the **Styx** as its tributary; the **Piras**, or **Achelous**, near Olenus; and the border streams of **Sythas** and **Larissus**, *Mana*, whose positions have been already noticed.

§ 8. The original inhabitants of Achaia, according to the Greek legends, were Pelasgians, named Ægialeis: the Ionians subsequently

---

[1] The grove was named after Molorchus, who is said to have entertained Hercules there on his expedition against the lion :—

    Cuncta mihi, Alpheum linquens lucosque Molorchi,
    Curalibus, et crudo decernet Græcia cæstu.—*Georg.* iii. 18.

    Dat Nemea comites, et quos in prœlia vires
    Sacra Cleoniæ cogunt *vincta Molorchi.*—STAT. *Theb.* iv. 159.

[2] The plain of Nemea is most abundantly watered, and well deserves the epithet of βαθυσθενος, which Pindar gives it :—

    Κα-
    ματωδεων δὲ πλαγᾶν
    Ἄκος ὑγιηρὸν ἐν
    Βαθυπεδίῳ Νεμέᾳ
    Τὸ καλλίνικον φέρει.   PIND. *Nem.* iii. 27.

[3] Αἰγιαλὸν τ' ἀνὰ πάντα, καὶ ἀμφ' Ἑλίκην εὐρεῖαν.—*Il.* ii. 575.

settled in it, and remained there until the time of the Dorian conquest, when the Achæans, having been ejected from Argos and Lacedæmonia, in turn ejected the Ionians, and gave the country its historical name of Achaia. There is some doubt, however, whether the Achæans were not really an undisturbed remnant of the old population. The Ionians are said to have lived in villages, and the cities to have been first built by the Achæans, who united several villages in each town. The Achæans formed a confederacy of 12 towns, each of which was an independent republic, but united with the others in concerns of common interest, whether political or religious. The list, as given by Herodotus, comprised the following towns from E. to W. :—Pellēne, Ægira, Ægæ, Bura, Helice, Ægium, Rhypes, Patræ, Pharæ, Olēnus, Dyme, and Tritæa. Polybius gives Leontium and Cerynia in the place of Rhypes and Ægæ, which had fallen into decay: Pausanias, on the other hand, retains the two latter, and substitutes Cerynia for Patræ. The meetings of the confederacy were held originally at Helice, and, after its destruction in b.c. 373, at Ægium. The Achæan towns were, almost without exception, well situated on elevated ground, more or less near the sea. None of them are known as commercial towns in the flourishing period of Greek history, though Ægium and Patræ possessed good harbours: the Romans constituted the latter their port-town, and rendered it the most important place on the W. coast. We shall describe the towns more at length, in order from E. to W.

Pellene was situated about 7 miles from the sea, upon a strongly fortified hill, the summit of which rose to a peak, dividing the city into two parts. It was a very ancient place, and appears in the Homeric Catalogue.[4] It was the first of the Achæan towns to join Sparta in the Peloponnesian War. In the wars of the Achæan League it was taken and retaken several times. The town possessed several fine buildings, particularly a temple of Minerva with a statue by Phidias. The ruins are at *Tzerkovi*. Near it was a village, also called Pellene, where the cloaks, which were given as prizes in the games of the city, were made.[5] Its port, named Aristonautæ, was probably at *Kamari*. A little to the E. near the coast was the fortress of *Olūrus*, which commanded the entrance to the plain at *Xylo-castro*. **Ægira** stood on an eminence near the river Crius, about a mile from the sea: it occupied the site of the Homeric Hyperesia, and possessed a port probably at *Mavra Litharia*, to the left of which are some vestiges of Ægira. The town contained numerous temples. In b.c. 220 it was surprised by some Ætolians, who were, however, soon driven out. **Ægæ**, at the mouth of the Crathis, is noticed by Homer, and was celebrated in the earliest times for the worship of Poseidon. It was early deserted by its inhabitants, who removed to Ægira. **Bura** occupied a height about 5 miles from the sea: it was destroyed by an earthquake in

---

[4] Πελλήνην τ' εἶχον, ἠδ' Αἴγιον ἀμφενέμοντο.—Il. ii. 574.
[5] καὶ ψυχρὰν ὑπὲρ' οἰκα-
νὸν φάρμακον αἰρεῖν
Πελλάνᾳ φέρει.   Pind. Olymp. ix. 146.

B.C. 373, but was rebuilt, and took part in the proceedings of the League in 275. Its ruins have been discovered near *Trupia*. **Helice**, on the coast between the rivers Selinus and Cerynites, was probably the most ancient of the Achaean towns, its foundations being ascribed to Ion, the progenitor of the Ionians. It possessed a celebrated temple of Poseidon[*] where the Ionians held their congress. The Achaeans continued to do the same until the destruction of the town by a tremendous earthquake in B.C. 373, by which the whole town was submerged by the sea:[†] a precisely similar disaster occurred at the same spot in A.D. 1817. **Cerynia** was situated on a lofty height S. of Helice and near the river Cerynites: it is mentioned as a member of the League on its revival in B.C. 280, and one of its generals became the first generalissimo of the League in 255. **Aegium** stood between two promontories in the corner of a bay which formed the best harbour next to Patrae. It appears in the Homeric Catalogue, and, after the destruction of Helice, became the chief town in the League. The meetings were held in the grove, named *Homagyrium* or *Homarium*, near the sea. The site of Aegium was on a hill E. of *Vostitza*. **Rhypes** was 30 stadia W. of Aegium on the right bank of the river *Thola*, and is only known as the birth-place of Myscellus, the founder of Croton. It fell early into decay, and its existence was terminated by Augustus, who removed its inhabitants to Patrae. **Patrae** stood on a spur of Panachaicus, which overhangs the coast W. of the promontory of Rhium: it was formed by the union of three villages. Patrae was the only Achaean town which joined Athens in the Peloponnesian War. After the death of Alexander, Cassander got possession of it for a short time, but in 314 his troops were driven out by the general of Antigonus; in 280 the Macedonians were expelled, and in 279 Patrae assisted the Aetolians. It suffered severely in the wars between the Romans and Achaeans, and for a while ceased to be of any importance except as a place of debarcation from Italy. It was restored by Augustus with the title of Col. Aug. Aroë Patrensis, and invested with the sovereignty not only of the adjacent district but even of Locris. Numerous buildings adorned it, particularly a temple of Artemis Laphria, and an Odeum, second only to that of Herodes at Athens. A manufactory of head-dresses and garments of byssus or flax was carried on there. The modern town of *Patras* occupies its site, and is one of the most important seaports in Greece. **Tritaea** was situated near the borders of Arcadia at *Kastritza*, and was one of the four cities which revived the League in B.C. 280: its territory was annexed to Patrae by Augustus. **Pharae** stood on the banks of the Pirus, near *Prevezo*, about 9 miles from the sea: its history is the same as that of Tritaea. **Olenus** stood at the mouth of the Pirus at *Kato*: it fell into a state of decay in the 2nd century B.C., its inhabitants having removed to Dyme. **Dyme** was situated near the coast at *Karavostasi*, about 3½ miles N. of the Larisus: it was formed by an union of 8 villages. It was one of the towns which revived the League in 280. In the Social War it suffered

---

[*] Homer refers to this temple :—

Οἱ δὲ ναι εἰς Ἑλίκην τε καὶ Αἰγὰς δῶρ' ἀνέγουσι
Πολλά τε καὶ χαρίεντα.   *Il. viii. 203.*

Ἱερὸν δ' εἰς Αἰγὰς, ὅθι οἱ κλυτὰ δώματ' ἔασιν.—*Od. v. 381.*

[†] Si quaeras Helicen et Buris Achaïdas urbes,
Invenies sub aquis, et adhuc ostendere nautae
Inclinata solent cum moenibus oppida mersis.—*Ov. Met. xv. 293.*

from the Eleans, who captured the fortress of Teichos near the promontory of Araxus.   Dyme joined Philip of Macedon against the Romans, and was consequently ruined by them.   Pompey made an attempt to settle some Cilician pirates there.

History.—The Achæans are seldom noticed in history until the time of Philip.   In 358 they joined the Athenians and Bœotians at Chæronea, and in 330 the Spartans at Mantinea, and on both occasions they suffered severely.   The Macedonians placed garrisons in their towns, but in 281 some of the cities rose against them, and in 280 the old League was revived by four cities and was subsequently joined by six more.   This League attained a national importance under Aratus of Sicyon in 251, who succeeded in uniting to it Corinth in 243, Megalopolis in 230, and Argos in 226, as well as other important towns, with a view of expelling the Macedonians from Peloponnesus.   Sparta became jealous, and war ensued between Cleomenes and Aratus in 227; the latter called in the aid of the Macedonians, who thus again recovered their supremacy over Achaia.   The Social War in 220 conduced to the same result, and the death of Aratus in 213 completed the prostration of the League.   It was regenerated by Philopœmen, who, under the patronage of the Romans, again united the cities of Peloponnesus: but the Romans soon crushed its real power, and adopted an imperious policy, which ended at length in the defiance of the Achæans, and in the subjection of Greece by Mummius in 146.

III. ELIS.

§ 9. The province of **Elis** extended along the coast of the Ionian Sea, from the river Larisus in the N., on the borders of Achaia, to the Neda in the S., on the borders of Messenia: on the E. it was bounded by the mountains of Arcadia.   Within these limits were included three districts: **Elis Proper** or **Hollow Elis** in the N., extending down to the promontory of Ichthys; **Pisatis**, thence to the river Alpheus; and **Triphylia** in the S.   The first of these was divided into two parts: the fertile plain of the Peneus, which was, properly speaking, the "Hollow" Elis; and the mountainous district of **Acroria** in the interior.   The former consists almost wholly of rich alluvial plains, separated from each other by sandy hills, and well watered by numerous mountain-streams.   These hills are the lower slopes of the Arcadian mountains,—the most prominent being **Scollis**, *Sandameriotiko*, on the borders of Achaia, identified by Strabo with the "Olenian Rock" of Homer;[*] **Pholoë**, in Pisatis, which forms the watershed between the basins of the Peneus and Alpheus; **Lapithas**, *Smerna*, and **Minthe**, *Alvena*, in Triphylia, between which the river Anigrus flows.   The latter is the loftiest mountain in Elis, and was one of the seats of the worship of Hades.

§ 10. The coast of Elis is a long and almost unbroken sandy level,

---

[*] Όφρ᾽ ἐπὶ Βουπρασίου πολυπύρου βήσωμεν ἵπποος
Πέτρης τ᾽ Ὠλενίης καὶ Ἀλησίου, ἔνθα κολώνη
Κέκληται.   *Il.* xi. 755.

varied by the promontories of Chelonátas, *C. Tornese*, a designation originally given to the whole peninsula, of which the promontory opposite Zacynthus forms part, from its supposed resemblance to a tortoise; and Ichthys, *Kutakolo*, so called from its resemblance to a fish. Between these two projecting points is the Sinus Chelonites, while to the N. of Chelonatas is the Sin. Cyllénes, and S. of Ichthys the great Sin. Cyparissius. The chief rivers are—the Penéus, *Gustuni*, which rises in Erymanthus, receives the Ladon (the Homeric Selleeis) as a tributary, and flows across the plain of Elis, joining the sea S. of Prom. Chelonatas [*]—the Alphéus,[1] *Rufia*, the lower course of which alone belongs to Elis; it flows by Olympia[*] into the Cyparissian Gulf, and has a wide gravelly bed, well filled in winter, but shallow in summer—the Anigrus, *Mauro-potamo*, the Minycius of Homer, in Triphylia, the waters of which had a remarkable fœtid smell—and the Neda, *Buzi*, on the S. border. The plain of Elis produced *byssus* or fine flax, wheat, hemp, and wine: its rich pastures were favourable to the rearing of cattle and horses, the latter being specially famous in antiquity.[*]

§ 11. The earliest inhabitants of Elis were Pelasgians, named Caucónes: these afterwards withdrew into the N. near Dyme, and to the mountains of Triphylia. The Phœnicians probably had factories on the coast, and introduced the growth of flax. In the Homeric age the people were named Epeans, a race connected with the Ætolians, and occupying not only Elis Proper, but Triphylia and the Echinades. The name of Eleans was restricted to the inhabitants of Elis Proper, and described the fusion of the Eleans and the Ætolians, who entered at the time of the Dorian invasion. Triphylia was so named probably as being occupied by the "three tribes" of the Epeans, Eleans, and Minyans, the latter of whom

---

[*] The Penæus appears to have formerly joined the sea north of the promontory.
[1] The Alpheus was believed to continue a submarine course, and to mingle with the fount of Arethusa in Sicily :—

　　　　'Αμπνευμα σεμνὸν 'Αλφεοῦ,
　　　　Κλεινᾶν Συρακοσσᾶν θάλος, 'Ορτυγία,
　　　　Δέμνιον 'Αρτέμιδος.　　　　Pind. *Nem.* l. 1.

Sicanio prætenta sinu jacet insula contra
Plemmyrium undosum : nomen dixere priores
Ortygiam. Alpheum fama est huc Elidis amnem
Occultas egisse vias subter mare, qui nunc
Ore, Arethusa, tuo Siculis confunditur undis.—*Æn.* III. 692.

Hence Ovid terms the nymph Arethusa, *Alpheias* :—

Tum caput Elois Alpheias extulit undis.—*Met.* v. 487.

[*] Aut Alpheum rotis prælabi flumina Pisæ,
Et Jovis in luco currus agitare volantem.—*Georg.* III. 180.

[a] 'Ηλιδ' ἐς εὐρύχορον διαβήμεναι, ἔνθα μοι ἵπποι
Δώδεκα θήλειαι,　　　　*Od.* I v. 635.
Οὐδ' ὅσσοι τρέφουσι πρὸς 'Ηλιδος ἱππόβοτον.—*Il.* XII. 347.

entered after their expulsion from Laconia by the Dorians. The towns of Elis were for the most part very ancient, many of them being noticed by Homer: few, however, attained to any historical celebrity. The great question which agitated this part of Greece was the presidency of the Olympian games. Pisa originally possessed this privilege; but on its destruction, in B.C. 572, Elis obtained undisputed supremacy, and became the capital of the whole country—a position to which its admirable site, and the fertility of its territory, predestined it. The most interesting place in Elis was Olympia; but this, it must be remembered, was only a collection of public buildings, and not in any sense a town. Most of the Elean towns occupied commanding positions, and were valuable in a strategetical point of view. The nature of the coast involved the absence of harbours, and consequently Elis never attained commercial importance. We shall describe the towns from N. to S.

Elis, the capital, was well situated on the banks of the Peneus just at the point where it emerges into the plain, and at the foot of a projecting hill of a peaked form about 500 ft. high, on which its acropolis was posted. In the time of Pausanias it was one of the finest cities in Greece, and possessed a magnificent gymnasium

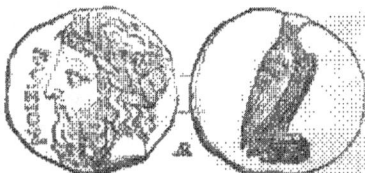

Coin of Elis.

named Xystus, an agora also used as an hippodrome, a building called Hellanodiceon, appropriated to the instruction of the presidents of the Olympic games, a theatre, and other buildings. The only remains are some masses of tile and mortar, a building square outside, but octagonal inside, and a few fragments of sculpture.[4] The site is occupied by two or three villages named *Palaeopoli*. Elis is noticed by Homer, but did not attain importance until after the Dorian Invasion, when it became the seat of government. After the Persian Wars the town spread from the acropolis, to which it was originally confined, over the subjacent plain. Pisa, the old capital of Pisatis, stood a little E. of Olympia, on the W. bank of a rivulet now named *Miraka*, near its junction with the Alpheus: it was celebrated in mythology as the residence of Œnomaus and Pelops: it had originally the presidency of the Olympian games, which led to frequent wars with Elis and to its utter destruction[5] in B.C. 572. Olympia was situated on a plain 3 miles long and 1

---

[4] The general disappearance of the buildings in Elis is attributable partly to the accumulation of the alluvial soil, and partly to the porous character of the stone.

[5] Even its existence has been doubted; but Pindar's testimony is conclusive on this point:—

'Ήτοι Πίσα μὲν Διὸς
'Ολυμπιάδα δ' ἔστα-
σεν 'Ηρακλέης,
'Ακρόθινα πολέμου.   Olymp. II. 4.

broad, open towards the W., but surrounded on other sides with hills,
among which **Mount Cronius** in the N., and **Typæus** in the S., are most

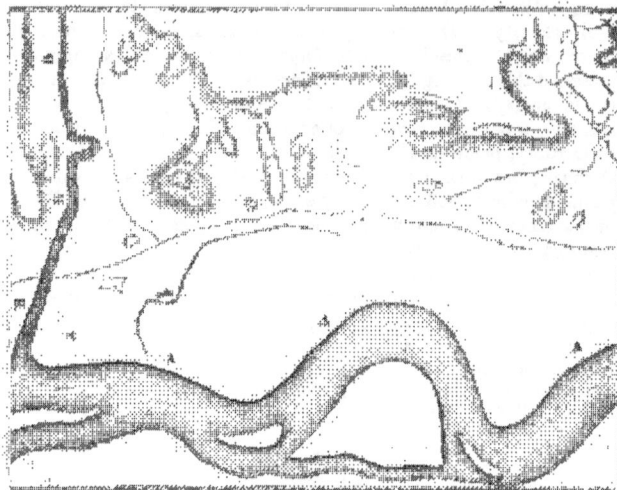

Plain of Olympia,

A A. Course of the Alpheus.   1. Site of Pisa.
B B. The Cladeus.   2. Mount Cronius.

conspicuous. The Alpheus flows between these ranges in a constantly
shifting course, and receives on its right bank a tributary from the N.
named **Cladeus**. Along the banks of this stream lay the Altis [4] or
Sacred Grove—a large enclosure, bounded on the S. and E. by a wall,
and elsewhere by hills, and adorned with trees, particularly a grove of
planes in its centre. Within it lay the most important buildings,
foremost among which we must notice the Olympiëum or temple of
Zeus Olympius near the S.W. corner, founded by the Eleans in B.C.
572, completed about 470, and decorated by Phidias about 435. The
date and cause of its destruction are unknown. Its foundations have
been laid bare in modern times, from which it appears that it was a
peripteral hexastyle building 230 feet long and 95 broad, of the Doric
order, with columns exceeding in size those of any other Greek building. The roof was covered with tiles of Pentelic marble ; the pediments were filled with sculpture, and their summits crowned with a

---

4 'Ο δ' ἐν Πίσᾳ ἔλσος ἄλσο τι στρατὸν
    Ἀσίαν τε πᾶσαν Διὸς ἄλσεσιν
    Υἱὸς ἐναλλάττο ζάθεον ἄλσος
    Πατρὶ μεγίστῳ· περὶ δὲ πάξαις
    Ἄλτιν μὲν ὅγ' ἐν καθαρῷ
    Διέκριτε.   PIND. Olymp. 2. 51.
    'Αλλ', ὦ Πίσας εὔδενδρον ἐπ' 'Αλφεῷ ἄλσος.—Id. viii. 12.

gilded statue of Victory. The colossal statue of Jupiter by Phidias, made of ivory and gold, was the most striking object inside; it existed until about A.D. 393, when it was carried off to Constantinople, and was burnt there in 476. The Heraeum, which comes next in importance, was also a Doric peripteral building; it contained the table on which the garlands for the victors were placed, and the celebrated chest of Cypselus. The great altar of Zeus, 22 feet high, was centrally situated. The thesauri, or treasuries, stood near the foot of Mount Cronius. The stadium and hippodrome appear to have formed a continuous area, the circular end of the former being at the back of Cronius, and the further end of the latter near the Alpheus. Various other temples were scattered over the intervening space, together with a large number of statues, computed by Pliny at 3000. The public games were said to have been originally instituted by Hercules; they were restored by Iphitus, king of Elis, in B.C. 884, and were celebrated every fourth year until A.D. 394; these periods were named Olympiads, and became a chronological era after B.C. 776. **Letrini** stood near the sea on the Sacred Way that connected Olympia with Elis; it joined Agis when he invaded Elis, and was made independent in B.C. 400; its site is at the village of *St. John*. **Lepreum**, the chief town of Triphylia, stood in the S. of the district, about 4½ miles from the sea, and appears from its ruins (near *Strovitsi*) to have been a place of considerable extent. It was the only Triphylian town which took part in the Persian Wars; it was also foremost in resisting the supremacy of Elis, from which it revolted in B.C. 421, and was formally freed in 400. Lepreum joined the Arcadian confederacy against Sparta about 370, and at a later period sided with Philip in his Ætolian War.

Of the less important towns we may notice — **Buprasium**, near the left bank of the Larissus, frequently noticed by Homer;[1] **Myrtuntium**, the Homeric Myrsinus, near the sea between Elis and Dyme; **Cyllene**, a seaport town usually identified with *Glarentza*, but more probably about midway between the promontories of Araxus and Chelonatas; it was burnt by the Corcyraeans in 435 and was the naval station of the Peloponnesian fleet in 429; **Hyrmine**, on the coast N. of Chelonatas at *Kunupeli*; **Pylus Eliacus**,[2] at the junction of the Ladon with the Peneus, where are the ruins of *Agrapidho-khori*; the only historical notices of it are its capture by the Spartans in 402, and its occupation by the exiles from Elis in 366; **Ephyra**,[3] the ancient capital of Augeas, on the Selleeis, or Ladon, about 14 miles S.E. of Elis; **Lasion**, the chief town of Acroria in the upper valley of the Ladon, and for a long

---

[1] The fertility of its district is remarked both by Homer and Theocritus:—

' Όφρ' έπι Βουπρασίου πολυπύρου βήσομεν ίππους.—Il. xi. 756.

Ου πάμαι βόσκονται ίαν βάσιν, ούθ' ένι χώρον
'Αλλ' αι μέν ρα νέμονται έν' όχθαις άμφ' Έλισούντος,
Αι δ' ιερόν θείοιο παρά ρόον 'Αλφειοίο,
Αι δ' έπι Βουπρασίου πολυβότρυος.　　　　Idyll. XXV. 8.

[2] This Pylus claimed to be Nestor's capital, on the strength of the following lines from the Iliad :—

γένος δ' ήν έν ποταμώ
'Αλφειού, δσ' εύρύ ρέει Πυλίων διά γαίης.　　　　v. 544.

The lines, however, only prove that the land or kingdom of Pylus extended to the north of Elis.

[3] Τήν άγετ' έξ 'Εφύρης ποταμού άπο Σελλήεντος.　　　　Il. ii. 659.

period in the occupation of the Arcadians; **Harpīna**, on the Alpheus near Olympia, said to have been named after the mother of Œnomaus; **Margīna**, in Pisatis, E. of Letrini; **Phea**, on the isthmus of Prom. Ichthys, with a port on the N. side of the isthmus which was visited by the Athenian fleet in in 431: the ruins of *Pontikokastro* are on its site; the Homeric stream of Iardanus[1] is probably the little torrent N. of Ichthys; **Epitalium**, *Agulenitza*, near the mouth of the Alpheus, and identified with the Homeric **Thryoessa**:[2] it commanded the coast road, and was hence garrisoned by Agis in 401, and taken by Philip in 218; **Scillus**, S. of Olympia, in the valley of the Selinus, destroyed by the Eleans in 572, and restored by the Lacedæmonians in 392, for 20 years the residence of Xenophon, who has left an interesting description of the place; **Hypāna**, in the interior of Triphylia, but of uncertain position; **Samicum**, *Khaiaffa*, on a hill near the coast midway between the Alpheus and Neda, identified with the Homeric **Arēne**;[3] it commanded the coast which here traverses a narrow pass; hence it was occupied by Polysperchon against the Arcadians, and taken by Philip in 219: near it was the temple of the Samian Poseidon, where the Triphylian cities held their congress; on either side of Samicum a large lagoon extends along the coast, into which the Anigrus flows: its water was efficacious in cutaneous diseases; **Macistus** or **Platanistus**, the chief town in Northern Triphylia, near Samicum, and not improbably the original name of the later town on the heights of *Khaiaffa*; some authorities place it more to the S.; **Phrixa**, on the left bank of the Alpheus, and on a hill now named *Paleofanaro*, founded by the Minyans; **Pylus Triphylīacus**,[4] N. of Lepreum, and in later times belonging to it; **Pyrgus** or **Pyrgi**, at the mouth of the Neda, an old settlement of the Minyas; and lastly, **Epēum**, the Homeric **Epy**,[5] so named from its lofty position, on the border of Arcadia, but of uncertain position.

History.—Elis, from its remote position, as well as from its privileged character as the Holy Land of Greece, took but a small part in the general history of the peninsula. We have already referred to the disputes for the supremacy between Pisa and Elis, in which the latter came off triumphant. A long period of peace ensued until in 421 Lepreum revolted, and a quarrel between Sparta and Elis resulted, which led ultimately to the invasions of Agis and the destruction of the supremacy of Elis in 400. An attempt to recover this supremacy after the battle of Leuctra in 371 led to an alliance between the Tri-

---

1　Φιὰς γὰρ τείχεσσιν, ᾽Ιαρδάνου ἀμφὶ ῥέεθρα.　　　　　*Il.* vii. 135.

2　᾽Εστι δέ τις Θρυόεσσα πόλις, αἰπεῖα κολώνη,
　Τηλοῦ ἐπ᾽ Ἀλφειῷ, νεάτη Πύλου ἠμαθόεντος.　　　　*Il.* xi. 710.

3　Οἳ δὲ Πύλον τ᾽ ἐνέμοντο, καὶ Ἀρήνην ἐρατεινήν.　　*Il.* II. 591.
　᾽Εστι δέ τις ποταμὸς Μινυήϊος εἰς ἅλα βάλλων
　᾽Εγγύθεν Ἀρήνης.　　　　　　　　　　　　　　*Il.* xi. 721.

4　The Triphylian Pylus was believed by Strabo to have been Nestor's capital, his main reason being that the account of Nestor's expedition against the Epeans (*Il.* xi. 676, *seq.*) implies a spot nearer than the Messenian Pylus, and that other passages (*Od.* iii. 423; xv. 199, *seq.*) are inconsistent with the idea of a seaport town. These objections are partly answered by the fact that Pylus applied to the *kingdom* as well as the city of Nestor. On the other hand, the account of the journeys of Telemachus from Sparta to Pylus *through Pherœ* (*Od.* iii. 485; xv. 182) is decisive for the Messenian town.

5　Καὶ Θρύον, Ἀλφειοῖο πόρον, καὶ ἐΰκτιτον Αἶπυ.—*Il.* ii. 592

phylian towns and the Arcadians, and to a war between the latter and
the Eleans, which lasted from 366 to 362 without any very decisive
result. The Eleans joined the Greeks in the Lamian War, and subsequently became members of the Ætolian League. They are not mentioned after this.

IV. MESSENIA.

§ 12. **Messenia** [4] lay in the S.W. of the Peloponnesus, bounded on
the N. by Elis and Arcadia, on the E. by Laconia,[7] and on the S.
and W. by the sea, viz. by the Messenian Gulf in the former, and
the Ionian Sea in the latter direction. The configuration of the
country is simple: on the N. frontier there is a band of mountains,
anciently named **Ira**, and now *Tetrazi*, forming the watershed of the
rivers Neda, Pamisus, and Alpheus; from this, ranges emanate towards the E. and W., the former named **Nomii Mts.**, *Makryplai*, the
latter **Elæum**, *Kuvela*, which is continued in a series of ranges skirting
the W. coast, named **Ægaleum**, between Cyparissia and Pylus,
Buphras and **Tomeus**, near Pylus, and **Tamathia**, *Lykodimo*, more to
the S., and terminates in the promontory of Acritas, *C. Gallo*. Returning to the N., the range of Nomii effects a junction, towards the
E., with **Taygetus**, which forms the general boundary on the side
of **Laconia** in the N.E., but runs into the latter country towards
the S. These mountains enclose an extensive plain, or rather series
of plains, watered by a river named, in its lower course, **Pamisus**,
Dhipotamo, and made up of the **Balyra**, the Amphitus, the Aris, and
other less important tributaries. The Pamisus falls into the Messenian Gulf, and is navigable for small boats. The basin of the
Pamisus is divided into two distinct parts by a ridge of mountains
crossing it in the neighbourhood of Ithome. The upper plain, named
Stenyclarus, is small, and of moderate fertility; the lower one, which
opens to the Messenian Gulf, is more extensive, and remarkably fertile, whence it was sometimes named **Macaria**, "the Blessed."[8]　The

---

[4] The Homeric form of the name is Messene :—
Τῷ δ᾽ ἐν Μεσσήνῃ ξυμβλήτην ἀλλήλοιϊν,
Οἴκῳ ἐν Ὀρσιλόχοια.　　　　　　　　　　Od. xxi. 15.

[7] The boundary on the side of Laconia varied at different times, Messenia
sometimes possessing and sometimes losing the border district, named Dentho-
liâtes Ager, which lay on the western slope of Taygetus, about Limnæ. This
was the cause of the first Messenian war ; it remained a subject of dispute under
the Romans; and even so late as A.D. 1835 it was transferred from the government of *Mistra* (Sparta) to that of *Kalamata*.

[8] It is, doubtless, to this district that Euripides refers in the following lines :—
Καλλίρρυτόν τε μυρίοισι νάμασι,
Καὶ βουσὶ καὶ ποίμναισιν εὐβοτωτάτην, .
Οὔτ᾽ ἐν πνοαῖσι χείματος δυσχείμερον,
Οὔτ᾽ αὖ τεθρίπποις ἡλίου θερμὴν ἄγαν.
　　　　　　　　　　　　Eurip. ap. Strab. viii. p. 366.
The climate of Messenia contrasts favourably with that of other parts of Greece,
in consequence of the lower elevation of the hills.

coast is tolerably regular, the most remarkable break being the deep bay of **Pylos**, *Navarino*, on the W. coast, which was 2¼ miles in diameter, bounded on the N. by the promontory of **Coryphasium**, and closed in front by the island of **Sphacteria**, *Sphagia*. More to the N. are the promontories of **Platamôdes**, near *Aia Kyriake*, and **Cyparissium**, which forms the southern limit of the **Cyparissius Sinus**.

§ 13. The earliest inhabitants of Messenia are said to have been Leleges. To these Æolians were added at an early period, whose chief settlement was at Pylus, the capital of Neleus. The Dorians conquered it, and remained the dominant race. It was divided by Cresphontes, the first Dorian king, into five parts, of which Stenyclêrus, Pylus, Rhium, Hyamia, and Mesôla, were the centres. The position of the two first is well known; Rhium was about the southern promontory, and Mesôla between Taygetus and the Pamisus; the position of Hyamia is unknown. The towns of Messenia were comparatively few. The earliest capitals were in the upper plain, Andania being that of the Messenian kings before the Dorians, and Stenyclerus that of the Dorians themselves. Pylus, on the W. coast, was the seat of an independent kingdom, which extended along the coast as far N. as the Alpheus. These towns fell into decay during the period when Messenia was subject to Sparta. The later capital, Messêne, was founded by Epaminondas, B.C. 369, and was advantageously placed between the two plains: it became one of the most important cities in Greece. Messenia possessed the harbours of Pylus and Methône on the W. coast, Asine and Corône on the E.: these do not appear, however, to have carried on an extensive trade. We shall describe the towns in order, commencing with those on the coast.

Pylus was the most important spot on the W. coast: the original town, Nestor's capital, was probably situated a little inland, with a port at Prom. Coryphasium: the later town, which was the scene of the operations in the Peloponnesian War, was on the coast itself, the inhabitants having at some early period moved thither from the old town. In the accompanying map, A marks the island of Sphacteria, B the town of Pylus on Prom. Coryphasium, C the modern *Navarino*, and D D the Bay of Pylus. Considerable changes have taken place in this locality since Thucydides wrote his account of it: the N. passage between the island and the mainland, which was formerly deep, and so narrow as to admit only two triremes abreast, is now 150 yards wide, and shallow, while the S. passage, which admitted only eight or nine triremes, is now 1400 yards wide. There is now a lagoon[*] at the back of the site of Pylus: in this direction Coryphasium is precipitous; but on the W. side it slopes down gently to the sea. It is covered with the foundations of Hellenic buildings, erected at the restoration of the town by

---

[*] This lagoon was probably a sandy plain in old times; hence the epithet which Homer applies to it:—

Πάσαι δ' ἐγγὺς ἄλὸς, νέαται Πύλου ἡμαθόεντος. *Il.* ix. 153.

Epaminondas. **Methône**, *Modon*, the Homeric **Pedāsus**, was situated at the extreme point of a rocky ridge, which runs into the sea N. of the Œnussæ Islands: it possessed an excellent harbour. It was held by the Messenians in the second war, and was afterwards given by the Spartans to the Nauplians. In 431 the Athenians vainly attempted to seize it. The Romans made it a free city. **Asine**, on the coast of the Messenian Gulf, was founded by the Dryopes, and was a place of considerable importance till the 6th century A.D.: its site is now occupied by *Koroni*, whence it appears to have received the population of **Corône**, which stood more to the N. at *Petalidhi*, where traces of the ancient mole and of the acropolis still exist. **Pharæ** was situated upon a hill, near the river Nedon, about a mile from the Messenian Gulf, occupying the site of *Kalamata*, the modern capital of Messenia. It is frequently noticed by Homer,[1] and

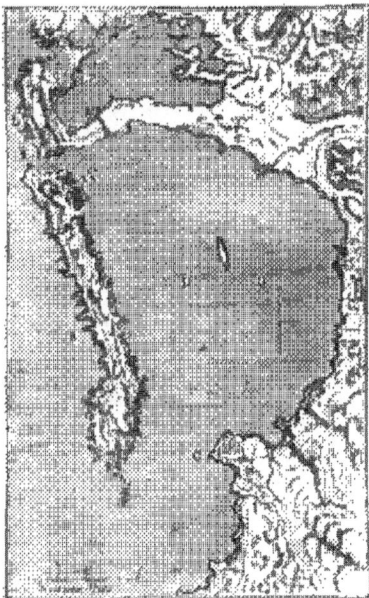

Map of the Bay of Pylus.

appears in his time to have been the chief town in the southern plain. It was annexed to Laconia by Augustus, but restored to Messenia by Tiberius. It possessed a roadstead, which was available only in the summer months. **Thuria**, on the river Aris, became one of the chief towns of the Lacedæmonian Perioeci after the subjugation of Messenia: it was identified with the Homeric **Anthêa**. The old town occupied the summit of a hill, now named *Paleokastro*; the later one was in the subjacent plain at *Palea Lutra*: remains of both exist. **Messêne**, the later capital of Messenia, built by Epaminondas in B.C. 369, was situated upon a rugged mountain which rises between the two Messenian plains, and which culminates in the heights of Ithôme and Eva, on the former of which the acropolis was posted, while the town lay in a hollow just W. of the ridge connecting the two summits. Ithome is 2631 feet high, with precipitous sides, and was connected by walls with the town. The circumference of the walls is about six miles, and the foundations still exist, together with the northern gate, called the Gate of Megalopolis, which has the appearance of a circular fortress,

---

[1] It was one of the 7 towns offered by Agamemnon to Achilles :—

Φηράς τε ζαθέας, ήδ' Άνθειαν βαθύλειμον.   *Il.* ix. 151.

The chief buildings in Messene were the Agora, near the village of Mauromati, containing a fountain in it named Arsinoë, and numerous temples; the stadium, some portions of which are still preserved; and the theatre, to the N. of it, of which there are also remains. The summit of Ithome is a small flat surface, extending from S.E. to N.W., and contained a temple of Zeus Ithomatas.

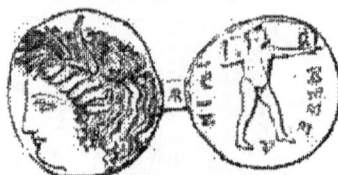

Coin of Messenia.

Messene was in vain attacked by Demetrius of Pharus, and by Nabis, the tyrant of Lacedæmon: it was, however, taken by Lycortas, the Achæan, in 182.

Of the less important places we may notice—Cyparissia, on the W. coast, possessing the best roadstead N. of Pylus, and well situated on an elevation; Abia, the Homeric Ira, on the sea-coast near the border of Laconia; Limnæ, more to the N., possessing a temple of Artemis, which was used jointly by the Messenians and Lacedæmonians, the ruins of which are at Bolimnos; Œchalia, in the plain of Stenyclarus, identified sometimes with Andania, the capital of the Lelegae, and the birth-place of Aristomenes—and sometimes with Carnasium, which stood a little to the N.E. of Andania, and possessed, in Pausanias's time, a sacred grove of cypresses, with statues of Apollo, Hermes, and Persephone; Stenyclarus, the capital of the Dorian conquerors, built by Cresphontes, in the plain which afterwards bore its name; and Ira, a fortress on the hill of the same name.

History.—The most important events in the early history of Messenia were the two wars with Sparta, the assigned dates of which are from B.C. 743 to 723, and from 685 to 668: after the second the whole of Messenia was incorporated with Sparta, the very name being superseded by that of Laconia. In 464 the Messenians rose against the Spartans, and the third war ensued, which terminated with the withdrawal of the Messenians to Naupactus in 455. The nationality was restored by Epaminondas in 369, when the Messenians returned from all directions, and rebuilt their old towns. After the fall of Thebes, the Messenians aided with Philip, and received in return Limnæ and other districts. They joined the Achæan League, but afterwards quarrelled with it, and were consequently engaged in war, which resulted in the secession of Abia, Thuria, and Pharæ, from the supremacy of Messene. Mummius restored these cities to it on the settlement of the affairs of Greece.

Islands.—Off the coast of Messenia are the following islands :—The Strophades, so named because the Boreadæ here *turned*[2] from the pursuit of the harpies: they are now named *Strofadia* and *Strivali*; Prote, which still retains its name. N. of Pylus; Sphacteria, *Sphagia*, opposite Pylus; the Œnussæ, a group, of which the two largest are now named *Cabrera* and *Sapienza*; and Thaganussæ, *Venetiko*, off the promontory of Acritas.

---

[1] Servatum ex undis Strophadum me littora primum
Accipiunt. Strophades Graio stant nomine dictæ
Insulæ Ionio in magno.   Virg. Æn. III. 209.

Gate of the Lions at Mycenæ.

CHAPTER XXIII.

PELOPONNESUS—*continued.*   LACONIA, ARGOLIS, ARCADIA.

V. LACONIA. § 1. Boundaries; Name. § 2. Mountains; Rivers.
§ 3. Inhabitants. § 4. Towns; History; Islands. VI. ARGOLIS,
with Cynuria. § 5. Boundaries; Name. § 6. Mountains; Rivers.
§ 7. Inhabitants; Towns; History. § 8. Cynuria. VII. ARCADIA.
§ 9. Boundaries. § 10. Mountains. § 11. Rivers. § 12. Inhabitants; Towns; History. § 13. SPORADES. § 14. CRETA. Mountains; Rivers. § 15. Inhabitants; Towns; History; St. Paul's
Travels.

V. LACONIA.

§ 1. **Laconia** occupied the S.E. portion of Peloponnesus, and was
bounded by Messenia on the W., Argolis and Arcadia on the N., and
in other directions by the sea. Its natural features are strongly
marked: it consists of a long valley,[1] surrounded on three sides
by mountains, and opening out towards the sea on the south
through the entire length of which the river Eurotas flows. The
approaches to it are difficult:[2] on the N. there are but two natural

---

[1] Hence the Homeric epithet " hollow " Lacedæmon :—

Οἳ δ' εἶχον κοίλην Λακεδαίμονα κητώεσσαν.   *Il.* II. 581.

The shape of the Laconian valley has been compared to that of an ancient
Stadium.

[2] This feature is forcibly described by Euripides :—

Πολὺν μὲν ἄροτον, ἱππονεῖς δ' οὐ ῥᾴδιον·
Κοίλη γάρ, ὄρεσι περίδρομος, τραχεῖά τε
Δυσείσβολός τε πολεμίοις.   *Ap.* Strab VIII. p. 366.

passes by which the plain of Sparta can be entered; on the W. the lofty masses of Taygetus present an almost insurmountable barrier; while on the E. the rocky character of the coast protects it from invasion by sea. The plain of Sparta is blessed with a fine climate and beautiful scenery,[3] but the soil is thin and poor, and adapted to the production of the olive rather than of grain crops.

Name.—The ancient name, as given by Homer, was Lacedæmon, which was occasionally used even in later times *(e. g.* Herod. vi. 58). The origin of the name was referred to a mythical hero, Laco, or Lacedæmon. Modern etymologists connect it with λάκος, *lacus, lacuna,* in reference to its being *deeply sunk* in the mountains.

§ 2. The chief mountain range of Laconia is **Taygetus**, which extends from the border of Arcadia in an almost unbroken line[4] for 70 miles to the promontory of **Tænarum**, *C. Matapan,* the extreme S. point both of Greece and of Europe. Taygetus attains its greatest elevation (7002 feet) near Sparta, in a hill named **Talêtum**, *St. Elias:* there are several other summits near Sparta, whence its modern name of *Pentedactylum,* " five fingers." Parallel to the central ridge is a lower one of less height bounding the plain of Sparta, which consists of huge projecting masses of precipitous rocks.[5] More to the S., it sends forth a lateral ridge, which forms the southern boundary of the Spartan plain. The sides of Taygetus are clothed with pine forests, which were in ancient times filled with game and wild beasts.[6] The southern part abounded in iron, marble,[7] and green porphyry; it also produced valuable whetstones. The range of **Parnon**, *Malevo,* which forms the boundary on the side of Argolis, consists of various detached mountains, the highest of which, attain-

---

[3] This portion of Laconia fully justifies the Homeric epithet " lovely :"—

Οὐδ᾽ ὅτε σε προτέρω Λακεδαίμονος ἐξ ἐραννῆς.—Il. iii. 443.

The climate is favourable to the complexion, and the present appearance of the Spartan women, as compared with the other Greeks, illustrates the other Homeric expression, Λακεδαίμονα καλλιγύναικα.

[4] The unbroken length of this range is well described by the epithet ἐρυμνότατον (see below, note [6]).

[5] The sides of Taygetus were much shattered by earthquakes, whence Laconia is described as " full of hollows :"—

Οἵ δ᾽ εἶχον κοίλην Λακεδαίμονα κητώεσσαν.   Il. II. 581.

[6] Hence it was one of the favourite haunts of Artemis :—

Οἵη δ᾽ Ἄρτεμις εἶσι κατ᾽ οὔρεος ἰοχέαιρα,
'Ἢ κατὰ Τηΰγετον περιμήκετον, ἢ Ἐρύμανθον,
Τερπομένη κάπροισι καὶ ὠκείῃς ἐλάφοισι.   Od. vi. 102.

For the same reason its dogs were celebrated :—

Voeat ingenti clamore Cithæron,
Taygetique canes, domitrixque Epidaurus equorum :
Et vox adsensu nemorum ingeminata remugit.—Georg. iii. 43.

Veloces Spartæ catulos.   Id. 405.

[7] Illic Taygeti virent metalla
Et crriant vario decore saxa.   Mart. vi. 42.

ing an elevation of 6355 feet, lies between the Eurotas and the sea.
On the W. Parnon sinks rapidly towards the valley of the Eurotas,
and breaks up into several hills, such as **Olympus** and **Evas**, near
Sellasia; **Thornax**, near the confluence of the Eurotas and Œnus; and
Menelaium, near Therapnæ.   The range continues towards the S. at
a less elevation, but again rises to a height of 3500 feet in Mount
Zarax on the E. coast, and terminates in Prom. Malea.   The ranges
of Parnon and Taygetus are connected in the N. by a rugged mountain district on the borders of Arcadia, named **Sciritis**.   The coast
is varied by the promontories of **Tænarum**,[1] *C. Matapan*, and **Malea**,[2]
C. Malio, on the S., and **Onugnathus** on the W. coast.   The only important river is the **Eurotas**,[1] *Busili-potamo*, which rises on the borders
of Arcadia, and flows towards the S.E. into the Laconian Gulf,

---

[1] Tænarum is more properly described as a circular peninsula, about 7 miles
in circumference, and connected with the range of Taygetus by an isthmus about
half a mile wide.   The peninsula was originally held to be sacred to the Sun:—

> Ἷξον, καὶ χῶρον περφωμβρότου Ἠελίοιο,
> Ταίναρον, ἵνθα τε μῆλα βαθύτριχα βόσκεται αἰεὶ
> Ἠελίοιο ἄνακτος, ἔχει δ᾽ ἐπιτερπέα χῶρον.
>
> Hom. *Hymn. in Apoll.* 411.

It was afterwards, however, sacred to Poseidon, who had a famous temple and
asylum there; reference is made to this in the line:—

> Ἱερόν τ᾽ ἀβραώστος Ταινάρου μένει λιμήν.—Eurip. *Cycl.* 292.

Near it was a cave, by which Hercules dragged Cerberus from the lower regions,
and which was hence regarded as one of the entrances to Hades:—

> τὰρ χθόνιον
> Ἄιδα στόμα, Ταίναρον εἰς ἱεράν.   Pind. *Pyth.* iv. 77.

> Tænarias etiam fauces, alta ostia Ditis,
> Et caligantem nigra formidine lucum
> Ingressus, Manesque adiit, regemque tremendum.—*Georg.* iv. 467.

The marble quarries of Tænarus were much valued:—

> Quidve domus prodest Phrygiis innixa columnis,
> Tænare, sive tuis, sive, Caryste, tuis?—Tibull. iii. 3, 13.

> Quod non Tænariis domus est mihi fulta columnis,
> Nec camera auratas inter eburna trabes.—Propert. iii. 2, 9.

[2] Malea was regarded with dread by ancient navigators:—

> Ἀλλά με κῦμα, ῥόος τε, περιγνάμπτοντα Μάλειαν,
> Καὶ βορέης ἀπέωσε, παρέπλαγξεν δὲ Κυθήρων.   *Od.* ix. 80.

> Nunc illas promitte vires,
> Nunc animos; quibus in Gætulis Syrtibus usi,
> Ionioque mari, Malemque aequacibus undis.—Virg. *Æn.* v. 191.

> Nec timeam vestros, curva Malea, sinus.—Ov. *Am.* ii. 16, 24.

[1] The banks of the Eurotas were in some parts overgrown with a profusion
of reeds:—

> Σπάρτην τ᾽ Εὐρώτα δονακοτρόφου ἀγλαὸν ἄστυ.—Theocr. 781.

Its groves were favourite haunts of the gods:—

> Qualis in Eurotæ ripis aut per juga Cynti
> Exercet Diana choros.   Virg. *Æn.* i. 498.

> Omnia quæ, Phœbo quondam meditante, beatus
> Audiit Eurotas, jussitque ediscere lauros.—*Ecl.* vi. 82.

receiving as tributaries, the Œnus on its left bank from the borders
of Argolis, and several lesser streams, of which the only ones that
received specific names were the Tiasa, below Sparta, and the
Phellias, which flows by Amyclæ. The mid-valley of the Eurotas,
below the junction of the Œnus, expands into a considerable plain.
More to the S. the river flows through a narrow gorge formed by
the advancing ranges of Taygetus : thence it emerges into the maritime plain of Helos, and flows through marshes and sandbanks into
the sea.

§ 3. Laconia is said to have been originally occupied by Leleges ;
then by Achæans ; and finally, by a mixed population, consisting
of (i.) the Spartans, or ruling caste of the Dorians ; (ii.) the Periœci,
"dwellers about the cities," who appear to have been partly Achæans
and partly Dorians of an inferior grade; and (iii.) the Helots, or
serfs, Achæans who had been taken captive in war. The number
of the Spartans at the time of the Persian wars was about 8000, and
of the Periœci probably 16,000 : the number of the Spartans diminished, and in B.C. 369 did not exceed 2000, and in 244 not more
than 700. The Helots were very numerous : at the battle of
Platæa there were 35,000 present. The towns were numerous, and
were situated partly in the valley of the Eurotas, and still more
numerously on the shores of the Laconian Gulf. In the Homeric
age Amyclæ was the chief town of the interior, and Helos the chief
maritime town : Phare, Sparta, and Bryscæ are also noticed as important cities of the vale ; Las, Œtylus, Messa, and Augiæ, or
Ægiæ, of the maritime district. Subsequently to the Dorian conquest, Sparta became the capital, with Gythium for its port-town.
With the exception of Sparta, the history of the Laconian towns is
comparatively uninteresting : they took little part in the general
affairs of Greece, and were rarely visited : indeed, without the
valuable work of Pausanias, we should have been devoid of any
description of them in their original condition.

§ 4. Sparta, or Lacedæmon, stood at the upper end of the mid-
valley of the Eurotas,[a] on the right bank of the river, and about two
miles E. of the modern *Mistra*. Like Rome, it was built partly on
some low hills, and partly on the adjacent plain. The names and
probable positions of the hills were as follows : Issorium, in the N. ;
Acropolis, more to the S., and divided from Issorium by a hollow
way communicating with a plain ; Colona, on the E., running

---

[a] The position of Sparta presents a striking contrast to that of Athens : the
former being inland, inaccessible by sea and land, remote from any great highway,
and possessing in her own territories all the necessaries of life—the latter, maritime, accessible, central, and dependent on other countries for her supplies. The
effect of geographical position may be traced in the history, policy, and institutions of each.

parallel to the Eurotas; and another to the S., on which New
Sparta is built. The town was made up of four villages—Pitāne, in

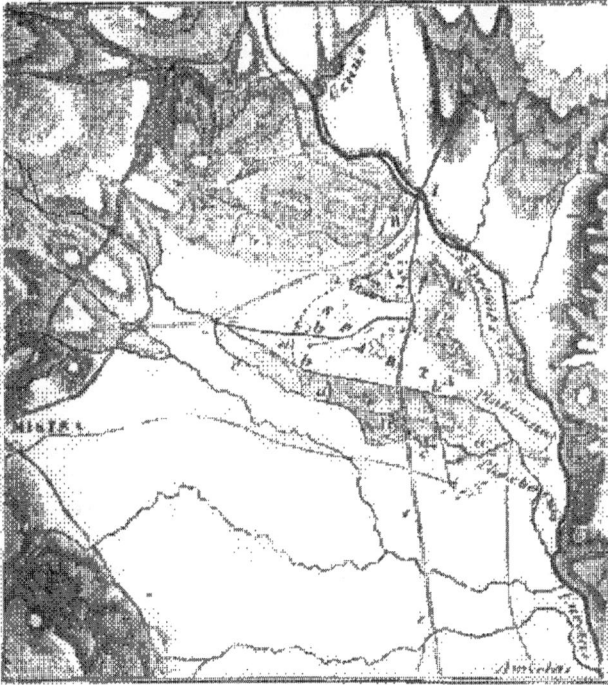

Sparta and its Environs.

A. Acropolis.	1. Theatre.	a a a. Circuit of Walls.
B. Mount Issorium	2. Agora.	b b. Osmyle.
C. Hill Colona.	3. Amphitheatre or Odeum.	c c. The Tomb.
D. New Sparta.	4. Bridge across the Eurotas.	d d. The Hyacinthian Road.
	5. Theræpne.	

the N., the residence of the wealthy; Limnæ, on the low marshy
ground near the Eurotas; Mesoa, in the S.E.; and. Cynosūra, in
the S.W. The town was not enclosed with walls until the Macedonian period: not a trace of them now remains. The general
appearance of the streets was poor, the houses being rude and
unadorned: there were, however, many fine public buildings, which
we shall notice in detail.

On the Acropolis stood the temple of Athena Chalciœcus, i. e. "of
the brazen house," so named from the bronze plates with which it was
adorned; the temples of Athena Ergane, of the Muses, and of Ares

Areia. **Below** the acropolis was the Agora, surrounded with colonnades, of which the most beautiful was the Persian stoa, so named as having been built out of the spoils of the Persian War, and representing the figures of Persians, particularly Mardonius and Artemisia. The agora contained the senate-house, the temple of Ophthalmitis, erected by Lycurgus on the spot where one of his eyes was struck out, and the Chorus, where the Spartan youths danced in honour of Apollo. W. of the Acropolis was the theatre, the centre being excavated out of the hill, and the wings being built up with enormous quadrangular stones, a large number of which still remain. S.E. of the agora was the Scias, a building used for public assemblies, though the name also applied to a street leading to the S.E. The Roman amphitheatre stood on the eastern hill; portions of its walls, 16 feet thick, remain: W. of it is a valley in the form of a horse-shoe which was probably a stadium. The part of the town in which these lay was named Dromus, from the gymnasia erected in it. To the S. of it was the Platanistas, a flat spot thickly planted with plane-trees and surrounded by streams: still more to the S., outside the city, was the district of Phoebæum. On the E. bank of the Eurotas, opposite Phoebæum, was the suburb of Therapne, or — ,¹ situated on Mount Mendælum (the Janiculum of Sparta), containing the temple of Menelaus, after which the hill was named, and the fountain of Messëis. According to the mythological account, Sparta was founded by Lacedæmon, a son of Zeus, who married Sparta the daughter of Eurotas. In the Homeric age it was subordinate to Argos, and the seat of the kingdom of Menelaus, the marriage of whose daughter Hermione with Orestes the son of Agamemnon, united these two kingdoms. On the Dorian conquest of Peloponnesus, Sparta became the capital. Its position secured it from attack until B.C. 390, when Epaminondas made an attempt on it from the side of Amyclæ. This was repeated in 362, when the Thebans penetrated into the agora. In 295 the town was surrounded with a ditch and palisade to withstand Demetrius Poliorcetes. In 218 Philip overran Laconia and passed the city twice without taking it. In 195 Q. Flamininus assaulted it, when it was held by Nabis, the tyrant, who had surrounded it with strong fortifications: he gained possession of the suburbs, but retired from the acropolis on the submission of the tyrant. In 192 it was again attacked by Philopœmen: its walls were then destroyed by the Achæan League, but restored by order of the Romans. In A.D. 396 it was taken by Alaric. In the 13th century it was still inhabited, but its inhabitants soon after removed to the fortress of *Mistra*, which became the chief place in the valley. The site of Sparta was occupied only by the villages of *Magula* and *Psychiko* until the present Greek government built *New Sparta*. In connexion with Sparta we may notice **Gythium**, which served as its port and arsenal: it was situated on the Laconian Gulf, about 30 miles from Sparta. In 455 it was burnt by the Athenians under Tolmidas; in 370 it was vainly besieged by Epaminondas; and in 195 it was taken by the Romans. Its fortifications were strong. Its ruins are found at *Paleopoli*, a little N. of *Marathonisi*: they belong to the Roman period, and consist of a theatre, sepulchres, &c.

¹ Τυνδαρίδαι δ', ἐν Ἀχαιοῖς ὀ-
φίνδον Θεράπνας αἰσίων Ῥος.
ἐν γυάλαις Θεράπνας,
Πότμον ἀμπικλάντες ὁμοῖον.

Pind. *Isthm.* i. 42.

Nem. x. 106.

Of the less important towns we may notice:—

(1.) *On the Coast.*—**Gerenia**, on the Messenian Gulf, originally some-what inland at *Zarnata*, afterwards at *Kitries* on the coast: it has been identified with the Homeric **Enōpe**: it was the reputed residence of Nestor in his youth, whence he was termed " Gerenian:" **Cardamŷle**, on a rocky height about a mile from the sea, near *Skardhamula*, one of the seven cities offered by Agamemnon to Achilles ; **Leuctrum**, ruins at *Leftro*, on the coast, said to have been founded by Pelops; **Thalāmæ**, on the minor Pamissus, probably at *Platza*, some distance from the coast, with a celebrated temple of Ino, where the future was revealed to those who slept in it; **Œtylus**, *Vitylo*, mentioned by Homer, with a temple of Sarapis, fragments of which still exist in the modern town ; **Messa**, on the W. coast of the Tænarian peninsula at *Mezapo*, where pigeons still abound ;[4] **Tænārum**, *Kypariseo*, about five miles N. of the Tænarian isthmus, named **Cænopōlis** by the maritime Laconians after they had thrown off the yoke of Sparta; **Psamāthus**, *Quaglio*, a harbour on the Tænarian promontory ; **Teuthrōne**, on the W. side of the Laconian Gulf at *Kotrones*, said to have been founded by the Athenian Teuthas; **Las**, about a mile from the W. shore of the Laconian Gulf: the town originally stood on the summit of a mount named Asia, *Passava*, but at a later time in a hollow between the three mountains, Asia, Ilium, and Cnacadium : it is noticed by Homer;[5] the name of **Asine**, given to it by Polybius and Strabo, is probably a mistake for Asia; **Helos**, E. of the mouth of the Eurotas, on a fertile though marshy plain: it was taken by the Dorians, and sunk into an insignifi-cant place; its site is probably at *Bizani*; **Epidaurus Limēra**, at the head of a spacious bay on the E. coast of Laconia, near which was the promontory of **Minōa**, now an island connected with the continent by a bridge: the ruins of Epidaurus are at *Old Monemvasia*, and consist of walls, terraces, &c.

(2.) *In the Interior.*—**Œum**, or **Ium**, in the district of Sciritis, com-manding the pass of *Klisura*, through which the road from Sparta to Tegea passed ; **Caryæ**, on the border of Arcadia, and originally an Arcadian town, but conquered by Sparta: it was celebrated for a temple of Artemis Caryatis, in which the Lacedæmonian virgins per-formed a peculiar dance at the time of the annual festival; from this dance the Greek artists gave the name of Caryatides to the female figures employed in architecture: Caryæ was probably situated on one of the side roads between Tegea and Sparta, near *Arakhova* ; **Sellasia**, on a mountain in the valley of the Œnus, just below the point where the roads from Argos and Tegea to Sparta unite: it was hence particu-larly exposed to attack; in B.C. 369 it was burnt by the Thebans: in 365 it was again destroyed by the Lacedæmonians: and again, in 221, after the famous battle between Cleomenes and Antigonus; the battle took place in the small plain of *Krevata*, which lies N. of the town between the mountains Olympus on the E., and Evas on the W., and through which the Œnus flows, receiving a small stream named Gorgylus from the W.; **Pellāna**, a fortress commanding the valley of the Eurotas, situated probably at *Mt. Burlaia*, about seven miles from Sparta; **Glyppia**, on the frontiers of Argolis, probably at *Lympiada*;

4　Φάρην τε, Σπάρτην τε, πολυτρήρωνά τε Μέσσην.—*Il.* II. 582.

5　Οἵ τ᾽ ἄρ᾽ Ἀμύκλας εἶχον, Ἕλος τ᾽, ἔφαλον πτολίεθρον.—*Il.* II. 584.
Πάρ δὲ Λακωνίδα γαῖαν, Ἕλος τ᾽, ἔφαλον πτολίεθρον.
Hom. *Hymn. in Apoll.* 410.

Geronthræ, *Gheraki*, on a height overlooking the valley of the Eurotas on the S.E, and famous for its prolonged resistance to the Dorian conquerors; **Bryseæ**, an old Homeric town S.W. of Sparta, with a temple of Dionysus which was accessible to women only; **Pharæ**, or **Pharis**, in the Spartan plain on the road to Gythium, an old Achæan town which maintained its independence until the reign of Teleclus: it was plundered by Aristomenes in the Second Messenian War: its site at *Baflo* is marked by a tumulus with an interior vault, which probably served as a treasury; **Amyclæ**, on the right bank of the Eurotas, two miles and a half from Sparta, in a remarkably fertile and beautiful district: it is said to have been the abode of Tyndarus and of Castor and Pollux:[*] it held out against the Dorians until the reign of Teleclus, after which it was chiefly famous for the festival of the Hyacinthia and for a temple and colossal statue of Apollo: its original site was probably at *Aghia-Kyriaki*, whence the population may have been removed into the plain nearer Sparta, the former spot being more than 20 stadia from Sparta; lastly, **Balemina**, or **Belbina**, on the N.W. frontier, originally an Arcadian town conquered by the Spartans, but restored to its former owners after the battle of Leuctra: the surrounding mountainous district, named Belminatis, was a constant source of contention between the Spartans and Achæans.

History.—At the Dorian conquest of the Peloponnesus, Laconia fell to the share of Eurysthenes and Procles, the sons of Aristodemus, who established themselves at Sparta. The Achæan cities were gradually subdued, and by the middle of the 8th century the Spartans were masters of all Laconia. Messenia was shortly after added to their territory, and by the time of the Persian Wars Sparta held the first place among the Greek powers. They retained this until B.C. 477, when the supremacy was transferred to Athens, and was not regained by Sparta until 404. The battle of Leuctra, in 371, deprived Sparta not only of her supremacy but also of the territories conquered from the neighbouring states. Attempts were made to recover her position during the Sacred War, and at a later period in the war with the Achæans; but the battle of Sellasia, in 221, completely frustrated the last of these attempts. The country now fell under the rule of tyrants, of whom Nabis was the most notorious: he was conquered by Flamininus, and, in 195, Sparta lost the maritime towns, which were placed under the Achæan League for a while, but were finally made independent by the Romans, with the title of Eleuthero-Lacones. There were originally twenty-four of these towns.

Off the S.E. extremity of Laconia lies the island of **Cythēra**, *Cerigo*, of an irregular oval shape, 20 miles long from N. to S., and 10 miles across in its widest part, very rocky, and containing only three towns; **Cythera**, on the E. coast at *Arionona*; an inland city also named **Cythera**, about three miles from the former; and **Scandea**, which appears to have been on the S. coast at *Kapsali*, though Pausanias seems to identify it with the seaport-town Cythera. The island was originally settled by Phœnicians, who carried on hence the purple fishery of the Laconian coast, and introduced the worship of Aphro-

---

[*] Castori Amyclæo et Amyclæo Polluci
Reddita Mopsopia Tænaris urbe soror.—Ov. *Heroid.* viii. 71.
Talis Amyclæi domitus Polluce habenis
Cyllarus. VIRG. *Georg.* III. 89.

dite.[7] It fell under the dominion first of the Argives, then of the Spartans, and was conquered by the Athenians under Nicias in B.C. 424, and under Conon in 393. Its chief productions were wine and honey.

VI. ARGOLIS, with CYNURIA.

§ 5. Argolis, in its most extensive sense, was bounded on the N. by Corinthia and Sicyonia; on the E. by the Saronic Gulf and the Myrtoan Sea; on the S. by the Hermionic and Argolic Gulfs, and Laconia; and on the W. by Arcadia. Within these limits are included the districts of Argolis Proper, i.e. the territory belonging to the city of Argos, and the peninsula between the Saronic and Argolic Gulfs, which was divided between the petty states of Epidaurus, Trœzen, and Hermiŏne. The former of these districts was by far the most important in ancient geography. The plain is enclosed on three sides by mountains,[8] and on the fourth lies open to the sea: it is from 10 to 12 miles long, and from 4 to 5 wide. Its fertility was great; and it was especially famous for its breed of horses.[9] The remainder of Argolis consisted of a broken, hilly district, with occasional plains by the sea-side.

Name.—The name Argos is said to have signified "plain" in the language of the Macedonians and Thessalians: it may be derived from the same root as the Latin "ager." In Homer, the name signifies both the town of Argos and the kingdom of Agamemnon, of which Mycenæ was the capital. The territory of Argos was most frequently termed by Greek writers Argeia, and occasionally Argolice and Argolis.

§ 6. The mountains of Argolis itself are not of much importance: they are connected with the great ranges on the borders of Arcadia, Parthenium and Artemisium. Arachnæum was the name of the ridge that separated the territories of Argos and Epidaurus: several lesser heights received specific names, which are, however, of no interest. The coast is irregular, and lined with islands: the most important promontories were on the Argolic Gulf—Buporthmus, *Muzaki*, on the S. coast; and Scyllæum, *Kavo-Skyli*, at the S.E. angle. On the N.E. coast is a considerable peninsula, connected by an isthmus, only 1000 feet broad, with the territory of Trœzen, and containing

---

[7] Est Amathus, est celsa mihi Paphus, atque Cythera,
Idaliæque domus.   *Æn.* x. 51.

Hunc ego sopitum somno, super alta Cythera,
Aut super Idalium, sacrata sede recondam.—*Id.* i. 680.

Mater Amoris
Nuda Cytheriacis edita fertur aquis.—Ov. *Heroid.* vii. 59.

[8] It is hence described by Sophocles as "hollow Argos:"—
Τὸ κοῖλον Ἄργος βὰς φυγὰς προσλαμβάνει.   *Œd. Col.* 378.

[9] The epithet "horse-feeding" is constantly applied to it by Homer:—
Ἐκθεῖσα τοι στείχοντες ἀπ' Ἄργεος ἱπποβότοια.—*Il.* II. 287.

a mountain, now named *Chelona*, above 2000 feet high: the peninsula was named after the town of Methana, which stood upon it.

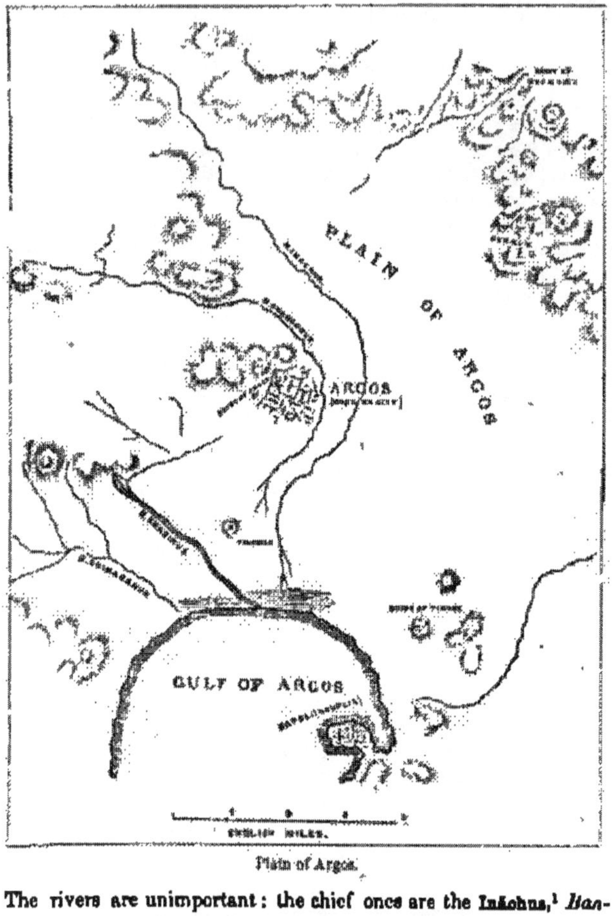

Plain of Argos.

The rivers are unimportant: the chief ones are the **Ināchus**,[1] *Bau-*

[1] The Inachus was regarded as the national stream of Argos; it was supposed to be connected by a subterraneous channel with the Amphilochian stream of the same name:—

Ὦ γῆς παλαιὸν Ἄργος, Ἰνάχου ῥοαί,
Ὅθεν ποτ᾽ ἄρας ναυσὶ χιλίαις Ἄρη
Ἐς γῆν ἔπλευσε Τρῳάδ᾽ Ἀγαμέμνων ἄναξ. Eurip. *Electr.* 1.

itza, and *Erasinus*,[1] *Kephalari*, in the plain of Argos—the former rising on the borders of Arcadia, and flowing towards the S.E. into the Argolic Gulf, receiving the Charadrus, *Xeria*, a little below Argos ; the latter issuing in several large streams from the rocks of Mount Chaon to the S.W. of Argos, and flowing in a short course across the plain into the gulf, receiving as a tributary the Phrixus shortly before its discharge. The celebrated Lake of Lerna lay at the S.W. extremity of the Argive plain, and was the centre of a marshy district[2] formed by numerous springs, and by the streams Pontinus and Amymóne,[4] which rise in the neighbouring hill of Pontinus: this district was drained in ancient times, and covered with sacred buildings, among which the temples of Demeter and Dionysus were most famous. The grove of Lerna lay between the rivers above named. The lake, which Pausanias names the Alcyonian Pool, was reputed to be unfathomable, and to be the entrance to the lower world : it is near the sea, and is a few hundred yards in circumference. Near it was the fountain of Amphiaraus, which can be no longer identified.

§ 7.   The population of Argolis was of a mixed character : the plain of Argos was originally held by Pelasgians, and afterwards by Achæans, while the coast districts of Trœzen and Epidaurus were held by Ionians. The Dorians subsequently entered as a conquering race and settled at Argos, and thenceforth the inhabitants of the Argolic plain were divided into three classes—the Dorians of the city ; the Periœci, or Achæan inhabitants ; and the Gymnesii, or bond-slaves, whose position resembled that of the Helots of Laconia. The towns may be divided into two classes—those of the plain

---

Ἴναχε γεννάτορ, καὶ κρηνῶν
Πατρὸς Ὠκεανοῦ, μέγα πρεσβύτων
Ἄργους τε γύας, Ἥρας τε πάγοις
Καὶ Τυρσηνοῖσι Πελασγοῖς.   SOPH. *Fragm.* 186.

Ørlslaque amnem fondcas pater Inachus urna.—VIRG. *Æn.* vii. 792.

[2] The Erasinus was universally believed to be the same as the river Stymphalus, which disappeared under Mount Apelauron. The distance between the two streams is so considerable as to make this opinion doubtful.

[3] The draining of the Lernæan Marsh by the Argives was the historical foundation of the legend of the victory of Hercules over the Hydra.

[4] Amymone is said to have been named after one of the daughters of Danaus whom Poseidon loved ; the stream gushed forth at the stroke of the god's trident :—

Ὅδ' ἐστὶν, αἰχμαλωτίδας
Ὃς δορὶ Θηβαίας Μυκηναίοισι
Ἀσγαλέᾳ τε διώσας Τριαίνῃ
Ποσειδωνίοις Ἀμυμωνίαις
Ὕδασι, δουλείαν περιβαλών.   EUR. *Phœn.* 188.

Testis Amymone, latices cum ferret in arvis,
Compressa, et Lerne puisa tridente palus.—PROPERT. II. 20, 47.

of Argos, of which the chief were Argos, Mycenæ, and Tiryns; and

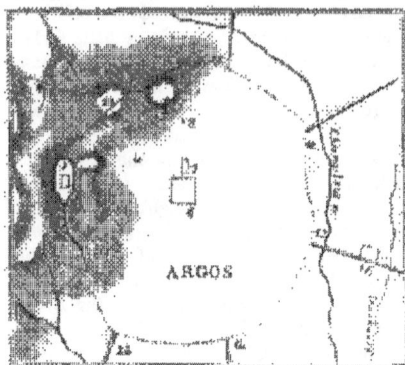

Plan of Argos.

those on the coast, Epidaurus, Trœzen, and Hermione. The former boasted of a remote antiquity, Argos being regarded as the most ancient city of Greece, and the others as hardly of later date. Mycenæ was the capital in the heroic age; Argos held that post subsequently to the Dorian conquest, and ultimately destroyed the other about B.C. 468. The remains of these cities afford remarkable specimens of the Cyclopean style of architecture. The towns of Epidaurus, Trœzen, and Hermione were well situated for purposes of trade, the two former facing the Saronic Gulf and Ægina, and the latter having a sheltered harbour on the S. coast. The secluded position of these towns enabled them to retain their independence, and they enjoyed at an early period a large amount of prosperity.

Argos or **Argi**, as the Romans usually termed it, was situated in the

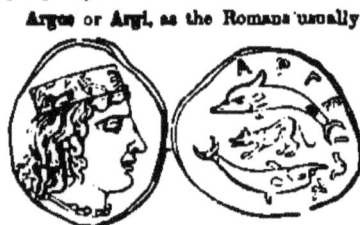

Coin of Argos.

plain named after it, about 3 miles from the sea and a little W. of the Charadrus. Its chief citadel,[b] Larissa (Map, 1), was built on an insulated conical hill, 900 feet high, on the W. side of the town. The second citadel stood on a lesser height named Aspis (Map, 3) in the N.W. of the city, and which was connected with Larissa by a ridge named Deiras (2). Argos was reputed the most ancient city of Greece, and was certainly one of the largest. It was founded by a Pelasgic chief named Phoroneus; and in the time of the Peloponnesian War it is computed to have had more than 16,000 citizens, and a total population of 110,000 in its territory. The city was surrounded by walls of Cyclopean structure, which extended over the acropolis and

---

[b] The present castle of Argos is a building of comparatively modern times, but contains some traces of Cyclopean masonry.

the adjacent hills, including the one named Aspis in the N.W., on which the second citadel stood.   The Agora (θ)[a] stood in the centre of the town.   The buildings in Argos were numerous: among them we may specify the temple of Apollo Lyceus (7) which stood near the agora; those of Zeus Larissæus and of Athena which crowned the summit of the Acropolis; two temples of Hera; the theatre (5) excavated out of the S. side of Larissa, remains of which still exist; and the monument of Pyrrhus in the agora.   Outside the town was the gymnasium (10), named Cylarabis, and about 5½ miles from it was the Heræum, or national temple of the tutelary goddess Hera, which was originally under the protection of the neighbouring town of Mycenæ, but afterwards under that of Argos.   It was well situated on a spur, overlooking the plain, and was adapted for the purposes of a fortress as well as of a temple.   The first temple was burnt down in B.C. 423, and a new one was erected in its place by Eupolemus.   The foundations of these temples have been discovered.   Argos was the seat of a famous school of statuary in which Phidias, Myron, and Polycletus were educated; music was also cultivated there, particularly under Sacadas; and in literature Argos produced the poetess Telesilla.   The remains of the town are few, and consist of traces of the walls, portions of the theatre, and of an aqueduct (9).   In connexion with Argos we may notice its port-town Nauplia[7] situated on a promontory running out into the Argolic Gulf about 6 miles from Argos, of which it became a dependency about the time of the second Messenian War: the modern town retains the ancient name.   Mycenæ was situated on a rugged height at the N.E. extremity of the Argive plain[8] near the village of Kharvati.   Its position gave it command of the roads between Argos and Corinth. The town was very ancient, its foundations being attributed to Perseus: it was the favourite residence of the Pelopidæ; and, under Agamemnon, was regarded as one of the chief towns of Greece.[9]   The town consisted of an Acropolis on the triangular summit of a steep hill, and a lower town on the S.W. side of the hill.   The Cyclopean walls[1] of the Acro-

---

[a] The temple of Apollo Lyceus stood on one side of the Agora; hence Sophocles says,

τοῦ Λυκοκτόνου θεοῦ

Ἀγορὰ Λύκειος.   Electr. 6.

[7] Ἥκει γὰρ ἐς γῆν Μενέλεως Τροίας ἄπο,
   Διμένα δὲ Ναυπλίειον ἐκπληρῶν πλάτῃ,
   Ἀκταῖσιν ὁρμεῖ, δαρὸν ἐκ Τροίας χρόνον
   Ἅλαισι πλαγχθείς.   Eurip. Orest. 53.

[8] It is hence described by Homer as being "in the corner" of the Argive land:—

Ἡμεῖσ· ὁ δ' εὔκηλος μυχῷ Ἄργεος ἱπποβότοιο.—Od. III. 263.

[9] Its wealth was proverbial:—

Ἢ αὐτὸν βασιλῆα πολυχρύσοιο Μυκήνης.   Il. vii. 180.

οἱ δ' ἱκανόμεν

Φάσκειν Μυκήνας τὰς πολυχρύσους ὁρᾶν.   Soph. Electr. 8.

Aptum dicet equis Argos, ditesque Mycenas.—Hor. Carm. i. 7, 9.

[1] The walls of Mycenæ excited the astonishment of the ancients, and were attributed to the Cyclopes; Homer gives the town the epithet "well-built:"—

Οἳ δὲ Μυκήνας εἶχον, ἐϋκτίμενον πτολίεθρον.   Il. il. 569.

Κυκλώπων βάθρα

Φοίνικι κανόσι καὶ τύποις ἡρμοσμένα.   Eurip. Here. Fur. 946.

 - Καλὴν πόλισμα Περσέως,

Κυκλωτίαν πόνον χερῶν;   In. Iph. in Aul. 1500.

polis still exist in a very perfect state, presenting good specimens both
of the polygonal and of the earlier style of that architecture: in some

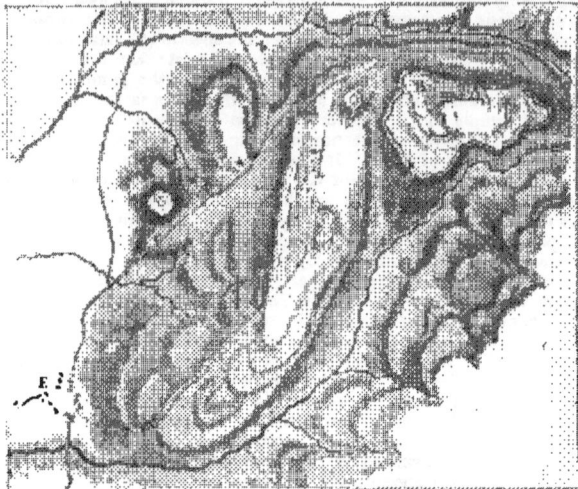

Plan of the Ruins of Mycenæ.

A. Acropolis.
B. Gate of Lions.
C. Subterraneous building usually called
D. Treasury of Atreus.
E. Subterraneous Building.
E. Village of Kharváti.

places they are from 15 to 20 feet high. One of the two gateways, by
which the Acropolis was entered, is also in existence, and is named
from the figures which crown the portal "the Gate of Lions." [1] The
lower town contained four subterraneous buildings, used either as
treasuries or perhaps rather as sepulchres (for they probably lay out-

Gallery at Tiryns.

side the walls): one of these
"the Treasury of Atreus"
still survives in a very perfect
state. Mycenæ sunk after the
occupation of Argos by the
Dorians, but it was not taken
by them until B.C. 468, when
it was destroyed. Thenceforth
it remained utterly desolate.
Tiryns was situated on an isolated hill, S.E. of Argos, and
about 1½ miles from Nauplia.
Its origin was traced back to
Proetus, whose house stood on
the highest part of the hill.

[1] The heads are now wanting: Pausanias is our authority for pronouncing the
animals to be lions. The column between the figures is conjectured to be the
symbol of Apollo Agyieus, whose aid is invoked in the Agamemnon of Æschylus
(1080, 1085), and in the Electra of Sophocles (1379).

Hercules resided there for some time.[3] Massive walls of Cyclopean structure surrounded it, and it was further defended by a citadel, named Licymna, the walls of which still exist, and are remarkable for their extreme strength, being in some places no less than 24 feet thick. The approaches of the citadel were defended by galleries of singular construction. Tiryns was conquered and destroyed by the Dorians of Argos in B.C. 468, and thenceforth remained desolate.[4] Epidaurus was the capital of a small district on the coast of the Saronic Gulf, consisting of a peninsula, on which the town itself stood, and a narrow, well-sheltered plain, on which the vine particularly flourished.[5] It derived its chief importance from the temple of Asclepius, 5 miles W. of the town, which was visited by patients from all parts of the Hellenic world, and which was, like the other celebrated fanes of Greece, surrounded by a grove and by numerous other buildings: extensive ruins cover the site, among which the theatre is the most important. The temple was plundered by Sulla. Epidaurus was reputed to have been founded by Carians, and afterwards colonized by Ionians, and conquered by the Dorians under Deiphontes: it was in early times a place of commercial importance, and sent colonies to Ægina, Cos, and other islands. It remained independent of the Argives, and was vainly attacked by them in 419. The name is preserved in that of the neighbouring village Pidhavro, but the remains are very scanty. Troezen was the capital of a small district in the S.E. angle of Argolis: it stood on a fertile maritime plain, about 2 miles from the sea, with Celenderis as its port-town on the Bay of Pogon, which offered a sheltered harbour. It was a very ancient city, and derived its name from a son of Pelops; it was the residence of Pittheus the grandfather of Theseus.[6] The Dorians settled there on their conquest of Peloponnesus, but the place retained its Ionic character. It became a powerful maritime state, and founded Halicarnassus and Myndus. It was allied with Athens until the time of the Peloponnesian War and afterwards with Sparta. The town was adorned with numerous fine buildings—consisting of the agora surrounded with colonnades; the temple of Artemis Lycia, with the stone upon which Orestes was

[3] Hercules is hence frequently termed "Tirynthian," e.g.:—
    Postquam Laurentia victor,
    Geryone extincto, *Tirynthius* attigit arva.—*Æn.* vii. 661.

The epithet is further applied to Herculaneum (Stat. *Silv.* ii. 2, 109), and Saguntum (Sil. Ital, ii. 509), as being founded by Hercules; and to the Fabian *gens*, as being descended from that god (Sil. Ital. vii. 33, vii. 218).

[4]   Suus excit in arma
    Antiquam Tiryntha Deus. Non fortibus illa
    Infecunda viris, famaeque immanis alumni
    Degeneras; sed lapsa situ fortuna, neque addunt
    Robur opes. *Raris cacumis habitator in arvis*
    Monstrat Cyclopum dextas auditibus acras.—STAT. *Theb.* iv. 146.

[5] Τροιζήν, 'Ηιόνας τε καὶ ἀμπελόεντ' 'Επίδαυρον.—*Il.* ii. 561.

It was also famous for its breed of horses:—
    Taygetique canes, domitrixque Epidaurus equorum.—*Georg.* iii. 44.

[6] The hero spent his youth at Troezen:—
    ἀλλὰ χαιρέτ', ὦ πόλις
    Καὶ γαῖ' 'Ερεχθέως· ὦ πέδον Τροιζήνιον,
    'Ως ἐγκαθηβῆς πόλλ' ἔχεις εὐδαίμονα,
    Χαῖρ'· ὕστατον γάρ σ' εἰσορῶν προσφθέγγομαι.—EUR. *Hipp.* 1097.

purified in front of it; the temple of Apollo Thearius, with the so-called tent of Orestes before it; the temple of Hippolytus; and the Acropolis, posted on a rugged and lofty hill: the ruins of Trœzen lie near *Dhamala*, and are insignificant. **Methana** stood on the W. coast of the peninsula of the same name N. of Trœzen, to which it belonged: the Athenians occupied the peninsula in 425, and fortified the isthmus. **Hermione** originally stood upon a promontory on the S. coast, but was afterwards removed about ¼ a mile inland to the slopes of a hill named Pron. It was founded by the Dryopes, and is noticed by Homer. It came under the power of Argos probably about B.C. 464, and was thenceforth a Doric city, but it regained its independence, and was allied to Sparta in the Peloponnesian War. The territory of Hermione extended over the S. angle of Argolis. Of the buildings in the town the most famous was the sanctuary of Demeter Chthonia on a height of Mt. Pron, which was an inviolable sanctuary.[7] The ruins of Hermione lie about *Kastri*.

Of the less important towns of Argolis we may notice—**Ornæ**, on the borders of Phliasia about 14 miles from Argos, a town which

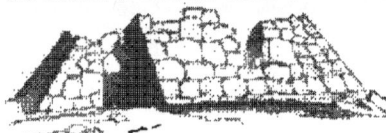

Ruins of a Pyramid in the Argeia.

retained its independence until B.C. 416, when it was destroyed by the Argives; **Œnoe**, on the Charadrus, W. of Argos, the scene of a victory gained by the Athenians and Argives over the Lacedæmonians; **Cenchreæ**, S. of Argos, near which were the sepulchral monuments of the Argives who fell at the battle of Hysiæ; a pyramid still existing, near the Erasinus, is probably one of these; **Hysiæ**, on an isolated hill below Mt. Parthenium; the scene of a battle between the Argives and Lacedæmonians in B.C. 669, destroyed by the Argives after the Persian War, and by the Lacedæmonians in 417; and **Asine**, on the coast near Nauplia, probably in the plain of *Iri*, founded by the Dryopes, and destroyed by the Argives in consequence of its having joined the Spartans against them; its inhabitants removed to Asine in Messenia.

Islands.—The coast of Argolis is fringed with islands, of which the most important are—**Tiparenus**, more probably Tricarenus, another form of **Tricarana**, *Trikkiri*, though frequently identified with *Spetzia*; **Hydrea**, *Hydra*, off the coast of Hermionis and Trœzenia; and **Calauria**, *Poro*, opposite Trœzenia, possessing an ancient temple of Poseidon, in which Demosthenes terminated his life.

History.—The authentic history of Argolis commences at the time of the Dorian invasion, when that country fell to the lot of Temenus, and Argos was constituted the Dorian capital. The conquest of the towns was gradual, and most of them retained their Achæan population. The sovereignty of Argos extended over the whole E. coast of Peloponnesus and even over Cythera, and she was the head of a league similar to the Amphictyonic, of which Phlius, Cleonæ, Sicyon, Epidaurus, Trœzen, Hermione, and Ægina, were members. Under Pheidon,

---

[7] Euripides refers to this :—

Θαρσει νιν ἀλσος, Ἑρμιονε τ᾽ ἐχει πολας.   *Herc. Fur.* 614.

B.C. 770-730, the power of Argos was at its highest, and an attempt was made to subject the whole of Peloponnesus. Subsequently, her power declined before that of Sparta, and the loss of Cynuria in 547 was followed by the decisive victory of Cleomenes near Tiryns. Argos took no part in the Persian Wars, but Tiryns and Mycenæ joined Sparta. These cities were destroyed by Argos about 466, and their population added to the capital, which thus regained its former supremacy. In the Peloponnesian War the Argives remained neutral for the first 10 years; in 421 they formed a league with the Corinthians and others against Sparta, which was dissolved in 418 by the battle of Mantinea. For a short period after this Argos joined Sparta, but soon withdrew from the alliance, and took an active part in the various combinations formed against that power. The subsequent history of Argos is unimportant : its towns fell under tyrants : it joined the Achæan League in 229, and yielded to the Romans in 146.

§ 8. The district of **Cynuria** was situated between Argolis and Laconia, and was debateable ground between the two states of Argos and Sparta, belonging alternately to each. The district consisted of a remarkably fertile plain, extending about six miles along the coast S. of Anigræa, bounded inland by the spurs of Parnon, and watered by two streams, named the **Tanus,**[*] Luku, and the **Charadrus,**[*] Kuni, which join the sea respectively N. and S. of the Thyreatic Gulf: the former was the boundary between the two states in the time of Euripides. The inhabitants were of Pelasgian origin, but were regarded as Ionians; they were a semi-barbarous and predatory tribe. There were five towns in the district—Thyrea, which may be regarded as the capital, and which is described as being situated about 10 stadia from the coast; Prasia, more to the S., on the coast; Anthēna, Neris, and Eva, in the interior. The exact position of these towns is undecided.

History.—Upon the conquest of Peloponnesus by the Dorians, Cynuria was subdued by Argos. As Sparta rose to power, there were numerous conflicts for it: Agis gained possession of it for Sparta about B.C. 1000, but Argos recovered it, and retained it until 547, when the dispute was decided in favour of Sparta by a pitched battle of 300 on each side. The Æginetans were settled there by the Spartans in 431, but were expelled by the Athenians in eight years. Philip, the father of Alexander the Great, restored Cynuria to Argos, which thenceforth retained it.

VII. ARCADIA.

§ 9. **Arcadia,** the central province of Peloponnesus, was bounded on the E. by Argolis, on the N. by Achaia, on the W. by Elis, and on the S. by Messenia and Laconia. Next to Laconia it was the

---

[*]   'Οι ἀμφὶ ποταμὸν Τάναον 'Αργείων ἔχουσι
Τάναοντα γαίαν.   EURIP. *Elect.* 410.

[*]   This name occurs only in Statius :—
Quæque pavet longa spumantem valle Charadrum
Neris.   *Theb.* iv. 46.

largest province in Peloponnesus, its greatest length being about 50 miles, and its breadth from 35 to 41. In its position it resembles a fortified camp, being surrounded on all sides by a natural wall of mountains which separate it from the other Peloponnesian states. The interior is broken up by irregular mountain-ranges, and the general appearance of the country justifies the name of "the Switzerland of Greece," which has been applied to it. The mountains vary in character and altitude in the E. and W.: in the latter they are wild, high, and bleak, with valleys of small extent and of little fertility; in the former they are of lower elevation, with small fertile plains embosomed in them, and so completely surrounded by hills that the streams can only escape by subterraneous outlets. These plains furnished the only attractive sites for towns, and we accordingly find all the chief places of Arcadia on this side of the country. Of the productions of the country, the best known were its asses, which were highly prized throughout Greece.

§ 10. The following were the principal mountains: in the N.E., Cyllene,[1] *Zyria*, 7788 feet high, reputed the loftiest in Peloponnesus, but in reality inferior to Taygetus—a massive, isolated peak, crowned with a temple of Hermes; Crathis and Aroanius, more to the W., forming the connecting links between Cyllene and the lofty and long range of Erymanthus[2] in the N.W.; Lampia and Pholoë, continuations of Erymanthus, separating Arcadia from Elis; Lycæus, *Dioforti*, in the S.W., in the district of Parrhasia, 4659 feet high, with a summit named Olympus, on which were situated the grove and altar of Zeus Lycæus,[3] together with a hippodrome and stadium

---

[1] It was celebrated as the birth-place of Hermes, or Mercury, in whose honour a temple was erected on the summit:—

Ἑρμῆν ὕμνει, Μοῦσα, Διὸς καὶ Μαιάδος υἱὸν,
Κυλλήνης μεδέοντα καὶ Ἀρκαδίης πολυμήλου.—Hom. *Hymn. in Merc.* 1.

Vobis Mercurius pater est, quem candida Maia
Cyllenæ gelido conceptum vertice fudit.—*Æn.* viii. 138.

He was hence termed Cyllenius by the poets:—
Ille primum paribus nitens Cyllenius alis
Constitit.   *Æn.* iv. 252.

[2] Erymanthus was covered with forests abounding with wild boasts, and was hence one of Diana's haunts and the fabled scene of Hercules's victory over the wild boar:—

Οἳ δ' Ἀρτεμις εἶσι κατ' οὔρεος ἰοχέαιρα,
ἢ κατὰ Τηΰγετον περιμήκετον, ἢ Ἐρύμανθον,
Τερπομένη κάπροισι καὶ ὠκείης ἐλάφοισι.   *Od.* vi. 102.

Ut Tegeæus aper cupressifero Erymantho
Incubet, et vasto pondere lædat humum.—Ov. *Heroid.* ix. 87.

Mœndriferumque Erymanthon.   STAT. *Theb.* iv. 298.

[3] Τὰ δὲ Παῤῥασίῳ στρατῷ
Θαυμαστὸς ἐὼν φάνη
Ζηνὸς ἀμφὶ πανάγυριν Λυκαίου·   PIND. *Olymp.* ix. 143.

for the celebration of the Lycæan games,[4] a temple of Pan,[5] and in the E. part of the mountain a sanctuary and grove of Apollo Parrhasius; Mænālus, in the interior, between the territories of Mantinea and Tegea, a well-wooded range rising to above 5000 feet in the summit of *Apano-Khrepa*, regarded as especially sacred to Pan;[6] and, lastly, Parthenium, Artemisium, and Lyrcēum, on the borders of Argolis.

§ 11. The chief river of Arcadia is the Alphēus, in its upper course named *Karüena*, in its lower *Rufia*, which rises in the S.E., on the borders of Laconia, near Phylace, and thence probably flowed in ancient times to the N.W.,[7] and disappeared in the Katavothrn of *Taki* : it then reappeared near Asea, and mixed with the Eurotas in the copious spring called *Frangovrysi* : the combined streams again disappear, and the Alpheus emerges at Pegæ, and flows towards the N.W., receiving the Helisson, on which Megalopolis was situated, then penetrating through a defile near Brenthe which separates the upper from the lower plain, and receiving, below Herœa, the Ladon,[8] *Rufia*, and the Erymanthus,[9] on the borders of

---

Ὄρη τε
'Αρκάδ' ἀνάσσων, μαρτυρήσω Λυκαίου βωμὸς ἀπαί.   PIND. *Olymp.* xiii. 162.

[4] These games resembled the Roman Lupercalia :—
Quid vetat Arcadio dictos a monte Lupercos?
Faunus in Arcadia templa Lycrœus habet.—Ov. *Fast.* II. 423.

[5] See quotations in next note.

[6] 'Ὁ Πὰν γὰρ, οὔτ' ἐσσὶ κατ' ὦρεα μακρὰ Λυκαίου,
Εἴτε τόγ' ἀμφιπολεῖς μέγα Μαίναλον.—THEOCR. *Idyll.* I. 123.
Pan, ovium custos, tua si tibi Mænala curæ,
Adsis, o Tegeæ, favens.   VIRG. *Georg.* I. 17.
Pialfer Illum etiam sola sub rupe jacentem
Mænalus, et gelidi fleverunt saxa Lycæi.—*Ecl.* x. 14.
Mænala transieram latebris horrenda ferarum,
Et cum Cyllene gelidi pineta Lycæi.—Ov. *Met.* I. 216.
Mænalius and Mænalis are frequently used by the Roman poets as equivalent to Arcadian :—
Pinigerum Fauni Mænalis ora caput—Ov. *Fast.* III. 84.
Sive fugæ comites, Mænall Nympha, tuæ.—*Id.* I. 634.

[7] It now flows to the N.E., and disappears in the katavothra of *Persova* at the foot of Mt. Parthenium ; its course is said to have been thus diverted in modern times.

[8] The Ladon is famed in mythology as the river into which Syrinx plunged when pursued by Pan :—
Donec arenosi placidam Ladonis ad amnem
Venerit ; hic illi cursum impedientibus undis,
Ut se mutarent, liquidas orasse sorores.—Ov. *Met.* I. 702
Its stream is described as being very rapid :—
Testis erit Pholoë, testes Stymphalides undæ ;
Quique citis Ladon in mare currit aquis.—Ov. *Fast.* II. 273.
Arcades hunc, Ladonque rapax, et Mænalos ingens
Rite colunt, Lunâ credita terra prior.   *Id.* v. 89.

[9] Ἀλθεὺς ἀλλ' οὔπω μέγας ἔρρυεν, οὐδ' Ἐρύμανθος
Λευκότατος ποταμῶν, ἔτι δ' ἄβροχος ἦεν ἅπασα
'Αρκαδίη.   CALLIM. *H. in Jov.* 19.
Et celer Ismenus cum Phocaico Erymantho.—Ov. *Met.* II. 244.

Elis. Of the numerous streams which rise in the E. district, the most important is the Stymphalus, which feeds the lake of the same name, and disappears in a *katavothra*, emerging (as it was universally believed) in the Argolic river Erasinus: the water of the Stymphalus was conveyed to Corinth by an aqueduct built by Hadrian.

§ 12. The inhabitants of Arcadia regarded themselves as the most ancient inhabitants of Greece,[1] and derived their name from Arcas, a son of Zeus. The Greeks described them as *autochthonous*, by which they understood that they were Pelasgians who had never changed their abode. They led a primitive and secluded life among their mountains, tending their flocks and herds, cultivating music with success,[2] but otherwise rather famed for stupidity,[3]—brave and hardy, and hence, like the Swiss, constantly employed as mercenaries. They lived for the most part in villages, in a state of political independence.[4] The country was divided into numerous districts, which were for the most part named after well-known towns in each. The exceptions are Parrhasia,[5] on the border of Messenia, which appears once to have possessed a town of the same name; Cynuria, to the N. of it; Eutresia, N. of Megalopolis; and Azania, which included numerous lesser districts in the N. of Arcadia. The towns were unimportant, with the exception of a few in the eastern district, particularly Tegea and the neighbouring Mantinea, which

---

[1] They termed themselves προσέληνοι, as having existed "even before the moon:"—

. Ἀρκάδες, οἳ καὶ πρόσθε Σεληναίης ἐδέοντο
Ζώειν, φηγὸν ἔδοντες ἐν οὔρεσιν· οὐδὲ Πελασγὶς
Χθὼν τότε κυδαλίμοισιν ἀνάσσετο Δευκαλίδῃσιν.—APOLL. *Argon.* iv. 264.

Orta prior Luna (de se si creditur ipsi)
A magno tellus Arcade nomen habet.—Ov. *Fast.* L 469.

[2] Hence "Arcades" became synonymous with pastoral poets :—
Ambo florentes ætatibus, Arcades ambo.—Virg. *Ecl.* vii. 4.
Tamen cantabitis, Arcades, inquit,
Montibus hæc vestris : soli cantare periti
Arcades. *Id.* x. 31.

[3] Arcadicus juvenis was tantamount to a " blockhead:"—
Nil salit Arcadico juveni. Juv. *Sat.* vii. 160.

[4] It is worthy of remark how the habits, mythology, and political condition of the Arcadians were influenced by the physical characteristics of their country. The poverty of the soil and severity of the climate necessitated a *pastoral* rather than an agricultural life; hence their love of music and their devotion to Pan, the inventor of the pipe, and Mercury, the god of the lyre. The great hydraulic works necessary to keep the eastern plains from inundation were ascribed to Hercules. The mountain-ranges which encircled and subdivided it precluded both external and internal union for political purposes.

[5] It is noticed by Homer (*Il.* ii. 608). The terms *Parrhasius* and *Parrhasis* are used by the Latin poets as equivalent to Arcadicus :—
Parrhasio dictum Panos de more Lupercæ.—Virg. *Æn.* viii. 344.
Cum Parrhasio Arcte.—Ov. *Met.* viii. 315.
So also *Æn.* xi. 31 ; *Fast.* L 618, iv. 377 ; *Trist.* ii. 190.

were exposed to inroads from the adjoining states of Sparta, Corinth, and Argos, and were not unfrequently rivals for the supremacy over each other. Megalopolis was founded at a comparatively late period, B.C. 370, and became the capital of the country. The towns fell into decay under the Roman dominion, and in the time of Strabo Tegea alone was inhabited.

Mantinea stood in the central portion of the plain of *Tripolitza*, and was the capital of a territory lying between the mountains Mænalus on the W. and Artemisium on the E., and separated by a low ridge from Orchomenia in the N., and by projecting spurs of the mountains already mentioned from Tegeatis on the S. The town itself was in nearly the lowest as well as the narrowest part of the plain. The small river Ophis[6] flowed originally through it, and afterwards just outside its walls, and disappeared in a katavothra to the N.W. of the town. The fortifications were regular; and the circuit of the walls, flanked with numerous towers, are still traceable on the site, now named *Paleopoli*. The position of Mantinea rendered it a place of great military importance: roads led from it to Orchomenus, Tegea, Pallantium, and Argos; and the character of the plain was adapted to the operations of an army. It was the scene of no less than five battles, of which the two first are of most historical importance; the first fought B.C. 418, in which the Argeans, Mantineans, and Athenians, were defeated by the Lacedæmonians under Agis, and the second in B.C. 362, in which the Lacedæmonians were defeated by Epaminondas, who perished in the battle. Both these battles were fought in the plain S. of the town, where it is contracted by the advancing ridge of Mænalus, named Scope.[7] Mantinea is said to have been so named after a son of Lycaon: it is noticed in Homer.[8] Originally it consisted of four or five villages, which were incorporated into one town. Its constitution was democratical, and hence it was hostile both to its neighbour Tegea and to Sparta. With the former it fought an indecisive battle in B.C. 423; by the latter it was defeated in the first great battle of Mantinea in 418, and again in 385, when the town capitulated, and its inhabitants were dispersed. The town was rebuilt in 371, and shortly after made an alliance with Sparta against the other Arcadian towns: this brought on the second great battle in 362, in which Epaminondas died. In 295 the Spartans were defeated near the town by Demetrius Poliorcetes, and in 242 by Aratus and the Achæans. In the Cleomenic War, Mantinea was taken in 226 by Aratus, and in 222 by Antigonus Doson, when it was plundered, and its name changed to Antigonia. In 207 the plain was the scene of a fifth great

---

[6] This stream rose in the territory of Tegea, and more than once was used as a weapon of offence in the Mantinean wars, the plain being so flat that the waters could be easily diverted from their usual channel, or wholly stopped by an embankment, in either of which cases the plain was inundated. This was done by Agesilaus in B.C. 385.

[7] This defile was the "narrow pass" in which Areithous was slain :—

Τὸν Λυκόοργος ἔπεφνε δόλῳ οὔτι κράτεΐ γε
Στεινωπῷ ἐν ὁδῷ.　　　　　　　*Il.* vii. 142.

[8] Καὶ Τεγέην εἶχον, καὶ Μαντινέην ἐρατεινήν.　　*Il.* ii. 607.

The epithet of "lovely," here applied, is now inappropriate to the plain, which is bare and covered to a great extent with stagnant water. In former times, however, forests of oaks and cork trees grew on it.

battle, in which the Achæans, under Philopœmen, defeated the Lacedæmonians. The old name of Mantinea was restored by Hadrian.
The only remains of it are traces of the walls and of the theatre.
Tegea stood in the southern part of the plain of *Tripolitza*, about 10
miles S. of Mantinea. Its territory extended over the surrounding
district, which was divided into the following portions:—The Tegeatic
plain to the N., extending to the hill Scope; the Manthyric to the
S.W.; and the Corythic to the E. The plain is watered by the upper
course of the Alpheus and its tributaries, as well as by the Garates:
these streams all disappeared in *katavothras*. The town was situated in
the lowest part of the plain, and hence the accumulation of soil has
entirely overlaid its site, leaving but a few buildings visible,—among
them the remains of a theatre, perhaps the one built by Antiochus
Epiphanes in 175, and of a temple of Athena Alea,[*] erected by Scopas
after the destruction of the former edifice in 394, and deemed the most
magnificent temple in the Peloponnesus. Tegea is noticed by Homer,

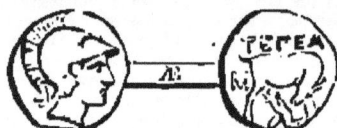

Coin of Tegea.

and was probably the most
celebrated of the Arcadian
towns in ancient times. Its
contiguity to Sparta brought
it into early conflict with that
state; and after numerous engagements it was obliged to
yield in about B.C. 560, though
it still retained its independ-
ence. War broke out again between them, and battles were fought in
479 and 464, on each of which occasions Tegea was unsuccessful.
Thenceforth there was a firm alliance between them until 371, when
Tegea joined the Arcadian confederacy, and fought against Sparta
and Mantinea in 362. It joined Sparta against the Achæan League,
and was hence taken by Antigonus Doson in 222, retaken in 218 by
Lycurgus the tyrant of Sparta, and subsequently by Machanidas, and
recovered by the Achæans after the death of the latter. The town
existed until the 4th century A.D. **Megalopolis**, "the Great City,"
was situated in the middle of a plain on the banks of Helisson,
about 2½ miles above its junction with the Alpheus: its ruins are
near *Sinanu*. It was founded in B.C. 370, as the capital of the Arcadian confederation; and it was peopled with the inhabitants of
forty townships, which thenceforth became desolate. The town itself
was 50 stadia in circumference, and its territory extended northwards for 23 miles, being the most extensive of all the Arcadian
states. Roads led in various directions to Messene, Sparta, Tegea,
Heræa, and other places. The most important buildings were the
theatre, on the S. side of the river, the largest in Greece; and the
agora on the N. side, which was on a magnificent scale, and was
adorned with colonnades, temples, and statues: the remains of the
theatre are extensive. Megalopolis was particularly exposed to the
enmity of the Spartans, not only from the object for which it was
founded, but also from its position. It hence allied itself first with
Thebes, and afterwards with Macedonia. It joined Cassander against
Polysperchon, and was besieged by the latter in 316. It was afterwards

---

[*]   Templumque Aleæ nemorale Minervæ.—STAT. *Theb.* iv. 288.
  The site of this temple is sometimes erroneously transferred to the town of
Alea in the N.E. of Arcadia.

governed by tyrants. In 222 Cleomenes III. reduced the greater part of it to ruins; it was soon rebuilt on its former grand scale, which had at all times been beyond the requirements of the population.[1] Megalopolis produced two eminent men—the general Philopœmen, and the historian Polybius. Herœa was the chief town in the lower plain of the Alpheus: it stood on the right bank of that river, about 2 miles above the junction of the Ladon. Its territory was fertile, and it lay on the high road between Olympia and Central Arcadia. It is said to have been founded by a son of Lycaon. About B.C. 580 it concluded a treaty with the Eleans, the original of which, on a bronze tablet, is in the British Museum. The town was enlarged by the Spartan king Cleombrotus, and was hence allied to Sparta. It became a member of the Achæan League, and was a place of some importance in the time of Pausanias: its ruins near *Aianni* are inconsiderable. Phigalia occupied the summit of a lofty hill in the S.W. corner of the country, on the right bank of the Neda. Its origin was traced back to Phigalus, a son of Lycaon. In B.C. 659 it was taken by the Spartans, and in 375 the place became notorious for the fierce disputes between its factions. In the wars between the Ætolians and Achæans it was occupied by the former. Phigalia possessed a beautiful temple of Apollo Epicurius, erected to commemorate the deliverance of the town from the plague in the Peloponnesian War: it stood at Bassæ, in a glen near the summit of Mt. Cotilium, and was the work of Ictinus, the architect of the Parthenon. It was a peripteral hexastyle building of the Doric order, 125 feet in length and 48 in breadth, with 15 columns on each side. It exists in a tolerably perfect state, and is altogether one of the most interesting ruins in Greece. Methydrium was situated on a lofty height "between the rivers" (whence its name) Malœtas and Mylaon, in the central district of Arcadia: its position is probably near *Nemnitza*. It was founded by Orchomenus, and destroyed at the foundation of Megalopolis. Orchomĕnus was situated N. of Mantinea, on a plain[2] which was bounded on the N. by the lofty chain of Oligyrtus, on the S. by the low ridge of Anchisia, and on the E. and W. by parallel chains, not distinguished by any special names, from which spurs project into the centre of the plain, dividing it into two parts. The acropolis stood on the western of these spurs, a lofty insulated hill, nearly 3000 feet high, commanding the two plains: this position was forsaken for a lower site at the foot of the hill. Orchomenus was one of the most powerful cities of Arcadia in ancient times: it was governed by kings, who, down to the time of the second Messenian War, exercised a supremacy over the whole country, and who continued to reign in their own territory until the Peloponnesian War. Orchomenus was generally on bad terms with Mantinea, but was unable to cope with it. It was taken by Cassander in 313, subsequently by Cleomenes in the Ætolian War, and retaken by Antigonus Doson. Some remains of temples and tumuli mark the site of the town at *Kalpaki*. Stymphălus lay on the S. side of the lake of the same name, where its ruins

---

[1] Its size was so excessive as to lead to the following *bon mot* of a comic poet :—

'Ερημία μεγάλη ἐστὶν ἡ Μεγαλόπολις.

[2] Οἱ θάνατόν τ' ἐνέμοντο, καὶ 'Ορχομενὸν πολύμηλον.—*Il.* II. 605.
Dives et Orchomenos pecorum.   STAT. *Theb.* IV. 293.

may still be seen. It is noticed by Homer and Pindar.[3] Its chief historical importance is due to its position on the road that leads into Arcadia from Argolis and Corinth. It possessed a temple of Artemis Stymphalia.

Of the less important towns we may notice—Pallantium, W. of Tegea, near Makri, a very ancient town, and the reputed residence of Evander, who transferred the name, together with a portion of its inhabitants, to the Palatine Hill at Rome;[4] Asea, about midway between Tegea and Megalopolis, near the joint source of the Eurotas and Alpheus; Lycosura, in Parrhasia, near Stala, reputed by Pausanias the most ancient city in Greece; Acacesium, in the same district, with a celebrated temple of Despœna in its neighbourhood; Alliphira, upon a steep and lofty hill, now named Nerovitza,.near the borders of Elis, with temples of Asclepius and Athena, and a celebrated bronze statue of the latter; Thelpûsa, on the Ladon,[5] N. of Herœa, taken by Antigonus Doson in 222; its ruins lie on the slope of a hill near Vanena; it possessed famous temples of Erynnys and Apollo, at a spot named Onceum; Psophis, Tripotamo, a very ancient town, situated on elevated ground at the junction of the Erymanthus and Aroanius, captured by Philip of Macedon in 219; Cleitor, ruins at Paleopoli, more to the E., situated on a brook of the same name, which falls into the Aroanius (not the river above mentioned), a tributary of the Ladon: its inhabitants were renowned for their love of liberty, and were frequently engaged in contests with the other Arcadian towns; a celebrated fountain was in its neighbourhood,[6] and the river Aroanius is said to have produced singing fishes; Cynætha, Kalavryta, on the N. side of the Arcadian mountains, destroyed in the Social War by the Ætolians; Nonacris, more to the E., famed for its vicinity to the river Styx,[7] which rises a short distance from the town, and descends perpendicularly over a precipice,[8] forming by far the highest waterfall in Greece; it falls into

---

[3] Στυμφηλόν τ' εἶχον, καὶ Παῤῥάσιον δονέοντα.   Il. II. 608.
  Οἰκοῦσιν οἶκαδ' ἀπὸ Στυμφαλίων
  Τειχέων ποτινισσόμενον,
  Ματέρ' εὐμήλοιο λείποντ' Ἀρκαδίας.   Pind. Olymp. vi. 167.

[4] Arcades his oris, genus a Pallante profectum,
  Qui regem Evandrum comites, qui signa secuti,
  Delegere locum, et posuere in montibus urbem,
  Pallantis proavi de nomine, Pallanteum.—Virg. Æn. viii. 51.

[5] Δίνῃ ἐδοῦσι τέῤῥοθος Τελφουσία
  Ἀδδωνος ἀμφὶ ῥεῖθρα νιίουσα σκύλαξ.   Lycophr. 1040.

[6] This spring was supposed to be a specific against the love of wine:—
  Clitorio quicunque sitim de fonte levarit,
  Vina fugit; gaudetque meris abstemius undis.—Ov. Met. xv. 322.

[7] Nonacris is used by Ovid as a synonym for Arcadius:—
  Et matri et vati paret Nonacrius heros (sc. Evander).—Fast. v. 97.
  Dum redit itque frequens, in virgine Nonacrina.—Met. ii. 409.

[8] It is correctly described by Homer and Hesiod:—
  Στυγὸς ὕδατος αἰπὰ ῥέεθρα.   Il. viii. 369.
  Καὶ τὸ κατειβόμενον Στυγὸς ὕδωρ.   Id. xv. 37.

                    ὕδωρ
  Ψυχρὸν δ' τ' ἐκ πετρῆς καταλείβεται ἠλιβάτοιο
  Ὑψηλῆς.   Theog. 785.
  Ὀγύγιον, τὸ δ' ἴησι καταστυφέλου διὰ χώρου.—Id. 805.

The description in Herodotus (vi. 74) is less correct. The old belief still holds good among the inhabitants of the neighbourhood; whence the modern names Mavro-Nero, "black waters," and Drako-Nero, "terrible waters."

the Crathis below Nonacris; its waters were believed to be poisonous, and hence the stream was transferred to the imagery of the nether world; **Pheneus**, *Fonia*, W. of Stymphalus, in a plain enclosed on every side by mountains, and watered by two streams, which disappear in a *katavothra*, and emerge as the sources of the Ladon : this outlet has occasionally become choked, and an inundation has ensued; a canal which was formed for the purpose of guiding

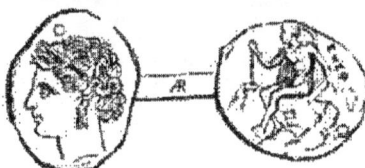

Coin of Pheneus.

the streams to the *katavothra* was ascribed to Hercules; the town is noticed by Homer, and is represented by Virgil as the residence of Evander;[*] lastly, **Caphyae**, N.W. of the lake of Orchomenus at *Khotussa*, the scene of a battle between the Ætolians and Achæans in 220; its territory was protected from inundation by embankments and trenches.

History.—The early history of Arcadia is unimportant. The people were divided into three separate bodies, named Azanes, Parrhasii, and Trapezuntii, governed by their separate kings. Homer notices only one Arcadian king, Agapenor. The Dorians did not conquer Arcadia on their first entrance into Peloponnesus, but the Spartans succeeded in gaining various districts adjacent to their frontier. The Arcadians were thus opposed to Sparta, and it was not until the defeat of the Tegeans in B.C. 560 that they changed their views, and became allies of that power. Between 479 and 464 they vainly endeavoured to shake off the supremacy. In the Peloponnesian War all the towns, except Mantinea, remained faithful to Sparta, and even Mantinea was obliged to succumb in 417. After the battle of Leuctra in 371, the Arcadians became independent, restored Mantinea, which had been destroyed in 385, and founded Megalopolis as the seat of a federal government. A battle, in which the Spartans were victorious in 367, and a war with the Eleans for the Olympian supremacy in 365, were the next events of importance: the latter led to disputes between Tegea and Mantinea, which were not settled until the battle of Mantinea in 362. The country subsequently joined the Achæan League, to which it belonged until the dissolution of the league by the Romans, when it became part of the province of Achaia.

§ 13. The islands of the Ægean Sea, which were not included in the Cyclades, were grouped together under the general name of the **Sporades**, "scattered." Some of these lie in close contiguity to the eastern and northern coasts of the Ægean, and have been already described in connexion with Asia Minor and Thrace. Another group is found between the coasts of Peloponnesus and Crete, consisting of Melos, Cimolos, Oliaros, Pholegandros, Sicinos, Ios, Thera, and Anaphe; while a third, lying E. of the Cyclades, included Amorgus, Astypalæa, and some lesser islands.

---

[*] Accersit et cupidus Phenei sub mœnia duxi.—*Æn.* viii. 165.

Melos, *Milo*, stands midway between Crete and Peloponnesus, 70 miles from the former, and 65 from the latter; it is about 15 miles long and eight broad, and resembles a bow in shape : it is mountainous, and of volcanic origin, and has warm springs: its chief productions were kids, sulphur, alum, pumice-stone, and a red pigment. A deep bay occurs on the N. coast, and served as the harbour of the chief town, which stood on its shore : remains of polygonal walls, of two theatres, and of the necropolis, still exist. Melos was originally occupied by Phœnicians, and afterwards by Lacedæmonians. It was cruelly ravaged by the Athenians in B.C. 416, when the population was exterminated, and Athenian settlers introduced. **Cimolos**, *Cimoli*, lies between Melos and Siphnus, in size 5 miles long by 3½ broad: it was particularly celebrated for its chalk [1] (*Cimolia creta*), used by fullers, and in medicine. The chief town stood opposite Melos on a rock, named *Daskalio*, which was formerly united to the island by an isthmus, but is now disjoined from it. **Oliaros**,[2] *Antiparo*, near Paros, is now famed for a stalactitic cavern, which appears to have been unknown to the ancients. **Pholegandros**, **Sicinos**, and **Ios**, lie in a line from W. to E., to the S. of Paros, and retain their names with but slight variation: Ios is celebrated as the burial-place of Homer; the alleged discovery of his tomb in 1771 is, however, problematical. **Thera**, *Santorin*, is the chief of the group, and lies nearest to Crete : it has the form of a crescent, with its horns elongated towards the W., and has a circumference of 30 miles, with a breadth nowhere exceeding three miles. It is said to have been first occupied by Phœnicians, but it was afterwards colonized by Spartans,[3] and itself colonized Cyrene, in Africa. Opposite the N. point of Thera is **Therasia**; and between this and the S. point is the inlet of *Aspronisi;* these three were originally united, and they form the walls of a vast crater, now a gulf of the sea, from the centre of which have arisen three peaks, named the *Kammenis*, the first of which made its appearance in B.C. 197, the second in A.D. 46, and the third in A.D. 1707. The volcanic eruptions in these islands have been very numerous and violent. There are remains of several ancient towns on Thera, particularly of one of considerable size on the summit of *Messa Vouno*. **Anaphe** lies E. of Thera, and contained a famous temple of Apollo Ægletes, said to have been founded by the Argonauts, of which considerable remains still exist: it has at all times abounded in partridges. **Astypalæa**, *Stampalia*, lies E. of Anaphe, and consists of two large rocky masses, united in the centre by an isthmus: two deep bays penetrate on the N. and S. coasts, and off the latter lie several desert islands.[4] It was colonized by Megarians, and is said to have been subdued by Minos: in B.C. 105 the Romans concluded a

---

[1]　　　Cretæque rara Cimoli.— Ov. *Met.* vii. 463.

[2] It is noticed by Virgil :—

　　Oleraon, nirsamque Paron, sparsasque per æquor.—*Æn.* III. 136.

[3] Its earliest name is said to have been Calliste :—

　　　　Καὶ, Λακεδαι-
　　μονίαν μιχθέντες ἀνδρῶν
　　Ἥθεσιν, ἐν ποτε Καλ-
　　λίσταν ἀπῴκησαν χρόνῳ
　　Νάσῳ.　　　　　　　Pind. *Pyth.* iv. 457.

[4] Ovid alludes to these in the line :—

　　Cinctaque piscosis Astypalæa vadis.—Ov. *Ar. Am.* II. 82.

treaty with it, and made it subsequently a "libera civitas." The town stood on the S. bay, and appears to have possessed handsome buildings. **Amorgos**, *Amorgo*, N.W. of Astypalæa, is chiefly celebrated as the birth-place of the poet Simonides, and for its linen fabrics. It was fertile, and was considered by the Romans as one of the most favourable places for banishment: it contained three towns. **Cinarus**, named after the artichoke (κίναρα) it produced, and **Lebinthus**, lie E. of Amorgos; **Lelandrus** and **Nicasia** N. of it; **Phacusa** and **Schœnus** W. of it.

§ 14. The large island of **Creta**, known to us under the name of *Candia*, but to its own inhabitants as *Kriti*, lies at the entrance of the Ægæan Sea, about 60 miles distant from the Peloponnesus, and double that distance from Asia Minor. Its length is about 160 miles, and its greatest breadth about 30. It is very mountainous and woody, and was celebrated in ancient times for its medicinal herbs (particularly the "dictamnon"), for its raisin-wine and honey, and its dogs. A chain of mountains traverses the whole length of the island: the central height, named **Ida**,[5] *Psiloriti*, terminates in three lofty peaks at an elevation of 7674 feet: the eastern prolongation was named **Dicta**, *Juktas*, and the western **Leuce**,[6] *Leuki*. The coast is irregular, and contains numerous promontories, of which we may notice, as most important—**Corycus**, *C. Grabusa*, in the N.W.; **Dictynnæum**, or **Psacum**, *C. Spadha*, a little to the E., the termination of a ridge of the same name, which was crowned with a temple of Dictynna; **Criumetōpon**, *C. Crio*, in the S.W.; **Matala**, *Matala*, on the S. coast; **Ampēlus**, *C. Xacro*, in the S.E.; and **Samonium**, the Salmone of Acts xxvii. 7, *C. St. Sidero*, in the N.E. The chief river, named **Lethæus**, *Maloyniti*, runs from E. to W. through the plain of Gortyna, joining the sea on the S. coast. The other streams derive their whole importance from poetical associations: they are the **Iardănus**,[7] *Platania*, on the N. coast, near which was the rock Lissa; and the **Oaxes**,[8] or **Axus**, flowing down from Ida to the N. coast, and still retaining its name.

§ 15. The earliest inhabitants of Crete were probably a mixed

---

[5] Ida, and particularly its summit, named Panacra, was regarded as especially sacred to Jupiter, where the bees nurtured him with their honey:—

> Κρήτη τιμήεσσα, Διὸς μεγάλοιο τιθήνη,
> Πολλή τε λιπαρή τε καὶ εὔβοτος ἦε ὑπὲρ Ἴδη,
> Ἴδη, καλλιεόμοισιν ὑπὸ δρυσὶ τηλεθάωσα.
> Καὶ τῆς τοι μέγεθος στριώσιον.   DION. PERIEG. 501.
> Γέντο γὰρ ἐξαπίναια Πανακρίδος ἔργα μελίσσης
> Ἰδαίοις ἐν ὄρεσσι, τά τε κλείουσι Πάνακρα.
>   CALLIM. Hymn. in Jov. 66.

[6] Leuca was well clothed with wood:—

> βαίνε δὲ κούρη
> Λευκὸν ἔσι, Κρηταῖον ὄρος, κεκομημένον ὕλη.—CALLIM. H. in Dian. 40.

[7] ᾗχι Κυδωνες ἔναιον, Ἰαρδάνου ἀμφὶ ῥέεθρα.   Od. III. 292.

[8] At nos hinc alii sitientes ibimus Afros,
Pars Scythiam, et rapidum Cretæ veniemus Oaxen.—VIRG. Ecl. I. 65.

race of Carians, Pelasgians, and Phœnicians. In the heroic age, Dorians were the dominant race, sharing the country with the Eteocretans, Cydonians, and other races.[6] The Cretans had a high reputation as light troops,[1] and served as mercenaries in Greek and barbarian armies.. They lived in separate communities, each town having its own senate, coins, &c., and only coalescing, or "syncretizing," when their common mother-country was threatened by a foreign foe. The towns are said to have been as many as 100.[r] Many of them were very ancient, and they existed until the invasion of the Romans under Q. Metellus. The most important were Cnossus, Gortyna, Cydonia, and, after the decay of the latter, Lyctus. The first two had a "hegemony," and were generally hostile to each other.

(1.) *On the Sea-Coast.*—Commencing in the N.W., the first important town we meet with is **Cydonia**, *Khania*, which existed in Homer's time,

Coin of Cydonia.

but was enlarged and adorned by the Samians under Polycrates. In the Peloponnesian War it was at war with the Gortynians and Athenians. It was besieged by Phalæcus the Phocian after the Sacred War, and again by the Roman general Metellus. The quince-tree derived its name from this place. **Itanus**, on the E. coast, near a promontory of the same name, was probably a Phœnician town. **Lebena**, *Leda*, on the S. coast, served as the port of Gortyna, and possessed a celebrated temple of Asclepius. **Phalasarna**, on the W. coast, a little S. of

6 Κρήτη τις γαῖ ἐστι, μέσῳ ἐνὶ οἴνοπι πόντῳ,
Καλὴ καὶ πίειρα, περίῤῥυτος· ἐν δ᾽ ἄνθρωποι
Πολλοί, ἀπειρέσιοι, καὶ ἐννήκοντα πόληες.
Ἄλλη δ᾽ ἄλλων γλῶσσα μεμιγμένη· ἐν μὲν Ἀχαιοί,
Ἐν δ᾽ Ἐτεόκρητες μεγαλήτορες, ἐν δὲ Κύδωνες,
Δωριέες τε τριχάϊκες, δῖοί τε Πελασγοί.   Od. xix. 172.

1 Their skill with the bow and arrow is frequently noticed :—
    Primævæ Teucer tela Cydonio
    Direxit arcu.   *   Hor. Carm. iv. 9, 17.
    Hastas et calami spicula Gnosii
    Vitabis.   Id. i. 15, 17.
            Libet Parthos torquere Cydonia cornu
    Spicula.   Virg. Ecl. x. 59.

r Ἄλλοι δ᾽, οἱ Κρήτην ἑκατόμπολιν ἀμφενέμοντο.—Il. ii. 649.
    Creta Jovis magni medio jacet insula ponto ;
    Mons Idæus ubi, et gentis cunabula nostræ.
    Centum urbes habitant magnas, uberrima regna.—Æn. iii. 104.
    Aut illo centum nobilem Cretam urbibus,
        Ventis iturus non tuis ;
    Exercitatas aut petit Syrtes Noto :
        Aut fertur incerto mari.   Hor. Epod. ix. 29.

Prom. Coryous, was the nearest port to Greece, and possessed a temple of Artemis. Remains of the walls, tombs, and of a singular chair cut out of the solid rock and destined for some deity, still exist.

(2.) *In the Interior.*—Polyrrhenia was the chief town in the N.W., and had Phalasarna as its port, from which it was distant about 7 miles : its war with Cnossus in B.C. 219 is the only historical event recorded: some walls near *Kisamo-Kasteli* mark its site. Lappa, or Lampa, possessed an extensive district, extending from sea to sea, and made use of Phœnix as its port. After its capture by Metellus it was made a free city by Augustus, and at a later period it became an episcopal see. Some ruins at *Polis* represent it. Gortyn, or Gortyna, stood S. of Ida, on a plain watered by the river Leth-œus, and possessed two har-bours, Leben and Metallum. It ranked next to Cnossus in importance, and in early times had leagued with that town for the purpose of sub-duing the whole of Crete, but afterwards was engaged in constant hostilities with

Coin of Gortyns.

it. In the Peloponnesian War it sided with Athens. Philopœmen was elected commander-in-chief of its army in B.C. 204, and, in 197, 500 Gortynians joined Quinctius Flamininus in Thessaly. Its site is uncertain ; It has been placed at *Haghios Dheka.* Cnossus, or Gnossus, the royal city[3] of Crete, was centrally situated near the N. coast, on the banks of a small stream named Cæratus,[4] after which it was originally named. It possessed two ports, Hera-clēum and Amnisus. Its foundation was attributed to Minos, who resided there. The locality abounded with

Coin of Cnossus.

mythological associations : Jupiter was believed to have been born and to have died there;[5] there Dædalus cultivated his art, and near it was the

---

[3] Τῇσι δ' ἐνὶ Κνωσσὸς, μεγάλη πόλις· ἔνθα τε Μίνως   *Od.* xix. 176.
'Εννέωρος βασίλευε Διὸς μεγάλου ὀαριστήν.

The whole island was occasionally named after it :—
Jupiter omnipotens ! utinam ne tempore primo
Gnossia Cecropiae tetigissent littora puppes ;
Indomito nec dira ferens stipendia tauro
Perfidus in Cretam religasset navita funem.—CATULL. lxiv. 171.

[4] Καῖρέ δι Καῖρατος ποταμὸς μέγα, χαῖρε δι Τηθύν.
                                                    CALLIM. *Hymn. in Dian.* 44.

[5] The Cretans pretended that they had his tomb, and hence obtained the cha-racter for lying attributed to them by Callimachus and Aratus, the latter of whom is quoted by St. Paul (*Tit.* i. 12) :—

Κρῆτες ἀεὶ ψεῦσται· καὶ γὰρ τάφον, ὦ ἄνα, σεῖο
Κρῆτες ἐτεκτήναντο, σὺ δ' οὐ θάνες, ἐσσὶ γὰρ αἰεί.
                                                    CALLIM. *Hymn. in Jov.* 8.

Κρῆτες ἀεὶ ψεῦσται, κακὰ θηρία, γαστέρες ἀργαί.

                                                                        Y

Labyrinth,[a] erected by him and inhabited by the Minotaur, a building which had no existence except in the imaginations of poets. Cnosus was colonised by Dorians, and became the leading town in Crete. The Romans made it a colony. Some rude masses of Roman brickwork and parts of a long wall, from which the site is now named *Makro-Taicho*, are the sole relics of it. Lyctus was situated in the interior,

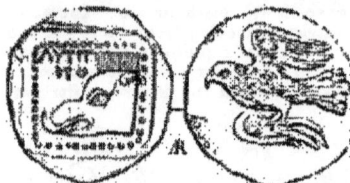

Coin of Lyctus.

S.E. of Cnossus: it was regarded as a colony from Sparta, and the worship of Apollo prevailed there. It was a constant rival of Cnossus. In 344 B.C. it was taken by Phalæcus, the Phocian, and an ally of Cnossus, and at a later period was utterly destroyed by the Cnossians : it was finally sacked by Metellus. Numerous remains of buildings, tombs, marbles, and particularly an immense arch of an aqueduct, exist at *Lytto*. Præsus stood under the N. slope of Mount Dicte and possessed a considerable territory, together with a famous temple of Dictæan Jupiter: its ruins still retain the name of *Præsus*.

History.—The history of Crete is somewhat bare of events. At the time of the Trojan War, Idomeneus, son of Deucalion and grandson of Minos, was king, and took part with the Greeks. After his return he was banished, and retired to Italy. The violent quarrels between the chief towns led to a reference to Philip IV. of Macedon as a mediator; but his intervention does not appear to have effected permanent good. In B.C. 67 Crete was reduced by Q. Metellus Creticus, and was annexed to Cyrene as a Roman province. This union remained in force until the time of Constantine, when they were constituted distinct provinces.

St. Paul's Travels.—In his disastrous voyage to Rome St. Paul visited the coasts of Crete. Sailing from Myra in Lycia with a N.W. wind, his vessel "ran under Crete over against Salmone," *i. e.* got under the lee of the island, easily rounding the cape, but afterwards with difficulty getting along the S. coast. Reaching the neighbourhood of Prom. Matala, whence it would have been necessary to cross the open sea, it was deemed prudent to put into a roadstead a few miles E. of the cape, named "Fair Havens," near which was a town named Lasæa, the ruins of which have been found five miles E. of the cape. Here the vessel remained some time; but, as the place was inconvenient for wintering, it was decided to go to Phœnice (the classical Phœnix) which lay more to the W., probably at *Lutro*, which is described as "looking toward the S.W. wind and N.W. wind," meaning probably the aspect which the place bore to one approaching it from the sea, in which case it would be sheltered from those winds. They set sail; but, after passing Cape Matala, they were blown off the shore of Crete by a N.E. wind, and carried by Clauda, the modern *Gozza*, a small island lying S.W. of Crete (Acts xxvii. 7-16).

---

a Ἐν δὲ χορὸν ποίκιλλε περικλυτὸς Ἀμφιγυήεις,
Τῷ ἴκελον, οἶόν ποτ' ἐνὶ Κνωσσῷ εὐρείῃ
Δαίδαλος ἤσκησεν καλλιπλοκάμῳ Ἀριάδνῃ.

Il. xviii. 590

Personification of the River Tiber.

CHAPTER XXIV.

ITALY.—VENETIA, ISTRIA, GALLIA CISALPINA, LIGURIA.

§ 1. THE peninsula of Italia was bounded on the N. by the Alps, on the E. by the Adriatic or Upper Sea, on the W. by the Tyrrhenian or Lower Sea, and on the S. by the open Mediterranean. The precise boundary on the N.E. and N.W. varied: in the latter direction it was originally fixed at Tropæa Augusti, where an advancing spur of the Maritime Alps formed a natural division, but by Augustus it was advanced westward to the river Varus, and thus included Nicæa; in the former direction the boundary originally stood at the river Formio, but was afterwards carried on to the Julian Alps and the river Arsia. The general direction of the peninsula is towards the S.E.; its extreme length, from the foot of the Alps to Prom. Leucopetra is about 700 miles; its width varies considerably, the northern portion spreading out into a broad expanse about 350 miles across, while the southern portion has an average width of about 100 miles; its area is estimated at 90,000 square miles.

Name.—The etymology of the name Italia is quite uncertain: the Greeks and Romans derived it from an eponymous hero, Italus;[1] others have connected it with an old Tyrrhenian word allied to *vitulus*, meaning "ox," in which case Italia would signify "the land of oxen." The name was originally applied only to the S. point of the peninsula, as far N. as the Scylletian Gulf. Thence it was extended, even in early times, to the whole tract along the shores of the Tarentine Gulf as high as Metapontum, and on the W. shore as high as the Gulf of Præstum, and in this sense it was co-extensive with Œnotria. At that time (about the 5th century, B.C.) the remaining portions of Italy were known by the names of Opica and Tyrrhenia. In the time of Pyrrhus it was extended northwards to the S. frontiers of Cisalpine Gaul and Liguria. In the later days of the Republic, when those countries were subjected to the arms of Rome, the name was extended in ordinary language to the foot of the Alps, though in official language the distinction between Italia and Cisalpine Gaul was still observed. Under the Emperors this distinction ceased, and Italy was carried to the natural limits of the peninsula, viz. the Alps. In the last ages of the Western Empire the name was applied exclusively to the northern provinces. We have further to notice the poetical names of Hesperia,[2] Ausonia,[3] and Saturnia.[4]

§ 2. The general features of the peninsula are the results of its physical structure. It consists of two great divisions: (i.) the alluvial plains of the *Po* in the N., lying between the Alps and the Apennines; and (ii.) the southern extension formed by the central ridge of the Apennines, which penetrates through the whole length of the peninsula, and reappears in the island of Sicily. Down to the head of the Bay of Tarentum this ridge is a single one: there it bifurcates, one of the branches continuing to the E. and forming the promontory of Iapygia, while the other descends first towards the S. and afterwards towards the S.W: hence arises the striking resemblance which the southern portion of the peninsula bears to a boot. The lateral ridges of the Apennines are generally of low elevation, and seldom reach the sea: hence the line of coast is generally regular. The rivers, with the exception of the *Po*, are necessarily of short course, the central chain forming an unbroken barrier throughout its whole length. The climate of Italy has in all ages been regarded as remarkably fine.[5] The peninsula lies between the

---

[1] Œnotri coluere viri; nunc fama, minores
   Italiam dixisse *ducis de nomine* gentem.—*Æn.* i. 532.
[2] Est locus, Hesperiam Graii cognomine dicunt,
   Terra antiqua, potens armis, atque ubere glebæ.—*Æn.* i. 530.
[3] Multi illam magno e Latio totaque petebant
   Ausonia.   *Æn.* vii. 54.
               Pertulit Ausonias ad urbes.—Hor. *Carm.* iv. 4, 56
[4] Salve, magna parens frugum, Saturnia tellus
   Magna virum.   *Georg.* ii. 173.
[5] The fine passage from Virgil (*Georg.* ii. 136, sq.) on this theme has been
already quoted (above, p. 323).

parallels of 38° and 46° N. lat., in the most favoured region of the
temperate zone, the natural heat due to its position being tempered
by the seas that bathe its coasts, and by the high ground of the
Apennines in the interior.   It was probably somewhat colder in
early times than at present.[*]   We have also reason to believe that
it was more healthy, the modern *malaria* being attributable in great
measure to want of population and cultivation.[†]   The soil was in
many parts very productive: Campania yielded corn in abundance,
while the olives of Messapia, Daunia, and the Sabines, and the vineyards of Etruria, the Falernian, and the Alban hills, were famed
throughout the ancient world.   The highlands of the Apennines
and the plains of Apulia afforded excellent pasturage for sheep,
horses, and cattle.   The plains of Lombardy, then covered with
forests, supported vast herds of swine.   The slopes of the Apennines
were clothed with magnificent forests.   Mineral productions were
not numerous:[‡] gold was at one time found in the Alpine streams;
copper was tolerably abundant; the island of Ilva yielded iron;
fine marble was found at Luna; and among the special productions
cinnabar and calamine are noticed.

§ 3. The mountains of Italy belong either to the chain of the Alps
or to that of the Apennines.   The general course of the former of
these chains has been already traced (p. 319).   It remains for us to
describe the divisions and principal heights known to the ancients,
which are as follows from W. to E. :—**Alpes Maritimæ**, from the

---

[*] Horace speaks of Soracte as white with snow, the Alban hills as covered
with it on the first approach of winter, and the rivers frozen :—

> Vides ut alta stet nive candidum
> Soracte, nec jam sustineant onus
> Silvæ laborantes, geluque
>   Flumina constiterint acuto.   *Carm.* I. 9, 1.

Quod si bruma nives Albanis illinet agris.—*Ep.* i. 7, 10.

Juvenal alludes to the Tiber being frozen, as if it were an ordinary occurrence :—

> Libernum fracta glacie deveradet in amnem,
> Ter matutino Tiberi mergetur.   *Sat.* vi. 522.

[†] Certain portions of the peninsula appear to have been unhealthy in early
times—the *Maremma* of Tuscany, for instance, and the neighbourhood of Ardea.
Even Rome itself was unhealthy in the summer and autumn, as the subjoined
lines from Horace show :—

> Frustra per autumnos nocentem
>   Corporibus metuemus Austrum.   *Carm.* II. 14, 15.

Auctumnusque gravis, Libitinæ quæstus acerbæ.—*Sat.* II. 6, 19.   _

[‡] The assertion of Virgil in the following lines partakes of poetical license :—

> Hæc eadem argenti rivos ærisque metalla
> Ostendit venis, atque auro plurima fluxit.—*Georg.* II. 165.

The gold mines were worked out in his day, and we have no specific statement
of the production of silver : the fact that the old coinage was of copper proves
that it was not abundant.

coast of Liguria to **M. Vesálus**,[1] *Monte Viso*, containing the sources
of the Po. A. Cottiæ, northwards to *Mont Cenis*, including **M. Matrōna**, *Mont Genévre*; they were named after a chieftain of eminence in these parts in the time of Augustus. A. Graiæ, from *Mont
Cenis* to *Mont Blanc*, including Cremonis Jugum, *Cramont*, and the
A. Centroniæ, about the *Little St. Bernard*. A. Pœninæ, from
Mont Blanc to *Monte Rosa*, including the *Great St. Bernard*; the
name is derived from the Celtic *Pen* or *Ben*, "summit." A. Rhætiæ, in the *Grisons* and *Tyrol*, including **M. Adula**, *St. Gothard*.
A. Carniæ or Venétæ, from the Atagis eastward, so named from the
tribes of the Carni and Veneti. And, lastly, **A. Juliæ**, extending
down to the coast of the Adriatic, and named after Julius Cæsar,
who reduced the mountain tribes to submission. The **Apenninus
Mons**[1] emanates from the Maritime Alps in the N.W. of Italy. At
first it runs parallel to the sea, and in close proximity to it, sweeping
round the head of the Ligurian Bay; it then almost crosses the peninsula to the Adriatic, in the neighbourhood of Ariminum; from
this point it turns to the S.S.E., and assumes a direction parallel to
the Adriatic down to the borders of Lucania. In the central portion of the peninsula the main range approaches nearer to the Adriatic than to the Tyrrhenian Sea, and leaves on the W. the plains of
Etruria and Latium; as it descends to the S., however, it approaches
the western coast, and leaves the plains of Apulia on the E. In the
S. of Samnium the chain presents the appearance of a confused knot
of mountains. More to the S. it resolves itself into a central range,
with numerous offshoots ramifying throughout the whole of Lucania.
In the N. of Bruttium there is a remarkable subsidence of the chain
between the Scylletian and Torinæan bays; in the S. it rises again
into a lofty and rugged mass to the height of about 7000 feet. The
highest summits of the Apennines are covered with snow during the
winter.[2] The sides were far more extensively covered with forests
formerly than they now are.[3]

§ 4. The line of coast contains the following bays and promontories from W. to E.:—**Ligusticus Sinus**, *G. of Genoa*, extending
along the coast of Liguria. **Lunæ Prom.**, on the borders of Liguria

---

[1] Ac velut ille canum morsu de montibus altis
Actus aper, multos Vesulus quem pinifer annos
Defendit. Æn. x. 707.

[1] Lucan (ii. 396, seq.) gives a correct description of the position which the
Apennines hold in the Italian peninsula.

[2] Hence the expression is strictly true:—
Gaudetque nivali
Vertice se attollens pater Apenninus ad auras.—Æn. xii. 702.

[3] The pine grows only on the loftier summits, as implied in the following
lines:—
Horrebat glacie saxa inter lubrica, summo
Piniferum cœlo miscens caput, Apenninus.—SIL. ITAL. iv. 743.

and Etruria. **Populonium Prom.**, opposite the isle of Ilva. **Circæum,**[4] *Monte Circeo*, in Latium, a bold and abrupt mass rising precipitously from the sea. **Misénum,**[5] *C. di Miseno*, in Campania, forming the northern limit of the **Sinus Cumánus**, *Bay of Naples*. **Prom. Minervæ**, *Punta della Campanella*, a bold rocky headland, forming the southern boundary of the *Bay of Naples*, and deriving its name from a temple of Minerva on its summit. **Pæstánus Sin.**, *G. of Salerno*, commencing at Prom. Minervæ in the N., and extending to the headland of **Posidium**, *Punta di Licosa*, in the S. **Palinūri Prom.**,[6] *C. Palinuro*, in Lucania, named after Palinurus, the pilot of Æneas, who is said to have been buried there; more to the S. a bay in Bruttium, known by the various names of **Sinus Hipponiātes, Lametinus, Terinæus, Vibonensis,** and **Napetinus**, after towns of similar names on its shore, and now the *Golfo di Santa Eufemia*. **Prom. Scyllæum**, *Scilla*, a projecting rocky headland jutting out boldly into the sea, at the entrance of the Sicilian Straits. **Leucopetra**, *C. dell' Armi*, the extreme S.W. point of Italy, and the termination of the Apennine range; its name refers to the white colour of the cliffs. **Prom. Herculis**, *C. Spartivento*, at the S.E. point of the peninsula. **Prom. Zephyrium**, *C. di Bruzzano*, a low headland on the coast of Bruttium, whence the Locrians were named Epizephyrii. **Sin. Scylletious**, *G. of Squillace*, named after the town of Scylletium. **Prom. Lacinium**, *C. delle Colonne*, a bold and rocky headland about 6 miles S. of Crotona, crowned in ancient times by a celebrated temple of Lacinian Juno.[7] **Sin. Tarentinus**, *Golfo di Taranto*, an extensive gulf between the two great peninsulas of Southern Italy, commencing at the Lacinian promontory in the W., and extending to the Iapygian in the E., named after the city of Tarentum. **Prom. Iapygium** or **Salentinum**, *C. di Leuca*, the extreme S.E. point of the *heel* of Italy, forming the E. boundary of the Tarentine Gulf. **Prom. Gargáni**, the N. point of the large projection occupied by Mt. Gar-

---

[4] The name was connected with the legend of Circe, though it does not appear why this promontory should be identified with the *island* of the Homeric myth (*Od.* xi. 135). Either the legend itself was of Italian origin, or perhaps the Cumæan Greeks identified some local deity with their own Circe. The popular belief is expressed by Virgil in the *Æneid*, vii. 10, *seq.*

[5] So named after Misenus, the trumpeter of Æneas, who was buried there :—
Monte sub aërio : qui nunc Misenus ab illo
Dicitur, æternumque tenet per sæcula nomen.—*Æn.* vi. 234.
Qua jacet et Trojæ tubicen Misenus arena.—PROPERT. iii. 18, 3.

[6] So named after the pilot of Æneas, who was buried at this spot :—
Et statuent tumulum, et tumulo sollemnia mittent ;
Æternumque loco Palinuri nomen habebit.—*Æn.* vi. 380.

[7] Hinc sinus Herculei, si vera est fama, Tarenti
Cernitur. Attollit se Diva Lacinia contra.—*Æn.* iii. 551.
Extenditque suas in templa Lacinia rupes.—LUC. ii. 434.

ganus, and, lastly, **aia. Targestinus**, *G. of Trieste*, at the N. end of the Adriatic Sea.

§ 5. The rivers of Italy derive their importance rather from historical and geographical associations than from their size. From this description we must however except the **Padus,**[*] *Po*, which deserves to be ranked among the chief rivers of Europe.[*] Rising in the Western Alps, it drains the wide basin of Northern Italy, receiving numerous tributaries from the Alps[1] on the N. and the Apennines on the S., and discharging itself into the Adriatic through several channels, the position and number of which has altered from time to time. Of these there were two principal ones, named Padoa and Olana, and five lesser ones: some of them were artificial; in others extensive embankments were raised to restrain the stream. The next important river in Northern Italy is the **Athisis,**[*] *Adige*, which in the lower part of its course flows parallel to the *Po*, and discharges itself into the Adriatic somewhat N. of it. In Central Italy we may notice the **Arnus**, *Arno*, which, rising on the western slopes of the Apennines, drains the northern portion of Etruria; and the **Tiberis,**[*] *Tiber*, which has its sources not far from the Arnus, and flows with a general southerly direction until it approaches the sea, when it turns towards the W.; its importance in the political geography of Italy is great, not only as being the river on which Rome itself stood, but as forming the boundary between Etruria on the W.,[*] and Umbria, the Sabini, and Samnium on the E. S. of

---

[*] The origin of the name Padus is uncertain; it comes probably from a Celtic root. The native Ligurian name was Bodencus. The Greeks identified it with the mythical Eridanus, and the Latin poets adopted the title from them.

[*] Virgil designates it very properly the "king" of the Italian rivers:—

> Proluit insano contorquens vortice silvas
> Fluviorum rex Eridanus, camposque per omnes
> Cum stabulis armenta tulit.   *Georg.* i. 481.

[1] As these streams were fed with the melted snow, the river has been at all times liable to heavy floods; hence we read in Virgil:— '

> Eridanus, quo non alius per pinguia culta
> In mare purpureum violentior effluit amnis.—*Georg.* iv. 372.

[*] Virgil couples it with the Po, and gives it the epithet of "pleasant:"—

> Quales aëriæ liquentia flumina circum
> Sive Padi ripis, Athesim seu propter amœnum,
> Consurgunt geminæ quercus.   *Æn.* ix. 679.

[*] The name was connected with that of a Tuscan prince, Tiberis or Thybris, who was said to have been drowned in it; its earlier name was Albula:—

> Tum reges, asperque immani corpore Tibris ,
> A quo post Itali fluvium cognomine Tibrim
> Diximus: amisit verum vetus Albula nomen.—*Æn.* viii. 330.
> Albula, quem Tiberim mersus Tiberinus in undis
> Reddidit.   Ov. *Fast.* ii. 389.

[1] Hence it is termed "Tuscan " by Virgil:—

> Di patrii Indigetes, et Romule, Vestaque mater,
> Quæ Tuscum Tiberim, et Romana palatia servas.—*Georg.* i. 198.

the Tiber are the Liris, *Garigliano*, which has its sources in the Central Apennines near the lake Fucinus,[a] and flows through the S.E. of Latium,[b] joining the sea at Minturnæ; and the Vulturnus, *Volturno*, which brings with it the collected waters of almost the whole of Samnium,[c] and in its lower course traverses the plain of Campania to the sea. Between Campania and Lucania is the Silarus,[d] *Sele*, which rises in the N.E. of Lucania, and flows into the Gulf of Prestum. On the E. of the Apennines the only noticeable river is the Aufidus, *Ofunto*, which rises in the S. of Samnium, and descends to the plains of Apulia, across which it flows with a gentle stream[e] to the Adriatic.

§ 6. The lakes of Italy form a conspicuous feature in that country. They may be arranged into three groups: (i.) those of Northern Italy, which are fed by the Alpine streams, and lie as it were in long, deep valleys; (ii.) those of Central Italy, which, with few exceptions, occupy the craters of extinct volcanoes, and are thus generally of circular or oval form, and of small size; (iii.) those few which do not fall under this description, but are simply basins surrounded by hills, whence the water has no natural outlet. 1. In the first of these classes we may enumerate—the Lacus Verbānus, *Lago Maggiore*, formed by the Ticinus; L. Larius,[f] *L. di Como*, by the

---

[a] Lucan is mistaken in placing its sources in the country of the Vestini:—
Umbrosæ Liris per regna Marica
Vestinis impulsus aquis.   II. 424.

[b] Its lower course crosses the plain of Campania with a slow gentle stream:—
Non rura, quæ Liris quieto
Mordet aqua *taciturnus* amnis.   Hor. *Carm.* i. 31, 7.

[c] Hence the Vulturnus is a rapid and turbid stream:—
Delabitur inde
Vulturnusque *celer*.   Luc. II. 422.
Multamque trahens sub gurgite arenam
Vulturnus.   Ov. *Met.* xv. 714.
Virgil characterises it as *vadosus*, referring apparently to the inequality of its stream:—
Amnisque vadosi
Accola Vulturni.   *Æn.* vii. 728.

[d] The Silarus is said to have possessed the quality of fossilising:—
Nunc Silarus quas nutrit aquis, quo gurgite tradunt
Duritiem lapidum mersis inolescere ramis.—Sil. Ital. viii. 582.

[e] The passages describing the rapidity of its stream apply only to its upper course, near which Horace lived (at Venusia), and to the period of the year when it was swollen by the mountain rains:—
Sic tauriformis volvitur Aufidus
Qui regna Dauni præfluit Appuli,
Cum sævit, horrendumque fluens
Diluviem meditatur agris.   *Carm.* iv. 14, 25.

[f] Virgil describes Larius as the *greatest* of the Italian lakes. Verbanus really holds this position, as its modern name implies; but he singularly omits all notice of this:—
Anne lacus tantos? te, Lari maxime, teque
Fluctibus et fremitu assurgens, Benace, marino.—*Georg.* II. 159.

Addua; L. Sabinus, L. d' Isco, by the Ollius; and L. Benacus, L. di Garda, by the Mincius. The L. di Lugano, between the two first lakes, though of large size, is not noticed by any writer earlier than the 6th century of our era. 2. In the second class are—L. Vulsiniensis. L. di Bolsena, in Southern Etruria, a basin of about 30 miles in circumference; L. Sabatinus, L. di Bracciano, and L. Ciminus, L. di Vico, in the same district; L. Albanus. L. d'Albuno, and L. Nemorensis, L. di Nemi, in Latium; and L. Avernus[2] in Campania. 3. In the third class are the two most important lakes of Central Italy—L. Trasimenus, L. di Perugia, in Etruria; and L. Fucinus,[2] L. Fucino, in the territory of the Marsi.

§ 7. The ethnography of Italy is still involved in much obscurity. The inhabitants may be divided into two classes: (I.) the occupants of the southern portion of the peninsula, who may be grouped under the following five heads—(1.) Pelasgians, (2.) Oscans, (3.) Sabellians, (4.) Umbrians, (5.) Etruscans; and (II.) the inhabitants of Northern Italy, who were either Celts—as the Gauls and the Carni, or of uncertain origin—as the Ligurians, Veneti, and Euganei. The former class alone call for detailed notice: (1.) The Pelasgi were in historical times confined to the S., where they existed under the following names:—the Messapians and Salentines in the Iapygian peninsula, and the Peucetians and Daunians in Apulia. The Siculi, who afterwards crossed into Sicily, belonged to the same stock; and at an early period a Tyrrhenian race prevailed in Campania and in Latium. Probably the inhabitants of Southern Etruria may be referred to the same class. (2.) The Oscans—whom we may identify with the Opicans and Ausonians of Greek writers, and with the Auruncans of Roman writers—were reputed the earliest inhabitants of Campania, and held Samnium before its occupation by the Sabines. The Volscians and Æquians belonged to this stock, and it also furnished an important element in the Latin nation.

---

[2] The mephitic exhalations arising from the lake and its neighbourhood suggested the idea that there was an entrance to the infernal regions here. To this circumstance its name was also referred, the Greek form 'Αορνος being derived from ά and όρνις, "the birdless lake:" the lines in Virgil, however, in which this is expressed, is probably interpolated:—

Spelunca alta fuit, vastoque immanis hiatu,
Scrupea, tuta lacu nigro nemorumque tenebris;
Quam super haud ullae poterant impune volantes
Tendere iter pennis: talis sese halitus atris
Faucibus effundens supera ad convexa ferebat.
[Unde locum Grall dixerunt nomine Aornon].—Æn. vi. 237.

[2] The "glassy" waters of this lake are noticed by the poets:—
Te nemus Angitiae, vitrea te Fucinus unda,
Te liquidi flevere lacus.   Æn. vii. 759.
        Vitreo quem Fucinus antro
Nutrierat, dederatque lacum transmittere nando.—Sil. Ital. iv. 346.

(3.) The Sabellians are said to have originally lived in the central Apennines and the upland valleys about Amiternum. Thence they spread southwards in a series of emigrations, defeating the Oscans, and occupying their territories as conquerors. To this class belonged the well-known nations of the Sabines and the Samnites; the Picēni, Peligni, Vestīni, and Marrucīni; probably the Marsi; the Frentāni and Hirpīni; the Lucanians and a portion of the Bruttians; and, lastly, the later masters of Campania, which country they seized between B.C. 440 and 420. The Sabellians in each case probably coalesced with the earlier Oscans, with whom they may have been allied in race and language. (4.) The Umbrians were regarded as the most ancient of the Italian races. At an early period they occupied not only the district which afterwards bore their name, but also Etruria and the plains on the Adriatic from Ravenna to Ancona: they were also allied to the Oscans and Sabellians. (5.) Of the Etruscans, or Tuscans proper, we can say nothing more than that they were entirely distinct from the surrounding nations, and that they were probably of Indo-European origin.

§ 8. The geographical divisions of Italy usually recognized had their origin in the names which the Romans found attached to the countries or their inhabitants at the period when they conquered them. No formal division of the country took place until the time of Augustus, who divided it into 11 regions, the limits of which were not in all instances coincident with that of the old provinces. The regions included the following countries: 1. Latium and Campania. 2. The Hirpini, Apulia and Calabria. 3. Lucania and Bruttium. 4. The Frentani, Marrucini, Peligni, Marsi, Vestini, Sabini and Samnium. 5. Picenum. 6. Umbria. 7. Etruria. 8. Gallia Cispadana. 9. Liguria. 10. The E. part of Gallia Transpadana, Venetia, and Istria. 11. The W. part of Gallia Transpadana. This division continued with but slight alteration to the time of Constantine, who added to Italy the provinces of Rhætia and Vindelicia, and the islands of Sicily, Sardinia, and Corsica, and arranged the whole into 17 provinces,—the northern being placed under the Vicarius Italiæ, and the southern under the Vicarius Urbis Romæ. This division survived into the Middle Ages.

I. ISTRIA AND VENETIA.

§ 9. The small district named Istria,[*] or Histria, lay in the extreme N.E. of Italy, on the borders of Illyricum, and consisted of the

---

[*] The name is derived both by Greek and Latin authors from the notion that a branch of the Ister or Danube flowed into the Adriatic. That notion, however, probably originated in the resemblance of the names Ister and Istri.

greater portion of the triangular peninsula which projects into the
Adriatic between the Sinus Tergestinus and the Sinus Flanaticus.
The river Arsia bounded it on the E., and the Formio on the N.,
where it adjoined Venetia. It was not a naturally fertile country,
but in the later ages it exported considerable quantities of corn,
wine, and oil to Ravenna. The Istrians were probably an Illyrian
race, but we know little of them. The towns are few, and, with
the exception of Pola, unimportant.

Pola, *Pola*, was situated near the S. extremity of the peninsula, on a
land-locked bay which formed an excellent port. Tradition assigned
to it a Colchian origin. We hear little of it until Augustus established
a colony there, with the name of Pietas Julia. There are considerable
remains, among which the amphitheatre, two temples, dedicated, the
one to Rome and Augustus, the other to Diana, and a triumphal arch,
named the Porta Aurea, are most famous. We may also notice—
Parentium, *Parenzo*, on the W. coast, about 30 miles N. of Pola, occupied by Romans, and raised to the rank of a colony by Trajan; and
Ægida, more to the N., also a Roman settlement, and restored by
Justin II. under the name of Justinopolis.

History.—The Istrians first appear in history as confederates of the
Illyrians in their piratical undertakings. Shortly before the second
Punic War they were reduced to submission by M. Minucius Rufus and
P. Cornelius. In B.C. 183 they were again attacked by M. Claudius
Marcellus; and in the years 178 and 177 they were finally subdued by
A. Manlius and C. Claudius.

§ 10. The boundaries of **Venetia** varied considerably at different
periods. In the later period of the Roman empire they were fixed
at the Athesis on the W., and the Formio in the E.; but in the
former direction, Verona, Brixia, and Cremona, and sometimes even
Bergomum, were included within its limits, while in the latter the
town of Tergesto was frequently regarded as belonging to Istria, in
which case the boundary would be placed at the Timavus. Sometimes Carnia was regarded as a distinct country from Venetia, and
again, previous to the time of the empire, both of these districts were
included in Cisalpine Gaul. The maritime district of Venetia consists of a broad and level plain, through which the Alpine streams
find their way in very broad beds, formed in the periods when they
are swollen by the melting of the snows. The coast itself in the
S.W. is fringed with lagunes, through which the rivers escape to
the sea by narrow outlets. The rivers are confined in their lower
courses by artificial barriers. The northern portion of Venetia is of
a mountainous character, being intersected with the spurs of the
Alps.

§ 11. The rivers of Venetia are numerous, and are the most
striking feature in the country. The **Athesis**, *Adige* (p. 488), is the
most important. The next in point of magnitude is the **Medoacus**,
or **Medoacus**. *Brenta*, which flows by Patavium, and receives as a

tributary the Meduacus Minor, *Bacchiglione*. Then follow, in order
from W. to E.—the **Silis**, *Sele*, a small stream flowing by Altinum ; the
Plavis, *Piave*, which enters the sea a few miles E. of Altinum ; the
Liquentia, *Livenza*; the **Tilaventus**, *Tagliamento*, the most important in the E. part of the province, having its sources in the high
ranges of the Alps above Julium Carnicum ; the **Turrus**, *Torre*,
Natiso, *Natisone*, and **Sontius**, *Isonzo*, which now unite their
streams, but which formerly flowed in other courses,—the Turrus
and Natiso under the walls of Aquileia, four miles W. of the present
channel, and the Sontius by an independent channel ; the **Frigidus**,
a tributary of the Sontius ; the **Timavus**, *Timao*, a river little more
than a mile long, but of great size and depth, being 50 yards broad
close to its source,[5] and deep enough to be navigable for vessels
of considerable size ; and the **Formio**, on the borders of Istria.

§ 12. The earliest inhabitants of Venetia were named **Euganei**, a
people of whom some traces remained in the valleys of the Alps
within the historical period, but of whose origin we know nothing.
The two chief races in later times were the **Veneti**, probably a
Slavonian tribe, who occupied the W. district from the Athesis to
the Plavis, and the **Carni**, probably a Celtic race, who occupied the
E. district. The towns of Venetia rose to high prosperity under the
Roman empire, not only from the fertility of the country, but
because they stood on the great high-road that communicated with
the E. To the latter circumstance they also owed their adversity :
for it was through Venetia that the barbarian hordes descended
into Italy. Aquileia ranked as the capital of the province, and from
its position near the head of the Adriatic, was the key of Italy, and
hence the scene of repeated contests for the possession of the Imperial power.

Tergeste, *Trieste*, was situated on the innermost bay of the Adriatic,
and on the confines of Istria. It appears to have been a Roman settlement as early as B.C. 51, when it was plundered by some barbarians:
in 32 it was fortified by Octavian, and it was made a colony by Augustus. It is seldom noticed, and never attained the importance which
its modern representative, *Trieste*, now enjoys. **Aquileia**, *Aquileia*, was
situated near the head of the Adriatic, between the rivers Alsa and
Natiso. It was founded by the Romans in B.C. 181, and named after

[5] The number of its sources is variously stated: Virgil makes them nine ;
modern travellers reduce them to four. There appears to have been formerly
some communication with the sea, by which some of the springs were rendered at
times brackish, and hence perhaps the term *pelagus* applied by Virgil ; this
phenomenon no longer exists :—

  Antenor potuit, mediis elapsus Achivis,
  Illyricos penetrare sinus atque intima tutus
  Regna Liburnorum, et fontem superare Timavi ;
  Unde per ora novem vasto cum murmure montis
  It mare proruptum, et pelago premit arva sonanti.—*Æn. I. 242.*

the accidental omen of an eagle appearing at the time of its foundation. It soon rose to importance, both as a place of trade and as a military station for the defence of the N.E. border.[a] In A.D. 238 it was besieged without effect by the tyrant Maximin; In 340, the younger Constantine was defeated and slain beneath its walls; in 388, it witnessed the defeat and death of the usurper Maximus by Theodosius the Great; and in 425, that of Joannes by the generals of Theodosius II. In 452 it was utterly destroyed by Attila. Forum Julii, *Cividale di Friuli*, lay about 25 miles N. of Aquileia, and nearly at the foot of the Julian Alps. It was probably founded by Julius Cæsar as a place of meeting for the Carni: but it did not rise to importance until the later ages of the Roman empire, and particularly after the fall of Aquileia, when it became the capital of Venetia. Julium Carnicum, *Zuglio*, was situated at the foot of the Julian Alps, and was probably founded at the same time as Forum Julii. Altinum, *Altino*, stood on the right bank of the Silis, and on the edge of a lagune, from which it is now two miles distant. It became a favourite residence of the wealthy Romans,[7] and was further known for its excellent wool[8] and its fish; it became a colony probably under Trajan. Patavium, *Padova*, was situated on the river Medoacus, about 30 miles from its mouth. Its mythical founder was Antenor.[9] The earliest historical notice of it is in B.C. 301, when it was attacked by the Lacedæmonian Cleonymus. In 174 it is again noticed, as seeking the interference of the Romans. Generally speaking, however, its history was uneventful, and it enjoyed a high degree of prosperity from its woollen manufactures,[1] which so enriched its citizens, that it was the only city of Italy, except Rome, able to produce 500 persons entitled to equestrian rank. It was the birthplace of the historian Livy. In A.D. 452 it was utterly destroyed by Attila. Near it were some celebrated mineral waters, at a place named Aponi Fons, *Bagni d'Abano*, situated at the foot of a singular volcanic group of hills named Euganeus Collis:[2] these waters were resorted to by patients from all parts of Italy. Lastly, Verona, *Verona*, though situated chiefly on the W. bank of the Athesis, may be regarded as a Venetian town, as it probably belonged to the Euganei. Of its early history we know nothing : it became under the Romans a colony, with the surname of Augusta, and was one of the finest cities in this part of Italy. The Campi Raudii, the scene of Marius's victory over the Cimbri,

---

[a] Ausonius places it ninth in his *Ordo Nobilium Urbium* :—

   Nona inter claras Aquileia cleberis urbes
   Itala ad Illyricos objecta colonia montes
   Moenibus et partu celeberrima.

[7] Æmula Balanis Altini littora villis.   MART. iv. 25.

[8] Velleribus primis Appulia, Parma secundis
   Nobilis; Altinum tertia laudat ovis.   MART. xiv. 155.

[9] Ille tamen ille (sc. Antenor) urbem Patavi, sedesque locavit
   Teucrorum, et genti nomen dedit; armaque fixit
   Troia.   Æn. L 247.

[1] Vellera cum sumant Patavinæ multa trilices
   Et pingues tunicas serra secare potest.   MART. xiv. 143.

[2] Euganeo, si vera fides memorantibus, augur,
   Collo sedens, Aponus terris ubi fumifer exit.   Luc. vii. 193.

[3] Tum Verona Athesi circumflua.   SIL. ITAL. viii. 597.

were near it. It was the birthplace of Catullus,[4] and the scene of some
interesting occurrences in the times of the later Roman empire. The
amphitheatre of Verona is in a good state of preservation: it was built
of marble, and was capable of holding 22,000 persons. There are also
remains of a theatre, of a gateway named *Porta de Borsari*, and of the
walls erected by Gallienus in A.D. 265.

Of the less important towns we may notice—Tarvisium, *Treviso*, on
the Silis, a considerable city after the fall of the western empire; Opitergium, *Oderzo*, between the rivers Plavis and Liquentia, a considerable town under the Romans, destroyed by the Quadi and Marcomanni
in A.D. 372, but afterwards restored; Ateste, *Este*, about 18 miles S.W.
of Patavium, a municipal town of some importance as early as B.C. 136,
and afterwards a Roman colony; and Vicentia, or Vicetia, *Vicenza*,
about 22 miles N.W. of Patavium, frequently noticed by Roman writers,
but evidently not a place of importance.

Roads.—Venetia was traversed by an important high-road, which
formed the chief line of communication between Mediolanum and the
Danube, and the provinces of the Eastern empire. It passed through
Aquileia, Altinum, Patavium, and Vicentia. From Patavium a branch
road joined the Æmilian Way at Mutina. The range of the Alps was
crossed at three points: (1.) by a road which led from Aquileia up the
valley of the Frigidus, and crossed Mount Ocra to Æmona in Pannonia;
(2.) by a road from Aquileia to Julium Carnicum, and thence across
the Alps to the valley of the *Gail* and the *Puster Thal*; and (3.) by a
route which left Opitergium and passed through the *Val Sugana* to
Tridentum, and there fell into the valley of the Athesis.

History.—The history of Venetia is unimportant: the Veneti concluded an alliance with Rome in B.C. 302 against the Gauls, and they
adhered to that alliance with great fidelity. The Carni were reduced
about B.C. 181. Before the close of the Republic, the Veneti had passed
from the condition of allies into that of subjects of Rome. They probably acquired the franchise in B.C. 49.

II. GALLIA CISALPINA.

§ 13. Gallia Cisalpina was bounded on the E. by the Athesis on
the side of Venetia and farther S. by the Adriatic Sea; on the S.
by the river Rubicon and the Apennines, separating it from Umbria
and Etruria respectively; on the W. by the Trebia on the side of
Liguria, and further N. by the Alps; and on the N. by the Alps
and Rhætia. This province may be described generally as consisting of the basin of the Po; for, with the exception of the portion
near the rise of that river which belonged to Liguria, the whole
course of the river falls within the limits of Gallia, which was unequally divided by it into two portions, named Transpadana and Cispadana. The basin is of a triangular form, the Adriatic Sea sup-
plying the base line, whence the sides of the valley gradually con-
tract towards the W. The greater portion of this district is an
alluvial plain, the length of which, from Augusta Taurinorum to

---

[4] Mantua Virgilio gaudet, Verona Catullo.   Ov. *Am* III. 15, 7.

the delta of the Po, is above 200 miles, while its width between
Bononia and Verona is about 70. Its soil was wonderfully fertile,
and the productions varied : we may particularly notice wool, swine,
flax, and every kind of grain.

Names.—Various designations were employed to distinguish the Gaul
of Italy from the northern country of that name. The most usual was
Cisalpina, *i. e.* "on this side of the Alps," as opposed to Transalpina ;
or Citerior, "nearer," as opposed to Ulterior, "further." The Greek
writers used the expressions "Gaul within the Alps," or "Gaul about
the Po ;" or, again, "the land of the Italian Gauls." After it had
become thoroughly Romanized, it was termed Gallia Togàta, in opposition to G. Braccàta or Comàta. Frequently it is termed simply
Gallia.

§ 14. The mountains that bound the basin of the Po are connected
either with the Alps or the Apennines ; only a few of them received
special designations. The rivers are for the most part tributaries of
the Po. Those on the left or N. bank are of considerable size and
length ; those on the S. bank are of less importance. This difference is due partly to the circumstance that the Po approaches the
Apennines more nearly than the Alps, and partly to the large
amount of snow that covers the latter range. The most important
of these tributaries, from W. to E., on its left bank were—the **Duria
Minor**, *Dora Riparia*, which joins it near Augusta Taurinorum ;
the **Stura**, *Stura* ; the **Orgus**, *Orcu* ; the **Duria Major**, *Dora Baltea*,
which has its sources in the Pennine and Graian Alps, and flows
through the valley of the Salassi by Augusta, *Aosta* ; the **Sesites**,
Sesia ; the **Ticinus**,[a] *Ticino*, flowing from the Lacus Verbanus, historically famous for the battle between Hannibal and Scipio, in
B.C. 218, as well as for engagements between the Alemanni and
Aurelian in A.D. 270, and between Magnentius and Constantius in
352 ; the **Addua**, *Adda*, from the Lacus Larius ; the **Ollius**, *Oglio*,
from the Lacus Sebinus ; and the **Mincius**,[b] *Mincio*, from the Lacus
Benacus, on whose banks Cornelius defeated the Insubres and Cenomani in B.C. 197. On the southern bank we have to notice in
Gallia, the **Trebia**, *Trebbia*, flowing by Placentia, and famed for the
victory gained by Hannibal over the Roman consul Sempronius, in

---

[a] Silius Italicus notices the remarkable clearness of its water :—

Cærulcus Ticinus aquas, et stagna vadoso
Perspicuus servat turbari nescia fundo,
Ac nitidum viridi lente trahit amne liquorem.—iv. 82.

[b] The Mincius, after it leaves lake Benacus, runs in a deep winding course,
and near Mantua spreads out into shallow lakes ; hence Virgil :—

Propter aquam, tardis ingens ubi *flexibus* errat
Mincius, et tenera prætexit arundine ripas.—*Georg.* III. 14.

B.C. 218; the **Scultenna**, *Panaro*, which flows not far from Mutina,
and which was the scene of a battle between the Ligurians and the
Romans under C. Claudius, in B.C. 177; and the **Rhenus**, *Reno*,
which flows near Bononia, and is celebrated for the interview between Antony, Octavian, and Lepidus, that took place on a small
island formed by its waters.   On the coast of the Adriatic were
several unimportant streams, one of which, the **Rubicon**, probably
the *Fiumicino*, has derived celebrity from its having formed the
boundary of Gallia Cisalpina ;[f] the passage of it by Cæsar was
therefore tantamount to a declaration of war.

§ 15. The original inhabitants of this district were Tuscans :
these were driven southwards by the Gauls, who crossed the Alps
at different periods in successive emigrations, commencing, according
to Livy, in the reign of Tarquinius Priscus.   The most important of
the Gaulish tribes, from E. to W., (1.) in Gallia Transpadana, were—
the **Cenomani**, between the Athesis and the Addua; the **Insubres**,
between the Addua and Ticinus; the **Levi** and **Libicii**, to the W. of
the Ticinus; the **Salassi**, to the N., in the valley of the Duria
Major; and the **Taurini**, a Ligurian tribe, in the Alpine valleys N.
of the Po. (2.) In Gallia Cispadana—the **Senones**, on the Adriatic,
between Ravenna and Ancona; the **Lingones**, more to the N., in
the low flat land E. of Mutina and Bononia; the **Boii**, between the
Po and the Apennines; and the **Ananes**, in the W., at the base of
the Apennines.   The towns of this province were in some instances
of Tuscan origin : this was certainly the case with Mantua, Adria,
and Bononia.   A few others, as Mediolanum and Brixia, were of
Gallic origin ; but, generally speaking, the Gauls lived in villages,
and the towns were erected by the Romans, in opposition to their
interests, as military posts to secure the conquest of the country.
The first that were thus established were Placentia on the S., and
Cremona on the N. side of the Po, in B.C. 219.   Subsequently to
the formation of the roads, the towns became wealthy and numerous.   The Æmilian Way, in particular, in Cispadana, was studded
with large and prosperous towns, such as Bononia, Mutina, Regium
Lepidi, and Parma.   In Transpadana there were two lines: one
running parallel to the Po, and marked by Mantua, Cremona, and
Ticinum; another at the foot of the Alps, by Brixia, Bergomum,
and Comum.   Between these, in the very centre of the country,
stood Mediolanum, the capital not only of Cisalpine Gaul, but at one
period of Italy itself.

(1.) *In Transpadana*, from E. to W.—**Mantua**, *Mantova*, was situated on the Mincius, about 12 miles above its confluence with the Po

---

[f] See Lucan L. 215.

Its antiquity was very great: it was founded by the Etruscans,[8] and retained much of its Etruscan character down to classical times. It is seldom noticed in history, and it derives its chief celebrity from Virgil[9] having been born either there or at Andes in its territory. **Brixia**, *Brescia*, lay at the foot of the Alps, about 18 miles W. of lake Benacus. It was probably founded by the Cenomani; it became under the Romans a thriving and opulent town, and was made a civic colony by Augustus with the title "Colonia Civica Augusta." It was plundered by the Huns in A.D. 452, but recovered the blow. The remains of antiquity are numerous and interesting. We may particularly notice a building called the Temple of Hercules (more probably a *basilica* than a temple), portions of the theatre, a bronze statue of Victory, and a large collection of inscriptions. **Cremōna**, *Cremona*, was situated on the N. bank of the Po, about six miles below the confluence of the Addua. It was colonised by the Romans in B.C. 219 with 6000 men. It suffered severely from the Gauls for its fidelity in the Second Punic War. In the Civil Wars it espoused the cause of Brutus, and suffered the confiscation of its territory in consequence.[1] In the Civil War of A.D. 69 it became the headquarters of the Vitellian forces; and, having been captured by Antonius, Vespasian's general, it was reduced to ashes. Though rebuilt, it never recovered its prosperity in ancient times. **Mediolānum**, *Milan*, was situated about midway between the rivers Ticinus and Addua, in a broad and fertile plain, about 28 miles from the foot of the Alps. It was founded by the Insubres, and was captured by the Romans in B.C. 222. We hear little of its early history: it probably submitted to Rome in 190, received the Latin franchise in 89, and the full Roman franchise in 49. Subsequently it became a place of literary distinction; but its ultimate greatness dates from the period when it became the Imperial residence, for which its central position in reference to Gaul, Germany, and Pannonia, particularly adapted it. Maximian (about A.D. 303) was the first to reside there permanently, and his successors followed his example down to Honorius in 404. It was taken and plundered by Attila in 452, but it retained its eminence, and became, in 476, the residence of the Gothic kings. It was adorned with many magnificent buildings, of which the only remains are sixteen columns of a portico formerly attached to the public baths. **Bergŏmum**, *Bergamo*, lay 33 miles N.E. of Mediolanum, between Brixia and the Lacus Larius. It is seldom noticed, but was, nevertheless, a considerable town: it derived its wealth chiefly from copper-mines in its territory. It was laid waste by Attila in 452. **Cŏmum**, *Como*, was situated at the S. extremity of the Lacus Larius. The earliest notice of it

[8] Virgil informs us of this, and further that it contained 12 peoples, wherein he probably refers to some internal divisions of the place:—

 Mantua, dives avis; sed non genus omnibus unum:
 Gens illi triplex, populi sub gente quaterni;
 Ipsa caput populis; Tusco de sanguine vires.—*Æn.* x. 201.

[9] The poet possessed an estate there, which was confiscated in the Civil Wars, but was restored to him by Augustus:—

 Fortunate senex, ergo tua rura manebunt;
 Et tibi magna satis, quamvis lapis omnia nudus,
 Limosoque palus obducat pascua junco.   *Ecl.* i. 47.

[1] Mantua was involved in this disaster; hence Virgil's exclamation:—
 Mantua væ miseræ nimium vicina Cremonæ!—*Ecl.* ix. 28.

occurs in B.C. 196, when it joined the Insubres against the Romans, and was consequently taken by them. It was several times furnished with Roman settlers; and, on the last of these occasions, when Julius Cæsar sent 5000 there, its name was changed to Novum Comum. The place is chiefly famous as the birthplace of the two Plinys, the younger of whom had several villas on the banks of the lake. Ticinum, *Pavia,* was situated on the Ticinus, about five miles above its confluence with the Po. It is not noticed until the time of Augustus, but it probably had risen to be a considerable place under the Republic. Its position on the extension of the Æmilian Way made it an important post. It was here that the troops of Vitellius rebelled, that Claudius II. was saluted with the imperial title, and that Constantius took leave of his nephew Julius. It was destroyed by Attila, but restored by the Gothic king Theodoric, and made one of the strongest fortresses of Northern Italy. From A.D. 570 to 774 it was the residence of the Lombard kings, who gave it the name of Papia, whence its modern name is derived. Vercellæ, *Vercelli,* the chief town of the Libicii, stood on the W. bank of the Sessites: it did not rise to importance until after Strabo's time. It was chiefly famous for its temple and grove of Apollo. Augusta Taurinōrum, *Turin,* the capital of the Taurini, was situated on the river Po at the junction of the Duria Minor. Its original name appears to have been Taurasia: its historical name dates from the time when Augustus planted a colony there. Its position was good, commanding the passage of the Cottian Alps, and at the head of the navigation of the Po. Augusta Prætoria, *Aosta,* in the valley of the Duria Major, was founded by Augustus with 3000 veterans, as a means of keeping the Salassi in subjection. It commanded the passes over the Pennine and Graian Alps, and was a place of considerable importance, as attested by its numerous remains, consisting of a triumphal arch, a gateway, a fine bridge, and some remains of an amphitheatre.

Of the less important towns we may notice—Adria, or Hadria. *Adria,* between the Po and the Athesis, formerly on the sea-coast but now 14 miles distant from it, an Etruscan town of early commercial importance, but insignificant under the Romans; Bedriacum, between Verona and Cremona, the scene of two important battles in A.D. 69 between the generals of Vitellius and those of Otho in the first instance, and of Vespasian in the second; Laus Pompeii, *Lodi Vecchio,* 16 miles S.E. of Mediolanum, probably so named in compliment to Pompeius Strabo, who conferred the Latin citizenship on the municipalities of these parts; Eporedia, *Ivrea,* on the Duria Major at the entrance of the valley of the Salassi, founded for the purpose of checking the Salassi, and, after the subjugation of this tribe, a place of wealth and importance; Novaria, *Novara,* between Mediolanum and Vercellæ, noticed as one of the cities which declared in favour of Vitellius in A.D. 69; and, lastly, Segusio, *Susa,* at the foot of the Cottian Alps in the valley of the Duria Minor, the capital of the chieftain Cottius, and of importance as commanding the passes over *Mont Genèvre* and *Mont Cenis.*

(2.) *In Gallia Cispadana.*—Ravenna, *Ravenna,* was situated near the coast of the Adriatic at the S. extremity of the long range of lagunes which stretch northwards as far as Altinum. It was originally an Umbrian town. No mention of it occurs until a late period of the Republic, nor is it known when it received a Roman colony. Its subsequent importance was due to Augustus, who constructed a port

named Portus Classis, or simply Classis, capable of holding 250 ships
of war, and made it the chief naval station on the Adriatic. The
town was very secure, being not only surrounded by lagunes, but
built on piles in a lagune like Venice,[1] and also being well fortified.
The later emperors frequently made it their military quarters, and
from the time of Honorius, in A.D. 404, it was selected from its great
security to be their permanent residence. The Gothic kings retained
it as their capital until 539, when it passed into the hands of the
Byzantines, and became the residence of the Byzantine exarchs. It
was captured by the Lombards in about 750. The sea-coast has now
receded more than four miles from the town. The only Roman
remains are a few basilicas and a sepulchral chapel. Bononia, *Bologna*,
lay at the foot of the Apennines on the river Rhenus. It was originally
an Etruscan town with the name of Felsina; it afterwards passed into
the hands of the Boian Gauls; and finally it became a Roman colony in
B.C. 189. It was centrally situated in reference to the lines of communication opened by the Romans. In B.C. 43 it was garrisoned by M.
Antonius, but was seized by Hirtius. It was under the patronage of
the Antonian family, and hence was not required to take up arms
against Antony in B.C. 32. Subsequently to the battle of Actium,
however, Octavian sent a colony thither. In A.D. 53 it was much
damaged by a fire, but it was restored by Claudius. Mutina, *Modena*,
lay 25 miles W. of Bononia, on the Via Æmilia. It fell into the hands
of the Romans probably in the Gaulish War, B.C. 225-222, and was
made a colony in 183. It played a conspicuous part in the Civil Wars.[2]
In 44 D. Brutus occupied it, and was besieged in it by M. Antonius,
who was defeated, however, outside the walls in two engagements in
43, and was obliged to raise the siege. In A.D. 452 its territory was laid
waste by Attila, and in about 600 it fell into decay. It was particularly famed for its wool.[3] Parma, *Parma*, between Mutina and Placentia, was established as a Roman colony in B.C. 183. It is seldom
noticed until the Civil Wars, when it sided against M. Antonius, and
was consequently taken and plundered in B.C. 43. Its territory was
celebrated for its fine wool.[4] It survived Attila's invasion, and was a
wealthy city after the Lombard conquest. Placentia, *Piacenza*, was
situated near the S. bank of the Po, near the confluence of the Trebia.
It was founded in B.C. 219 by the Romans, and supplied with 6000
colonists. In B.C. 200 it was captured by a sudden attack of the Gauls,
and for some years was liable to their incursions, so much so that in
190 a fresh body of 3000 colonists were sent there. Thenceforward it
prospered, and under Augustus it is noticed as one of the most flourishing cities of Cispadana.

---

[1] All the allusions to Ravenna bear upon its " watery " character :—
    Quique gravi remo limosa regniter undis
    Lenta *paludosæ* proscindunt stagna Ravennæ.— Sil. Ital. viii. 602.
    Sit cisterna mihi, quam vinea, malo Ravennæ ;
    Quum possim multo vendere pluris aquam.—Mart. III. 56.
[2] Illa Cæsar, Perusina fames, Mutinæque labores
    Accedant.   Luc. I. 41.
[3] Sutor cerdo dedit tibi, culta Bononia, munus;
    Fullo dedit Mutinæ.   Mart. III. 59.
[4] Velleribus primis Appulia, Parma secundis
    Nobilis.   Mart. xiv. 155

Of the less important towns we may notice—**Faventia**, *Faenza*, on the Via Æmilia, famed for its vines and its manufacture of linen, and noted in history as the place where Carbo and Norbanus were defeated by Metellus in B.C. 82; **Forum Cornelii**, *Imola*, 10 miles W. of Faventia, said to have been named after the dictator Sulla, the residence of Martial at one period of his life; **Claterna**, on the Via Æmilia, the scene of some military operations during the Civil War in B.C. 43, and almost the only town on the Via Æmilia which has ceased to exist in modern times; **Brixellum**, *Bresello*, on the S. bank of the Po, chiefly celebrated as the place where the Emperor Otho put an end to his life; **Regium Lepidi**, *Reggio*, 17 miles W. of Mutina, deriving its surname probably from Æmilius Lepidus, the constructor of the great road, a place frequently mentioned in the Civil War with M. Antonius; and, lastly, **Clastidium**, *Casteggio*, on the borders of Liguria, 7 miles S. of the Po, chiefly celebrated for the victory gained there in B.C. 222, by Marcellus over the Insubrians, and a place apparently of some importance until the end of the Second Punic War.

Roads.—We have frequently mentioned the **Via Æmilia** in the preceding pages. It was constructed in B.C. 187, by M. Æmilius Lepidus, to connect Placentia with Ariminum. It runs in a direct line for 180 miles through a level plain, and is still the great high road of that district. So great was its importance that its name was transferred to the provinces through which it passed.[*] From Placentia the road was continued to Mediolanum, probably after the complete subjugation of the Transpadan Gauls. From Mediolanum branch-roads led to Augusta Prætoria in the W. and to Aquileia in the N.E. There were also branch-roads from Mutina to Patavium, and from Placentia to Ticinum and Augusta Taurinorum, and so on to the Cottian Alps. There were five important passes over the Alps in this province:—(1.) Across the Rhætian Alps, between Verona and Augusta Vindelicorum, by way of Tridentum, the valleys of the Athesis and Atagis, and the pass of the *Brenner*. (2.) Between the Lacus Larius and Brigantia, on the *Lake of Constance*, either by the *Splügen* or by the *Septimer*, both of which passes are noticed in the Itineraries. (3.) Across the Pennine Alps, between Augusta Prætoria and Octodurus, *Martigny*, by the *Great St. Bernard*. (4.) Across the Graian Alps, between Augusta Prætoria and the valley of the Isara, by the *Little St. Bernard*. (5.) Across the Cottian Alps, between Augusta Taurinorum and Brigantio, *Briançon*, in Gaul, by the pass of *Mont Genèvre*. Lastly, the Apennines were crossed by a road between Bononia and Arretium.

History.—The Gauls became first known to the Romans by the formidable incursions undertaken by them towards the S., in one of which, in B.C. 390, the city of Rome itself was taken and in part destroyed. The first tribe on whose territory the Romans obtained a permanent footing were the Senones, who lived in the extreme S.E. and in Umbria: this occurred in 282. It was not until fifty years later that the great Gallic War took place in consequence of the distribution of the "Gallicus ager." In this the Romans gradually subdued all the Gaulish tribes; Placentia and Cremona were occupied as colonies in 219; the Boii, in Cispadana, yielded in 191; and the Gauls of Transpadana, among whom the Insubres were most conspicuous for their re-

---

[*] This usage appears to have commenced at a very early period :
Romam vade, liber. Si, veneris unde, requiret,
Æmiliæ dices de regione viæ.   MART. III. 4.

sistance to Rome, yielded about the same time. Of the history of
Gallia Cisalpina, as a Roman province, we know little, except that in
B.C. 89 the Jus Latii was conferred on the towns N. of the Po, in re-
ward for the fidelity of the Gauls in the Social War.

III. LIGURIA.

§ 16. The province of **Liguria** extended along the N. coast of the
Tyrrhenian Sea, from the river Varus on the W., separating it from
Gaul, to the Macra on the E., separating it from Etruria; northwards it extended inland to the river Padus, the right bank of which,
down to the confluence of the Trebia, formed the boundary. This
district is throughout of a mountainous and rugged character, being
intersected in all directions by the ridges of the Apennines. The
chief exports were timber, cattle, hides, and honey. Certain portions of the country were adapted to agriculture, but the majority
of the inhabitants subsisted on the produce of their herds and flocks.
Among the special productions may be noticed a breed of dwarf
horses and mules, and a mineral resembling amber, called *ligurium*.
The coast is steep, and affords few natural ports. The rivers on
the S. of the Apennines are small, and call for no special notice : on
the N. of them there are several important tributaries of the Padus,
particularly the **Tanarus**, *Tanaro*, with its confluent the **Stura**.

§ 17. The inhabitants of Liguria (the Ligyes and Ligystini of the
Greeks, the Ligüres of the Romans) were a wild and hardy race,
chiefly noted for their excellence as light-armed troops. They were
divided into a number of independent tribes, which coalesced only
on occasions of public danger. Of these tribes the most important
were—the **Apuani**, in the valley of the Macra; the **Ingauni**, on the
W. coast; the **Intemelii**, on the borders of Gaul; the **Vagienni**, in
the mountainous district N. of the Apennines to the sources of the
Padus; and the **Taurini**, who occupied the country on both sides of
the Padus, but whose capital (*Turin*) was on the left bank of the
river. The Ligurians lived for the most part in villages and mountain fastnesses, and even under the Romans the towns along the sea-
coasts were few. Genua served as the chief port, and Lunæ Portus
in the E. was also a place of trade. In the interior there were
several flourishing places under the Romans, situated at the points
where the mountains declined towards the plain, such as Augusta
Vagiennorum, Alba Pompeia, Asta, and Dertona. These are seldom
noticed in history, but nevertheless appear to have been of importance. We shall describe the towns in order from W. to E., taking
first those on the sea-coast, and afterwards those in the interior.

(1.) *On the Coast.*—**Nicæa**, *Nice*, was situated at the foot of the
Maritime Alps, and on the borders of Gaul. It was a colony of Massilia, and was therefore not a Ligurian possession. In B.C. 154 it was

attacked by the Ligurians. In the later period of the Roman empire it was attached to Gaul. Herculis Monœci Portus, *Monaco*, was also a Massilian colony, and derived its name from a temple of Hercules. It possessed a small harbour, which was frequently resorted to by vessels trading to Spain. Album Intemelium, *Vintimiglia*, the capital of the tribe of the Intemelii, was situated at the mouth of the Rutuba, and derived its name Album from its proximity to the Maritime Alps. Album Ingaunum, *Albenga*, the capital of the Ingauni, on the coast more to the E., became a municipal town of importance under the Romans. Genua, *Genoa*, stood at the head of the Ligurian Gulf, and was the chief town in Liguria, an eminence which it owed partly to its excellent port, and partly as being the point whence the valley of the Po was most easily accessible,—a road crossing the Apennines at this point. Hence it was visited by Scipio and by Mago in the Second Punic War. By the latter it was destroyed in B.C. 205, but was rebuilt by the Romans in 203. It is seldom mentioned afterwards.

(2.) *In the Interior.*—Augusta Vagiennorum, the capital of the Vagienni, stood between the Stura and Tanarus, probably at a place near *Bene*, where considerable ruins exist, comprising remains of an aqueduct, amphitheatre, baths, &c. Pollentia, *Polenza*, was situated near the confluence of the Stura and Tanarus. Its chief celebrity is due to the great battle fought there between Stilicho and the Goths under Alaric, in A.D. 403. Its pottery and its dark-coloured wool are noticed. Alba Pompeia, *Alba*, on the Tanarus, owed its distinctive appellation to Cn. Pompeius Strabo, who conferred many privileges on the towns of this district. It was the birth place of the emperor Pertinax. Asta, *Asti*, on the Tanarus, became a Roman colony, probably under the Emperor Trajan. It was noted for its pottery. Aquæ Statiellæ, *Acqui*, was the chief town of the Statielli, and owed its name to the mineral springs there. Some remains of the ancient baths and numerous other antiquities still exist. Dertona, *Tortona*, was founded by the Romans under the republic, and recolonised by Augustus. It stood on the road leading from Genua to Placentia, and was a convenient station for troops. Cemenelium, *Cimiez*, near Nicæa, the resort of wealthy Romans under the later empire, on account of its mild air. Vada Sabata, *Vado*, possessing one of the best roadsteads on the Ligurian coast, and the point whence a road crossed the Apennines.

Roads.—The position of Liguria made it the greatest thoroughfare between Rome and Gaul. The maritime road was a continuation of the Via Aurelia, and was constructed as far as Vada Sabata by Æmilius Scaurus in B.C. 109. It was not until the time of Augustus, in B.C. 14, that it was carried on to Gaul. This was a work of some difficulty, the road requiring to be cut in the face of the mountain in certain places. At the head of the pass Augustus erected a massive trophy or monument, named Tropæa Augusti, the remains of which may be seen at *Turbia*.

History.—We have some few notices of the Ligurians in early Greek writers, from which we conclude that they were a more powerful and widely-spread nation in early than in late times. The Romans first entered into warfare with them in B.C. 237, and continued a series of wars for above eighty years. The progress of their arms was very slow. The Apuani were removed in a body to Samnium in 180; the Ingauni and Intemelii were conquered in 181, and the Statielli in 173; but the Ligurians were not really reduced to peaceable subjection until the construction of the roads just described, in the years 109 and 14.

Natural Bridge in the Neighbourhood of Veii.

CHAPTER XXV.

IV. ETRURIA.

§ 1. **Etruria** (the **Tyrrhenia** of the Greeks) was bounded on the
N.W. by the river Macra, separating it from Liguria; on the N. by
the Apennines; on the E. by the Tiber, separating it from Umbria,
the Sabini, and Latium; and on the W. by the Tyrrhenian Sea.
This province is varied in character; the N. and N.E. is very mountainous, being intersected with lofty and rugged spurs belonging
to the central chain of the Apennines; the central district, though

still of a mountainous character, has ridges of less height intermixed
with valleys of considerable width and fertility, such as are those
of the Arnus and the Clanis; the maritime district, now called the
Maremma, is a plain of varying width, according as the ridges
approach to or recede from the coast. The general direction of the
ranges in the central region is parallel to that of the Apennines,
i. e. from N.W. to S.E.; and the rivers find outlets to the sea at
places where the ranges are interrupted. Near the coast the hills
strike out at right angles to their former course, and in some
instances descend to the very coast itself. In the S.E. there is a
volcanic region of some extent, connected with that of the Roman
Campagna. The volcanoes have not, however, been active in historical times, the craters having been transformed into lakes. Certain
portions of Etruria were remarkably fertile, particularly the plain
of the Arnus, the valleys of the Clanis and the Umbro, and the
maritime plain. The coast-line is broken at certain points by the
protrusion of the ranges, but still there is a deficiency of good
harbours.

§ 2. Few of the Etrurian mountains are known to us by special
names; we may, however, specify Argentarius, *Argentaro,* a remarkable mountain, forming a promontory on the coast; Soracte,[1] *Monte
S. Oreste,* near the Tiber, a bold and abrupt mass, rising out of the
Roman plain on the N., and hence a conspicuous object from Rome
itself; and Ciminius Mons, *Monte Cimino,* a range that stretches
away in a S.W. direction from the Tiber to the sea-coast, and forms
the boundary of the great plain of the *Campagna* on the N. The
two chief rivers of Etruria are the Arnus and the Tiberis (p. 488).
Of the affluents of the Arnus the only one whose ancient name has
come down to us is the Auser, *Serchio,* which flowed by Luca, and
formerly joined the Arnus, but now reaches the sea by an independent channel. Of the affluents of the Tiber, we have to notice the
Clanis, *Chiana,* which drains a valley between the Arnus and the
Tiber of such remarkable flatness that the waters can be carried off
in either direction: in ancient times the outlet was to the Tiber; at
present there are two channels, one into the Arnus, the other into
the Tiber; and the Cremĕra, *Fosso di Valca,* a small and generally
sluggish brook,[2] flowing through a deep valley from Veii to the Tiber,

---

[1] It is referred to by Horace in the well-known ode:—
  Vides, ut alta stet nive candidum
  Soracte.   *Carm.* 1. 9, 1.
On its summit were a temple and grove of Apollo:—
  Summe deûm, sancti custos Soractis, Apollo.   *Æn.* xi. 785.
[2] It is only after heavy rains that its stream is violent:—
  Ut celeri passu Cremeram tetigere rapacem
    (Turbidus hibernis ille fluebat aquis),
  Castra loco ponunt.   Ov. *Fast.* ii. 205.

and celebrated for the defeat of the Fabii in B.C. 476. On the coast
between the Arnus and Tiber we meet with the *Cæcina, Cecina*,
which watered the territory of Volaterræ; the *Umbro, Ombrone*,
which flowed beneath the walls of Russellæ; and the *Minio, Mig-
none*, a small stream noticed by Virgil. The chief lakes of Etruria
have been already noticed: two of them were historically famous—
the **Lacus Trasimēnus**, for the great victory obtained there by Hannibal over the Roman consul, C. Flaminius, in B.C. 217; and the
Lacus Vadimōnis, a mere pool near the Tiber, for two successive
defeats of the Etruscans by the Romans. The **Lacus Clusīnus** was
a stagnant accumulation of water connected with the river Clanis.

§ 3. The origin of the Etruscans [3] is still wrapped in obscurity.
The ancients, from Herodotus downwards, believed them to be
Lydians.[4] The probability is that they were a mixed people, containing three distinct elements: the Pelasgi, who supplied the bulk
of the population; the Rasenna, or proper Etruscans, who entered
from the N. as a conquering race, and subdued the Pelasgi; and the
Umbrians, who were regarded as the aboriginal population of
Central Italy. The Etrurians were the most refined of all the
inhabitants of Italy, and were particularly skilful in various kinds
of handicraft. Their architecture resembled the Cyclopean style of
the Greeks, the walls being built of large irregular blocks, rudely
squared, and laid, without cement, in horizontal courses. They
were skilful in the construction of sewers, and in the laying
out of streets; in the erection of sepulchres, and the adornment of
the interior walls with paintings; in the manufacture of earthenware vases and domestic utensils; in the sculpture of sarcophagi
and sepulchral urns; and in the casting of figures in bronze. They
were not united under a single government, but formed a confederacy of twelve cities, each of which was an independent state, and
united with the others only in matters of common interest. The
following nine were unquestionably members of the league—Tarquinii, Veii, Volsinii, Clusium, Volaterræ, Vetulonium, Perusia,
Cortōna, and Arretium: to these may probably be added, Cære and
Falerii, though the claims of Fæsulæ, Rusellæ, Pisæ, and Volci are
nearly equally strong. Some of the Etruscan towns were of very
great antiquity: Perusia, Cortona, and a few others, traced back
their existence to the time when the Umbrians occupied the coun-

---

[3] The people were named by Latin writers either Etrusci or Tusci, both of
which are modifications of the same original name Turaci.

[4] Hence the epithets " Lydian " and " Mæonian " are used as equivalent to
Tuscan:—

　　　　　　　　ubi Lydius arva
Inter opima virûm leni fluit agmine Tibria.　　　Æn. ii. 781.

　　O Mæoniæ delecta Juventus.　　　Id. viii. 499.

try : others claimed a Pelasgic origin, as Cære (under its older name of Agylla), Falerii, and Pisæ; others again were of a purely Etruscan origin, as Tarquinii, Volaterræ, and many others; and, lastly, a few, as Sena Julia, Saturnia, and Florentia, dated only from the Roman period. The Etruscan towns occupied remarkable positions, being generally erected on the summits of precipitous hills. The walls which surrounded them were of the most massive character. Possessed of this double security, they appear to have passed a tolerably peaceable existence subsequently to the time of the Roman conquest. We shall describe them in order from N. to S.

Luna, *Luni*, was situated on the left bank of the Macra, on the borders of Liguria. At the time the Romans first knew it, the Ligurians had gained possession of it from its old masters, the Etruscans. The Romans colonized it, first in B.C. 177, and again under the Second Triumvirate; but it never rose to any eminence.[6] Its territory was famous for its wine and its cheeses,[6] and still more for its quarries of white *Carrara* marble, which was used both for building and for statuary.[7] About five miles from the town there is a magnificent gulf called Portus Lunæ,[8] now the *G. of Spezia*: a range of rocky hills intervenes between the town and the bay, so that it does not appear now it could have served as the port of Luna. Luca, *Lucca*, was situated in a plain at the foot of the Apennines, near the left bank of the Auser, and about 12 miles from the sea. It was rather a Ligurian than an Etruscan town, and was included within the limits of Liguria by Augustus. It was colonized in 177, and became a municipium in 49. Cæsar, while in charge of the province of Gaul, frequently appointed it as a rendezvous for his political friends. There are remains of an amphitheatre visible. Pisæ, *Pisa*, was situated on the right bank of the Arnus, at a distance formerly of 2½, but now of 6, miles from its mouth. Most ancient writers connected it with Pisa in Elis,[9] and supposed it to be founded by Peloponnesians after the Trojan War. It appears probable that it was a Pelasgic settlement; but it afterwards passed into the hands of the Etruscans, and became one of their chief cities. Its position rendered it an important frontier town in the wars of the Romans with the Ligurians. A Roman colony was planted there

---

[6] It was deserted even in Lucan's time :—
   Aruns incoluit desertae moenia Lunae.   L 586.
[6] Casens Etruscm signatus imagine Lunae,
   Praestabit pueris prandia mille mis.   MART. xiii, 30.
[7] Anno metallifera repetit jam moenia Lunae ?—STAT. Silv. iv. 4, 23.
   Lunaque portandis tantum suffecta columnis.   Id. iv. 2, 29.
[8] Tunc quos a nivea exegit Luna metallis
   Insignis portu ; quo non spatiosior alter
   Innumeras cepisse rates, et claudere pontum.—SIL. ITAL. viii. 482.
[9] Hence the epithet of " Alphean " :—
   Hos parere jubent Alpheae ab origine Pisae :
   Urbs Etrusca solo.   Æn. x. 179.
   Nec Alphea capiunt navalia Pisae.   CLAUD. B. Gild. 483.

in 180, at the request of the Pisans themselves, and again by Augustus. Its territory was fertile, producing a fine kind of wheat and excellent wine. Its port was situated at a point between the mouth of the Arnus and *Leghorn*. **Faesulæ**, *Fiesole*, was situated on a hill about three miles N. of the Arnus. It is noticed in the great Gaulish War in B.C. 225, and in the Second Punic War, as it stood on the route which the invading hosts followed. It was destroyed by Sulla, and restored by a colony of his party, who afterwards rendered it the head-quarters of Catiline.[1] The circuit of the walls, the remains of a theatre, a curious reservoir, and other objects, have been found on its site. **Florentia**, *Florence*, on the Arnus, probably derived its origin as a town from a Roman colony planted here, originally perhaps by Sulla, but renewed by the triumvirs after the death of Cæsar. From the latter of these periods it became a flourishing town, though seldom noticed in history. There are some remains of an amphitheatre there. **Arretium**, *Arezzo*, was situated in the upper valley of the Arnus. It became in the Gaulish Wars a military post[2] of the highest import-ance, as commanding the communications between Cisalpine Gaul and Etruria. In the civil wars of Sulla and Marius it aided with the latter, and suffered severely in consequence. Cæsar occupied it in B.C. 49, at the commencement of the Civil War; but subsequently to this its name is scarcely mentioned in history. It was celebrated for its pottery of a bright red hue,[3] many specimens of which are still extant. Numerous works in bronze have also been discovered there. Mæcenas was probably a native of this place. **Cortona**. *Cortona*, stood on a lofty hill, S. of Arretium, and about 9 miles N. of the Lacus Trasimenus. It was reputed a very ancient city, having been founded by the Um-brians, then occupied by Pelasgians under the name of Corythus,[4] and finally by Etruscans. It received a Roman colony, probably in Sulla's time. Its walls may still be traced, and present some of the finest specimens of Cyclopean architecture to be seen in all Italy. **Sena Julia**, *Sienna*, was situated nearly in the centre of Etruria, and appears to have been founded by Julius Cæsar: it is seldom noticed. **Volaterræ**, *Volterra*, stood about 5 miles N. of the river Cæcina, and 15 from the sea. Its position was fine, the height of the hill on which it stood being about 1700 feet. It was a city of the highest antiquity, and one of the twelve chief towns of Etruria. In the civil wars between Sulla and Marius, it became the last stronghold of the party of the latter, and was besieged for two years by Sulla himself, and, after its capture, suffered various losses. It received a fresh colony under the Trium-virate, but is not subsequently mentioned. The ancient walls may be traced throughout their whole circuit, and in some places are in a high

---

[1] Its inhabitants were noted for their skill in divination :—
  Adfuit et sacris interpres fulminis alis
  Faesula. SIL. ITAL. viii. 478.

[2] An, Corrine, sedet, clausum se consul inerti
  Ut tenent vallo ; Pœnas nunc occupet altos
  Arreti muros. ID. v. 121.

[3] Aretina nimis ne spernas vasa, monemus :
  Lautus erat Tuscis Porsena fictilibus. MART. xiv. 98.

[4] The Latin poets have borrowed this name from them :—
  Surge age, et hæc lætus longævo dicta parenti
  Haud dubitanda refer. Corythum, terrasque require
  Ausonias. Æn. iii. 169.

state of preservation: two of the ancient gateways, probably of the
Roman period, also remain. The sepulchres are numerous, and have
yielded a large collection of urns, many of which are adorned with
sculptures and bas-reliefs. Clusium, *Chiusi*, was situated on a gentle
hill rising above the valley of the Clanis, and near the lake named
after it. Its antiquity was believed to be very great, and Virgil represents it as aiding Æneas against Turnus.[3] It was one of the cities that
joined in the war against Tarquinius Priscus. The invasion of the
Gauls in B.C. 391 resulted (it was said) from an internal dissension at
Clusium ; in 295 the Senones cut to pieces a Roman legion stationed
there; and again in 225 the Gauls once more appeared under its walls.
In the civil wars of Sulla and Marius, two battles were fought in its
neighbourhood, in both of which Sulla's party were successful. Portions of the walls are visible, and the sepulchres are very numerous and
rich in urns, pottery, bronzes, and other objects. The district of
Clusium was famous for its wheat and spelt, and also possessed sulphureous springs. Perusia, *Perugia*, stood on a lofty hill on the right
bank of the Tiber, overlooking the Trasimene Lake, and thus near
the borders of Umbria. No notice occurs of the time when it yielded
to Rome; but in the Second Punic War it comes prominently forward
as an ally of that power. In the civil war between Octavian and L. Antonius in 41, the latter threw himself into Perusia: Octavian besieged it,
and, on its capture, gave it up to plunder, and put its chief citizens to
death.[4] The town was accidentally burnt at that time, but it was
restored by Augustus. Portions of the walls and two gateways survive,
the latter belonging to the Roman period. The sepulchres are numerous
and interesting : a specimen of the Etruscan language was found in one
of them. Volsinii, *Bolsena*, was situated on the shore of the lake
named after it. The old Etruscan town stood on a hill; the later
Roman one in the plain by the lake. After numerous wars with Rome,
it was finally subdued in 280. The old town was then destroyed, and
the new one built: some remains of the latter exist, the most remarkable
being those of a temple. It was the birth-place of Sejanus, the favourite
of Tiberius. Cosa, *Ansedonia*, stood on a height near the sea-coast, somewhat S. of Mons Argentarius. Its name first appears in B.C. 273, when
a Roman colony was planted there: Virgil, however, assigns to it a
higher antiquity.[7] In the Second Punic War it is noticed among the
allies of Rome, and in 196 a new colony was sent thither, apparently
from losses sustained in that war. Its port was a convenient point of
embarkation for Corsica and Sardinia, and to this it owed its chief
importance. The walls of Cosa still exist, but are probably of the
Roman period. Tarquinii, near *Corneto*, was situated about four miles
from the coast, near the left bank of the river Marta. It was reputed
the most ancient of the Etruscan cities, its origin being attributed to
Tarchon, son of the Lydian Tyrrhenus.[8] Its proximity to Rome brought

[3] Massicus armat princeps acrat æquora Tigri ;
    Sub quo mille manus juvenum ; qui mœnia Clusi,
    Quique urbem liquere Cosa.   *Æn*. x. 166.
[4] Ille Cæsar, *Perusina fames* Mutinæque labores
    Accedant fatis.   Luc. i. 41.
[7] See note [3] above, where it appears as one of the allies of Æneas.
[8] Ipse oratorea ad me regnique coronam
    Cum sceptro misit, mandatque insignia Tarcho :
    Succedam castris, Tyrrhenaque regna capessam.—*Æn*. viii. 503.

it into early connexion with that town, and it was reputed the native town of the two Tarquins, whose father, Demaratus, had emigrated from Corinth to Tarquinii. From B.C. 398 to 309 Tarquinii was engaged in wars with Rome at intervals; but subsequently to the great battle at Lake Vadimo it fell into a state of dependency, and is seldom noticed afterwards. The circuit of the ancient walls may be traced at *Turchina*, about 1½ mile from *Corneto*: there is also a very extensive necropolis, containing some tombs adorned with paintings: the paintings themselves are of Greek character, but the subjects are purely Etruscan. *Falerii, Sta. Maria di Falleri*, stood N. of Mt. Soracte, a few miles W. of the Tiber. It was of Pelasgic origin, and retained much of its Pelasgic character after its conquest by the Etruscans. It is first noticed in B.C. 437, as joining the Veientes against Rome. After the fall of Veii it came to terms with Rome, but contests were from time to time renewed until B.C. 241, when their city was destroyed, and rebuilt on a new site of less natural strength. The position of the old Etruscan town was at *Civita Castellana*, and of the later Roman town at *Sta. Maria di Falleri*, a deserted spot where the ancient walls are still visible. The surrounding territory was very fertile, and Falerii was much famed for its sausages.[*] Its inhabitants were named Falisci, and sometimes Æqui Falisci, *i. e.* "Faliscans of the Plain." Veii stood about 12 miles N. of Rome, at *Isola Farnese*. It was a powerful city at the time of the foundation of Rome, and possessed a territory extending along the right bank of the Tiber, from Soracte down to the mouth of the river. The Veientes first engaged in war with the Romans for the recovery of Fidenæ: they were defeated by Romulus, and lost a portion of their territory near Rome, known as Septem Pagi. War was renewed in the reigns of Tullus Hostilius, Ancus Marcius, L. Tarquinius, and Servius Tullius, and on every occasion with an unfavourable result for Veii. After the expulsion of the second Tarquin, the Veientines, with the aid of Porsena of Clusium, recovered their territory for a brief space; and thenceforward the war was of a more serious character, as the Veientes obtained the assistance of the Etruscans. The slaughter of the Fabii in B.C. 470, who had gone out to check the incursions of the Veientes, and the capture of Veii itself by Camillus, after a ten years' siege, in 390, are the most striking incidents in these wars. After its capture it fell gradually into decay,[1] but continued to exist till a late period. There are remains of the ancient walls, and numerous sepulchres, on its site. *Cære, Cervetri*, was situated a few miles from the coast, on a small stream formerly named Curetanus Amnis,[2] and now *Vaccina*. Its ancient name was Agylla,[3] and its founders were Pelasgi. It was conquered by the

* It was the birthplace of Ovid's wife :—
     Cum mihi pomiferis conjux foret orta Faliscis,
     Mœnia contiginus victa, Camille, tibi.   Ov. *Am.* iii. 13, 1.

[1] Lucan speaks of it as utterly desolate :—
                      Tunc omne Latinum
     Fabula nomen erit : Gabios, Veiosque, Coramque
     Pulvere vix tectæ poterunt monstrare ruinæ.   vii. 391.

[2] It is the *Cæritis amnis* of Virgil :—
     Est ingens gelidum lucus prope Cæritis amnem.—*Æn.* viii. 597.

[3] Haud procul hinc saxo incolitur fundata vetusto
     Urbis Agyllinæ sedes : ubi Lydia quondam
     Gens, bello præclara, jugis insedit Etruscis.   *Id.* viii. 478.

Etruscans, but, like Falerii, it probably retained much of its Pelasgic character. It is first noticed by Herodotus as joining in an expedition against the Corsican Phocæans, and it appears to have been an important maritime town at that time. It engaged in war with Rome under the elder Tarquin, and was the place whither the second king of that name first retired into exile. In B.C. 353 the Cærites again took up arms against Rome to no effect; and it was probably on this occasion that they received the Roman citizenship without the right of suffrage—a political condition which was tantamount to disfranchisement, and which gave rise to the expression, "in tabulas Cæritum referre."

Of the less important towns we may notice—1. *On the Coast.*—
Vetulonium, *Magliano*, one of the twelve confederate cities, reputed to be the place where the Etruscan insignia of magistracy (lictors, toga prætexta, sella curulis, &c.) were first used.[4] Populonium,[5] *Populonia*, on the promontory of the same name, opposite the island of Ilva, the chief maritime town of Etruria, and the only city which possessed a silver coinage of its own. Rusellæ, *Roselle*, about 14 miles from the sea, and 4 from the right bank of the river Umbro, the scene of a battle between the Romans, under Valerius Maximus, and the Etruscans in B.C. 301, and afterwards captured by Megellus in 294. Telamon, *Telamone*, on a promontory between Mons Argentarius and the Umbro, noticed in B.C. 225 as the scene of a great battle between the Romans and Gauls, and in 87 as the spot where Marius landed on his return from exile. Volci, near *Ponte della Badia* on the river Armina, about 8 miles from its mouth, a place seldom noticed in history, but known to be a large town from the extent of its necropolis, which was discovered in 1828, and in which no less than 6000 tombs have been opened, yielding a vast number of painted vases, bronzes, &c. Saturnia, *Saturnia*, a little N. of Volci, so named by the Romans when they sent a colony thither in B.C. 183, the former Etruscan name having been Aurinia. Graviscæ, on the sea-coast, probably at *S. Clementino*, about a mile S. of the Marta, colonized in B.C. 181, but owing to the unhealthiness[6] of its situation a poor place. Centumcellæ, *Civita Vecchia*, on the sea-coast, 47 miles from Rome, a town which owed its existence to the magnificent port which Trajan constructed there. Castrum Novum, *Torre di Chiaruccia*, about 5 miles S. of Centumcellæ, colonized by the Romans in B.C. 191. Pyrgi, *Santa Severa*, on the

Coin of Populonium.

Obverse: Gorgon's head of much the reverse is plain, with an type or legend.

[5] Mœniaque decus quondam Vetulonia gentis.
Bissenos hæc prima dedit præcedere fasces,
Et junxit totidem tacito terrore secures:
Hæc alias eboris decoravit honore curules,
Et princeps Tyrio vestem prætexuit ostro:
Hæc eadem pugnas accendere protulit ære.   SIL. ITAL. VIII. 485.

[4] It was one of the cities that assisted Æneas:—
Una torvus Abas: huic totum insignibus armis
Agmen, et aurato fulgebat Apolline puppis.
Sexcentos illi dederat Populonia mater
Expertos belli juvenes.   Æn. x. 169.

[6] Et Pyrgi veteres, intempestæque Graviscæ.   Id. x. 184.

coast, 34 miles from Rome, probably a Pelasgian[f] town, and the seat of a celebrated temple of Eileithyia, which was plundered in B.C. 354 by Dionysius of Syracuse. **Alsium**, *Palo*, on the sea-coast, colonized in B.C. 245, and a favourite residence of the wealthy Romans under the empire. **Fregenæ**, *Torre di Maccarese*, between Alsium and the mouth of the Tiber, colonized probably in B.C. 245, and situated in an unhealthy position.* 2. *In the Interior.*—**Pistoria**, *Pistoja*, under the Apennines, between Luca and Fæsulæ, the scene of Catiline's final defeat in B.C. 62. **Ferentinum**, *Ferento*, N. of the Ciminian range, and about 5 miles from the Tiber, the birthplace of the Emperor Otho, and a place of consideration under the empire; the theatre is still in a high state of preservation. **Sutrium**, *Sutri*, on an isolated hill 32 miles N. of Rome, a place frequently noticed in the wars of the Romans and Etruscans; its amphitheatre remains, excavated in the tufo rock. **Fescennium**, S.E. of Falerii, of which it was a dependency, a place of small importance, and chiefly notorious as having given name to a rude kind of dramatic entertainment styled "Fescennini Versus," which afterwards degenerated into mere licentious songs. **Capena**, about 8 miles S. of Soracte, an ally of Veii in her Roman wars, and consequently reduced by the Romans after the fall of that town; its territory was remarkably fertile, and was further noted for the grove and temple of Feronia* situated at the foot of Soracte. Lastly, **Nepête**, *Nepi*, between Falerii and Veii, and probably a dependency of the latter; it is first mentioned in B.C. 386 as an ally of Rome, and it received a colony in 363.

Roads.—Three great high-roads traversed Etruria in its whole extent:—The **Via Aurelia**, which led from Rome to Alsium, and thence along the sea-coast to Pisæ and Luna; the **Via Cassia**, from Rome, through the heart of the province by Sutrium and Clusium, to Arretium, and thence by Florentia across the Apennines; and the **Via Clodia**, which took an intermediate line by Saturnia, Rusellæ, and Sena to Florentia, where it joined the Via Cassia. The dates of the construction of these roads are quite uncertain. The **Via Flaminia** skirted the S.E. border of Etruria, entering it by the Milvian bridge, about 3 miles from Rome, and striking to the N. under Soracte to Ocriculum in Umbria.

Islands.—Off the coast of Etruria there are several islands, the most important of which, named **Ilva** by the Latins, **Æthalia** by the Greeks, and *Elba* by ourselves, was only about 6 miles distant from the mainland, and was particularly famous for its iron-mines.[i] The ore was originally smelted on the island itself, whence its Greek name (from αἴθαλη, "soot"); but in later times, when fuel had run short, it was brought over to Populonium for that purpose.

History.—The Etruscans were once widely spread over Central and

---

Virgil refers to its antiquity; see previous note.

     [g] Alsium et obscuro campo squalente Fregenæ.   Sil. Ital. viii. 477.

     [h]   Itur in agros
Dives ubi ante omnes colitur Feronia luco,
Et sacer humectat fluvialia rura Capenas.   Id. xiii. 83.

     [i]   Ilva trecentos,
Insula inexhaustis Chalybum generosa metallis.—Æn. x. 173.
Non totidem Ilva viros, sed lætos cingere ferrum,
Armatæ patris, quæ nutrit bella, metallo.   Sil. Ital. viii. 616.

Northern Italy, occupying not only Etruria, but a portion of Gallia
Cisalpina in the N. and Campania in the S. They possessed from an
early period great naval power, and engaged in maritime war with
the Phocæans of Alalia in B.C. 538, with Hiero of Syracuse in 474, and
with other cities. They also founded colonies in Corsica. Their maritime supremacy waned, however, about the time of the capture of Veii.
Their territorial influence was at its highest about 620-500 B.C., and was
coincident with the rule of the Tarquins at Rome. At a subsequent
period constant wars occurred between Rome and Veii, which terminated only with the destruction of the latter in 396. Thenceforward
the Romans advanced northwards, reaching Sutrium in 390, crossing
the Ciminian forest in 310, defeating the Etruscans at Lake Vadimo
in 309, at Sentinum in Umbria in 295, and again at Lake Vadimo in
293, and reducing the Volsinienses in 265. The Roman conquest does
not appear to have interfered with the Etruscan nationality: colonies
were founded in the S., and at Pisæ and Luca in the N., but elsewhere
the population remained unchanged. The Etruscans received the
Roman franchise in 89. In the civil wars of Marius and Sulla they
sided with the former, and were severely handled by Sulla at the completion of the war: they again suffered from the Catiline War. Finally,
Cæsar established a number of military colonies throughout the land.

V. UMBRIA.

§ 4. Umbria, in its most extensive sense, was bounded on the W.
by the Tiber, from its source to a point below Ocriculum; on the E.
by the river Nar, separating it from the land of the Sabines, and by
the Æsis, separating it from Picenum; on the N.E. by the Adriatic
Sea; and on the N. by the Rubico, separating it from Gallia Cisalpina. Within the limits specified are contained (1) Umbria Proper,
which lay on the W. of the Apennines, and (2) the district of the
Senones, or, as the Romans termed it, the Gallicus Ager, on the E.
of the range. Umbria is generally mountainous, being intersected by the Apennines, which, though neither so lofty nor yet so
rugged as they become more to the S., are very extensive, occupying, with their lateral ridges, a space varying from 30 to 50 miles
in width. On the W. the lateral ridges extend to the valley of
the Tiber, but between them and the central range is a fertile and
delightful district, watered by the Tinia and Clitumnus, and renowned for its rich pastures. On the E. of the central range the
country is broken up by a vast number of parallel ridges, which
strike out at right angles to the main range, and subside gradually
as they approach the sea.

§ 5. The rivers of Umbria were numerous, but not of any great
size. Of the tributaries of the Tiber, which may be considered as
in part an Umbrian river, the most important is the Nar, Nera,
which rises in the country of the Sabines, and in its lower course,
from Interamna to the Tiber, flowed entirely through Umbria. The

Clitumnus,[3] *Clitumno*, or Tinia (as it was called in its lower course), was a small stream which flowed through a tract of great fertility by the town of Mevania. The streams which flow into the Adriatic are—the *Æsis*, *Esino*, which formed the limit on the side of Picenum ; the *Sena*,[1] *Nevola*, which flowed under the walls of Sena Gallica ; the *Metaurus*, *Metauro*, which joins the sea at Fanum Fortunæ, and is celebrated in history for the great battle,[4] in B.C. 207, between Hasdrubal and the Romans ; the *Pisaurus*, *Foglia*, which gave name to the city of Pisaurum ; and the *Ariminus*, *Marecchia*, which flowed by Ariminum.

§ 6. The Umbrians at one period occupied a very extensive region in the northern part of Central Italy, spreading on each side of the Apennines from sea to sea. We know nothing of their character beyond the fact that they were reputed brave and hardy warriors. They were not united under one government, but lived in separate tribes, each of which followed its own line of policy. The towns were numerous, but not of any great importance. Several of them received Roman colonies after the country was conquered, as Narnia, Spoletium, Sena, Ariminum, and Pisaurum. The towns in the E. district were situated on the sea-coast, at the mouth of the rivers ; those in the western district were in the fertile valleys of the Tiber, the Nar, and the Clitumnus. We shall describe these in order from N. to S., beginning with those on the W. of the Apennines.

Mevania, *Bevagna*, was situated on the Tinia, in the midst of the luxuriant pastures[5] for which that stream was so celebrated. It was an important town under the Umbrians, and was their head-quarters in B.C. 308. Its chief fame, however, rests upon its claim to be considered the birth-place of the poet Propertius.[6] **Tudar**, *Todi*, was

---

[1] The waters of this river were supposed to impart the white colour for which the cattle that fed on its banks were famous :—

    Hinc albi, Clitumne, greges, et maxima taurus
    Victima, sæpe tuo perfusi flumine sacro,
    Romanos ad templa deûm duxere triumphos.   *Georg.* II. 146.
    Qua formosa suo Clitumnus flumina luco
    Integit, et niveos abluit unda boves.   *Propert.* II. 19, 25

[2] Et Clanis, et Rubico, et Senonum de nomine Sena.—Sil. Ital. viii. 455.

[3] Quid dobeas, O Roma, Neronibus,
    Testis Metaurum flumen, et Hasdrubal
    Devictus.   Hor. Carm. iv. 4, 37.

[4] Illa urbes Arna et lætis Mevania campis. Sil. Ital. viii. 456.
    Tauriferis ubi se Mevania campis
    Explicat.   Luc. i. 478.

[5] The passage on which this claim is grounded is of an ambiguous character :—
    Umbria te notis antiqua penatibus edit.
    Mentior ! an patriæ tangitur ara tuæ !
    Qua nebulosa cavo rurat Mevania campo,
    Et lacus æstivis intepet Umber aquis.   Propert. iv. 1, 121.

situated on a lofty hill,[r] rising above the left bank of the Tiber. It received a colony under Augustus, and, though seldom mentioned in history, appears to have been a considerable town under the Roman Empire. The walls of the city, partly of an early Etruscan and partly of a later Roman character, still remain, as also do portions of a building (probably a basilica) called the "Temple of Mars."[s] Numerous coins and bronzes have also been found there. **Spoletium**, *Spoleto*, was situated near the sources of the Clitumnus. We have no notice of its existence before B.C. 240, when a Roman colony was planted there. It was attacked by Hannibal, in 217, without success. A battle was fought beneath its walls in 82, between the generals of Sulla, and Carrinas, the lieutenant of Carbo, and the town suffered severely in consequence of having received the latter after his defeat. An arch, named *Porta d'Annibale*, some remains of an ancient theatre and of two or three temples, still exist. **Narnia**, *Narni*, was strongly situated on a lofty hill[t] on the left bank of the Nar, about 8 miles above its confluence with the Tiber. Previous to the Roman conquest it was named Nequinum: it was taken and colonized in 209. For some time it appears to have been in a depressed condition, and in 190 it received a fresh colony, but afterwards its position on the Flaminian Road secured to it a high degree of prosperity. The Emperor Nerva was born there. The chief remains of antiquity are one of the arches and the two other piers of a magnificent bridge which Augustus constructed for the Flaminian Road. **Ariminum**, *Rimini*, lay on the sea-coast about 9 miles S. of the Rubicon. It is first noticed in B.C. 268, when the Romans established a colony there, which became a military post of the highest importance, and was justly considered the key of Cisalpine Gaul. It was strongly occupied by the Romans in the Gaulish War in 225, in the Second Punic War in 218, and again in 200. It suffered severely from Sulla's troops in the Civil War with Marius. Cæsar occupied it in his war against Pompey, and we have it mentioned in several subsequent wars. The most striking remains of antiquity are a splendid marble bridge of five arches over the Ariminus, commenced by Augustus and finished by Tiberius; and a triumphal arch, erected in honour of Augustus. **Fanum Fortunæ**, *Fano*, stood on the left bank of the Metaurus, at the point where the Flaminian Road fell upon the sea-coast. Its name is due to a temple of Fortune that stood there. It was occupied by Cæsar in B.C. 49, and by the generals of Vespasian in A.D. 69, and was undoubtedly of importance as a military post. A triumphal arch, erected in honour of Augustus, is the only important relic of antiquity.

Of the less important towns we may notice in the same order:—

---

[r]　　　Excelso summum qua vertice montis
　　Devexum lateri pendet Tuder.　　　　　SIL. ITAL. vi. 645.

[s] This name has been assigned to it from the fact that Mars was worshipped at Tuder :—

　　Et gradivicolam celso de colle Tudertem.　　SIL. ITAL. iv. 222.
　　Haud parci Martem colpisse Tudertes.　　　ID. viii. 464.

[t]　　　　Duro monti per saxa recumbens
　　Narnia.　　　　　　　　　　　　　　ID. viii. 459.

　　Narnia, sulphureo quam gurgite candidus amnis
　　Circuit, ancipiti vix adeunda jugo.　　　MART. vii. 93.

1. *W. of the Apennines.*—Iguvium,[1] *Gubbio,* strongly situated on the W. slope of the Apennines, the place where the Illyrian king Gentius and his sons were confined, but more celebrated for the seven tables with inscriptions in the old Umbrian tongue, which were found about 8 miles off, on the site of a temple of Jupiter Apenninus. Hispallum, *Spello,* N. of Mevania, colonized under Augustus and again under Vespasian, and regarded by some critics as the birthplace of Propertius. Ameria,[2] *Amelia,* the most ancient of the Umbrian towns, situated on a hill between the Tiber and the Nar. Interamna, *Terni,* "between the branches" of the river Nar, which here divides and forms an island, a municipal town of some importance, and generally regarded as the birth-place of the historian Tacitus. Ocriculum, *Otricoli,* the southernmost town of Umbria, near the Tiber, and on the Flaminian Road, which leads to frequent incidental notices of it; it became a favourite residence of the wealthy Romans; and, from the remains discovered by excavating in 1780, it appears to have been a splendid town. 2. *E. of the Apennines.*—Sarsina, *Sarsina,* in the extreme N., chiefly famed for having given birth to Plautus. Urbinum, surnamed *Hortense, Urbino,* situated on a hill between the valleys of the Metaurus and Pisaurus, the place where Fabius Valens was put to death in A.D. 69. Pisaurum, *Pesaro,* at the mouth of a river of the same name, colonized by the Romans in B.C. 184, again by M. Antonius, and a third time by Augustus, having been destroyed by an earthquake in B.C. 31. Sena, surnamed Gallica, to distinguish it from the Etrurian city of the same name, founded by the Romans in B.C. 289 after their conquest of the Senones, and situated on the coast S. of Fanum Fortunæ: the name has been corrupted into *Sinigaglia.* Sentinum, *Sentino,* near the sources of the Æsis, celebrated as the spot where Q. Fabius defeated the Samnites and Gauls in B.C. 295, and itself a strong town, besieged by Octavian in the Perusian War without success. Camerinum, *Camerino,* in the Apennines near the frontiers of Picenum, the old capital of the Camertes, and occupied as a stronghold on several occasions in the Roman Civil Wars.

Roads.—Umbria was traversed in its whole length by the celebrated Via Flaminia, constructed by the censor C. Flaminius, in B.C. 220, as a means of communication with Cisalpine Gaul. It entered the province at Ocriculum, passed by Narnia, and thence either by Mevania or by a more circuitous route by Spoletium to Fulginium, and across the Apennines to Fanum Fortunæ on the Adriatic. A branch road left at Nuceria for Ancona, whence a road was carried along the coast by Sena Gallica to Fanum Fortunæ.

History.—The early history of the Umbrians is almost unknown. They were expelled from the maritime district by the Senonian Gauls. They made common cause with the Etruscans against the Romans, and suffered in consequence several defeats, the last of which, near Mevania in B.C. 308, was a decisive blow. They passed into the condition of a subject state, and remained, with few exceptions, faithful to their allegiance. Augustus retained the name for the sixth region in his division, but it was subsequently united to Etruria.

---

[1]   Infretum nebulis humentibus olim
    Iguvium.   SIL. ITAL. viii. 459.
[2] Its orders are noticed by Virgil :—
    Atque Ameriae parant lentæ retinacula viti.   *Georg.* i. 265.

VI. PICENUM.

§ 7. Picenum extended along the coast of the Adriatic from the river Æsis, which separated it from Umbria, to the Matrinus, which separated it from the territory of the Vestini; inland, on the W., it was bounded by the central ridge of the Apennines. It is a district of great fertility and beauty, the greater part of it being occupied by the secondary ridges of the Apennines, which in their upper regions were clothed with extensive forests, while the lower slopes produced abundance of fruit, especially apples* and olives, as well as good corn and wine. The rivers are numerous, but of short course: the most important is the Truentus, *Trento,* which flowed by Asculum.

§ 8. The inhabitants of this district, named Picentes, are generally regarded as a branch of the Sabine race.[1] The Prætutii, who lived in the S., were to some extent a distinct people, as also were the inhabitants of Ancona, who were Syracusan Greeks. The towns of Picenum were numerous, and many of them of considerable size, but they did not attain to any historical celebrity. With the exception of Ancona, which alone possessed a good port, the most important cities were situated inland on hills of considerable elevation, and were thus so many natural fortresses. Asculum ranked as the capital. We shall describe these towns in order from N. to S., commencing with those on the sea-coast.

Ancōna, or Ancon, *Ancona,* was so named from its being on an "elbow" (ἀγκών) or bend of the coast between two promontories, a peculiarity of position which furnished the town with a device for its coins. It was founded by some dissatisfied Syracusans in B.C. 392; and it became, under the Romans, one of the most important seaport towns on the Adriatic, and the chief entrepôt for the trade with Illyria. Trajan constructed an excellent harbour there, by the formation of a mole,

Coin of Ancona belonging to the Greek colony

Obverse, head of Venus. Reverse, a bent arm, or elbow, in allusion to its name.

---

* Picenis cedant poma Tiburtia suco.   Hor. *Sat.* ii. 4, 70
  Quid quam Picenis excerpens semina pomis
  Gaudes?   *Id.* ii. 3, 272.
              De corbibus isdem
  Æmula Picenis, et odoris mala recentis.   *Juv.* xi. 73.

[1] The name was usually derived from *picus* "a wood-pecker," which guided the emigrants on their road. Silius Italicus, however, refers it to an Italian divinity of that name:—
  Hoc Picus, quondam nomen memorabile ab alto
  Saturno, statuit genitor, quem carmine Circe
  Exutum formæ volitare per æthera jussit,
  Et sparsit plumis croceum fugientis honorem.   viii. 441.

which still remains, and is adorned with a triumphal arch of white marble, erected in honour of that emperor. The town possessed a celebrated temple of Venus,[4] and was also noted for its purple dye. The surrounding district yielded large crops of wheat. The population was very large, the number of citizens at the time of the Roman conquest having been 360,000, according to Pliny. **Firmum**, *Fermo*, was situated about 6 miles from the coast, on which it possessed a port or emporium called Castellum Firmanum. The Romans colonized it at the beginning of the First Punic War. It was strongly placed, and was occupied on several occasions by Roman generals. **Castrum Novum** was founded by the Romans at the same time as Firmum: it probably occupied the site of the deserted town of S. Flariano. **Hadria**, or **Adria**, *Atri*, stood between the rivers Vomanus

Coin of Adria.

The coin belongs to the class commonly known as Æs grave.

and Matrinus, about 5 miles from the coast, on which it possessed a port named Matrinum : it was occupied by a Roman colony in B.C. 282, and was recolonized by Hadrian whose family originally belonged to this place. The coins of Adria are remarkable for their great weight. Great part of the circuit of the walls and other ancient remains exist there. **Auximum**, *Osimo*, the most northerly town in the interior, stood on a lofty hill about 12 miles S.W. of Ancona ; from the strength of its position, it was occupied by Pompey in his wars against Sulla and Cæsar, but it declared in favour of the latter.[7] It did not become a colony until B.C. 157, though it was fortified by the Romans some twenty years earlier. **Urbs Salvia**, *Urbisaglia*, was situated in the upper valley of the Flusor, and was a municipal town. **Asculum**, *Ascoli*, stood on the banks of the Truentus.[8] It bore an important part in the Social War, which commenced in that town. It was hence besieged by Pompeius Strabo, and not reduced till after a long and obstinate defence.

Of the smaller towns we may notice— **Potentia**, at the mouth of the river of the same name, colonized by the Romans in B.C. 184; **Cupra Maritima**, 8 miles N. of the Truentus, the site of an ancient temple of Cupra (Juno), founded by the Etruscans;[9] **Cingulum**, *Cingoli*, W. of Auximum, a place of great strength,[1] noticed in the Civil War between

---

[4] It is noticed by Juvenal :—
　Ante domum Veneris, quam Dorica sustinet Ancon.—iv. 40.

[5] Stat fucare colus nec Sidone villor Ancon
　Murice nec Libyco.　　　　　　SIL. ITAL. viii. 438.

[7] Lucan refers to this in the line—
　Varus, ut admotæ pulsarunt Auximon alæ, &c.　ii. 466

[8] The natural strength of its position was remarkable, and it was further fortified by art :—
　Et inclemens hirsuti signifer Ascli.　　　SIL. ITAL. viii. 440.

[9] Et queis littoreæ fumant altaria Cupræ—　　Id. viii. 434.

[1] 　　　　　Celsis Lablenum Cingula saxa
　Miserunt muris.　　　　　　　　ID. x. 34.

Cæsar and Pompey; Truentum, or Castrum Truentinum, at the mouth
of the Truentus, one of the places occupied by Cæsar in the Civil
Wars; and, lastly, Interamna, *Teramo*, the capital of the Prætutii,
whose name was subsequently applied to the town under the form of
Aprutium, whence the modern name of the province *Abruzzo*.

Roads.—Picenum was reached from Rome by the **Via Salaria**, which
crossed the Apennines to Asculum and thence descended to the
Adriatic. Another road followed the line of coast from Ancona to
Aternum, where it united with the Via Valeria. A third left Ancona
and Auximum for Nuceria, where it fell into the Via Flaminia.

History.—The history of Picenum is unimportant: it was reduced
by the Romans in a single campaign in B.C. 268: it suffered severely
from the ravages of the Second Punic War. The Social War took its
rise in this province in B.C. 90, and led to the siege of Asculum. Cæsar
occupied it at the commencement of the Civil War.

VII. The Sabini, Marsi, Vestini, Marrucini, and Peligni.

§ 9. The country of the **Sabini** was a narrow strip, extending
about 85 miles in length, from the sources of the Nar in the N. to
the junction of the Tiber and Anio in the S. It was bounded on the
N. and W. by the Umbrians and Etruscans; on the N.E. by
Picenum; on the E. by the Vestini, Marsi, and Æquiculi; and on
the S. by Latium. This country is generally rugged and mountainous: but the valleys are fertile, and the sides of the hills and
lower slopes of the mountains are adapted to the growth of the vine
and the olive. The lower valley of the Velinus, about Reate, was
particularly celebrated for its fertility. The country produced large
quantities of oil and wine, though not of the best quality.[1] The
savin, which was used instead of incense,[2] derives its name from the
Sabine hills, where it was found in abundance. The neighbourhood
of Reate was famous for its mules and horses, and the mountains
afforded excellent pasturage for sheep.

§ 10. The Apennines attain their greatest elevation in this part
of their course. A few of the prominent points received special
names, as Tetrica and Sevărus,[3] but it is difficult to identify them.
Of the lesser heights we may notice **Mons Lucretilis**,[4] *Monte Gennaro*,

---

[1] Deprome quadrimum Sabinâ,
    O Thaliarche, merum diotâ.   Hor. *Carm.* i. 9, 7.
    Vile potabis modicis Sabinum
    Cantharis.   *Id.* i. 20, 1.

[2] Ara dabat fumos herbis contenta Sabinis.   Ov. *Fast.* i. 343.

[3] Qui Tetricae horrentes rupes, montemque Severum.—*Æn.* vii. 713.

[4] Horace's villa was situated near it; hence the allusion:—
    Velox amoenum sæpe Lucretilem
    Mutat Lycæo Faunus; et igneam
    Defendit æstatem capellis
    Usque meis, pluviosque ventos.   *Carm.* i. 17, 1.

which rises on the borders of the Roman *Campagna*. The chief
rivers were the **Nar**, the **Tiber**, and the **Anio**. The two former have
been already noticed : the Anio belongs more properly to Latium.
Among the tributaries of these rivers we may specially notice the
Velinus, *Velino*, which rises in the Apennines N. of Interocrea, and
flows in the upper part of its course from N. to S., then to the W.,
and finally to the N.W., discharging itself into the Nar about 8
miles above Interamna. The Tolênus, *Turano*, is a small tributary
of the Velinus, joining it a few miles below Reate. We may also
notice the small stream **Digentia**, *Licenza*, a tributary of the Anio,
on the banks of which Horace had a farm;[5] and the still smaller
Allia, also a tributary of the Anio, and probably to be identified with
the *Scolo del Casale*, 12 miles from Rome, memorable for the defeat
sustained by the Romans from the Gauls under Brennus in B.C. 390.[7]

§ 11. The Sabines were members of a race which was widely
spread throughout Central and Southern Italy, and which may be
divided into three great classes :—the Sabini, with whom we are now
more immediately concerned ; the Sabelli, including the various
lesser tribes of the Vestini, Marsi, &c. ; and the Samnites, who were
the most important of all. The earliest abode of the race appears
to have been about Amiternum, at the foot of the Apennines :
thence they issued in a series of migrations founded on a peculiar
custom called *Ver Sacrum*, which consisted in the dedication of a
whole generation to some god under the pressure of any great calamity. The Sabines were a frugal[8] and hardy race, deeply imbued
with religious feelings, and skilled in augury and magical rites.
They dwelt principally in villages, and the towns were accordingly
very few. Reate ranked as the capital, and Amiternum was a place
of some importance.

Amiternum was situated in the upper valley of the Aternus. We
have already stated that it was the cradle of the Sabine race. It suffered severely in the Social and Civil Wars, but subsequently became a
place of much importance, as the ruins at *San Vittorino* testify. It was
the birth-place of the historian Sallust. **Reate**, *Rieti*, was situated on
the Via Salaria, 48 miles from Rome, and on the banks of the Velinus.
The surrounding district was one of the most fertile and beautiful in
the whole of Italy; the plains that intervened between the town and the

---

[5] Me quotiens reficit gelidus Digentia rivus,
    Quem Mandela bibit, rugosus frigore pagus.   Hor. *Ep.* i. 18, 104.

[7] This disaster is frequently referred to by the Roman poets :—
    Quosque secans Infaustum interluit Allia nomen.—*Æn.* vii. 717.

[8]   Cedant feralia nomina Cannæ
    Et damnata die Romanis Allia festis.   Luc. vii. 408.

[9] Vel Gabiis, vel cum *rigidis* æquosa Sabinis.   Hor. *Ep.* II. 1, 25.
    Translatus subito ad Marsos mensamque Sabellam
    Contentusque Illic veneto duroque cucullo.   Juv. III. 169.

Lacus Velinus were known as the Roseæ Campi,[*] and the valley is termed by Cicero the "Reatine Tempe." The plain was however liable to inundation from the blocking up of the channel of the Velinus, and disputes occurred between Reate and Interamna on this subject. **Nursia**, *Norcia*, was situated in the upper valley of the Nar at a great elevation, and consequently enjoyed a very cold climate.[1] It is noticed in B.C. 205, along with Reate and Amiternum, as aiding Scipio with volunteers. It was also the birth-place of Vespasian's mother. We may further notice—**Falacrinum**, on the Via Salaria, the birth-place of the Emperor Vespasian; **Interocrea**, between Reate and Amiternum, deriving its name from its position between two rugged mountains ; **Cutilia**, between Reate and Interocrea, with a lake in its neighbourhood famed for the phenomenon of a floating isle, and also possessing medicinal springs of great repute, which were visited by Vespasian ; **Cures**, *Correse*, about 3 miles from the Tiber and 24 from Rome, the birth-place of Numa[2] and the city of Tatius, but afterwards a poor decayed village; and **Eretum**, *Grotta Marozza*, about 18 miles from Rome, at the junction of the Via Nomentana with the Via Salaria, and from its position frequently mentioned in connexion with the wars between the Sabines and Romans.

Roads.—The territory of the Sabini was traversed throughout its whole length by the **Via Salaria**, which proceeded from Rome by Reate and Interocrea across the Apennines to Picenum.

History.—The Sabines occupy a prominent place in the early history of Rome. They established themselves on the Quirinal Hill, and became a constituent element in the Roman population. Wars nevertheless ensued between the two nations, and were continued down to B.C. 290, when the Sabines were subdued by M. Curius Dentatus. The most signal event in the course of these wars was the decisive victory gained in B.C. 449 by M. Horatius. They are seldom mentioned after their incorporation with the Roman state.

§ 12. The **Marsi** occupied a mountainous district around the basin of Lake Fucinus, having to the N. of them the Sabines, to the E. the Peligni, and to the W. and S. the Æqui, Hernici, and Volsci. Their territory lies at an elevation of more than 2000 feet above the sea : hence the climate is severe, and ill adapted to the growth of corn ; fruit, however, abounded, and wine of an inferior quality was produced there. In addition to the basin about the lake, the Marsi also possessed the upper valley of the Liris. The **Fucinus Lacus** has been already briefly noticed : we may here add that it is about 29

---

[*] Qui Nomentum urbem, qui rosea rura Velini
Casperiamque colunt.   *Æn.* vii. 712.

[1] Qui Tiberim Fabarimque bibunt, quos *frigida* misit
Nursia.   *Id.* vii. 715.
                    Necnon habitata pruinis
Nursia.   *Sil. Ital.* viii. 418.

[2]   Nosco crines incanaque menta
Regis Romani ; primus qui legibus urbem
Fundabit, Curibus *parvis* et paupere terra
Missus in imperium magnum.   *Æn.* vi. 809.
To Tatius, *parvique* Cures, Caninaque sensit.   Ov. *Fast.* II. 135.

miles in circumference, of oval shape, and so completely shut in by mountains that there was no natural passage for its waters; these were originally carried off by subterranean channels, and the waters were supposed to reappear at the sources of the Aqua Marcia,[3] in the valley of the Anio, though the grounds for such belief are very insufficient. An artificial duct was made with immense labour by the Emperor Claudius, through the solid limestone rock, to the valley of the Liris; and by this means the inundations, to which the country of the Marsi was liable, were for a while checked. The duct is now closed. The Marsi were a Sabellian race, and resembled the Sabines in character. They possessed the art of charming venomous reptiles.[4] Their principal and indeed only town was Marruvium.[5]

Marruvium lay on the E. shore of the Fucine Lake, and evidently derived its name from the Marsi, whose capital it was. Under the Romans it became a flourishing municipal town. Portions of the walls and of an amphitheatre still remain at a spot now named *S. Benedetto*. We may further notice **Lucus Angitiæ**, *Luco*, a place which grew up about the grove and sanctuary of the goddess Angitia, on the W. bank of the lake; and **Cerfennia**, on the Via Valeria, at the foot of the pass (the *Forca di Caruso*) leading across to the valley of the Peligni.

Road.—The Marsian district was traversed by the **Via Valeria**, which was originally constructed from Tibur to the Fucine Lake and Cerfennia, but was afterwards, in the reign of Claudius, carried over Mons Imeus to the valley of the Aternus and the Adriatic.

History.—The Marsi are first noticed in B.C. 340 as being on friendly terms with Rome. In 308, however, they joined the Samnites against the Romans; and in 301 they appear to have undertaken war with them single-handed, and were consequently reduced with ease. At a later period they took a prominent part in the Social, or, as it was more usually termed, the Marsic War; and, even after the other tribes had yielded, they maintained an unequal struggle, which terminated in their complete subjection.

§ 13. The **Vestini** occupied a mountainous tract between the Pyrenees and the Adriatic, bounded by the Matrinus on the N.W., and by the Aternus on the S.E. Within these limits are two distinct regions: the upper valley of the Aternus, a bleak and cold upland tract lying at the back of the *Monte Corno*; and the district that lies between that range and the Adriatic, which, though hilly, enjoys a tolerably fine climate. The mountains were the haunts of wild animals to a late period. The upland pastures were good, and

---

[3] Hence Statius speaks of the aqueduct as—
   Marsasque nives et frigora ducens.   *Silv.* I. 5, 26.

[4]   At Marsica pubes
   Et bellare manu, et chelydris cantare soporem,
   Vipereumque herbis habetare, et carmine dentem.—SIL. ITAL. viii. 497.
See also VIRG. Æn. vii. 750.

[5] Marruvium, veteris celebratum nomine Marri,
   Urbibus est illis caput.   IN. viii. 507.

from them an excellent kind of cheese was produced. The Apennines here attain their greatest elevation in the group now called *Monte Corno*, which may perhaps represent the *Mons Fiscellus* of the ancients. The only river worthy of notice is the *Aternus*, *Pescara*, which rises near Amiternum, and in its upper course flows from N. to S. through a broad valley, some 2000 feet above the sea, and, after passing through a gorge between two masses of mountains, descends in a N.E. direction to the sea. The inhabitants of this district were a Sabellian race, and participated in the Sabine character. Their chief towns were Pinna in the interior, and Aternum on the sea-coast.

Pinna, *Penne*, was situated on the E. slope of the Apennines, about 15 miles from the sea. The only historical notice of it is in the Social War, when it stood firm to the Roman allegiance. **Aternum,** *Pescara,* stood at the mouth of the Aternus, and was a place of considerable trade. It joined the cause of Hannibal, and was consequently besieged and taken by the Romans in B.C. 213. It afterwards became a municipium, and its port was improved by the Emperor Tiberius.

History.—The Vestini are first mentioned in B.C. 324, when they joined the Samnites against Rome; they were defeated by the consul D. Junius Brutus. In 301 they concluded a treaty with the Romans. They joined in the Social War, and were again conquered by Pompeius Strabo in 89. They were generally in league with the Marrucini and Peligni, and the histories of all these tribes are almost identical.

§ 14. The **Marrucini** occupied a narrow strip of territory on the S. bank of the Aternus, between the Adriatic and the Apennines. On the W. they adjoined the Peligni, from whom they were separated by the lofty ranges now named *Majella* and *Morrone*; on the S. the *Foro*, 7 miles from the Aternus, appears to have been their boundary on the side of the Frentani. Their district was fertile, and produced corn, wine, oil, and especially excellent fruit and vegetables. It appears to have been subject to earthquakes.[6] The people were a Sabellian race, and their name is only another form of Marsi. The only town of consequence was the capital, Teate.

Teate, *Chieti*, was situated on a hill about 3 miles from the Aternus, and 8 from the Adriatic. Though the capital of the district, and described by Silius Italicus[7] as the "great" and "illustrious," it is not mentioned in history. It was the native place of Asinius Pollio the orator.

§ 15. The **Peligni** occupied a small inland district in the very heart of the Apennines, between the Marrucini on the E., the Marsi

---

6   Procul ista tuis sint fata Teate
    Nec Marrucinos agat hæc insania montes.   Stat. *Sil.* iv. 4, 85.

7   Marrucina simul Frentanis æmula pubes
    Corfini populos, magnumque Teate trahebat.   Sil. *Ital.* viii. 521.
              Cui nobile nomen
    Marrucina domus, clarumque Teate ferebat.   Id. xvii. 453.

on the W., and the Vestini on the N. Their district consisted of
the valley of the *Gizio*, which runs northwards into the Aternus;
in this direction alone did it lie open; elsewhere it was surrounded
on all sides by lofty mountains. The climate was proverbially
severe[a] from the elevation of the land; still the valley of the *Gizio*
was sufficiently fertile in corn and wine, and even produced the olive
in some places. The people were a Sabellian race, and resembled the
other branches of that race in character. They possessed three principal towns: Corfinium, Sulmo, and Superæqueum.

Corfinium was situated in the valley of the Aternus, near the point
where that river makes its great bend to the E. It is not noticed
earlier than in the Social War, B.C. 90, when its position led to its
being selected by the allied nations as the site of their capital. It was
occupied by L. Domitius in the Civil War between Cæsar and Pompey,
and held out for a time against the former. The ruins of the city are
found at *S. Pelino*. Sulmo, *Sulmona*, stood seven miles S. of Corfinium, in the valley of the *Gizio*, and is chiefly celebrated for its having
been the birthplace of Ovid.[b] It is noticed in B.C. 211, as suffering
from the ravages of Hannibal's army; and, like Corfinium, it was occupied by L. Domitius in the Civil War. Superæqueum stood on the
right bank of the Aternus, about four miles from the Via Valeria: it
was a municipal town, but without any historical interest: the name
Subequo still attaches to its site.

Roads.—The territory of the Peligni was centrally situated in reference to the lines of communication of Central Italy. The Via Valeria
traversed it between the Marsi and Marrucini, entering the district by
the pass of Imsus, and leaving it by the gorge of the Aternus. In
another direction the valley of the Aternus opened a natural route to
Reate and the valley of the Tiber; and in the opposite direction a
practicable pass crossed the Apennines into the valley of the Sagrus.

VIII. SAMNIUM, WITH THE FRENTANI.

§ 10. Samnium was an extensive district in the centre of Italy,
bounded on the N. by the Marsi, Peligni, and Marrucini; on the
W. by Latium and Campania; on the S. by Lucania; and on the
E. by the Frentani and Apulia. The whole of this district is of a
mountainous character, and is broken up by lofty ranges emanating
from the Apennines, which in this part of their course cease to be a
regular chain, and resolve themselves into distinct and broken
masses. The most important of these masses, now named *Monte
Matese*, lies S.W. of Bovianum, and separates the basins of the
Tifernus and Vulturnus: a portion of it, containing the sources of
the former river, was named **Mons Tifernus**. The next most impor-

---

[a] Quæ præbente domum, et quota,
   Pelignis curream frigoribus, tacce   Hor. *Carm.* iii. 19, 7.
[b] Sulmo mihi patria est, gelidis uberrimus undis.—*Trist.* iv. 10, 3.
   Pars me Sulmo tenet, Peligni tertia ruris;
   Parva, sed irriguis ora salubris aquis.   *Am.* ii. 16, 1.

tant group was that named Mons Taburnus,[1] separated from *Matese*
by the valley of the Calor, and forming the boundary of the Campanian plain : the W. extremity of this ridge is the Mons Tifata, so
celebrated in the campaigns of Hannibal.   Several chains strike out
on the E. side of the Apennines, forming distinct and parallel
valleys, through which the rivers seek the Adriatic.   On the W.
side there are two extensive valleys—the northern one, in which
the Vulturnus flows in a direction from N.W. to S.E. ; the southern,
in which its tributary, the Calor, flows in an opposite direction,
having its upper course in an extensive basin lying at the back of
the groups of *Matese* and Taburnus.   As Samnium thus includes
the whole breadth of the Apennines, the rivers which belong to it
seek both the Adriatic and the Mediterranean Seas.   In the former
direction run the Sagrus, *Sangro*, which rises S. of the Fucino lake,
and flows through a broad upland valley by the walls of Aufidena ;
the Trinius, *Trigno* ; the Tifernus, *Biferno*, which rises near Do-
vianum in *Monte Matese* ; the Frento, *Fortore* ; and, lastly, the
Aufidus, *Ofanto*, in the extreme S.   In the latter direction runs
the Vulturnus, *Volturno*, which rises about five miles S. of Aufidena,
and pursues a S.E. course until its junction with the Calor, *Calore*,
which rises on the borders of Lucania and flows by Beneventum,
receiving in its course the tributary waters of the Sabatus and
Tamarus.

§ 17.   The country we are now describing was originally held by
the Opicans, or Oscans.   The Samnites were a Sabine race, who
entered as an invading host and conquered the Opicans, coalescing
with them afterwards, and adopting their language.   They were
divided into four tribes, the most important of which were the
Caudini and Pentri, who lived respectively S. and N. of the *Matese*,
while the less important were the Caracēni, in the valley of the
Sagrus, and the Hirpini, in the upper valleys of the Calor and its
tributaries.   The Samnites were a brave and frugal race, leading a
rude, pastoral life, and superstitious.   They lived for the most part
in villages, but they possessed some towns—as Æsernia and Bovianum,—which were strongly fortified.   These, and all the Samnite
towns, were utterly destroyed by Sulla after the Marian War ; nor
did any of them, although supplied with colonists from Rome, rise

---

[1] This mountain forms a very conspicuous object from the Campanian plain :
its upper regions are described by Virgil as being clothed with forests, while on
its lower slopes the olive flourished :—

Ac velut ingenti Sila, summove Taburno
Cum duo conversis inimica in proelia tauri
Frontibus incurrunt.   *Æn.* xii. 715.

Neu segnes jaceant terrae.   Juvat Ismara Baccho
Conserere, atque olea magnum vestire Taburnum.—*Georg.* ll. 37.

again to importance, with the exception of Beneventum, which was centrally situated on the Via Appia.

Æsernia, *Isernia*, was situated on a tributary of the Vulturnus, in the upper valley of that river. It was captured by the Romans in B.C.

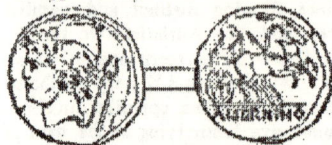

Coin of Æsernia.

293, and was colonized by them in 264. After its destruction by Sulla, colonies were sent to it by Cæsar, Augustus, and Nero; and it became a municipal town of importance in the time of Trajan and the Antonines: there are remains of an aqueduct and of a fine bridge of this period. Bovianum, *Bojano*, was situated close to the sources of the Tifernus, amidst lofty mountains. It was the capital of the Pentri, and hence figures in the Second Samnite War. It was besieged without success in B.C. 314, but was taken in 311, again in 305, and a third time in 298. In the Social War, it became the head-quarters of the allies after the fall of Corfinium; it never recovered its destruction by Sulla. Some portions of its ancient walls, of a very massive order, are still visible. Beneventum, *Benevento*, was situated on the banks of the

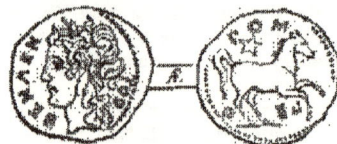

Coin of Beneventum.

Calor, and on the Via Appia.[1] It was a very ancient town, and its foundation was attributed to Diomedes. Its original name was Maleventum, which the Romans deemed of ill omen, and therefore changed it to Beneventum, in B.C. 268, when they planted a colony there. Its strength and the centrality of its position lead to frequent notices of it. Several colonies were sent there by the Roman emperors, and it was visited by Nero, Trajan, and Septimius Severus. A triumphal arch in honour of Trajan still remains. Caudium, the capital of the Caudini, stood on the Via Appia between Beneventum and Capua. It is noticed in the history of the Samnite Wars, and is particularly memorable for the disastrous defeat of the Romans in B.C. 321, which took place at a pass called Furculæ Caudinæ, "the Caudine Forks," the position of which is near *Arpaja*, between *Sta. Agata* and *Moirano*.

Of the less important towns we may notice—Aufidena, *Alfidena*, the capital of the Caraceni, in the upper valley of the Sagrus, a fortress of great strength; Allifæ, *Alifa*, in the valley of the Vulturnus, on the borders of Campania, the scene of several military events, and a place of importance under the empire; Calatia, *Caiazzo*, about a mile N. of the Vulturnus, and ten miles N.E. of Capua, the town at which the Romans were encamped before their disaster at the Caudine Forks; Saticula,[2] S. of the Vulturnus, and probably in the valley at the back

---

[1] Hence the well-known notice in Horace in his journey to Brundusium :—
Tendimus hinc recta Beneventum, &c. *Sat.* i. 5, 71.

[2] Virgil adopts the ethnic form *Saticulus* for Saticulanus :—
Accola Volturni, pariterque Saticulus asper. *Æn.* vii. 729.

of Mount Tifata, besieged and taken by the Romans in B.C. 315; **Equus Tuticus**, S. *Eleuterio*, in the district of the Hirpini, on the Via Trajana; **Trivicum**, *Trevico*, on the Via Appia,[4] but not on the line of road followed in later times; **Romulea**,[5] on the same road at *Bisaccia*, noticed as a large town at the time of its capture by the Romans in B.C. 297, but not mentioned subsequently; **Compsa**, *Conza*, on the borders of Lucania, the place where Hannibal deposited his baggage in B.C. 218, and subsequently taken by the Romans in 214; and, lastly, **Abellinum**, *Avellino*, near the Campanian frontier, a place of wealth and importance under the Empire.

Roads.—Samnium was traversed by several high-roads. The **Via Appia** entered it from Capua, and passed through the S. part of the province, by Beneventum and the valley of the Calor, to Venusia in Apulia. A branch-road struck off from this at Beneventum, which joined the Via Egnatia at Æcæ in Apulia: this was named **Via Trajana**, having been constructed by the Emperor Trajan. Another road, also starting from Beneventum, followed the valley of the Vulturnus to Venafrum and Æsernia, whence it crossed the ridge to Aufidena, in the valley of the Sagrus. Another crossed from Æsernia to Bovianum, and thence followed the valley of the Tifernus in one direction; and in another crossed to Equus Tuticus, where it fell into the Via Trajana.

History.—The Samnites are first noticed in B.C. 354, as concluding a treaty with Rome. Subsequently war broke out between the two peoples, in consequence of the Samnite invasion of Campania. These wars continued, with a few interruptions, for fifty-three years (from 343 to 290), when the Samnites were completely subdued. They joined the allies in the Social War in 90, and continued the struggle after the others had given way. In the Civil War between Sulla and Marius they again broke out; but they were defeated by Sulla, in 82, before the gates of Rome, and suffered severely from his revenge, the whole country being reduced to a state of utter desolation, from which it never recovered.

§ 18. The **Frentani** occupied a maritime district between Samnium and the Adriatic Sea, from the border of the Marrucini in the N.W. to Apulia in the S.E., from which it was separated by the Tifernus. It is for the most part hilly, but fertile, and well watered

---

[4] Incipit ex illo montes Appulia notos
Ostentare mihi, quos torret Atabulus; et quos
Nunquam erepsemus, nisi nos vicina Trivici
Villa recepisset.   Hor. Sat. L 5, 77.

[5] Between this and Beneventum lie the valley and lake of Ampsanctus, which Virgil describes. The spot is now named *Le Mofete*, and the sulphureous vapours are remarkably strong. The woods which formerly surrounded it have been cut down.

Est locus Italiae medio sub montibus altis
Nobilis, et fama multis memoratus in oris,
Amsancti valles: densis hunc frondibus atrum
Urget utrimque latus nemoris, medioque fragosus
Dat sonitum saxis et torto vertice torrens:
Hic specus horrendum, saevi spiracula Ditis
Monstrantur, ruptoque ingens Acheronte vorago
Pestiferas aperit fauces; quis condita Erinnys,
Invisum numen, terras coelumque levabat.   Æn. vii. 503.

by the lower courses of the rivers **Tifernus, Trinius,** and other streams which take their rise in the mountains of Samnium. The Frentani were a Samnite race. The towns of importance on the sea-coast were **Ortona,** *Ortona,* **Histonium,** and **Buca,** probably at *Termoli,* none of which have any historical associations: Histonium appears to have ranked as the capital under the Roman empire; there are extensive remains of it at *Il Vasto.* **Anxanum,** *Lanciano,* in the interior, may also be noticed as a municipal town of some size.

History.—The Frentani are first noticed in B.C. 319, when they were at war with Rome, and were speedily reduced. In 304 they concluded peace with the Romans, and they remained faithful to them, even after the battle of Cannæ. They joined in the Social War without taking any prominent part in it.

Beneventum.

Alban Hills and Remains of Roman Aqueduct.

CHAPTER XXVI.

ITALY—*continued.* LATIUM.

IX. LATIUM. § 1. Boundaries, and General Description. § 2. Mountains. § 3. Rivers. § 4. Inhabitants. § 5. Rome. § 6. Remaining Towns of Latium. Roads. Islands. History.

IX. LATIUM.

§ 1. In fixing the boundaries of Latium,[1] care must be taken to distinguish between Latium in the original and *historical* sense, and Latium in its later *geographical* sense. The former was a small country, bounded on the N. by the Tiber and the Anio (with the exception of a small district N. of the Anio, at the confluence of these rivers, which was included in Latium) ; on the E. by the lower ranges of the Apennines, a little E. of Tibur and Præneste ; on the S. by a line drawn from the latter town to the promontory of Circeii ; and on the W. by the Tyrrhenian Sea. The latter comprehended, in addition to the territory just described, the districts of the Æqui and Hernici in the E., and the Volsci and Aurunci in the

---

[1] The origin of the name "Latium" is unknown : the Romans themselves connected it with *lateo* because Saturn had there *latu hid* from Jupiter :—

Composuit legesque dedit, Latiumque vocari

Maluit, his quoniam *latuisset* tutus in oris.   Æn. viii. 322.

The name is undoubtedly connected with Lavinium and Lavinus, and probably the oldest form was Latvinus. It should be observed that the name Latium was derived from Latini, and not *vice versâ.*

S., so that it bordered in the former direction on Samnium, and in the
latter on Campania, the point of separation being just S. of Sinuessa.
The greater portion of Latium consists of a broad undulating plain,
now called the *Campagna*, extending from the sea to the advanced
ridges of the Apennines, and interrupted only by the isolated group
of the Alban hills : this plain, though apparently level, is intersected
by ravines which the streams have worn, for themselves, and which
generally have rugged, precipitous sides, particularly in the E. portion of it. The eastern part of Latium, occupied by the Æqui and
Hernici, is hilly ; and the southern district again, occupied by the
Volsci, is intersected by an extensive range, similar in character
to the Apennines, but separated from them by the valleys of the
Trerus and Liris. The districts vary in regard to the fertility of
the soil : the *Campagna* and the Alban hills are of volcanic origin ;
the former, though at present utterly desolate, was well cultivated
in ancient times, and produced considerable quantities of corn. The
slopes of the hills have been in all ages well adapted to the growth
of the vine, the olive, and other fruit-trees ; and among the special
products of the country, we may specify the wine of the Alban hills,[1]
the figs of Tusculum, the hazel-nuts of Præneste, and the pears of
Crustumerium and Tibur.

§ 2. Of all the hills of Latium the most important and conspicuous
is the group of the Alban hills, the central height of which is the
Albanus Mons[2] of the ancients and the *Monte Cavo* of modern times.
The name does not appear to have been extended to the general group,
though modern usage has effected this. The Alban hills are a nearly
circular mass, about 40 miles in circumference, of volcanic origin, and
forming apparently at one time a single great crater, the edge of which
has been broken up into numerous summits, while from the lower
slopes numerous spurs project into the plain, affording admirable sites
for towns. The summit of Albanus Mons was crowned with the
temple of Jupiter Latiaris, in which the Latins held their congress.
In the N.E. quarter **Algidus**[4] was a name applied either to a

[1] Horace classes it with the Falernian :—
, Ille horras, Albanum, Mæcenas, sive Falernum,
Te magis appositis delectat ; haberaus utramque.—*Sat.* II. 8, 16.

[2] This summit commands a magnificent view of the *Campagna* ; hence Virgil
represents Juno as observing from this point the combat between the Trojans and
Latins :—
At Juno ex summo, qui nunc Albanus habetur,
Prospiciens tumulo, campum spectabat.		*Æn.* xII. 134.

[4] The sides of this hill were covered, in the time of Horace, with dense
forests :—
Nam, quæ nivali pascitur Algido
Devota, quercus inter et ilices,
Aut crescit Albanis in herbis
Victima.		*Carm.* III. 23, 9.

single summit or to that portion of the group; the plain which intervenes between it and Tusculum was the scene of frequent engagements between the Romans and the Æquians. The Volscian hills, now known as the *Monti Lepini*, received no special name in ancient times. They rise immediately S. of the Pontine Marshes, and fill up the whole intervening space (from 12 to 16 miles in breadth) between them and the valley of the Trerus; they descend to the coast between Tarracina and the Liris, and form a succession of headlands. We must also notice the small **Mons Sacer**[5] which overlooks the Anio at a distance of about 3 miles from Rome, and is memorable as the spot whither the Plebeians seceded in B.C. 494 and 449[6].

§ 3. The chief river in Latium is the **Tiber**[7], the lower course of which falls within the limits of this province; about 2 miles above Rome it receives an important tributary in the **Anio**[8], *Teverone*, which rises in the Apennines near Treba, and descends rapidly through the Æquian hills to Tibur, where it forms a remarkable waterfall[9], and

Doris ut ilex tonsa bipennibus
Nigræ feraci frondis in Algido.   *Carm*. iv. 4, 57.

At a later period the wealthy Romans had villas there, and its character was changed :—

Nec Tusculanos Algidosve secessus
Prænestis nec sic Antiumve miratur.   *Mart*. x. 30.

                    nec amœno retinentant
Algida.   Sil. Ital. xii. 536.

[5] The name is derived from the *Lex Sacrata* passed there in B.C. 494.

[6] Plebs vetus et nullis etiam nunc tuta Tribunis,
Fugit; et in Sacri vertice montis abit.   Ov. *Fast*. iii. 663.

[7] The yellow hue and turbid character of its stream are frequently noticed by the poets :—

Vidimus *flavum* Tiberim, retortis
Littore Etrusco violenter undis.   Hor. *Carm*. i. 2, 13.

In fluvium dedit : ille suo cum gurgite *flavo*
Accepit venientem ac mollibus extulit undis.   *Æn*. ix. 816.

            Hunc inter fluvio Tiberinus amœno
Verticibus rapidis et multa flavus arena
In mare prorumpit.   *Id*. vii. 30.

The river is frequently called Albula by the Roman poets, from a tradition that such was its earliest name, its later designation having been derived from a king named Tibris, according to Virgil (*Æn*. viii. 330), or from an Alban king, Tiberinus, according to Livy (i. 3).

[8] The oblique cases of this name come from a more ancient form, Anien, which is itself used by some of the later poets (Stat. *Silv*. i. 3, 20).

[9] The present cascade is artificial, having been constructed in 1834; but there was always a considerable fall there, as the subjoined passages imply :—

Et *præceps* Anio.   Hor. *Carm*. i. 7, 13.

Et *cadit* in patulos Nymphæ Aniena lacus.   Propert. iii. 16, 4.

Aut ingens in stagna *cadit* vitreæque natatu
Plaudit aqua.   Stat. *Silv*. i. 3, 73.

It appears from the last two passages that the fall was broken towards its lower part by projecting ledges, which caused it to form small pools.

thence pursues a winding course through the *Campagna*; its water was very jaire and it was one of the sources whence Rome drew its supply. The Liris, *Garigliano* (p. 489), is the chief river in the southern district; it receives the Trerus, *Sacco*, from the neighbourhood of Præneste, a stream which, though itself important and flowing through a wide valley, is unnoticed by the historians and poets of ancient times. Of the lesser streams which crossed the plain, we may notice the Numicius[1], *Rio Torto*, on the banks of which Æneas was burial; the Astūra[2], or Storas, which rises at the foot of the Alban hills, and on the banks of which was fought the last great battle between the Romans and Latins in B.C. 338; the Amasēnus[3], *Amaseno*, which rises in the Volscian hills and descends through the Pontine Marshes to the sea near Tarracina; and the Ufens, *Ufente*, a sluggish stream which now joins the Amisenus in the Pontine Marshes[4]. There were numerous small lakes in Latium, the chief of which was Albānus Lacus, *Lago di Albano*, beneath the mountain of the same name, 6 miles in circumference, undoubtedly occupying the crater of an extinct volcano, and so entirely surrounded by mountains that there was no natural outlet for the surplus waters; these were carried off by an artificial emissary pierced through the solid rock, constructed in B.C. 397 and still existing, which conducts the waters by a stream named the *Rivo Albano* to the Tiber. We may also notice L. Nemorensis, *Lago di Nemi*, near Aricia, also a volcanic crater, of small size but remarkable for its picturesque appearance, and famed in antiquity for the sanctuary of Diana (Nemus Dianæ), to which it owed its name; and L. Regillus, at the foot of the Tusculan hills, the scene of the great battle between the Romans and

---

[1] He was here worshipped under the title of Jupiter Indiges :—
  Ille sanctus eris, quum te veneranda Numici
    Unda Deum cœlo miserit Indigetem.   TIBULL. ll. 5, 42.
There was also on its banks a grove sacred to the nymph Anna Perenna :—
  Corniger hanc cupidis rapuisse Numicius undis
  Creditur, et stagnis occuluisse suis
  .   .   .   .
  Ipsa loqui visa est, placidi sum nympha Numici;
  Amne perenne latens Anna Perenna vocor.—Ov. *Fast.* lll. 647.
[2] At its mouth was a small islet, now converted into a peninsula by an artificial causeway: it was a favourite residence of the Romans and, among others, of Cicero.
[3] Virgil describes it as swollen to a large stream in his account of the escape of Metabus:
  Ecce, fugæ medio, summis Amasenus abundans
  Spumabat ripis; tantus se nubibus imber
  Ruperat.   *Æn.* xI. 547.
[4] Et quos pestifera Pomptini uligine campi,
  Qua Saturæ nebulosa palus restagnat, et atro
  Liventes cœno per squalida turbidus arva
  Cogit aquas Ufens, atque influit æquora limo.   SIL. ITAL. viii. 381.

Latins in B.C. 496; it probably occupied a small crater at *Cornufelle* which has since been drained of its waters. The **Pomptinæ Paludes** form an important feature in the S. of Latium; they occupy an extensive tract between the Volscian mountains and the sea, about 30 miles in length by 7 or 8 in breadth, and are the results of a considerable depression of the land, in which the waters of the Amasenus and other streams stagnate. The Via Appia was carried across them in B.C. 312, and a canal formed by its side between Forum Appii and Tarracina. Fruitless attempts were made to drain the marshes by Cornelius Cethegus in 160, and subsequently by Cæsar, Augustus, and Trajan.

§ 4. The inhabitants of Latium consisted of several distinct peoples. The **Latini**[5] occupied Latium proper; the limits of their territory on the side of the Volscians were fluctuating; on the one hand several towns in the Volscian mountains, as Velitræ, Cora, Norba and Setia, belonged to the Latins, and on the other hand Antium belonged to the Volscians. The **Volsci** spread over the greater part of the southern district from the seacoast to the borders of Samnium; they thus held the Pontine Marshes, the Volscian hills (*Monti Lepini*), and the valley of the Liris. The **Aurunci** were a petty nation on the left bank of the Liris and on the borders of Campania[6]; and the **Ausones**, who were originally identical with the Aurunci[7], lived in later times on the right bank of the Liris between the sea and the Volscian mountains. The **Hernici**[8] held the upper valley of the Trerus, and the hill country adjacent to it. The **Æqui** occupied the mountainous district in the upper valley of the Anio, between the Sabini on the W. and the Marsi on the E. The towns of Latium were numerous and remarkable for the natural strength of their position, furnishing a complete illustration of Virgil's line:

"Tot congesta manu præruptis oppida saxis."[9]

---

[5] The origin of the term "Prisci Latini," which occurs in Roman history subsequent to the fall of Alba, is uncertain: perhaps it represented a league of a portion of the Latin cities formed at that time, who set themselves up as the "old Latins."

[6] Their capital, Aurunca, stood about five miles N. of Suessa, on a spur of *Monte di Sta. Croce*: to this Virgil alludes:

Mille rapit populos: vertunt felicia Baccho
Massica qui rastris; et quos de *collibus altis*
Aurunci misere patres.   *Æn.* vii. 725.

[7] The names are in fact the same, the *r* being changed into *s*, as is common in Latin. The distinction between the two tribes first appears in the 4th cent. B.C. The name is probably derived from the same root as Oscus.

[8] The name is said to have been derived from a Sabine word, *herna* "a rock:" if so, it was truly appropriate to the district which the Hernicans occupied, which Virgil describes as,—

Roscida rivis
Hernica saxa.   *Æn.* vii. 683.

[9] *Georg.* ii. 156.

Not only do the Alban hills abound with sites of remarkable strength
overlooking the plain from a great height, but the *Campagna* itself,
furrowed as it is with deep channels formed by the streams in the
tufo rock, afforded admirable positions for ancient towns. These natural advantages were improved by art, and walls of great strength
in the Cyclopean style were erected on the brows of the cliffs, specimens of which remain to this day at Signia, Cora, and other places.
The Latins possessed a confederacy of thirty towns, at the head of
which stood Alba. The brilliant period of the Latin towns generally
was anterior to the rise of the supremacy of Rome. They subsequently became little else than suburbs of the great metropolis, and
derived their prosperity from the patronage of the wealthy Romans
who erected their villas wherever the scenery or the fine air attracted
them. The towns on the Appian Way, however, retained some importance as places of trade.

§ 5. **Rome**, the metropolis not only of Italy but of the ancient
world, was situated on the Tiber, about 15 miles from its mouth. The
chief part of the town lay on the left bank, where the ground is
broken by a group of hills, and the river winds about with a treble
curve. Of the seven hills which formed the site of the city, three are
isolated, and the other four connected at their bases. Of the former
the Capitoline stands about 300 paces from the river at its most
easterly point, and is the hill to which all the others seem to point;
it is of a saddle-back shape, depressed in the centre and rising towards
its N. and S. extremities. To the S.E. lies the Palatine, a little inferior in point of height, and of a lozenge shape; and still more to the
S. is the Aventine, closely bordering on the Tiber. The four connected
hills[1] are, from S. to N., the Caelian, the largest of the whole group,
lying opposite the Aventine; the Esquiline, which divides at its extremity into two tongues, named Cispius and Oppius; the Viminal,
a small hill almost enclosed between the Esquiline and Quirinal; and
the Quirinal, which curves round in a hooked shape towards the
Esquiline. Still further to the N., but outside the walls, is the Pincian hill, while on the opposite side of the Tiber lie the Janiculan, a
ridge which runs in a direct line between the two curves of the Tiber,
and the Vatican yet more to the N. Rome is said to have been
founded in B.C. 753; the original city of Romulus stood on the Palatine[2], while a Sabine town occupied the Quirinal and Capitoline, and
Etruscans were settled on the Caelian and Esquiline. The Sabine

---

[1] This part of Rome has been compared to the back of a man's hand when
slightly bent and held with the fingers open, the latter representing the Esquiline, Quirinal, and Viminal (Arnold's *Rome*, i. 31).

[2] Inde petens dextram, porta est, ait, ista Palati;
Ille Stator, hoc primum condita Roma loco est.—Ov. *Trist.* iii. 1, 31.

and Roman towns were incorporated in the reign of Romulus, and the Etruscans were removed from their settlement to the plain between the Cælian and Esquiline. Ancus Martius added the Aventine, and built a fortress on the Janiculan. Tarquinius Priscus drained the low ground between the Palatine and Capitol, and planned the Circus Maximus and Forum. Finally Servius Tullius added the Viminal and Esquiline, and surrounded the seven hills with walls extending about 7 miles in circumference. In course of time the city outgrew these limits, and in the reign of Vespasian reached a circumference of 13 miles, at which period it is computed to have contained a population of nearly two millions. Subsequently its size was somewhat diminished, the walls of Aurelian having a circumference of only 11 miles. The

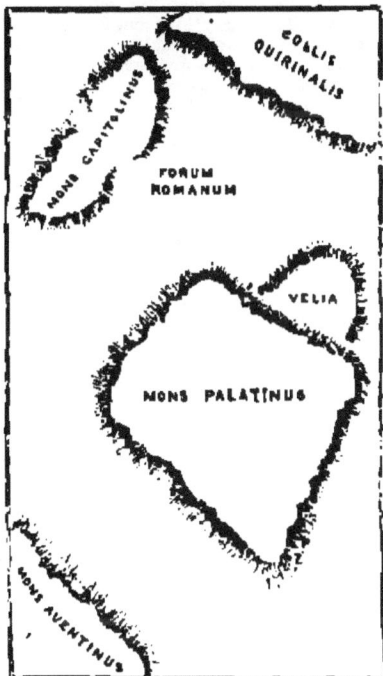

Plan of the City of Romulus.

general appearance of the city was for a long period but poor; after its destruction by the Gauls in B.C. 890, it was rebuilt in haste with narrow crooked streets, and these remained down to the time of Nero, when two-thirds of the town were burnt down (A.D. 64), and were rebuilt with wide and regular streets. The houses were of two classes, called *domus* and *insulæ*, the former being the private houses of the wealthy, the latter the residences of the middle and lower classes, who occupied *flats* or portions of houses, which were carried to the unsafe height of 60 or 70 feet. There were 46,602 of the latter, and 1,790 of the former.

I. *Divisions of the City.*—Servius Tullius divided the town into four regions—Suburana, Esquilina, Collina, and Palatina—corresponding to the number of the city tribes: these were subdivided into 27 Sacella

Argæorum. This division held good until the time of Augustus, who
rearranged the whole city in fourteen Regions, named as follows:—
(1) Porta Capena; (2) Cælimontium; (3) Isis et Serapis; (4) Via
Sacra; (5) Esquiliæ cum Viminali; (6) Alta Semita; (7) Via Lata;
(8) Forum Romanum; (9) Circus Flaminius; (10) Palatium; (11) Circus
Maximus; (12) Piscina Publica; (13) Aventinus; (14) Trans Tiberim.
The localities of these divisions are in several instances pointed out
by the names which correspond to those of the hills and well-known
quarters of the city: it will suffice to add that Isis and Serapis was
at the back of the Esquiline, Alta Semita on the Quirinal and Pincian,
Via Lata on the E. of the Campus Martius, and Piscina Publica S. of
the Aventine.

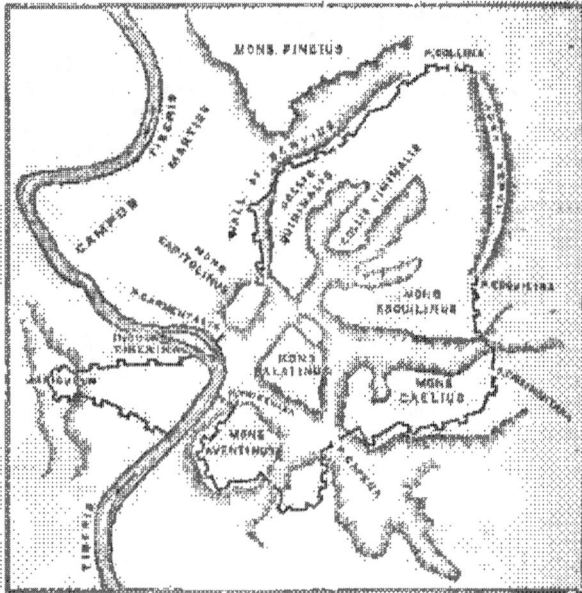

- Map of Rome, showing the Servian Walls and the Seven Hills.

II. *Walls and Gates.*—The Wall of Servius, which was built of stone,
surrounded the whole city, with the exception of the Capitoline Hill
and the portion adjacent to the Tiber, which were both defended by
nature. On the E. side of the town a portion of the *agger* still remains
at the back of the Esquiline and Quirinal hills.[3] In other directions

---

* Recurring to the comparison already made (note ¹), the position of the walls
of Servius would be represented by a line drawn across the *knuckles*; those of
Aurelian by a line drawn across the *wrist.*

its course may be traced by means of the gates, of which no less than twenty are enumerated, the most important being the Porta Collina, at the N. extremity of the Quirinal; Ratumēna, beneath the N. point of the Capitoline Hill; Carmentalis, at the S. foot of the Capitoline; Trigemina, near the Tiber at the foot of the Aventine; Capéna, at the foot of the Cælian; Cælimontāna, on the Cælian; Esquilina and Viminalis, at the back of the hills of the same name. These gates remained to a late period, but the wall fell into decay, nor was there any necessity to rebuild it until the German hordes threatened the city. Aurelian commenced a new wall in A.D. 271, which was completed by Probus and repaired by Honorius: it is substantially the same as now exists. It enclosed a much larger area than that of Servius, including the Piucian Hill and the Campus Martius on the N., the Janiculum on the W. of the Tiber, and a considerable district S. of the Aventine, and at the back of the Esquiline and Quirinal. It had 14 principal and several lesser gates.

Temple of Jupiter Capitolinus restored.

III. *The Capitol.*—The Capitoline Hill rose to a double summit at its N.E. and S.W. extremities, as already noticed. On the former probably stood the Temple of Capitolina Jupiter, founded by Tarquinius Priscus; the Temple of Jupiter Feretrius,[4] in which the *spolia opima* were dedicated; and a Temple of Fides. On the S.W. summit stood the Arx; the Temple of Jupiter Tonans,[5] erected by Augustus; the Temple of Juno Moneta, erected by Camillus in 345 and used as a public mint; and the Temple of Honos and Virtus, built by C. Marius. Between the two summits lay the Asylum of Romulus: this name was afterwards transferred to a spot on the N.E. summit. The Rupes

[4] Nunc spolia in templo tria condita : causa Feretri
    Omine quod certo dux ferit ense ducem.
    Seu quia victa suis humeris huc arma ferebant
    Hulo Feretri dicta est ara superba Jovis.   PROPERT. iv. 10, 45.

[5]   O magne qui mœnia prospicis urbis
    Tarpeia de rupe Tonans !   LUC. i. 195.

Tarpeia[4] was probably on the E. side, facing the Forum, though the name *Rupe Tarpea* is now assigned to a cliff on the W. side.

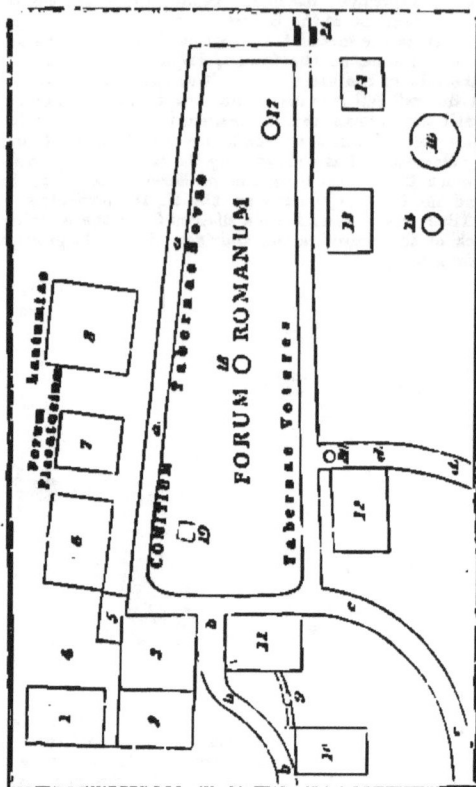

Plan of the Forum during the Republic.

The Forum and its Environs.—The Forum, the great centre of Roman life and business, was situated in a deep hollow between the Capitoline and Palatine hills. It was of an oblong shape, 671 feet long, and diminishing in breadth from 202 feet at the W. end to 117 at the E. It was bounded on the N. by the Via Sacra,[7] (see Plan, *aa*)

[5] From this criminals were executed by being hurled down :—

Tunc Hyrī, Damæ, aut Dionysi filius, audes
Dejicere e saxo cives, aut tradere Cadmo.   Hor. Sat. i. 6, 38.

[7] The Via Sacra was the route by which the processions of victorious generals ascended to the Capitol ; the name was more particularly applied to a portion of

which led from the Colosseum to the Capitoline. Two parallel streets led out towards the S., the Vicus Jugarius (Plan, *cc*) from its W. end, and the Vicus Tuscus (Plan, *dd*), the best shopping street in Rome, from the centre. The Comitium, where public business was transacted, occupied the E. end of the Forum. The Forum was surrounded with porticoes and shops, those on the N. side being named Tabernæ Novæ, and those on the S. side Tabernæ Veteres.

The Forum itself contained the following buildings and objects:—the Rostra (Plan, 19), or stage, in front of the Curia, and so named from the *beaks* of the vessels taken from the Antiates in 537, with which it was adorned; the Lacus Curtius (Plan, 18) in the very centre of the Forum, which was drained by Tarquinius Priscus, the site of it being subsequently marked by a depression; the Jani, the chief resort of the money-lenders, in front of the Basilica Æmilia on the N. side; the Tribunal of the Prætor, at the E. end of the Forum; the Puteal Libonis (Plan, 17), near it, so called from the resemblance it bore to the top of a well; the Temple of Divus Julius, erected on the spot where the

Puteal Libonis or Scribonianum.

body of Cæsar was burnt, also at the E. end of the Forum; the Rostra Julia, in front of it; the Milliarium Aureum, or gilt mile-

---

the street which formed the ascent of the Velia, and which was otherwise called " Sacer Clivus " :—

Quandoque trahet feroces
Per sacrum clivum, merita decorus
    Fronde Sicambros.　　　　　　Hor. *Carm.* iv, 2, 34.

Intactus aut Britannus ut descenderet
    Sacra catenatus *Via.*　　　　　Id. *Epod.* vii. 7.

At the summit of the ascent, called Summa Sacra Via, a market was held for the sale of fruit and toys, and the street was generally a lounge for idlers : —

Ibam forte Via Sacra, sicut meus est mos,
    Nescio quid meditans nugarum.　　Hor. *Sat.* i. 9, 1.

⁸ Scents, frankincense, silks, &c., were sold there :—

Deferar in vicum vendentem thus et odores
Et piper, et quicquid chartis amicitur ineptis.　　Id. *Ep.* ii. 1, 269.

Nec nisi prima velit de Tusco servica vico.　　Mart. xi. 27.

⁹ Curtius ille lacus, siccas qui sustinet aras
    Nunc solida est tellus, sed lacus ante fuit.　　Ov. *Fast.* vi. 403.

¹ There were probably two of them, and when Horace speaks of the *middle* Janus, he means the middle of the street :—

Postquam omnis res mea Janum
Ad medium fracta est.　　　　　Sat. ii. 3, 18.

²　　　　　　　　　　　Ante secundam
Roscius orabat sibi adesse ad Puteal cras.　　Hor. *Sat.* ii. 6, 34.

³ Ovid describes it as facing the Capitol : —

Ut semper Capitolia nostra forumque
Divus ab excelsa prospectat Julius æde.　　Ov. *Met.* xv. 841.

stone, erected by Augustus; the statue of Marsyas[4]—the resort of lawyers and courtezans—and numerous other statues; the Columna Mænia, commemorative of the victory of Mænius over the Latins, in 338; and the Columna Rostrāta, adorned with the beaks of the ships taken by Duilius from the Carthaginians ir 260.

Adjacent to the Forum we may note the Temple of Vesta[5] (Plan,

Temple of Vesta. (From a Coin.)

16), at its S.E. end, erected by Numa Pompilius; the Temple of Castor and Pollux[6] (Plan, 13), just under the Palatine, vowed by Postumius in the Latin War, and dedicated by his son in 484, of which three columns still remain; the Basilica Julia, between the Vicus Tuscus and Jugarius, erected by Cæsar for the accommodation of the law-courts; the Temple of Saturn (Plan, 11), at the W. end of the Forum under the Capitoline Hill, dedicated in 497, and of which eight columns remain; the Temple of Concordia (Plan 2), erected by L. Opimius, in 121, at the N.W. end of the Forum and on the rise of the Capitoline; the Senaculum (Plan, 3), an elevated area between the Temple of Concord and the Forum, where the senators met before entering the Curia; the Tullianum, or lower dungeon of the Mamertine prison, erected by Servius Tullius, and still in existence; the Curia (Plan, 6),

Arch of Septimius Severus.

or Senate-House, on the N. side of the Forum, at its W. end; the Græcostāsis (Plan, 5), adjacent to it at its S.W. angle, a place set apart as a waiting-room for foreign ambassadors; the Basilica Porcia (Plan, 7), on the E. of the Curia, erected in 184, by Porcius Cato, for the assemblies of the tribunes of the people; the Basilica Æmilia (Plan, 8), originally erected in 179 by M. Æmilius Lepidus;

and, lastly, the Arch of Severus, erected in A.D. 203, at the N.W. angle of the Forum, and still in a good state of preservation.

V. *The Imperial Fora.*—As Rome increased in size the old Forum

---

[4]   Obeundus Marsya, qui se
Vultum ferre negat Noviorum posse minoris.   Hor. *Sat.* l. 6, 120.
Ipse potest fieri Marsya candidus.   Mart. II. 64.

[5] Illo loco est Vestæ: qui Pallada servat et ignem.—Ov. *Trist.* III. 1, 29.

[6] At quæ venturas præcedet nixta Kalendas
Hæc sunt Lederis templa dicata Deis.
Fratribus illa Deis fratres de gente Deorum
Circa Juturnæ composuere lacus.   Id. *Fast.* I. 705.

was found insufficient for the transaction of law business; and hence numerous fora were erected by the emperors in the ground intervening between the Forum and the Quirinal, and in a line diverging to the N.W. from the old Forum. The chief of these imperial fora were the **Forum Julium**, founded by Cæsar and finished by Augustus, which was situated at the back of the Basilica Æmilia; the **Forum Augusti**, to the N. of the Forum Julii, enclosing a temple of **Mars Ultor**,[1] of which three columns still remain ; the **Forum Transitorium**, commenced by Domitian and completed by Nerva, and containing a temple of **Minerva**, situated E. of the Forum Julium ; the **Forum Trajani**, the most magnificent of them all, situated between the Quirinal and Capitoline, and containing, in addition to the Forum itself, the **Basilica Ulpia**, at the W. end of which stands the famous **Column of Trajan**,

Temple of Trajan.

commemorating the wars of that emperor with Decebalus ; and, lastly, W. of the Basilica, completing the range of buildings, the Temple of **Divus Trajanus**, erected by Hadrian.

VI. *The Palatine and Velia.*—After the Capitol and Forum, the Palatine Hill is the most interesting spot in Rome, both as having been the cradle of the eternal city and the later residence of the emperors in the time of its highest glory. The declivity towards the Capitoline was called **Germalus**, or **Cermalus**, and contained the **Lupercal**, a grotto sacred to Pan;[2] the **Ficus Ruminalis**, the fig-tree under which Romulus and Remus were suckled by the wolf; and the **Casa Romuli**,[3] a hut in which Romulus was nurtured. These objects were probably at the W. angle of the hill, near the Circus. Among the illustrious Romans who had houses on the Palatine, we may notice Vitruvius Vaccus, whose house was pulled down in B.C. 335, Fulvius Flaccus, who perished in the sedition of Gracchus, Cicero, who lived on the N.E. side of the hill, Catiline, Antonius, and Scaurus. Augustus was born in this quarter, and adorned it with a splendid Temple of Apollo, surrounded with a portico containing the **Bibliothæca Græca et Latina**: the temple itself was built of solid white marble, and contained statues of the god and of Augustus himself; the columns of the portico were of African marble and between them stood statues of the fifty daughters of Danaus:[4] its exact position is not known. The Palace of Augustus

---

[1] It was vowed by Augustus in the civil war undertaken to avenge his father's death :—

Mars, ades, et satia scelerato sanguine ferrum :
Sitque favor causa pro meliore tua.
Templa feram et, me victore, vocaberis Ultor.   Ov. *Fast.* v. 575.

[2] Hinc lucum ingentem, quem Romulus acer Asylum
Rettulit, et gelida monstrat sub rupe Lupercal,
Parrhasio dictum Panos de more Lycæi.   *Æn.* viii. 342.

[3] Romuleoque recens horrebat regia culmo.   *Id.* viii. 654.

[4] Quæris cur veniam tibi tardior ? aurea Phœbi
Porticus a magno Cæsare aperta fuit :   Tota

appears to have stood on the N.E. side of the hill, and the Palace of Tiberius near the N.W. corner. The two palaces of Nero, named Domus Transitoria and Domus Aurea, probably covered the whole of the hill. The Velia was the rising ground between the valley of the Forum on the one side and the Colosseum on the other. It contained the following objects: — the Ædes Penatium, an ancient fane in which the images of the household gods brought from Troy were preserved; the Temple of Peace, erected by Vespasian after his triumph over Jerusalem, with the spoils of which it was adorned; the Basilica Constantini, erected by Maxentius in honour of Constantine, of which three massive arches still remain; the splendid Temple of Roma and Venus, built by Hadrian, considerable remains of which exist behind the convent of S. Francesca Romana; the Arch of Titus, which spanned the Via Sacra at the very summit of the Velian ridge, adorned with beautiful reliefs illustrating the Jewish triumphs of Titus, and still existing; the Arch of Constantina, at

Arch of Titus restored.

Arch of Constantine.

the N.E. corner of the Palatine, erected in honour of Constantine's victory over Maxentius, and still in a good state of preservation; and

Tota erat in speciem Penis digesta columnis
  inter quas Danai femina turba senis.   PROPERT. ii. 31, 1.

Horace alludes to the Library :—
  Scripta Palatinus quaecunque recepit Apollo.   Ep L 3, 17.

the Meta Sudans, a fountain erected by Domitian, of which there are some remains.

VII. *The Aventine.*—The Aventine was regarded as ill omened in the early days of Rome: it contained, nevertheless, several famous spots, such as the Altar of Evander, the Cave of Cacus,[2] and the Temple of Jupiter Inventor, dedicated by Hercules after he had found his cattle. The Temple of Diana, erected by Servius Tullius as the sanctuary of the cities of the Latin League, stood on the side of the hill facing the Circus, while at its N. extremity, near the Porta Trigemina, stood the famous Temple of Juno Regina, built by Camillus after the conquest of Veii. A portion of the summit, probably about the centre of it, named Saxum,[3] was the spot where Remus was reputed to have taken his auguries: a Temple of the Bona Dea,[4] was afterwards erected there. There was also a Temple of Luna,[5] probably on the side next the Circus, and one of Libertas, founded by T. Sempronius Gracchus. We have notice of houses of Sura, of Trajan before he became emperor, and of Ennius the poet, on this hill. The strip of ground between the Aventine and the Tiber was one of the busiest parts of the city, as it contained the emporium or quays for the discharge of the cargoes of ships, and the principal corn-market. L. Æmilius Lepidus and L. Æmilius Paulus founded a regular Emporium and a portico named after them Porticus Æmilia. The broad level space to the S. of the hill was probably the site of large warehouses for storing goods. The *Monte Testaccio*, which is in the same district, is an artificial hill of potsherds, 153 feet high, the origin of which is shrouded in mystery.

VIII. *The Velabrum, Forum Boarium, and Circus Maximus.*—Between the Palatine, Aventine, and Tiber, the level ground was occupied by two districts named the Velabrum and the Forum Boarium, while between the two hills was the Circus Maximus. The Velabrum was originally a marsh[6] and afterwards a quarter of the town at the head of the Vicus Tuscus; its name is preserved in that of the modern church of *S. Giorgio in Velabro*, near which still stand two ancient monuments, the Arcus Argentarius, built by the silversmiths in honour of Septimius Severus, and a square building named Janus Quadrifrons. The F. Boarium was a large unenclosed space extending from the Velabrum to the ascent of the Aventine, and from the Tiber to the Circus.[7] It probably derived its name from having been an old cattle-market: it was rich in temples and monuments, particularly a Temple

---

[2] Ille spelunca fuit, vasto summota recessu,
Semihominis Caci facies quam dira tenebat
Solis inaccessam radiis, *Æn.* viii. 193.

[3] Interea Diva canenda Bona est.
Est moles nativa, loco res nomina fecit.
Appellant Saxum ; pars bona montis ea est.—Ov. *Fast.* v. 148.

[4] Templa Patres illic, oculos exosa viriles,
Leniter acclivi constituere jugo. Id. *Fast.* v. 153.

[6] Luna regit menses; hujus quoque tempora mensis
Finit Aventino Luna colenda jugo. Id. *Fast.* III. 883.

[6] At qua Velabri regio patet, ire solebat
Exiguus pulso per vada linter aqua. Tibull. II. 5, 33.

[7] Pontibus et Magno juncta est celeberrima Circo
Area, quae pomilo da bove nomen habet. Ov. *Fast.* vi. 477.

of Hercules, covering the altar said to have been built by Evander; another round temple of the same god, possibly represented by the remains still existing at

Temple of Hercules.

the church of *S. Maria del Sole*: temples of **Fortuna** and **Mater Matúta**, both of them built by Servius Tullius and of uncertain position;[2] and a temple of **Pudicitia Patricia**, which may perhaps be represented by the elegant remains now forming the Armenian church of *S. Maria Egiziaca*. The **Cloaca Maxima** discharges itself into the Tiber in this district, and its mouth is visible when the river is low. The **Circus Maximus** was nearly half a mile long and was the principal racecourse in Rome:

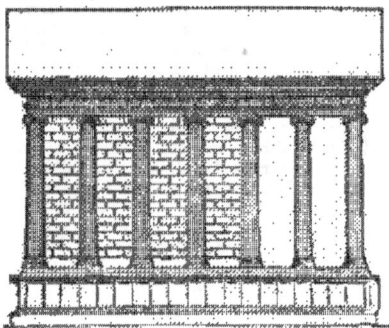

Temple of Pudicitia Patricia.

it was founded by Tarquinius Priscus, but it remained in a rude state until the time of Julius Cæsar, who placed permanent seats, the lower ones of stone and the upper of wood. It was further improved by Augustus, Claudius, and Trajan. It was probably capable of containing about 385,000 spectators.

IX. *The Cælian Hill.* —The Cælian Hill was not much frequented in early times. The only public buildings on it worthy of notice were—a little temple of **Minerva Capta** on the declivity of the hill;[3] a temple of **Divus Claudius**, begun by Agrippina, destroyed by Nero, and restored by Vespasian; and the **Arch of Dolabella**, erected in the consulship of Dolabella, A.D. 10, and probably designed as an entrance to some public place. In the imperial times many illus-

---

[2] They are referred to by Ovid:—

Lux eadem, Fortuna, tua est, auctorque, locusque,
Sed superinjectis quis latet æde tugis?
Servius est.　　　　　　　　　　　　*Fast.* vi. 569.
Hac tibi loco ferunt Matutæ mora parenti
Scriptriferas Servi templa dedisse manus.　　*Id.* vi. 479.

[3] Cælius ex alto qua Mons descendit in æquum;
Hic ubi non plana est, sed prope plana via est,
Parva licet videas Captæ delubra Minervæ.　　Ov. *Fast.* iii. 835.

trious Romans had fine houses here, particularly Mamurra, Annius Verus the grandfather of Marcus Aurelius, and the Laterani, whose house appears to have been confiscated after the treason of Plautius Lateranus in Nero's reign.

X. *The District S. of the Cælian.*—To the S. of the Cælian were the 1st and 12th regions of Augustus, named *Porta Capena* and *Piscina Publica.* In the former of these lay the **Porta Capena** itself;[1] the Valley of **Egeria**,[2] watered by the small stream Almo,[3] and the traditional scene of Numa's interviews with the nymph; and the **Thermæ Antoninæ** or **Caracalla**, on the right of the Appian Way, remains of which are still in existence. For several miles the tombs of eminent Romans skirt the Via Appia, commencing immediately outside the P. Capena. The most interesting of these is the Tomb of the **Scipios**, about 400 paces within the *P. S. Sebastiano;* while the mausoleum of **Septimius Severus** and that of **Cæcilia Metella** deserve notice, though the latter lies beyond the limits of the city.

Tomb of Cæcilia Metella.

XI. *The Esquiline and its Neighbourhood.*—The Esquiline was originally covered with a thick wood, to which its name may be referred. On the larger and more southerly of the two tongues into which the

---

[1] A branch of the Aqua Marcia passed over this gate, and kept it in a dripping state :—

Substitit ad veteres arcus, *madidamque* Capenam.—Juv. iii. 11.

Capena grandi porta, qua pluit gutta.   Mart. iii. 47.

[2] In vallem Egeriæ descendimus et speluncas
Dissimiles veris.   Juv. iii. 17.

[3] The waters of this stream were sacred to Cybele :—
Et parvo lotam revocant Almone Cybeben.   Luc. i. 600.

hill is divided, viz. Mons Oppius, was situated the district named
Carinæ, extending down from the extremity of the hill into the subjacent valleys. In the valley between this and the Cælian lay the
gigantic Amphitheatrum Flavium, more commonly known as the
Colosseum, probably from a colossal statue of Nero. It was com-

Colosseum.

menced by Vespasian, was completed by Domitian, and was capable of
holding 87,000 spectators. On the hill above the Colosseum were
the Thermæ Titi, of which there are still considerable remains; and
near them the Thermæ Trajani. The Vicus Cyprius ran along the N.
base of Mons Oppius, under the Carinæ, and ascended the hill at the
head of the valley between the Oppian and Cispian mounts by the
Clivus Urbius, near which point the palace of Servius Tullius stood.
In the valley between the extremities of the Quirinal, Viminal, and
Esquiline, lay the populous region of Suburra, the resort of hucksters,
prostitutes, and the dregs of the population.[c] During the republic a
part of the Esquiline outside the walls, named Campus Esquilinus, was
used as a burying-ground for paupers and slaves. Mæcenas converted
this into a public garden or park, the celebrated Horti Mæcenatis,[b] extending to the Agger of Servius Tullius, which then became the resort
of fortune-tellers.[d] In the same part of the town were the Horti Lamiani,

---

          Latrent Suburranæ canes.   Hor. Epod. v. 57.

    Dum tu forsitan inquietus erras
    Clamosa, Juvenalis, in Suburra.   Mart. xii. 18.

    Ego vel Prochytam præpono Suburræ.   Juv. iii. 5.

    [b] Nunc licet Esquiliis habitare salubribus atque
    Aggere in aprico spatiari, quo modo tristes
    Albis informem spectabant ossibus agrum.   Hor. Sat. i. 8, 14

    [d] Plebeium in circo positum est et in aggere farum.—Juv. vi. 588.

belonging perhaps to Ælius Lamia, and the Horti Palantii, founded
apparently by Pallas, the freedman of Claudius. It was also the residence of the poets Virgil and Propertius, and a favourite resort of
Horace. Pliny the younger also had a house there. There were numerous temples, the most important of which was the Templum Telluris.

XII. *The Colles, or the Viminal, Quirinal, and Pincian Hills.*—The
Viminal is separated from the Esquiline by a valley through which ran
the Vicus Patricius, and from the Quirinal by a valley the N. part of
which was named Vallis Quirini.[1] The Viminal was chiefly inhabited
by the lower classes, the only remarkable building being the palace of
C. Aquilius. The Quirinal was separated from the Pincian on the N.
by a deep valley, and skirted the Campus Martius on the W. It was
the most ancient quarter of the town, and abounded in fanes and
temples, the most famous of which was the Temple of Quirinus, originally erected by Numa to Romulus after his apotheosis. Numa resided on the Quirinal: his capitol probably stood on the W. side of
the hill, and contained a temple to Jupiter, Juno, and Minerva. Near
it was the Temple of Flora, and the house of the poet Martial. The part
adjacent to the Porta Salutaris was named Collis Salutaris, after an
ancient shrine of Salus. Between the temples of Salus and Flora
stood the shrine of Semo Sanctus or Dius Fidius, an old Sabine deity,
said to have been founded by Tatius. We may also notice the Horti
Sallustiani, formed by Sallust the historian, in the valley between the
Quirinal and Pincian, the subsequent residence of the emperors Vespasian, Nerva, and Aurelian; the Thermæ Diocletiani, the largest of all
the Roman baths, but now in a very ruined state; the Campus Scele-
ratus, where Vestal virgins convicted of unchastity were buried alive;
the Templum Gentis Flaviæ, a magnificent mausoleum[2] erected by
Domitian for his family; and the Prætorian Camp, established in the
reign of Tiberius outside the Porta Collina. The Pincian Hill was so
named from a magnificent palace of the Pincian family on it: previously it had been called Collis Hortorum, from the gardens which
covered it. The only place to be noticed on it was the Gardens of
Lucullus, the scene of Messalina's infamous marriage with Silius, and
of her death by the order of Claudius.

XIII.—*The Campus Martius, Circus Flaminius, and Via Lata.*—The
Campus Martius was the plain lying between the Pincian, Quirinal, and
Capitoline hills on the E., and the Tiber on the W. It was intersected
in its whole length by the Via Flaminia. The S. portion of the plain
between the road and the river constituted the 9th region of Augustus,
under the name of Circus Flaminius; and the S. portion, on the other
side of the road, between it and the hills, formed the 7th region, with
the name of Via Lata. The temples and public buildings in this district were very numerous. The Circus Flaminius contained the Temple
of Pietas, dedicated by the son of M. Acilius Glabrio, in B.C. 180; the

---

Some of the tombs remained in this part of the grounds, as alluded to by Horace
in describing the magical rites of Canidia :—

        Lunamque rubentem,
Ne foret his testis, post magna latere sepulchra.—*Sat.* I. 8, 35.

[1]   Officium cras
Primo sole mihi peragendum in valle Quirini.   Juv. II. 132.

[2] Jam vicina jubent nos vivere Mausolea
     Quum doceant ipsos posse perire deos.   Mart. v. 64.

Temple of Janus; the Theatre of Marcellus; the Temple of Apollo,
dedicated in U.C. 430; the Temple of Bellona, said to have been built
in pursuance of a vow made by Appius Claudius Cæcus, in the battle
against the Etruscans in B.C. 207, and the place where the assemblies
of the Senate met outside the *pomœrium*; the Circus Flaminius, under
the Capitol, extending in a westerly direction towards the river; the
Porticus Octavia, erected by Augustus in honour of his sister, containing a library, and Temples of Jupiter Stator and Juno; the Porticus
Philippi,[9] enclosing a Temple of Herculas Musarum, built by M. Fulvius
Nobilior, and rebuilt by L. Marcius Philippus, the stepfather of Augustus; the Theatre of Pompey, with a portico, adjoining the scena; a
Curia, or large hall in the portico, used both for scenic purposes and
for the assemblies of the Senate, with a statue of Pompey in it, before
which Cæsar was assassinated; and another portico, named Hecato-
stylon, from its having 100 columns.[1] The Campus Martius itself was
originally nothing more than an open plain used for gymnastic and
warlike exercises,[2] and also for large public assemblies of the people.
Subsequently to the 6th century of the city, temples began to be built
there; and gradually it was almost covered with important edifices,
among which the most conspicuous were—the Septa Julia, a marble

building commenced
by Cæsar, and com-
pleted after his death
for the purpose of
holding the assemblies
of the Comitia Cen-
turiata; the Villa Pub-
lica, adjoining the
Septa Julia on the
S., used by the con-
suls for the levying
of troops, and for the
reception of foreign
ambassadors; the
Pantheon of M. Vip-
sanius Agrippa, in the
very centre of the
Campus, and still in
a very good state of

Pantheon of Agrippa.

preservation; the Thermæ of Agrippa, adjoining the Pantheon on the
S.; the Diribitorium, also adjoining it, a large building erected by
Agrippa, and used for the scrutiny of the voting tablets used in the
Comitia; the Porticus Argonautarum,[3] erected in commemoration of
Agrippa's naval victories, and named after a picture of the Argonauts,

---

[9] Vites ærneo porticum Philippi :
     Si te viderit Hercules, periisti.   MART. v. 49.

[1] Inde petit centum pendentia tecta columnis;
     Illinc Pompeii dona, nemusque duplex.   ID. IL 14.

[2] Tunc ego me memini ludos in gramine campi
     Adspicere; et didici, lubrice Tibri, tuos.   OV. *Fast*. vi. 237.

     Quamvis non alias flectere equum sciens
     Æque conspicitur gramine Martio.   HOR. *Carm*. iii. 7, 25.

[3] An spatia carpit lentus Argonautarum !   MART. iii. 20.

with which it was adorned; the Mausoleum of Augustus in the northern
angle of the Campus, between the Via Flaminia and the river, wherein
were deposited the ashes of Marcellus,[4] Agrippa, Octavia, Drusus, Augustus, and other illustrious personages; the Thermæ Neronianæ,[5]
erected by Nero close to the baths of Agrippa; the Temples of Isis and
Serapis,[6] in the same quarter, restored by Domitian after the fire in the
reign of Titus; and the Temple and Column erected in honour of M.
Aurelius Antoninus, the latter of which (named Columna Cochlis, from
the spiral staircase inside it) was erected by M. Aurelius and L. Verus,
and now stands in the Piazza di Monte Citorio. The Via Lata contained
the Campus Agrippæ, used, as the Campus Martius was, for gymnastic
exercises and amusement, the buildings about it having been erected
by Vipsanius Agrippa for that purpose; the Triumphal Arches of
Claudius and M. Aurelius; and the Forum Suarium or pork-market.

XIV.—The Transtiberine District.—The district beyond the Tiber
was never regarded as a portion of the Urbs, properly so called, although it formed one of Augustus's regións, and was included within
the walls of Aurelian. It may be divided into three parts: the Insula
Tiberina, said to have been formed by the corn of the Tarquins thrown
into the river, and on which stood a Temple of Æsculapius, much
visited by sick persons; the Janiculum,[7] enclosed between a ridge
running due S. from the point where the Tiber takes its first great
bend and the river itself, a considerable space, chiefly occupied by the
lower classes, but containing the Horti Cæsaris,[8] which Cæsar bequeathed to the Roman people, and two Naumachiæ, constructed by
Augustus[9] and Domitian; and the Mons Vaticanus,[1] a little N.W. of
the Mons Janiculus, not included in the walls of Aurelian, and noted
for its unhealthy air and its execrable wine. The only building of note
between this hill and the river was the Mausoleum or Moles Hadriana,
erected by Hadrian, and the tomb of himself and the succeeding emperors until the time of Commodus, and now known as the Castle of
St. Angelo.

XV. Bridges.—The Tiber was crossed by seven bridges, which may
be enumerated in the following order from N. to S.:—Pons Ælius,
built by Hadrian to connect his mausoleum with the city. P. Neroni-

---

[4]   Quæ, Tiberine, videbis
Funera, quæm tumulum præterlabere recentem.—Æn. vi. 874.

[5]   Quid Nerone pejus!
Quid Thermis melius Neronianis.   MART. vii. 34.

[6] A Meroë portabis aquas, ut spargat in æde
Isidis, antiquo quæ proxima surgit ovili.   JUV. vi. 528.

[7] The name was derived from Janus :—
Hanc Janus pater, hanc Saturnus condidit arcem :
Janiculum huic, illi fuerat Saturnia nomen.   Æn. viii. 357.

[8] Trans Tiberim longe cubat is, prope Cæsaris hortos.—HOR. Sat. i. 9, 18.

[9] The lake of this one remained for a long period :—
Continuo dextras flavi pete Tibridis oras,
Lydia quæ penitus stagnum navale exercet
Ripa, suburbanisque vadum prætexitur hortis.—STAT. Silv. iv. 4, 5.

[1]   Simul et jocosa
Redderet laudes tibi Vaticani
Montis imago.   HOR. Carm. i. 20, 5.

Mole of Hadrian restored.

anus or Vaticanus, leading from the Campus Martius to the Vatican and the Gardens of Nero; the remains of its piers are still visible. P. Aurelius, on the site of the *Ponte Sisto*, leading to Janiculum. P. Fabricius[2] and P. Cestius, the former connecting the Insula Tiberina with the city, the latter with the Janiculum; they still exist under the names of *Ponte Quattro Capi* and *Ponte S. Bartolommeo*. P. Senatorius or Palatinus, opposite the Palatine Hill; and P. Sublicius,[3] the oldest of all, said to have been erected by Ancus Martius, and named after the "wooden beams" (*sublices*) of which it was built. We may also notice the P. Milvius or Mulvius, the present *Ponte Molle*, 2 miles N. of the city at the point where the Flaminian Way crossed the river.

XVI. *Aqueducts.*—Rome was supplied with water by fourteen aqueducts, the first of which was constructed in B.C. 313 by the Censor Appius Claudius Cæcus, and was named after him Aqua Appia. Of the others we may notice the Anio Vetus, constructed in 273, which derived its supply from the Anio above Tibur, and was 43 miles in length; the Aqua Marcia, built in 144 by the Prætor Q. Marcius Rex, and which was reputed to bring the most wholesome water of all; the Aqua Julia, built by Agrippa in his ædileship in 33, a very magnificent work; the Aqua Claudia, begun by Caligula, and dedicated by Claudius; and the

---

[2] It was the favourite bridge for suicides :—

Jussit sapientem pascere barbam
Atque a Fabricio non tristem ponte reverti.   Hor. Sat. ii. 3, 35.

[3] A stone bridge was erected by the side of the old wooden one: it was called Pons Æmilius, and is noticed in the following line :—

Cum ipsi vicinam se probat Æmilius pons ?   Juv. vi. 32.

Insula Tiberina, with the Pons Fabricius and Pons Cestius.

Anio Novus, also completed by Claudius, 59 miles in length, and with arches occasionally 109 feet high. The two last were the most gigantic of all the Roman aqueducts.

§ 6. The remaining towns of Latium were as follows:—

Ostia. *Ostia*, was situated at the mouth (as its name implies) of the river Tiber[4] on its left bank, and was the original port of Rome. It was founded by Ancus Martius, and in the time of the Second Punic War was important both as a commercial and naval station. It suffered severely in the Civil Wars of Sulla and Marius, and was destroyed by the latter in B.C. 87. As the coast had advanced considerably through the alluvial deposit of the Tiber, it was found necessary to make a new port; and this was effected by Claudius, who constructed a basin about 2 miles N. of Ostia,[5] which he connected with the river by means of a canal. This was designated Portus Augusti, and was further enlarged by the addition of an inner dock by Trajan, which was named after him Portus Trajani. The canal was enlarged, and henceforth known as Fossa Trajana, and now as the *Fiumicino;* and an extensive town named Portus Ostiensis, or simply *Portus,* grew up about the place. The remains of this town still retain the name of *Porto,* and the outline of the mole and dock may be traced. It became blocked up by sand in the 10th century, and the trade returned to the old channel. The ruins of Ostia itself are extensive, but uninteresting : the statues and other objects discovered there prove it to have been a place of con-

---

[4] Ostia contigerat, qua se Tiberinus in altum
Dividit, et campo liberiore natat. *Ov. Fast.* iv. 291.

[5] Non ita Tyrrhenus stupet Ioniusque magister,
Qui portus, Tiberine, tuos, clarumque serena
Arce Pharon praeceps subiit : nusquam Ostia, nusquam
Ausoniam videt. *Val. Flac.* vii. 83.

alderable wealth. Antium, *Porto d'Anzo*, was situated on a promontory about 38 miles from Rome. It was in the early age of Roman history the resort of Tyrrhenian pirates. In B.C. 468 it was captured and colo-

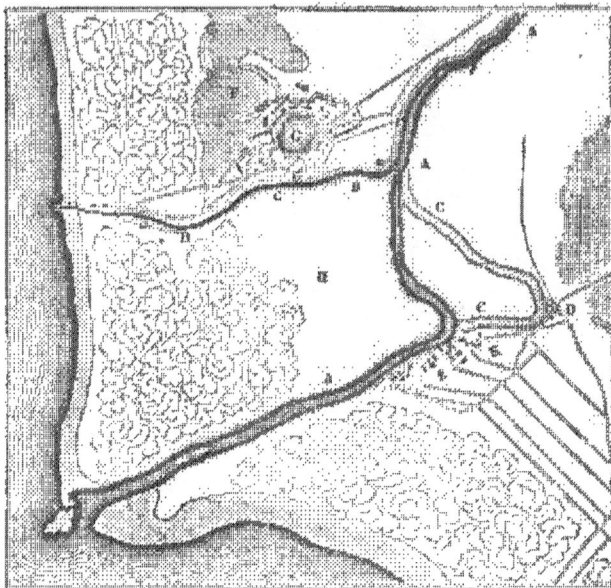

Plan of Ostia.

A A. Main Channel of the Tiber. B. Right arm of ditto, the Fossa Trojana, now called Fiumicino. C. Dry bed of ancient course of Tiber. D. Modern village of Ostia. E. Ruins of Ancient Ostia. F. Portus Augusti. G. Portus Trajani.

nised by the Romans; in 459 it revolted, and remained independent for 120 years, during which it waged several wars with Rome. Thenceforth its history is unimportant; but it remained a very flourishing place, and was the residence of Cicero and the birth-place of Caligula and Nero. It possessed a celebrated Temple of Fortune,[*] and another of Æsculapius. On the site of the old town numerous works of art have been discovered, particularly the statues of the Apollo Belvedere and the Fighting Gladiator. Circeii lay at the foot of Mons Circeius, on its N. side, and not far from the sea. It was founded by Tarquinius Superbus, and rose to such a state of commercial prosperity that it appears among the towns with which Carthage concluded a treaty. In B.C. 340 it was a member of the Latin League, having revolted from

---

[*] Hence Horace addresses Fortuna as—

O Diva, gratum quæ regis Antium.   *Carm.* L 35, 1.

Rome ; and thenceforth its name seldom appears in history.   It became
a favourite residence of the wealthy Romans, and was the occasional
abode of the Emperors Tiberius and Domitian.   Its chief fame, however, is due to its excellent oysters.[7]   A few polygonal blocks of
masonry are all that remains of it.   Tarracina, *Terracina*, was situated
on the summit of a white cliff,[8] about 10 miles S. of Circell, and at the
extremity of the Pontine Marshes.   It was also called Anxur, a name
familiar to us from its being constantly used by the poets.   In B.C. 509
Tarracina appears in the Carthaginian treaty as a dependent of Rome ;
in 406 it was under the Volscians, and was attacked and taken by M.
Fabius Ambustus ; in 402 it was again under the Volscians, and in 400
was recaptured by the Romans ; finally, in 329, a colony was sent there
by them.   Its position on the Appian Way rendered it always a place
of importance and of resort.   Considerable portions of the walls remain, as well as some tombs.   It possessed an artificial port, which is
noticed in B.C. 210, and was subsequently improved under the emperors.   Formiæ, *Mola di Gaëta*, was situated on the innermost point
of the Sinus Caietanus and on the Appian Way.   It is first noticed in
B.C. 338 as being on friendly terms with Rome, and as receiving the
Roman citizenship in reward for its services.   From the beauty of its
position it became a favourite resort of the wealthy Romans,[9] and,
among others, of Cicero, who perished there in B.C. 43.   The ruins of
villas and sepulchres line the coast and the Appian Way for some miles
E. of Formiæ.   The hills at the back of the town produced a good
kind of wine.[1]   Caiëta,[2] *Gaëta*, was situated on a projecting headland
on the S. side of the bay named after it, and about 4 miles from Formiæ.
The town itself was poor, but the port was frequented from the earliest
ages, and is spoken of by Cicero[3] as " portus celeberrimus et plenissi-
mus navium."   Antoninus Pius had a villa there, which the younger
Faustina frequented.   Among the ancient remains we may notice the
sepulchre of L. Munatius Plancus, and portions of a temple of Serapis

[7]   Circæis nata forent, an
Lucrinum ad saxum, Rutupinove edita fundo
Ostrea, callebat primo deprendere morsu.   Juv. iv. 140.

[8] Millia tum prænoi tria repimus ; atque sublimus
Impositum saxis late candentibus Anxur.   Hor. *Sat.* i. 5, 25.
Sive salutiferis candidus Anxur aquis.   Mart. v. 1.
                    Scopulosi verticis Anxur.   Sil. Ital. viii. 392.

[9] Martial enlarges on its many recommendations in the poem commencing, –
O temperatæ dulce Formiæ littus,
Vos, quum acveri fugit oppidum Martis,
Et inquietas fessus exolt curas,
Apollinaris omnibus locis præfert.   x. 30.
The wealthy Mamurra resided there ; hence the allusion in Horace.
In Mamurrarum lassi deinde urbe manemus.   *Sat.* L 5, 37.

[1]   Mea nec Falerno
Temperant vites, neque Formiani
                Pocula colles.   Hor. *Carm.* i. 20, 10.

[2] It is said to have been named after the nurse of Æneas :—
Tu quoque litoribus nostris, Æneïa nutrix,
Æternam moriens famam, Caieta, dedisti :
Et nunc servat honos sedem tuum.   *Æn.* vii. 1.

[3] Pro *Leg. Manil.* 12.

and of an aqueduct. **Minturnæ** was situated on the right bank of the
Liris, about 3 miles from the sea, and on the Appian Way. It was
originally an Ausonian town, but was colonised by the Romans in
B.C. 296. Its position on the Appian Way secured its prosperity, in
spite of the unhealthiness of the locality. The only interesting event
connected with it is the capture of C. Marius in 88 in the neighbouring
marshes,[4] and his subsequent release. Extensive ruins of an amphitheatre, of an aqueduct, and of other buildings, mark its site. Near it
were the celebrated grove and temple of the goddess Marica.[5] **Sinuessa**,
the most southerly town of Latium, stood on the shore of the Sinus
Caiotanus, about 6 miles N. of the river Vulturnus, and on the Appian
Way.[6] It was colonised at the same time as Minturnæ, the object of
this step being the protection of the Roman border against the Samnites. In its neighbourhood was produced the famous Massic wine ;[7]
and near it there were some much-frequented baths named Aquæ
Sinuessanæ, and now *I Bagni*. The ruins of Sinuessa lie just below
the hill of *Mondragone*, and consist of the remains of a triumphal arch,
an aqueduct, and other buildings.

2. *In the Interior.*—**Tibur**, *Tivoli*, was situated on the banks of the
Anio, just above the spot where that river makes its descent into the
Campagna. It thus appeared from one side to stand on the summit of
a lofty cliff.[8] The town was very ancient, and was believed to have
been founded by the Siculi. It is first noticed in B.C. 446 as the
place whither M. Claudius retired in exile. In 357 it was engaged in
disputes with Rome; and for the next twenty years frequent wars took
place between them, ending in the capture of Tibur by L. Furius
Camillus in 335. It enjoyed the privileges of an asylum,[9] and was the
place of exile of M. Claudius in 446, of Cinna after the murder of Cæsar,
of Syphax king of Numidia, and of the beautiful Zenobia. It possessed
a very famous temple of Hercules[1] Victor Tibur, with a library, a
treasury, and an oracle attached. It became, from the beauty of its
scenery, a favourite resort of the wealthy Romans. Mæcenas, Catullus,

---

[4] Exsilium, et career, Minturnarumque paludes	
Hinc causas habuere.	JUV. x. 276.
[5] Et umbrosæ Liris per regna Maricæ.—LUC. II. 424.	
Cæruleus nos Liris amat, quem silva Maricæ	
Protegit.	MART. xIII. 83.
[6] Postera lux oritur multo gratissima : namque	
Plotius et Varius Sinuessæ, Virgiliusque	
Occurrunt.	HOR. *Sat.* L. 3, 39.
[7] Quocunque lectum nomine Massicum	
Servas, moveri digna bono die.	ID. *Carm.* III. 21, 5.
Uviferis late florebat Massicus arvis.	SIL. ITAL. VII. 207.
[8] Hence Horace's epithet :—	
Præneste, seu Tibur supinum.	*Carm.* III. 4, 23.
[9] Quid referam veteres Romanæ gentis, apud quos	
Exsilium tellus ultima Tibur erat.	OV. *ex Pont.* L. 3, 81.
Hence the epithet " Herculeus " was applied to it :—	
Itur ad Herculei gelidas qua Tiboris arces.	MART. L. 12.
Venit in Herculeos colles : quid Tiburis alti	
Aura valet !	ID. vii. 13.

Tivoli, the ancient Tibur.

Horace,[2] Sallust, Vopiscus, and Quinctilius Varus had villas there; and about 2 miles S. of the town the emperor Hadrian erected a magnificent palace with an immense number of buildings, such as a lyceum, an academy, &c., and extensive pleasure-grounds. Considerable remains of the buildings are still visible. The chief remains of Tibur are a circular, peripteral temple, reputed to be dedicated to the sibyl Albunea, with ten out of the original eighteen columns still existing; an oblong temple, supposed to be of Vesta; part of a temple which stood in the ancient forum; together with remains of two bridges, and the villas of Mæcenas, Varus, &c. The surrounding country was celebrated for its fruit, and for its extensive quarries, which supplied Rome with the *travertino* used in the Colosseum and the basilica of St. Peter. *Præneste, Palestrina,* stood on a projecting spur[3] of the Apennines, directly opposite the Alban Hills, and 23 miles E. of Rome. Various accounts were given of its origin, not one of which is trustworthy. It

---

[2]   Mihi jam non regia Roma
Sed vacuum Tibur placet.   Hor. *Ep.* L 7, 44.

Sed quæ Tibur aquæ fertile præfluunt
Et spissæ nemorum comæ,
Fingent Æolio carmine nobilem.   Id. *Carm.* iv. 8, 10.

[3] Quique *altum* Præneste viri, quique arva Gabinæ.—*Æn.* vii. 682.

was a member of the Latin League; in B.C. 499 it seceded and joined the Romans; in 389 it commenced hostilities against them; in 380 it was captured by T. Quinctius Cincinnatus after the defeat of its army in the open field; in 340 it took a prominent part in the great Latin War; and in 338 it shared in the defeat at Pedum. In the Civil War between Sulla and Marius it was occupied by the latter, who put an end to his life there. The city was subsequently destroyed by Sulla, and its site removed from the hill to the subjacent plain. Its elevated position and bracing air[4] made it a favourite retreat of the Romans during the summer months; and it was the occasional abode of Augustus, Horace,[5] Hadrian, and M. Aurelius. It also possessed a celebrated shrine of Fortune,[6] of which the terraces still remain, and the temple itself existed until the 13th century. There are also extensive remains of Hadrian's villa. Tusculum, *Frascati*, stood on a spur of the Alban Hills, about 15 miles S.E. of Rome, with its citadel posted on a very lofty peak on the E. of the town. Its foundation was attributed to Telegonus,[7] the son of Ulysses and Circe. It first appears in history as the abode of Octavius Mamilius, the son-in-law of Tarquinius Superbus, who took refuge there on his expulsion from Rome, and thence headed the Latins against the Romans at the battle of Lake Regillus. Thenceforward the Tusculans appear as the steady allies of Rome. They nevertheless joined in the great Latin War against Rome, but were favourably treated in the settlement that took place in 335. Many of the Tusculan families were distinguished at Rome, particularly the gens Mamilia, the Porcia, the Fulvia, &c. Among the eminent Romans who had villas there, we may notice Lucullus, Cato, Marcus Brutus, L. Crassus, Mæcenas,[8] and particularly Cicero, who there composed most of his philosophical works, one of which, the 'Tusculan Disputations,' derives its name from the place: his abode is probably identical with the ruins of *Villa Rufinella*. The chief relics of the town are portions of the walls, of a piscina, and of two theatres. Aricia, *La Riccia*, was situated on the Appian Way, at the foot of the Alban Mount and on the Appian Road,[9] 16 miles from Rome. It was a member of the Latin League, and appears to have been one of the most powerful in the time of Tarquinius Superbus. It took part in the great Latin War, and subsequently received the full rights of

---

[4]   Seu mihi *frigidum*
    Præneste, seu Tibur supinum,
        Seu liquidæ placuere Baiæ.   Hor. *Carm.* iii. 4, 23.
Quis timet aut timuit gelida Præneste ruinam.—Juv. iii. 190.

[5] Dum tu declamas Romæ, Præneste relegi.   Hor. *Ep.* i. 2.

                    Sacriaque dicatum
Fortunæ Præneste jugis.   Sil. *Ital.* viii. 366.

[7] Inter Ariciam, Albanaque tempora constant
    Factaque Telegoni mœnia celsa manu.
Quid petis .Exi mœnia Telegoni ?   Ov. *Fast.* iii. 91.
                                                     Propert. ii. 32, 3.

[8] Nec ut superni villa candens Tusculi
    Circæa tanget mœnia.   Hor. *Epod.* i. 29.

[9] Ne semper udum Tibur, et Æmiæ
    Declive contempleris arvum, et
Telegoni juga parricidæ.   Id. *Carm.* iii. 29, 6.
Egressum magna me accepit Aricia Roma.   Id. *Sat.* i. 5, 1.

Roman citizenship. **Anagnia,**[1] *Anagni,* was situated on a hill to the left of the Via Latina, 41 miles S.E. of Rome. It appears to have been the capital of the Hernican cities, but its history is devoid of interest. Its position on the Via Latina exposed it to the ravages of invading armies, and it suffered both from Pyrrhus and from Hannibal. Its territory was remarkably fertile, and the city abounded in temples and sanctuaries.

Of the less important towns we may notice—(1.) *On the Coast.*— **Laurentum,** *Torre di Paterno,* about 16 miles from Rome, the ancient capital of Latinus, with marshes about it,[2] and a very extensive forest, in which the laurel was common, and was supposed to have given name to the place;[3] **Lavinium,** *Pratica,* S. of Laurentum (said to have been founded by Æneas, and named after his wife Lavinia), the sacred metropolis of the Latin League, but an insignificant place in the later days of the republic, and finally (probably in the reign of Trajan) re-colonised and united with Laurentum under the name of Lauro-Lavinium; **Ardea,** *Ardea,* 24 miles S. of Rome, and about 4 miles from the sea-coast, a city of great antiquity, said to have been founded by Danaë[4] the mother of Perseus, the capital of the Rutuli and royal abode of Turnus, but in later times a poor decayed place,[5] probably from the unhealthiness of the neighbourhood; **Lautulæ,** a spot between Tar-racina and Fundi, where a narrow pass (the *Passo di Portella*) occurs, through which the Appian Way passed, the scene of the insurrection of the Roman army under C. Marcius Rutilus in B.C. 342, and of a battle between the Romans and Samnites in 315; **Fundi,**[6] *Fondi,* on the Appian Way, between Tarracina and Formiæ, and near a considerable lake named Lacus Fundanus, *Lago di Fondi,* which intervened between it and the sea, a town of no pretensions, but noted for the excellence of the wine, particularly the Cæcuban,[7] produced in its territory; and, lastly, **Amyclæ,** on the shores of the bay named after it, Sinus Amy-

---

[1]　　　　Surgit suspensa tumenti
Dorso frugiferis Cerealis Anagnia glebis.　　　SIL. ITAL. xll. 332.

[2] Nam Laurens malus est, ulvis et arundine pinguis.—HOR. Sat. II. 4, 42.

These marshes were the haunts of wild boars:—
Inter quæ rari Laurentem ponderis aprum
Misimus, Ætola de Calydone putes.　　　MART. ix. 49.

[3] Ipse ferebatur Phœbo sacrasse Latinus;
Laurentisque ab ea (sc. lauro) nomen posuisse colonis.—Æn. vii. 62.

[4]　　　　Quam dicitur urbem
Acrisioneis Danae fundasse colonis,
Præcipiti delata Noto, locus Ardea quondam
Dictus avis; et nunc magnum manet Ardea nomen.—Æn. vii. 400.

[5] Magnanimis regnata viris, nunc Ardea nomen.　　　SIL. ITAL. l. 291.

[6] The pompousness of the "mayor" of this town was the object of Horace's ridicule:—
Fundos Aufidio Lusco prætore libenter
Linquimus, insani ridentes præmia scribæ,
Prætextam, et latum clavum, prunæque batillum.—Sat. l. 5, 34.

[7] Cæcuba Fundanis generosa coquuntur Amyclis.　　　MART. xiii. 115.
Absumet hæres Cæcuba dignior
Servata centum clavibus.　　　HOR. Carm. ii. 14, 25.

olanus, a place which had altogether disappeared in the time of Pliny.[*]

(2.) *In the Interior.*—Corİŏlİ, supposed to have been situated on the most westerly of the Alban Hills, chiefly celebrated for its connexion with the legend of C. Marcius Coriolanus. Alba Longa, situated on a long narrow ridge between the Alban Mount and Lake, the ancient capital of the Latin cities, said to have been founded by Ascanius[*] the

Gateway of Signia.

son of Æneas, and destroyed by Tullus Hostilius. Lanuvium,[*] *Civita Lavinia,* on a southern spur of the Alban Hills, about 20 miles from Rome, a member of the Latin League, but still more famed for its temple of Juno Sospita,[*] and as the birth-place of Antoninus Pius, who made it his occasional residence. Velitræ, *Velletri,* on a southern spur of the Alban Hills, overlooking the Pontine Marshes, probably a member of the Latin League, though otherwise regarded as a Volscian town, and an active opponent of Rome in the Latin Wars, subsequently an ordinary

municipal town, and the native place of the Octavian family, from which the Emperor Augustus was descended. Signia, *Segni,* on a lofty hill at the N.W. angle of the Volscian Hills, founded by Tarquinius Superbus, and, with few exceptions, a faithful dependant of Rome, chiefly noted in later times for its astringent wine[*] used for medicinal purposes, its pears and vegetables, and a kind of cement known as "opus Signinum:" its Cyclopean walls may still be traced,

---

[*] It is said to have fallen through a law imposing silence on its inhabitants in reference to any report of an enemy approaching :—

Magnanimo Volscente satum, ditissimus agri
Qui fuit Ausonidum, et *tacitis* regnavit Amyclis.—*Æn.* x. 563.

[*] The name was connected with the tradition of a white sow appearing to Æneas :—

Ex quo ter denis urbem redeuntibus annis
Ascanius clari condet cognomini Albam. *Id.* viii. 47.

[*] The names Lanuvium and Lavinium are constantly interchanged in early Roman history; the modern name affords a further illustration of this.

[*] Lanuvio generate, inquit, quem Sospita Juno
Dat nobis, Milo, Gradivi cape victor honorem. *Sil. Ital.* xiii. 364.

[*] Quos Cora, quos spumans *inmiti* Signia musto.—*Id.* viii. 380.

Potabis liquidum Signina morantia ventrem;
Ne nimium sistant, sit tibi parca sitis. *Mart.* xiii. 116.

and there is a remarkable gateway in the same style. **Cora,**[4] *Cori* on a bold hill S.E. of Velitræ, at a very early period one of the first cities of Latium, for a time conquered by the Volscians, but regained by the Latins, now remarkable for the remains of its ancient walls, and a bridge thrown over a deep ravine. **Suessa Pometia,**[5] on the borders of the Pontine Marshes, which were supposed to be named after it, a place of great wealth at the time of its capture by Tarquinius Superbus, but not mentioned after B.C. 495, and utterly extinct in Pliny's time. **Setia,** *Sezze,* on a lofty hill overlooking the Pontine Marshes, about 5 miles to the left of the Appian Way, a Latin city, but at one period subject to the Volscians, the place where the Carthaginian hostages were deposited at the close of the Second Punic War, and celebrated under the empire for its superior wine. **Privernum,** *Piperno Vecchio,* on the E. slope of the Volscian Hills, overlooking the valley of the Amisenus, an important town of the Volscians, engaged in hostilities with Rome in B.C. 358 and 327, and under the empire noted for its wine. **Fregellæ,** on the left bank of the Liris, near its junction with the Trerus, a Volscian city, destroyed by the Samnites, but rebuilt by the Romans in B.C. 328, and subsequently signalized for its fidelity to Rome in the Second Punic War, and for its defection from that power in 125, when it was utterly destroyed. **Arpinum,** *Arpino,* on a hill in the upper valley of the Liris, originally belonging to the Volscians, then to the Samnites, and captured by the Romans in B.C. 305, chiefly famous, however, as the birth-place of Cicero and C. Marius,[7] the former of whom possessed a patrimonial estate there, and now remarkable for the remains of its Cyclopean walls and an old gateway. **Sora,** *Sora,* about 6 miles higher up the river, a Volscian town, captured by the Romans in B.C. 345, and colonised by them: under the empire a cheap, retired country town. **Frusino,** *Frosinone,* on the Via Latina, belonging originally to the Volscians, but in close connexion with the Hernicans, and at a later period having the same character as Sora. **Ferentinum,** *Ferentino,* on the Via Latina, between Frusino and Anagnia, a Hernican town, but subject to the Volscians about B.C. 413, actively engaged in the war against Rome in 361, a severe sufferer from the ravages of Hannibal's army in 211, and now famous for the remains of its Cyclopean walls. **Pedum,** *Gallicano,* between Tibur and Præneste, a member of the Latin League, and an active participator in the wars with Rome, particularly in the last great war, when it became the centre of hostilities, and was captured by Camillus. **Labicum**[8] or **Lavicum,** *La*

[1] Virgil (Æn. vi. 775) reckons it among the colonies of Alba :—
Pometios, Castrumque Inui, Bolamque, Coramque.

[3] See previous note.

[4] Nec faelli pretio, sed quo contenta Palerni
Testa sit, aut cellis Setia cara suis.			MART. x. 36.
Tunc illa time, cum pocula sumes
Gemmata, et lato Setinum ardebit in auro.			Juv. x. 26.

[7] Juvenal contrasts these two great men in the passages commencing—
Hic novus Arpinas ignobilis, et modo Romæ
Municipalis eques, &c.
Arpinas alios Volscorum in monte solebat
Poscere mercedes, &c.			Sat. viii. 237, 245.

[8] It is noticed by Virgil as one of the towns allied to Turnus :—
Et Sacranæ acies, et picti scuta Labici.			Æn. vii. 796.

Colonna, at the N.E. foot of the Alban Hills, and about 15 miles from Rome, a member of the Latin League, frequently mentioned in the history of the Æquian wars, but in after times a poor decayed place. Gabii, between Rome and Præneste, a colony of Alba[9] and a member of the Latin League, captured by stratagem by Tarquinius Superbus, and thenceforward a place rarely montioned in history, having sunk gradually to a state of decay[1] until a temporary revival of it took place under the emperors, probably on account of its cold sulphureous springs. Fidēnæ, *Castel Giubileo*, on a steep hill overlooking the Tiber, 5 miles from Rome, founded by Alba,[2] conquered and colonised by Romulus, and engaged in constant feuds with Rome until B.C. 438, when it was destroyed, and thenceforward remained a poor deserted place,[3] notorious only for a terrible disaster which happened in the time of Tiberius, when 50,000 persons were either killed or hurt by the fall of a wooden amphitheatre. **Ficulæa**, *Cesarini*, between Rome and Nomentum, about 9 miles from Rome, said to have been founded by the Aborigines, conquered by Tarquinius Priscus. **Crustumerium**, on the borders of the Sabine territory, and at one time regarded as a Sabine town, captured by Romulus, and again by Tarquinius Priscus, but subsequently unnoticed in history. **Nomentum**, *Mentana*, on the Sabine frontier N. of the Anio, and 14½ miles from Rome, a colony of Alba, and frequently noticed as a Latin town, and as taking part in the wars against Rome, the abode in later times of Seneca, Martial, Q. Ovidius, and Nepos.

Roads.—As Latium contained the metropolis of Italy, it was naturally the point to which all the great roads converged : we shall therefore consider ourselves as stationed at Rome, and describe the roads that issued from it. 1. The **Via Latina**, which we mention first as being probably the most ancient of all the Italian roads, issued from the Porta Capena, and led through Ferentinum, Frusino, Aquinum, and Teanum, to Casilinum in Campania, where it fell into the Via Appia. It skirted the Alban Hills near Tusculum, and followed the valleys of the Trerus and Liris to the borders of Campania. 2. The **Via Appia**, the great southern road of Italy, also issued from the Porta Capena, and made in a straight line for Tarracina on the sea-coast ; thence it went by Fundi to Formiæ, and then followed the sea-coast to Sinuessa, whence it struck inland to Capua, Beneventum, and ultimately to Brundisium. It was constructed as far as Capua in B.C. 312, by the Censor Appius Claudius. Between Rome and the Alban Hills this road was bordered with tombs and other buildings, the remains of which render it, even at the present day, one of the most remarkable objects in the neighbourhood of Rome. 3. The **Via Ostiensis** originally passed through the Porta Trigemina, but afterwards through the Porta Ostiensis, and followed the left bank of the Tiber to Ostia. 4. The **Via Portuensis** issued from the Porta Portuensis in the walls of Aurelian, and followed the right bank of the

---

[9] Ei tibi Nomentum, et Gabios, urbemque Fidenam,
Illi Collatinas imponent mœnibus arces.   *Æn.* vi. 773.

[1] Scis Lebedus quid sit ; Gabiis desertior atque
Fidenis vicus.   *Hor. Ep.* i. 11, 7.

Gabios, Velorque, Coramque
Pulvere vix tectæ poterant monstrare ruinæ.   *Luc.* vii. 392.

[2] See note [9] above.

[3] See quotation from Horace in note [1].

Tiber to Portus Trajani. 5. The **Via Labicana** passed out by the Porta Esquilina, and, passing by Labicum, fell into the Via Latina at Bivium, 30 miles from Rome. 6. The **Via Prænestina**, or, as it was originally called, **Via Gabina**, issued from the Porta Esquilina and led to Præneste; a branch thence communicated with the Via Latina near Anagnia. 7. The **Via Tiburtina** issued from the Porta Esquilina, crossed the Anio by a bridge 4 miles from Rome, and re-crossed it at the foot of the hill on which Tibur stood; it was thence continued, under the name of Via Valeria, to Corfinium and the Adriatic. 8. The **Via Nomentana** left by the Porta Collina, crossed the Anio just under the Mons Sacer, and thence reached Nomentum; a branch road from this point led to Eretum, where it fell into the Via Salaria. 9. The **Via Salaria** also issued from the Porta Collina, struck into the heart of the Sabine country by Reate, and thence was carried across the Apennines to Picenum and the Adriatic. 10. The **Via Flaminia**, the great northern road of Italy, crossed the Campus Martius and issued from the Porta Flaminia, crossed the Tiber by the Pons Milvius, 3 miles from Rome, into Etruria, where its course has been already described (pp. 512, 516). It was constructed by the censor C. Flaminius in B.C. 220. 11. The **Via Aurelia**, the Great Coast Road, issued from the Porta Janiculensis, and struck off towards the W. for the coast, which it reached at Alsium, whence it followed the line of coast throughout Etruria and Liguria (see pp. 512, 503).

Islands.—Off the coast of Latium lies a group of islands of **volcanic** origin, of which **Pontia**, *Ponza*, was the most considerable; it was colonized by the Romans in B.C. 313, and became under the emperors a place of confinement for state prisoners. The others were named **Palmaria**, *Palmarula*, **Sinonia**, *Zannone*, and **Pandataria**, *Vandotena*, also used as a State prison.

History.—The extension of the Roman supremacy over Latium was a long and gradual process. We find the kings waging successful war with the Latin cities (Alba itself being destroyed by Tullus Hostilius), and shortly after taking the supremacy of the Latin league, as appears from the treaty concluded with Carthage in B.C. 509. Upon the expulsion of the kings, however, the Latins regained their independence, and in 493 they concluded a treaty with Rome, the object of which appears to have been to counteract the growing power of the Volscians and Æquians. For the next 100 years little occurred to break this arrangement; some small wars were then waged with the Prænestines and others, which were but a prelude to the great struggle for independence in the war of 341-338, when the Latins combined with the Volscians, Æquians, and Hernicans against Rome. The battles of Vesuvius, Pedum, and Astura, decided the struggle in favour of the latter power. The Latins were subdued in 338, the Hernicans in 306, and the Æquians in 304. The period of the final subjection of the Volscians is not so certainly fixed; they were subjected, however, before 326.

Ruins of Capua.

CHAPTER XXVII.

ITALY—*continued.* CAMPANIA, APULIA, CALABRIA, LUCANIA, THE BRUTTII.

X. CAMPANIA.

§ 1. Campania was bounded on the N. by Latium, on the E. by Samnium, on the S. by Lucania, from which it was separated by the river Silarus, and on the W. by the Tyrrhenian Sea. These limits include the district of the Picentini in the S. The chief portion of this province consists, as its name (from *campus*) implies, of an extensive plain extending from the sea to the Apennines, and broken only by a group of volcanic hills between Cumæ and Neapolis, and by the isolated mountain of Vesuvius. This plain was bounded on the S. by a lateral ridge which strikes off from the Apennines at right angles to the general direction of the range, and protrudes into the sea at Prom. Minervæ, forming the southern termination of the Sinus Cumanus. On the other side of this range follows the hilly country of the Picentini. The soil of this plain is of volcanic origin, and has been celebrated in all ages for its extraordinary fertility.[1] It produced three and even four crops in the year, and was particularly famous for its sheep, its wine,[2] and its oil.[3] The genial mildness of the climate, combined with the beauty of the scenery, and the numerous thermal springs it possessed, rendered it highly attractive to the luxurious and wealthy Romans.

§ 2. The most conspicuous feature in the Campanian plain is the volcanic mountain Vesuvius, which rises in an isolated conical mass to the height of 4,020 feet, to the E. of Neapolis. No eruption is recorded before the terrible one in A.D. 79, which overwhelmed Herculaneum and Pompeii, and in which the elder Pliny perished;[4] two subsequent eruptions are recorded in ancient times, in A.D. 203 and 472. The summit of the mountain is described by Strabo as nearly level, and probably the present central cone first came into existence in A.D. 79. The volcanic group to the W. of Naples culminated in **Mons Gaurus**, *Monte Barbaro*, about 3 miles N. E. of Cumæ, famed for its excellent wines.[5] The plains to the N. of this were denominated by the Greeks of Cumæ the **Campi Phlegræi**, from the evident

---

[1]   Illa ubi laeta intexet vitibus ulmos ;
       Illa ferax oleo est : illam exsperiere colendo
       Et facilem pecori, et pallentem vomeris usu.
       Talem dives arat Capua et vicina Vesevo
       Ora jugo, et vacuis Clanius non æquus Acerris.—*Georg.* II. 221.

[2] The Massic, Falernian, Gaurian, and Surrentine, were the most celebrated kinds.

[3] The oil of Venafrum was particularly prized :—
                      Insuper addes
       Pressa Venafranæ quod bacca remisit olivæ.   Hor. *Sat.* II. 4, 68.
       Hos tibi Campani sudavit bacca Venafri.   Mart. xiii. 101.

[4] Previous to this, the fertility of the soil about Vesuvius was famed (see *Georg.* II. 221, above quoted). Martial contrasts with this the desolation that reigned there in his time :—

       Hic est pampineis viridis modo Vesbius umbris ;
       Presserat hic madidos nobilis uva lacus.
           .   .   .   .
       Cuncta jacent flammis et tristi mersa favilla :
       Nec superi vellent hoc licuisse sibi.   iv. 44.
[5]   frondentia læto
       Palmite devastat Nysæa cacumina Gauri.   Sil. Ital. xii. 160.

signs of volcanic action apparent on them:[4] they were also called
Campi Laborini, a designation preserved in the modern *Terra di La-
voro*, now applied to the whole district. On the borders of Samnium,
the ranges which overlook the plain, and which stand forth as the advanced guard of the central Apennines, were named **Tifata**, *Monte di
Maddaloni*, near Capua, and **Taburnus**, *Taburno*, S. of the Via Appia.
The range which we have already noticed as bounding the plain on
the S. was named **Lactarius**, *Monte S. Angelo*, from the excellent milk
produced from its pastures. Between the projecting points of **Prom.
Minerva** and **Misenum** lies the deep and beautiful *Bay of Naples*, to
which the ancients gave the name of **Crater** from its *cup-like* form,
though it was also otherwise named after the towns of Cumae and
Puteoli. The rivers of Campania are unimportant, with the exception
of the **Vulturnus**, previously described (p. 489); we may notice the
Savo, *Savone*, a small sluggish[7] stream N. of the Vulturnus; the
Clanius, to the S. of it, now converted into the canal of *Lagno*; the
Sebethus, which flows under the walls of Neapolis; the **Sarnus**, *Sarno*,
which waters the plain to the S. of Vesuvius; and the **Silarus**, *Sele*,
on the S. border. Campania possessed a few small lakes, one of
which, **Avernus**, has been previously noticed (p. 400), while another
hardly less famous was known by the name of **Lucrinus** *Lacus*; this
lay at the head of the Sinus Baianus, and was separated from the sea
only by a narrow barrier of sand: it was shallow, and hence peculiarly adapted for oyster-beds.[8] Agrippa constructed a port, named
Julius Portus, by opening communications between the Lucrine Lake
and the sea on one side and Lake Avernus on the other; at the same
time he constructed a mole of great strength outside the barrier of sand.[9]
This project turned out a failure. A large portion of the Lucrine Lake is
now occupied by the *Monte Nuovo*, a hill some 400 feet high, which
was thrown up by volcanic action in 1538.

§ 3. The original inhabitants of Campania were an Oscan or Opican

---

[4]　　　　　　　Tum sulphure et igni
Semper anhelantes coctaeque bitumine campos
Ostentant. Tellus, atro exundante vapore
Suspirans, ustisque diu calefacta medullis
Æstuat, et Stygias exhalat in aëra fistus.　　　Sil. Ital. xii. 133.

[7] Statius (*Silv.* iv. 3, 66) describes it as "piger Savo."

[8] Non me Lucrina juverint conchylia.　　　Hor. *Epod.* ii. 49.
Murice Baiano melior Lucrina peloris.　　　Id. *Sat.* ii. 4, 32.

[9] An memorem portus, Lucrinoque addita claustra,
Atque indignatum magnis stridoribus aequor,
Julia qua ponto longe sonat unda refuso,
Tyrrhenusque fretis immittitur aestus Averni?—*Georg.* ii. 161.
Debemur morti nos nostraque: sive receptus
Terra Neptunus classes aquilonibus arcet,
Regis opus.　　　Hor. *Art. Poet.* 63.

race. They were subdued by the Etruscans, and the date of this occurrence is variously fixed at B.C. 471 and 771. Finally the Samnites entered as a conquering race, and established themselves in the neighbourhood of Capua about B.C. 440. Throughout all these changes, however, the Oscan element remained the basis of the population, and imposed its language upon the conquerors. We have yet to notice the Greek settlers on the coast, who exercised a material influence in works of art. The Campanians were reputed generally a soft and luxurious race; at the same time they are noticed in history as serving as mercenaries in the Carthaginian armies. The towns of Campania rose at different periods of its history: the earliest settlement of which we hear was the Greek colony of Cumæ, founded (according to tradition) in B.C. 1050; this in turn founded the other Greek cities on the coast, Dicæarchia, Palæpolis, and Neapolis, and, according to some writers, Nola and Abella in the interior. The Etruscans are said to have had a confederacy of twelve cities in Campania, as they had in Etruria and Gallia Cisalpina, at the head of which stood Capua. This remained the chief town under the Samnites also, and was the place with which the Romans came into contact in the 4th century B.C. Under the Roman empire the towns on the Campanian coast rose to wealth and celebrity as the fashionable watering places of Italy; new towns sprang up at Baiæ and Bauli on the N. coast of the *Bay of Naples*; the whole circuit of the bay was studded with villas and palaces, and Neapolis, Pompeii and Surrentum were much frequented. The terrible disaster in A.D. 79 gave a temporary check to this prosperity; but the country soon recovered the blow, and remained one of the most flourishing and populous provinces of Italy down to the very close of the Western Empire. We shall describe the towns in their order from N. to S., taking first those on the sea-coast, and then those in the interior.

1. *On the Coast.*—CUMÆ, one of the most ancient and celebrated Greek colonies in Italy, stood on the summit of a cliff, 6 miles N. of Prom. Misenum. It was founded jointly by Chalcidians of Eubœa,[1] under Megasthenes, and Cymæans of Æolis, under Hippocles; and, according to agreement, it received the name of the one town and ranked as the colony of the other. The assigned date of its foundation (B.C. 1050) is too early to be accepted. It soon rose to commercial wealth and power, and founded several colonies in the neighbourhood. Its fall may be attributed to its internal dissensions, which led to the establishment of a despotism under Aristodemus, in 505, during whose rule Tarquinius Superbus took refuge and died there in 496. It suffered from the growing power of the Etruscans, who attacked it in

---

[1] Hence the epithet of Euboic, commonly applied to it:

Et tandem Euboicis Cumarum allabitur oris.   *Æn.* vi. 2.

Redibus Euboicam Stygiis emergit in urbem
Troius Æneas.   Ov. *Met.* xiv. 155.

474, and were only resisted by the aid of Hieron of Syracuse; and it was finally crushed by the Samnites, who captured it in 420. Under the Romans it became a *municipium* and a colony, but never regained its importance.[2] It was noted for its red earthenware and its flax. The chief celebrity of Cumæ is, however, derived from its being the reputed residence of the Sibyl, whose cave [3] existed in historical times, probably on the E. side of the cliff. The remains

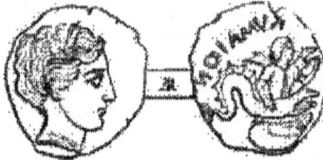

Coin of Cumæ.

of Cumæ are inconsiderable, but valuable works of art (statues, vases, &c.) have been discovered on its site. **Misenum**, on the promontory of the same name, first rose to importance under Augustus as the station of a fleet for the defence of the Tyrrhenian Sea, and is memorable as the scene of an interview between Octavian, Antony, and Sextus Pompeius. Lucullus had a magnificent villa there, which the Emperor Tiberius [4] subsequently acquired, and in which he died. Several interesting inscriptions have been found on the site of Misenum. **Baiæ**, *Baja*, was situated W. of Misenum and on the S.W. side of a bay, named after it, which penetrates inland between Misenum and Puteoli. Its port was frequented in early times; but the town rose, under the patronage of the Romans, towards the end of the Republic, and became one of the most popular watering-places on this coast.[5] Among the illustrious men who had villas there, we may notice Cicero, Lucullus, C. Marius, Pompey, Cæsar, Nero, Caligula, Hadrian (who died there), and Alexander Severus. Many of the villas were built on piles actually in the sea.[6] The chief relic of antiquity is the so-called Temple of Venus, near the sea-coast. **Puteoli**, *Pozzuoli*, was situated on the promontory which forms the E. boundary of the Sinus Baianus. It was founded by Greeks of Cumæ, in B.C. 521, and was originally named **Dicæarchia**. This was exchanged for Puteoli when the Romans got possession of it in the Second Punic War, the new name being

[2] Juvenal speaks of it as quite deserted :—
    Laudo tamen vacuis quod sedem figere Cumis
    Destinet, atque unum civem donare Sibyllæ.   *Sat.* III. 2.

[3] Excisum Euboicæ latus ingens rupis in antrum;
    Quo lati ducunt aditus centum, ostia centum,
    Unde ruunt totidem voces, responsa Sibyllæ.   *Æn.* vi. 42.

[4] Cæsar Tiberius quum petens Neapolim
    In Misenensem villam venisset suam,
    Quæ monte summo posita Luculli manu,
    Prospectat Siculum et prospicit Tuscum mare.—PHÆDR. ii. 5, 7.

[5] Nullus in orbe sinus Baiis prælucet amoenis.   HOR. *Ep.* i. 1, 83.
    Littus beatæ Veneris aureum Baiæ,
    Baiæ superbæ blanda dona naturæ,
    Ut mille laudem, Flacce, versibus Baias,
    Laudabo digne non satis tamen Baias.   MART. xi. 80.

[6] To this Horace alludes :—
    Mariæque Baiis obstrepentis urges
    Summovere littora
    Parum locuples continente ripa.   *Carm.* ii. 18, 20

derived either from the stench of the sulphureous springs,[7] or from the
wells (*putei*) of a volcanic origin about it. It was colonized by the
Romans in 194. It possessed an excellent harbour, which was further
improved by a mole, and which became the most frequented port for
Egyptian, Tyrian, and Spanish traffic. It was also frequented by the
wealthy Romans, and Cicero possessed a villa there, at which Hadrian
was afterwards buried. Caligula established a temporary bridge, two
miles long, between Baiæ and Puteoli. The remains are extensive, the
most important being those of the amphitheatre, of the mole, and of the
so-called temple of Serapis, probably used as a bath-house, and interesting from the proof which it affords of extensive changes in the level
of the soil on which it stands. **Neapolis**, *Naples*, was situated on the
W. slope of Mt. Vesuvius and on the banks of the small stream

Sebethus. It was founded by
Greeks of Cumæ,[8] and was
named Neapolis, "New City,"
in contradistinction to Palæpolis, "Old City," which had
been previously established, probably on the hill of Pausilypus.
The name of Parthenope appears
to have originally belonged to
Palæpolis, but was subsequently

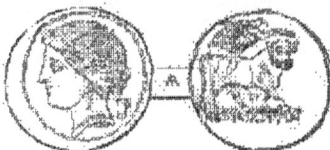

Coin of Neapolis.

transferred to Neapolis.[9] Neapolis was conquered by the Samnites
in B.C. 327, and passed into the hands of the Romans in 290: it
retained its Greek character under them, and hence became a favourite
resort[1] of the Romans before the end of the Republic. It was subsequently made a *municipium*, and finally a colony, though the
date of this latter change is uncertain. Of the Roman villas about
Neapolis that of Vedius Pollio, on the ridge named by him Pausilypus, and now *Posilippo*, was the most famous. The Emperors
Claudius and Nero had villas there, as also had the poets Virgil
(who was buried there), Statius, and Silius Italicus. The only remains of the town are two arches, part of an aqueduct, and the
ruins of a temple of Castor and Pollux. The tomb of Virgil[2] also
survives. **Pompeii** stood at the mouth of the river Sarnus and on the S.
side of Vesuvius. The line of the coast has been carried out two miles

---

[7] Near Puteoli was a spot called Forum Vulcani, now *Solfatara*, from the
number of holes whence issued sulphureous vapours.

[8] Hence the epithets of Euboic and Chalcidian given to it :—
  Anne quod Euboicos fessus remeare penates
  Auguror. STAT. *Silv.* iii. 5, 12.
  Omnia Chalcidicas turres obverm asistant. *Id.* ii. 2, 94.

[9] This is the name usually adopted by Statius and Silius Italicus.

[1]   In *otia natam*
  Parthenopen. Ov. *Met.* xv. 711.
  Et otiosa credidit Neapolis. Hor. *Epod.* v. 43.

Many literary men settled there; hence the epithet *docta* :—
  Et quas docta Neapolis creavit. MART. v. 78.

[2] Statius refers to it as being near Neapolis :—
    Maroneique sedens in margine templi
  Sumo animum, et magni tumulis adcanto magistri.—*Silv.* iv. 4, 51.

from the site of the town by the changes produced by the catastrophe in A.D. 79. The town was a very ancient one, and belonged successively to the Oscans and Etruscans; it served as the port of Nola, Nuceria, and other inland towns. It became a favourite abode of the

Temple of Venus at Pompeii.

Romans; and, among others, Cicero had a villa there. It was partly destroyed by an earthquake in A.D. 63, and utterly by the eruption of 79, which buried it beneath a vast shower of ashes and other volcanic substances. So completely did the town disappear, that even its site was unknown: it was discovered accidentally in 1689, and

Street of the Tombs at Pompeii.

excavations were commenced in 1755, which have been carried on at intervals to the present day, so that about half the town is now exposed to view. The most remarkable buildings are found in the Forum, and consist of the Temples of Jupiter, Venus, and Mercury, a Basilica, Baths, a Pantheon, &c. Outside the gate leading to Herculaneum lies the Street of the Tombs. The light which has been thrown on the private life of the ancients by these discoveries is invaluable. **Surrentum,** *Sorrento,* stood on the S. coast of the *Bay of Naples,* about 7 miles N.E. of Prom. Minervæ. It was reputed a Greek town, but this, as well as its remaining history, is a matter of uncertainty. It was chiefly famed for the wine grown on the neighbouring hills,[3] and for its pottery. Pollius Felix, the friend of Statius, had a villa there, of which extensive ruins still remain. **Salernum,** *Salerno,* was situated in the territory of the Picentini on the N. shore of the Sinus Pæstanus. We know nothing of it previous to the settlement made by the Romans there, in B.C. 194, for the purpose of holding the Picentini in check. It thenceforward became the chief town in this part of Campania.[4]

2. *In the Interior.*—**Teanum,** surnamed **Sidicinum,** to distinguish it from the Apulian town of the same name, stood on the Via Latina in the extreme N.E. of the province. It was originally the capital of the Sidicini, and its position on the Via Latina made it important as a military post. It received a colony under Augustus, and remained a large and populous town under the Empire. Remains of an amphitheatre and of a theatre

Coin of Teanum Sidicinum.

exist on its site. **Capua,**[5] *Sta. Maria di Capua,* was situated about two miles S. of the Vulturnus and one from the foot of Mount Tifata. It was called Vulturnum under the Etruscans; it was either founded or colonized by the Etruscans, but the date of this event is quite uncertain. The Samnites captured it in B.C. 423; its first intercourse with Rome was in 343, when it obtained aid against the Samnites; in 216 it joined the cause of Hannibal, and in 211 was severely punished by Rome for this defection. It was placed under a Roman Præfectus, was made a colony by Cæsar in 59, and was re-colonized by Nero. The luxury and refinement of the Capuans became proverbial. The town, being built on a plain, was of great extent; it was surrounded by walls, and had seven gates. In the neighbourhood the famous

---

[3] Inde legit Capreas, promontoriumque Minervæ,
   Et Surrentino generosos palmite colles.   Ov. *Met.* xv. 709.
   Carasque non molli jugra Surrentina Lyæo.   Stat. *Silv.* iii. 5, 102.

[4] It was visited by Horace for the improvement of his health :—
   Quæ sit hiems Veliæ, quod cœlum, Vala, Salerni,
   Quorum hominum regio, et qualis via ?   *Ep.* i. 15, 1.

[5] The origin of the name is uncertain ; Virgil derives it from Capys :—
   Et Capys : hinc nomen Campana ducitur urbi.—*Æn.* x. 145.

It is probably connected with Campus on account of its situation on a plain.

Falernian wine was produced. Some portions of the ancient walls, of an amphitheatre, and of a triumphal arch remain. The town was destroyed A.D. 840, and was rebuilt on the site of Casilinum, 3 miles distant, which has hence inherited the name of *Capua*. Nola, *Nola*, stood 21 miles S.E. of Capua, between Vesuvius and the Apennines: it was a town of great antiquity, founded by the Ausonians, colonized by the Greeks of Cumæ,[6] occupied successively by the Etruscans and Samnites, and, finally, conquered by the Romans in B.C. 313. It was signalized for its fidelity to Rome after the battle of Cannæ, in reward for which it was allowed to retain its constitution; it withstood Hannibal on no less than three occasions in the Second Punic War.[7] It bore a conspicuous part in the Social War, having been occupied by the allies, and subsequently captured and destroyed by Sulla. It was rebuilt, and received colonies under Augustus and Vespasian. Augustus died there. Numerous inscriptions in the Oscan language and a vast number of Greek painted vases have been found at Nola. Nuceria, *Nocera*, surnamed Alfaterna, to distinguish it from other towns of the same name, stood on the Sarnus, about 9 miles from its mouth, and on the Appia Via. Its early history is unknown. In B.C. 315 it is noticed as joining the Samnites against Rome, and in 308 it was taken by the consul Fabius. In 216 it was taken by Hannibal, and its inhabitants were subsequently re-settled at Atella. Nuceria was, however, rebuilt and received colonies under Augustus and Nero.

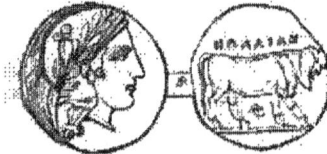

Of the less important towns we may notice:—

(1.) *On the Coast.*—Volturnum, *Castel Volturno*, at the mouth of the Volturnus, originally only a fort erected by the Romans in the Second Punic War, but subsequently colonized in B.C. 194; Liternum, *Tor di Patria*, on the verge of a marsh or lagoon called the Literna Palus,[8] a place famous as the retreat of Scipio Africanus, who died and, according to one account, was buried there; Bauli, between Baiæ and

Coin of Capua.

Coin of Nola.

---

[6] Hence it is termed Chalcidian:—

    Illinc ad Chalcidicam transfert citus agmina Nolam.—SIL. ITAL. xii. 161.

[7] Campo Nola ardet crebris circumdata in orbem
    Turribus, et celso facilem tutatur adiri
    Planitiem vallo.   SIL. ITAL. xii. 162.

[8] Illinc calidi fontes, lentisciferumque tenentur
    Liternum.   Ov. Met. xv. 713.

Prom. Misenum, a favourite resort of the Romans, and, among others, of Hortensius and of Nero, who here planned the death of Agrippina;[9] Herculaneum, *Ercolano*, at the foot of Vesuvius, founded by the Oscans, occupied by the Etruscans, and subsequently by Greeks, captured by the Romans in the Social War, and finally buried to a depth of from 70 to 100 feet beneath the ground by the same catastrophe which destroyed Pompeii; it was discovered in 1738, and partly explored, the chief buildings found being a theatre capable of seating 10,000 persons, portions of two temples, and other buildings; Stabiæ, *Castell-a-Mare di Stabia*, 4 miles S. of Pompeii, destroyed by Sulla in the Social War, subsequently the residence of several Romans, and, among others, of Pomponianus, the friend of the elder Pliny, who perished here in the overwhelming catastrophe of A.D. 79; and, lastly, Picentia, *Vicenza*, the chief town of the Picentini.

(2.) *In the Interior.*—Cales, *Calvi*, on the Via Latina, S.E. of Teanum, originally the capital of the Ausonian tribe named Caleni, subsequently taken and colonized by the Romans in B.C. 335, and especially famed for its fine wine;[1] Casilinum, *Capua*, on the Vulturnus, famed for the noble stand made there by 1000 Roman troops against the whole army of Hannibal in B.C. 216; Atella, midway between Capua and Neapolis, historically famous only for the severe punishment inflicted on it by the Romans in B.C. 211 for its defection to Hannibal, and otherwise better known for the dramatic representations, named "Fabulæ Atellanæ," which originated there; and, lastly, Acerræ, *Acerra*, 8 miles N.E. of Neapolis, which received the Roman franchise in B.C. 332, was destroyed by Hannibal in 216, and rebuilt in 210.[2]

Roads.—Campania was traversed by the Via Appia, which entered it at Sinuessa, struck inland to Casilinum and Capua, and quitted it for Caudium and Beneventum; this portion of the road could not have been constructed before the end of the Samnite Wars. The Via Latina entered Campania near Teanum and passed by Cales to Casilinum, where it fell into the Appian Way. Other roads, the names of which are unknown, led from Capua by Nola and Nuceria to Salernum, and so on to Rhegium, and again from Sinuessa along the coast to Cumæ and Neapolis.

Islands.—Off the coast of Campania lie the following islands:— Prochyta, *Procida*, off Prom. Misenum, from which it is distant about 3 miles, a flat and comparatively low island, and, though now thickly populated, formerly uninhabited:[4] Ænaria; *Ischia*—the Pithecusa of

---

[9] Dum petit a Baulis mater Cærelia Baiæ,
   Occidit insani crimine mersa freti.   MART. IV. 63.

[1] Cæcubum et præso domitam Caleno
   Tu bibes uvam.   HOR. *Carm.* i. 20, 9.
   Premant Calena falce, quibus dedit
   Fortuna vitem.   *Id.* i. 31, 9.

[2] It appears to have been a poor, forsaken place:—
   Et racui Clanius non æquus Acerris.—VIRG. *Georg.* ii. 225.
   Allifæ, et Clanio contemptæ semper Acerræ.   SIL. ITAL. viii. 537.

[3] Virgil's epithet "alta" is incorrect:—
   Tum sonitu Prochyta *alta* tremit, durumque cubile
   Inarime Jovis imperiis imposta Typhœo.   Æn. ix. 715.

[4] Ego vel Prochytam præpono Suburræ.   JUV. iii. 5.

the Greeks, and the Inarime[b] of the Latin poets — a little W. of Prochyta, of volcanic origin, and hence both fertile and provided with thermal springs; and Capreæ,[c] *Capri*, off Prom. Minervæ and at the S. extremity of the Bay of Naples, a lofty and almost inaccessible mass of limestone rock, which became the imperial abode, occasionally of Augustus and permanently of Tiberius,[f] during the last ten years of his life.

History.—We have already stated that the Oscans, the Etruscans, and the Samnites became the successive masters of the rich plains of Campania. It remains for us to narrate the circumstances of the Roman conquest. Capua, having been attacked afresh by the Samnites, in B.C. 343, solicited the aid of Rome, which was accorded, and resulted in the victories of Valerius Corvus at Mt. Gaurus and Suessula, and the expulsion of the Samnites. The Campanians, *i. e.* the Capuans, thus became the nominal subjects of Rome: nevertheless, they joined in the Latin War, in 340, and were defeated at the foot of Mt. Vesuvius by the consuls T. Manlius and P. Decius. The submission of the other towns of Campania shortly afterwards followed, viz., of Neapolis, in 326, of Nola, in 313, and of Nuceria, in 308, and at the end of the Second Samnite War, in 304, Rome was master of all the province. In the Second Punic War, when Campania was one of the chief seats of war, Capua and some of the smaller towns espoused the cause of Hannibal, while Casilinum, Nola, and Neapolis, remained faithful. The capture of Capua by the Romans, in 212, re-established their supremacy.

XL APULIA.

§ 4. **Apulia** was situated on the E. coast of Italy, and was bounded on the N. by the Tifernus, dividing it from Picenum; on the W. by Samnium; on the S. by Lucania and Calabria, from the former of which it was separated by the river Bradanus, and from the latter by a line drawn across the Messapian peninsula from the head of the Tarentine bay to a point between Egnatia and Brundusium; and on the E. by the Adriatic Sea. The N. portion, from the Tifernus

---

[b] The name Inarime appears to be derived from the Homeric Ἄριμοι, the fable of Typhœus having been transferred from Asia to Italy. Ovid incorrectly distinguishes Inarime and Pithecusa :—

> Orbataque præside pinus
> Iuarimen, Prochytenque legit, sterilique locatas
> Colle Pithecusas, habitantum nomine dictas.   *Met.* xiv. 88.

[c] The original occupants of this island are said to have been named Teleboæ, a people whom we only know as occupying the Echinades, off the W. coast of Greece :—

> Œbale, quem generasse Telon Sebethidis nympha
> Fertur, Teleboum Capreas quum regna teneret.—*Æn.* vii. 734.

[f] Juvenal speaks of him as—

> Principis, angusta Caprearum in rupe sedentis.—*Sat.* x. 93.

Statius applies to it the epithet "dites," apparently in reference to the palaces erected by Tiberius :—

> *dites* Capreæ viridesque resultant
> Taurubulæ, et terris ingens redit æquoris echo.—*Silv.* iii. 1, 138.

Plain of Cannæ.

to the Aufidus, consists almost wholly of a great plain sloping down
from the Pyrenees to the sea, the only exception being the isolated
mass of Garganus, the "spur" of Italy, on the sea-coast. The S.
portion is for the most part covered with barren hills, which emanate
from the Apennines near Venusia, and extend in a broad chain towards Brundusium : between these and the sea is a narrow strip of
land of great fertility. The northern plains afford pasture for vast
numbers of horses and sheep during the winter months; in the
summer they become parched in consequence of the calcareous
nature of the soil, and at this period the flocks are removed to the
highlands of Samnium, which are then rich, but are covered with
snow in the winter. A constant interchange thus takes place between these two districts, and has done so from the earliest ages :
the Romans imposed a tax on all flocks and herds thus migrating.
The only mountains that received special designations were Gar-
ginus, which projects above 30 miles into the sea, forming a
vast promontory,[2] of which Mons Matinus[3] was the most southerly

----

[2] The forests, for which it was formerly so famous, have now disappeared :—
        Aquilonibus
    Querceta Gargani laborant.   Hor. Carm. ii. 9, 7.
    Garganum mugire putes nemus, aut mare Tuscum.—In. Ep. ii. 1, 202.
[3] This and all the other heights of Garganus are covered with aromatic herbs,
and produce excellent honey :—
        Ego apis Matinæ
    More modoque
    Grata carpentis thyma per laborem
    .   .   .   .
    Carmina fingo.   _   Hor. Carm. iv. 2, 27.

offshoot; and **Vultur**, *Monte Voltore*, an isolated hill of volcanic origin on the borders of Lucania and Samnium. The rivers are—the **Tifernus**, *Biferno*, on the N. boundary; the **Frento**, *Fortore*, N. of Garganus; the **Cerbalus**, *Cervaro*, S. of that mountain; the **Aufidus** (p. 489); and the **Bradanus**, *Braduno*, on the borders of Lucania, falling into the Tarentine Gulf. These rivers are small in summer, but exceedingly violent in winter, and at this season they not unfrequently inundate the plains.

§ 5. The inhabitants of Apulia were a mixed race, consisting of the three following elements :—(1.) The **Apuli**, probably an Oscan race; (2.) the **Daunii**, a Pelasgian race; and (3.) the **Peucetii** or **Pœdiculi**, also of Pelasgian origin. The two former races were fused into one people in historical times, and occupied the plains of Northern Apulia; the third lived separately in the hilly country of the S. The Apulians were not united under one government at the time the Romans came in contact with them, but each town formed an independent community. Of these, Arpi, Canusium, Luceria, and Teanum, appear to have been most prominent. These towns are frequently mentioned in the Second Samnite, the Second Punic, and the Social Wars, but subsequently became historically unimportant. Their chief interest is derived from the large amount of Hellenic influence which was infused into them by Tarentum and the other Greek towns in those parts, and which is manifest both in their coins and in the numerous works of art, particularly painted vases, discovered on their sites. We shall describe first those in the interior, then those on the coast.

(1.) *In the Interior.*—**Larinum**, *Larino Vecchio*, was situated 14 miles from the coast, a little S. of the Tifernus. It is sometimes regarded as belonging to the Frentani; it did not originally belong to either, but formed a separate and independant state. In Augustus's [1] division, however, it was included in Apulia. During the Second Punic War its territory was the scene of several operations between the Roman and Carthaginian armies; the town itself is seldom noticed. **Arpi**, *Arpa*, the **Argyripa** of the poets, [2] stood in the centre of the great Apulian plain, 20 miles from the sea. Its foundation was attributed to Diomede, but without any solid reason. Its extent and population were very large at the time of the Second Punic War. In this it was originally friendly to Rome, but after the battle of Cannæ it joined Hannibal, and was in consequence severely punished by the Romans in B.C. 213 ; from that time it

[1] Horace seems to refer to its position as partly in and partly out of Apulia, when he says :—

> Me fabulosæ Vulture in *Appulo*
> Altricis *extra limen* Apullæ.   *Carm.* III. 4, 9.

[2] The name first appears in Lycophron : it was adopted from the Greeks by the Latins :—

> Ille urbem Argyripam, patriæ cognomine gentis,
> Victor Gargani condebat Iapygis arva.   *Æn.* xi. 246.

sank. Canusium, *Canosa*, stood near the right bank of the Aufidus, about 15 miles from its mouth. Its origin was attributed to Diomede, and it certainly had a strong infusion of the Greek element in it,[3] but there are no grounds for supposing it to be a Greek colony. It was conquered by the Romans in B.C. 318, and is memorable for the hospitality afforded to the Roman army after the defeat at Cannæ. It received a colony under M. Aurelius. It possessed a splendid aqueduct, made by Herodes Atticus, to supply its natural deficiency of water.[4] Its remains, consisting of portions of the aqueduct, of an amphitheatre, and a gateway, belong to the Roman era. Luceria, *Lucera*, was situated about 12 miles W. of Arpi; it was probably of Oscan origin. It first appears in history as friendly to Rome in the Second Samnite War, then as captured by the Samnites, and recovered by the Romans in B.C. 320, recaptured by the Samnites, and again recovered in 314, and finally besieged by the Samnites in 294. In the Second Punic War it was the head-quarters of the Romans in Apulia. It subsequently became a colony, and remained a considerable town.[5] Venusia, *Venosa*, lay on

the frontiers of Lucania,[6] and on the Appia Via. It was captured by the Romans in B.C. 262, and shortly afterwards was colonized by them. It became the Roman head-quarters after the battle of Cannæ. In the Social War it was the stronghold of the allies in these parts. Its position

Coin of Venusia.

on the Appian road secured its subsequent prosperity, and it is well known to us as the birth-place of Horace. (2.) *On the Coast.*—Sipontum,[7] *Sta. Maria di Siponto*, stood immediately S. of Garganus, and was reputed to have been founded by Diomede. It was captured by Alexander of Epirus,

---

[3] That the Greek tongue prevailed here to a great extent, appears from Horace's allusion :—

  Canusini more bilinguis.   *Sat.* I. 10, 30.

[4] To this Horace alludes :—

  Nam Canusi lapidosus ; aquæ non ditior urna :
  Qui locus a forti Diomede est conditus olim.   *Sat.* i. 5, 91.

The gritty quality of the bread, to which "lapidosus" refers, is still noticed by travellers, and arises probably from defective millstones.

[5] Its wool was famous :—

  Te lanæ prope nobilem
  Tonsæ Lucerism, non citharæ, decent.   Hor. *Carm.* III. 15, 13

[6] Hence Horace speaks of himself as—

   Lucanus an Appulus, anceps,
  Nam Venusinus arat finem sub utramque colonus.—*Sat.* ii. 1, 34.

[7] The poets adopted the Greek form of the name, Sipus :—

  Quæsivit Calaber, subducta luce repente
  Immensis tenebris, et terram et littora Sipus.   Sil. Ital. viii. 631.

  Quas recipit Salapina palus, et subdita Sipus
  Montibus.   Luc. v. 377.

in B.C. 330, was colonized by the Romans in 194 and again at a later period, and became a place of considerable trade in corn. **Salapia,** *Salpi,* lay more to the S. on a lagoon named Salapina Palus,[a] which formerly had a natural, but now has only an artificial outlet to the sea. It was the head-quarters of Hannibal in B.C. 214, was captured by the Romans in 210, and again attacked by the Carthaginians in 208. It was destroyed by the Romans in the Social War, and never recovered its prosperity.

Of the less important towns we may notice **Teanum,** surnamed **Apulum,** *Civitate,* on the Frento, about 12 miles from its mouth, noticed as being conquered by the Romans in B.C. 318, and the head-quarters of M. Junius Pera in the Second Punic War; **Herdonia,** *Ordona,* on the Via Egnatina, the scene of the Roman defeats by Hannibal in B.C. 212 and 210; **Asculum,** *Ascoli,* 10 miles S. of Herdonia, the scene of the great battle between Pyrrhus and the Romans in B.C. 279; **Cannæ,**[b] *Canne,* on the Aufidus, 6 miles from its mouth, celebrated for the memorable defeat of the Romans by Hannibal in B.C. 216, which took place on the N. side of the river (see note at end of chapter, BATTLE OF CANNÆ); **Barium,** *Bari,* on the coast, about 36 miles S. of the Aufidus, on the Via Trajana, noticed by Horace as a fishing town;[1] and **Egnatia,**[c] or **Gnatia,** at the point where the Appia Via came upon the coast.

Roads.—Apulia was traversed by the two great branches of the Appian Way—the **Via Trajana,** which passed through Herdonia, Canusium, and Barium to Brundusium, and the **Via Appia,** properly so called, which passed through Venusia to Tarentum.

History.—Apulia first comes into notice in the Second Samnite War, as in alliance with Rome, with the exception of a few towns which joined the Samnites. Pyrrhus reduced several of its cities in B.C. 279, but did not shake the fidelity of the province generally. In the second Punic War it was for several successive years the winter quarters of Hannibal, and, after the battle of Cannæ, many of the cities joined his cause. The punishment inflicted subsequently by the Romans was very severe. In the Social War the Apulians embraced the side of the allies, and the renewed punishment then inflicted on them by the Romans proved fatal to the prosperity of the province.

XII. CALABRIA.

§ 6. **Calabria** was the name given to the peninsula which runs out to the S.E. of Tarentum, and which is commonly known as the

---

[a] See Luc. v. 377, in previous note.

[b]   Ut ventum ad Cannas, urbis vestigia prisco,
      Defigunt diro signa infellcia vallo.   SIL. ITAL. viii. 624.

[1]   Postera tempestas melior, via pejor, adusque
      Bari mœnia piscosi.   HOR. Sat. i. 5, 96.

[c]   Horace seems to describe its water as bad ("lymphis iratis"), but it is now celebrated for the abundance and excellence of its water. The pretended miracle which he witnessed is also noticed by Pliny (ii. 111).

           Dehinc Gnatia, Lymphis
      Iratis exstructa, dedit risusque jocosque,
      Dum flamma sine thure liquescere limine sacro
      Persuadere cupit.   Sat. i. 5, 97.

Brundusium.

"heel" of Italy. The Greeks named it Messapia and Iapygia—terms which are used with varying significance by different writers. The whole of this peninsula is occupied by broad and gently undulating hills of small elevation. The soil is dry, being of a calcareous nature : it was nevertheless famed for its fertility, and particularly for its growth of olives. The province was also famous for its horses, wines, fruit, honey, and wool, and, in another sense, for its venomous serpents. It possesses no stream of any size. The inhabitants of Calabria were divided into two tribes—the Messapii or Calabri proper, who occupied the E., and the Sallentini, who occupied the W. and S. coasts. These tribes belonged to the Pelasgian stock, and were not originally distinct. They appear to have attained a certain degree of culture before the appearance of the Greek settlers, and they possessed the towns of Hydruntum and Hyria. The foundation of Tarentum, about 708 B.C., formed an era in the history of this province. It was the metropolis of this part of Italy until the period when the Romans established their ascendency. Under them Brundusium rose to importance as the terminus of the Appian Way, and the chief port for communication with Greece.

Brundusium or Brundisium, *Brindisi*, was situated on a small enclosed bay, which communicated with the sea by a narrow channel, and ter-

---

* The Sallentini were traditionally believed to be of Cretan origin :—

Et Sallentinos obsedit milite campos

Lyctius Idomeneus.   *Æn.* III. 400.

minated Inland in two arms, giving it a general resemblance to a stag's head, from which it is said to have derived its name. This bay formed an admirable port, about which the Sallentini[4] built a town, and which the Romans acquired in B.C. 267 and colonized in 244. It was the scene of many interesting events ; of Sulla's landing from the Mithridatic war in 83, of Cicero's return from his exile, of the blockade of the fleet of Pompey by Cæsar, of the death of Virgil, and of Agrippina's landing with the ashes of Germanicus. Its name is familiar to us from the visit of Horace, who went thither with Mæcenas and Cocceius, when

the place was threatened by Antony in 41. Hydruntum, *Otranto*, the Hydrus of the Greeks, was situated S.E. of Brundusium, and was the nearest point to Greece. It was a customary port of embarkation for the East as early as 191 B.C., and ultimately, in the 4th century A.D., supplanted Brundusium as the principal port in that district. Tarentum, *Taranto*, was situated on a peninsula at the entrance of an extensive but shallow bay, which runs inland for some 6 miles from the head of the Tarentine Gulf. This bay served as its port, being connected with the sea by a channel so narrow that a bridge is now thrown across it. The surrounding country was remarkably fertile, and its climate luxuriously soft. It was founded by a colony from Sparta,[5] led by Phalanthus[6] in B.C. 708. For the first two centuries of its existence we hear little of it, but it was growing in wealth and commercial greatness. A terrible defeat sustained by the Tarentines from the Messapians in 473 is the first event of importance in their history. In 432 they were engaged in war with the Thurians, which ended in the joint foundation of Heraclea. In 346

Plan of Brundusium.

A A. Inner Harbour.   B. Outer Harbour.   C. Spot where Cæsar tried to block up the entrance of the Inner Harbour.   D. Modern city of *Brindisi*.   E. Islands of S. Andrea, the ancient Barra.

---

[4] Hence its foundation is assigned by Lucan (ii. 610) to the Cretans.

[5] Hence the epithet of "Lacedæmonian," and the name Œbalia, an ancient name of Laconia, are applied to it :—

Navigat Ionium, Lacedæmoniumque Tarentum.—Ov. *Met.* xv. 50.
  Aut Lacedæmonium Tarentum.   Hor. *Carm.* III. 5, 56.

Namque mb Œbaliæ memini me turribus altis
Qua niger humectat flaventia culta Galæsus,
Corycium vidisse senem.   *Georg.* iv. 125.

[6] Dulce pellitis ovibus Galesi
Flumen, et regnata petam Laconi
  Rura Phalanto.   Hor. *Carm.* II. 6, 10.

they were involved in a more serious struggle with the Lucanians and
Messapians, and they were obliged to call in the aid first of the Spartans,
whose leader, Archidamus, fell in battle in 338, and afterwards of Alexander of Epirus, who finished the war with the Lucanians, and then
himself became the enemy of the Tarentines. In 302 they came for
the first time into collision with the Romans in consequence of an attack
made on ships that had passed the stipulated boundary, viz., the Lacinian cape. The Tarentines called in the aid of Pyrrhus in 281, after
whose withdrawal in 274 resistance became futile, and their city was
taken in 272. The only other important events are the revolt of Tarentum to Hannibal in 212, and its recovery by the Romans in 207,
when it was most severely treated. It then fell into a state of decay,
but was subsequently revived by a colony sent there in 123, and it
became a naval station of importance under the empire. The general
form of the city was triangular, having the citadel at the apex,
adjoining the mouth of the harbour. Hardly any remains of it exist.
The chief productions of its territory were honey, olives, wine,[7] wool
of the very finest description,[8] horses, fruit, and shellfish, which were
used both as an article of diet, and for the preparation of the famous
purple dye.[9] The Tarentines were reputed a luxurious and enervated
race.[1]

Of the less important places we may notice: Castra Minervæ, between
Hydruntum and the Iapygian promontory, named after a temple of
Minerva which occupied a conspicuous position on a cliff;[2] Manduria,
Manduria, 24 miles E. of Tarentum, the scene of the great battle in
which Archidamus perished; Uria or Hyria, midway between Brun-
dusium and Tarentum, the ancient metropolis of the Messapians; And
Callipolis, *Gallipoli*, on the W. coast, a Lacedæmonian colony with an
excellent port, which is, however, unnoticed in ancient times.

Roads.—There were three roads in Calabria—one a continuation of
the Via Trajana, which led from Brundusium to the Iapygian promontory; another from Tarentum to the same point; and a third from
Tarentum to Brundusium.

History.—The history of Calabria may be disposed of in a few words,
In spite of the great defeat which the Tarentines received in B.C. 473,
as already related, they succeeded in establishing a supremacy over the
tribes of the peninsula. The fall of Tarentum into the power of the
Romans involved almost as a matter of course the submission of the
whole peninsula, which was obtained in a single campaign.

---

[7] The best kind was grown on a hill named Aulon, as we learn from the
passage in which Horace expatiates on the fertility of the Tarentine territory :—

　　　Amicus Aulon
　Fertili Baccho minimum Falernis
　Invidet uvis.　　　　　　　　　Hor. *Carm.* II. 6, 18.

[8] The pastures about the small stream Galæsus produced the best (see notes [7]
and [9] above).

　　[9] Lana Tarentino violas imitata veneno.　　　ID. *Ep.* II. 1, 207.

　　[1] Pectinibus patulis jactat se molle Tarentum.　　　ID. *Sat.* II. 4, 34.
　　Sed vacuum Tibur placet, aut imbelle Tarentum.—ID. *Ep.* I. 7, 45.

　[2] Virgil represents this as the first object which met the eye of Æneas as he
approached the Italian coast :—

　　Crebrescunt optatæ auræ : portusque patescit
　　Jam propior, templumque apparet in arce Minervæ.—Æn. III. 530.

XIII. Lucania.

§ 7. **Lucania** was bounded on the N. by an irregular line crossing from the Silarus on the Tyrrhenian coast to the Bradanus on the Tarentine Bay; in this direction it was contiguous to Campania, Samnium, and Apulia; on the S. it was separated from the land of the Bruttii by the rivers Latis and Crathis; on the E. and W. it was bordered by the sea. The province is traversed in its whole length by the Apennines, which approach more nearly to the W. than the E. coast, and descend on the former side in lofty and rugged chains almost to the coast itself, while on the latter they slope gradually off, and leave a broad and remarkably fertile strip of plain between the mouths of the Bradanus and the Siris. S. of the Siris the mountains approach the W. coast, but again recede and leave a considerable plain about the Crathis. The interior of Lucania was and still is one of the wildest regions of Italy, most of it being covered with immense forests which gave support to vast herds of swine, as well as to wild boars and bears. The only mountain with whose name we are acquainted is **Alburnus**,[3] *Monte Alburno*, S. of the river Silarus. The rivers, though numerous, are unimportant: on the E. coast we may notice, from N. to S., the **Bradanus**, *Bradano*, on the borders of Apulia; the **Casuentus**, *Basiento*, which runs parallel to it and joins the sea at Metapontum; the **Aciris**, *Agri*, and **Siris**,[4] *Sinno*, which join the sea at no great distance from each other; the **Sybaris**, *Coscile*, a small stream flowing by the town of the same name; and the **Crathis**[5] on the S. frontier. On the W. coast the chief stream is the **Silarus**, *Sele*, with its tributaries the **Tanager**, *Tanagro*, and the **Calor**, *Calore*.

§ 8. The earliest inhabitants of this country were a Pelasgic race, named Œnotrians: they seem to have been an unwarlike people, and were gradually driven into the interior by the Greeks, who settled on the coast and gave to it and the coast of the adjacent province of Bruttium the title of **Magna Græcia**. The Lucanians were a branch of the Samnite nation, who pressed down southward probably about B.C. 400, subdued the Greek cities, and spread over the

---

[3] It is noticed by Virgil, *Georg.* iii. 147.

[4] The beauty of the district about the Siris, called Siritis, is noticed by Archilochus:—

Οὐ γάρ τι καλὸς χῶρος, οὐδ᾽ ἐφίμερος
Οὐδ᾽ ἐρατός, οἷος ἀμφὶ Σίριος ῥοάς.   *Ap. Athen.* xii. p. 523.

[5] The waters of the Crathis were reputed to turn the hair to a golden hue:—

Ὁ ξανθὰ χαίτας Φυσσπίνων
Κράθις.   *Euarr. Trœzd.* 223.

Crathis et huic Sybaris, nostris conterminus arvis
Electro similes faciunt auroque capillos.   *Ov. Met.* xv. 315.

whole of the interior. The towns of Lucania may be divided into two classes: those on the coast, which were of Greek origin; and those in the interior, which were either native Lucanian towns or Roman colonies of a later date. The former class comprises some of the most important towns of Magna Græcia, such as Heraclea, Sybaris, Velia, and Pæstum. In the latter class we may specially notice Grumentum on the Aciris. We shall describe these towns in order, commencing with those on the E. coast, from N. to S.

Metapontum was situated on the coast between the rivers Bradanus and Casuentus, about 24 miles from Tarentum. It was founded by Achæans under Leucippus, probably about 700–690 B.C., on the site (as it was said) of an earlier town. The philosopher Pythagoras retired and died there. In 415 the Metapontines joined the Athenians in their Sicilian expedition. In 332 they aided Alexander of Epirus against the Lucanians, but

Coin of Metapontum.

in 303 they refused the alliance of Cleonymus, and suffered in consequence. In the Second Punic War Metapontum was occupied by Hannibal in the years 212–207, and after his withdrawal it was forsaken by its inhabitants, and the place ceased to be of any importance. The remains consist of the ruins of a Doric temple, of which 15 columns are standing, and some portions of another temple; they lie near *Torre di Mari*. Heraclea was situated between the rivers Aciris and Siris. It was founded in B.C. 432 by a joint colony of Thurians and Tarentines. It soon rose to importance and became the place of congress for the Italiot Greeks. It was taken by Alexander of Epirus, and was the scene of a battle between the Romans and Pyrrhus in 280. It was

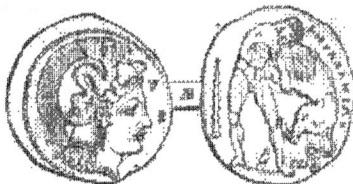

Coin of Heraclea.

partly destroyed in the Social War. Large heaps of ruins near a farm, named *Policoro*, mark its site; in these have been found coins, bronzes, &c., and particularly two tables, known as the Tabulæ Heraclienses, containing much information relating to municipal law. Zeuxis, the painter, was probably born at this Heraclea. Siris stood at the mouth of the river of the same name. It was a place of great antiquity and was reputed a Trojan colony, but was more probably a city of the Chones. Ionians from Colophon settled there between 690 and 660 B.C., and made it a flourishing Greek town. Of its history we know nothing: it probably perished between 550 and 510. Sybaris was situated between the rivers Crathis and Sybaris, its exact position being unknown. It was founded by Achæans and Trœzenians in B.C. 720, and soon rose to a state of the highest prosperity from the extensive trade it prosecuted with Asia Minor and other countries. The town itself was about 6 miles in circumference ; its power was extended over 25 cities, and it could muster an army of 300,000 men. The wealth and luxuriousness of its inhabitants became proverbial. Internal dissensions proved its ruin ; the Trœzenians, having been ejected by the Achæans, sought the aid of Croton, and in the war that ensued the Sybarites were defeated in 510 on the banks of the Crathis, and their town was destroyed by a diversion of the stream against it. A desolate swamp now covers its site. The inhabitants took refuge in Laus and Scidrus; they returned 58 years after, and attempted to rebuild the town, but the opposition of the Crotoniats defeated this plan, and they ultimately joined a mixed body of Greeks, more especially of Athenians, in the foundation of Thurii, at a little distance from the site of the old town, and

Coin of Thurii.

probably to the N. of the river Sybaris, though its site has not yet been identified. The foundation of Thurii is variously assigned to the years 446 and 443 B.C.; Herodotus and the orator Lysias were in the number of the original colonists. The Sybarites were expelled, and fresh colonists introduced from Greece. The town rose to a state of the greatest prosperity, and carried on independent wars against the Lucanians and Tarentines, from the former of whom the Thurians received a severe defeat in 390. The Romans subsequently aided them against these enemies about 286, and thenceforth the town became subject to Rome. In the Second Punic War it revolted to Hannibal, who nevertheless plundered it and removed its inhabitants to Crotona on his withdrawal in 204. It was revived by a Roman colony in 194, under the name of Copiæ, and remained the most important town in these parts until a late period.

Buxentum, *Policastro*, the **Pyxus** of the Greeks, was situated on the W. coast, some distance N. of the Laus. Its foundation is attributed to the Rhegians under Micythus in B.C. 470, but there was certainly an earlier town, probably a colony from Siris, on the spot. The Romans sent colonies there in 194 and again in 196. **Elea** or **Velia**, *Castell a Mare della Brucca*, stood midway between Buxentum and Pæstum. It was founded by the fugitive Phocæans about 540 B.C. Though it became undoubtedly a prosperous place, we know nothing of its history. Its chief celebrity is due to the philosophical school planted there by Xenophanes of Colophon, and carried on by Parmenides and Zeno. Cicero frequently visited Velia, and it appears to have been noted for its healthiness.[*] It possessed a famous temple of Ceres. **Pæstum**, *Pesto*, the **Posidonia** of the Greeks, was situated about 5 miles S. of the Silarus. It was a colony from Sybaris, founded probably by the expelled Trœzenians of that place. We know nothing of its early history ; it was captured by the Lucanians some time before B.C. 390, and

---

[*] Horace refers to this when he writes—

Quæ sit hiems Veliæ, quod cœlum, Vala, Salerni.—*Ep.* i. 15, 1.

passed along with the rest of Lucania into the hands of the Romans, who sent a colony there in 273, and changed its name to Pæstum. It remained a considerable place, though of no historical importance. Its chief celebrity in ancient times arose from its roses,[1] which flowered twice a year, a quality which they still retain. The ruins of Pæstum consist of the circuit of the walls and three temples, the finest of which (commonly known as the Temple of Neptune) is of the Doric order, 195 feet long by 79 wide, and in a remarkably perfect state ; the second is 180 feet long by 80 wide, and appears from its construction to have been two temples in one ; the

Plan of Pæstum.

A. Temple of Neptune.	D. Amphitheatre.
B. Temple, commonly called Basilica.	E. Other ruins of Roman time.
C. Smaller Temple of Ceres or Vesta.	F F Gates of the City.
	G. River Salso.

third (known as the Temple of Ceres or Vesta) is much smaller ; there are also remains of an amphitheatre and of an aqueduct. About 5 miles from Pæstum, at the mouth of the Silarus, was a famous temple of Juno. **Grumentum**, *Saponara*, was situated in the interior on the Aciris, and was a native Lucanian town. It is first mentioned in B.C. 215, when Hanno was defeated there by the Romans. In the Social War the Roman prætor Licinius Crassus took refuge there after his defeat by the Lucanians. It afterwards became a *municipium*.

Of the less important towns we may notice—**Blanda**, 12 miles S.E. of Buxentum, noticed among the towns which revolted to Hannibal, and were recovered by Fabius in 214 ; **Laüs**, on the borders of the Bruttian territory near Scalea, a colony of Sybaris, and the place whither the expatriated Sybarites retired in B.C. 510 ; the scene also of a great defeat sustained by the Greeks from the Lucanians ; **Nerulum**, to the S.E. of Blanda, captured by Æmilius Barbula in 317 ; **Numistro**, on the borders of Apulia, the scene of a battle between Hannibal and Marcellus in 210 ; **Potentia**, near *Potenza*, on the Casuentus, a considerable town, though historically unnoticed ; and **Volceium** or **Volcentum**, *Buccino*, W. of Potentia, the chief town of the Volcentes, who are noticed as revolting to Hannibal, but returning to their allegiance in 209.

Roads.—The principal road in Lucania was the **Via Popilia**, which traversed the province in its whole length on its way between Capua

[1] Forsitan et, pingues hortos quæ cura colendi
Ornaret, canerem, *biferique* rosaria Pæsti.   *Georg.* iv. 118.

Vidi ego odorati victura rosaria Pæsti
Sub matutino coela jacere noto.   PROPERT. iv. 3, 39.

Lecoeniamque petit, tepidique rosaria Pæsti.   OV. *Met.* xr. 708.

and Rhegium ; it followed the valley of the Tanager. Roads followed the coasts between Pæstum, Velia, and Buxentum on the W., and between Thurii and Metapontum on the E.

History.—The history of Lucania, as distinct from that of the Greek cities on its coasts, commences with the entrance of the Lucanians towards the end of the 5th century B.C. In 393 a league was formed against them by the Greeks, but this was crushed by the defeat sustained by the latter near Laüs in 390. The Lucanians then became masters of the whole country, and were at the height of their power about 350. The wars which they subsequently waged against the Tarentines and their allies, Archidamus and Alexander, appear to have shaken their power by the end of the 4th century. In 326 the Lucanians entered into an alliance with Rome, which they shortly after gave up, and were severely handled in 317 in consequence. In 288 their attack on Thurii again drew on them the vengeance of Rome. In 281 they joined Pyrrhus, and in 272 were again reduced to submission. In 216 they declared in favour of Hannibal, and in 209 they returned to their allegiance. In the Social War they again revolted, and in the Civil War between Sulla and Marius they joined the latter, and suffered severely at the hands of Sulla.

XIV. The Bruttii.

§ 9. The land of the Bruttii * occupied the S. extremity of the Italian peninsula from the borders of Lucania. This region is cor-

Coin of the Bruttii.

rectly described by Strabo as a "peninsula including a peninsula within it." The first or larger peninsula is formed by the approach of the Tarentine and Terinæan gulfs on the borders of Lucania; the second or included peninsula by the approach of the Scyllacian and Hipponian gulfs, more to the S. The general configuration of the country thus resembles a boot, of which the heel is formed by the Lacinian promontory, and the toe by Leucopetra. It is traversed through its whole length by the Apennines, which in the N. district approach very close to the Tyrrhenian Sea, leaving room on the E. for the extensive outlying mass now named *Sila*; the range sinks at the point where the Hipponian and Scyllacian bays approach, and rises again more to the S. in the rugged masses anciently named Sila,* and now *Aspromonte*. These mountains have been always covered with dense forests, which supplied the Romans with timber

---

* The name " Bruttium," given to the country by modern writers on ancient geography, is not found in any classical author.

    § Ac velut ingenti Sila, summove Taburno
    Cum duo conversis inimica in prœlia tauri
    Frontibus incurrunt, pavidi cessere magistri.—*Æn.* xii. 715.

and pitch. Along the coasts there are alluvial plains of great fertility but small in extent, skirting the bays. The rivers are numerous, but unimportant: we may notice, on the E. coast, the Crathis, on the borders of Lucania; the Næthus, *Neto*, the largest of them all, joining the sea about 10 miles N. of Crotona; and, on the W. coast, the Medma. *Mesima.*

§ 10. The province we are describing was originally occupied by the Œnotrians, who were divided into two tribes named Chones and Morgètes. The Greeks subsequently became the virtual owners of the land, occupying the whole of the valuable sea-coast, and leaving the interior to the Œnotrians. The period of their supremacy lasted from about 700 B.C. to 390, when the Lucanians overran the country, and established their dominion over the interior. These were succeeded, in 356, by the people called Bruttii, who are represented as having been an heterogeneous collection of revolted slaves and bandits, but who nevertheless were strong enough to dispossess the Lucanians of their supremacy, and to enter upon war with the Greek cities. The towns may be divided into two classes :—(1.) The Greek colonies on the coast, of which the most important were Crotòna, Caulonia, Locri, Rhegium, Medma, Hipponium, and Terìna; and (2.) the proper Bruttian cities, of which the most considerable were Clampetia and Tempsa on the coast, and Consentia in the interior. We shall commence with those on the E. coast, from N. to S.

Croton or Crotòna, *Cotrone*, was situated about 6 miles N. of Prom. Lacinium, at the mouth of the little river Æsarus. It was founded by Achæans under Myscellus in B.C. 710, and at an early period of its existence attained a high pitch of power. Its walls were 12 miles in circumference, its authority extended to the other side of the peninsula, and it could bring into the field 100,000 men. Pythagoras

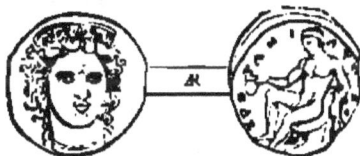

Coin of Croton.

established himself there about 540, and introduced great changes of a political and social character. War occurred between Croton and Sybaris in 510, and terminated in the destruction of the latter city. The battle of the Sagras, in which the Crotoniats were defeated with heavy loss by the Locrians and Rhegians, took place probably after 510. It suffered severely in the wars waged by the Syracusan tyrants, being captured by Dionysius in 389, and by Agathocles in 299. It became subject to Rome in 277, while it was under the power of Pyrrhus. Its ruin was completed in the Second Punic War, when it was held for three years by Hannibal, and, in spite of a Roman colony sent there in 194, it sank into insignificance. The healthiness of Crotona and the fertility of the pastures about the Æsarus are much praised. Scylacium or Scylletium, *Squillace*, stood near the inmost recess of the bay named after it. There are traditions as to its being a Greek city, but they are not trustworthy.

We first hear of it as a dependency of Crotona. In B.C. 124 the Romans sent a colony there, and from this time it became a considerable town, and remained such under the empire. Caulon or **Caulonia** was a colony of Achæan origin, its founders being partly natives of Crotona, and partly from the mother country. Its early history is lost to us. It was destroyed by Dionysius of Syracuse in 389, and again, during the war with Pyrrhus, by some Campanian mercenaries. On each occasion it was rebuilt, and it is again noticed in the Second Punic War as revolting to Hannibal, after which it probably fell into decay. Its site is still unknown.[1] Locri, surnamed **Epizephyrii**, to distinguish it from the cities of the same name in Greece, was situated 15 miles N. of Prom. Zephyrium, from which its surname was derived. It was founded by Locrians[2] in B.C. 683, or even earlier, and was originally built on the promontory itself. Its early history is unknown, and its chief celebrity is due to the excellence of its laws, which were drawn up by Zaleucus[3] probably about B.C. 660. It took part in the battle against Crotona at the Sagras. It maintained a close alliance with Syracuse, and an enmity against Rhegium. In the Second Punic War it revolted to Hannibal in 216, and was not recovered by the Romans until 205, after which we hear little of it. The ruins of Locri are about 5 miles from *Gerace*, and consist of the circuit of the walls and the basement of a Doric temple. A celebrated temple of Persephone belonged to it. **Rhegium,**[4] *Reggio*, was situated on the E. side of the Sicilian

Coin of Caulonia.

[1] It appears to have stood on an elevation :—

    Attollit se diva Lacinia contra
    Cauloniaque arces, et navifragum Scylaceum.   Æn. III. 552.

[2] They were supposed to be of the Opuntian branch; whence the epithet " Narycian " is applied to them :—

    Hinc et Narycii posuerunt mœnia Locri.   Æn. III. 399.
    Naryciæque picis lucos.   Georg. II. 438.

[3] Pindar eulogizes the character of the Locrians :—

    Νέμει γὰρ Ἀτρέκεια πόλιν Λοκρῶν
    Ζεφυρίων· μέλει τέ σφισι Καλλιόπα
    Καὶ χάλκεος Ἄρης.   Olymp. x. 17.

[4] The name Rhegium was commonly derived from ῥήγνυμι, " to break," in allusion to the idea that the shores of Italy and Sicily were broken asunder by an earthquake :—

        Hæc loca, vi quondam, et vasta convulsa ruina
        (Tantum ævi longinqua valet mutare vetustas)
        Dissiluisse ferunt : cum protinus utraque tellus
        Una foret ; venit medio vi pontus, et undis
        Hesperium Siculo latus abscidit, arvaque et urbes
        Littore diductas angusto interluit æstu.   Æn. III. 414.

            Zancle quoque juncta fuisse
        Dicitur Italiæ : donec confinia pontus
        Abstulit ; et media tellurem reppulit unda.   Ov. Met. xv. 290.

Straits, almost directly opposite to Messana in Sicily. It was founded probably about 740 by a joint colony of Chalcidians and Messenians, the latter having left their country after the First Messenian War. A fresh band of Messenians was added in 668 at the close of the Second Messenian War. Its government was originally oligarchical, but in 494 Anaxilaus made himself tyrant, and was succeeded in 476 by his sons, who, however, were expelled in 468. Dionysius the elder carried on a series of wars with Rhegium. It received a colony in the time of Augustus, and was named Julium. Its position, at the termination of the great line of communication with Sicily, secured its prosperity under the empire; the point where the transit was effected was, however, not at Rhegium itself, but 9 miles N. of it, at Columna Rhegina. Rhegium gave birth to the poet Ibycus, the historian Lycus, and the sculptor Pythagoras. Medma or Mesma stood on the W. coast between Hipponium and the mouth of the Metaurus, its exact position being unknown. It was a colony of the Epizephyrian Locrians, and is always noticed among the Greek cities of Italy, but its history is wholly lost to us. Hipponium or Hippo, otherwise known by its Latin names of Vibo[1] and Vibo Valentia, *Bivona*, was situated on the shore of the bay named after it, now the *Gulf of St. Eufemia*. It was also a colony of Locri, and is historically unknown until the time of its capture and destruction by Dionysius of Syracuse in B.C. 389. In 192 it received a Roman colony with the name of Valentia, and became important as the place where timber was exported and ships were built. The plains about it were celebrated for beautiful flowers, and a temple of Proserpine was appropriately erected there. Temesa or Tempsa was situated a little N. of the Gulf of Hipponium. It is said to have been an Ausonian town, and it subsequently became hellenised, though no Greek colony is known to have been planted there. Between 480 and 460 it was under the power of the Locrians, from whom it passed to the Bruttians, and ultimately to the Romans, who sent a colony there in 194. Its copper mines are frequently noticed.[2] In the Servile War it was seized and held by a body of the slaves. It afterwards disappeared, and even its site is unknown. Clampetia or Lampetia stood more to the N., probably at *Amantea*. The only notice of it is its recovery by the Romans during the Second Punic War.

Of the less important towns we may notice—Terina on the Terinæus Sinus, a colony of Crotona, and, as we may conjecture from the character of its coinage, a place of wealth and importance; Petella or Petilia, *Strongoli*, about 12 miles N. of Crotona, and 3 miles from the coast, the metropolis of the Lucanians, and otherwise famous for

Coin of Rhegium.

---

[1] Vibo is the Bruttian or Oscan form of Hippo, and was probably the original name of the town.

[2] Et cui se tollis Temese dedit hausta metalla.—STAT. *Silv.* l. 1, 42.

Evincitque fretum, Siculique angusta Pelori
Hippotadæque domos regis Temeseque metalla.—Ov. *Met.* xv. 705.

the long siege it sustained from the Carthaginians and Bruttians in
B.C. 216 ; Pandosia, an old Œnotrian town, somewhere between Thurii
and Consentia, afterwards a colony of Crotona, famous as being the
place near which Alexander of Epirus was slain in 326; and, lastly,
Consentia, Cosenza, in the mountains near the sources of the Crathis,
the metropolis of the Bruttians, noticed in the Second Punic War as
being taken by Himileo in 216, and by the Romans in 204, and in the
Servile War as being besieged by Sextus Pompeius without success.

Roads.—This province was traversed by the Via Popilia, which passed
up the valley of the Crathis to Consentia, thence descended to the
shores of the Gulf of Hipponium, and followed the line of coast to
Rhegium. A second road, constructed by Trajan, followed the E.
coast, and a third followed the W. coast from Blanda to Hipponium
where it fell into the Via Popilia.

History.—The rise of the Bruttii has been already traced. They appear to have attained their highest prosperity about 300 B.C., after their
wars with Alexander of Epirus and Agathocles were concluded, and
before the contest with Rome began. In 282 they joined the Lucanians
against Rome ; they are again numbered among the allies of Pyrrhus,
after whose defeat they were attacked and subdued by C. Fabricius and
L. Papirius. In the Second Punic War the cities in some cases revolted
to Hannibal, in other cases were subdued by him, and for four successive years he maintained himself in this province. After his retreat
the Romans effectually subdued the Bruttians, and they disappear, as a
people, from history.

BATTLE OF CANNÆ.

The scene of the battle of Cannæ has been controverted, some writers assuming
that it took place on the S. side of the Aufidus. The following observations,
bearing upon the point, lead
to the opposite conclusion.
Two days before the battle
the Romans had established
themselves at a camp about
50 stadia distant from the
enemy (Plan, A). The next
day they advanced, and
formed two camps; the
larger one on the S. side of
the river (B), and the
smaller one on the N. side
(c); Hannibal was also
encamped on the S. side
(D). On the day of the
battle Varro crossed the
river (a k) from the larger
camp and drew up his
forces in a line facing the
S. Hannibal also crossed,
and drew up opposite him.
The battle was fought at a
spot (s) where the Aufidus

Plan of Cannæ.

takes a sudden bend ; and hence we can understand how the Roman army had its
left wing on the bank of the river, and still faced the S. The town of Cannæ
was on the S. side, at r ; Canusium, at o ; and the bridge of Canusium, at s.

Nuraghe in Sardinia.

CHAPTER XXVIII.

SICILY, SARDINIA, CORSICA, AND THE ADJACENT ISLANDS.

I. SICILY. § 1. General description. § 2. Mountains and rivers. § 3. Inhabitants; towns; lesser islands; history. § 4. Melita. II. SARDINIA. § 5. General description; mountains and rivers. § 6. Inhabitants; towns; history. III. CORSICA. § 7. General description; towns; history.

I. SICILIA.

§ 1. The important island of Sicilia lies off the southern extremity of the peninsula of Italy, from which it is divided by a narrow strait formerly called Fretum Siculum, and now the *Straits of Messina.* At its W. extremity it approaches within 80 geographical miles of the continent of Africa near Carthage, and it forms the great barrier between the eastern and western basins of the Mediterranean. Its form is triangular,[1] the E. side representing the base, and the W. angle the apex. It is for the most part mountainous, being traversed through its whole length by a range which may be regarded as a continuation of the Apennines, and which sends out an important offshoot to the

---

[1] The names "Trinacria" and "Triquetra" have direct reference to its shape :—

Terra tribus scopulis vastum procurrit in æquor
Trinacris, a positu nomen adepta loci.   Ov. *Fast.* iv. 419.

Insula quem Triquetris terrarum gessit in oris :
Quam fluitans circum magnis anfractibus æquo·
Ionium glauca aspergit virus ab unda :
Angustoque freta rapidum mare dividit undis
Italiæ terraï oras a finibus ejus.   Lucret. L 716.

          Militibus promissa Triquetra
Prædia Cæsar ; an est Itala tellure datorus ?   Hor. *Sat.* ii. 6, 55.

S.E. angle of the island, communicating to it its peculiar configuration. The space between these limbs is filled up on the E. coast by the volcanic mountain of Ætna, and on the S.W. coast by a range of inferior height. The fertility of the soil of Sicily has been in all ages the theme of admiration;[3] though it possesses few plains, its well-watered valleys and the slopes of the mountains admit of the most perfect cultivation. It was believed to be the native country of wheat; and it was celebrated for its honey and saffron, its sheep and cattle, and particularly for its horses, those of Agrigentum[3] being the most famous. The climate appears to have been more healthy in ancient than in modern times: the temperature varies considerably in different parts of the island, on the N. coast resembling that of Italy, on the S. that of Africa.

§ 2. The general name for the range, which runs parallel to the N. shore, appears to have been **Nebrôdes Mons**,[4] though this may have been also more particularly applied to the central and highest portion of the chain, now named *Monte Madonia*. Distinct names were given to portions of the chain, among which we may notice **Neptunius Ms**, in the immediate vicinity of Messana; **Herei Mts** near Enna, and **Cratas** to the S. of Panormus, in the W. portion of the island. This range is, however, far inferior in height to **Ætna**, which attains an elevation of nearly 11,000 feet, and covers with its base a space not less than 90 miles in circumference. The volcanic character[4] of this mountain was known to the Greeks at an early

---

[3] Multa solo virtus : jam reddere foenus aratris
Jam montes umbrare olea, dare nomina Baccho
Cornipedemque citum lituis generasse ferendis,
Nectare Cecropias Hyblaeo accedere cerna.   SIL. ITAL. xiv. 23.

[3] Arduus inde Acragas ostentat maxima longe
Moenia, magnanimûm quondam generator equorum.—Æn. iii. 703.

[4] Nebrodes gemini nutrit divortia fontis
      Quo mons Sicania non surgit ditior umbrae.   SIL. ITAL. xiv. 236.

[1] The eruptions were ascribed by the poets to the struggles of the giant Typhoeus, or (according to Virgil) of Enceladus, who was buried under the mountain by Zeus after the defeat of the giants :—

Καὶ νῦν ἀχρεῖον καὶ παρήορον δέμας
Κεῖται στενωποῦ πλησίον θαλασσίου
Ἰπούμενος ῥίζαισιν Αἰτναίαις ὕπο·
Κορυφαῖς δ' ἐν ἄκραις ἥμενος μυδροκτυπεῖ
Ἥφαιστος, ἔνθεν ἐκραγήσονταί ποτε
Ποταμοὶ πυρὸς δάπτοντες ἀγρίαις γνάθοις
Τῆς καλλικάρπου Σικελίας λευροὺς γύας·
Τοιάνδε Τυφὼς ἐξαναζέσει χόλον
Θερμοῖς ἀπλήστου βέλεσι πυρπνόου ζάλης,
Καίπερ κεραυνῷ Ζηνὸς ἠνθρακωμένος.   ÆSCH. PROM. 363.

Fama est, Enceladi semiustum fulmine corpus
Urgeri mole hac, ingentemque insuper Ætnam
Impositam, rupis flammam exspirare caminis ;
Et, fessum quoties mutet latus, intremere omnem
Murmure Trinacriam et coelum subtexere fumo.—Æn. iii. 578.

The

period : the date of the first eruption which they witnessed is not
known ; the second occurred in B.C. 475, and is noticed by Pindar
and Æschylus ; the third in 425 : many eruptions are subsequently
recorded.   At the other extremity of the island lies a mountain of
considerable fame in antiquity, named **Eryx**, *Monte S. Giuliano*, an
isolated peak, rising out of a low tract, and hence apparently higher
than it really is.[6] Its summit was crowned with a famous temple of
Venus,[7] said to have been founded by Æneas.   The three promontories,[8] which form the salient points of the island, are **Pelorus**, *Capo*

The snow-clad summit of the mountain is frequently referred to, as well as the
contrast exhibited between the perpetual fire and the perpetual snow :—

Νιφόεσσ' Αἴτνα, πάντες
Χίονος ὀξείας τιθήνα·
Τᾶς ἐρεύγονται μὲν ἀελδ-
του πυρὸς ἁγνόταται
Ἐκ μυχῶν παγαί.   PIND. *Pyth.* L. 38.

At Ætna eructat tremefactis cautibus ignis
Inclusi gemitus, pelagique imitata furorem
Murmure per cæcos tonat irrequieta fragorre
Nocte dieque simul : fonte e Phlegethontis ut atro
Flammarum exundat torrens, piceaque procella
Semiambusta rotat liquefactis mixta cavernis.
Sed quanquam largo flammarum exæstuet intus
Turbine, et assiduæ submænens profluat ignis,
Summo cana jugo cohibet (mirabile dictu)
Vicinam flammis glaciem, æternoque rigore
Ardentes horrent scopuli : stat vertice celsi
Collis hiems, calidaque nivem tegit atra favilla.—SIL. ITAL. xiv. 38.

Virgil's well-known description of an eruption supplied Silius Italicus with
many of his ideas :—

Portus ab accessu ventorum immotus, et ingens
Ipse ; sed horrificis juxta tonat Ætna ruinis,
Interdumque atram prorumpit ad æthera nubem,
Turbine fumantem piceo et candente favilla ;
Attollitque globos flammarum, et sidera lambit :
Interdum scopulos avulsaque viscera montis
Erigit eructans, liquefactaque saxa sub auras
Cum gemitu glomerat, fundoque exæstuat imo.—Æn. III. 570.

[6] Hence the poets class it with the loftiest mountains in the world :—
Quantus Athos, aut quantus Eryx aut ipse coruscis
Cum fremit ilicibus, quantus, gaudetque nivali
Vertice se attollens pater Apenninus ad auras.—Æn. xii. 701.
Magnus Eryx, deferre velint quem vallibus imbres.
VAL. FLACC. ii. 528.

[7] Tum vicina astris Erycino in vertice sedes
Fundatur Veneri Idaliæ.   Æn. v. 759.
Hence Venus is termed Erycina :—
Sive tu mavis, Erycina ridens.   HOR. Carm. l. 2, 83.
Tu quoque, quæ montes celebras, Erycina, Sicanos.
OV. Heroid. xv. 57.

[8] The position of these is well described by Ovid :—
Tribus hæc excurrit in æquora linguis.
E quibus imbriferos obversa Pachynos ad Austros ;

di Faro,[a] in the N.E., immediately opposite the Italian coast, and hence important as a naval station; **Pachynus.**[1] *C. Passaro*, in the S.E., and the most southerly point of the island; and **Lilybæum,** *C. Boeo*, in the W., a low, rocky point with reefs about it, which rendered navigation dangerous. The rivers of Sicily are generally little more than mountain torrents, swollen in winter, and nearly dry in summer. The most important are—the **Symæthus,**[2] *Giaretta*, which flows by the roots of Ætna, and falls into the sea S. of Catana, receiving in its course the **Chrysas,** *Dittaino*, and the **Cyamosorus,** *Fiume Salso*; the **Himera,** *Fiume Salso*, which rises on the S. side of Nebrodes, only about 15 miles from the N. coast, and traverses the whole breadth of Sicily, falling into the sea W. of Gela; the **Halycus,** *Platani*, which rises not far from the Himera and enters the sea at Heraclea Minoa; and the **Hypsas,** *Belici*, also on the S. coast, a few miles E. of Selinus. The lakes of Sicily are unimportant; we may notice, however, **Palicorum Lacus,** a deep pool of volcanic origin, about 15 miles W. of Leontini, the waters of which were set in commotion by jets of volcanic gas;[3] and **Pergus,**[4] near Enna, which is also still in existence.

§ 3. The most ancient inhabitants of Sicily of whom we hear are the **Sicani,** who claimed to be autochthons, and who, in historical times, occupied the W. and N.W. of the island. A second and more widely-spread race were the **Siculi** or **Sicelli,** after whom the island was named, and who occupied the greater part of the interior;

Mollibus expositum Zephyris Lilybæon: at Arcton
Æquoris expertem special Boreanque Peloros.   *Met.* xiii. 734.
Jamque Peloriaden, Lilybæaque, Jamque Pachynon
Lustrarat, terræ cornua prima suæ.   *Fast.* iv. 479.

[a] The modern name is derived from a lighthouse (Pharos) which once stood on it, as also did a temple of Neptune. The position of this promontory in the Sicilian straits is well described by Virgil's expression "*angusti claustra Pelori*" (*Æn.* iii. 411).

[1] It is correctly described by Virgil as formed by bold projecting rocks:—
Hinc alta saxa *projectaque saxa* Pachyni
Radimus.   *Æn.* iii. 699.

[2] Rapidique colunt vada flava Symæthi.   Sil. Ital. xiv. 231.
Quaeque Symæthæas accipit æquor aquas.   Ov. *Fast.* iv. 472.

[2] The pool is now called *Lago di Naftia* from the naphtha with which it is impregnated. Formerly there appear to have been two separate pools or craters; there is now but one. The spot was consecrated to the indigenous deities, called Palici; hence Virgil speaks of the son of Arcens as—
Eductum matris luco, Symæthia circum
Flumina: pinguis ubi et placabilis ora Palici.   *Æn.* ix. 584.
The pool is described by Ovid:—
Perque lacus altos, et olentia sulfure fertur
Stagna Palicorum, rupta ferventia terra.   *Met.* v. 40

[4] Haud procul Hennæis lacus est a mœnibus altæ,
Nomine Pergus, aquæ, &c.   Ov. *Met.* v. 385.

they were a Pelasgic race, and crossed over into Sicily from Italy within historical times. The Elymi, in the N.W. corner of the island, were a distinct people of no great importance. In addition to these, which we may term the indigenous races of Sicily, numerous foreign settlements were made on the coasts by the Phœnicians and Greeks, by the former merely for trading purposes, by the latter as permanent colonies. The most important towns of Sicily were founded by the Greeks between 750 and 600 B.C. : Naxos was the first in point of time, in 735; then followed in rapid succession Syracuse in 734, Messana, of uncertain date, Leontini and Catâna about 730, Megara Hyblæa about 726, Gela in 690, Selinus in 626, and Agrigentum in 580, all of which rose to eminence, and some became the parents of fresh colonies. Naxus, Leontini, and Catana, were of Ionian origin; the rest were Dorian. The Phœnicians were gradually driven to the W. by the Greeks, and were at last confined to three towns at the N.W. corner of the island, viz., Motya, Panormus, and Solœis. These fell under the dominion of Carthage, probably about the time when Phœnicia itself became subject to the Persian empire. The Carthaginians themselves founded several important towns about the W. extremity of the island, particularly Lilybæum and Drepânum. Several important towns owed their origin to the elder Dionysius, 405–368, as Tauromenium, which arose in the place of Naxos, Tyndaris, and Alœsa on the N. coast. The flourishing period of the Greek towns lasted until the time of the Roman conquest of Sicily in 241. A long series of wars, and still more the exactions of Roman governors, proved fatal to them, and in Strabo's time many were in actual ruins, and others in a declining state. We shall describe them in order, commencing with the E. coast.

(1.) *Towns on the E. coast from N. to S.*—**Messâna,** *Messina,* stood on the Sicilian straits opposite Rhegium ;[a] it owed its chief importance partly to its position in reference to Italy, and partly to the excellence of its port, formed by a projecting spit of sand, which curves round in the shape of a sickle[b] (whence its older name of Zancle), and which constitutes a natural mole. Immediately behind the town, which encircles the harbour, rises the range of Neptunius. Messana was first colonized by Chalcidians of Eretria, having been previously occupied by the native Siceli. In 494 it was seized by Samians and Milesians, who had emigrated from Asia Minor after the fall of Miletus. These were driven out by Anaxilas, a Messenian, who crossed with a body of his countrymen from Rhegium, and changed the name from Zancle to·

[a] Liquerat et Zanclen, adversaque mœnia Rhegi.—Ov. *Met.* xiv. 5.
Incumbens Messana ferto minimumque revelas
Discreta Italia atque Osco memorabilis arta.   SIL. ITAL. xiv. 194.
[b] Quique locus curvæ nomina falcis habet.   Ov. *Fast.* iv. 474.

Messana. At the commencement of the fourth century B.C., it was one of the most important cities in Sicily. Having been destroyed in 396 by the Carthaginians, it was restored by Dionysius, and regained its prosperity. It fell from time to time under the dominion of tyrants, and was conquered by Agathocles of Syracuse in 312, who introduced into it the Mamertini from Campania. After the death of Agathocles in 282, these Mamertini seized the town and massacred all the males; thenceforth it was named Mamertina. These bandits were attacked in 271 by Hiero of Syracuse, against whom they called in the aid, first of the Carthaginians, and afterwards of the Romans, who entered Sicily as the allies of Messana in 264, and were immediately engaged in the First Punic War. Messana was constituted a *fœderata civitas*, and it became one of the finest and wealthiest of the Sicilian cities. Near it was the famous, and, in early times, much dreaded whirlpool named Charybdis.[7] Naxos was situated on a low rocky headland at the mouth of the river Acesines; it ranked as the oldest of all the Greek cities in Sicily, having been founded by Chalcidians in B.C. 735. Its early history is not known to us; it was taken by Hippocrates of Gela, about 493, was depopulated by Hieron in 476, and was restored about 461. It fell under the enmity of Syracuse, in consequence of its having espoused the cause of Athens in 415; and in 403 it was utterly destroyed by Dionysius,[8] and its inhabitants expatriated. The Siculi, to whom the territory was then given, erected

Coin of Messana.

Coin of Naxos.

---

[7] The earliest notice of this occurs in Homer, who describes it as opposite to Scylla, though it is really some ten miles distant. Scylla offers no particular risks to the navigator: Charybdis, on the other hand, might well be dreaded by the ancients, whose vessels were small and undecked; even at the present day larger vessels are sometimes endangered by it. It is formed by the meeting of opposite currents, which are much affected by certain winds. The following passages illustrate the above remarks:—

Τῇ δ' ὑπὸ δῖα Χάρυβδις ἀναρροιβδεῖ μέλαν ὕδωρ·
Τρὶς μὲν γάρ τ' ἀνίησιν ἐπ' ἥματι, τρὶς δ' ἀναροιβδεῖ
Δεινόν.　　　　　　　　　　　　　　Hom. Od. xii. 104.

Dextrum Scylla latus, lævum implacata Charybdis
Obsidet: atque imo barathri ter gurgite vastos
Sorbet in abruptum fluctus, rursusque sub auras
Erigit alternos, et sidera verberat unda.　　　　Æn. iii. 420.

Scylla latus dextrum, lævum irrequieta Charybdis
Infestat.　　　　　　　　　　　　　　Ov. Met. xiii. 730.

Nec Scyllæ sævo conterruit impetus ore
Nec violenta suo consumsit in orbe Charybdis.—Tibull. iv. 1, 71.

a new town about three miles from Naxos, on the slope of Taurus,[1] which they named **Tauromenium**, and which is still called *Taormina*. To this place the old Naxian exiles were brought back in 358 by Andromachus, and it was henceforth regarded as the representative of the old town. It appears subsequently to have fallen under the power of Syracuse, and ultimately passed with the rest into the hands of the Romans, who made it a *fœderata civitas*, and afterwards a colony. The remains of Tauromenium are numerous, and consist of a theatre in a very perfect state, and, in point of size, second only to that of Syracuse, a building styled a naumachia, parts of the ancient walls, reservoirs, sepulchres, tessellated pavements, &c. The position of this town was remarkably strong; it stood on a projecting ridge some 900 feet above the sea, and was backed by an inaccessible rock some 500 feet higher, on which its citadel was posted. **Catina** or **Catina**, *Catania*, was situated midway between Tauromenium and Syracuse, and almost immediately at the foot of Ætna. It was founded about B.C. 730 by Naxos, and it remained independent until 476, when it was taken by Hiero I., its inhabitants removed to Leontini, and fresh settlers from Syracuse and Peloponnesus introduced in their stead. In 461 the old inhabitants returned, and the place subsequently attained a high degree of prosperity. In the Athenian invasion, Catana was seized and occupied by the Athenians. In 403 it was conquered by Dionysius of Syracuse, and was held by a body of Campanian mercenaries until 396. It was afterwards governed by tyrants. In 263 it yielded to Rome, and was prosperous until the time of Sextus Pompeius, from whom it suffered much: it was colonized by Augustus. It was the birth-place of the philosopher Charondas, and the residence of the poet Stesichorus. From its proximity to Ætna,[2] it suffered from the eruptions, especially in B.C. 121, when much of its territory was overwhelmed. The remains of Catana belong to the Roman period, and consist of the ruins of a theatre, of an odeum, of baths, and of an aqueduct. **Leontini**, *Leontini*, was situated on the small river Lissus, about eight miles from the sea.

It stood on a hill, which divides into two summits with an intervening valley, and was surrounded by a district of extraordinary fertility. It was founded by Naxians in B.C. 730, and retained its independence until 498, when it fell under the yoke of Hippocrates of Gela. In 476 it was subject to Hiero of Syracuse, but in 466 it was again independent, and at its highest prosperity.

Coin of Leontini.

Subsequently it became entangled in disputes with its powerful neighbour Syracuse, and from 427 down to the time of the Roman conquest, it was either subject to or at war with that state. Under the Romans it sunk into a state of decay. It was the birth-place of the orator Gorgias. **Megara**, surnamed **Hyblæa**, to distinguish it from the town

---

[1] Its elevated position is implied in the following line :—
  Tauromenitana cernunt de sede Charybdim.   Sil. Ital. xiv. 256.
[2] Tum Catane, nimium ardenti vicina Typhæo.   Id. xiv. 196.

In Greece, was situated on a deep bay between Catana and Syracuse, probably at *Agosta*. It was founded by colonists from Megara in Greece, on the site of an older town named Hybla, about B.C. 726, and it became the parent of Selinus. In 481 it was destroyed by Gelon, and it was not rebuilt until 415, when a new town arose at the mouth of the river Alabus, *Cantaro*, sometimes called Megara, and sometimes Hybla, which was held by the Syracusans, and was captured by Marcellus in 214. The neighbouring hills produced excellent honey.[1] Syracuse, the most powerful of all the Sicilian cities, was situated on a triangular plateau, which projects into the sea between two bays, that on the S. being small, and forming the great harbour of Syracuse, while that on the N. stretches out as far as Thapsus. The extremity of the hill is about 2½ miles broad; inland it narrows gradually till it terminates in a ridge which connects with the table-land of the interior. The plateau is divided into two portions by a depression running N. and S., about a mile from the sea. Opposite the S.E. angle of the plateau is the island of Ortygia, between which and the plateau itself a low level tract intervenes. S. of the great harbour rises a peninsular promontory named Plemmyrium. The town, which was founded in B.C. 734 by Corinthians and other Dorians under the guidance of Archias, was originally built on Ortygia: subsequently, by the time of the Peloponnesian War, it had been extended to the mainland, and the extremity of the hill, as far back as the depression already noticed, was built over and described as the "outer city" in contradistinction to the "inner city," or acropolis on Ortygia. At this period there appears to have been no suburb outside the walls with the exception of Temenitis on the S. side of the plateau: the whole of the triangular space at the back of the "outer city" was then named Epipolae. Subsequently, however, to this period, an extensive suburb, named Tyche, grew up immediately W. of the "outer city," or as it was afterwards called Achradina : Temenitis was also enlarged, and its name changed to Neapolis: the low ground between the "outer" and "inner" cities was built over : and finally the whole of the triangular space was enclosed within walls by Dionysius I. The city was thus composed of five towns, viz. Ortygia, Achradina, Tyche, Epipolae, and Neapolis. 1. Ortygia[3] was an island of oblong shape, about a mile in length, stretching across the mouth of the great harbour. It was joined to the mainland in the first instance by a causeway, but in the Roman period by a bridge. It contained the famous fountain of Arethusa,[3] the citadel, a magnificent temple of Minerva, of which there

---

[1] Florida quam multas Hybla tuetur apes.   Ov. *Trist.* v. 6, 38.
     Hyblæis apibus florem depasta salicti.   VIRG. *Ecl.* i. 55.

[2] Ortygia was held sacred to Diana, and is hence described by Pindar as "the couch of Artemis," and the "sister of Delos" :—

          Ἄμπνευμα σεμνὸν Ἀλφεοῦ,
          Κλεινᾶν Συρακοσσᾶν θάλος, Ὀρτυγία,
          Δέμνιον Ἀρτέμιδος,
          Δάλου κασιγνήτα.   *Nem.* i. 1.

[3] Arethusa was supposed to be connected by a submarine current with the Alpheus in Elis :—

          Alpheum fama est huc, Elidis amnem,
          Occultas egisse vias subter mare; qui nunc
          Ore, Arethusa, tuo Siculis confunditur undis.   *Æn.* III. 694.

are considerable remains built into the church of *Santa Maria delle Colonne*, a temple of Diana, the palace of Hiero, and other edifices. 2. **Achradina**, "the *outer* city" of Thucydides, contained the forum, the temple of Jupiter Olympius, a theatre, and the catacombs. 3. **Tycha**, so named after an ancient temple of Fortune, became one of the most populous parts of Syracuse, subsequently to the time of the Athenian expedition. 4. **Neapolis**, "the new city," contained the theatre, capable of holding 24,000 spectators, an amphitheatre, several temples, and the Lautumiæ, or quarries. 5. **Epipolæ**, which, in the time of Thucydides, was applied to the whole of the plateau W. of Achradina, was afterwards restricted to the most inland and highest portion of it. This contained the fort of Euryalus, now called *Mongibellisi*, erected probably by Dionysius, and enlarged by Hiero II.

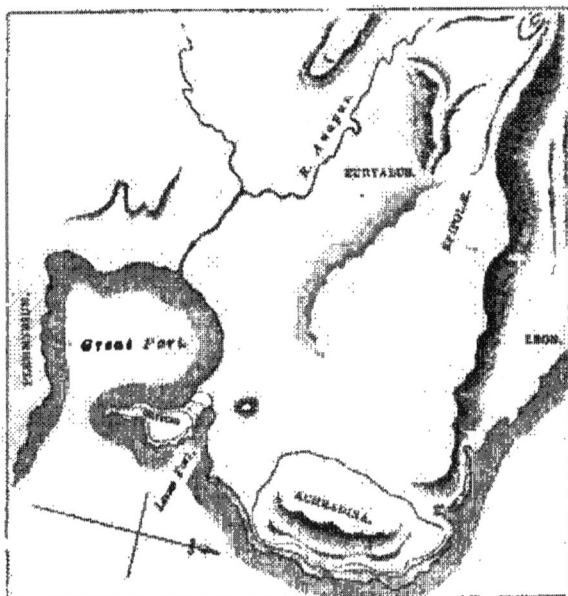

Map of Syracuse at the time of the Peloponnesian War.

Syracuse possessed two ports, the great harbour, the entrance to which was on the S. side of Ortygia, a land-locked bay, 13 miles in circumference, and the small harbour between Ortygia and Achradina. A fine

Extremum hunc, Arethusa, mihi concede laborem,
Pauca meo Gallo, sed quæ legat ipsa Lycoris,
Carmina sunt dicenda : neget quis carmina Gallo?
Sic tibi, cum fluctus subterlabere Sicanos,
Doris amara suam non intermisceat undam.   VIRG. *Ecl.* x. 1.

aqueduct, constructed by Gelon, and improved by Hiero, supplied the town with water. About 1½ miles from Neapolis, and on the S. side of the Anapus, stood the Olympiêum, or temple of Olympian Jove, about which a village named Polichne grew up, and which was important as a military post, commanding the bridge over the Anapus, which discharges itself into the great harbour. Syracuse was originally governed by an aristocracy: this was superseded by a democracy in about 486, and this by a tyranny in the person of Gelon in 485. Under the reigns of Gelon (485-478,, and Hiero (478-467), Syracuse became wealthy and prosperous: Hiero's successor, Thrasybulus, was expelled after a brief reign on account of his cruelty, and a democracy was established. In 415 the Athenians appeared before Syracuse; in 414 the siege of the town was commenced, and ended in the following year in the total defeat of the Athenians. In 405 the democracy was succeeded by a tyranny in the person of the elder Dionysius, who had a long and prosperous reign, and was followed, in 367, by his son, Dionysius the younger, whose reign was quite of a different character, and who was

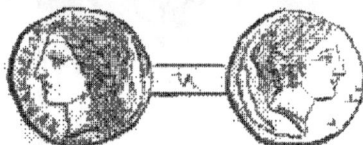

Coin of Syracuse.

expelled by Timoleon in 343. For about 26 years a republic prevailed: but, in 317, Agathocles re-established the tyranny. He reigned until 289, and then followed an interval of anarchy and dissension until 270, when the Syracusans elected Hiero II.

as their king. During his reign the town was peaceable and prosperous, mainly through the wise policy which he adopted towards Rome. His successor, Hieronymus, adopted another line, and joined the Carthaginians; this resulted in the siege of the town by Marcellus, prolonged through the skill of Archimedes for two years, but ending in its capture in 212. The modern *Syracuse* is a comparatively small town confined to the island of Ortygia.

(2). *On the S. Coast.*—Camarina, *Camarana*, was situated at the mouth of the little river Hipparis, about 40 miles W. of Prom. Pachynus. It was founded by Syracuse in B.C. 599, and in 46 years it was strong enough to attempt a revolt against its parent city, which, however, proved unsuccessful, and resulted in the destruction of the town in 552. In 495 it was rebuilt by Hippocrates of Gela, and in 485 was again destroyed by the removal of its inhabitants. In 461 it was for a third time rebuilt, and for the next 50 years reached a high degree of prosperity, which was terminated in 405 by the invasion of its territory by the Carthaginians, and the temporary withdrawal of its inhabitants. In 258 it was betrayed to the Carthaginians, but was speedily recovered by the Romans. In 255 the Roman fleet was wrecked near it. Adjacent to the town was a marsh, which rendered the air unhealthy: the citizens drained this, in opposition to the warning of an oracle, and, in so doing, they exposed their walls to their enemies: hence arose a proverbial saying.[4] Gela, *Terranova*, was situated at the mouth

---

[4] Μὴ κίνει Καμάριναν ἀκίνητος γὰρ ἀμείνων.

Nunquam concessa moveri

Apparet Camarina procul.　　　　　　　　　　　Æn. III. 700.

Et cui non licitum fatis, Camarina, moveri.　　　Sil. Ital. xiv. 198.

of a river of the same name,[5] between Camarina and Agrigentum. It was founded by a joint colony of Rhodians and Cretans in B.C. 690, and in 582 it was sufficiently strong to found Agrigentum. Its constitution was originally oligarchical; but in 505 Cleander established a tyranny, and was succeeded in it by Hippocrates in 498, under whom it attained a very high pitch of power, and by Gelon, in 491, who succeeded in making himself master of Syracuse itself, and removed much of the population thither. These returned to their native city in 466, and a period of prosperity followed until 406, when the town was besieged, and in the next year taken by the Carthaginians. After various fortunes, its final ruin was effected by the removal of its inhabitants to Phintias, the city founded by the tyrant of Agrigentum. To the W. of the town are the broad plains named Campi Geloi, celebrated for their extreme fertility. Gela was the birthplace of Apollodorus, a comic poet, and the place to which Æschylus retired, and where he ended his days. **Agrigentum,** *Girgenti,* the **Acragas**[6] of the Greeks and of the Latin poets, was situated about

midway between Gela and Selinus. It stood on a hill between 2 and 3 miles from the sea, at the base of which flowed the small river Acragas. It was founded by Gela in B.C. 582. It soon fell under the power of despots, of whom Phalaris (about 570) was the first, and who was succeeded by Alcamenes, Alcander, Theron, who de-

Coin of Agrigentum.

feated the Carthaginians in 480, and Thrasydaeus in 472. A democracy followed, and under it Agrigentum spent 60 years of the greatest prosperity, during which its population is computed to have amounted to 200,000. This happy period was terminated by the destruction of the city in 405, by the Carthaginians. It was rebuilt by Timoleon in 340, and again attained a high pitch of power. In 309 it took the lead in the war against Agathocles, but without success. After his death Phintias became despot of the city. In the First Punic War it was held by the Carthaginians, and was consequently besieged by the Romans, who took it after 7 months, in 262. It was again taken and destroyed by the Carthaginians in 255, retaken and held by them in the Second Punic War, and finally recovered by Rome in 210. Under the Romans it still flourished, though not again historically famous. Its ruins are extensive and beautiful;[7] among them we may notice the so-called

---

[5] Immanisque Gela, fluvii cognomine dicta.   Æn. iii. 702.

The river Gela is at times an impetuous torrent; hence Ovid—

    Et te, vorticibus non adeunde Gela.   Fast. iv. 470.

[6] Ovid adopts the Greek form in the line,—

    Himeraque et Didymen, Acragantaque Tauromenonque.—*Fast. iv. 475.*

[7] These justify the encomium which Pindar passes on it as "the fairest of mortal cities:"—

    Αἶτέω σε, φιλάγλαε, καλ-
    λίστα βροτεᾶν πολίων,

temples of Juno Lacinia and of Concord, both of the Doric order, the basement and some fragments of the great temple of Olympian Jove, and the foundation walls of several other temples. Agrigentum was the birth-place of Empedocles and other famous men : it was celebrated for the luxury, the hospitality, and the lavish expenditure of its citizens, the last of which qualities was specially manifested in their sepulchral monuments. Heraclea, surnamed Minôa, stood at the mouth of the river Halycus, between Agrigentum and Selinus. Its surname was attributed traditionally to its having been founded by Minos, king of Crete. In historical times it appears first as a colony of Selinus ; it was subsequently, in B.C. 510, seized by Spartans, who gave it the name of Heraclea. It was soon after destroyed by the Carthaginians, but was rebuilt, and remained in their hands, with but few intervals, until the time of the Roman conquest. During this period it derived importance from the circumstance that the Halycus formed the boundary between the Carthaginian and Greek districts. Selinus was situated at the mouth of the river of the same name, in the S.W. part of the island. It was founded by the Sicilian Megara about B.C. 628, and probably derived its name from the abundance of parsley (σέλινον) found there.[a] It was the most westerly of the Greek cities, and was consequently exposed to the attacks of the Carthaginians, who destroyed it on two occasions, viz. in 409, when no less than 16,000 of its citizens were killed, and 5000 taken captive, and in 250 when its inhabitants were removed to Lilybæum. Near it were some sulphureous springs, called Thermæ Selinuntiæ, which were much frequented by the Romans. The circuit of the walls, the remains of 3 large and 1 small Doric temples within the walls, and 3 temples of yet larger dimensions outside the walls, of the largest of which 3 columns are still standing, mark the site of the town at Torre dei Pulci. Lilybæum, Marsala, was situated on the promontory of the same name in the extreme W. of the island. It was founded by Carthaginians about B.C. 397, and became their stronghold in Sicily, being the nearest point to the African continent.[b] In 250 it was increased by the addition of the population of Selinus, and in the same year commenced the siege of it by the Romans, which lasted for 10 years, and was brought to a close by the peace at the conclusion of the First Punic War. Thenceforth it remained in the hands of the Romans, under whom it became the chief port for African commerce, and the residence of one of the 2 quæstors of Sicily. Numerous vases, sculptures, and coins, have been found on its site: the latter are of a Greek character, a circumstance which shows the predominating influence of the Greeks in Sicily.

(3). *On the N. Coast.*—Eryx, *S. Giuliano,* was situated on the W. slope of the hill of the same name,[c] about 2 miles from the sea-coast. Both

---

Φερσεφόνας ἴδοι, ἅ-
τ' ὄχθαις ἔπι μαλοβότον
Ναίεις 'Ακράγαντος εὐ-
δμάτου κολώναν.   *Pyth.* xii. 1.

[a] It seems to have been yet more famous for its palm-trees :—
    Teque datis linquo ventis, *palmosa* Selinus.   *Æn.* iii. 705.
    Audax Hybla favis, palmisque arbusta Selinus.—Sil. Ital. xiv. 200.

[b] The entrance to the harbour was dangerous from shoals and reefs :—
    Et vada dura lego saxis Lilybeïa cæcia.   *Æn.* iii. 706.

[c] See note 7, p. 591.

the town and the famous temple of Venus appear to have been of Pelasgic origin, nor do the Greeks ever appear to have settled here. It passed into the hands of the Carthaginians, and remained under them until its capture by Pyrrhus in B.C. 278. It was destroyed by the Carthaginians in 260, and its inhabitants removed to Drepanum. It appears to have been partly rebuilt, and it was again the scene of operations between the Romans and the Carthaginians in the First Punic War. Drepanum, or Drepana, *Trapani*, was situated about 6 miles from Eryx, immediately opposite to the Ægates. It derived its name from the promontory on which it stood, which resembled a sickle (δρέπανον) in shape.[?] It was founded by the Carthaginian general Hamilcar in B.C. 260, and was peopled with the inhabitants of Eryx; it was retained by Carthage until the end of the First Punic War, when it was besieged by Lutatius Catulus, and taken after the battle of the Ægates in 241. Segesta, the Egesta or Ægesta of the Greeks, was situated on a hill about 6 miles from the sea-coast, and 3 miles N.W. of *Calatafimi*. Its origin was mythically ascribed to the Trojans,[?] and it appears to have been neither a Greek nor a native Sicanian town. It was engaged in perpetual hostilities with the neighbouring town of Selinus, and is historically famous as having given occasion to the Athenian expedition against Sicily. In 409 it was taken and destroyed by the Carthaginians, was rebuilt, and captured in 307 by Agathocles, who destroyed its citizens, changed its name to Dicæopolis, and peopled it with fugitives from all quarters. It was, however, reoccupied by its old inhabitants, and fell under the power of the Carthaginians until 264, when it was taken by the Romans. Its site is marked by the ruins of a temple and theatre, the former of which is in a very perfect state, and is one of the most striking ruins in Sicily : it is of the Doric order, and has 6 columns in front and 14 on each side. Panormus, *Palermo*, stood on an extensive bay, now named the *Gulf of Palermo*, about 50 miles from the W. extremity of the island. It was of Phœnician origin, and was originally called Machanath "a camp," but received its historical name from the Greeks, who named it from its spacious bay, Panormus, or "all-port." The Carthaginians made it one of their chief naval stations,

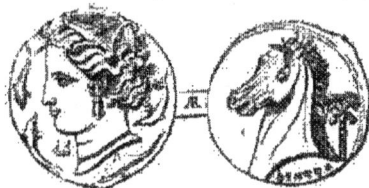

Coin of Panormus.

and, with the exception of a short time when Pyrrhus became master of it in 276, they held it until 254, when it was taken by the Romans. Under its walls the Carthaginians were defeated by C. Metellus in 250. Under the Romans it became a flourishing town, and received

---

[?] Virgil makes it the scene of the death of Anchises :—
　　Hinc Drepani me portus et illætabilis ora
　　Accipit. Hic, pelagi tot tempestatibus actus,
　　Heu genitorem, omnis curæ casusque levamen,
　　Amitto Anchisen.　　　　　　　　　　Æn. III. 707.

[?] Virgil attributes its foundation to Acestes and calls the town Acesta :—
　　Urbem appellabant permisso nomine Acestam.— Æn. v. 718.
Silius Italicus (xiv. 220) describes it as *Trojana Acesta*.

several special privileges. It also received colonies under Augustus, Vespasian, and Hadrian. The town consisted of an inner and outer city, each with its separate inclosure of walls. Numerous inscriptions and coins have been found on its site. Himéra was situated some distance E. of Panormus, near *Termini*. It was founded by Chalcidians

Coin of Himera.

of Zancle, mixed with Syracusans, in B.C. 648. The earliest notice of it is in 560, when it was under the power of Phalaris of Agrigentum. In 490 it received Scythes, the tyrant of Zancle, and shortly after itself became subject to a tyrant named Terillus, and it was at his invitation that the Carthaginians made their first great expedition into Sicily, which ended in their total defeat by Theron of Agrigentum and Gelon of Syracuse in 480. The town then became subject to Theron, who placed his son Thrasydæus in charge of it. In 476 a large number of disaffected citizens were put to death and exiled, and the town was replenished with Dorian settlers. On the death of Theron in 472 Himera became independent, and enjoyed a high state of prosperity until 408, when it was taken and destroyed by the Carthaginians. In 405 the surviving inhabitants founded a new town, named Thermæ, from some hot springs; this appears to have become an important town, and a Roman colony under Augustus. The baths were much frequented by the Romans, and still exist under the name of *Bagni di S. Calogero*. The old town was probably situated about 8 miles to the W. at *Torre di Bonfornello*, where vases, bronzes, &c., have been found. Himera was the birth-place of the poet Stesichorus,[4] and Thermæ of the tyrant Agathocles. Mylæ, *Milazzo*, was situated on a promontory, opposite to the Liparæan Islands. It was founded by Zanclæans some time before B.C. 648, and always remained a dependency of Messana. In 427 it was attacked by the Athenians under Laches; in 315 it was captured by Agathocles; and in 270 it was the scene of the defeat of the Mamertines by Hiero of Syracuse. It sank into insignificance under the Romans.[5] The bay, which lies E. of the promontory, was the scene of the defeats of the Carthaginian fleet by Duilius in 260, and of the fleet of Sextus Pompeius by Agrippa in 36. Near Mylæ stood a famous temple of Diana.[5]

(4). *In the Interior.*—Centuripa, *Centorbi*, stood on a lofty hill, S.W. of Mount Ætna, and appears in the first instance as a stronghold of the Siculi, and as generally preserving its independence, though occasionally under tyrants, and at one time subject to Agathocles. In the First Punic War it was taken by the Romans, and it became subsequently one of the most important cities of Sicily, being situated in the midst of a remarkably fertile corn-producing district. Hybla, surnamed

---

[4] Littora Thermarum, prisca dotata Camœnæ,
Armavere anos, qua mergitur Himera ponto
Æolio.　　　　　　　　　　　　　Sil. Ital. xiv. 232.

[5] Et justi quondam portus, nunc littore solo
Subsidium insidam fugientibus æquora, Mylæ.　　In. xiv. 201.
Mille Thoantæ ædes Phacelina Dianæ.　　　　　In. xiv. 260.

Major, was situated S. of Ætna, and near the Symæthus, probably at *Paterno.* It was a city of the Siculi, and became in later times dependent on Catana. Its history is unimportant, and much confusion exists in the notices of this and of the other Hybla. **Enna,** or *Henna, Castro Giovanni,* was situated nearly in the centre of the island, where it occupied a position of remarkable strength, on the level summit of a gigantic hill, the sides of which are precipitous. It was a Siculian town, and retained its independence until the time of Dionysius of Syracuse, who gained possession of it by treachery. In 214 its citizens were massacred by the Romans, and in the Servile War in 134-132 it became the head-quarters of the insurgents. Enna was celebrated in mythology as the place where Pluto carried off Proserpine :[7] it possessed a very famous temple of Ceres.

Coin of Enna.

Of the less important towns we may notice—(1). *On the E. Coast*—**Callipolis,** a colony of Naxos, N. of Tauromenium, destroyed at an early period, probably by Hippocrates of Gela ; and **Helōrum,** or **Helōrus,** at the mouth of a river of the same name,[8] about 25 miles S. of Syracuse, of which it was a dependency, and probably a colony. *On the S. Coast*—**Motya,** between Lilybæum and Drepanum, a Phœnician colony, captured by Dionysius of Syracuse in 397, after a desperate defence, but recovered by Himilco in 396, who, however, removed its inhabitants to Lilybæum—**Selus,** or **Soluntum,** *Solanto,* about 12 miles east of Panormus, a Phœnician colony, and one of their last positions in the island, subsequently in the hands of the Carthaginians, with whom it remained until the First Punic War—**Cephalœdium,**[9] *Cefalu,* E. of Himera, origi-

---

[7] This event is said to have taken place at a small lake, fringed with flowery meadows, and surrounded by lofty mountains, with a cavern near it, whence Pluto issued. The place is still shown, but the flowers have disappeared. Ovid calls the lake Pergus (see p. 592). The myth is told at length in *Met.* v. 385-408, and more briefly by Silius Italicus :—

    Enna deûm lucis sacras dedit ardua dextras,
    Illo specus, ingentem laxans telluris hiatum,
    Cæcum iter ad manes tenebroso limite pandit,
    Qua novus ignotas Hymenæus venit in oras,
    Has Stygius quondam, stimulants Cupidine, rector
    Ausus adire diem, mæstoque Acheronta relicto
    Egit in illicitas currum per inania terras.
    Tum rapta præceps Ennæa virgine ficult
    Attonitos cœli visu lucemque paventes
    In Styga rursus equos, et prædam condidit umbris. —xiv. 239.

[8] This river, now the *Abisso,* stagnates about its mouth, but in its upper course is a brawling impetuous torrent: the following descriptions are equally correct of its different parts :—

    Exaupero præpingue solum stagnantis Helori.—*Æn.* iii 698.
    Unde clamosus Helorus.   Sil. Ital. xiv. 269.

[9] Quæque procelloso Cephalœdias ora profundo
    Cæruleis horret exmpis pascentia cete.   Ib. xiv. 252.

nally only a fortress on a lofty rock belonging to the Himeraeans, but
afterwards a town, first noticed in 396, and captured by treachery by
the Romans in 254—Haléea, or Alæsa, near Tusa, on the N. coast, a
Sicilian town, founded in B.C. 403 by citizens of Herbita and others,
and under the Romans one of the chief towns of Sicily, until ruined
by the exactions of Verres—Calacta,[1] Caronia, situated E. of Halesa, on
a portion of the coast which, for its beauty and fertility, was named
" the fair coast," a name which was subsequently affixed to a town
founded by Sicilians and others about B.C. 400—Aluntium, San Marco,
E. of Calacta, a place which suffered severely from the exactions of
Verres—Tyndáris, Tindaro, W. of Mylæ, founded by the elder Dionysius in B.C. 395, and peopled with Messenians, the head-quarters of
Agrippa in the war against Sextus Pompeius—and Abacænum, between
Tyndaris and Mylæ, about 4 miles from the N. coast, a city of the
Siculi, and at one time a place of importance, but from the time of
Hiero insignificant.

(2). *In the Interior.*—Ætna, at the S. foot of the mountain of the
same name, originally a Sicilian town with the name of Inessa, but
afterwards occupied by the colonists whom Hiero had sent to Catana,
and who changed its name to Ætna ; it was a strongly situated place,
vainly attacked by Laches in 426, seized by Dionysius in 403, and
peopled by him with Campanian mercenaries, who held it till 339.
Agyrium, S. Filippo d'Argiro, on the summit of a lofty hill, between
Centuripa and Enna, a Sicilian town, first noticed in B.C. 404 as the
residence of a powerful prince, named Agyris, under the Romans a
place of wealth and importance from the fertility of its territory in
corn, also known as the birth-place of the historian Diodorus Siculus.
Engyum, Gangi Veteri, S. of Halesa, celebrated for its temple of the
Magna Mater, which was plundered by Verres. Halicyæ, Salemi, 10
miles S. of Segesta, a town which, in the First Punic War, joined the
Romans at an early period, and was rewarded with immunity from
taxes and other privileges. Entella, Rocca d'Entella, on the left bank
of the Hypsas, said to have been founded by Acestes, first noticed in
B.C. 404 as being seized by the Campanian mercenaries, and held by
them until about 345, when the Carthaginians obtained possession of
it. Herbíta, Nicosia, 10 miles N.W. of Agyrium, first noticed in B.C.
445, as under the rule of a tyrant named Archonides, who held out
against Dionysius of Syracuse ; it is better known in connexion with
the exactions of Verres. Morgantia, S.W. of Catana, a Sicilian town,
first noticed in B.C. 459, as being taken by Ducetius, and repeatedly
mentioned during the Second Punic War. Menænum, Mineo, about 18
miles W. of Leontini, a Sicilian city, founded by Ducetius in B.C. 459,
conquered by Dionysius in 396, and mentioned by Cicero as one of the
flourishing towns of Sicily at that time. Acræ, Palazzolo, on a lofty
hill, 24 miles W. of Syracuse, of which it was a colony, planted in B.C.
663, and to which it was valuable as a military post. Casmînæ,
founded by Syracusans in B.C. 643, and noticed by Herodotus as the
place whither the exiled Gamori retired.

Off the coast of Sicily lie two groups of islands—the Ægatæ Insulæ,
off the W. angle, containing three islands, named Hiéra, Ægusa, and
Phorbantia, and historically famous for the victory obtained by Lutatius Catulus over the Carthaginians in B.C. 241, which put an end to

---

¹ Littus placens Calacte.   Sil. Ital. xiv. 231.

the first Punic War.[2] : and the volcanic group variously named *Æoliæ*,
from the Homeric island *Æolus*,[3] *Vulcaniæ* or *Hephæstiæ*, from their
volcanic character,[4] and *Liparæ*, after *Lipara*, the largest of the group,
a name which they still retain as the *Lipari Islands*. There are 7
larger and several smaller islands : of these, **Hiera**, *Vulcano*, the most
southerly, and **Strongyle**, *Stromboli*, the most northerly, were active
volcanoes : *Lipara* was the only one that possessed any considerable
population, together with a town, founded by Dorians from Cnidus in
B.C. 627, and a place of some historical importance : **Didyme**, *Salina*,
derived its name from the *twin* conical mountains on it ; *Phœnicūssa*,
Felicudi, from its palms (φοίνικες); *Ericūsa*, *Alicudi*, from its heath
(ἐρείκη), and **Euonymus**, *Panaria*, from the circumstance of its lying
on the *left hand*, as one sailed from Lipara to Sicily.

History.—The history of Sicily resolves itself very much into those
of its several towns. These have been already related, but it may be
useful to give a connected statement of the states which held the predominant power at different periods. During the 6th cent. B.C. Gela
and Agrigentum were the most powerful cities. Syracuse first rose
under Gelon in 485, and attained the ascendency over the Greek towns,
both under him and under his successor Hiero. On the expulsion of
Thrasybulus in 467, most of the towns adopted a democratic government, and from 461 to 409 they retained their independence of Syracuse,
and enjoyed the highest degree of prosperity. The Carthaginians, who
had failed in their first endeavour to obtain a footing in Sicily in B.C.
480, were more successful in 409, when they took Selinus, Himera,
and Agrigentum, and established themselves firmly in the W. of the
island. To counteract this power, the Greek cities threw themselves
more under the authority of Syracuse, which was raised by Dionysius I.
to the sovereignty of all eastern Sicily. Internal dissensions followed,
and at length, by the aid of Timoleon in 343, the cities were restored to
liberty. Again Syracuse became predominant under Agathocles from
317 to 289. Agrigentum had now revived, and was the second town in
Sicily. Under Hiero II. Syracuse was flourishing, and the other cities,

---

[2]   Illa super, ævi
    Flore virens, avet Ægates abolere, parentum
    Dedecus, ac Siculo demergere fœdera ponto.   SIL. ITAL. l. 60.

[3] It was the fabled residence of Æolus, the god of the winds :—
    Nimborum in patriam, loca fœta furentibus Austris,
    Æoliam venit. Ille vasto rex Æolus antro
    Luctantes ventos tempestatesque sonoras
    Imperio premit, ac vinclis et carcere frœnat.
    Illi indignantes, magno cum murmure montis,
    Circum claustra fremunt. Celsa sedet Æolus arce,
    Sceptra tenens ; mollitque animos, et temperat iras.— *Æn.* i. 31.

[4] Or as being (according to the mythical account) the workshop of Vulcan :—
         Jam silicato nectare turgens
    Brachia Vulcanus Liparæa nigra *taberna*.   JUV. xlii. 44.

    Insula Sicanium juxta latus Æoliamque
    Erigitur Liparen, fumantibus ardua saxis :
    Quam subter specus et Cyclopum exesa caminis
    Antra Ætnæa tonant, validique incudibus ictus
    Auditi referunt gemitum, striduntque cavernis
    Stricturæ Chalybum, et fornacibus ignis anhelat ;
    Vulcani domus, et Vulcania nomine tellus.   *Æn.* viii. 416.

which adopted the side of Carthage in the First Punic War, were reduced
by the Romans. In the Second Punic War, Syracuse fell in 212, and
the whole island was reduced to the condition of a Roman province.
It suffered severely from the Servile wars in 135-132, and 103-100,
from the exactions of Verres, and subsequently from those of Sextus
Pompeius. It was originally governed by a prætor and two quæstors,
but it was placed by Augustus under a proconsul.

§ 4. Melita, *Malta*, lies about 50 miles S. of Sicily; it is about
17 miles long, and 9½ broad, and is separated only by a narrow

Coin of Melita.

channel from the island
of Gaulos, *Gozo*. Melita
was conveniently situated
as a trading station, and
was from an early period
occupied by a Phœnician
settlement. It passed into
the hands of the Cartha-
ginians, who held it until
the Second Punic War,
when it was taken by Tib. Sempronius, in B.C. 218. It was famous
for its wool,[1] and for the manufacture of a fine cotton fabric,
known at Rome as "vestis Melitensis." It derives its chief interest
from the shipwreck of St. Paul on its coasts (Acts, xxviii.): the
memory of this event is preserved in the title of *St. Paul's Bay*, on
the N E. coast of the island. W. of Melita lies the small and barren
isle of Cosyra,[2] *Pantellaria*.

§ 5. The large island of Sardinia, the *Sardo* of the Greeks, lies S.
of Corsica, and N.W. of Sicily, and is distant only 120 geographical
miles from the coast of Africa. Its form resembles an oblong parallelogram;[3] its length is above 140 geographical miles, and its
average breadth about 60. It is traversed by a chain of mountains

---

[1]   Telaque superba
      Lanigera Melite.   Sil. Ital. xiv. 250

[2] Ovid contrasts the barrenness of Cosyra with the fertility of Malta: the
contrast does not hold good as regards the latter island, which is rocky and
dry :—
            Fertilis est Melite, sterili vicina Cosyrae
      Insula, quam Libyci verberat unda freti.   Fast. iii. 567.

[3] It resembles somewhat the print of a man's foot, and hence was named
Ichnusa by the Greeks :—
            Insula, fluctisono circumvallata profundo,
            Castigatur aquis, compressaque gurgite terras
            Enormes cohibet nudae sub imagine plantae.
            Inde Ichnusa prius Graiis nemorata colonis,
            Mox Libyci Sardus generoso sanguine fidens
            Herculis, ex sese mutavit nomina terrae.
            Affluxere etiam, et sedes posuere coactas
            Disperei pelago, post eruta Pergama, Teucri.   Sil. Ital. xii. 355.

from N. to S., a portion of which in the N. was named **Insani Montes**, from the violent storms which sailors encountered off that part of the island. There are several plains of considerable extent in the S. and W. parts. The rivers are small, but numerous: the chief are the **Thyrsus**, *Tirso*, and the **Sacer Fluvius**, *R. di Pabillonis*, on the W. coast; the **Termus**, *Temo*, on the N.; and the **Cædrius**, *Fiume dei Orosei*, on the E. coast. The climate of Sardinia has been in all ages unhealthy :[8] the soil was fertile,[9] and yielded large quantities of corn, and among the special productions of the island may be noticed a poisonous plant of extreme bitterness,[1] which, from the contortions it produced in the countenance, gave rise to the expression "Sardonicus risus." Wool was abundant, and Sardinia also possessed mines of silver and iron.

§ 6. The population of Sardinia was of a very mixed character: three native tribes are noticed—the Iolaï or Iolaenses, who (according to tradition) were of Trojan origin,[2] but more probably were Tyrrhenians; the Balari, probably of Iberian extraction; and the Corsi, from the neighbouring island of Corsica. The Greeks were acquainted with the island, and some of the towns have Greek names, but we have no record of their ever having settled on it. The Phœnicians, and in later times the Carthaginians, had stations on it. The Sardinians enjoyed an ill fame for general worthlessness of character. The towns were but few: the most important were founded by the Phœnicians, viz. Caralis, Nora, and Sulci. Of the antiquities of the country we may notice the peculiar towers named *Nuraghe*, built very massively, and containing one or two vaulted chambers. The number of these is very great, but both their use and their origin is unknown.

Caralis, *Cagliari*, was situated on the S. coast, and was said to have been founded by the Carthaginians. From the time of the Second Punic War, it became the chief naval station of the Romans, and the residence of the prætor. There are remains of an amphitheatre and of an aqueduct. **Sulci** was situated on a small island in the S.W. corner of the island. It was undoubtedly founded by the Carthaginians, and it reached a high degree of prosperity, both under them and under the Romans. **Neapolis**, *Nabui*, on the W. coast, would seem, from its name, to have been founded by the Greeks. **Olbia**, *Terranova*, was situated near the N.E. extremity of the island. Its name also is Greek, and tradition assigned to it a Greek origin. It was the ordinary place of communication with Italy, and hence rose to importance under the Romans. In 259 it was the scene of warlike operations between the

[8] Silius Italicus describes it as—
        Tristis cœlo et multa vitiata palude.   xii. 371.

[9]   Optimæ
        Sardiniæ segetes feracis.   Hor. *Carm.* i. 31, 3.

[1] Immo ego Sardois videar tibi amarior herbis.   Virg. *Ecl.* vii. 41.

[2] See note 7 above.

Romans under Cornelius and the Carthaginians. We may further
notice as considerable towns—**Nora**, on a promontory, about 20 miles S.
of Caralis, now named *Capo di Pula*, where are remains of a theatre,
an aqueduct, and quays—**Tharras**, on a promontory on the W. coast
now named *Capo del Sevo*, a Phœnician settlement—**Cornus**, on the W.
coast, about 16 miles N. of Tharras, the head-quarters of the revolted
Sardinian tribes in the Second Punic War—**Bosa**, *Bosa*, at the mouth of
the Temus—**Turris Libysõnis**, *Porto Torres*, a Roman colony on the
N. coast—and **Tibula**, near the extreme N. point, the port of communication with Corsica.

History.—The Carthaginians conquered Sardinia about 500-480 B.C.,
and it was held by them until 238, when the Romans got possession of
it. In 215 the natives rose in rebellion, and again a portion of them in
181 and in 114; but on all these occasions they were easily put down.
Sardinia was united with Corsica as a province under a proconsul. It
became a place of exile for political offenders under the Empire.

§ 7. **Corsica** (the **Cyrnus** of the Greeks) lies N. of Sardinia, from
which it is separated only by a narrow strait. Its size was unduly
magnified by the ancients: its length is really 126 miles, and its
greatest breadth about 51. Almost the whole of it is occupied by
lofty and rugged mountains, whose sides were clothed with the
finest timber. The central mass was named **Aureus Mons**, now
Monte Rotondo. The principal rivers are the **Rhotanus**, *Tavignano*,
and the **Tuola**, *Golo*, which enter the sea on the E. coast. Honey
and wax[3] are noted among the productions of the island, but the
former had a very bitter flavour,[4] from the number of yew trees
on the island. The earliest inhabitants were probably Ligurians:
Greeks settled at Alalia, in B.C. 664; and subsequently Tyrrhenians
and Carthaginians. The **Corsi** were reputed a wild and barbarous
race; they lived chiefly on the produce of their herds. The most
important towns were Mariana and Aleria.

Mariana stood on the E. coast, and was founded by and named after
C. Marius: it probably occupied the site of an earlier town **Nicæa**,
whose name bespeaks a Greek origin. **Aleria** (the *Alalia* of the
Greeks) also stood on the E. coast, near the mouth of the Rhotanus.
It was founded by Phocæans, in 564, but was abandoned by them about
540, in consequence of a severe defeat they sustained from the Tyrrhe-
nians and Carthaginians. It was captured by the Romans under
L. Scipio, in 259, and subsequently received a colony under Sulla.

History.—Corsica, like Sardinia, was under the power of Carthage at
the time of the First Punic War. The capture of Aleria was followed
by the nominal subjection of the island to Rome. It was not, however,
until the time of Sulla that it was really brought into a state of peaceable submission. It was made a place of banishment by the Romans,
and, among others, Seneca spent some time there.

---

[3] Ite hinc difficiles, funebria ligna, tabellæ :
Tuque negaturis cera referta notis.
Quam, puto, de longæ collectam flure cicutæ
Melle sub infami Corsica misit apis, Ov. *Am.* I. 12, 7.

[4] Sic tua Cyrneas fugiant examina taxos. Virg. *Ecl.* ix. 30.

Remains of the Great Theatre, Saguntum, Spain.

CHAPTER XXIX.

HISPANIA.

§ 1. Boundaries. § 2. Mountains. § 3. Rivers. § 4. Bays and Promontories. § 5. Climate and Productions. § 6. Inhabitants. § 7. Divisions. I. BÆTICA. § 8. Boundaries, &c. § 9. Inhabitants, Towns, &c. II. LUSITANIA. § 10. Boundaries; Rivers. § 11. Inhabitants; Towns. III. TARRACONENSIS. § 12. Boundaries; Rivers. § 13. Tribes and Towns on the Mediterranean. § 14. Tribes and Towns near the Pyrenees. § 15. Tribes on the N. Coast. § 16. Tribes and Towns of the Interior; Islands; History.

§ 1. Hispania, *Spain*, has been already noticed as the most westerly of the three southern peninsulas of the continent of Europe. It is bounded on the E. and S.E. by the Mare Internum, on the S.W. and W. by that portion of the Atlantic Ocean which was called Oceanus Gaditanus, and on the N. by the Mare Cantabricum, *Bay of Biscay*, and the Pyrenæi Montes, which stretch across the greater portion of the isthmus, connecting it with the continent. Its form is neither a quadrangle, as Strabo supposed, nor yet a triangle, as others describe it, but a trapezium. It lies between 36° 1' and 43° 45' N. lat., and between 3° 20' E. and 9° 21' W. long., its greatest length from N. to S. being about 460

miles, its greatest breadth from E. to W. about 570, and its area,
including the Balearic Isles, about 171,300 square miles. The
greater part of the peninsula is an elevated table-land, sustained by
lofty mountain ranges, sloping down gradually to the W. coast, and
terminated eastwards by the ranges which bound the valley of the
Iberus.

Names.—The name "Hispania" came into use when the Romans
became connected with the country: its original form was *Span*, or
Sapan, supposed to be derived from a Phoenician root signifying
"rabbit," in reference to the number of those animals in the country:
it has also been derived from the Basque *Espana*, "margin," in reference
to its position on the shores of the ocean. The Greeks termed it
"Iberia," from the river Iberus, and "Hesperia," as the most westerly
portion of the known world, to which the Latins added the epithet
"Ultima." The interior of the country was occasionally termed
"Celtica" from its Celtic population; and the S. portion. outside the
straits, "Tartessis," the same as the scriptural Tarshish. The ethnic
forms were "Iber," and in the plural "Ibêres," or "Iberi," and
"Hispânus," or "Spanus:" the adjective forms were "Ibericus,"
"Ibêrus," or "Iberiacus," and "Hispaniensis."

§ 2. The chief mountain range is the **Pyrenæi Montes**, already
described as crossing the isthmus between the Mediterranean Sea
and the *Bay of Biscay*. The great table-land of Spain is bounded
on the N. by the continuations of the Pyrenean range, under the
names of **Vasconum Saltus** and **Vindius Ms.**; on the E. by a range
which strikes off from the eastern extremity of the latter towards
the S.E. and S., under the names of **Idubêda**, *Sierras de Oca* and
de Lorenzo, and **Orospêda** or **Ortospêda**, *Sierra Molina;* on the S.
by the **Mariânus Ms.**, *Sierra Morena*; while towards the W. it sinks
down gradually towards the Atlantic. The table-land itself is crossed
by two chains which spring out of Idubeda, and run towards the
S.W., neither of which received specific names in ancient geography,
with the exception of the W. portion of the northerly one, which
was called **Herminius**, *Sierra de Estrella*. An important range, now
Sierra Nevada, runs parallel to the Mediterranean Sea, portions of
which were named **Solorius** and **Ilipûla**. This was connected with
Ortospeda by cross ranges, named **Castulonensis Saltus** and **Argentarius**, which closed in the head of the valley of the Bætis.

§ 3. The great rivers of Spain have their basins clearly defined by
the chains just described—the **Ibêrus**, *Ebro*, draining the large triangular space enclosed by the Pyrenees on the N. and Idubeda on
the W., and opening out to the Mediterranean Sea on the S.E.; the
Bætis, *Guadalquivir*, between the ranges of Ilipula and Marianus;
the **Anas**, *Guadiana*, between Marianus and the southerly of the two
ranges that cross the table-land; the **Tagus**, between the two central
ranges and the **Durius**, *Douro*, between the northern one and Ms.

Vindius. Of these rivers the three last belong mainly to the central table-land, the two first to the surrounding district. The valleys of the Iberus and Bætis, together with the intervening maritime district, were the most important portions of the peninsula in ancient times, that of the Iberus lying conveniently open to the Mediterranean, and that of the Bætis being so enclosed with mountain ranges as to be almost a distinct country.

§ 4. The line of coast presents the following promontories and bays :—**Pyrénæ** or **Vanaris Prom.**, *C. Creus*, the E. extremity of the Pyrenean range; **Dianium**, *C. St. Martin*, which forms the S. extremity of the **Sucronensis Sinus**, *G. of Valencia*; **Saturni Prom.**, *C. de Palos*, which encloses on the S., as Dianium on the N., the **Illicitanus Sinus**, *B. of Alicante*; **Charidémi Prom.**, *C. de Gata*, between which and Saturni lies the **Mamilánus Sinus**; **Calpe**, *Gibraltar*, at the E. end of the **Fretum Gaditanum**, *Straits of Gibraltar*; **Junónis Prom.**, *C. Trafalgar*, outside the W. entrance of the Straits; **Cuneus**, *C. de Santa Maria*, and near it **Sacrum Prom.**, *C. St. Vincent*, at the S.W. extremity of the peninsula; **Barbarium Prom.**, *C. Espichel*, and **Magnum Prom.**, *C. da Roca*, respectively S. and N. of the estuary of the Tagus; **Celticum** or **Nerium Prom.**, *C. de Finisterre*, at the N.W. extremity; and **Corn** or **Trileucum Prom.**, *C. Ortegal*, at the extreme N.

§ 5. The climate of Spain varies with the varying altitude of the districts. In the central table-land the cold is very severe for a considerable portion of the year; the southern maritime districts have an almost tropical heat. Equally various are the soil and productions : while large portions of the centre are barren, and others only adapted for hardy productions, such as wheat, the valleys of Bætica are suited to the growth of the palm-tree and other tropical plants. The latter region was therefore most attractive for early colonization : it produced corn, wine, oil, and figs, in abundance. Lusitania was famed for its fine-wooled sheep; Celtiberia for its asses; the fields of Carthago Nova and other plains for its *spartum*, out of which cordage was made; and Cantabria for its pigs. The most valuable productions, however, were minerals: silver was abundant, and one of the mountains we have noticed, Argentarius, was named after its valuable mines of this metal; tin was found in Lusitania, Gallæcia, and Bætica; lead in Saltus Castulonensis; iron and copper in many places, the latter especially at Cotinæ.

§ 6. The population of Spain consisted mainly of Iberians, the progenitors of the modern *Basques*; another very important, though less numerous element was supplied by the Celts. These two coalesced to a certain extent, and formed a mixed race named Celtiberian, which occupied the centre of the country as well as parts of Lusitania and of the N. coast. In other parts they lived distinct—

the Iberians in the Pyrenees and along the coast-districts, the Celts
on both sides of the Anas and in the extreme N.W. of the peninsula
about Prom. Nerium. Lastly, there was a large admixture of Phœnicians in Bætica; colonies were established on the S.E. coast by
the Carthaginians, and by various Greek states; and at a later period
there was also a large influx of Romans. The tribes were very
numerous, and differed materially in character and acquirements.
The Cantabrians and the peoples of the N. coast were the wildest
and rudest; the Celtiberians had a higher character, but were hardly
more civilized; the Vaccæi were (under the Romans at least) highly
civilized; while the Turdetani cultivated science, and had a literature of their own. In some respects the Iberians [1] contrasted favourably with the civilized nations of antiquity, particularly in the
higher position assigned to women in their social system; but on the
other hand they were cunning, mischievous, and dishonest. Under
the Romans the country was thoroughly civilized: many very considerable colonies were planted, and were adorned with magnificent
productions of Roman architecture, some of which remain to this
day, while vast numbers have been barbarously demolished for the
sake of the materials alone. Roads were constructed through every
part of the country, and so completely was the Roman influence infused into it, that in Bætica the natives had forgotten even their
own language. The degree of culture may to a certain extent be
inferred by the numerous illustrious men who were born in Spain—
the Emperors Trajan and Hadrian; the poets Silius Italicus, Lucan,
Martial, Prudentius, and Columella; the two Senecas; the geographer Pomponius Mela; the rhetorician Quinctilian; and many
others.

§ 7. The earliest political division of Spain dates from the time
when the Romans gained a footing in the country. In B.C. 205 it
was divided into two parts—Citerior and Ulterior, respectively E.
and W. of the river Iberus, which formed the original line of demarcation between the Roman and Carthaginian possessions. Other
designations were occasionally employed, as Celtiberia for the E. and
Iberia for the W. by Polybius. As the Roman conquests advanced
into the country, Citerior advanced with them until it embraced
the whole country as far as the borders of the later Bætica. A new
arrangement was introduced by Augustus by the division of Ulterior
into two provinces, named Bætica and Lusitania, and the alteration
of the name Citerior into Tarraconensis. He further subdivided the

---

[1] The general bearing of the ancient Iberian was strikingly similar to that of
the modern Spaniard; he was temperate and sober, indolent and yet spirited,
successful in guerilla warfare, and stubborn to the last degree in the defence of
towns, but deficient in the higher military qualities requisite for pitched battles
or other operations in the field.

whole country into 14 *conventus juridici*.   Constantine divided
Spain, with its islands and part of Mauretania, into 7 provinces.
We shall adopt the division of Augustus in the following pages.

I. BÆTICA.

§ 8. Bætica was bounded on the N. and W. by the river Anas,
on the E. by a line drawn from the upper valley of the Anas across
that of the Bætis to the sea near Prom. Charidemi, and on the S. by
the sea : it thus corresponds to the modern *Andalusia*.   It derived
its name from the chief river in the district, the Bætis,[3] *Guadalquivir*, which rises in Mt. Argentarius near Castulo, and flows towards the S.W., reaching the sea a little W. of Gades ; it receives
only one important tributary, the Singilis, *Xenil*, which rises in
Ilipula, and flows towards the N.W., joining it in its mid-course.
The Bætis was navigable for small boats as far as Corduba, and for
light vessels as far as Hispalis.   Bætica was the portion of Spain
which, from its fertility and its contiguity to the *Straits of Gibraltar*,
became first known to the commercial nations of antiquity.   The
Phœnicians carried on an extensive trade with Tartessus,[3] the
Tarshish of Scripture, which appears to have been the name both
of a town and of a district W. of the Columns of Hercules about the
mouth of the Bætis ; they planted the colonies of Gades and Carteia
there.   It was visited by Samians about B.C. 650, and by Phocæans
in 630 ; and at this period its trade extended to Britain and Africa.

§ 9. The principal tribes were—the Bastuli on the S. coast, from
Calpe on the W. to the E. border ; the Turduli and Turdetani, two
tribes closely connected together, occupying the lower valley of the
Bætis ; and a tribe of Celtici in the district of Bæturia, which lay
between Ms. Marianus and the Anas.   Bætica possessed some of the
finest towns of Spain : Corduba, on the right bank of the Bætis,
ranked as its capital, being the residence of the Roman governor,

---

[3] The indigenous name was Certis or Perces ; the early Greeks described it as
the Tartessus : the modern Arabic name signifies the " Great River."   The name
was used by the poets as equivalent to the country which it watered :—

> Bætis olivifera crinem redimite corona ;
> Aurea qui nitidis vellera tingis aquis.   MART. xll. 100.
>
> An Tartessiacus stabuli nutritor Iberi
> Bætis in Hesperia te quoque lavit aqua ?   ib. vlli. 28.

[5] Tartessus became a synonymous term for the West among the Latin
poets, e. g. :—

> Premerat occiduus Tartessia littora Phœbus.   OV. *Met.* xiv. 416.
>
> Armat Tartessos, stabulanti conscia Phœbo.   SIL. ITAL. iii. 399.

And sometimes for Spain :—

> mcoque subibat
> Germano devexa jugum Tartessia tellus.   IB. xiii. 673.

and the seat of a *conventus.* Three other towns were the seats of
conventus, viz. Gades- on the sea-coast, Astigi on the Singulis, and
Hispalis on the left bank of the Bætis. The whole number of towns
under the Romans was 175, of which 9 were *coloniæ,* 8 *municipia,*
29 endowed with the Latin franchise, 6 free, 3 allied, and 120 *stipendiaria.*

(1.) *Towns along the Coast from W. to E.* —Onōba Œstuaria stood
near the mouth of the Luxia and near an island named Herculis Insula,
Saltes: it had a mint. There are a few Roman remains of it, particularly an aqueduct, at *Huelva.* Asta⁴ stood on an estuary of the *Gulf of
Cadiz,* about 12 miles from Gades: it was the ancient seat of congress
for the people of that neighbourhood, and, under the Romans, became
a colony: its ruins are called *Mesa de Asta.* Gades,⁵ *Cadiz,* one of the
most famous cities of Spain, was situated on a small island now named
Isla de Leon, separated from the mainland by a narrow strait, the
River of St. Peter, over which a bridge was thrown. It was founded
by the Phœnicians at a very early period. Originally the town, which
was very small, stood on the W. side of the island: under the Romans
it was enlarged by the building of the "New City," and, even then,
it did not exceed 2½ miles in circumference, as the more wealthy
citizens had their villas outside the town, either on the mainland or
on the isle of *Trocadero.* The territory of the city was but small, its
great wealth and power being wholly derived from its commerce. It
entered into alliance with Rome in B.C. 212, and this alliance was
confirmed in 78: it was visited by Julius Cæsar in 49, when the
civitas of Rome was conferred upon its citizens. Under Augustus it
became a *municipium.* Gades possessed famous temples of the Phœnician Saturn and Hercules, the latter of which stood on *St. Peter's Isle,*
and had an oracle. The wealth of Gades induced habits of luxury
and immorality.⁶ Belon stood at the mouth of the river *Barbate,* W.
of *Tarifa,* and was the usual place of embarkation for Tingis on the
opposite side of the straits: its ruins are at *Belonia.* Cartela⁷ was an

---

⁴ The root *Ast,* which appears in this and various other Spanish names, is supposed to mean "hill-fortress."
⁵ The Phœnician form of the name was Gadir, or, with the article, Hagadir,
which is the usual inscription on the coins. The Greeks called it Gadeira. Its
meaning is thus explained by Avienus:—

　　　　Gaddir hic est oppidum :
　　Nam Punicorum lingua conseptum locum
　　Gaddir vocabat.　　　　　　　　　　*Ora Marit.* 267.

The Greeks and Romans regarded it as the extreme W. point of the world :—

　　Omnibus in terris quæ sunt a Gadibus usque
　　Auroram et Gangem.　　　　　　　　Juv. x. 1.

⁶ Forsitan exspectes, ut Gaditana canoro
　　Incipiat prurire choro.　　　　　　Id. xi. 162.

　　Gaudent jocosæ Canio suo Gades :
　　Emerita Deciana meſ.　　　　　　Mart. l. 61.

　　Nec de Gadibus improbis puellæ
　　Vibrabunt.　　　　　　　　　　Id. v. 78.

⁷ Cartela is probably identical with Calpe, which was one of the Greek forms of
the name, the others being *Carpia, Carpeia, Calpis;* it may also be identical with
Tartessus, which was sometimes described as Carpessus. The same root lies at

ancient Phœnician town, situated on the Bay of Gibraltar, at *Rocadillo*, about half way between *Algesiras* and *Gibraltar*, where the remains of an amphitheatre still exist. In the Punic War it was an important naval station, and the scene of a sea-fight in which Lælius defeated Adherbal, in B.C. 206: in 171 it became a colony, and was peopled with the offspring of Roman soldiers and Spanish women. Cu. Pompeius took refuge there after his defeat at *Munda*. **Malaca**, *Maluga*, was an important town, situated on a river of the same name, now the *Guadal-medina*, E. of Calpe: it was probably of Phœnician origin: under the Romans it became a *fœderata civitas* and had extensive establishments for salting fish.

(2.) *Towns in the Interior.*—Illiturgis was situated on a steep rock on the N. side of the Bætis, near *Andujar*. In the Second Punic War it joined the Romans, and was twice besieged by the Carthaginians: it afterwards revolted, and was stormed and destroyed by Publius Scipio in B.C. 206, and again in 196. Under the Roman empire it was a considerable town with the surname of Forum Julium. **Munda** probably stood, not on the site of the present *Monda*, but near *Martos* to the S.E. of Corduba, where are the remains of an ancient town: it was the scene of two great battles, the first in B.C. 216, when Cn. Scipio defeated the Carthaginians,[a] the second in 45, when Julius Cæsar defeated the sons of Pompey. **Astigi**, *Ecija*, stood on the plain S. of the Bætis. Though a considerable town, it possesses no historical associations. **Hispalis**,[b] *Seville*, stood on the left bank of the Bætis, and from its position gradually rose to the highest eminence, being styled *metropolis* by Ptolemy: as a Roman colony it bore the titles of Julia Romula and Colonia Romulensis. **Italica**, *Old Seville*, on the opposite side of the river, was founded by Scipio Africanus, in B.C. 207, as a settlement for his disabled veterans: it was a *municipium*, and the native place of the Emperors Trajan, Hadrian, and Theodosius, and, as some say, of the poet Silius Italicus: its inhabitants migrated to *Seville* in the Middle Ages: the ruins of an amphitheatre and of some reservoirs alone remain. **Sisapon**, *Almaden*, was the chief town in Bæturia, and derived its importance from its silver mines and veins of cinnabar. **Corduba**, *Cordova*, on the right bank of the Bætis,[c] is said to have been founded by Marcellus, who made it his head-quarters in the Celtiberian War. Its population was a mixture of Romans and natives, and it was the first Roman colony in those parts: it suffered severely in the great Civil War, and was taken by Cæsar in B.C. 45, when 22,000 of its inhabitants were slaughtered. It became the capital of the province,[d] and had the surname of Patricia from the

the bottom of all these words, and appears in the Phœnician name of Hercules, *Mel-Carth*.

[a] Pœni saturentur sanguine manes;
     Ultima funesta concurrant prœlia Munda.   Luc. L 39.
     Non Utica Libyæ clades, Hispania Munda
     Plesset.   Id. vi. 306.

[b] The tide reached up to Hispalis :—
     Et celebre Oceano atque alternis æstibus Hispal.—Sil. Ital. iii. 393.

[c] In Tartessiacis domus est notissima terris
     Qua dives placidum Corduba Bætin amat.   Mart. ix. 62.

[d] The bright colour of the wool in this neighbourhood is often noticed :—
     Uncto Corduba lætior Venafro,
     Illinc nec minus absoluta testa.

number of patricians among its colonists. It was also the birthplace
of Lucan and the two Senecas.[3]

We may further briefly notice—**Illibbris**, the original of *Granada*,
noticed by Hecatæus under the form of Elibyrge; **Urso**, *Osuna*, in the
mountains S.E. of Hispalis, the last resort of the Pompeians, and a
Roman colony with the name Genua Urbanorum; **Carmo**, *Carmona*, a
strongly-fortified town N.E. of Hispalis, one of the head-quarters of the
rebellion in Bæturia, B.C. 197, and in the Julian Civil War described as
the strongest city in Hispania Ulterior: **Astapa**, in an open plain on the
S. margin of the valley of the Bætis, celebrated for its devoted attach-
ment to the Carthaginians, and for the consequent self-destruction of
its inhabitants when besieged by the Romans in the Second Punic
War; **Acinipo**, about 6 miles N. of *Ronda*, worthy of notice for the
ruins of an aqueduct and theatre on its site; **Aurinx**, or **Oringis**, near
Munda, the head-quarters of Hasdrubal in B.C. 207, and also wealthy
from its silver-mines and the fertility of its territory; **Calentum**,
Casalla, famous for the manufacture of a very light kind of tile;
Ilipa, on the right bank of the Bætis, with great silver-mines in its
neighbourhood, and just at the point where the river ceased to be
navigable for vessels: its ruins are near *Peñaflor*; and **Intibili**, near Illiturgis, the scene of a victory gained by the Romans over the Carthaginians in the Second Punic War.

II. LUSITANIA.

§ 10. **Lusitania** was bounded on the W. and S. by the Atlantic
Ocean, on the N. by the river Durius, and on the E. by the Anas
as far as above Metellinum, and by a line drawn thence to the
Durius, at a point below the junction of the river Pisoraca. It
corresponds generally to the kingdom of *Portugal*, but while on the
one hand it was less extensive than that kingdom in the N. (for
Portugal extends to the *Minho*), it was more extensive towards the
E., and included the N. part of Spanish *Estremadura*, and the S.
part of *Leon*. The country is generally lofty and rugged on the
E. side, but more level as it approaches the sea. It is divided into
two portions by the range of **Herminius**, which separates the basins
of the Tagus and Durius. The chief rivers are the **Tagus**,[4] which

Albi quæ superas oves Galesi,
Nolle murice, nec cruore mendax,
Sed tincta gregibus colore vivo.   Maur. xii. 64.

    Qua dives placidum Corduba Bætin amat ;
Vellera nativo pallent ubi flava metello,
Et linit Hesperium bractea viva pecus.   In. ix. 62.

[*] Docuque Senecas unicumque Lucanum
    Facunda loquitur Corduba.   Jw. i. 62.

[·] It was famed for its fish and oysters :—
    Sed quærcunque tamen feretur illic
    Piscosi calamo Tagi notata
    Macrum pagina nostra nominabit.   In. x. 78.

Likewise for its gold sand, of which at the present time the quantity is very
small : —

falls into the ocean near Olisipo; the **Durius**, *Douro*, on the N. border; and the smaller streams of the **Callipus**, *Sadao*, S.E. of the Tagus, and **Vacua**, *Vouga*, between the Tagus and Durius.

§ 11. Lusitania was occupied by five chief tribes—the **Lusitani** on the W. coast, between the Tagus and Durius; the **Vettönes**, E. of them, between the Durius and Anas; the **Turduli Veteres**, on the banks of the Anas; the **Turdetani**, between the lower course of the Anas and the S. and W. coasts; and the **Celtici** in various positions, chiefly S.E. of the Lower Tagus, and on the S. coast in the district named **Cuneus**, where they bore the distinctive name of **Conii**. Of the towns we have not much information; Olisipo, *Lisbon*, was the old capital of the Lusitani, and Emerita Augusta, in the S.E. of the province, the later capital of the Romans, while Pax Julia, near the Callipus, and Scallabis on the Tagus, were, with Emerita, the seats of the three *conventus* into which the province was divided. The total number of towns was 46, of which 5 were *coloniæ*, 1 a *municipium*, 3 with the Latin franchise, and 37 *stipendiaria*.

Towns from S. to N.—**Balsa**, *Tavira*, stood on the coast W. of the Anas, and was a *municipium*, with the title of Felix. **Myrtilis**, *Mertola*, on the river Anas, had the *jus Latii*, with the surname of Julia. **Pax Julia**, *Beja*, lay on a hill to the N., and was a Roman colony, and the seat of a conventus: it was probably the same as Pax Augusta. **Salacia**, *Alacer do Sal*, to the N.W., was celebrated for its manufacture of fine woollen cloths. **Ebora** was an important town and a *municipium*, with the surname of Liberalitas Julia: there are fine ruins at *Evora*, especially of an aqueduct and a temple of Diana. **Augusta Emerita**, *Merida*, on the Anas,[a] was built, in B.C. 23, by Publius Carisius, the legate of Augustus, and was colonised with the veterans of the 5th and 10th legions. It was a *colonia* from the first, and had also the *jus Italicum*, was the residence of the prætor, and the seat of a conventus. The ruins of the town are magnificent; the circus is nearly perfect; the great aqueduct is one of the grandest remains of antiquity in the world, and the old Roman bridge remained uninjured until A.D. 1812, when some of the arches were blown up: in respect to its ruins it has been termed "the Rome of Spain." **Metellinum**, *Medellin*, was higher up the Anas; the modern town stands on the S. side of the river, and would thus have fallen within the limits of Bætica; it was a *colonia*. **Olisipo**, *Lisbon*, stood on the right bank of the Tagus, near its

Cedat et auriferi ripa beata Tagi.   Ov. *Am.* l. 15, 34.

 Tanti tibi non sit opaci
Omnis arena Tagi, quodque in mare volvitur aurum.— Juv. III. 54.

Æstus serenos aureo franges Tago,
 Obscurus umbris arborum.   Mart. L 50.

[a] Clara mihi post has memorabere, nomen Iberum,
Emerita æquoreus quam præterlabitur amnis,
Submittit cui tota suos Hispania fasces.
Corduba non, non arce potens tibi Tarraco certat
Quæque sinu pelagi jactat se Bracara dives.
                              Ausonius, *Ord. Nob. Urb.* 13.

mouth : its territory was celebrated for a remarkably swift breed of
horses: the name is also given as Ulysaipo, from a mistaken idea that
the legend of a town founded by Ulysses applied to it. **Norba Cæsaria,**
Alcantara, lay on the left bank of the Tagus, N.W. of Emerita : a
magnificent bridge over the river, built by Trajan, still remains.
Scalabis, *Santarem,* lay between Olisipo and Emerita : it was a colony,
with the surname Præsidium Julium, and one of the three *conventus.*
Salmantica, *Salamanca,* also called Helmantica and Hermandica, stood
S. of the Durius : the piers of a bridge of 27 arches over the *Tormes,*
built by Trajan, are still in existence.

III. TARRACONENSIS.

§ 12. **Tarraconensis** was bounded on the E. by the Mare Internum;
on the N. by the Pyrenees, which separated it from Gallia, and
further W. by the Mare Cantabricum ; on the W. by the Atlantic
Ocean, as far S. as the Durius, and below that point by the province
of Lusitania ; and on the S. by the provinces of Lusitania and
Bætica, from the former of which it was separated by the Durius,
from the latter by Ms. Marianus. It thus embraced the modern
provinces of *Murcia, Valencia, Catalonia, Arragon, Navarre,
Biscay, Asturias, Gallicia,* the N. parts of *Portugal* and of *Leon,*
nearly all the *Castilles,* and part of *Andalusia.* This extensive
district contains within its limits the upper courses of all the large
rivers already noticed, the **Bætis, Anas, Tagus,** and **Durius,** together
with the whole course of the **Iberus,** which was historically the
most important river of Spain, and which received as tributaries,
on its left bank, the **Gallicus,** *Gallego,* and the **Sicoris,**[*] *Segre,* and on
its right, the **Salo,** *Xalon.* In addition to these we may notice the
following important rivers which flow into the Mediterranean : the
Rubricatus, *Llobregat,* joining the sea a little W. of Barcino ; the
Turia, *Guadalaviar,* near Valentia, famed for a battle fought on its
banks between Pompey and Sertorius ; the **Sucro,** *Xucar,* more to
the S.; and the **Tader,** *Segura,* N. of Carthago Nova. On the W.
coast, the **Minius,** *Minho,* which rises in the mountains of Gallæcia,
is an important river : it is said to have been so named from the
minium, or vermilion, carried down by its waters. We may also
notice the following tributaries of the Durius : on its right bank the
Pisoriea, *Pisuerga,* and the **Astura,** *Ezla,* and on its left the **Cuda,**
Coa. In describing Tarraconensis we shall adopt a fourfold division
of the tribes, as follows : (1) those along the coast of the Mediterranean ; (2) those at the foot of the Pyrenees ; (3) those along the
N. coast ; (4) those in the interior. It only remains for us here to
observe that Tarraconensis was divided into seven *conventus juridici,*

Inter
Stagnantem Sicorim et rapidum depressus Iberum.—LUCAN, iv. 331.

containing 472 towns and villages, of which 12 were *coloniæ*, 13 *oppida civium Romanorum*, 18 *Latinorum veterum*, 1 *fœderatorum*, and 135 *stipendiaria*.

§ 13. The tribes on the coast of the Mediterranean from S.W. to N.E. were—the **Bastetáni**, on the borders of Bætica, sometimes identified with the Hastuli ; the **Contestáni**, on the coast from the borders of Bætica to the river Sucro ; the **Edetáni**, or **Sedetáni**, between the Sucro and the Iberus ; the **Deresónes**, in that portion of the sea-coast of Edetania which lies S.W. of the Iberus ; the **Cosetáni**, from the mouth of the Iberus northwards to near the Rubricatus ; the **Lasetáni**, or **Laletáni**, thence to the territory of the **Indigétes**, who lived on the bay of Emporiæ in the extreme N.E. This district contained the most important towns in Spain—Carthágo Nova, the Punic capital, in the territory of the Contestani ; Tarráco, the Roman capital, on the coast N. of the Iberus ; Cæsaraugusta, the chief town in the upper valley of the Iberus ; Barcino and Emporiæ, flourishing sea-ports between the Iberus and the Pyrenees. The origin of many of these towns is unknown : Carthago Nova was founded, within historical times, by the Carthaginians ; the names of Tarraco and Barcino also bespeak a Punic origin. Saguntum and Emporiæ, on the other hand, were attributed to the Greeks, the latter having an undoubted Greek name, and the former being regarded as a modification of Zacynthus. The inland towns belonged to the Iberians, their names being either Latinized forms of the original ones, or, as in the case of Cæsaraugusta, new names assigned to them by the Romans.

Carthágo Nova, *Cartagena,* stood a little W. of Prom. Saturni, at the bottom of a bay, which, having its entrance nearly closed by the isle of Scombraria, was thus converted into a sheltered harbour. The site of the town was an elevated tongue of land projecting into the bay, surrounded on the E. and S. by the sea, and on the W. and partly on the N. by a lake communicating with the sea, the isthmus between the lake and the sea being only 250 paces wide. A range of hills encircled the town on the land side. Carthago Nova was a colony of Carthage, planted by Hasdrubal in B.C. 242, the site being selected partly from the excellence of its harbour and its central position in reference to the coast of Spain and the opposite coast of Africa, and partly from its proximity to the richest silver mines of Spain. It became the Carthaginian capital of the country—at once the treasury, the arsenal, and the base of operations. It was surprised by P. Scipio in 210, and became thenceforward the rival of Tarraco. It was made a colony, with the title of Col. Victrix Julia Nova Carthago, and was the seat of a *conventus*. It remained an important place of commerce even after its size was much diminished. It sometimes received the surname of Spartana, from the valuable plant (a kind of broom) which grew in its neighbourhood. **Sætábis,** *Jativa,* was a Roman *municipium*, and the seat of a large flax[f] and linen manu-

---

[f] Sætabis et telas Arabum spreveris superba
Et Pelusiaco filum componere lino.   SIL. ITAL. III. 374.

facture : it lay upon an eminence S. of the Sucro. **Valentia** belonged to the Edetani, and was situated on the Turia, about 3 miles from its

mouth: it became a colony, and was peopled with the soldiers of Viriathus: it was destroyed by Pompey, but was restored; it still exists, with the same name. **Saguntum** was seated on an eminence[6] on the banks of the river Pallantias, between the Sucro and Tarraco, and not far from

Coin of Valentia in Spain.

the sea. It was said to have been founded by Zacynthians,[7] with whom, according to some accounts, Rutuli from Ardea were mixed. It lay in a fertile district, and attained to great wealth by its commerce. Its capture by Hannibal, in B.C. 218, after a long resistance, was the cause of the Second Punic War. It was recovered by the Romans in 210, and made a Roman colony. A manufacture of earthenware[1] cups (*calices Saguntini*) was carried on there. The modern town is named *Murviedro*, from the *muri veteres* of the old town; the remains of them, however, are now insignificant: the framework of the theatre exists, and there are portions of the walls of the Circus Maximus. **Caesaraugusta**, *Zaragosa*, *Saragossa*, stood on the right bank of the Iberus, and was the central point whence all the great roads of Spain radiated. Its original name, as a town of the Edetani, was Salduba, which was changed in honour of Augustus, who colonised it in B.C. 25; it was a *colonia immunis*, and the seat of a *conventus*. The first Christian poet, Aurelius Prudentius, is said to have been born there in A.D. 348. **Dertosa**, *Tortosa*, stood on the left bank of the Iberus, not far above the delta of the river, in the territory of the Ilercaones: it became a colony. **Tarraco**,[2] *Tarragona*, was finely seated on a rock,[3] between 700 and 800 feet high, overhanging a bay of the Mediterranean sea: it possessed no harbour. It was fortified by the Scipios, who converted it into a fortress against the Carthaginians; subsequently it became the capital of the province and the seat of a *conventus*. Augustus wintered there after his Cantabrian

---

Nam sudaria Setaba ex Iberis
Miserunt tibi muneri.   CATULL. xli. 14.

[6] Haud procul Herculei tollunt se littore muri,
Clementer crescente jugo, quis nobile nomen
Conditus excelso sacravit colle Zacynthos.   SIL. ITAL. i. 273.

[7] Mox profugi ducente Noto advertere coloni,
Insula quos genuit Graio circumflua ponto,
Atque auxit quondam Laertia regna, Zacynthos:
Firmavit tenues ortus mox Daunia pubes,
Sedis inops, misit largo quam dives alumno,
Magnanimis regnata viris, nunc Ardea nomen.   ID. i. 288.

[1] Sume Saguntino pocula ficta luto.   MART. xiv. 108.

Pugna Saguntina fervet commissa lagena.   JUV. v. 29.

[2] The name seems to imply a Phoenician foundation: it comes from *Tarchon*, "citadel."

[3] Hispanae pete Tarraconis arces.   MART. x. 104.

campaign. Its fertile plain and sunny shores are celebrated by Martial [1] and other poets; and its neighbourhood produced good wine [1] and flax. There are numerous remains of the old town, particularly the so-called palace of Augustus, now used as a prison, some cyclopean foundations near it, with traces of the circus and amphitheatre: near the town is a magnificent aqueduct, 700 feet long, with two tiers of arches, the loftiest of which are 96 feet high, and a sepulchre called the "Tower of the Scipios." Barcino, *Barcelona*, was a city of the Laletani, and stood on the coast, a little N. of the Rubricatus: it is said to have been founded by Hercules 400 years before Rome, and to have been rebuilt by Hamilcar Barcas, who gave it the name of his family. Under the Romans it became a colony, with the surname of Faventia. It possessed an excellent harbour and a beautiful situation, [4] and so attained a state of high prosperity. Emporiæ, or Emporium, *Ampurias*, was on the small gulf which lies below the E. extremity of the Pyrenees and at the mouth of the river Clodianus, which forms its port. It was the natural landing-place from Gaul, and is said to have been colonized by Phoceans [7] of Massalia, who originally occupied a small island, and thence passed over to the main land.

§ 14. The tribes at the foot of the Pyrenees from E. to W. were—the Ausetani, W.-of the Indigetes and Lasetani; the Castellani; the Cerretani, [8] in the upper valley of the Sicoris; the Laeetani, N. of the Laletani, and not improbably but another form of the same name; the Jaccetani, [9] below the central portion of the chain; the Ilergetes, to the S., along the N. bank of the Iberus, from the Gallicus in the W. to the Sicoris in the E.; and the Vascones, [1] between the upper valley of the Iberus and the Pyrenees. The towns of this district were generally unimportant: Ilerda, on the Sicoris, the capital of the Ilergetes, Pompelo, the capital of the Vascones, and Calagurris, also in their territory, on the Iberus, deserve notice as important military positions.

Pompelo, *Pamplona*, stood at the foot of the Pyrenees, in one of the lateral valleys of the *Ebro*, and on one of the roads leading into Gallia.

---

[1] *Aprica repetes Tarraconis littora.*   MART. I. 30.

[1] Tarraco, Campano tantum erasura Lyæo,
   Hæc genuit Tuscis æmula vina cadis.   Ir. xiii. 118.

[4] Et Barcinonem amœna sedes ditium.   AVIEN. *Or. Mar.* 520.

[7] *Phocæicas dant Emporiæ, dat Tarraco pubem.*   SIL. ITAL. iii. 369

[8] They were very famous for their hams:—
   Cerretana mihi fiet vel missa licebit
   De Menapis: lauti de petasone vorent.   MART. xiii. 54.

They are also noticed by Silius Italicus:—
   Nec Cerretani, quondam Tiryuthia castra.   iii. 357.

[9] The territory of the Jaccetani formed a part of the theatre of war in the contests between Sertorius and Pompey, and between Julius Cæsar and Pompey's legates.

[1] The name of the Vascones is preserved in that of the *Basques*; they went to battle bareheaded:—
   Aut Vasco, *insuetus galeæ.*   SIL. ITAL. iii. 358.

Calagurris, *Calahorra*, stood on a rocky hill[2] near the right bank of the Iberus: it is first noticed in the Celtiberian War, B.C. 186, but it obtained a horrible celebrity afterwards for its resistance in the Sertorian War to Pompey's legate Afranius, when its defenders consumed the flesh of their own wives and children.[3] It afterwards became a *municipium* with the *Civitas Romana*; it was surnamed Nassica, to distinguish it from Calagurris Fibularia, *Loarre*, N. of the Ebro. It was the birth-place of the rhetorician Quinctilian. Ausa, the capital of the Ausetani, stood on a tributary of the Alba at *Vique*: and lower down the river was their other town, Gerunda, *Gerona*. The only town belonging to the Cerretani was Julia Libyca, near *Puigcerda.* Jaca, the capital of the Jaccetani, is still named *Jaca.* Ilerda, *Lerida*, the
capital of the Ilergetes,

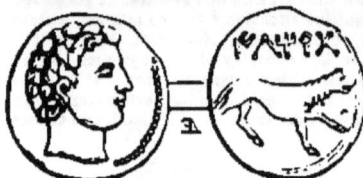

Coin of Ilerda.

stood upon an eminence[4] on the right bank of the Sicoris, and from its position, commanding the great road from Tarraco to the N.W. of Spain, which here crossed the Sicoris, it was a place of great importance. Afranius and Petreius[4] occupied it in the first year of the Civil War, B.C. 49, and were besieged by Cæsar, whose successful operations have made Ilerda the *Badajoz* of antiquity: under the Romans

Coin of Osca.

it became at first a flourishing place, but afterwards fell into decay. Osca, *Huesca*, N.E. of Cæsaraugusta, is chiefly known as the place where Sertorius died: it was a Roman colony, and had a mint, to which perhaps the expression *Argentum Oscense* refers. Celsa, on the Iberus, was a Roman colony, with the surname Victrix Julia: the river was here crossed by a bridge, the ruins of which remain at *Velilla*.

§ 15. The tribes on the N. coast from E. to W. were the Vardúli,

---

[2] hærens scopulis Calagurris.   Auson. *Ep.* xxv. 57.

[3] Vascones, hæc fama est, alimentis talibus olim
Produxere animas: sed res diversa, sed illic
Fortunæ invidia est, bellorumque ultima, casus
Extremi, longæ dira obsidionis egestas.   Juv. *Sat.* xv. 93.

[4] Its position is thus described by Lucan:—
Colle tumet modico, lenique excrevit in altum
Pingue solum tumulo: super hunc fundata vetusta
Surgit Ilerda manu: placidis præabitur undis
Hesperios inter Sicoris non ultimus amnes,
Saxeus ingenti quem pons amplectitur arcu.   iv. 11.

Its *remoteness* is the point in Horace's line:—
Aut fugies Uticam, aut vinctus mitteris Ilerdam.—*Ep.* I. 20, 13.

[4]   Postquam omnia fatis
Cæsaris ire vides, celsam Petreius Ilerdam
Deserit.   Lcc. iv. 143.

W. of the Vascones, reaching from the upper *Ebro* to the S.E.
corner of the *Bay of Biscay*; the **Caristi**; the **Autrigônes**, from the
upper valley of the *Ebro* to the sea-coast about the mouth of the
Nerva; the **Cantăbri**,[6] an important tribe occupying the mountains
about the sources of the *Ebro* and the maritime district to the N.
of them; they offered an obstinate resistance to the Romans, having
been first subjugated by Augustus in B.C. 25, and again by Agrippa
in 19: the *Basques* are their genuine descendants: they were
divided into seven tribes, of which the **Concăni** were the most
notorious for their savage habits;[7] the **Astûres**, between the upper
Durius and the sea, in a country abounding in gold, and also famed
for a breed of horses, the small ambling jennet now named *Asturco*;[8]
the N. part of their country (the modern *Asturias*) is the " Wales " of
Spain, and has always been the stronghold of Spanish independence;
the people were a wild and warlike[9] race, and were defeated by the
Romans on the banks of the Astura in B.C. 25;[1] the **Gallaeci**, or
Callaïci, who were divided into two great tribes, the **Lucénses** in the
N., from the coast in the neighbourhood of the river Navia to the
Minius; and the **Bracări** in the S., from the Minius to the Durius,
a warlike but barbarous race, who imposed ordinary labour upon their
women; the Bracari were subdued by Decimus Brutus in B.C. 136;
the Lucenses yielded to Augustus along with the other northern
tribes; lastly, the **Artăbri** in the extreme N.W. The towns along
the coast of the *Bay of Biscay* were unimportant, but in the interior
there were some places which were occupied as military stations by

---

[6] The name was sometimes loosely applied to the inhabitants of all the mountainous districts on the N. coast: *e. g.* in Juvenal, xv. 108, compared with 93. Their hardihood and bravery are frequently noticed :—

Cantaber ante omnes, hiemisque æstusque famisque	
Invictus, palmamque ex omni ferre labore.	SIL. ITAL. iII. 326.
Septimi, Gades aditure mecum, et	
Cantabrum indoctum juga ferre nostra.	HOR. *Carm.* II. 6, l.
Quid bellicosus Cantaber.	*Id.* II. 11, 1.
[7] Et lætum equino sanguine Concanum.	*Id.* III. 4, 34.
Nec qui Massageten monstrans feritate parentem	
Cornipedis fusa satiaris, Concane, vena.	SIL. ITAL. III. 360.
[8] Meruerit Asturii scrutator pallidus auri.	LUC. iv. 298.
Ille brevis, ad numerum rapidos qui colligit ungues,	
Venit ab auriferis gentibus, Astur equus.	MART. xiv. 199.
[9] Exercitus Astur.	SIL. ITAL. I. 252.
Belliger Astur.	ID. xII. 748.
[1] Gold was abundant in their country :—	
Astur avarus	
Visceribus laceræ telluris mergitur imis,	
Et redit infelix effuso concolor auro.	SIL. ITAL. L. 231.
Accipe Callaïci quidquid fodit Astur in arvis.	MART. x. 16.

the Romans, and were thus raised to great prosperity : these are
still important towns, and retain, with but slight alteration, their
modern names : we may instance Asturica, *Astorga*, and Legio VII.
Gemina, *Leon*, in the country of the Asturs, Lucus Augusti, *Lugo*,
and Bracara Augusta, *Braga*, in the districts of the Lucenses and
Bracari.

Asturica Augusta² stood in a lateral valley of the N.W. mountains
of Asturia, on the upper course of one of the tributaries of the Astura.
It obtained its surname probably after the Cantabrian War, and it
became the seat of a conventus. Pliny describes it as *urbs magnifica*,
and the modern *Astorga* gives a perfect idea of a Roman fortified town.
Legio VII. Gemina was admirably situated at the confluence of two
tributaries of the Astura, at the foot of the Asturian mountains. It
was the station of the new seventh legion which was raised by the
Emperor Galba in Spain, and which was named Gemina from its amal-
gamation by Vespasian with one of the German legions. Brigantium
was an important seaport town of the Callaïci Lucenses, variously identified with *El Ferrol* and with *Corunna*. Lucus Augusti, *Lugo*, stood
on one of the upper branches of the Minius : it was originally the chief
town of a small tribe named the Capori, but under the Romans it
became the seat of a *conventus*, and the capital of the Callaïci Lucenses.
Bracara Augusta, *Braga*, stood between the Durius and Minius, near
the river Nebis, and was the seat of a *conventus*; among its ruins are
the remains of an aqueduct and amphitheatre.

§ 10. Tribes of the interior from W. to E. : the Vaccaei, between
the Cantabri on the N. and the river Durius on the S. ; the
Celtiberi,³ a very important race occupying the whole central plateau
from the borders of Lusitania in the W. to the mountains that
bound the valley of the Ebro in the E. ; they were subdivided into
four tribes, of whom the Arevacae, in the N., were the most powerful,
while the Pelendones lived more to the E., the Berones, between Idubeda and the Iberus, and the Lusones, about the sources of the Tagus ;
the Carpetani or Carpesii,⁴ one of the most numerous and most
powerful in the whole peninsula, occupying the great valley of the
upper Tagus and the intervening district to the Anas in the S. ; and
the Oretani, more to the S., on the borders of Baetica. The only
famous town in this district was Numantia.

Pallantia, *Palencia*, the capital of the Vaccaei, stood on a tributary
of the Durius. Clunia stood on the summit of an isolated hill sur-

---

² The Asturians attributed its foundation to Astur, son of Memnon :—
  Armiger Eoï non felix Memnonis Astur.   Sil. Ital. iii. 334.
³ The origin of the name has been already referred to; it is thus expressed by
Lucan :—
        Profugique a gente vetusta
  Gallorum Celtae miscentes nomen Iberis.   iv. 9.
⁴ Their name appears to be connected with that of Calpe and Carpessus, or
Tartessus ; they may, therefore, have once stretched down to the Mediterranean
coast.

rounded with rocks, somewhat N. of the Durius: it belonged to the Arevacæ, and is described by Pliny as *Celtiberiæ finis:* under the Romans it became a colony, and the seat of a *conventus.* Numantia, the capital of the Arevacæ, stood on a moderately high but steep hill near the Durius, and was accessible only from one side, in which direction it was strongly defended:[5] it was besieged and destroyed by Scipio Africanus in B.C. 134:[6] the ruins at *Puente de Don Guarray* are supposed to mark its site. Bilbilis, *Bambola,* the second city of the Celtiberi, stood on a rocky height overhanging the river Salo:[7] it was the birthplace of the poet Martial. It was famed for its manufacture of steel, the water of the Salo being remarkably adapted to tempering the metal;[8] gold was also found there.[9] Under the Romans it became a *municipium,* with the surname of Augusta. The neighbourhood was for some time the scene of the war between Sertorius and Metellus. Segobriga, the capital of the Celtiberi, lay S.W. of Cæsaraugusta, near *Priego;* the surrounding district was celebrated for its talc. Contrebia, one of the chief cities of Celtiberia, lay S.E. of Cæsaraugusta, probably near *Albarracin:* it was besieged by Sertorius, and held out for forty-four days. Toletum, *Toledo,* the capital of the Carpetani, was situated on the Tagus: it was a very strong town, and famed for its manufacture of arms and steel-ware; there are numerous remains of Roman antiquities, especially the ruins of a circus. Castulo, *Carlona,* was on the upper course[1] of the Bætis, near the E. border of Bætica: it was the chief city of the Oretani, and one of the most important towns in the S. of Spain, having very rich copper and lead mines[2] in its neigh-

---

[5] Nolis *longa ferro* bella Numantiæ.　　　　Hor. *Carm.* II. 12, 1.

[6] Hence named Numantinus:—

Ille Numantina traxit ab urbe notam.　　　　Ov. *Fast.* L 596.

Afra Numantinos regna loquuntur avos.　　　　Propert. iv. 11, 30.

[7] Municipes, Augusta mihi quos Bilbilis acri
Monte creat, *rapidus quem Salo cingit aquis;*
Ecquid lætæ juvat ventri vos gloria vatis?
Nam decus et nomen, famaque vestra sumus.—Mart. x. 103.

citatus

Aliam Bilbilin, et tuum Salonem
Quinto forsitan exardo videbis.　　　　In. x. 104.

[8] Sevo Bilbilin optimam metallo,
Quæ vincit Chalybasque, Noricosque,
Et ferro Platram suo sonantem,
Quam Sucin tenui, sed inquieto
Armorum Salo temperator ambit.　　　　Id. iv. 55.

[9] Me multos repetita post Decembres
Accepit mea, rusticumque fecit
Auro Bilbilis, et *superba ferro.*　　　　Io. xii. 18.

[1] The valley in which Castulo stood has some resemblance to that above Delphi; hence the allusion in Silius Italicus:—

Fulget præcipuis *Parnasia* Castulo signis.　　　　iii. 391.

At contra Cirrhæi sanguis Imilce
Castulli.　　　　iii. 97.

[2] These mines are still productive; the well-known mine of *Linares,* the property of an English company, is near Castulo; and perhaps the mine whence Hannibal's wife drew her wealth is the one N. of *Linares,* named *Los Pozos de Anibal.*

bourhood: Himilce, the rich wife of Hannibal, was a native of Castulo. In the Second Punic War it revolted from the Carthaginians to the Romans, and became the head-quarters of P. Scipio; it afterwards returned to the Punic alliance, but was obliged to yield to Rome in 206: under the Romans it became a municipium with the *Jus Latinum*.

Islands.—Off the E. coast of Spain lies an important group of islands, consisting of the Baleáres,[3] or Gymnesiæ, and the Pityusæ. The former contained two chief islands, named, from their respective sizes, Major, *Majorca*, and Minor, *Minorca*: the latter also contained two, Ebúsus, *Iviza*, and Colubraria, or Ophiúsa, *Formentera*. The Baleares had numerous excellent harbours, and were extremely fertile in all produce, except wine and olive-oil. They were celebrated for their cattle, and especially for the mules of the lesser island. Their chief mineral product was the red earth named *sinopa*. The inhabitants were famous for their skill as slingers:[4] they were quiet and inoffensive. The Carthaginians originally colonized these islands; after the fall of Carthage they were independent until B.C. 123, when they were subdued by the Romans under Cæcilius Metellus. The chief towns of *Majorca* were Palma, on the S.W., and Pollentia on the N.E. coast, both of which still retain their names; and of *Minorca*, Jamna, *Ciudadela*, on the W., and Mago, *Port Mahon*, on the E. coast, both of them Phœnician colonies.

History.—The earliest notices of Spain are connected with the commerce of the Phœnicians: the Tyrians are described by Ezekiel as trading to Tarshish for silver, iron, tin, and lead; and the extent to which this commerce was carried is incidentally proved by the Biblical expression "ships of Tarshish," meaning large, sea-going merchantmen. The Phœnicians settled chiefly on the S. coast and in Bætica, but did not endeavour to found a dominion in Spain until B.C. 237, when Hamilcar formed the design of establishing a new Carthaginian empire there, partly as a counterpoise for the loss of Sicily and Sardinia, and partly perhaps as an asylum for himself, should he be expelled from Carthage. His plan was successful, and the rights of the Carthaginians were so far recognized by the Romans that a treaty was concluded with Hasdrubal in 228, by which the Iberus was fixed as the boundary between the two states, with a special stipulation in favour of Saguntum, as an ally of Rome. The infraction of this stipulation led to the Second Punic War, when the contest was transferred by Scipio to Spain itself in 210, and the Carthaginians were wholly expelled in 206. The subsequent progress of the Roman arms has been already traced in Chap. iv.

---

[3] The name Baleares was derived by the Greeks from Βάλλω, in reference to this distinguishing feature of the inhabitants; it is, however, derived from the Phœnician root *Bal*. The Greek name Gymnesiæ may have reference to the practice of slinging, as usual among light-armed troops (γυμνῆτες).

[4] Stuppea torquentem Balearis verbera fundæ.   Virg. *Georg.* L. 309.

Non secus exarsit, quam cum Balearica plumbum
Funda jacit.   Ov. *Met.* II. 727.

                ductor
Impiger et torto Balearis verbere fundæ
Ocior.   Lrc. L. 229.

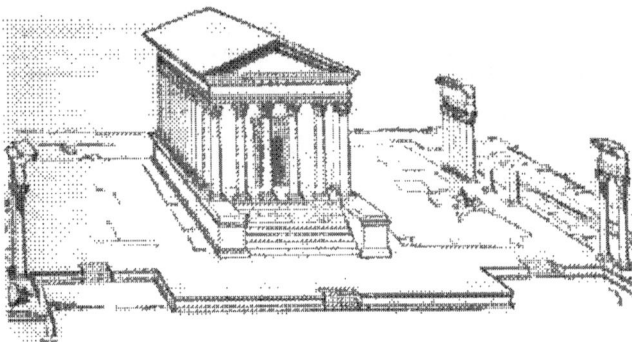

Temple at Nemausus, now called the *Maison Carrée*.

CHAPTER XXX.

GALLIA.

§ 1. Boundaries. § 2. Mountains and rivers. § 3. Inhabitants. § 4. Divisions. I. AQUITANIA. § 5. Boundaries; rivers. § 6. Tribes; towns. II. NARBONENSIS. § 7. Boundaries; rivers. § 8. Tribes; towns; roads; Hannibal's march. III. LUGDUNENSIS. § 9. Boundaries; rivers. § 10. Tribes; towns. IV. BELGICA. § 11. Boundaries; rivers. § 12. Tribes; towns; history.

§ 1. The boundaries of Gallia coincided with those of modern *France* on three sides, viz.: on the N., W., and S.—the Mare Britannicum, the Atlantic, and the Mediterranean, with the Pyrenees, forming the natural limits in these directions. On the E. there is a considerable difference, as the ancient Gallia was carried forward to the *Rhine* in its lower and middle course, and thus included the greater part of *Switzerland*,[1] the *Duchy of Luxemburg*, *Germany* W. of the Rhine, *Belgium*, and part of the *Netherlands*. The soil was fertile, and the climate good : corn, wine, and oil were produced in various districts, and fruits of all kinds ripened. Cattle, pigs, and horses were abundant, and of good quality. Iron, lead, silver, and even gold, are enumerated among its mineral productions ; and its rock salt and brine springs were well known.

---

[1] The eastern part of *Switzerland* was not in Gallia. The provinces S. of the *Lake of Geneva* and of the upper *Rhone* were not included among Cæsar's Helvetii, and must therefore have been regarded as a border country between Gaul and Italy. In the extreme S. the French border until recently coincided with that of later Gallia, the Varus being regarded as the limit : the addition of *Nice* to France has once more reinstated the maritime Alps as the boundary.

Names.—Gallia proper was commonly described as **Transalpina,** and occasionally as **Ulterior,** to distinguish it from the Italian Gallia. It was also described as G. **Comāta,** from the fashion of letting the hair grow, which prevailed among all the Gauls except the Narbonenses; while Narbonensis itself was named **Braccāta,** from the *braccæ* or "breeches" worn in that part. The Greeks termed it originally Celtice, then Galatia, and finally Gallia.

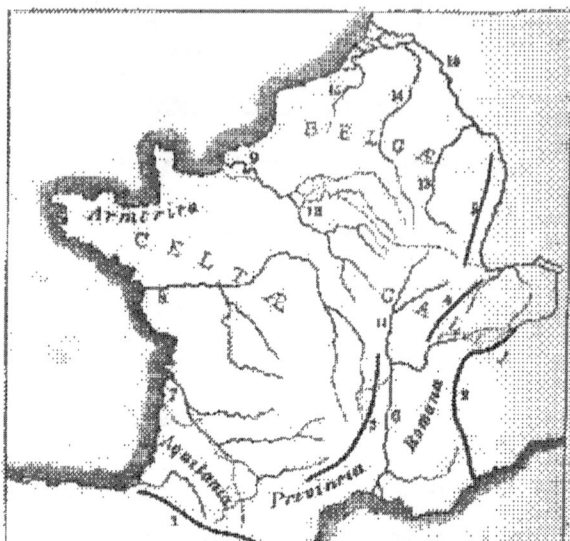

Sketch Map of the physical features of Gallia and the political divisions in Cæsar's time

1. Pyrenæus Ms.   2. Alpes.   3. Cebenna.   4. Jura.   5. Vosegus.   6. Rhodanus.   7. Garumna.   8. Liger 9. Sequana.   10. Elaver.   11. Arar.   12. Matrona.   13. Mosella.   14. Mosa.   15. Scaldis.

§ 2. The chief mountain ranges of Gaul (exclusive of the **Alpes,** on the borders of Italy, and the **Pyrenæi Mts,** on the borders of Spain) are the **Cebenna,** *Cevennes,* extending in a S. and S.W. direction between the basins of the Rhone on the E. and the Liger and Garumna on the W.; **Jura,** *Jura,* between the Rhone near *Geneva*

---

[1] Et nunc tonse Liger, quondam per colla decora
    Crinibus effusis toti prælate Comatæ.   Luc. i. 442.

[2] The *Cevennes* culminate in *Mt. Mezenc* at a height of 5820 ft. When Cæsar crossed this range the snow lay 6 ft. deep on the summit of the pass. Hence Lucan's description:
         qua montibus ardua summis
  Gens habitat cana pendentes rupe Cebennas.   i. 434.

and the Rhine near *Basle* ; and **Vosigus**, or **Vogesus**,[4] *Vosges*, running parallel to the left bank of the Upper Rhine for above 170 miles. A high wooded district between the Rhine and the Mosa, in the N.E. of the country, was named **Arduenna Silva**,[5] the *Ardennes*. The most important rivers of Gallia are the **Rhenus**, on the borders of Germania, rising in the Alps, and flowing northwards into the German Ocean ; the **Rhodanus**, *Rhone*, rising in the same range, and flowing southwards to the Mediterranean ; the **Garumna**, *Garonne*, in the S.W., flowing into the Atlantic ; the **Liger**, *Loire*, which traverses an extensive district in central Gaul, having a circuitous course, first towards the N., and then towards the W. into the Atlantic ; and the **Sequana**, *Seine*, the chief river on the N. coast, flowing into the Mare Britannicum.[6] Of the numerous lakes in *Switzerland*, only the **Lacus Lemannus**, *L. of Geneva*, is spoken of by ancient writers : the Venetus Lacus, *L. of Constance*, was outside the limits of Gaul.

§ 3. The inhabitants of Gallia belonged to various stocks : the proper Galli, who supplied the bulk of the population, were Celts ; in the S.W., between the Garonne and the Pyrenees, were an Iberian race, named Aquitani ; and in the N.E. were numerous Germanic and semi-Germanic tribes. In addition to these, Greek settlers occupied at an early period some spots on the S. coast ; and at a later period Romans were dispersed in great numbers over the whole country. The Celts appear to have been divided into two great branches—the Galli, whose name survives in the present *Gael* of Scotland ; and the non-Galli, corresponding to the modern *Cymry* of Wales. The latter class occupied the N. and N.W. districts, and have preserved their language to the present day in *Brittany* : the Belgae appear to have been substantially *Cymry*, but were in many instances intermixed with Germans. The Gauls are described as a fine, stalwart race, with fair complexions, blue eyes, and light hair. The prominent features in their character were desperate courage, skill in war, fickle temper, and great ingenuity. When the Romans

---

[4] This form appears in Lucan :—

Castraque quo Vogesi curvam super ardua rupem
Pugnaces pictis cohibebant Lingonas armis.   L. 397.

[5] The extent of this tract is over-estimated by Cæsar (*B. G.* vi. 29), unless the present reading be (as is probable) a mistake of the copyists. The text states it as 500 miles, whereas the whole distance from Coblentz to the German Ocean does not exceed 300. The name is probably significant of a "forest," and reminds us of our "Arden" in Warwickshire.

[6] These rivers exercised an important office as the commercial routes of ancient Gaul. The Rhone, the Arar or *Saône*, and the Sequana, formed the links in the chain of communication between the Mediterranean and the British Channel ; the Rhone and the Liger between the Mediterranean and the *Bay of Biscay* ; and again the Atax and the Garumna in the S.W.

first entered the country, their social and political condition were
low : drunkenness and many barbarous practices prevailed : the poor
were in a state of servitude, and the nobles engaged in constant
feuds. Their religion was a form of Druidism. Great improvements
took place under the Romans : universities were established ; the
Latin language and Roman law were introduced ; and the religion
was modified by an infusion of the Roman tenets. The towns were
beautified with temples and other public buildings, roads and aqueducts were formed, and the remains of these magnificent structures
prove, better than anything else, the advance of wealth and civilization. Literature was cultivated, and the Gauls were noted for their
skill in rhetoric even as early as the days of Juvenal.[1]

§ 4. The first political division in Gaul dates from the time that
the Romans entered the country, when they named their conquests
in the S.E. Provincia, in contradistinction to the rest of Gaul, which
was independent. Cæsar divided Gallia (by which he means Gallia
exclusive of Provincia) into three portions, corresponding to the main
elements in the population, viz. : Aquitani, between the Garumna
and the Pyrenees ; Celtæ, between the Garumna, the Atlantic, the
Sequana, and the limits of Provincia ; and Belgæ, between the
Sequana and the Rhine. Augustus, who first organized the country,
modified these divisions by substituting the name of Narbonensis for
that of Provincia, enlarging Aquitania by the addition of an
extensive district N. of the Garumna, and assigning the name of
Lugdunensis to the remainder of Cæsar's Celtica. These divisions
were retained until the 4th century A.D., when the whole was
re-arranged into 17 provinces, which were collectively described as
"Galliæ et septem Provinciæ," the former term including Lugdunensis in four provinces, Belgica in five, and a part of Narbonensis,
bordering on the Alps, named Alpes Penninæ ; the latter, including
the remainder of Narbonensis in four, and Aquitania in three provinces. We shall adopt the division of Augustus in the following
pages.

I. AQUITANIA.

§ 5. Aquitania was bounded on the W. by the Atlantic Ocean, on
the S. by the Pyrenees and the Mediterranean, on the E. by the
lower course of the Rhone and the Cevenna, and on the N. by the
Liger. This district contained within it the northern slopes of the
Pyrenæi Mts., and the whole range of Cebenna.[2] The rivers which

[1] Nunc totus Graias nostrasque habet orbis Athenas.
  Gallia causidicos docuit facunda Britannos.—XV. 110.
[2] The name survives in the corrupted form *Guienne :* it has been conjectured
that the original name was derived from the numerous springs (aquæ) in this
district ; but this etymology is doubtful. The Romans undoubtedly were acquainted

fall within it are—the **Atūrus**, *Adour*, which rises in the Pyrenees and enters the *B. of Biscay* near its S.E. corner; the **Garumna**,[*] which rises in the Pyrenees, and flows towards the N.W., into a large estuary of the *B. of Biscay*, receiving in its course as tributaries, on its right bank, the **Tarnis**, *Tarn*, the **Oltis**, *Lot*, and the **Duranius**, *Dordogne*; the **Carantōnus**, *Charente*, which joins the sea more to the N., flowing through the country of the Santones; and the **Liger**,[1] the border stream on the side of Lugdunensis, receiving on its left bank as tributaries the **Elāver**, *Allier*, which joins it at Noviodunum, and the **Caris**, *Cher*, which joins it at Cæsarodunum.

§ 6. The tribes[2] of Aquitania were the **Tarbelli**,[3] along the coast in the extreme S.W.; the **Convēnæ**,[4] N. of the Pyrenees, on the upper course of the Garumna; the **Ausci** (*Auch*), N. of the Convenæ; the **Elusātes** (*Eause*), N. of the Aturus; the **Vasātes** (*Bazas*), N.W. of the Elusates; the **Bituriges Vivisci**, about the estuary of the Garumna; the **Petrocorii** (*Perigord*), N. of the Duranius; the **Nitiobriges**, on the middle course of the Garumna; the **Cadurci** (*Cahors*), more to the E., along the course of the Oltis; the **Ruteni**[5] (*Rodez*), extending along the base of Cebenna, in the valleys of the Tarnis and its tributaries; the **Gabali** (*Javols*), on the range of Cebenna, somewhat N.

---

[*] with the mineral springs; for we have notices of Aquæ Tarbellicæ, *Dax*; Aquæ Convenarum, *Bagnères* in *Cominges*; Aquensis Vicus, *Bagnères de Bigorre*; Aquæ Calidæ, *Vichy*; Aquæ Bormänis, *Bourbonne-les-Bains*; and Aquæ Siccæ, perhaps *Sèches*.

[*] The gender of Garumna is dubious. Tibullus (l. 7, 11) calls it "magnus Garumna;" but Ausonius (*Mosella*, v. 463), "æquorem Garumnæ." The tide enters the *Garonne* with great violence:—

> Quosque rigat retro pernicior unda Garumnæ,
> Oceani pleno quoties impellitur æstu.   Claud. *in Ruf.* ii. 113.

[1] The proper Greek form is Λείγηρ: hence the first vowel would naturally be long. The Romans, however, made it short, as in the lines interpolated in Lucan (l. 438):—

> In nebulis, Mednana, tuis marcere perosus
> Andus jam placida Ligeris recreatur ab unda.

And in Tibullus:—

> Testis Arar, Rhodanusque celer, magnusque Garumna,
> Carnuti et flavi cærula lympha Liger.   i. 7, 11.

[2] The names of almost all the Gallic tribes correspond to the modern names either of districts or towns—generally the latter—to which they were transferred in the later Roman era. As these towns serve to identify the position of the ancient tribes, we have added them in the text.

[3] They extended down to the Aturis and the Pyrenees:—

> Qui tenet et ripas Aturi, qua littore curvo
> Molliter admissum claudit Tarbellicus æquor.   Luc. l. 420.

> Tarbella Pyrene
> Testis, et oceani littora Santonici.   Tibull. i. 7, 9.

[4] The Convenæ were (as their name implies) a mixed race.

[5] Solvuntur flavi longa stationa Ruteni.   Luc. l. 402.

of the sources of the Tarnis; the Arverni [a] (*Auvergne*), in the valley of the Elaver and the adjacent highlands; the Bituriges Cubi (*Bourges*), along the course of the Liger from the Elaver to the Caris; the Lemovices (*Limoges*), to the W. of the Arverni; the Santones (*Saintes*), N. and E. of the estuary of the Garumna; and the Pictones, or Pictavi (*Poitiers*), along the left bank of the Liger. Of the above-mentioned tribes only the Tarbelli, Convenæ, Ausci, Elusates, and Vasates, were proper Aquitani, i.e. of the stock allied to the Iberians. The others were Celtæ, whom Augustus added to the Aquitani when he extended the borders of the country from the Garumna to the Liger. Of the towns in Aquitania we know little more than the names: Burdigala, the ancient representative of *Bourdeaux*, appears to have been the most important: and the Roman remains at Mediolanum, *Saintes*, and Limonum, *Poitiers*, prove them to have been large towns. It may be said generally that almost every place of present importance was in existence in the Roman era, the names in most instances corresponding to those of the ancient tribes.[f]

Lugdunum,[g] the chief town of the Convenæ and a Roman colony, stood on an isolated hill by the Garumna; it is now named *St. Bertrand de Cominges*. Elusa,[h] the capital of the Elusates, stood at *Civitas* near *Eause*. Burdigala,[i] *Bordeaux*, on the left bank of the Garumna, was the port of the Bituriges Vivisci, and a place of great commerce under the empire: it became the metropolis of Aquitania II., and was also the seat of an university. The only Roman building still existing is the amphitheatre, called the *Arènes*, now in a much shattered state. Vesunna, *Perigueux*, the capital of the Petrocorii, was on a branch of the Duranius: the Roman remains are extensive, consisting of several bridges, the ruins of an amphitheatre, and of the citadel, and a round building named the *Tour de Vesone*, about 200 ft. in circumference; there

[a] The Arverni claimed descent from the Trojans:—
    Arvernique ausi Latio se fingere fratres
    Sanguine ab Iliaco populi.   Luc. L. 427.

[f] Note [f] above.
[g] The terminations of very many of the Gallic names of towns were significant; e.g. -dunum = "hill;" -durum (compare the Welsh dwr) = "water;"
-ritum = "ford;" -bona = "boundary;" -brica = "bridge;" -magus = "field."
These Celtic terminations were combined by the Romans with Latin prefixes in
many cases; e.g. Augustobona, Juliomagus, &c.

[h] It is noticed by Claudian (in *Ruf.* 1. 137):—
    Invadit muros Elusæ, notissima dudum
    Tecta petens.

[i] The pronunciation of the name is decided by Ausonius, who was a native of
Burdigala, and describes the place at length in his *Ordo Nobilium Urbium*:—
    Burdigala est natale solum, clementia cœli
    Mitis ubi et riguæ largo indulgentia terræ.   xiv. 5.
    Diligo Burdigalam: Romam colo. Civis in hac sum,
    Consul in ambabus. Cunæ hic, ibi sella curulis. *Id.* 39.

are several Roman camps about the town. **Divǒna**,[7] *Cahors*, the capital of the Cadurci, stood on the Oltis: it was supplied with water by an aqueduct about 19 miles in length, a magnificent work, some remains of which are still extant: ruins of the baths and of the theatre have also been discovered. **Segodǔnum**, *Rodez*, the capital of the Rutoni, was on a tributary of the Tarnis. **Anderitum**, the capital of the Gabali, has been variously identified with *Javols* and *Anterrieux*. **Gergovia**, a town of the Arverni, was situated on a mountain, still named *Gergois*, about 4 miles S. of *Clermont*, and W. of the Elaver ; in front of the town is a lower hill named *Puy de Jussat*: this place was the scene of some important operations in the Gallic War, when Vercingetorix was attacked by Cæsar: the former was encamped on the plateau of Gergovia ; the latter seized the *Puy de Jussat*, and brought it into communication with his camp: he then assaulted Gergovia from the S. side, and at the same time diverted the enemy's attention by a feigned attack on the N.W. ; the troops succeeded in getting on the plateau, but were afterwards driven back. **Augustonemǎtum**, *Clermont*, the capital of the Arverni, was on the Elaver : the modern name is derived from the *Clarus Mons* of the Middle Ages. **Avaricum**, *Bourges*, the capital of the Dituriges Cubi, stood on a branch of the Caris : its walls are particularly described by Cæsar ' *B. G.* vii. 23), by whom it was besieged and taken in B.C. 52. **Augustoritum**, *Limoges*, was the capital of the Lemovices. **Mediolǎnum**, *Saintes*, the capital of the Santones, stood on the Carantonus : the remains still existing of an aqueduct and an amphitheatre prove it to have been an important town: there is an arch in honour of Germanicus Cæsar, singularly placed in the middle of the *Charents*. **Limǒnum**, *Poitiers*, the capital of the Pictones, was situated on a tributary of the *Vienne*; there are remains of a huge amphitheatre, capable of holding 20,000 persons ; the walls are 7 French feet thick.

II. NARBONENSIS.

§ 7. **Narbonensis**,[8] or, as it was originally termed, **Provincia**, extended along the Mediterranean Sea from the Alps to the Pyrenees, and inland to the Rhone on the N., and Ms. Cebenna on the W. With the exception of the three chains already noticed as forming its limits, there were no other mountains in this portion of Gallia. The chief river was the **Rhodǎnus**, which enters the province at the Lacus Lemannus, and runs first to the W., as far as Lugdunum, then S. to the Mediterranean, where it forms a delta : it receives as tributaries,

---

[7] The name is derived by Ausonius from *di*, "god," and *von*, "water":—
Divona Celtarum lingua, Fons addita Divis.   *Clar. Urb.* (*Burdig.*) 32.

[8] Its limits are thus described by Ausonius :—
Insinuant qua se Sequanis Allobroges oris,
Excluduntque Italos Alpina cacumina fines ;
Qua Pyrenaicis nivibus dirimuntur Iberi ;
Qua rapitur præceps Rhodanus genitore Lemano,
Interiusque premant Aquitanica rura Cebennæ,
Usque in Tectosagos primævo nomine Volcas,
Totum Narbo fuit.   *Ord. Nob. Urb.* xiii. 4.

on its left bank, the Isára,[4] *Isère,* which rises in the Alps, and, flowing by Gratianopolis, *Grenoble,* joins the main stream a little N. of Valentia; the Sulgas, *Sorgue,* which joins at Vindalium; and the Druentia, *Durance,* which rises in the Cottian Alps, and rushes down with a violent course to the Rhone at Avenio. The other rivers which flow into the Mediterranean are—the Varus, *Var,* which in its lower course forms the boundary on the side of Italy; the Arauris, *Herault,* rising in the Cebenna, and entering the sea near Agatha; the Atax,[5] Attagus, or Narbo, *Aude,* rising in the Pyrenees, and falling into the Sinus Gallicus to the E. of Narbo: and lastly the Telis or Ruscino, *Tet,* near the border of Spain.

§ 8. The chief tribes from S.W. to N.E. were—the Sardones, at the foot of the Pyrenees and on the adjacent sea-coast; the Volcæ divided into two branches, the Tectosages and the Arecomici, who occupied the whole country between the Garonne and the Rhone, the former W., the latter E. of the range of Cebenna; the Salyes, or Salluvii, E. of the Rhone from the Druentia to the Mediterranean; the Cavares, N. of the Druentia about Avenio; the Vocontii,[6] more to the E., at the foot of the Alps from the Druentia to the Isara; and lastly, the Allobroges,[7] between the Rhone, the Isara, and the Lake Leman. Narbonensis contained, as might be inferred from its proximity to the Italian frontier, some of the most important towns of Gaul. In the interior were Aquæ Sextiæ, the first Roman colony in the country; Narbo, the earliest colony W. of the Rhone, and the future capital of the province, Arelate, commanding the valley of the Rhone, Nemausus on the road between Arelate and Narbo, and Vienna on the E. bank of the Rhone, S. of Lugdunum. These towns were adorned with magnificent buildings, some of which rank among the finest specimens of Roman architecture. On the coast we meet with the old Greek colony of Massalia, which attained a high pitch of commercial prosperity, and planted several colonies along the coast; and Forum Julii, a Roman colony, and the chief naval station on this coast.

Illíbèris, *Elne,* was the nearest town to the Spanish frontier on the coast-road from Narbo: Hannibal passed through it on his advance to Italy. Ruscino lay on the same route and on the river of the same name:

---

[4] Hannibal followed the course of this stream, " Arar," in Livy, xxi. 31, being a corrupt reading for Isara. The insula of which he speaks was at the junction of the rivers.

[5] Mitis Atax Latias gaudet non ferre carinas,
Finis et Hesperiæ, promoto limite, Varus.   Luc. l. 403.

[6] Hannibal's route lay through their territory :—
Jam faciles campos, jam rura Vocontia carpit.   Sil. Ital. iii. 467.

[7] Æmula nec virtus Capuæ, nec Spartacus acer,
Novisque rebus infidelis Allobrox.   Hor. Epod. xvi. 5.

its name has been transformed into *Roussillon*, and the site of the town is at *Castel Roussillon*. **Tolosa**, *Toulouse*, a town of the Tectosages, stood on the right bank of the Garumna: it was enriched with the gold and silver found in the surrounding district, and which was kept in the temples as a sacred deposit. The plunder of these treasures by Cæpio, followed as it was by his defeat by the Cimbri, led to the proverb "Aurum Tolosanum," as a warning against sacrilege. It afterwards became a *colonia*, and appears to have been a seat of art and literature.[a] The important town of **Narbo**, or **Narbona**, *Narbonne*, which the Romans elevated into the capital of the province, stood on the river Atax: it belonged originally to the Volcæ Arecomici, and was first occupied by a Roman colony in B.C. 118, and surnamed "Martius" or Marcius, probably after a consul who was engaged in a contest with a Ligurian tribe in that year. It was at all times an important commercial town, the Atax being navigable up to it; but its chief importance was due to its position in reference to Spain and Aquitania. It was adorned with public buildings,[b] none of which are now in existence, though numerous antiquities have been discovered. The adjacent coast was famous for its oysters. **Bæterræ**, *Béziers*,[1] was on the Orbis, E. of Narbo, in the midst of a wine-producing district; there are vestiges of an amphitheatre and of an aqueduct. **Nemausus**, *Nîmes*, the chief town of the Volcæ Arecomici, stood a little W. of the Rhone on the road between Arelate and Narbo. The town was itself large, and contained twenty-four villages in its territory. The remains of the old town are very fine: the amphitheatre, which is tolerably perfect, was 437 feet in diameter, and could hold 17,000 persons; the present height of the walls is 70 feet: there is also a beautiful temple dedicated to M. Aurelius and L. Verus, now used as a museum, and named *Maison Carrée*, 76 feet long, and 40 wide, with 30 Corinthian fluted pillars. The famous fountain, noticed by Ausonius,[2] still exists, but the chief supply of water was obtained from some springs near *Uzès*, and conveyed by a splendid aqueduct: a portion

Aqueduct of Nemausus, now called the *Pont du Gard*.

---

[a] Hence Martial (ix. 100) terms it Palladia :—

    Te sibi Palladiæ antetulit toga docta Tolosæ.

[b] Quam pulcherrimas jam redire Narbo.   MART. viii. 72.

[1] Festus Avienus (589) furnishes us with a link between the ancient and modern names :—

                Dehinc
Besaram statione fama cæca tradidit.

[2] Non Aponus potu, vitrea non luce Nemausus
    Purior.   Ord. Nob. Urb. xiv. 32.

of this work remains across the valley of Vardo, and is named the *Pont du Gard*: it has three tiers of arches; the lowest containing six arches, the next eleven, and the upper one thirty-five; the total height is about 155 feet, and the length on the top about 870. Arelâte, *Arles*, a town of the Salyes, was situated on the left bank of the Rhone at the point where it bifurcates. It became a Roman colony in the time of Augustus, with the surname of Sextani, and was a place of considerable trade. It was improved by Constantine, and a new town [3] added on the other bank of the river at *Trinquetaille.* The amphitheatre, of which there are remains, was capable of holding 20,000 spectators: it is not in so perfect a state as that of Nemausus. An Egyptian obelisk and some ancient tombs are the other most interesting monuments. Aquæ Sextiæ, *Aix*, the first Roman colony planted in Gaul, B.C. 122, stood about 18 Roman miles N. of Massilia. Its name indicates both the presence of mineral waters, and that it was founded by Sextius Calvinus. The great battle, in B.C. 102, between Marius and the German tribes of the Cimbri and Teutones, was probably fought at *Meiragues*, two leagues from Massilia, the modern name being a corruption of *Marii Ager.* Massilia, or Massalia, as the Greeks wrote it, *Marseilles*, stood on a bay some distance E. of the mouth of the Rhone, in the midst of a rather sterile district.[4] The accounts of its foundation are somewhat conflicting, but they agree in asserting that Phocæans settled there about 600 B.C.[5] It was built on rocky ground: the harbour, named Lacydon, faced the S., and lay beneath a rock in the form of a theatre. Both the harbour and city were well walled, and the town was of considerable extent, but contained few buildings worthy of notice except the Ephesium, or temple of Ephesian Artemis, and the temple of Delphinian Apollo, both of which stood on the citadel. Massalia became an ally of Rome in the Second Punic War, and was aided by her, in B.C. 154, against the Ligurian tribes of the Oxybii and Deceates. In B.C. 49, it sided with Pompey in the Civil War, and was taken after a long siege by C. Trebonius, Cæsar's *legatus.* The constitution of the town was aristocratic, and its institutions were generally good. The habits of the people were simple and temperate: literature and medicinal science were cultivated to a certain extent. Its commerce was extensive, and it planted colonies on the shores of Gaul and Spain. Its prosperity declined after the planting of a Roman colony at Narbo. Forum Julii, *Fréjus*, was the chief naval station of the Romans, and held the same position which *Toulon* (the ancient Telo Martius) now holds on this coast. It lay considerably E. of Massalia, at the bottom of a small bay, which was partly enclosed by

---

[3] It is hence termed by Ausonius *duplex:* it also received the name of Constantina:—

> Pande, duplex Arelate, tuos, blanda hospita, portus,
> Gallula Roma Arelas: quam Narbo Martius, et quam
> Accolit Alpinis opulenta Vienna colonis.   *Ord. Clar. Urb.* viii. 1

[4] It produced the vine:—

> Cum tua centenos expugnet sportula cives,
> Fumea Massilliæ ponere vina potes.   MART. xiii. 123.

[5] Aristotle names Euxenus, and Plutarch Protos, as its founder. There is a romantic story that one of these two was chosen as husband by the daughter of Nannus, king of the country, her choice being signified by the presentation of a cup of water, or of wine and water.

two moles: the entrance of the bay has been choked up by the deposits of the river Argenteus, and the entrance to the·port is now 3000 feet from the sea.   The place was probably named after Julius Cæsar, but it first became a station in the time of Augustus.   It had various surnames, such as Classica, from its being the station of the fleet, and Octavanorum, probably from the 8th legion being settled there.   It was the birth-place of Cn. Agricola, and was further known for the manufacture of the sauce named *garum.*   A triumphal arch, the ruins of the amphitheatre, an old gateway, and parts of the aqueduct still remain. **Antipolis,** *Antibes,* further E. on the coast, was a colony of Massalia, and under the Romans a *municipium:* it was rather famous for its pickle: there are remains of a theatre and a few other buildings there.[1] **Avenio,** *Avignon,* stood at the junction of the Druentia with the Rhone: it was reputed a colony·of Massalia.   **Arausio,** *Orange,* was in the territory of the Cavares, near the E. bank of the Rhone; it became a colony with the additional title of Secundanorum.   The Roman remains are numerous, the most remarkable being a triumphal arch, about 80 feet high, with three archways, inscribed " Mario," but of a later period than the Marius who defeated the Teutones ; and the remains of an aqueduct near the town.   **Ebrodunum,** *Embrun,* was situated on the upper course of the Druentia under the Cottian Alps: it became the capital of Alpes Maritimæ.   **Brigantium,** *Briançon,* was the first town in Gaul on the road from Segusio over *Mont Genèvre:* at this point the road branched off W. to the valley of the Isara, and S.W. to that of the Druentia.   **Vienna,** *Vienne,* lay on the E. bank of the Rhone, in the country of the Allobroges.   Under the Roman empire it became a *colonia,* and a great place, even rivalling Lugdunum.[7]   The foundations of the massive Roman walls, 20 feet thick, still remain; there are also some arcades which probably served as the entrance to the thermæ, a well preserved temple of the Corinthian order, dedicated to Augustus and Livia, now used as a museum, and the remains of an amphitheatre, and of four large aqueducts, chiefly constructed under ground.   Pilate is said to have been banished to Vienna: an unfinished pyramid on a quadrangular base, of a total height of 52 feet, is called, without any good reason, " Pontius Pilate's Tomb."

Roman Roads.—The **Via Aurelia** was carried on under Augustus from Vada Sabbata in Liguria to Arelate on the Rhone, passing through Antipolis, Forum Julii, and Aquæ Sextiæ.   From Arelate the chief line of communication with Spain commenced, passing through Nemausus and Narbo.   A road sometimes named **Via Domitia** ran along the E. bank of the Rhone through Vienna to Lugdunum.   From Vienna roads led to the Alpis Graia, *Little St. Bernard,* and to the Alpis Cottia, *Mont Genèvre.*

---

[1] Antipolitani, fateor, sum filia thynni
    Esaem si scombri, non tibi missa forem.   MART. xiii. 103.

[7] Its beauty is referred to by Martial, and its state of culture may be inferred from the circumstance that both his own and Pliny's works were to be had at the booksellers' shops there :—

    Fertur habere meos, si vera est fama, libellos,
    Inter delicias, pulchra Vienna, suas.   ID. vii. 88.

It was also famous for its wine :—

    Hæc de vitifera veni ce picata Vienna
    Ne dubites : misit Romulus ipse mihi.   ID. xiii. 107.

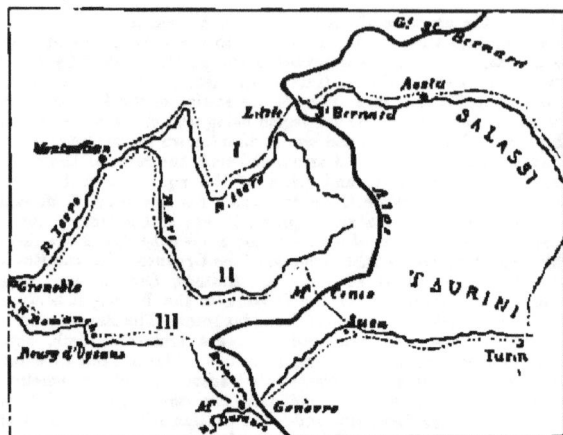

The Passes of the Alps, to illustrate Hannibal's Route.

Hannibal's March.—The route pursued by Hannibal in his celebrated expedition from Spain to Italy, lay wholly through the portion of Gaul we have been describing. He entered it by the E. extremity of the Pyrenees, and thence followed the coast-road by Ruscino, Narbo, and Nemausus, reaching the Rhone a little above Avenio. Having crossed the river, he followed up the left bank to the Isara, and thence along the latter stream to the point where it emerges from the lower ridges of the Alps near *Grenoble*. From this point his route is uncertain: according to some authorities he pursued the route marked I. in the accompanying plan, which follows the Isara, and crosses the *Little St. Bernard* into the valley of *Aosta*, and thence down to *Turin*: according to others he pursued route II., which follows the *Arc* over *Mt. Cenis*, and thence straight down to *Susa* and Turin: lastly, he may have pursued route III., following the *Romanche* by *Bourg d'Oysans* and across *Mt. Genèvre*. The objections to route I. are its length, and the fact that the valley of the *Dora* was occupied by a very warlike tribe, the *Salassi*, who would not have permitted Hannibal's army to pass unopposed. Between II. and III. there is not much to choose: but the latter was probably the one: at all events the *Mont Genèvre* route was the more frequented route in the Roman period; it was probably the one explored by Pompey in B.C. 77, and was certainly followed by Caesar in his expedition against the Helvetians. The two stations Ad Martis and Brigantio are the modern *Oulx* and *Briançon*.

Many of the villages on the road to St. Bernard derive their names from the Roman miles measured from Vienna, as *Septème* (7), *Oytier* (8), and *Diémos* (10).

III.—LUGDUNENSIS.

§ 9. Lugdunensis was separated from Aquitania on the S.W. by the Liger, and from Narbonensis on the S.E. by the Rhodanus: on

the E., where it was contiguous to Belgica, there was no natural
boundary, but the limit between them would be coincident with a
line leaving the Rhine near its great bend at *Basle*, and striking
across to the *British Channel* at the point where the 50th parallel
falls on it. The mountain range of *Jura* lies wholly in Lugdunensis,
and the chief rivers are the border streams of the *Liger* and the
Rhodanus, the former of which receives on its right bank at Juliomagus the **Meduana**, *Mayenne*, while the latter has an important
tributary in the **Arar**, *Saône*,[*] which rises in Vosegus, and flows
with a slow current to the S., receiving the **Dubis**, *Doubs*, on its left
bank, and joining the main stream[*] at Lugdunum. We have
further to notice the **Sequana**, *Seine*, which rises in the high lands
S. of *Langres*, and flows to the N.W. into the *British Channel*: it
receives on its right bank the **Matrôna**, *Marne*, and the **Isâra**, *Oise*,
with its tributary the **Axôna**, *Aisne;* and on its left bank the
Icaunus, *Yonne*, which is known to us only from inscriptions.

§ 10. The nations occupying Lugdunensis from S.E. to N.W.
were—the **Segusiani**, between the Liger and the Rhodanus, and, in
Cæsar's time at all events, in the angle formed by the Rhodanus
and the Arar; the **Ædui**, between the Liger and the Arar; the
Lingônes,[1] *Langres*, about the sources of the *Marne* and *Seine*, N. of
the Ædui; the **Senônes**, *Sens*, N.W. of the Ædui to the Sequana
near *Paris*; the **Carnûtes**,[2] *Chartres*, between the Sequana below
Paris, and the Liger, and even beyond the Liger to the Elaver;
the **Aulerci**, between the Sequana in its lower course and the Liger,
divided into two great branches, the **Eburovices**,[3] *Evreux*, in the N.,
and the **Cenomani**, *Mans*, in the S.; the **Namnêtes**, *Nantes*, on the
right bank of the Liger near its mouth; the **Armorici**,[4] a general
name for the *maritime* tribes between the mouths of the Liger and
of the Sequana, of which the most important were the **Venêti**,

---

[*] The modern name is derived from *Sauoona*, which appears to have been the
true Gallic name of the river.

[*]   Qua Rhodanus raptum velocibus undis
    In mare fert Ararim.   Luc. l. 433.

[1] The Lingones are described as a warlike race by Lucan :—
    Castraque quæ Vogesi curvam super ardua rupem
    Pugnaces pictis cohibebant Lingonas armis.   l. 397.

[2] They are noticed by Tibullus (l. 7, 12) under the form of Carnuti :—
    Carnuti et flavi cærula lympha Liger.

[3] In Cæsar (*B.G.* iii. 17) the text has *Eburones* instead of *Eburovices*. The
reading in vii. 75, "**Brannovii**," as a branch of the Aulerci, is probably an
interpolation; the **Brannovices** noticed in the same passage must have been a
distinct tribe, as they lived S. of the Ædui; the **Diablintes**, N.W. of the
Cenomani, are noticed as a branch of the Aulerci by Ptolemy.

[4] The name Armorica is derived from the Celtic words *ar*, "on," and *mor*,
"sea."

Vannes, on the coast W. of the Namnetes, a sea-faring race, who
carried on trade with Britain, and who, from the character of their
coast, broken up by numerous promontories or *lingulæ* surrounded
with shallow water, enjoyed great security ; the Osismii, in the extremity of *Bretagne*; and the Uselli, in the peninsula of *Cotantin*.
Lugdunensis contained comparatively few towns of importance :
Lugdûnum, the capital of the province, stood opposite the point of
junction of the Rhone with the Arar. Augustodûnum, near the
Liger, is proved, by its extensive remains, to have been a fine
town ; and the position of Genâbum, in command of the passage
across the Liger, rendered it a valuable military station. The
modern capital of *France* is represented by Lutetia, which appears
to have been a small place, but valuable from its safe position on
an island in the *Seine*, whence either bank was accessible to its
inhabitants.

The Roman colony of Lugdûnum was planted by L. Munatius Plancus
in B.C. 43, and peopled with the inhabitants of Vienna. It stood on the
right bank of the Arar on the slope of a hill named *Fourvière*. The
modern town of *Lyons* originally occupied the same site, that portion of
the city which lies between the two rivers Arar and Rhone being a
modern addition. The position of Lugdunum, as a place of trade
and a central spot of communication, secured to it a large amount of
prosperity. It was destroyed by fire in Seneca's time, and restored
by the Emperor Nero. It was again burnt by the soldiers of Septimius
Severus in A.D. 197. Between the two rivers stood the Ara Augusti,[a]
dedicated to the emperor by the sixty states of Gaul, each of which was
represented by a figure. A church was planted at Lugdunum at an
early period, which suffered a furious persecution in the time of Marcus
Aurelius in A.D. 172 or 177 : Irenæus was one of its bishops. The
Roman remains are small : there are traces of a theatre on the *Place des
Minimes*, and of a camp on the W. side of the *Saône:* some of the
arches of the great aqueduct (50 miles long) are preserved at *Cham-
ponost:* there were two other aqueducts of great length. Cabillônum,
Châlon, was a town of the Ædui on the Arar: the Romans kept a fleet
of some kind there, and it appears to have been a place of commercial
importance. Bibracte, or, as it was afterwards called, Augustodûnum,
whence the modern *Autun*, was the chief town of the Ædui, and stood
on a tributary of the Liger: it was the chief place of education for the
noble youths of Gaul, and was altogether a very important town. Near
it Cæsar defeated the Helvetii in a pitched battle: it was seized by
Sacrovir in A.D. 21, was taken by Tetricus in the time of Gallienus, and
is said to have been destroyed by Attila. The Roman remains at
Autun are numerous, consisting of the circuit of the walls, with two of
the main entrances, *Porte d'Arroux*, 50 feet high and 60 broad, and
Porte St. André, 60 feet high and 40 broad, the ruins of a theatre, traces
of an amphitheatre with a naumachia near it supplied by an aqueduct
from three large ponds outside the town, and the remains of a magnificent temple of Janus: the names *Monjeu* (Mons Jovis) and *Chaumar*

---

[a] Aut Lugdunensem rhetor dicturus ad aram.   Juv. I. 44.

(Campus Martius) are vestiges of the Roman era. Alesia, *Alise*, a town of the Mandubii, was situated on a lofty hill between the streams *Lose* and *Lozerain*, tributaries of the *Yonne*. It was here that the Gauls, under Vercingetorix, made their final stand in B.C. 52. Agendicum, *Sens*, was the chief town of the Senones, and under the later Roman empire became the capital of Lugdunensis IV. Lutetia, *Paris*, the capital of the Parisii, stood on the Sequana, and was originally confined to an island forming a portion of *La Cité* (derived from *civitas*), the original isle having been increased since the Roman period, by the addition of two other small islands. It was never a large place under the Romans, though it may have occupied some ground on one or both of the banks of the river, with which the island was connected by bridges. The place was threatened by Labienus, in B.C. 52, without effect. Julian spent a winter, and was proclaimed Augustus there, A.D. 358. Some sculptured stones, and a portion of a subterranean aqueduct, are the only Roman remains. Rotomagus, which was afterwards contracted into Rotomum, and this into *Rouen*, was the chief town of the Vellocasses on the Sequana. Genabum, or, as it was afterwards called, Aurelianl, and hence *Orléans*, was an emporium of the Carnutes on the Liger : it was the focus of the great insurrection in B.C. 52, and was taken and destroyed by Cæsar : its later name is supposed to have been given after the Emperor Aurelian, in whose reign the walls, of which there are some traces, may have been built. Cæsarodûnum, *Tours*, the chief town of the Turones (whence the modern name), was on the S. bank of the Liger. Juliomagus, the capital of the Andecavi, from whom its modern name *Angers* comes, was on the Meduana, a short distance above its junction with the Liger.

Islands.—Off the coast of Lugdunensis were two groups of islands— Venetiæ Insulæ, off the W. coast, of which Vindilis, *Belle-Isle*, is the largest ; and Cæsarèa, *Jersey*, Sarnia, *Guernsey*, and Ridûna, *Alderney*, off the N. coast.

IV.—BELGICA.

§ 11. **Belgica** was bounded on the W. by the rivers Sequana and Matrona ; on the N. by the Fretum Gallicum, *Straits of Dover*, and the German Ocean ; on the E. by the Rhine ; and on the S. by the Rhone and the Alps. The mountain ranges of Jura, Vosgus, and the Pennine Alps, fall within these limits, together with the following rivers—the border stream of the Rhenus, which rises on the W. side of Ms. Adûla, passes through the Lacus Venètua, *L. of Constance*, in its upper course, receives as tributaries on its left bank, the Nava, *Nahe*, at Bingium, *Bingen*, and the Mosella,[*] *Moselle*,

---

[*] The Mosella is undoubtedly noticed in Cæsar (*B. G.* iv. 15) in the words "ad confluentem Mosæ et Rheni." Whether the river was called Mosa as well as Mosella, or whether there is a mistake of the author or his copyists, is uncertain. The banks of the Moselle presented very much the same appearance in the 4th century A.D. as at present, being well clad with vines :—

Quæ sublimia apex longo super ardua tractu
Et rupes et apricæ jugi, flexusque sinusque
Vitibus adsurgunt naturalique theatro.   AUSON. *Idyl.* x. 154.

at Confluentes, *Coblentz*, and finally discharged its waters through two main streams,[?] of which the western, uniting with the Mosa, received the name of **Vahalis**, *Waal*, while the eastern retained the name of the original stream; the **Mosa**, *Meuse* or *Maas*, which rises about 48° N. lat., and flows towards the N., receiving, as above noticed, a branch of the Rhine, before its discharge; the **Scaldis**, *Schelde*, more to the W., which is described by Cæsar (*B. G.* vi. 33) as flowing into the Mosa; and, lastly, the **Samara**, *Somme*, which falls into the Fretum Gallicum in the W. part of the province.

§ 12. The most important tribes[a] were located in the following manner—the **Helvetii**,[b] in the plains of *Switzerland*, between Jura on the W., the Rhone on the S., and the Rhine on the N. and E.; they were divided into four *pagi*, or cantons, of which two are named, viz.: Urbigenus, or Verbigenus, which is supposed to have reached

---

Its cheerful aspect is noticed:—

Haud aliter *placida* subter vada *lenta* Mosellæ
Deteg't admixtos non concolor herba lapillos.—AUSON. *Idyl.* x. 78.

[?] Pliny notices, in addition to the two already specified, a third, named Flevum, which flowed towards the N. into the lakes (*Zuider Zee*). This was probably identical with the artificial channel, **Fossa Drusiana**, of which Tacitus speaks (*Ann.* ii. 6). Ptolemy notices three outlets, all of them N. of the Mosa. In the midst of these somewhat conflicting statements it is clear that the Rhine Proper, which deviates from the *Waal* at *Pannerden* and enters the sea near *Leyden*, was the boundary between Gaul and Germany.

[a] The ethnology of Belgica is involved in considerable difficulty: generally speaking it will be found that the divisions of this province represent the two main elements of the population, i.e. that the tribes in the two Germanies were Germans, and those in the two Belgicæ were Belgians. But it must be remarked that many of the tribes on the border of Belgica were to a certain extent Germans. We may instance the Menapii, Nervii, and Treveri.

[b] The Helvetii come prominently forward in the history of Cæsar's wars in

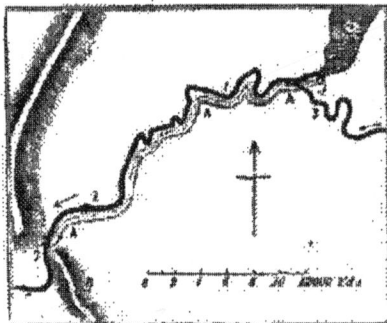

B.C. 58. They formed the plan of migrating in a body from their own territory into the heart of Gallia. Cæsar prevented them from entering Provincia by throwing up a wall, probably of earth, 19 miles long and 16 feet high, marked A A in the accompanying plan, along the S. bank of the Rhone (1) from the point (4) where it issues from L. Leman (2) across the Arve (3) to where the *Mt. aux Vaches* (6) presses the S. side of the river. The Helvetii were compelled therefore to go through the pass of *Fort l'Ecluse* on the N. side, and thus to follow

Map showing the position of Cæsar's Murus.

the right bank of the Rhone to the Arar. They were met by Cæsar and utterly defeated near Bibracte, and only 110,000 returned home out of 300,000.

from Salodurum, *Solothurn*, as far as Aquæ Helveticæ, *Baden*, near
the *Aar*; and Tigurinus more to the S., between *L. Morat* on the N.,
Jura on the W., and the *L. of Geneva* on the S., its limits on the
E. not being known; the other two pagi are not named, but may
have been the Tugëni, between *L. Zürich* and *L. of Constanz*, and the
Ambrönes to the S. of the two first; the **Rauraci**, along the Rhine
in the neighbourhood of *Bâle*; the **Sequani**,[1] between the upper Arar
in the W., Jura in the E., and the Rhone, near Geneva, in the S.;
the **Leuci**,[2] in the valley of the *Upper Moselle*; the **Remi**, *Rheims*,
between the *Marne* and the *Meuse*; the German tribes of the **Triboeci**,
Nemëtes, and **Vangiönes**,[3] along the Upper Rhine; the **Mediomatrici**,
Metz, N. of the Leuci on the course of the *Moselle*, and at one time
reaching E. to the Rhine, but subsequently restricted to the W. of
the *Vosges* by the German immigrants; the **Treviri** or **Treveri**,
Trèves, on the *Lower Moselle*, from the *Meuse* in the W. to the
Rhine in the E., though their position on the course of the latter
river is by no means well defined; the **Ubii**, a German tribe, who
in Cæsar's time lived E. of the Rhine opposite the Treveri, but in
the time of Augustus crossed the river and occupied a district
between the Treveri and the Gugerni, in the middle of which stood
Cologne; the **Gugerni** to the N.; the **Batavi**,[5] a branch of the Chatti
who left their country and settled, before Cæsar's time, in the island
(Batavorum insula) formed by the two great branches of the Rhine
on the N. and S. and the sea on the W.; the **Menapii**, in Cæsar's
time, on both sides of the Rhine, and along the coast as far W. as
the Morini (the German tribes of the **Usipëtes** and **Tenethëri** crossed
the Rhine and settled in their territory); the **Nervii**[6] on the right

---

[1] The Sequani appear to have been skilful weavers :—

Hanc tibi Sequanicæ pinguem textricis alumnam,
Quæ Lacedæmonium barbara nomen habet;
Sordida, sed gelido non aspernanda Decembri
Dona, peregrinam mittimus endromida.   MART. iv. 19.

[2] The Leuci and Remi are noticed by Lucan as skilful spearmen :—

Optimus excusso Leucus Rhemusque lacerto.   L. 424.

[3] Et qui te laxis imitantur, Sarmata, braccis
Vangiones.   LUC. i. 430.

[4] Tu quoque, lætatus converti prœlia, Trevir.   ID. i. 441.

[5] The Batavi are described as a fierce race, of large size, with light or red
hair :—

———— Batavique truces, quos ære recurvo
Stridentes acuere tubæ.   LUC. i. 431.

Rie petit Euphraten juvenis domitique Batavi
Custodes aquilas, armis industrius.   JUV. viii. 51.

Sum figuli lusus, rufi persona Batavi.   MART. xiv. 176.

Jam puer auricomæ præformidate Batavæ.   SIL. ITAL. iii. 608.

[6] The Nervii offered a most determined resistance to the Romans: they were

bank of the upper Scaldis ; the Morini [1] along the sea-coast from the mouth of the Scaldis in the E. to the territory of the Ambiani in . the W. ; the Bellovàci, *Beauvais*, between the upper Samara and the Sequana, reputed the first of the Belgic tribes in numbers and influence ; and the Calèti, *Caux*, on the sea-coast E. of the Sequana. The towns of Belgica rose to importance at a comparatively late date. The dangers that threatened the Roman empire on the side of Germany necessitated a number of garrisons along the course of the Rhine, commencing with Argentorätum, and extending down to Lugdunum Batavörum, between which points we have Mogontiàcum, Bingium, Bonna, Colonia Agrippina, Asciburgium, Castra Vetèra, and other less important towns. Augusta Trevirörum, on the Mosella, was the finest town in this part of the country, and the general residence of the Roman Emperors in their visits to northern Gaul. Divodûrum, on the same river, and Durocortörum, on a branch of the Isara, are also proved by their remains to have been important and fine cities.

Colonia Equestris Noviodunum, *Nyon*, was in the country of the Helvetii on the *L. of Geneva:* the name of *Equestris* is said still to attach to the neighbourhood of *Nyon*. Aventicum, *Avenches*, the capital of the Helvetii, stood N.E. of *Geneva:* it became a Roman colony with the name Pia Flavia Constans Emerita : there are remains of its amphitheatre and aqueduct, and part of its wall. Salodûrum, *Solothurn*, was another town of the Helvetii, of which some ancient remains are still extant. Vindonissa, *Windisch*, near the Aar, was a considerable place, and the station of the 21st Legion in A.D. 71 : there are traces of an amphitheatre, and various other Roman remains on its site. Augusta Rauracörum, *Augst*, 6 miles E. of *Basle*, was the chief town of the Rauraci : a Roman colony was planted there in the time of Augustus by L. Munatius Plancus. Vesontio, *Besançon*, the chief town of the Sequani, stood on the Dubis, *Doubs*, a tributary of the Arar : the position of the town is correctly described by Cæsar as being on a peninsula surrounded by the Dubis ; but he is wrong in stating the width of the neck of land which connects it with the adjacent country as 600 Roman feet, its width really being 1500. Vesontio suffered severely from the Alemanni, Huns, and others ; a triumphal arch and a part of the aqueduct are all the remains of the old town. Tullum, *Toul*, was the chief city of the Leuci. Catalauni, or, as the name is otherwise given, Durocatalaunum, *Châlons-sur-Marne*, in the territory of the Remi, was famous for the defeat of Attila and his Huns by the Roman Aëtius in A.D. 451 : the name implies a people as well as a town. Durocortörum, *Reims*, was the capital of the Remi, and the centre where

---

cut up by Cæsar on the banks of the Sabis in B.C. 57 ; they revolted in 54, and were again defeated by Cæsar in 53.

———————— nimiumque rebellis -

Nervius, et cæsi pollutus sanguine Cottæ.   Luc. L. 426.

[1] Their name, from *mor*, " the sea," bespeaks a Celtic origin : they are noticed by Virgil as the most distant of the Continental nations :—

Extremíque hominum Morini.   *Æn.* v iii. 737.

numerous roads met; it also possessed a school of rhetoric: it contained numerous Roman edifices, of which a triumphal arch with three gateways and eight Corinthian columns, and some traces of the Thermæ, are the only remains. **Divodūrum**, *Metz*, probably derived its name from being situated at the junction (*divo* = "two") of the *Moselle* and *Seille*: it was the chief town of the Mediomatrici, and became from its position an important place. In A.D. 70, 4000 of its inhabitants were massacred by the soldiers of Vitellius: it was destroyed by the Huns in the fifth century. The town was supplied with water by a magnificent aqueduct six French leagues in length; of this, five arches remain on the left bank of the *Moselle*, and seventeen on its right bank at *Jouy*, one of which is 64 feet high. **Argentarātum**, afterwards Stratisburgium, whence its modern name *Strasburg*, was the chief town of the Tribocci on the Rhine. The Romans had a manufactory of arms, and Julian defeated the Alemanni there. **Noviomāgus**, *Speier*, lower down the course of the Rhine, was the capital of the Nemetes. **Mogontiācum**, *Mainz*, on the Rhine, was a *municipium*, and is noted as the spot where a monument was erected in honour of Drusus, father of Germanicus. **Bingium**, *Bingen*, at the junction of the Nava and the Rhine, was a Roman station, and is noticed by Tacitus in connection with the war of Civilis. **Augusta Trevirūrum**, *Trier* or *Trèves*, was a Roman colony, planted probably by Augustus, on the right bank of the *Moselle*: it was connected with the other side of the river by a bridge, and it appears to have been walled from the time of its erection. Ausonius places Treviri fourth in his list of "nobiles urbes:" it appears to have been the regular imperial residence in this part of Gaul in the fourth century. It was one of the sixty cities taken by the Franks and the Alemanni after the death of Aurelian, and recovered by Probus. Constantine the Great frequently resided there, and restored the place, and Eumenius the rhetorician speaks of the great circus, the basilica, the forum, and the walls, as the works of that emperor. The piers of the bridge, the remains of the amphitheatre, and a gigantic gate—a quadrangular construction, 115 feet long, 91 high, and 67 deep—are the most striking Roman monuments. **Bonna**, *Bonn*, was a town of the Ubii, on the Rhine: it was here that Drusus made his bridge of boats across the river in B.C. 12 or 11. It became a military station of the Roman legions, which were attacked here in A.D. 70, by the Batavi and Canninefates. It was probably taken by the Alemanni, as the walls were repaired by Julian in A.D. 359. **Colonia Agrippina**, *Cologne* on the left bank of the Rhine, was originally called Oppidum Ubiorum, as being the chief town of the Ubii: the change of name was effected, in A.D. 51, by Claudius, at the request of his wife Agrippina, who was born there, and at the same time a colony of veterans was planted there. The town was well situated at the chief place of transit between the E. and W. sides of the Rhine, and the inhabitants soon became enriched with the tolls they levied on the merchandize that crossed there, as well as probably on that which passed down the river. It became the chief town of Germania Secunda, and enjoyed the *jus Italicum*. Aulus Vitellius was proclaimed emperor by the soldiers in A.D. 69, and Trajan assumed the imperial insignia there in 98. The place was taken by the Franks, but recovered by Julian about A.D. 356. The Roman remains consist of a gateway, the *Pfaffen-porta*, supposed to be the Porta Claudia, and portions of the walls, with numerous antiquities: the name *Cologne* is a modification of Colonia. **Asciburgium**, *Asburg*, on the lower Rhine, was a Roman station in A.D. 70. **Castra Vetēra**, *Xanten*,

was an important Roman station on an elevation near the Rhine, formed
in the time of Augustus: Civilis blockaded and captured some Roman
legions there in A.D. 70.   **Lugdūnum Batavōrum**, *Leyden*, was the chief
town in the Batavian Isle : the name itself is Celtic, and leads to the
inference that the Celts had occupied this district before the entrance
of the Batavi.   **Gesoriǎcum** or **Bononia**, *Boulogne*, was the chief port of
the Morini, and the place whence Claudius crossed into Britain : it is
described by Pliny as Portus Morinorum Britannicus, and the distance
across (probably to Rutupiæ) is estimated by him at 50 M. P. : there are
no Roman remains at *Boulogne*.   The **Itius Portus**, whence Cæsar sailed
certainly in his second expedition, and probably in his first, is more to
the E. at *Wissant*, where there is no port strictly speaking, but a wide,
sheltered, sandy bay : the Ulterior Portus of which he speaks would
thus be *Sangaits*.   **Castellum**, *Cassel*, near *Dunkerque*, was a Roman
station, as also was **Tarusnna**, *Térouenne*, both in the territory of the
Morini.   **Samarobriva**, *Amiens*, a town of the Ambiani, was situated (as
its name implies) on the Samara.   **Augusta Suessiōnum**, the capital of
the Suessiones, is the present *Soissons;* and **Julleböna**, the capital of the
Caleti, is *Lillebonne*, where are the remains of a theatre, and tombs,
together with other antiquities.

History.—The history of Gallia commences with the settlement of
Massalia by the Phocæans of Asia Minor, about B C. 600, who introduced
the vine, and taught the Galli the use of letters.   We hear little of the
country until the time that the Romans entered it in 125, as allies
of the Massaliots against the Salyes.   In this and the two following
years the Salyes were attacked, and finally subdued ; and in 122 the
colony of Aquæ Sextiæ was planted.   The gradual progress of the
Roman arms has been already traced,[*] and need not be repeated here.
After the completion of Cæsar's conquests, various colonies were planted
throughout the country, but no regular government was introduced
until B.C. 27, when Augustus established the fourfold division to which
we have referred.

_____

[*] See pp. 52, 53.

Coin of Nemausus.

Remains of Roman Wall.

CHAPTER XXXI.

BRITANNICÆ INSULÆ. GERMANIA.

I. BRITANNICÆ INSULÆ. § 1. Names and divisions of the islands. § 2. Rivers, &c. of Britannia Romana. § 3. Climate and productions. § 4. Inhabitants. § 5. Roman divisions; towns; roads; walls; history. § 6. Britannia Barbara. § 7. Hibernia. II. GERMANIA. § 8. Boundaries and general description. § 9. Mountains; rivers. § 10. Tribes; towns; history.

I. BRITANICÆ INSULÆ.

§ 1. The term Britannicæ Insulæ was employed by Greek writers to describe the whole group of the British Isles, but more especially the two largest of them, *Great Britain* and *Ireland.* Subsequently to the time of Cæsar these two were distinguished, the former as Britannia,[1] or Albion,[2] the latter as Hibernia, or Ierna. At

---

[1] The Greeks generally wrote the name Βρεττανία, with a double *t*; the Latins used the single *t*. Lucretius alone lengthens the *i* in the line—

Nam quid Britannis cœlum differre putamus.   vi. 1105.

The origin of the name is not known; it is usually referred to the Celtic *brit*, "painted."

[2] This name is generally regarded as derived from *albus*, in reference to the "white" cliffs on the S. coast. It is more probably connected with the Celtic *Alben*, signifying "height."

a later period Britannia was applied to *England* as distinguished from *Scotland*. The position of the group in relation to the continent of Europe, and particularly to Gaul, was well known to the later Romans;[3] very inaccurate views, however, prevailed, even down to the time of Ptolemy, as to the form and relative positions of the islands themselves. The seas which surround them are—the Mare Britannicum on the S., the Mare Germanicum on the E., the Oceanus Atlanticus on the W., and the Mare Cronium or Pigrum on the N. Britannia itself was divided into two portions, **Romana** and **Barbara**, corresponding generally to the modern *England* and *Scotland*, though Romana was sometimes carried into *Scotland* as far as the *Firths of Forth* and *Clyde*, and was sometimes restricted to Hadrian's Wall. As the latter appears to have been the proper boundary of the Roman province, we shall regard it as the limit of Britannia Romana in the following pages.

§ 2. The names of the physical features of Britannia Romana are known to us, partly from the writings of Cæsar and Tacitus, and partly from the description of Ptolemy. The notices of the two former writers are few, but are the only ones that possess any historical interest: Cæsar mentions the **Prom. Cantium**, *North Foreland*, and the river **Tamésis**, *Thames;* and Tacitus the river **Sabrina**, *Severn*, and the **Antona**, for which we should probably read **Aufona**, *Avon.* The description in Ptolemy is sufficiently full, but consists of names alone without any associations. These are valuable, as proving the identity[4] of the modern and ancient names, and occasionally as affording indications of the ancient British language. We give them

---

[3] The remoteness of Britain is noticed by Virgil and Horace:—

     Et penitus toto divisos orbe Britannos.   VIRG. *Ecl.* L 67.

     Serves Iturum Cæsarem in ultimos
     Orbis Britannos.   HOR. *Carm.* i. 35, 29.

     Pervemque, a populo, principe Cæsare, in
     Persas atque Britannos
         Vestra motus aget preca.   *Id.* i. 21, 14.

     Te belluosus, qui remotis
     Obstrepit, Oceanus, Britannis.   *Id.* iv. 14, 47.

[4] The identity is not indeed universal, but it holds good in many instances where there is an apparent discrepancy; *e. g.* Idumania and *Blackwater* probably have the same meaning, the latter being a translation of the former: *Garrhuenus* is radically the same as *Yare;* while Ituna, *Solway Firth*, no doubt has reference to the *Eden*, which flows into it. The orthography of the classical names is very doubtful: we have, for instance, three forms for the ancient name of the *Thames*, viz., Tamesis in Cæsar, Tamesa in Dion Cassius and Tacitus, and Iamesa in Ptolemy, the latter being probably an error of a copyist. So also of the tribes; *e. g.*, Trinobantes and Trinoantes, Damnonii and Dumnonii, Demetæ and Dimetæ, &c. And so still more of the towns; *e. g.*, Camalodunum and Camulodunum, Verolanium and Verulamium, Luguvallium, Luguvallium, and Lugubalium, &c. We deem it unnecessary to specify all these variations in the text.

therefore in brief, taking a survey of the coasts, beginning with the N.E.

(1.) *On the E. coast.* Rivers—**Vedra**, *Weur;* **Abus**, *Humber;* **Gabranuenus**, *Yare;* **Sturius**, *Stour;* **Idumania**, *Blackwater;* and **Tamesa**, *Thames.* *Estuaries and Bays.*—**Dunum Sinus**, *Dunsley Bay,* near *Whitby;* **Gabrantuicōrum Sin.**, *Filey Bay;* **Metăris Æstuarium**, the *Wash;* and. **Tamēsa Æst.**, the mouth of the *Thames. Promontories.—* **Ocellum**, *Flamboro' Head;* and **Cantium**, *North Foreland.*

(2.) *On the S. coast.* Rivers.—**Trisanton**, probably the *Aron;* **Alaunus**, perhaps the *Axe;* **Isāca**, *Exe;* and **Tamărus**, *Tamar. Promontories.—***Damnonium** or **Ocrinum**, the *Lizard;* and **Antivestæum** or **Bolerium**, *Land's End.*

(3.) *On the W. coast.* Rivers.—**Sabrina**, *Severn;* **Rhatostathybius**, *Taff;* **Tobius**, *Towey;* **Tuerōbis**, *Teify;* **Stucia**, *Dovey;* and **Tuesōbis**, *Conway. Æstuaries.*—**Sabrina Æst.**, *Bristol Channel;* **Setëia Æst.**, mouth of the *Dee;* **Belisama Æst.**, mouth of the *Ribble;* **Moricambe Æst.**, *Morecambe Bay;* and **Itūna Æst.**, *Solway Firth. Promontories.—***Hercūlis Prom.**, *Hartland Point;* **Octapitārum**, *St. David's Head;* and **Ganganōrum Prom.**, *Braich-y-Pwll.*

§ 3. The climate and productions of Britain are described by several writers. The former is characterised as humid and foggy, but otherwise temperate. A large amount of the country was covered with forests and morasses, which rendered it more moist than it now is. The soil was regarded as fertile: in Cæsar's time a very small portion of it was cultivated, but in the later times of the Empire a large amount of corn was exported for the use of the Roman troops in Germany.[5] The greater part of the island was given up to pasture, and the native British lived mostly on the produce of their flocks and herds. The country was rich in minerals: the tin-mines of Cornwall were probably worked by the Phœnicians from a very early period,[6] and led to the application of the name Cassiterides to the S.W. coast and the *Scilly Isles.* In addition to this we have notices of lead, iron, silver, and even gold.[7] The dogs[8] of Britain were particularly prized, and the oysters of Rutupiæ[9] were well known at Rome. Pearls were found in con-

---

[5] About A.D. 360 Julian had 600 vessels built for the express purpose of importing corn to the provinces bordering on the Rhine.

[6] This however has been denied by many modern writers, as no Phœnician coins have been found nor any other evidence of their having settled in Britain. It has been supposed that the tin was carried across Gaul to Massilia and other Greek colonies, and then sold to the Phœnician merchants.

[7] Specimens of these metals, as produced by the Romans, are still in existence. Blocks of tin are rare; those of lead are more common, and bear inscriptions giving the name of the emperor in whose reign they were smelted. A square ingot of silver has also been found with a Latin inscription; and there are undoubted proofs that the Romans crushed quartz for gold in the neighbourhood of *Llampeter* in Wales.

[8] They are noticed by Claudian as a very powerful breed:—

Magnaque taurorum fracturæ colla Britannæ.—*De Laud. Ital.* iii. 301.

[9] ―――― Rutupinove edita fundo
Ostrea.   Juv. iv. 141

siderable numbers, but of poor quality. We have also evidence that
there were abundance of sheep, pigs, goats, deer, oxen, and horses
on the island.[16] The seas about the shores of Britain were reputed
to abound with a kind of whale.[1]

§ 4. The inhabitants of Britannia Romana were Celts of the
Cymry branch, and are described as similar to the Gauls in person
and manners. They had attained but a low degree of civilisation
at the time the Romans became acquainted with them : their clothing was made of skins, and they were in the habit of staining and
tattooing their bodies.[2] They were warlike,[3] and fought without
armour, but were acquainted with the use of the war-chariot.
They were divided into numerous tribes, which lived independently
of each other under their own chieftains. Their religion was
Druidism, and the priests exercised considerable influence in the
state, as the depositaries of learning and the administrators of
justice. Their towns were little else than stockaded villages. The
introduction of Roman civilisation effected without doubt a considerable improvement in their condition, though we have not much
information on this subject. It appears, however, that they acquired
the art of coining money. The chief memorials of the ancient
British people consist of "cromlechs," barrows, and circles of stones,
all of which are connected with their sepulchres, camps, traces of
villages, and above all the mysterious construction at *Stonehenge*.
The articles discovered in the sepulchres consist chiefly of urns,
sometimes rudely ornamented, and instruments of stone and bronze,
such as "celts" or chisels, arrow-heads, and the heads of axes and
hammers.

---

[16] Proofs of the existence of these animals are found in the Roman rubbish-pits,
where their bones exist in great quantities, showing that they were largely eaten.
From this source we learn that there was a very large breed of oxen then in the
island, described by naturalists as *bos longifrons*.

[1] Quanto delphinis balæna Britannica major.   Juv. x. 14.

    —— *belluosus* qui remotis
    Obstrepit Oceanus remotis.   Hor. Carm. iv. 14, 47.

[2] This custom is frequently noticed by the Latin poets :—

    Claudia *cæruleis* cum sit Rufina Britannis
    Edita.   Mart. xi. 53.

    Barbara de *pictis* veni bascauda Britannis.   Id. xiv. 99.

    Nunc etiam *infectos* demens imitare Britannos,
    Ludis et *externo tincta nitore* caput.   Propert. ii. 14, 25.

    Sed Scythiam, Cilicasque feros, *viridesque* Britannos.—Ov. Am. ii. 16, 39.

[3] Visam Britannos hospitibus feros.   Hor. Carm. iii. 4, 33.

    Qua nec terribiles Cimbri nec Britones unquam
    Sauromatæve truces aut immanes Agathyrsi.   Juv. xv. 124.

    Gallicum Rhenum, horribilesque ulti-
        mosque Britannos.   Catull. xi. 11.

The native tribes of Britain were arranged as follows :—(1.) *S. of the Thames*—the Cantii in *Kent*; the Regni in *Surrey* and *Sussex*; the Belgæ in *Wilts, Hants,* and *Somersetshire*; the Durotriges in *Dorsetshire*; the Atrebatii in *Berks*; and the Damnonii in *Devon* and *Cornwall.* (2.) Between the *Thames,* the *Severn,* and the *Humber*—the Trinobantes in *Middlesex, Essex,* and the .*S.* of *Suffolk*; the Dobuni in *Oxfordshire* and *Gloucestershire,* with the Catuellāni as a subdivision; the Catyenchlāni in *Northamptonshire, Beds, Hunts,* and *Rutland*; the Cenimagni in the *N.* of *Suffolk*; the Icēni in *Norfolk*; the Coritāni in *Lincolnshire* and *Leicestershire*; and the Cornavii in *Cheshire* and parts of *Staffordshire* and *Shropshire.* (3.) *W.* of the *Severn*—the Silūres in *Monmouthshire* and the *E.* of *S. Wales*; the Dimētæ in the three W counties of *S. Wales*; and the Ordovices in *Shropshire* and *N. Wales.* (4.) Between the *Humber* and Hadrian's Wall—the Brigantes, with the Setantii as a subordinate tribe on the banks of the *Ribble,* and the Parisii just *N.* of the *Humber.* The position of the Cangi, noticed by Tacitus, is quite uncertain.

§ 5. The Romans first entered Britain in B.C. 55, under Cæsar; but they did not permanently occupy it until about one hundred years later, when Claudius subdued the tribes S. of the Thames (A.D. 43). That emperor constituted Britain a province under the government of a consular legatus and a procurator. It remained in this state until A.D. 197, when it was divided into two provinces, Superior and Inferior, the latter being in the S., each under a separate Præses. It was subsequently, probably under Constantine, subdivided into four provinces named as follows: Britannia Prima, S. of the *Thames*; Brit. Secunda, W. of the *Severn*; Maxima Cæsariensis, between the *Thames* and the *Humber*; and Flavia Cæsariensis, N. of the *Humber.* Our information with regard to the political and social state of Britain under the Romans is unfortunately scanty : the sources whence it is derived may be classed under three heads,—(1) historical documents; (2) itineraries and geographers, particularly Ptolemy; (3) existing remains. 1. From the first of these sources we learn somewhat of the topography of the country and of the political status of the towns ; the classical writers notice the capital Londinium, *London,* Camalodūnum, *Colchester,* the first Roman colony, Verulamium, *St. Alban's,* the capital of Cassivelaunus, and Rutupiæ, *Richborough,* the chief port for communication with the continent ; later writers (Dio Cassius, Eutropius, &c.) notice Eborăcum, *York,* the great station of the Romans in the later period of their occupancy : and a very much later authority, Richard of Cirencester,* who, however, probably drew his information from original sources,

---

* Richard of Cirencester flourished in the 14th century. Among other works he composed a treatise, " De Situ Britanniæ," which was not known to the world until 1747, when it was discovered by Dr. Bertram of Copenhagen. The manuscript has been lost, and it is doubtful whether Bertram has given his author with fidelity. There seems, however, to be no doubt that Richard of Cirencester's treatise contained local information not found in the Itineraries.

informs us that there were in Britain 2 *municipia*, viz., Verulamium and Eboracum; 9 *coloniæ*, viz., Londinium, Camaloduoum,
Rutupiæ, *Richborough*, Aquæ Solis, *Bath*, Isca, *Caerleon*, Deva,
Chester, Glevum, *Gloucester*, Liudum, *Lincoln*, and Camboricum,
Cambridge; 10 cities *Latio jure donatæ*, of which we may notice
Durobrivæ, *Custor*, Luguballium, *Carlisle*, and Corinium, *Cirencester*; and 12 *stipendiariæ*, of which we may notice Venta Bel-
garum, *Winchester*, Segontium, *Carnarvon*, Mariduoum, *Carmarthen*, Ratæ, *Leicester*, Cantiopolis, *Canterbury*, Durinum,
Dorchester, Isca, *Exeter*, and Durobrivæ, *Rochester*. 2. From the
Itineraries we obtain information with regard to the roads constructed by the Romans, and the numerous towns which lined
them. No less than fifteen routes are given in the Itinerary of Antonine, and eighteen in that of Richard of Cirencester. These routes de-
monstrate how completely the Romans had opened up the country, and
how great was the communication carried on between the different
districts. Ptolemy also mentions numerous towns. 3. From the
third source of information we obtain a vivid idea of the extent to
which the country was Romanised, and the high pitch of wealth and
refinement that prevailed, through all parts. We learn, for instance,
from this source, that the towns were inclosed within strong walls [5]—
that every one of any size possessed its *basilica* or court-house, and
its public baths—that magnificent temples were erected [6]—that
many of them had amphitheatres [7]—and that all were furnished
with large cemeteries outside the walls. We further learn that
villas were dispersed all over the land, and that in the southern
counties they were almost as numerous as gentlemen's seats in the
present day—that these villas were of vast extent [8] and of great
magnificence, furnished with "hypocausts" for the purpose of
warming the rooms, and with baths, and adorned with painted walls
and mosaic floors with elaborate designs. We further learn that the
Romans carried on extensive manufactories of pottery [9] and of
iron, [10] and that, as we have previously noticed, they worked and
smelted other metals. We further learn that there was the usual
amount of refinement in matters of personal appearance: among the

[5] Specimens of Roman walls and gates are found at *Richborough, Burgh* in
Suffolk, Lymne, York, Lincoln, Chichester, Pevensey, and other places.
[6] We know of the existence of a temple of Minerva at *Bath*, a temple of Neptune
and Minerva at *Chichester*, and a temple of Minerva at Cocelum, *Ribchester.*
[7] As at *Dorchester, Cirencester, Caerleon, Richborough, Colchester,* and *Silchester.*
[8] The most perfect remains of villas are found at *Bignor* in *Sussex*, and at
Woodchester in *Gloucestershire.*
[9] Remains of potteries have been found at *Upchurch Marshes* on the *Medway,*
and at *Castor* in *Northamptonshire.*
[10] The *Forest of Dean* was the main seat of the iron-works: the heaps of scoriæ
may still be seen there in vast numbers. Iron also appears to have been made in
the *Weald* of *Sussex.*

articles which have been discovered, are *fibulæ* or buckles, bone and bronze hair-pins, metal *specula* or looking-glasses, gold *torques* or collars, bracelets, needles, *styli* or pens, spoons, &c. Lastly, the vast number of coins which are discovered amid Roman ruins, extending over the whole period of their occupation of the country, affords no slight indication of the extent to which Roman [1] influence prevailed in the transactions of daily life. From this brief review of the state of Britain under the Romans we now revert to the notices of the towns.

Towns.—**Londinium**, the capital of Roman Britain, originally stood wholly on the N. side of the Thames; but in the time of Hadrian and Antoninus Pius it had extended to the S. bank (where *Southwark* now stands), and is hence described by Ptolemy as a town of the Cantii. It is first noticed by Tacitus, who speaks of it as a place of great trade; it was plundered by the Britons at the time of Boudicea's revolt. It bore at a later date the surname of Augusta, and became the terminus of the great roads of Britain. The remains that have been discovered, show the extent and magnificence of the town. The walls enclosed the same circuit as those of mediæval London; they were 12 feet thick, and were furnished with at least seven gates. Numerous tessellated pavements and fragments of statuary and sculpture have been discovered at depths varying from 12 to 20 feet below the present level of the soil. There was a mint at London, the coins struck in it belonging chiefly to Carausius, Allectus, and Constantinus. **Verulamium**, *Old Verulam*, near *St. Alban's*, was probably the residence of Cassivelaunus, which was taken by Cæsar: it was afterwards the capital of a prince named Tasciovanus, some of whose coins still exist: it was plundered at the time of Boadicea's revolt. It subsequently became a *municipium*, and one of the chief Roman stations in the island. The abbey church of *St. Alban's* is built to a great extent of Roman tiles taken from the old town. **Camalodunum** was the chief town in the country of the Trinobantes. It was the residence of Prince Cunobelinus in the reign of Tiberius, and was taken by Claudius in A.D. 43, and converted into a Roman colony under the name of Col. Camalodunensis Victricensis. Tacitus (*Ann.* xii. 32, 33) states that this was done for the repression of the Silures; but this is clearly erroneous. He also informs us (*Ann.* xiv. 31, 32) that it possessed a temple of Claudius, a curia, and a theatre. It was taken and destroyed by the Britons before Boadicea's revolt. Some doubt exists as to whether it is to be identified with *Maldon* or *Colchester*: the general opinion is that the Roman Colonia and Camalodunum were the same place, in which case it would be *Colchester*, where a vast number of Roman remains have been

---

[1] It is important to observe that the Romans of Britain were not all of them Italians. With regard to the civilians, indeed, we know little or nothing; but the legionary troops who were stationed in the island were drawn from the most remote and widely-separated districts. There were, for instance, Gauls stationed at *Lymne;* Spaniards at Anderida, *Procursacy;* Dalmatians at Branodunum, *Branconater;* Thracians at Gabrosentum, *Drumburgh,* and Dacians at Amboglanna, *Birdoswald.* These nations introduced various kinds of religious worship; and hence we find altars not only of Jupiter and the other Roman gods, but of deities whose names even are unknown to us.

discovered. **Venta**, the capital of the Iceni, and hence surnamed *Icenôrum*, to distinguish it from the other towns of the same name, probably stood at *Caistor*, a little S. of *N. rwich*. **Lindum** was an important town in the district of the Coritani, and a colony, as its modern name *Lincoln*, from "Linli colonia," implies. The Roman remains are very important, and consist of a gateway, named *Newport*, still in use, a sewer, a wall now known as the "Mint Wall," numerous inscriptions; coins, &c. **Eboracum**, *York*, was situated in the country of the Brigantes, and became from its northerly position the chief military station of the Romans in the later period of their residence in the island. It was the station of the 6th Legion, surnamed *Victriz*. The emperors Severus and Constantius Chlorus died there; and Constantine the Great is said (but on insufficient authority) to have been born there. The foundations of the old Roman walls on three sides have been discovered, together with the remains of one of the gates, probably the Prætorian, facing the N. The town appears to have been of rectangular form, 650 yards long by 550 broad, and to have been protected by a wall, with a rampart on the inside and a fosse on the outside. Outside these limits were suburbs of considerable extent. The remains of private dwellings, baths, tessellated pavements, and votive tablets, particularly two to Serapis and Mithras, are very numerous. **Luguvallum**, *Carlisle*, appears to have been an important place, though the notices of it are very scanty. It stood near the W. extremity of Hadrian's wall, and on one of the roads leading into Caledonia. **Deva**, *Chester*,[2] was so named from the river on which it was built. It was an important military station, and the head-quarters of the 20th Legion, surnamed *Valeria Victriz*. The Roman remains are numerous, consisting of the foundations of the walls, a postern now called *Shipgate*, altars, and baths, statues, particularly one of Mithras with a Phrygian bonnet, vases, &c. **Uriconium**, *Wroxeter*, was situated on the main road between Deva and Londinium, and in the territory of the Cornavii. The explorations which have been made here prove that it was a very important town. The buildings as yet discovered consist of a *basilica*, *thermæ*, a forum, and numerous other objects. **Isca**, *Caerleon*, in the country of the Silures, was an important military post for keeping that nation in order, and was at one time the station of the 2nd Legion, surnamed *Augusta*. Numerous antiquities have been discovered there, particularly an amphitheatre, the remains of a Roman villa, with specimens of Samian ware and bronze ornaments, tessellated pavements, and inscriptions. In the same neighbourhood stood **Venta**, surnamed *Silurum*, *Caerwent*, where are traces of the Roman walls. **Corinium**, or **Durocornovium**,[3] *Cirencester*, was centrally situated at the

---

[2] In many instances, where the ancient differ from the modern names, the former still exist in reference to other objects; *e.g.* we may compare Deva with the river *Dee*; Uriconium with the mountain *Wrekin*; Segontium with the river *Seiont*; and Isca with the *Usk*. In other cases the ancient names are modified by the addition of the word *castra* in different forms. The Saxons turned this into *chester* or *cester*, and the Danes into *castor* or *caster*, while the British used the form *caer* or *car* as a prefix. Hence we have the names *Glow-cester* as equivalent to "Glevi castra," *Don-caster* to Danii castra, *Carlisle* to Castra Luguvalli, *Caer-leon* to Castra Legionis, *Carmarthen* to Castra Maridani; and in some instances we have simply Castra, as in *Chester* and *Caistor*.

[3] The prefix *Duro*, which appears in numerous instances, is equivalent to the Welsh *dwr*, "water," and expresses the position of the town by a river. The two

junction of three Roman roads, and in the midst of a well-occupied district. Many villas have been discovered in and about the town, and it appears to have been one of the most fashionable towns of Roman Britain. Aquæ Solis, *Bath*, was the favourite watering-place of the Romans. "Solis" may be a corruption of "Sulis," a British goddess, whose name appears on an altar found there. Remains of the baths and of a temple of Minerva have been discovered there, together with inscriptions which prove that it was much frequented. Durnovaria, *Dorchester*, was one of the chief towns on the S. coast: the walls have been traced, and an amphitheatre is still in existence. Venta Belgārum, *Winchester*, and Sorbiodūnum were the chief towns of the Belgæ; the walls of the latter have been traced at *Old Sarum* near *Salisbury*, and numerous coins have been found there. Calleva, the chief town of the Atrebates, is represented by *Silchester*, where walls three miles in circuit mark the site of the old town. Finally, in *Kent* we have to notice Durobrivæ, *Rochester*, where coins, *fibulæ*, and pottery have been found; Durovernum, *Canterbury*; Regulbium, *Reculver*, a fort, of which some walls still exist, commanding the entrance of the channel that separated the isle of *Thanet* from the mainland; Rutupiæ, *Richborough*, its port being named Portus Rutupensis (Trutulensis in Tac. *Agric.* 38); it was evidently a town of great magnificence; portions of its walls still exist to the height of between 20 and 30 feet, as well as the foundations of its amphitheatre, and a vast number of smaller objects, such as *fibulæ*, pottery, coins, &c.; we have already noticed Rutupiæ as the chief port for the Continental traffic; there were also ports at Dubris, *Dover*, where is a tower supposed to have been a lighthouse; and at Portus Lemanis, *Lymne*, where one of the gates has been discovered as well as the old walls: both *Dover* and *Lymne* were stations for the marines (*Classiarii Britannici*).

Roads.—The Roman roads were constructed in a most substantial manner, and may still be traced in many parts of the country. The most remarkable feature about them is the undeviating directness of their course. The original names have not come down to us, with the exception perhaps of the Via Julia along the coast of S. *Wales*: in their place, we have the names given to some of them by the Saxons. Five main routes traversed the country in various directions, as follows:—1. *Watling Street*, from Rutupiæ through Durobrivæ to Londinium (where the name is still applied to an important street), and thence by Verulamium, Venonæ, *High Cross* in *Leicestershire*, and Etocetum, *Wall* in *Staffordshire*, to Uriconium, where it divided, one branch going through *Wales* to Segontium, *Carnarvon*, while another went northwards to Deva and Mancunium, *Manchester*, whence it was carried on by Caractonium, *Catterick*, to Cortnopitum, *Corbridge* on the *Tyne*, and thence into *Scotland*. 2. *Ermine Street*, or the great north road, which appears to have started from Anderida, *Pevensey*, on the S. coast, and passed through Londinium, by Durolipons, *Godmanchester* in *Huntingdonshire*, Durobrivæ, *Castor*, and Causennæ, *Ancaster*, to

forms Corinium and Duro-cornovium differ mainly through the addition of the prefix in the latter case, and the same root lies at the bottom both of these and of the modern *Ciren-cester*, all of them having reference to the river *Churn*. So again Durobrivæ and *Rochester* may be identified by the rejection of the prefix *Duro* in the ancient, and the affix *chester* in the modern names, the connecting links between the remaining—brivæ and Ro—being found in the forms "Civitas Ribi," and the Saxon *Hrofe-ceaster*.

Lindum, whence it was continued in one direction to the *Humber*, in another to Danum, *Doncaster*, and Eboracum. 3. *Ikenild Street*, from Venta Icenorum by Camboricum, *Cambridge*, Sorbiodunum, and laca Damnoniorum, *Exeter*, to the extremity of *Cornwall*. 4. *Fosse Way*, from Lindum in a S.W. direction by Ratæ, *Leicester*, Corinium, Aquæ Solis, and Ischalis, *Ilchester*, to Moridunum, probably *Seaton* near *Honiton*. 5. *Rykniield Street*, from Hadrian's wall near *Tynemouth*, in a S.W. direction to Glevum, *Gloucester*, and thence along the coast of *S. Wales* by Nidum, *Neath*, to Maridunum, *Carmarthen*. Important roads also led from Londinium to the eastern counties by Cæsaromagus, *Chelmsford*, to Camalodunum and Venta Icenorum; and again to the W. by a route which crossed the Thames at Pontes, *Staines*, and thence by Callova and Spinæ, *Speen* in *Berks*, to Corinium in one direction, and Aquæ Solis in another: from the latter place it was continued across the *Bristol Channel* (where the old Roman name for the passage, Augusti Trajectus, is still preserved in the form *Aust*) to Venta Silurum, Burrium, *Usk*, Gobannium, *Abergavenny*, Luentinum, in *Cardiganshire*, and thence in a line parallel to the coast to Conovium, *Conway*: this road is now called *Sarn Helen* in Wales.

Roman Walls.—Among the monuments which survive to tell of the presence of the Romans, none are more striking than the lines of defence erected by them on the N. frontier. The first in point of time was erected by Agricola in A.D. 81, between the Firths of *Clyde* and *Forth*, and consisted of a chain of forts, of which there are said to have been nineteen in all, though the sites of only thirteen have been discovered. This line of defence was completed in A.D. 144, by the addition of a rampart and ditch, constructed by Lollius Urbicus, the lieutenant of Antoninus Pius, and named, after the emperor, **Vallum Antonini**. It began near *Old Kirkpatrick* on the *Clyde* and terminated between *Abercorn* and *Borrowstoness* on the *Forth*: its course can still be traced in some parts. Another and more important line of defence was erected between the *Tyne* and *Solway Firth*, consisting of a wall of stone, and a vallum or rampart of earth running parallel to it on the S. side, with an interval of space between the two generally of 60 to 70 yards, but sometimes as much as half a mile, and sometimes only a few yards. It has been generally assumed that the two lines were erected at different periods, the *Vallum* by Hadrian in A.D. 120, and the wall by Severus in 208-211. It is, however, far more probable that they were both erected by Hadrian, and were subsequently repaired by Severus. The wall was probably from 18 to 20 feet high, and from 6 to 9½ feet thick. It was protected on the outside by a fosse, in some places 40 feet wide and 20 deep. Between the wall and the rampart were stations at intervals of four miles, eighteen of them on the wall, the others on either side of it. These stations enclosed areas of from three to six acres, and one of them, named Borcovicus, *Housesteads*, even fifteen acres. In addition to these there were *Castella*, or forts, about 60 ft. square, at intervals of a mile.

History.—The first expedition of Cæsar took place in B.C. 55: starting from *Portius Itius* he crossed the channel to the neighbourhood of *Dover*, and thence coasted along probably to *Deal*.[4] He defeated the

Britons, but did not advance far from the coast. In 54 he again
invaded the island, defeated the Britons, probably on the banks of
the *Stour*, crossed the Thames near *Chertsey*, and took the capital
of Cassivellaunus, which stood probably on the site of Verulamium.
Having received the homage of most of the southern tribes, he retired. The permanent conquest of Britain was commenced by Claudius, who sent over Aulus Plautius in A.D. 43, and shortly after
followed himself, and took Camalodunum, the capital of Cunobeline.
Plautius was succeeded in 50 by Ostorius Scapula, who advanced
the Roman frontier to the banks of the *Severn*, defeated the Iceni
of *Norfolk*, the Brigantes of *Yorkshire*, and the Silures of *S. Wales*,
under their king Caractacus. Didius, who succeeded Ostorius, was
again engaged in war with the Silures. He was succeeded in 57 by
Veranius, and he by Paulinus Suetonius, who attacked the isle of
Mona, but was summoned thence to quell the insurrection of the Iceni
under Boadicea. The next important event was the reduction of the
Brigantes by Petilius Cerealis in the reign of Vespasian. Julius Fron-
tinus succeeded as propraetor, and defeated the Silures; but the final
conquest of Britain was achieved by Julius Agricola, who became governor in 78, defeated the Ordovices of *N. Wales*, reduced Mona,
adopted various measures for civilising the tribes, and in 80 crossed
the frontier of *Scotland*, and succeeded in extending the Roman dominion as far as the *Firths of Forth* and *Clyde*, between which he erected
the line of forts already described: beyond this he advanced in 84 to
the foot of the Grampians, and defeated the Caledonians under Galgacus in a pitched battle, believed to have taken place on *Ardoch Moor*
in *Perthshire*. In the reign of Hadrian these conquests are said to
have been given up, and the boundary was fixed at the *Tyne* and the
Solway. Antoninus Pius again advanced the border, and established
the *vallum* parallel to Agricola's chain of forts in A.D. 144. The remaining facts in the history of Britain are—the death of the emperor
Severus at York, in A.D. 211; the revolts headed by Carausius and
Allectus; the appearance of the Picts in the reign of Diocletian, and
of the Attacotti and Scots in that of Julian A.D. 360. Britain was
abandoned by the Romans early in the 5th century in consequence of
the difficulties under which the empire laboured; shortly afterwards
the Angli and Saxones made their appearance and subdued it.

own words, " ventum et metum uno tempore nactus secundum, circiter millia
passuum vii. ab eo loco progressus, aperto ac plano littore naves constituit"
(*B. G.* iv. 23). As low water occurs at 2 P.M. on that day, it was inferred by Dr.
Halley that Cæsar was carried by the flowing tide to the N. and landed at Deal.
Mr. Airy, the Astronomer Royal, has stated that the stream off Dover does not
turn at the time of high water, but runs westward for 7 hours, commencing
with the 4th hour after high water, and that consequently Cæsar was carried
westward. The accuracy of this statement has been in turn disputed by Dr.
Cardwell, who has ascertained that there is a difference in the currents of the mid-
channel and the in-shore water, the change taking place in the latter from one to
two hours earlier than in the former. Moreover the westward set of the mid-
channel current commences at half ebb and continues until half flood, whereas the
Astronomer Royal's computation adds one hour to the former and two to the
latter. Allowing for these differences, Dr. Cardwell thinks it more than probable that Cæsar was carried northward by the *in-shore* current, which would
commence on the day in question at 3 P.M. (See *Archæol. Cantian.* vol. III.)
Those who have adopted the Astronomer Royal's view, have placed the landing
either at Romney Marsh, W. of Hythe, at Rye, or even at Pevensey.

Islands.—Off the coast of Britannia were the islands—**Vectis,** *I. of Wight,* which was conquered by Vespasian in the reign of Claudius; **Ictis,** *St. Michael's Mount,* whither (according to Diodorus) the Britons conveyed their tin in waggons when the tide was out; **Mictis** (apparently one of the *Scilly Isles*), noticed by Pliny as a place where tin was found, and which the natives reached in coracles; **Silura,** or **Sylina,** the former appearing in Solinus, the latter in Sulpicius Severus, probably one of the *Scilly Isles;* **Mona,** *Anglesey,* the head-quarters of the Druids in the time of the Romans, and hence attacked by Paulinus in A.D. 61, and again by Agricola in A.D. 78; and lastly, **Monapia,** or **Monarina,** *Isle of Man,* which is also named **Mona** by Cæsar (*B. G.* v. 13).

§ 6. Britannia Barbara embraces the whole of Britain N. of the great rampart between the *Solway* and the *Tyne:* it corresponds generally to the **Caledonia** [5] of the ancients in its extended sense, and to the modern *Scotland.* The Romans were very slightly acquainted with this district, at all events with that portion of it which lies N. of the *Firths* of *Clyde* and *Forth.* The names of the tribes and localities are chiefly valuable to the ethnologist as indicative of the races to which the inhabitants belonged. The occurrence, for instance, of the names Cantæ and Cornubii in N. Britain, which are almost identical with the Cantium and Cornubii of S. Britain, and, again, the appearance of the element *Car* in many of the names, leads to the inference that the population of Scotland was originally British rather than Gaelic.[6] This is further supported by the probable etymology of the name Caledonii. The names of Picti [7] and Scoti appear only in late writers: the latter were undoubtedly a Gaelic race who immigrated into the N. of *Scotland* from *Ireland,* and subdued the occupants of the whole district N. of the *Clyde;* the former, the Picti, appear to have been identical with the Caledonii, the name being a mere translation of the term *brit,* "painted,"

---

[5] The name Caledonia first appears in Pliny; it occurs frequently in Tacitus's *Agricola* as applicable to all the populations N. of the rampart, while in Ptolemy the Caledonii are a tribe resident in the W. of Scotland. It appears again in the *Oceanus Deu-caledonius* of the same writer, and in the *Di-calidones,* one of the two *gentes* into which the Picts are divided by Ammianus Marcellinus. It is probably derived from the Welsh *celeddon,* "wooded district." A comparison of the passages in which it occurs leads to the inference that until the invasion of Agricola the term was restricted to the residence of the Caledonii or Di-caledonii between *Loch Fyne* and the *Murray Firth,* and that Agricola, having become first acquainted with this people as living immediately N. of his rampart, extended the term to all the tribes of Scotland.

[6] The limit between the British and Gaelic Celts is marked by the prevalence of the prefix *aber* in the former, and *inver* in the latter. This line runs obliquely from *Loch Fyne* on the W. coast to the *Spey* on the E. On the N. of it are the names *Inver-ness, Inver-ary,* &c.; on the S. *Aber-deen, Aber-dour,* &c.

[7] Ille leves Mauros, nec falso nomine Pictos
Edomuit, Scotumque vago mucrone secutus,
Fregit Hyperboreas remis audacibus undas.

CLAUDIAN. *de* III. *Cons. Honor.* 54.

See also note [5] below.

which is supposed to be at the root of the name Briton. The nationality of the Picts is, however, a subject of much mystery.

Physical Features.—There is but one mountain range named by ancient writers, viz. **Grampias Ms.**, which evidently answers in name to the *Grampians*, the scene of Galgacus's resistance to the Roman arms. There is also a forest, **Caledonia Silva**,[5] noticed by Ptolemy; the position of this could not, from the geological character of the country, have been further N. than the *Clyde* on the W. and the *Dee* on the E. coast. The chief promontories, from the S.W. round to the S.E. are—Prom. **Novantarum**, *Corsill Point*; Prom. **Epidium**, *Mull of Cantyre*; Prom. **Tarvēdum** or **Orcas**, *Dunnet Head*; **Verubium**, *Noss Head*; and **Taexalōrum Prom.**, *Kinnaird's Head*. The rivers and estuaries are—the **Novius**, *Nith*; **Deva**, *Dee*; **Iēna Est.**, *Wigton Bay*; **Rerigonius Sin.**, *Loch Ryan*; **Clota Est.**, *Firth of Clyde*; **Lelannonius Sin.**, *L. Linnhe*; **Volsas Sin.**, *Loch Broom*; **Varar Est.**, *Firth of Cromarty*; **Tuaesis Est.**, *Murray Firth*; **Tava Est.**, *Firth of Tay*; and **Boderia Est.**, *Firth of Forth*.

Tribes.—(1.) In Valentia, from S. to N., the **Selgōvæ** in *Dumfriesshire*; the **Novantæ** in *Wigtonshire*; the **Gadēni** in *Roxburghshire*: the **Otadini** in *Northumberland* and *Berwickshire*; and the **Damnii** or **Dumnonii** in *Peebles, Selkirk, Lanark, Edinburgh, Linlithgow, Renfrew,* and *Stirling.* (2.) To the N. of the *Clyde*, from S. to N.: in the W., the **Epidii, Cerones, Vacomagi, Carnonacæ,** and **Careni**; in the E., the **Venicontes, Texali, Decantæ, Meretæ, Lugi,** and **Cornavii.**

Towns.—**Blatum Bulgium**, *Middleby*, in *Dumfriesshire*, where there are Roman remains; **Brumenium**, a town of the Otadini, variously identified with *Brampton, Riechester,* and *Newcastle*; **Colania** and **Coria**, towns of the Damnii, identified with *Carstairs* and *Crawfurd* respectively; **Vanduara** or **Vandogara**, *Paisley,* and **Victoria**, either on *Inchkeith Island* or *Abernethy* near *Perth*, also towns of the Damnii; and **Alāta Castra** near *Inverness*, the northernmost station of the Romans, probably raised by Lollius Urbicus in A.D. 139, but soon abandoned.

Islands.—Off the W. coast of *Scotland* lie the **Hebudæ** or **Ebudæ**, *Hebrides*, which are noticed by Pliny and Solinus; and off the N. coast the **Orcādes**,[9] the *Orkney* and *Shetland Isles*, which are noticed by several writers. We may here notice **Thule**,[10] which Pytheas, its dis-

---

[9] Martial implies that bears were imported at Rome from the wilds of Scotland:—

　　　Nuda Caledonio sic pectora præbuit urso.　　　*De Spectac.* vii. 3.

　———————— Arma quidem ultra
Littora Juverdae promovimus, et modo captas
Oreadas, et minima contentos nocte Britannos.
　　　　　　　　　　　　　　　　　　　　Juv. ii. 159.

Quid rigor æternus cœli ? quid sidera prosunt ?
Ignotumque fretum ? maduerunt Saxone fuso
Orcades : incaluit Pictorum sanguine Thule :
Scotorum cumulos flevit glacialis Ierne.—CLAUDIAN. *de IV. Cons. Honor.* 30.

[10] Thule was always regarded as the farthest point of the known world; and this is supposed to be expressed in the name itself, the Gothic *tiel* or *tiule* denoting the remotest land :—

　———————— Ubi servial ultima Thule.—VIRG. *Georg.* i. 30.

We seem to have some reference to the frozen waters of the arctic seas in the following lines of Claudian :—

coverer, places at six days' sail from the Orcades, and thus leads us to identify it with *Iceland*, while Ptolemy places it more to the S., in the latitude of the *Shetlands*, so that we may identify it with *Mainland*.

§ 7. The ancient accounts of *Ireland*[1] are chiefly interesting as illustrative of the progress of geographical knowledge: they also, to a certain extent, assist the ethnologist. The oldest form of the name is **Ierne**, which appears in Aristotle, and which most nearly approximates to the native name *Eri*. Diodorus Siculus calls it **Iris**; Strabo, **Ierne**; Mela, **Iverna**; Pliny, **Hybernia**; Solinus, **Hibernia**; and Ptolemy, **Ivernia**. The statements of these writers are somewhat fabulous. The people were cannibals, according to Diodorus; and the country was so cold as to be barely habitable, according to Strabo. Ptolemy alone gives any details as to the geography, and his description of it is fuller even than that of Britain. It may be observed that many of the rivers and places retain their ancient names at the present time. The population was substantially Gaelic. The occurrence of the German names Cauci and Menapii, and of the British name Brigantes, suggests the probability of colonies having been planted on the E. coast from Germany and Britain. The Scoti, who migrated to Scotland, are not noticed by Ptolemy, but appear in Claudian.[2]

Physical Features.—The rivers noticed are—the **Bargus**, *Barrow*; **Senus**, *Shannon*; **Libdus**, *Liffy*; **Oboca**, *Avoca*; and **Isrnus**, probably the *Kenmare*. The promontories are—**Sacrum**, *Carnsore Point*, at the S.E.; **Isamnium**, *St. John's Point*; **Robogdium**, *Fair Head*, at the N.E.; **Borvum**, *Malin Head*; and **Notium**, *Mizen Head*, on the S.W. angle.

Tribes.—The **Brigantes** and **Coriondi** on the S. coast; the **Velleböri**, **Gangäni**, **Autini**, **Nagnätæ**, **Erdini**, and **Vennienii**, along the W. coast from S. to N.; the **Darini** and **Robogdii**, along the N. coast; the **Voluntii**, **Ebläni**, **Cauci**, and **Manapii**, along the E. coast from N. to S.

Towns.—The situations of the towns noticed by Ptolemy are problematical. **Ebläna** represents *Dublin*; **Nagnata**, described as an im-

Facta tui numerabat avi, quem litus Aduatæ
Horrescit Libyæ, ratibusque impervia Thule.—*De III. Cons. Honor.* 52.

[1] It is difficult to decide the date of the earliest notice of *Ireland*. If the Orphic poem on the Argonautic expedition were composed by Onomacritus, we should carry it back to the reign of Darius I. The form of the name is the old one :—

νήσοισιν Ἰέρνισιν ἄγχου ἰαμμαι.   Orpheus, 1184.

The knowledge of Avienus was derived from the Carthaginians, perhaps from the account of Hanno's expedition: he describes it as the "sacred isle," from the similarity of the name to ἱερα :—

Ast in duobus in *Sacram*, sic insulam
Dixere prisci, solibus cursus rata est.
Hæc inter undas multa cæspitem jacit
Eamque late genus Hibernorum colit.   *Or. Marit.* 109.

[2] ———— totam quum Scotus Iernen
Movit.   *In I. Cons. Stilich. ii.* 251

portant town, was probably on *Sligo Bay*; **Managia** may be *Wexford*. In addition to these, six inland towns are enumerated, proving that the country was well occupied: their names were Rhæba, Laverus, Dunum (a well-known Celtic termination), Macolicum, perhaps *Millick* on the *Shannon*, and two named Rhegia.

II. GERMANIA.

§ 8. The boundaries of **Germania** were the Rhine on the W., the Danube on the S., the Sarmatian Mountains and the Vistula on the E., and the Mare Suevicum, *Baltic*, and Mare Germanicum on the N. Sometimes indeed the peninsula of Scandia was regarded as a part of Germany, in which case the N. boundary was carried on to the Oceanus Septentrionalis. Taken at its fullest extent, it would include, in addition to the greatest part of *Germany, Holland,* the W. of *Poland, Denmark, Norway,* and *Sweden*. The greater portion of this extensive district was unknown even to the Romans: the parts with which they were best acquainted were in the W. and S. It is described as a wild and inhospitable[3] country, covered with forests and marshes, excessively cold, and much infested with wild beasts. Its soil was generally unfertile, yet it produced, in certain parts, wheat, barley, oats, flax, and various edible roots. The vine was not introduced until the 6th century of our era: the ordinary drink of the country was a kind of beer. The country supported a large number of pigs, together with a fair amount of sheep and goats, valuable hounds, strong but small horses, and short-horned cattle. Numerous kinds of wild beasts are mentioned, particularly elks (*alces*) and wild oxen (*uri*).

Name.—The name was regarded by many ancient writers as derived from the Latin *germani*, and as intended to describe the "brotherhood" supposed to exist between the Gauls and Germans. Tacitus, however, regarded it as originally the name of a particular tribe, the Tungri. It has also been derived from the Persian tribe of the same name, noticed by Herodotus (i. 125). Most probably it is of Celtic origin, and came into use among the Celts in Gaul before the time of Cæsar. It has been referred to a Gaelic root *gair*, "to cry out," giving it the sense of the Homeric βοὴν ἀγαθός, a fierce warrior. The indigenous name has always been *Deutsch*, which appears in the classic form *Teutones*. Germany proper was named Germania *Magna, Transrhenana,* or *Barbara,* in contradistinction to the Germania on the W. of the Rhine.

§ 9. The mountain ranges of Germany received for the most part specific designations. The **Hercynia**[4] **Silva** has been already noticed

---

[3] Quis Parthum pavent? quis gelidum Scythen?
Quis, Germania quos *horrida* parturit
Fœtus, incolumi Cæsare?   Hor. *Carm.* iv. 5, 25.

[4] The name is of Celtic origin, signifying a "wooded mountain;" it still survives in the modern *Harz*.

(p. 320). The other ranges are—**Taunus**, in the angle between the Rhine and the *Mœnus, Maine;* **Rhetico**, of uncertain position, in the same neighbourhood; and **Saltus Teutoburgiensis** in the N., between the *Lippe* and *Weser.* The only promontory noticed is **Cimbrōrum Prom.,** *Skagen,* the N. point of *Denmark.* Several great forests[3] are noticed, as **Casia Silva,** between the rivers *Lippe* and *Yssel;* **Baduhennæ Lucus,** *Holtpade* in *West Friesland;* **Harculla Silva,** *Suntelgebirge,* W. of *Minden;* **Semnōnum Silva,** between the *Elster* and *Spree;* and **Naharvalōrum Silva,** between the *Oder* and *Vistula.* The chief rivers are—the border stream of the **Rhenus,** *Rhine,* which receives on its right bank the tributary waters of the **Nicer,** *Neckar,* **Mœnus,** *Maine,* and **Luppia,** *Lippe,* with others of less consequence; the **Amisia,** *Ems,* flowing into the German Ocean, and historically known for a battle fought on its banks in B.C. 12 between Drusus and the Bructeri; the **Visurgis,** *Weser,* reaching the ocean in the district of the Chauci; the **Albis,** *Elbe,* the most easterly river reached by the Romans, having been crossed by Domitius Ahenobarbus in B.C. 3; the **Viadus,** *Oder,* which flows into the Mare Suevicum in the land of the Rugii; the **Vistūla** on the E. border; and the **Danubius,** which has its sources in Abnoba Ms., and receives numerous tributaries on its left bank, of which the **Marus,** *March,* is the most important. In the N.W. of Germany a large lake is noticed under the name of **Flevo Lacus,** now the *Zuider Zee.* This was connected with the Rhine by a canal cut by Drusus, and named after him **Fossa Drusiana,** which commences below the separation of the Rhine and *Waal,* and joins the *Yssel* near *Doesburg:* this new outlet for the Rhine was named Flevum Ostium.

§ 10. The Germans are said to have regarded themselves as an autochthonous race, and they certainly have preserved no tradition of their Asiatic origin. In physical appearance they were tall and handsome, with blue eyes[4] and fair or red hair.[7] They subsisted chiefly on the cattle they reared, and on the proceeds of the chase and war. They enjoyed a character for independence and faithfulness combined with cunning and falsehood. The various tribes were classified by Tacitus in three groups: the **Ingævōnes** on the ocean, the **Hermiōnes**

---

[3] The forests of Germany were in many cases sacred to certain gods, as in the case of the Semnonum and Baduhennæ groves :—

Ut procul Hercyniæ per vasta silentia silvæ
Venari tuto liceat, lucosque relusta
Religione truces ——   CLAUDIAN. in 1. Stil. l. 228.

[4] Nec fera cœrules domuit Germania pube.   HOR. Epod. xvi. 7.

[7] They had a custom of heightening the red colour of their hair by artificial means :—

Caustica Teutonicos accendit spuma capillos;
Captivis potaris cultior esse comis.   MART. xiv. 26.

in the interior, and the **Istævônæs** in the E. and S. To these we may add the inhabitants of the Scandinavian peninsula, who bore the general name of **Hillevionæs**. The chief tribes belonging to these groups were located in the following manner :—

(1.) On the coast.—The **Frisii**, about Lake Flevo, between the *Rhine* and *Ems*, divided into two clans, Majores and Minores, the former living probably W. of the canal of Drusus in *N. Holland*, the latter E. of it, in *Friesland*, which still retains the ancient name. The **Chauci**, between the *Ems* and the *Elbe*, in *Oldenburg* and *Hanover*, also divided into Majores and Minores, living respectively W. and E. of the *Weser*; they were skilful navigators, and much addicted to piracy. The **Saxones**, E. of the *Elbe* in *Holstein*, a people whose name does not appear in history until A.D. 287, but who may have occupied that district in the days of Pliny and Tacitus. The **Cimbri**, in the Chersonesus Cimbrica,[8] *Jutland*, in all probability a Celtic race, as the ancients themselves believed, their name bearing a close resemblance to that of *Kymri*, and their armour and customs differing from those of the Germans ; the **Varini**, between the Chalusus, *Trave*, and the Suebus, *Warne*; the **Teutones**, also between the *Trave* and the *Warne*, the representatives of the original tribe which sent forth the mighty horde whom the Romans defeated in B.C. 102 ; the **Sidöni**, between the Suebus and the Viadus, *Oder*; and lastly the **Rugii**, between the *Oder* and *Vistula*, and on the island which still bears the name of *Rugen*.

(2.) South of these, from E. to W., lived — the **Helvecônæs**, below the Rugii. The **Burgundiônes**,[9] a Gothic race, between the Vistula and Viadus; in later times (A.D. 289) a people of the same name appear in the S.W. of Germany, and in the early part of the 5th century these crossed the Rhine and established themselves in *Burgundy*. The **Vandali**, a powerful race, of which the Burgundiones were regarded as a tribe, and whose settlements were frequently shifted: we first hear of them as seated on the Palus Mæotis, then (in Pliny's time) between the Vistula and Viadus, next in the country N. of *Bohemia*, about the *Riesengebirge*, which were named Vandalici Mts. after them; in the reign of Constantine in *Moravia*, whence they were transplanted by that emperor into Pannonia ; in the reign of Probus in Dacia ; in A.D. 400 ravaging Gaul ; in 409 in Spain ; in 429 across the Straits of Gibraltar in Africa, where they established themselves for above one hundred years, when Belisarius succeeded in destroying their power, A.D. 534 ; they have been variously regarded as a German or a Slavonic race. The **Semnones**, a Suevic[1] tribe between the Viadus and Albis, and between

---

[8] ———— latisque paludibus exit
Cimber.							CLAUDIAN. *de* IV. *Cons. Hon.* 451.

[9] The name is explained by Ammianus Marcellinus as meaning those who lived in "townships" (*burgi*). It is uncertain whether the later Burgundians were the same race as those of the N.E., but they probably were so.

[1] Suevi appears to have been a general designation, embracing a great number of the tribes of Central Germany. By Cæsar they are placed on the E. bank of the Rhine in *Baden*; by Tacitus to the N. and E. of that district; by Strabo between the *Rhine* and *Elbe*. The Suevi of Cæsar were true Germans; those of Tacitus and Strabo contained Celtic or Slavonian elements. About A.D. 250 a people calling themselves Suevi, though they appear to have belonged to various tribes, settled in *Suabia*, which still retains their name. Their general position is indicated by Lucan :—

Potsdam in the N. and the hills of *Lusatia* in the S.; they are mentioned after the time of M. Aurelius. The Langobardi,[2] a Suevic tribe,
first met with on the left bank of the *Elbe*, N. of the junction of the
Sala; then on the right bank, having been probably driven across the
Elbe by Tiberius in the reign of Augustus; and again, in Ptolemy's
time, between the Rhine and the *Weser*; a people of the same name,
and probably of the same tribe, are next heard of in Pannonia, and late
in the 5th century A.D. on the right bank of the Danube in *Hungary*,
whence they extended their sway along the Danube into Dacia, and
finally crossed into Italy in A.D. 568, and settled in the country which
still bears their name, viz. *Lombardy*. The Anglii or Angli, a Suevic
tribe, occupying, according to Ptolemy, an extensive district on the left
bank of the *Elbe*, whence they subsequently migrated to Britain; the
Angrivarii, on both sides of the *Weser*, but mainly between that river
and the *Elbe*; and the Bructeri,[3] between the Rhine and the *Ems*, divided by the river Luppia into two branches, the Majores to the N.,
and Minores S. of that stream.

(3,) Tribes yet more to the S., from W. to E.—The Usipetes,[4] originally belonging to the interior; then settled on the right bank of the
Luppia, after their defeat by Cæsar; and afterwards, as it appears,
more to the S., in the neighbourhood of the Marsi. The Tencteri, a
companion tribe to the Usipetes; they apparently emigrated from the
interior, crossed the Rhine in Cæsar's time, were defeated and almost
cut to pieces by him, and finally settled on the right bank of the Rhine,
between the *Ruhr* and the *Sieg*. The Sicambri, originally on the right
bank of the Rhine, between the *Sieg*[5] and the *Lippe*; afterwards, when
they had received the Usipetes and the Tencteri into their territory,
they were transplanted to Gaul by Tiberius, and settled between the
Meuse and Rhine, with the exception of a section which remained in
Germany about Mons Rhetico. The Catti[6] or Chatti, E. of the Tencteri,

Fundat ab extremo flavos Aquilone Suevos
Albis, et indomitum Rheni caput.　　　　　　　　　　II. 51.

[2] The name has been generally understood to mean "long-bearded;" but more
probably it is derived from the *lange Börde*, "the plain by the side of the river"
Elbe, where they are first found, and where the name still attaches to a district
near *Magdeburg*.
　　　　　　　　　Venit accola silvæ
Bructerus Hercyniæ.　　　　　　　　Claudian. de IV. Cons. Hon. 450.

[4] Rem factam Pompilius habet, Faustine: legetur,
　　Et nomen toto sparget in orbe suum.
Sic leve flavorum valeat genus Usiplorum,
　　Quisquis et Ausonium non amat imperium.　　Mart. vi. 60.

[5] Their name is generally derived from this river; but this is doubtful. In
B.C. 17 they invaded Gaul, but at the approach of Augustus retired to their own
territory. To this Horace alludes in the following lines, which also indicate the
reputed character of this people:—
　　　　　　　—— quandoque trahet *feroces*
Per sacrum clivum, merita decorus
　　Fronde, Sicambros.　　　　　　　　Hor. *Carm.* iv. 2, 34.
Te cæde gaudentes Sicambri
　　Compositis venerantur armis.　　　　Id. iv. 14, 51.

[6] The Catti obtained great celebrity for their resistance to the Romans:—
Traxerat attonitos et federuaro coactos,
Tanquam de Cattis aliquid torvisque Sicambris
Dicturus.　　　　　　　　　　Juv. iv. 146.

between the *Saale* in the E., the *Maine* in the S., and the upper course of the *Weser* in the N., thus occupying the country which still retains their name, *Hesse*; in Ptolemy's time they appear to have lived more to the E. The **Mattiaci**, probably a branch of the Chatti, occupying the present *Nassau*, on the right bank of the Rhine. The **Tubantes**, originally between the Rhine and the *Yssel*, but in the time of Germanicus S. of the *Lippe*, in the former territory of the Sicambri, and in Ptolemy's time still more to the S., near the *Thüringer-Wald*. The **Cherusci**, an important tribe between the *Weser* in the W., the *Elbe'* in the E., Melibocus Ms. in the N., and the Sala in the S.; after their conquest by the Chatti they dwindled down to a small tribe, which in the time of Ptolemy lived in the *Harz* Mountains. And, lastly, the **Lygii**, a widely-spread nation, containing a number of tribes, settled between the *Vistula* and *Oder*; they were probably Slavonians who had been subdued by the Suevi.

(4.) Tribes along the course of the Danube from E. to W.—The **Quadi**, in *Moravia*, the N.W. of *Hungary*, and the E. of *Bohemia*; they were regarded by Tacitus as Germans, but they may have been Sarmatians; their name disappears towards the end of the 4th century of our era. The **Marcomanni**, *i.e.* "march-men," or "borderers," a tribe who first appear on the Rhenish frontier about the lower course of the *Maine*, as having crossed thence into Gaul, and being driven back by Cæsar in B.C. 58; hence they migrated into the territory of the Celtic Boii, *Bohemia*, where they organised a powerful kingdom about A.D. 6; they came prominently forward in their wars with the Romans, A.D. 166-180, and made inroads into Italy; they are last mentioned as forming a portion of Attila's army. The **Hermunduri**, between the mountains in the N.W. of *Bohemia* and the Roman wall in the S.W., which bounded the Agri Decumates; they were a Suevic race, and first appear in history at the time of Domitius Ahenobarbus, who settled them between the *Maius* and the Danube, whence they spread out in a N.E. direction. Lastly, within the limits of the **Agri Decumates**, *i.e.* "tithe-lands," which lay in the S.W. of Germany, and were separated from the interior by a wall from *Ratisbon* on the Danube to *Lorch*, and thence by an earthwork to the Rhine near *Cologne*, were located various immigrant bands of Gauls and Germans, to whom were subsequently added colonies of veterans for the defence of the border; this district was incorporated with the empire, as a part of the province of Rhætia, but it was wholly lost about A.D. 284.

The distinctive names of the German tribes appear to have fallen into disuse about the end of the 3rd century of our era, and the whole nation was classified under two broad appellations, **Alemanni** and **Franci**, the first applying to the tribes that lived on or about the Upper Rhine, the second to those on the Lower Rhine. Alemanni was (as the word itself implies, being derived from *Alle Männer*, "all men") a confederacy of many tribes, chiefly of the Suevic race. It first appears in the history of Dion Cassius, about A.D. 400; and it is preserved in the modern French name of Germany, *Allemagne*. The chief seat of the contest between them and the Romans was in the Agri Decumates. The Franci, *i.e.* "free men," are first mentioned in A.D. 240, and were also a confederacy of which the Sicambri were the most influential member. They conquered the N. of Gaul, and, having there adopted the civilisation of the Romanised Celts, they acquired such power that they were enabled, in A.D. 496, to return and subdue their German kinsmen.

Towns.—Of the towns which were scattered over the extensive districts above referred to, we know little else than the names. It is interesting, however, to observe that the much-frequented watering-places in the neighbourhood of the Rhine were not unknown in ancient times, *Baden* being described as Aquæ Aureliæ, and *Wiesbaden* as Aquæ Mattiacæ. Mattium, the capital of the Chatti, which was burnt down, A.D. 15, in the war with Germanicus, was at *Maden*, on the right bank of the *Eder*. The only district bearing marks of Roman occupation is the Agri Decumates, where not only roads, but walls, inscriptions, and numerous antiquities, have been discovered in many places: we may instance the remains of Samulocenæ at *Sülchen*, of Cana at *Cannstadt*, of Clarenna at *Köngen*, all of them on the *Neckar*. The position of Solicinium, in the same district, rendered famous by the victory gained by Valentinian over the Alemanni in A.D. 369, is uncertain. In the territory occupied by the Quadi the names of several towns (such as Eburodunum, Mellodunum, &c.) indicate a prior occupation of that country by the Celts.

Islands.—The ancients not unnaturally regarded the Scandinavian peninsula as an island or collection of islands. Pliny names two of those islands Scandia and Scandinavia, the latter being the largest in the whole group. Ptolemy speaks of four under the general name of Scandiæ Insulæ, of which the largest was Scandia. Tacitus does not mention Scandia, but the tribes of the Sitones and Suiones must undoubtedly be placed there : the latter name is the original of *Swedes*, and the southern part of *Sweden* still bears a name not unlike Scandia, *Scania*, *Scone*, or *Schonen*. Pliny also speaks-of an island named Nerigos, whence people used to sail for Thule : this has been identified with *Norway*; in which case his Bergi may represent *Bergen*, and Dumna Dunoes : this is, however, uncertain.

History.—We have no connected history of the German nations until the time of Julius Cæsar, who in his Gallic campaigns came in contact with and defeated Ariovistus. Cæsar himself crossed the Rhine twice, in B.C. 55 and 54, but he did not attempt to maintain himself in Germany. In B.C. 37 Agrippa transplanted the Ubii to the W. bank of the Rhine, as a barrier on the side of the German border. This plan, however, did not fully succeed ; and hence Nero Claudius Drusus undertook a series of expeditions against the Germans from the Insula Batavorum. He advanced as far as the *Elbe*; and on his death, in B.C. 9, the operations were carried on by Tiberius and Domitius Ahenobarbus, who subdued for a while the tribes between the Rhine and the *Weser*; but in A.D. 9, Arminius, king of the Cherusci, defeated the Romans in the Teutoburg forest, and terminated their supremacy in the N., while the resistance of Maroboduus, the Marcomannian, on the Middle Rhine, checked them in that direction. In the latter district Germanicus gained some advantages, but was unable to re-establish a permanent ascendancy. The Romans then withdrew within the Agri Decumates, which they fortified between A.D. 16 and 69. The great revolt of the Batavi, in A.D. 70 and 71, was followed by repeated wars with several German tribes, until in the reign of M. Antoninus the great Marcomannic war broke out on the Danube, resulting in the surrender of the Roman forts along the course of that river in A.D. 180. Soon afterwards the German tribes began to pour over the Rhine ; and towards the end of the 5th century they had subdued Gaul, Spain, and Italy, and had even crossed over into Africa.

The Court-yard of Diocletian's Palace at Salonæ (Spalato).

CHAPTER XXXII.

THE DANUBIAN PROVINCES, ILLYRICUM, MŒSIA, DACIA, AND
SARMATIA.

I. The DANUBIAN PROVINCES. § 1. Vindelicia. § 2. Rhætia. § 3.
Noricum. § 4. Pannonia. § 5. Its inhabitants and towns. II.
ILLYRICUM. § 6. Boundaries. § 7. Mountains and rivers. § 8.
Inhabitants; Towns; Roads; History. III. MŒSIA. § 9. Bounda-
ries; Rivers. § 10. Inhabitants; Towns. IV. DACIA. § 11. Boun
daries; Mountains; Rivers. § 12. Inhabitants; Towns. § 13. The
Jazyges Metanastæ. V. SARMATIA EUROPÆA. § 14. Boundaries:
Tribes; Towns.

§ 1. **Vindelicia**,[1] the most westerly of the four Danubian provinces, was bounded on the N. by the Danube and the Vallum
Hadriani, on the W. by the territory of the Helvetii, on the S. by

---

[1] This name contains the root *Vind*, which occurs in other Celtic names, such
as Vindobona, Vindomagna, &c.

Rhætia, the ridge of the Rhætian Alps forming the limit, and on the E. by the river Ænus, separating it from Noricum. It embraced the N.E. of *Switzerland*, the S.E. of *Baden*, the S. of *Würtemburg* and *Bavaria*, and the N. of *Tyrol*. The country is for the most part flat, but spurs of the Rhætian Alps traverse the S. district. The chief river is the **Danubius**, which receives numerous tributaries on its right bank, of which the **Ænus**, *Inn*, is the most important. The **Brigantinus Lacus**, *L. of Constanz*, belonged to this country. The inhabitants were in the time of Augustus a Celtic race, and were divided into numerous tribes. They were subdued by Drusus and Tiberius[*] in B.C. 15, and their country was formed into a separate province. About the end of the first century after Christ, it was united with Rhætia, but subsequently was separated from it with the title of Rhætia Secunda.

The towns possess no historical associations: the capital **Augusta Vindelicōrum**, *Augsburg*, was founded by Augustus about A.D. 14, at the junction of the rivers Licus and Virdo. The other important towns were—**Brigantium**, *Bregenz*, on the lake named after it; **Campodūnum**, *Kempten* on the *Iller*; **Reginum**, *Ratisbon*, on the Danube; and **Veldidēna** on the Ænus.

§ 2. **Rhætia**, or, more properly, **Rætia**, was bounded on the N. by Vindelicia, on the W. by the territory of the Helvetii, on the S. by the Alps from Mons Adula to M. Ocra, and on the E. by Noricum and Venetia. It comprised the modern *Grisons*, the *Tyrol*, and a portion of *Lombardy*. It is throughout a mountainous country, being traversed by the ranges of the Rhætian Alps. The valleys were fertile, and produced a wine[*] not inferior to that of Italy; the inhabitants depended on their flocks rather than on agriculture: wax, honey, pitch, and cheese were largely exported. The chief rivers are the **Ænus**, which flows northwards to the Danube; and the **Athěsis**, *Adige*, with its tributary the **Atǎgis**, *Eisach*, which flows S. into the Adriatic. In addition to these the upper streams of many of the Alpine streams, such as the **Addua**, **Sarius**, **Ollius**, and **Mincius**, fall within the limits of Rhætia. The inhabitants of this

---

[*] The expedition of Drusus is commemorated by Horace :—
Videre Rhæti bella sub Alpibus
Drusum gerentem Vindelici.   *Carm.* iv. 4, 17.
The expedition of Tiberius, which took place at a later period of the same year, is commemorated in the following lines :—
Quem legis expertes Latinæ
Vindelici didicere nuper
Quid Marte posses.   *Id.* iv. 14, 7.
[*] ———— et quæ te carmine dicam,
Rhætica! nec cellis ideo contende Falernis.   Virg. *Georg.* ii. 95.
Si non ignota est docti tibi terra Catulli,
Potasti testa Rhætica vina mea.   Mart. xiv. 100.

province in the time of Augustus were mainly a Celtic race.[4] They were a wild, cunning, and rapacious mountain people, ardent in their love of freedom, and fierce in their defence of it. They were conquered by the Romans under Drusus and Tiberius[5] in B.C. 15, and their country was reduced to a province. The chief tribes were the Lepontii who inhabited the valleys on the S. side of the Alps about the head of the lakes of *Como* and *Maggiore*; the Tridentini in the valley of the Athesis; and the Euganei,[6] who at one time occupied the whole tract from the Alps to the Adriatic, but were driven by the Veneti into the Alpine valleys; they were a distinct race from the Rhætians, but their ethnological position is quite unknown.

The only important town in Rhætia was Tridentum, *Trent*, on the Athesis, which appears to have been made a Roman colony: it stood on the road which the Romans constructed between Verona and Augusta Vindelicorum. Another road,[7] between the latter town and Comum, passed through Rhætia.

§ 3. Noricum[8] was bounded on the W. by Rhætia and Vindelicia, on the N. by the Danube, on the E. by Mons Cetius, which separated it from Pannonia, and on the S. by the Savus, the Alpes Carnicæ, and Mount Ocra. It comprised portions of *Austria*, the greater part of *Styria*, *Carinthia*, and portions of *Carniola*, *Bavaria*, and *Tyrol*. It is a mountainous country, intersected by numerous

---

[4] An opinion prevailed among the ancients that the Rhætians were Etruscans who had been driven into the Alps from Lombardy by the Gauls. This view has been adopted by some eminent scholars in modern times, who have discovered in some remote districts (the *Grödnerthal* and the valley of the *Engadine*) names of places, peculiar words, and a few monuments, all of which bear some resemblance to those found in Etruria. This question does not affect the statement that in the time of Augustus the Rhætians were essentially Celts.

[5] The Genauni lived between the lakes *Maggiore* and *Como*:—

————————— Milite nam tuo
Drusus Genaunos, implacidum genus,
    Breunosque veloces, et arces
    Alpibus impositas tremendis
Dejecit acer plus vice simplici;
Major Neronum mox grave prœlium
Commisit, immanesque Rhætos
    Auspiciis pepulit secundis.   *Hor. Carm.* iv. 14, 9.

[6] They left a memorial of their former residence in the Euganeus Collis and the Euganei Lacus, and in the modern *Colli Euganei*, the volcanic group near *Padua*.
[7] This second route crossed the *Splügen* to Curia, *Coire*; it is described by Claudian:—

Protinus, umbrosa qua vestit littus oliva
Larius, et dulci mentitur Nereu fluctu,
Parva puppe lacum prætervolat. Ocius inde
Scandit inaccessus brumali sidere montes.   *Bell. Get.* 319.

[8] The name is probably derived from that of the town Noreia; its use dates from the time that the Romans became acquainted with the country.

valleys opening out towards the Danube, along the course of which
there are some plains. The climate was rough and cold, and the
soil unfertile. The wealth of the country consisted in its iron
mines,[9] which were extensively worked by the Romans. Salt was
also abundant. The chief range of mountains is the **Alpes Noricæ**,
which traverses the country from E. to W. **Cetius Mons**, *Kahlen-
berg*, lies on the borders of Pannonia; **Ocra** was the name given to
the lowest part of the Carnic Alps between Aquileia and Æmona.
The chief rivers are the **Danubius**, the **Ænus** with its tributary the
Jovávus, *Salzach*, and the upper courses of the **Dravus**, *Drave*, and
Sávus, *Save*, which rise, the former in the Norican, the latter in the
Carnic Alps, and flow in an easterly direction with nearly parallel
courses through the S. part of the province. The Norici were a
Celtic race whose ancient name was Taurisci; about B.C. 58, the
kindred race of the Boii immigrated into the northern part of the
country. The Noricans offered an obstinate resistance to the
Romans, but were subdued about B.C. 13 by Tiberius, Drusus, and
P. Silius, and their country was formed into a province, which was
subdivided in the later division of the empire into two, Noricum
Ripense about the Danube, and N. Mediterraneum in the S. The
Romans were obliged to keep a strong military force in it as a safe-
guard partly against the inhabitants themselves, partly against the
Trans-Danubian tribes; they also maintained three fleets on the
Danube, named Classes Comaginensia, Arlapensis, and Laureacensis,
for the latter purpose.

The capital **Noreia**, *Neumarkt*, was situated S. of the river Murius,
and formed the central point for the gold and iron trade: it is celebrated for the defeat there sustained by C. Carbo against the Cimbri
in B.C. 113, and for its siege by the Boii, about B.C. 58. The other
important towns were—**Boiodūrum**, *Innstadt*, at the mouth of the
Ænus, a town of the Boii, as its name indicates; **Ovilāba**, *Wels*, a
Roman colony, to the S.W. of Boiodurum; **Lauriācum**, *Lorch* near
Ens, at the junction of the river Anisius with the Danube, the head-
quarters of the third legion, a fleet station, an arsenal, and probably
a Roman colony; **Juvávum**, *Salzburg*, on the left bank of the river
Jovavus, the station of a cohort, the residence of the governor of the
province, and in early times probably the residence of the native kings;
Virūnum, an important town on the road from Aquileia to Lauriacum,
the ruins of which are found at *Mariasaal* near *Klagenfurt*; **Celeia**,
Cilly, in the S.E. corner of the country, a fine town, as its remains
testify; and **Teurnia**, on the Upper Dravus near *Spital*.

---

9 ———— quas neque *Noricas*
  Deterret ensis.   HOR. *Carm.* l. 16, 9.

Voles modo altis desilire turribus
Modo enes pectus *Norico* recludere.   ID. *Epod.* xvii. 70.

Sæve Bilbilin optimam metallo,
Quæ vincit Chalybasque *Noricosque*.   MART. iv. 55.

§ 4. **Pannonia** was bounded on the N. and E. by the Danube, on the S. by Illyricum and Mœsia, the valley of the *Savo* forming the limit in this direction, and on the W. by Noricum and Italy. It comprehends the E. portions of *Austria, Carinthia, Carniola,* the S.W. of *Hungary, Slavonia,* and parts of *Croatia* and *Bosnia.* It is a vast plain, enclosed on the W. and S. by lofty mountains, but elsewhere traversed by hills of only moderate height. The climate is described as severe, and the soil unproductive; but this is not the present character of the country. The vine and olive were not introduced until the time of the Emperor Probus; previously the beverage of the country was a kind of beer, named Sabaia. The mines do not appear to have been known to the ancients; timber was the most important production.[1] The mountains were described by the general name of **Pannonicae Alpes,** the special names being **Cetius** and **Carvancas** for the ranges on the side of Noricum, and **Albii** or **Albani Mts.** on the side of Illyricum. The chief rivers are—the **Danubius,** which in this part of its course deviates from its usual easterly course by a southerly bend; the **Dravus** and **Savus,** which flow in parallel courses to the Danube, and receive as tributaries, the former the **Murius,** *Muhr,* on its left bank; the latter the **Drinus,** *Drina,* and several less important streams on its right bank. The Danube receives also the **Arrabo,** *Raab,* previous to taking its southerly bend. A large lake named **Pelso,** *Plattensee,* lies in the N. part of the province.

§ 5. The Pannonians were generally reputed an Illyrian race; the Greek writers, however, identified them with the Pæonians of Thrace. Whatever their origin may have been, it is certain that there was a large admixture of Celts among them.[2] They are described as a brave and warlike people, faithless and cunning, and, previous to their subjection to the Romans, rude and uncivilized. They were conquered in the first instance by Octavianus in B.C. 35 and completely subdued by Tiberius in A.D. 8, and again by Drusus when they had broken out after the death of Augustus. The country was then divided into two portions, Pannonia Superior and P. Inferior, the boundary being formed by a line drawn from Arrabona in the N. to Servitium in the S., Superior lying W. of the

---

[1] Among the animals of Pannonia we have notice of bears, an unknown animal named *catta*, hounds, and the *sharas* or black-cock:—

Pannonis haud aliter post ictum saevior ursa
Se rotat in vulnus.   Luc. vi. 220.

Pannonicas nobis nunquam dedit Umbria cattaa.—Mart. xiii. 69

[2] —————— testis quoque *fallax*
Pannonius gelidas passim di-jectas in Alpes.   Tibull. iv. 1, 108.

Hunc quoque perque novem timuit Pamphylia menses
Pannoninaque ferox.   Stat. *Sile.* i. 4, 77.

line. In the 4th century, Galerius subdivided Inferior by taking
away the part N. of the Dravus, and constituting it a province with
the name of Valeria. Finally, Constantine the Great equalized the
size of the provinces by adding to Inferior the S. part of Superior.
Under the Romans the people became thoroughly civilized ; colonies
and *municipia* were established, and fortresses were built for its
protection ; military roads were constructed, of which we may
especially notice those from Æmona, where the road from Aquileia
in Italy emerges from the Julian Alps, down the Savus and across
to the Danube at Vindobona, another along the course of the Danube,
and again one through the central district from Vindobona to
Sirmium. The chief towns were situated on the Danube, and on
the course of the Savus, with some few on the cross roads. They
were all strongly fortified, but of their history we know little.

(1.) *In P. Superior.*—**Vindobona**, *Vienna*, on the Danube, was
originally a Celtic town: the Romans made it a municipium with the
name of Juliobona, and it became their most important military position as the station of the Danubian fleet and of the Legio X. Gemina.
Carnuntum, near *Haimburg*, on the Danube, was a place of the
greatest importance as the station of the fleet after its transfer from
Vindobona, and as the head-quarters of a legion. M. Aurelius made it
the base of his operations against the Marcomanni and Quadi: Severus
was here when he was proclaimed emperor, and, though destroyed
by the Germans in the fourth century, it was restored and was the
centre of Valentinian's operations against the Quadi. **Petovio**, *Pettau*,
on the Dravus, was a Roman colony with the surname of Ulpia, and
was probably founded either by Trajan or Hadrian: it was the station
of a legion, and an imperial palace existed outside its walls. **Emona**,
Laybach, on the Savus, was a strongly-fortified town and a place of
considerable trade : it became a Roman colony with the title of Julia
Augusta. **Siscia**, or **Segesta**, *Sissek*, stood on an island formed by
the junction of the rivers Colapis and Odra with the Savus, together
with an artificial canal dug by Tiberius: it was from the first a strong
fortress, and after its capture by Tiberius it became one of the most
important places in Pannonia, being centrally situated on the great
road from Æmona to Sirmium. It was made a colony, possessed a
mint, and was the station of a small fleet on the Save : it sunk with
the rise of Sirmium.

(2.) *In P. Inferior.*—**Sirmium**, *Mitrovitz*, stood on the left bank of
the Savus, and was the point at which several roads centered: it was
hence selected as an arsenal by the Romans in their wars against the
Danubian tribes and as the residence of the admiral of the first Flavian
fleet on the Danube: it contained a large manufactory of arms, an
imperial palace, and other public buildings. **Taurunum**, *Semlin*, was
a strong fortress at the junction of the Savus with the Danube, and
the station of a small fleet. **Cibalis** stood near lake Hiulcas, between
the Savus and Dravus, its exact position not being known: it was the
birthplace of the Emperor Valentinian, and in its vicinity Constantine
defeated Licinianus in A.D. 314. **Mursa**, *Essek*, on the Dravus, was
made a colony by Hadrian with the surname of Ælia: it was the
residence of the Roman governor of P. Inferior, and near it Gallienus

defeated Iugebus. Aquincum, or Acincum, Alt-Buda, a strong fortress
on the Danube, was the centre of the Roman operations against the
Jazyges, and possessed a manufactory of bucklers. Bregetium, E. of
Comorn, on the Danube, was another very strong fortress: the Emperor Valentinian died there.

II. ILLYRICUM.

§ 6. The country which the Greeks named Illyris (very rarely
Illyria), and the Latins Illyricum,[3] lay along the eastern shore of the
Adriatic (in this part termed the Illyrian Sea[4]), from the river Arsia
in the N.W., dividing it from Istria, to the Ceraunian Mountains in
the S., on the borders of Epirus; on the E. it was contiguous to
Moesia and Macedonia; and on the N. to Pannonia. It was divided
by the river Drilo into two portions, Illyris Romana or Barbara,
which included the modern districts of Dalmatia, Herzegovina,
and Monte-Negro, with parts of Croatia, Bosnia, and Albania, and
I. Graeca, answering to nearly the whole of Albania. The former
was the proper province of Illyricum; the latter was annexed to
Macedonia by Philip of Macedon, and formed a portion of the
Roman province of Macedonia. The country is generally wild and
mountainous, and, with the exception of the coast of the southern
district, unproductive.

§ 7. The ranges which traverse Illyricum in a direction parallel
to the sea-coast from N.W. to S.E. are the connecting links between
the Italian Alps and the systems of the Thracian Haemus and the
Greek peninsula. They were but little known to the ancients:
the most northerly range was named Albanus Ms., which was
followed by Ardius Ms., the Bebii Mts. on the borders of Moesia,
Scardus and Candavia Mts. on the borders of Macedonia, and the
Ceraunii Mts. on the borders of Epirus. The chief rivers from N.
to S. are: in Barbara, the Naro, Narenta, which waters the central
district, and which is described as navigable for a distance of 80
stadia; the Barbana, Bojana, which flows through lake Labeatis;
and the Drilo, Drin, rising in lake Lychnitis. In Graeca, the
Genusus, Tjerma, rising on the borders of Macedonia; the Apsus,[5]

---

[3] The name was occasionally applied in a broader sense to the countries S. of
the Danube. It may have been used in this indefinite sense by St. Paul (Rom. xv.
19). After the subjection of the Dalmatae by the Romans the province was
officially named Dalmatia; and henceforward Illyricum and Dalmatia became con-
vertible terms. It is thus that the term is used by St. Paul (2 Tim. iv. 10).

[4] Tu mihi, seu magni superas jam saxa Timavi:
Rive oram Illyrici legis aequoris.   Virg. Ecl. viii. 6.
Antenor potuit, mediis elapsus Achivis,
Illyricos penetrare sinus.   Æn. l. 242.

[5] Both the Apsus and Genusus, particularly the former, are mentioned in connexion with the campaign of Caesar and Pompey :—

Beratinos, which rises in the Candavian range, and receives an important tributary in the Eordaïcus, *Devol*; and the **Aöus**,[6] *Voyussa*, which rises in Mount Lacmon and flows generally to the N.W., reaching the sea near Apollonia; in its midcourse it takes a sudden turn for 12 miles to the S.W., passing between lofty cliffs which formed the Fauces Antigonenses of the ancients (so named from the neighbouring town of Antigonia), where Philip V. engaged the Roman consul Flaminius. There are several large lakes in Illyricum, particularly **Labeātis**, *Scutari*, and **Lychnitis**, *Okridha*, both of which abound with fish. The sea-coast is extremely irregular and, in the northern district, is fringed with islands. The only important bays are the **Sinus Flanaticus**, *G. di Quarnero*, in the extreme N., and the land-locked **Sin. Rhisonicus**, *B. of Cattaro*, near Epidaurus.

§ 8. The Illyrians were regarded by the ancients as a separate race, distinct both from the Thracians and the Epirots; they are undoubtedly the progenitors of the modern *Albanians*, who have now spread southwards over Epirus under the pressure of the Slavonian tribes. They were a warlike and, previously to the Roman conquest, a thoroughly uncivilized race. Like the Thracians they tattooed their bodies, and offered human sacrifices. The northern tribes, particularly the Liburnians, were skilful sailors and built peculiarly swift vessels[7] (*Liburnicæ naves*). They were much devoted to piracy, for the prosecution of which their coast offered great advantages.[8] They were divided into numerous tribes,[9] of

---

Prima duces junctis vidit consistere castris
Tellus, quam volucer Genusus, quam mollior Apsus
Circumeunt ripis. Apso gestare carinas
Causa palus, leni quam fallens exerit unda.
At Genusam nunc sole nives, nunc imbre solutæ
Præcipitant. Neuter longo se gurgite lassat,
Sed minimum terræ, vicino littore novit. Luc. v. 461.

[6] Lucan's description is hardly appropriate to the Aoüs, which is a considerable stream:—

Purus in occasus, parvi sed gurgitis, Æas
Ionio fluit inde mari. vi. 361.

[7] Ibit Liburnis inter alta navium,
Amice, propugnacula;
Paratus omne Cæsaris periculum
Subire, Mæcenas, tuo. Hor. Epod. i. 1.

[8] Hence Virgil's description:—
————— intima tutus
Regna Liburnorum. Æn. i. 243.

[9] The Liburnians appear to have been numerous at Rome, where they acted as attendants in menial offices:—

Procul horridus Liburnus, et querulus cliens;
Imperia viduarum procul. Mart. i. 50.

Primus, clamante Liburno,
Currite! jam sedit! rapta properabat abolla
Pegasus, attonitæ positus modo villicus urbi. Juv. iv. 75.

which the most important were—the **Iapydes** in the N. in the interior; the **Liburni** on the adjacent sea-coast, from the extreme N. of the Adriatic southwards; and the **Dalmatæ** in the central district. The country was divided by the Romans into three parts, named, after the above tribes, Iapydia, Liburnia, and Dalmatia.

The following towns are described in order from N. to S.:—

(1.) *In Barbara.*—**Metūlum**, the capital of the Iapydes, was situated on the frontier of Pannonia either at *Mottling* or *Metlica.* **Iadēra**, *Zara,* the capital of Liburnia, was made a Roman colony by Augustus. **Scardōna** stood on the estuary of the Titius, somewhat W. of the modern *Scardona;* as one of the three "conventus" of Dalmatia it must have been an important place. **Tragurium**, *Trau,* celebrated for its marble, stood on an island cut off from the mainland by an artificial canal. **Salōna**, more correctly **Salōnæ**, *Salona,* the capital of Dalmatia, stood on the banks of the river Iader,[1] which falls into a small inlet of the Adriatic. It was the head-quarters of Metellus in B.C. 117, and was again besieged and taken by Cosconius in 78: in the Civil War it was vainly attacked by the Pompeian fleet under M. Octavius: it was again taken by Asinius Pollio in 39, and from that time became the great bulwark of Roman power on this side of the Adriatic. All the great roads met here, and it became one of the three "conventus" of Dalmatia. Its neighbourhood was selected by Diocletian as the place of his retirement: he built about 3 miles from the town a magnificent palace covering no less a space than eight acres and containing temples dedicated to Jupiter and Æsculapius, the former of which is now named the *Duomo,* while the latter is a baptistery of *St. John:* the modern name *Spalato* is a corruption of Salonæ Palatium. **Narōna** stood on the river Naro, about 2½ miles from its mouth at *Vido,* and was a Roman colony and a "conventus;" the Romans made it their head-quarters in the Dalmatian war. **Epidaurus**, *Ragusa-Vecchia,* is first noticed as being besieged by M. Octavius in the Civil War. It afterwards became a Roman colony. **Scodra**, *Scutari,* was a very strong place at the outlet of lake Labeatis: Gentius was defeated under its walls in B.C. 168. **Lissus**, *Lesch,* at the mouth of the Drilo, was founded by Dionysius the elder in B.C. 385, and was the limit appointed by the Romans for Illyrian commerce: Philip of Macedon captured it in 211.

(2.) *In Illyris Græca.*—**Epidamnus**, or **Dyrrhachium**, the latter name being descriptive of the ruggedness of its situation, was founded by a mixed colony of Corcyræans and Corinthians about 627 B.C. It stood on the isthmus of a peninsula,[2] and from its favourable position rose to commercial importance at an early period. The dispute relative to it between Corcyra and Corinth led to the Peloponnesian War: from 312 it was much exposed to attacks from the Illyrians until it obtained the protection of the Romans. It was the scene of the

[1] Qua maris Hadriaci longas ferit unda Salonas
Et tepidum in molles Zephyros excurrit Iader.   Luc. iv. 404.

[2] Its position is thus described by Lucan:—
Sed munimen habet nullo quassabile ferro,
Naturam, sedemque loci. Nam classe profundo
Undique, et illisum scopulis revomentibus æquor,
Exiguo debet, quod non est insula, colli.   vi. 22.
2 o 2

contest between Cæsar and Pompey, and during the last Civil Wars
it sided with M. Antonius. Its inhabitants, whose patron deity was
Venus, were an immoral race:[3] it is still, as *Durazzo*, an important
town. Apollonia, *Pollina*, a colony of Corcyræans and Corinthians,
stood about 10 stadia from the right bank of the Aous and 60 from
the sea. Under the Romans it became the seat of a flourishing
university, and in the Civil Wars between Cæsar and Pompey it was
an important military post. Lychnidus, on the E. shore of lake
Lychnitis near its S. extremity, was, from its position on the frontier,
an important point in the Macedonian Wars of the Romans: it was
on the Egnatia Via. Oricus, or Oricum, *Ericho*, was a harbour[4] not
far S. of the mouth of the Aous, much frequented by the Romans in
their communication with Greece. It was taken by Philip V. in
B.C. 214, but afterwards fell into the hands of the Romans. Here
Æmilius Paulus embarked his army for Italy in 167; and here Cæsar
laid up his fleet in his war with Pompey. The place was famous for
its turpentine.[5]

Roads.—The great thoroughfare between Rome and the East, the
Via Egnatia, crossed the southern part of Illyricum, where it received
the special name of Candavia from the ridge[6] which it crossed on the
border of Macedonia. There were two branches of it, one starting
from Dyrrhachium, the other from Apollonia: these united at Clodiana
on the Genusus, and passed round the head of lake Lychnitis to Lychnidus, and thence to Heraclea in Macedonia.

History.—The Illyrians first encountered the Greeks under Brasidas
and Perdiccas in the Peloponnesian War. They were defeated and
their country partly conquered by Philip of Macedon about B.C. 360.
Their piratical practices led to the interference of the Romans in 233,
when an honourable peace was concluded, and again, in 219, when the
whole country was subdued. Various wars followed: the Liburnians
yielded to Rome in 176; the Dalmatæ, though defeated by L. Cæcilius
Metellus in 119, were not incorporated into the Roman Empire until
the year 23; the Iapydes were defeated in 129 by D. Junius Brutus,
and were united with the Liburni in a province by Augustus, but

---

[3] Nam ita est hæc hominum natio Epidamnia,
Voluptarii atque potatores maximi:
Tum sycophantæ et palpatores plurimi,
In urbe hac habitant: tum meretrices mulieres
Nusquam perhibentur blandiores gentium.
Propterea huic urbi nomen Epidamno inditum est,
Quia nemo ferme huc sine damno divertitur.   PLAUT. *Menæch.* II. 1

[4] Ille Notis actus ad Oricum
Post insana Capræ sidera, frigidus
Noctes non sine multis
Insomnis lachrimis agit.   HOR. *Carm.* III. 7, 3.

Ut te felici provecta Ceraunia remo
Accipiat placidis Oricos æquoribus.   PROPERT. L. 8, 19.

[5] ——————— quale per artem
Inclusum buxo, aut Oricia terebintho,
Lucet ebur.   ÆN. x. 135.

[6] ——————— sic fatus, in arma
Phœbeam convertit iter, terræque secutos
Devia, qua vastos aperit Candavia saltus,
Contigit Emathiam, bello quam fata parabant.   LUC. vi. 329.

were not finally conquered until 34 by Octavianus. The province of Illyricum embraced the northern district as far as the Drilo. In Constantine's division, Illyricum Occidentale was a diocese of the Prefecture of Italy, and included Dalmatia, Noricum, Pannonia, and other provinces, while Illyricum Orientale embraced Illyris Græca and a large number of provinces out of Illyricum proper.

Islands.—Off the coast of Illyris Romana lie from N. to S.:—The **Absyrtides**, *Cherso, Osero*, and others, said to have been named after Absyrtus, brother of Medea ; and the **Liburnides**, the chief of which are *Lissa, Grossa, Brattia, Brazza.* Pharus, *Lesina,* Corcyra Nigra, *Curzola,* Melita, *Melada,* and Issa, *Lissa,* on which Dionysius the elder planted a colony in B.C. 387 ; the attacks on it by Agron and Teuta brought on the first Illyrian War in B.C. 229. Its inhabitants were skilful sailors, and the "Lembi Issaici" did the Romans good service in their war with Philip of Macedon. These islands (Issa excepted) fringe the coast in a parallel direction from N.W. to S.E., and are uniformly long and narrow: the channels between them are deep and give ships a secure passage between them off the coast of Illyris Græca. The small island of **Saso**,[7] *Sasso*, N. of the Acroceraunian promontory, was a station for pirates: the approach to it was deemed very dangerous.

III. MŒSIA.

§ 0. **Mœsia**[8] was bounded on the W. by Ms. Scordus and the rivers Drinus and Savus, separating it from Illyricum and Pannonia ; on the S. by Ms. Hæmus on the side of Thrace, and Orbelus and Scordus on the side of Macedonia ; on the E. by the Euxine Sea ; and on the N. by the Danube, separating it from Dacia. It corresponds to the present *Servia* and *Bulgaria*. It was an irregular country, intersected by the various offsets of the lofty ranges which surround it, viz.: **Hæmus** in the S.E., **Orbelus** and **Scordus** in the S.W. and W. The rivers are all tributaries of the great border stream of the **Danubius**, which in this country resumes its easterly course and retains it until it approaches the Euxine, when it turns northwards for a while, and then to the S.E. entering the sea by several channels,[9] some of which enclosed the triangular isle of **Peuce**.[1] Its chief tributaries are—the **Savus**, of which only a

---

[7] ————— cum totas Hadria vires
Movit, et in nubes abiere Ceraunia, cumque
Spumose Calaber perfunditur æquore Sason.
Non humilem Sasona vadis, non littora curvæ
Thessaliæ saxosæ. Ib. v. 650.
Hadriaci fugite infaustas Sasonis arenas. SIL. ITAL. vii. 480.
Luc. ii. 625.

[8] The Greek form of the name was Μυσία, sometimes with the addition of ἡ ἐν Εὐρώπῃ, to distinguish it from the country of the same name in Asia.

[9] Multifidi Peucen unum caput adluit Istri. Luc. iii. 202.

[1] Martial describes it as a Getic, Valerius Flaccus as a Sarmatian isle :—
I, liber, ad Geticam Peucen, Istrumque tacentem,—MART. vii. 84.
Insula Sarmatica Peuce stat nomine Nymphæ,
Torvus ubi, et ripa semper metuendus utraque
In freta per æquos Ister descendit Alanos. VAL. FLAC. viii. 217.

small portion belongs to Mœsia; the **Drinus**, a feeder of the Savus, rising in M. Scordus; and the **Margus**, *Morawa*, which rises in Orbelus and joins the Danube W. of Viminacium.

§ 10. The inhabitants were reputed to be a Thracian race, allied to the Mysians of Asia Minor. Among them were settled a Celtic tribe, named Scordisci, who entered under Brennus in B.C. 277. The Romans subdued Mœsia in B.C. 29 under the generalship of M. Licinius Crassus, and kept military possession of it as a frontier province. It was originally organized as a single province, but early in Trajan's reign was divided into two provinces, separated from each other by the river Ciabrus, **Mœsia Superior** to the W., and **M. Inferior** to the E. When Aurelian withdrew from Dacia, he formed a settlement in the heart of Mœsia which was named after him **Dacia Aureliani**. The most important of the tribes were— the **Mœsi** proper on the river Ciabrus; the **Triballi** to the W. in the valley of the Margus; the **Peucini** on the Isle of Peuce at the mouth of the Danube; and the **Crobyzi** near the frontiers of Thrace.[9] The towns of Mœsia may be divided into three classes: (1) the Greek commercial towns on the shores of the Euxine, which were colonies of Miletus, such as Istropolis, Tomi, Callātis, and Odessus; (2) the Roman fortresses along the course of the Danube, such as Singidūnum, Ratiaria, and others, which became of great importance after the Romans had withdrawn from Dacia; and (3) the towns of the interior, which were comparatively few and little known. The names of many towns in the second class betoken a Celtic origin, e.g. Singi-*dunum*, Duro-storum, and Novio-*dunum*. The historical associations are very scanty. The Danubian towns were mostly destroyed by Attila and his Huns, and restored by Justinian. Mœsia gave three emperors to Rome, Constantine the Great, Maximian, and Justinian.

(1.) *Towns along the course of the Danube from W. to E.*—**Singidūnum**, *Belgrade*, at the spot where the Savus falls into the Danube; **Margum**, at the junction of the Margus, known as the scene of Diocletian's victory over Carinus; **Viminacium**, somewhat E. of the Margus, either at *Rum* or *Kostolacz*, the head-quarters of the Legio VII. Claudia; **Egēta**, near Trajan's bridge over the Danube; **Ratiaria**, *Arzar-Palanca*, the head-quarters of a legion and the station of a fleet on the Danube; **Œscus**, *Orexovitz*, near the mouth of the river of the same name; **Durostōrum**, celebrated as the birth-place of Aetius; and **Noviodūnum**, *Isaczi*, a little above the point where the Danube divides; near it Valens constructed a bridge over the river.

(2.) *In the Interior.*—**Naīssus**, *Nissa*, upon a tributary of the Margus, the birth-place of Constantine the Great, and also known for

---

[9] In addition to these tribes a number of Goths settled in the country in A.D. 395, and were thenceforward named Mœso-Goths. They were converted to Christianity, and for their use Ulphilas made a translation of the Scriptures, parts of which still exist.

a victory obtained by Claudius II., in A.D. 269, in its neighbourhood; **Serdica**, or **Sardica**, the later capital of Dacia Interior, situated in a fruitful plain at the spot where the sources of the Œscus unite, and from the time of Aurelian surnamed Ulpia; the Emperor Maximilian was born near there; **Scupi**, *Uschkūb*, a most important point as commanding the passes into Illyricum: near it was Tauresium, the birthplace of Justinian; **Marcianopolis**, near *Pravadi*, founded by Trajan and named after his sister Marciana; near it Claudius II. defeated the Goths in several battles.

(3.) *On the Euxine from N. to S.*—**Istropolis**, situated at the S. end of lake Halmyris and a place of considerable trade; **Tomi**, *Tomisrar*, some 40 miles to the S., the reputed spot where Medea cut up her brother's body,[3] but still better known as the place to which Ovid was banished; **Callātis**, *Collat*, originally colonized by Miletus and afterwards replenished with settlers from Heraclea; lastly, **Odessus**, *Varna*, which appears to have presided over the Greek towns on this coast: its coins bear devices relating to the worship of Serapis, the god imported from the shores of Pontus to Alexandria by Pompey.

IV. DACIA, WITH THE COUNTRY OF THE JAZYGES METANASTÆ.

§ 11. **Dacia** under the Romans was bounded on the S. by the Danube; on the E. by the river Hierasus; on the N. by M. Carpātes; and on the W. by the river Tysia, separating it from the country of the Jazyges. It thus contains the *Banat of Temesvar*, *Hungary* E. of the *Theiss*, *Transylvania*, the *Bukowina*, the S. point of *Gallicia*, *Moldavia* W. of the *Pruth*, and *Wallachia*. The only range of mountains noticed by ancient writers is **Carpates Mons** described by Ptolemy as an insulated range lying between the sources of the Tibiscus and the Tyras. It thus answers to the *W. Carpathians*. The rivers are all tributaries of the **Danubius**; they are, as follows, from W. to E.: the **Tysa**, or **Tisiānus**, *Theiss*, with its tributaries the **Gerāsus**, *Körös*, and the **Marisus**, *Maroxch*; the **Tibiscus** or **Pathissus**, *Temes*; the **Alūtas**, *Aluta*; and the **Hierāsus**, *Sereth*.

§ 12. The inhabitants of Dacia belonged to the Thracian group of nations. Their original name was **Getæ**,[4] which was subsequently changed to **Daci**, though the date and the causes of this change are quite unknown. The position of this people varied at different historical periods,[5] but at the time they became known to the

[3] This legend probably arose from a fancied derivation of the name from τέμνω "to cut."

[4] The resemblance of the names Getæ and Goths has occasionally led to a mistaken idea that the two races were identical. The names Geta, Dacus, and Davus, are the generic titles of slaves in the plays of Aristophanes and Terence. This originated in the number of captives made by the Gauls when they invaded Eastern Europe, and sold as slaves to the Athenians.

[5] Herodotus and Thucydides describe them as living between the Ister and Mt. Hæmus. When Philip invaded Scythia they had been displaced from these

Romans,* they occupied the district we have above described. The Romans first entered the country under Lentulus in B.C. 10, but they did not subdue the country until Trajan's expeditions' (A.D. 101-105), when a large number of the inhabitants migrated to the banks of the Borysthenes, where they were known as Tyragetæ. The country was now reduced to a province, and remained an integral portion of the Roman empire until the time of Aurelian (A.D. 270-275), when the Roman settlers withdrew to the S. of the Danube and settled in Dacia Aureliani, leaving Dacia Proper to the Goths.* It remained for a long time a barrier against the barbarian tribes of the north, but it was at length overrun by Attila and his Huns about A.D. 376. The conqueror of Dacia, Trajan, connected Dacia with Mœsia by a magnificent bridge,* and constructed three important roads, connected with the Via Trajana, which ran along the S. side of the Danube, partly cut in the rock and partly supported by wooden beams set up against the perpendicular wall of rock above the water of the river.' The first

quarters by the Triballi, and had been driven N. of the Ister. Here they were attacked by Alexander in B.C. 335, and by Lysimachus in B.C. 292.

' The Daci were regarded by the Romans as a formidable race: they served under Antony as mercenaries at Actium, to which Horace alludes in the following passages:—

Pæne occupatam seditionibus
Delevit urbem Dacus et Æthiops;
Ille classe formidatus, ille
    Missilibus melior sagittis.   Carm. III. 6, 13.
Frigidus a rostris manat per compita rumor;
Quicunque obvius est, me consulit: O bone (nam tu
Scire, deos quoniam propius contingis, oportet)!
Num quid de Dacis audisti?   Sat. II. 6, 50.

They were in consequence attacked by Lentulus about B.C. 25, to which the same poet refers in Carm. III. 8, 18:—

Occidit Daci Cotisonis agmen.

' In his first campaign Trajan passed through Pannonia, crossed the Theiss, and followed the course of the Marosch into Transylvania: his first great battle was fought on the Crossfield near Thorda, which still retains the name of Prat de Trajan (Pratum Trajani). In his second campaign he crossed the Danube below the Iron Gate, where his bridge was afterwards built, and, sending one part of his army along the Aluta, he himself followed the valley, which leads from Orsova by Mehadia (through the Iron Gate pass) to the capital, Sarmisegethusa, which the inhabitants set on fire.

* Though the Roman dominion lasted only about 170 years in Dacia, yet in no country has it left more unequivocal traces in the language of the people. The Wallachian is a Romance language, derived from the Latin, like the Italian, Spanish, and French.

* This bridge was situated at the point where the river makes a double bend near Severin. It was built by Apollodorus, and consisted of twenty piers, 150 ft. high, 60 thick, and 170 distant from each other. It was destroyed by Hadrian about A.D. 120. All that now remains of it is a solid mass of masonry about 30 ft. high on each bank, and the foundations of the piers, some of which are visible when the river is low.

' The sockets in which the beams were inserted to support this road are visible in many places. The road was in fact nothing but a wooden shelf.

road ran between Viminacium and Tibiscum; the second between Pons Trajani and Parolissum, by the banks of the *Temes* (through the narrow gorge of the Iron Gate) into the valley of the *Marosch*, and so on into *Transylvania*; and the third between Trajan's bridge by the valley of the *Aluta* to Apula, where it fell into the last-mentioned road. The so-called wall of Trajan, which ran through a great part of Dacia from the S.W. to N.E., and of which the remains may still be found, belongs to a later period. Of the towns we know but little. **Sarmizegethūsa**, the old Dacian capital and the chief garrison of the Romans, stood about five Roman miles N. of the *Vulkan* Pass at *Varhely* on the river *Strel* or *Strey*. It became a colony, and possessed an aqueduct and baths.

The other important towns were:—**Tibiscum** or **Tivisoum**, *Kavaran*, on the Tibiscus; **Tiarna**, on the Danube, at the mouth of a river of the same name; **Apōla**, *Weissenburg*, a Roman colony on the Marisus; and **Parolissum**, a *municipium* more to the N., the position of which is not well ascertained.

§ 13. The **Jazyges Metanastæ** were a Sarmatian race, whose original settlements were on the Palus Mæotis. Thence they wandered to the banks of the Lower Danube, and in A.D. 50 a portion of them transferred their residence to the country between the *Theiss* and the *Danube*, where they received the surname of Metanastæ, *i.e.* "transplanted," to distinguish them from the rest of the race.[1] They were a wild, nomad race, living in tents and waggons, and perpetually at war with the Romans. They called themselves Sarmatæ Limigantes, and were divided into two classes, slaves and freemen. The towns in this district were founded by the slaves who preceded the Jazyges. We know nothing of them beyond their names.

V. SARMATIA EUROPÆA.

§ 14. The extensive district which lies E. of the Vistula and N. of Dacia was comprised under the general name of **Sarmatia**; northwards it extended to the *Baltic*, and eastwards to the *Tanais*, which formed the boundary between Europe and Asia. It thus included parts of *Poland* and *Gallicia*, *Lithuania*, *Esthonia*, and *Western Russia*. The only portion of this enormous extent of country really known to the ancients was that which was adjacent to the coasts of the Euxine, answering to the Scythia of Herodotus. Of the rest we have a description by Ptolemy, consisting of nu-

---

[1] This was their position in Ovid's time:—
Jazyges, et Colchi, Metareaque turba, Getæque,
Danubii mediis vix prohibentur aquis.   *Trist.* ll. 191.

merous names of tribes and mountains, of which the former are interesting to the ethnologist, while the latter are so vaguely described as to be beyond the reach of identification. We have already noticed the chief rivers that discharge themselves into the Euxine, in connexion with the geography of Herodotus. It only remains for us to notice the most important tribes and towns known to the ancients.

§ 15. The chief tribes were the **Tauri** in the Chersonesus Taurica, Crimea, probably the remains of the Cimmerians, who were driven out of the Chersonese by the Scythians. They were a rude, savage people, much addicted to piracy.[1] The **Roxolani**, a Sarmatian race, who first appear in history about 100 B.C., when they occupied the steppes between the *Dnieper* and the *Don*. They waged war with Mithridates, and were defeated by his general Diophantus. They were also defeated by the Romans in Otho's reign. The **Jazyges**, whom we have lately referred to, and who once lived between the *Dnieper* and the *Sea of Azov*. The **Bastarnæ**, a powerful tribe, generally supposed to be of German extraction, whose earliest settlements seem to have been in the highlands between the *Theiss* and *Marosch*, whence they pressed down the course of the Danube to its mouth, where a portion of them settled in the Isle of Peuce under the name of **Peucini**. They are afterwards found between the *Dniester* and *Dnieper*. The **Alani**, a branch of the Asiatic race of the same name, a wandering horde that issued from the steppes between the Euxine and the Caspian. The **Hamaxobii**, on the

---

[1] The Taurians worshipped Diana, or, according to their own statement, Iphigenia :—

Ἐν δ' εἶπας ἐλθεῖν Ταυρικῆν μ' ὅρους χθονὸς,
Ἔνθ' Ἀρτέμις, σῇ σύγγονος, βωμοὺς ἔχει,
Λαβεῖν τ' ἄγαλμα θεᾶς, ὅ φασιν ἐνθάδε
Ἐς τούσδε ναοὺς οὐρανοῦ πεσεῖν ἄπο.   EURIP. *Iph. in Taur.* 85.
Ἥλθες ἀπὸ Σκυθίας, ἀπὸ δ' εἶπαο τείχεα Ταύρων.   CALLIM. *Hymn. in Dian.* 174.

Ovid refers to their barbarous custom of immolating human victims in honour of Diana Tauropolis :—

Est locus in Scythia, Tauros dixere priores,
Qui Getica longe non ita distat humo.
Hac ego sum terra (patrio nec pœnitet) ortus.
Consortvm Phœbi gens colit illa deam.
Templa manent hodie vastis innixa columnis ;
Perque quater denos itur in illa gradus.
Fama refert illic signum cœleste fuisse.
Quoque minus dubites, stat basis orba dea ;
Ataque, quæ fuerat natura candida saxi,
Decolor affuso tincta cruore rubet.   *Ex Pont.* iii. 2, 45.
Nec procul a nobis locus est, ubi Taurica dira
Cæde pharetratæ pascitur ara deæ.   *Trist.* iv. 4, 63.

There was a famous temple of this goddess near Chersonesus, Sebastopol ; but its exact position is undecided.

banks of the *Volga*, also a nomad race, as their name ("livers in waggons") implies. The **Agathyrsi,**[*] located in the time of Herodotus on the banks of the *Theiss*, afterwards in the Palus Mæotis, and again more to the N.; and the **Venēdæ,** on the shores of the Sinus Venedicus, *Gulf of Riga*. The only towns which we shall notice are the Greek colonies on the mainland and in the Tauric Chersonese.

(1.) *Towns on the Mainland.*—**Tyras** was a Milesian colony near the mouth of the river of the same name, probably at *Ackermann.* **Olbia,** or **Borysthenes,** stood on the right bank of the Hypanis, about 25 miles from its mouth; it was founded by Milesians in B.C. 655, and became a most important place of trade, and also produced some literary men of distinction: it appears to have been destroyed by the Getæ about B.C. 50, but was afterwards restored; its ruins are at *Stomogil.* **Carcina** stood at the entrance of the *Crimea* on a river which has been identified with the *Kalantchak.*

(2.) *Towns in the Tauric Chersonese.*— **Chersonēsus** was founded by the Dorians of Heraclea in Pontus, probably in the 5th century B.C., at the S.W. extremity of the peninsula. The original town stood close to *C. Fanari:* this was destroyed, and its successor occupied a portion of the site of the famous *Sebastopol.* A wall was constructed for the defence of this place from the head of the harbour to **Symbōlon,** *Balaclava:* the remains of the wall and town were considerable until the Russians erected *Sebastopol.* Near it was **Eupatorium,** generally identified with the now famous *Inkermann.* **Theodosia,** *Caffa,* a colony of the Milesians, stood on the S.E. coast, and was a place of considerable trade, particularly in corn: its native name was Ardabka, "town of the seven gods." **Nymphæa** was also a Milesian town with a harbour, the ruins of which are at the S. point of the *Lake of Tchourbache.* **Panticapæum,** *Kertch,* stood at the W. side of the Cimmerian Bosporus: the date of its foundation is not certain, but it must have been about 500 B.C.: it was the capital of the kings of Bosporus, and hence was itself occasionally called B.sporus. The old town occupied the eminence at the foot of which *Kertch* stands: numerous tumuli have been discovered about it, from which antiquities of all sorts have been extracted. The kingdom of Bosporus existed under various dynasties from about B.C. 500 to about A.D. 350. The events of chief interest connected with it are its conquest by Mithridates the Great, King of Pontus, and its subsequent submission to the Romans, who appointed Pharnaces king.

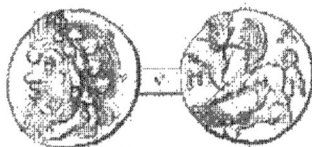

Coin of Panticapæum.

---

[*] They practised the art of tattooing:—
　Cretesque Dryopesque fremunt, *pictique* Agathyrsi.—*Æn.* iv. 146.

Arch of Volaterra.

INDEX.

ABBREVIATIONS.

Fl. = Flumen.	L. = Lacus.	Pr. = Promontorium.
Fret. = Fretum.	Ms. = Mons.	S. = Sinus.
I. = Insula or -æ.	Mts. = Montes.	

BATAVORUM.	BRITANNIA.	COMMAGENORUM.	CANTABRI.
Batavórum, L. 641	Bīocium, 332	Britannia Secunda, 641	Cæsaromágus, 646
Bathys Portus, 404	Bongrius, Fl., 391	Britannicæ, L. 637 ff.	Casia Sylva, 662
Batne, 709	Bous, Fl., 87	Britannicum Mare,	Caicus, Fl., 93
Bæti, 530	Bodencus, Fl., 488	317	Calcin, 333
Bautinus, Fl., 76	Boderia, Æst., 639	Brixellum, 591	Calabria, 476 ff.
Bæzium, Pr., 254	Bobe, 162	Brixia, 498	Calacte, 604
Bebryces, 89	Borbēto, L. 361	Brongus, Fl., 32	Caiagurris Fibularia,
Bechires, 159	Bæotia, 226 ff.	Bructeri, 664	622
Bedriacum, 499	Boii (Gall. Cis.), 497	Brundusium, 475	Calagurris Nassica,
Beeroth, 188	Boii (Germ.), 505	Brutii, 589	622
Beersheba, 184	Boiodūrum, 670	Bryges, 89	Calah, 12
Begorra, 344	Boium, 387	Brysæ, 409	Calatiæ, 476
Begorrita, L., 319	Bollo, 113	Buana, 344	Calauria, L. 468
Belbina, 460	Belbitine, 272	Bubassus, Sin., 115	Calbis, 115
Belbina, L. 411	Bolerium, Pr., 649	Bubassus, 271	Caledonia, 648
Belemina, 460	Bomi, Mts., 381	Bura, 478	Calentum, 516
Belgæ, 651	Bomienses, 381	Bucephala, 249	Cales, 333
Belgica, 641	Bonna, 644	Budini, 35	Caleti, 641
Beltinma, Æst., 649	Bononia (Gall.), 646	Budorum, Pr., 421	Callinge, 250
Belus, Fl., 208	Bononia (Ital.), 500	Bulla Regia, 556	Callaici, 517
Bellovaci, 644	Borconvicus, 656	Bumādus, Fl., 227	Callatis, 629
Belo, 614	Boreium, Pr. (Cyren.),	Buphras, Mts., 449	Calleva, 655
Bembina, 440	291	Buporthmos, Mts., 461	Callichrōma, Mts., 259
Benācus, L., 490	Borgium, Pr. (Hibern.),	Bupsasium, 447	Callicha, 281
Heneventum, 476	662	Bura, 441	Callicuscæ, 281
Benjamin, 187	Borsippa, 331	Burdigala, 642	Calligerum, Pr., 250
Berecyothus, 89	Borysthénes, 662	Burgundiónes, 661	Callipolis (Calabr.), 479
Herrædær, 278	Borysthénes, Fl., 32	Burrium, 655	Callipolis (Cicil.), 602
Berenice (Cyren.), 291	Bosa, 608	Busiris, 272	Callipolis (Thrac.), 631
Berenice Epideiros,	Bosporus, 682	Butheōtum, 371	Calliphae, 34
288	Bosporus Cimmerius,	Buto, L., 272	Callipus, Fl., 517
Berenice Panchrysus,	70	Buxentum, 591	Callirhoe, 416
288	Bosporus Thracius, 70	Hyblas, 170	Callisthenes, 41
Hergomum, 498	Bostra, 204	Byhatora, 147	Callium, 384
Hermius, Mts., 117	Bostrenus, Fl., 169	Byzacium, 554	Calneh, 12
Berœa (Maced.), 346	Bottiæa, 341	Byzantes, 299	Calor, Fl., 475
Berœa (Syr.), 165	Bovillanni, 575	Byzantium, 631	Colpe, 614
Berœa (Thrac.), 331	Bozrah (Arab.), 19		Calycadnus, Fl., 114
Berothai, 10	Bozrah (Peræa), 204		Calydon, 384
Berytus, 170	Bracara Augusta, 524		Calymnus, L. 333
Besa, 118	Brachōdes, Pr., 299	C.	Calynda, 122
Bethany, 188	Brodānus, Fl., 574		Camalodūnum, 651
Bethel, 188	Branchidæ, 119	Cabalis, 15, 147	Camarina, 498
Bethsaida, 193	Brannovices or Bran-	Cabīliōnum, 640	Cambūnii, Mts., 338
Beth-horon, 187	novii, 639	Cabira, 160	Cambysæ, Fl., 342
Bethlehem, 189	Bratia, L. 677	Cabolicæ, 243	Camerinum, 336
Bethsaida, 193	Brauron, 419	Cabura, 249	Camirus, 123
Bethsan, 186	Brigetium, 671	Cadmein, 401	Campania, 561 ff.
Beth-shemesh, 186	Bremenium, 659	Cadmus, 16	Campi Laborini, 562
B-tijm, Ms., 150	Beurthe, 472	Cadmus, Ms., 115	Campi Phlegræi, 562
Bezitha, 190	Brigiæi, 89	Cadurci, 641	Campodūnum, 668
Bibracte, 640	Brigantes (Brit.), 651	Cadytis, 17	Campus Esquilinus,
Bibilis, 623	Brigantes (Hibern.),	Cæcilia, Fl., 586	546
Idlierus, Fl., 151	662	Cæsbbus Ager, 857	Campus Martius, 547
Bingium, 643	Brigantinus, L. 668	Cædrius, Fl., 607	Cana, 193
Blandia, 140	Brigantium (Gall.),	Cœlius, Ms., 524	Canaria, L. 113
Bistones, 118	637	Cænæ, 337	Canastræum, Pr., 338
Bistonis, L., 117	Brigantium (Hisp.),	Cænopolis, 459	Candavia, Ms. 671
Bithroi, 151	624	Cære, 510	Candidum, Pr., 299
Bithynia, 192 ff.	Brigantium (Vindel.),	Cæsar, Commentaries	Cane, Fr., 92
Bisturiges, 641	668	of, 53	Cane, 199
Bisturiges Cubi, 641	Brigæ, 89	Cæsaraugusta, 620	Cingamórum, Pr., 649
Biseda, 581	Brilessus, Ms., 406	Cæsaria (Cappadoc.),	Cangi, 651
Bisium Belgium, 650	Britain, Discovery of,	162	Canoe, 575
Blaundus, 119	44	Cæsarea (Cilic.), 136	Canopus, 269
Blemmyes, 284	Britannia, 637	Cæsarea, L. 641	Cantaber, Oceanus, 522
	Britannia Barbara, 658	Cæsarea (Mauret.), 559	Cantabri, 621
	Britannia Prima, 641	Cæsarea (Palest.), 194	
		Cæsarea Philippi, 193	
		Cæsarodūnum, 641	

I. J.

2 R 2

2 I

MURRAY'S STUDENTS' MANUALS
FOR ADVANCED SCHOLARS.

"This series of 'STUDENTS' MANUALS' edited for the most part by Dr. WM. SMITH, possess several distinctive features which render them singularly valuable as educational works. While there is an utter absence of flippancy in them, there is thought, in every page, which cannot fail to excite thought in those who study them, and we are glad of an opportunity of directing the attention of such teachers as are not familiar with them to *these admirable school-books.*"—*The Museum.*

I.—ENGLAND.

THE STUDENT'S HUME: A HISTORY OF ENGLAND from the Earliest Times to the Revolution of 1688. By DAVID HUME; corrected and continued to 1868. Woodcuts. Post 8vo, 7s. 6d.

II.—FRANCE.

THE STUDENT'S HISTORY OF FRANCE. From the Earliest Times to the Establishment of the Second Empire, 1852. By W. H. PEARSON, M.A. Woodcuts. Post 8vo. 7s. 6d.

III.—GREECE.

THE STUDENT'S HISTORY OF GREECE. From the Earliest Times to the Roman Conquest. By Dr. WM. SMITH. Woodcuts. Post 8vo. 7s. 6d.

IV.—ROME.

(1) *The Republic.*
THE STUDENT'S HISTORY OF ROME. From the Earliest Times to the Establishment of the Empire. By DEAN LIDDELL. Woodcuts. Post 8vo. 7s. 6d.

(2) *The Empire.*
THE STUDENT'S GIBBON: An Epitome of the History of the Decline and Fall of the Roman Empire. By EDWARD GIBBON. Woodcuts. Post 8vo. 7s. 6d.

V.—GEOGRAPHY.

THE STUDENT'S MANUAL OF ANCIENT GEO-GRAPHY. By Rev. W. L. BEVAN. Woodcuts. Post 8vo. 7s. 6d.

VI.—SCRIPTURE HISTORY.

THE OLD TESTAMENT HISTORY: FROM THE CREATION to the Return of the Jews from Captivity. Edited by WM. SMITH, LL.D. Maps and Woodcuts. Post 8vo. 7s. 6d.

THE NEW TESTAMENT HISTORY. With an Introduction, Connecting the History of the Old and New Testaments. Maps and Woodcuts. Post 8vo. 7s. 6d.

[continued.]

www.ingramcontent.com/pod-product-compliance
Lightning Source LLC
Chambersburg PA
CBHW031932220326
41598CB00062BA/1622